Antioxidant and Redox Regulation of Genes

Antioxidant and Redox Regulation of Genes

Edited by

Chandan K. Sen
Lawrence Berkeley National Laboratory
University of California
Berkeley, California
and
University of Kuopio
Kuopio, Finland

Helmut Sies
Heinrich-Heine-Universität
Düsseldorf, Germany

Patrick A. Baeuerle
Micromet GmbH
Martinsried, Germany

Academic Press
San Diego   London   Boston   New York   Sydney   Tokyo   Toronto

Front cover photograph: NF-κB p65 homodimer. A redox-sensitive transcription factor.
Courtesy of Drs. Y. F. Wang and A. Sarai, RIKEN Gene Bank, Japan.

This book is printed on acid-free paper. ∞

Copyright © 2000 by ACADEMIC PRESS

All Rights Reserved.
No part of this publication may be reproduced or transmitted in any form or by any means, electronic or mechanical, including photocopy, recording, or any information storage and retrieval system, without permission in writing from the publisher.

Academic Press
A Division of Harcourt, Inc.
525 B Street, Suite 1900, San Diego, California 92101-4495, USA
http://www.apnet.com

Academic Press
24-28 Oval Road, London NW1 7DX
http://www.hbuk.co.uk/ap/

Library of Congress Catalog Card Number: 99-63956
International Standard Book Number: 0-12-636670-5

PRINTED IN THE UNITED STATES OF AMERICA
99 00 01 02 03 04 MM 9 8 7 6 5 4 3 2 1

Contents

Contributors xv

Foreword xxi
  Luc Montagnier

Preface xxiii

I Reactive Species as Intracellular Messengers

1 Signaling by Singlet Oxygen in Biological Systems
Lars-Oliver Klotz, Karlis Briviba, and Helmut Sies

  I. Singlet Oxygen in Biological Systems 3
  II. Effects of Singlet Oxygen on Gene Expression 6
  III. Comparison of Effects of Singlet Oxygen and Other Oxidants on the AP-1 Pathway 7
  IV. Induction of Transcription Factors by Singlet Oxygen: Mechanism 9
     References 14

2 Redox Regulation of Gene Expression
Dana R. Crawford, Toshihide Suzuki, and Kelvin J. A. Davies

    I. Introduction 21
   II. Methods of Analysis 22
  III. Modulation of Nuclear Gene Expression 24
  IV. Modulation of Mitochondrial Gene Expression 28
   V. Stable versus Transient Gene Expression 29
  VI. Modes of Regulation 31
 VII. Oxidant Stress-Related Pathologies 33
VIII. Concluding Remarks 37
      References 39

3 Rac, Superoxide, and Signal Transduction
Hamdy H. Hassanain and Pascal J. Goldschmidt-Clermont

    I. Introduction 48
   II. Activation of Rac Proteins 50
  III. Rac and Respiratory Burst in Phagocytes 52
  IV. Rac in Nonphagocytic Cells 54
   V. Rac and ROS Production 57
  VI. Rac and the Cytoskeleton 63
 VII. Rac Gene in Development 66
VIII. Oxidative Burst in Plants 67
      References 69

4 Redox Regulation of Ion Channels
Suneil K. Koliwad, Anna K. Brzezinska, and Stephen J. Elliott

  I. Ion Channels 81
 II. The Patch Clamp 85
III. Redox Regulation of Ion Channels 87
    References 101

5 Cell Ca²⁺ in Signal Transduction: Modulation in Oxidative Stress

Julio Girón-Calle and Henry Jay Forman

   I. Ca^{2+} in Signal Transduction and Cell Death   106
  II. Lethal Alterations of the Intracellular Ca^{2+} Concentration: Necrosis, Apoptosis, and Oxidative Stress   111
 III. Modulation of Ca^{2+} Signaling by Oxidative Stress   114
  IV. Concluding Remarks   122
      References   122

6 Induction of Protein Tyrosine Phosphorylation by Oxidative Stress and Its Implications

Gary L. Schieven

   I. Introduction   129
  II. Tyrosine Phosphorylation Signal Pathways   130
 III. Ionizing Radiation   133
  IV. Ultraviolet Radiation   135
   V. Chemical Oxidants   136
  VI. Dimerization Hypothesis   141
      References   142

7 Regulation of Signal Transduction and Gene Expression by Reactive Nitrogen Species

Ami A. Deora and Harry M. Lander

   I. Introduction   147
  II. Signal Transduction by Reactive Nitrogen Species   149
 III. Gene Expression by Reactive Nitrogen Species   163
      References   170

II Redox-Sensitive Molecular Processes and Cellular Responses

8 Reactive Oxygen Species as Costimulatory Signals of Cytokine-Induced NF-κB Activation Pathways

Patrick A. Baeuerle

    I. Functions of Transcription Factor NF-κB   181
   II. How TNF and IL-1 Activate NF-κB   184
  III. NF-κB Activation by Prooxidant Conditions   187
  IV. Inhibitory Effect of Antioxidants and Redox-Regulating Proteins on NF-κB Activation by Cytokines   190
   V. How Can Antioxidants and Enzymes Affect NF-κB Activation by Cytokines?   191
  VI. Why Should ROS Modulate Proinflammatory Cytokine Signaling?   195
 VII. Conclusion: ROS May Act as Costimulatory Signals   197
      References   198

9 Redox Regulation of NF-κB

Takashi Okamoto, Toshifumi Tetsuka, Sinichi Yoshida, and Takumi Kawabe

    I. Introduction: Transcription Factor NF-κB and Its Biological Functions   203
   II. Involvement of NF-κB in Various Pathologies   205
  III. Signal Transduction Pathway to Activate NF-κB   207
  IV. Conclusion: Redox Regulation of NF-κB   214
      References   215

10 Oxidants and Antioxidants in Apoptosis: Role of Bcl-2

Chandan K. Sen

    I. Introduction   221
   II. Induction of Apoptosis by Reactive Oxygen Species   222

III. Antiapoptotic Properties of Antioxidants   225
IV. Proapoptotic Properties of Antioxidants   228
V. Bcl-2   229
VI. Bcl-2 Functions via an Antioxidant Mechanism?   231
VII. Summary   236
References   237

11 Role of Reactive Oxygen Species in Tumor Necrosis Factor Toxicity

Vera Goossens, Kurt De Vos, Dominique Vercammen, Margino Steemans, Katia Vancompernolle, Walter Fiers, Peter Vandenabeele, and Johan Grooten

I. Introduction   245
II. Reactive Oxygen Species as Mediators of Cell Death by TNF   248
III. Concluding Remarks   259
References   260

12 Redox Regulation of Cell Adhesion Processes

Sashwati Roy, Chandan K. Sen, Alexia Gozin, Valèrie Andrieu, and Catherine Pasquier

I. Introduction   266
II. Cell Adhesion Molecules   266
III. Multistep Model of Cell Adhesion   270
IV. Direct Activation of Cell Adhesion Processes by Reactive Oxygen Species   271
V. Ionizing Radiations and Ultraviolet Radiation   276
VI. Cigarette Smoke, Heavy Metals, and Other Oxidative Environmental Pollutants   278
VII. Nitric Oxide   278
VIII. Antioxidant Regulation of Agonist-Induced Cell Adhesion Process   278
IX. Cell Adhesion Processes and Pathologies with Redox Imbalances   283
References   287

13 Role of Thioredoxin and Redox Regulation in Oxidative Stress Response and Signaling

Hiroshi Masutani, Masaya Ueno, Shugo Ueda, and Junji Yodoi

    I. Introduction   298
   II. Thioredoxin and Related Molecules   298
  III. Thioredoxin Induction by Oxidative Stress   299
  IV. Cytoprotective Action of Thioredoxin   300
   V. Oxidative Stress and Apoptosis   301
  VI. Redox Regulation of p53   302
 VII. Nuclear Translocation of Thioredoxin and Gene Regulation   303
VIII. Conclusion   305
      References   305

14 Redox Regulation of p21, Role of Reactive Oxygen and Nitrogen Species in Cell Cycle Progression

Axel H. Schönthal, Sebastian Mueller, and Enrique Cadenas

   I. Introduction   312
  II. p21 Expression and Cell Cycle Control   313
 III. Induction of p21: p53-Dependent and -Independent Pathways   316
 IV. Role of Free Radicals in the Redox Regulation of p21   318
  V. Concluding Remarks   330
     References   331

III Clinical Implications of Redox Signaling and Antioxidant Therapy

15 Effects of Lipoxygenases on Gene Expression in Mammalian Cells

Helena Viita and Seppo Ylä-Herttuala

  I. Introduction   339
 II. Effects of Lipoxygenases on Gene Expression   343
III. Conclusions   352
    References   352

16 α-Tocopherol in the Pathogenesis of Atherosclerosis: One Molecule, Several Functions

Angelo Azzi

I. Vitamin E as an Antioxidant   360
II. Vitamin E Function at the Level of Protein Kinase C: Smooth Muscle Cell Proliferation, Monocyte Adhesion, Thrombocyte Aggregation and Adhesion, and Neutrophil Activation   363
III. Consumption of Vitamin E and the Development of Atherosclerotic Modifications of the Arterial Wall   365
IV. Other Nonantioxidant Properties or Effects of Vitamin E   374
References   374

17 Peroxiredoxins in Cell Signaling and HIV Infection

Dong-Yan Jin and Kaun-Teh Jeang

I. Peroxiredoxins: A Large Family of Antioxidant Enzymes   382
II. Peroxiredoxins: Structure and Function   391
III. Peroxiredoxins in Cell Signaling   394
IV. Peroxiredoxins in HIV Infection   399
V. Future Directions   400
References   401

18 Enhanced Activity of an Oxidation Product of Lycopene Found in Tomato Products and Human Serum Relevant to Cancer Prevention

John S. Bertram, Timothy King, Laurie Fukishima, and Frederick Khachik

I. Introduction   410
II. Materials and Methods   411
III. Results   413
IV. Discussion   419
References   421

19 Antioxidant Genes and Reactive Oxygen Species in Down's Syndrome

Cécile Bladier, Judy B. de Haan, and Ismail Kola

- I. Introduction 425
- II. Premature Aging 427
- III. Alzheimer's Disease 432
- IV. Other Neuropathologies 435
- V. Immunological Deficiency 436
- VI. Inflammation 437
- VII. Other Antioxidant Genes or Genes Regulated by an Altered Redox State on Chromosome 21 440
- VIII. Conclusion 441
    - References 443

20 Oxidant-Mediated Repression of Mitochondrial DNA Transcription

Bruce S. Kristal and Byung P. Yu

- I. Introduction 452
- II. Mitochondrial Genome 453
- III. mtDNA Expression and Replication 454
- IV. Mitochondria–Oxidant Interactions 457
- V. Oxidant-Mediated Repression of mtDNA Transcription 458
- VI. Summary 471
    - References 472

21 Suppression of Insulin Gene Promoter Activity by Oxidative Stress

Yoshitaka Kajimoto, Yoshimitsu Yamasaki, Taka-aki Matsuoka, Hideaki Kaneto, and Masatsugu Hori

- I. Introduction 479
- II. Chronic Hyperglycemia Induces Glycation Reaction in Pancreatic β Cells 482
- III. *Cis*- and *Trans*-Acting Factors Regulating Insulin Gene Transcription 482

    IV. Sensitivity of Insulin Gene Promoter to Glycation and Oxidative Stress   484
    V. Glycation Reduces Insulin mRNA Amount and Insulin Content   485
    VI. Glycation-Dependent Reduction of DNA-Binding Activity of PDX-1   486
    VII. Conclusions   488
        References   488

22 Redox Regulation of Ischemic Adaptation
Nilanjana Maulik and Dipak K. Das

    I. Introduction   492
    II. Redox Regulation of Ischemia–Reperfusion   493
    III. Redox Regulation of Ischemic Adaptation   496
    IV. Activation of Multiple Transcription Factors   503
    V. Programed Cell Death in Ischemia–Reperfusion   505
    VI. Ischemic Adaptation and Apoptosis   507
    VII. Summary and Conclusion   508
        References   511

23 Oxidative Stress as a Governing Factor in Physiological Aging
William C. Orr and Rajindar S. Sohal

    I. Introduction   517
    II. Reformulation of the Oxidative Stress Hypothesis of Aging   518
    III. Genetic Manipulations to Attenuate Oxidative Damage   519
    IV. Factors Affecting the Generation of Reactive Oxygen Species   523
    V. What Are the Targets of Oxidative Stress?   524
    VI. Concluding Remarks   527
        References   527

24 Antioxidants in Senescence and Wasting
Wulf Dröge, Volker Hack, Raoul Breitkreutz, Eggert Holm, Stefanie Holm, and Ralf Kinscherf

    I. Introduction: Loss of Skeletal Muscle Mass and Muscle Function as a Correlate of Senescence and Wasting   531
   II. Redox State as a Target for Therapeutic Intervention   532
  III. What Determines the Plasma Cystine Level?   539
  IV. Diseases and Conditions Associated with Relatively Low Plasma Cystine Levels (Low CG Syndrome)   545
   V. Plasma Glutamate and Intracellular Glutathione Levels   549
  VI. Summary and Conclusions   549
      References   551

Index   557

Contributors

Numbers in parentheses indicate the pages on which the authors' contributions begin.

Valérie Andrieu (265), INSERM U 479, CHU Xavier Bichat, 75870 Paris, France

Angelo Azzi (359), Institute of Biochemistry and Molecular Biology, University of Bern, 3012 Bern, Switzerland

Patrick A. Baeuerle (181), Micromet GmbH, D-82152 Martinsried, Germany

John S. Bertram (409), Cancer Research Center of Hawaii, University of Hawaii, Honolulu, Hawaii 96813

Cécile Bladier (425), Centre for Functional Genomics and Human Disease, Institute of Reproduction and Development, Monash Medical Centre, Monash University, Victoria 3168, Australia

Raoul Breitkreutz (531), Deutsches Krebsforschungszentrum, Division of Immunochemistry, D-69120, Heidelberg, Germany

Karlis Briviba (3), Institut für Physiologische Chemie I, Heinrich-Heine-Universität, D-40001, Dusseldorf, Germany

Anna K. Brzezinska (81), Department of Physiology, Medical College of Wisconsin, Milwaukee, Wisconsin 53226

Enrique Cadenas (311), Department of Molecular Pharmacology and Toxicology, School of Pharmacy, University of Southern California, Los Angeles, California 90089

Dana R. Crawford (21), Department of Biochemistry and Molecular Biology, Albany Medical College, Albany, New York 12208

Kelvin J. A. Davies (21), Ethel Percy Andrus Gerontology Center, University of Southern California, Los Angeles, California 90089

Dipak K. Das (491), University of Connecticut School of Medicine, Farmington, Connecticut 06032

Judy B. de Haan (425), Centre for Functional Genomics and Human Disease, Institute of Reproduction and Development, Monash Medical Centre, Monash University, Victoria 3168, Australia

Ami A. Deora (147), Department of Biochemistry, Cornell University Medical College, New York, New York 10021

Kurt De Vos (245), Flanders Interuniversity Institute for Biotechnology, University of Gent, B-9000 Gent, Belgium

Wulf Dröge (531), Deutsches Krebsforschungszentrum, Division of Immunochemistry, D-69120 Heidelberg, Germany

Stephen J. Elliott (81), Doctors Medical Center, Modesto, California 95350

Walter Fiers (245), Flanders Interuniversity Institute Biotechnology, University of Gent, B-9000 Gent, Belgium

Henry Jay Forman (105), Department of Environmental Health Sciences, School of Public Health, University of Alabama at Birmingham, Birmingham, Alabama 35294

Laurie Fukishima (409), Cancer Research Center of Hawaii, University of Hawaii, Honolulu, Hawaii 96813

Julio Girón-Calle (105), Department of Environmental Health Sciences, School of Public Health, University of Alabama at Birmingham, Birmingham, Alabama 35294

Pascal J. Goldschmidt-Clermont (47), Heart and Lung Institute, Ohio State University, Columbus, Ohio 43210

Vera Goossens (245), Flanders Interuniversity Institute for Biotechnology, University of Gent, B-9000 Gent, Belgium

Alexia Gozin (265), INSERM U 479, CHU Xavier Bichat, 75870 Paris, France

Johan Grooten (245), Flanders Interuniversity Institute for Biotechnology, University of Gent, B-9000 Gent, Belgium

Volker Hack (531), Deutsches Krebsforschungszentrum, Division of Immunochemistry, D-69120 Heidelberg, Germany

Hamdy H. Hassanain (47), Heart and Lung Institute, Ohio State University, Columbus, Ohio 43210

Eggert Holm (531), Medical Clinic I, 68167 Mannheim, Germany

Stefanie Holm (531), Medical Clinic I, 68167 Mannheim, Germany

Masatsugu Hori (479), Department of Internal Medicine and Therapeutics, Osaka University Graduate School of Medicine, Suita 565-0871, Japan

Kuan-Teh Jeang (381), Laboratory of Molecular Microbiology, National Institute of Allergy and Infectious Diseases, Bethesda, Maryland 20892

Dong-Yan Jin (381), Laboratory of Molecular Microbiology, National Institute of Allergy and Infectious Diseases, Bethesda, Maryland 20892

Yoshitaka Kajimoto (479), Department of Internal Medicine and Therapeutics, Osaka University Graduate School of Medicine, Suita 565-0871, Japan

Hideaki Kaneto (479), Department of Internal Medicine and Therapeutics, Osaka University Graduate School of Medicine, Suita 565-0871, Japan

Takumi Kawabe (203), Department of Molecular Genetics, Nagoya City University Medical School, Nagoya 467, Japan

Frederick Khachik (409), Department of Chemistry and Biochemistry, Joint Institute for Food Safety and Applied Nutrition, University of Maryland, College Park, Maryland 20742

Timothy King (409), Cancer Research Center of Hawaii, University of Hawaii, Honolulu, Hawaii 96813

Ralf Kinscherf (531), Deutsches Krebsforschungszentrum, Division of Immunochemistry, D-69120 Heidelberg, Germany

Lars-Oliver Klotz (3), Institut für Physiologische Chemie I, Heinrich-Heine-Universität, D-40001 Düsseldorf, Germany

Ismail Kola (425), Centre for Functional Genomics and Human Disease, Institute of Reproduction and Development, Monash Medical Center, Monash University, Victoria 3168, Australia

Suneil K. Koliwad (81), Department of Pediatrics, Baylor College of Medicine, Houston, Texas 77030

Bruce S. Kristal (451), Dementia Research Service, Burke Medical Research Institute, White Plains, New York 10605, and Department of Biochemistry, Cornell University Medical College, New York, New York 10021

Harry M. Lander (147), Department of Biochemistry, Cornell University Medical College, New York, New York 10021

Hiroshi Masutani (297), Institute for Virus Research, Kyoto University, Sakyo, Kyoto 606-8397, Japan

Taka-aki Matsuoka (479), Department of Internal Medicine and Therapeutics, Osaka University Graduate School of Medicine, Suita 565-0871, Japan

Nilanjana Maulik (491), University of Connecticut School of Medicine, Farmington, Connecticut 06032

Sebastian Mueller (311), Department of Internal Medicine IV, University of Heidelberg, 69115 Heidelberg, Germany, and Department of Molecular Pharmacology and Toxicology, School of Pharmacy, University of Southern California, Los Angeles, California 90089

Takashi Okamoto (203), Department of Molecular Genetics, Nagoya City University Medical School, Nagoya 467, Japan

William C. Orr (517), Department of Biological Sciences, Southern Methodist University, Dallas, Texas 75275

Catherine Pasquier (265), INSERM U 479, CHU Xavier Bichat, 75870 Paris, France

Sashwati Roy (265), Department of Molecular and Cell Biology, University of California, Berkeley, California 94720

Gary L. Schieven (129), Bristol-Myers Squibb Pharmaceutical Research Institute, Princeton, New Jersey 08453

Axel H. Schönthal (311), Department of Molecular Microbiology and Immunology, School of Medicine, University of Southern California, Los Angeles, California 90089

Chandan K. Sen (221, 265), Lawrence Berkeley National Laboratory, University of California, Berkeley, California 94720, and Department of Physiology, University of Kuopio, Kuopio, Finland

Helmut Sies (3), Institut für Physiologische Chemie I, Heinrich-Heine-Universität, Düsseldorf, D-40001, Postfach 101007, Germany

Rajindar S. Sohal (517), Department of Biological Sciences, Southern Methodist University, Dallas, Texas 75275

Margino Steemans (245), Flanders Interuniversity Institute for Biotechnology, University of Gent, B-9000 Gent, Belgium

Toshihide Suzuki (21), Department of Biochemistry and Molecular Biology, Albany Medical College, Albany, New York 12208

Toshifumi Tetsuka (203), Department of Molecular Genetics, Nagoya City University Medical School, Nagoya 467, Japan

Shugo Ueda (279), Institute for Virus Research, Kyoto University, Sakyo, Kyoto 606-8397, Japan

Masaya Ueno (297), Institute for Virus Research, Kyoto University, Sakyo, Kyoto 606-8397, Japan

Katia Vancompernolle (245), Flanders Interuniversity Institute for Biotechnology, University of Gent, B-9000 Gent, Belgium

Peter Vandenabeele (245), Flanders Interuniversity Institute for Biotechnology, University of Gent, B-9000 Gent, Belgium

Dominique Vercammen (245), Flanders Interuniversity Institute for Biotechnology, University of Gent, B-9000 Gent, Belgium

Helena Viita (339), A.I. Virtanen Institute, University of Kuopio, FIN-70211 Kuopio, Finland

Yoshimitsu Yamasaki (479), Department of Internal Medicine and Therapeutic, Osaka University Graduate School of Medicine, Suita 565-0871, Japan

Seppo Ylä-Herttuala (339), A.I. Virtanen Institute, University of Kuopio, FIN-70211 Kuopio, Finland

Junji Yodoi (297), Institute for Virus Research, Kyoto University, Sakyo, Kyoto 606-8397, Japan

Sinichi Yoshida (203), Department of Molecular Genetics, Nagoya City University Medical School, Nagoya 467, Japan

Byung P. Yu (451), Department of Physiology, University of Texas Health Science Center, San Antonio, Texas 78284

More than anyone else we know, Lester Packer has created and maintained contacts among the international community of scientists active in free radical research in an exponentially growing fashion. When his 70th birthday approached, we thought it would be appropriate to dedicate a book to him. Rather than collecting contributions from all of his friends and colleagues (which would have made a mega volume), we chose a topic that Les has entered in recent years and which has developed at a rapid pace: antioxidant and redox regulation of signal transduction. In the name of the contributing authors and of his many friends around the globe, we extend our heartfelt wishes to Les in dedicating this book to him.

tion, activation of transcription factors, regulation of cellular cycle, and apoptosis which are all dependent on changes in redox equilibrium. Their role in physiological and pathological mechanisms and in senescence are also well analyzed.

We thank the editors and contributors for having dedicated this book to Lester Packer. Through the high quality of his scientific work, his communicative dynamism, his talent as an organizer, no one has contributed more to the development of this field and made it a highlight of modern biology. It is with great pleasure that I join this collective homage to Lester.

Luc Montagnier
Department of Retrovirus
Institut Pasteur
Paris, France

Preface

Reactive oxygen species and reactive nitrogen species have been implicated in a wide variety of physiological as well as pathophysiological processes including immune defense, vasorelaxation, aging, and a multitude of disease processes. Early free radical research led to the hypothesis that the primary mechanism by which reactive oxygen species influence biological processes is oxidative stress, a state wherein oxidants overwhelm antioxidant defense mechanisms leading to elevation in the levels of oxidative damage markers. As a result, antioxidants were mostly studied for their ability to prevent or minimize oxidative stress. Work done during the past decade, particularly during the past five years or so, has illuminated yet another major biological function of reactive oxygen and nitrogen species: the regulation of cellular signal transduction processes. Certain forms of these reactive species appear to be able to specifically regulate the function of discrete molecules in the complex signaling network. Evidence to support this hypothesis is being published at a rapidly growing pace, and the study of oxidation–reduction or redox-dependent regulation of molecular processes has gained momentum. The regulation of gene expression by oxidants, antioxidants, and the cellular redox status has emerged as a novel subdiscipline that integrates basic and clinical research in medicine.

This volume examines the molecular basis of oxidant and antioxidant action. The twenty-four chapters have been collated into three sections addressing the role of reactive species as intracellular messengers, redox regulation of cellular responses, and clinical implications of redox signaling and antioxidant therapy. Leading experts have provided novel insight, making this treatise a focal point in an emerging field of interest.

Developing this volume has been a pleasant experience, primarily because of the enthusiasm shared among the authors and the editors. We hope that this volume will contribute to the further development of this important field of research.

Chandan K. Sen
Helmut Sies
Patrick A. Baeuerle

I Reactive Species as Intracellular Messengers

1 Signaling by Singlet Oxygen in Biological Systems

Lars-Oliver Klotz, Karlis Briviba, and Helmut Sies
Institut für Physiologische Chemie I
Heinrich-Heine-Universität
D-40001 Düsseldorf, Germany

Singlet oxygen can be produced photochemically as a result of the irradiation of endogenous or exogenously applied photosensitizers with visible or ultraviolet light or in dark reactions, e.g., by stimulated phagocytes during the so-called oxidative burst. There is increasing evidence that singlet oxygen, apart from inflicting damage to biomolecules, has pronounced effects on cellular signaling events leading to the induced expression of a variety of proteins. Novel observations are the activation of transcription factor AP-2 and cellular signaling cascades comprising the activation of c-Jun-N-terminal kinases (JNK), p38-MAPK, and the NF-κB system. This chapter attempts to delineate mechanisms of action of singlet oxygen leading to the activation of signaling processes.

I. SINGLET OXYGEN IN BIOLOGICAL SYSTEMS

Singlet oxygen ($^1O_2$) is an electronically excited form of molecular ground state triplet oxygen. The energy needed to transfer molecular oxygen from its triplet ground state to the physiologically relevant first singlet state $^1\Delta_g$ ($^1\Delta_g$ here is implied when the terms "singlet oxygen" or $^1O_2$ are used) is 94.3 kJ/mol and may be generated chemically or provided from type II photochemical reactions.

There are two major sources of $^1O_2$ in biological systems: the inflammatory process and photosensitization in light-exposed areas. In fact, singlet oxygen has been shown to be generated by stimulated neutrophils employing a chemical trap, 9,10-diphenylanthracene (Steinbeck et al., 1992). $^1O_2$ seems to be produced in a series of reactions involving myeloperoxidase (MPO) (reactions 1–4). From experiments with stimulated macrophages that express NADPH oxidase but lack myeloperoxidase, Steinbeck et al. (1993) deduced that additional singlet oxygen may be derived from the spontaneous rather than the enzymatic dismutation of superoxide formed in the NADPH oxidase reaction [reactions (1) and (2a)].

$$2\ O_2 + NADPH \xrightarrow{\text{NADPH oxidase}} 2\ O_2^{\cdot-} + NADP^+ + H^+ \quad (1)$$

$$2\ O_2^{\cdot-} + 2\ H^+ \xrightarrow{\text{spontaneous}} H_2O_2 + {}^1O_2 \quad (2a)$$

$$2\ O_2^{\cdot-} + 2\ H^+ \xrightarrow{\text{SOD}} H_2O_2 + O_2 \quad (2b)$$

$$H_2O_2 + Cl^- \xrightarrow{\text{MPO}} H_2O + OCl^- \quad (3)$$

$$H_2O_2 + OCl^- \rightarrow {}^1O_2 + H_2O + Cl^- \quad (4)$$

A series of probably biologically relevant chemical reactions forming singlet oxygen has been worked out, such as the disproportionation of hydrogen peroxide reacting with peroxynitrite (Di Mascio et al., 1994) or hypohalites [reaction (4); for review, see Kanofsky, 1989] or the reaction of hydroperoxides with peroxynitrite (Di Mascio et al., 1997). Singlet oxygen may be formed from triplet-excited carbonyl species generated in the dark by chemiexcitation (Cilento, 1982; Schulte-Herbrüggen and Sies, 1989; Briviba et al., 1996). Singlet oxygen is also produced in type II (energy transfer) photodynamic reactions with endogenous or exogenous sensitizers (Fig. 1). UVA (320–400 nm) irradiation of endogenous sensitizers such as porphyrins, flavins, and quinones serves to generate singlet oxygen, a mediator of UVA-induced effects (Tyrrell, 1991; Herrmann et al., 1996). In photodynamic therapy (PDT) sensitizers such as phthalocyanines, porphyrins, or metabolic precursors of porphyrins, e.g., δ-aminolevulinic acid (Bonnett, 1995; Fritsch et al., 1997), are applied topically or systemically followed by irradiaton with light. Topical irradiation of (tumor) tissue preexposed to a sensitizer then leads to necrosis via the photodynamic production of oxidative species, predominantly singlet oxygen (Bonnett, 1995).

Singlet oxygen can damage biomolecules and thus exert toxicity (for reviews, see Sies, 1986; Piette, 1991; Sies and Menck, 1992). Damage to DNA consists primarily of the oxidation of guanine residues, mainly to 7-

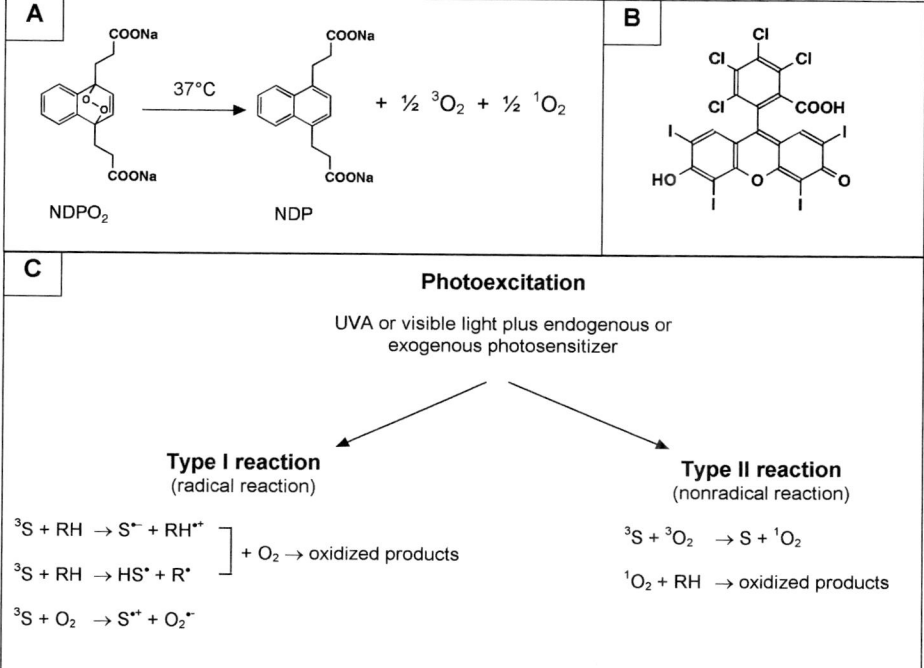

Figure 1 (A) Decomposition of $NDPO_2$ to generate singlet oxygen. The half-life of $NDPO_2$ at 37°C is 23 min (Di Mascio and Sies, 1989; Pierlot *et al.*, 1996). (B) Structure of Rose Bengal used as a photosensitizer to generate singlet oxygen on irradiation with white light. (C) Scheme outlining type I (electron transfer) and type II (energy transfer) photooxygenation reactions on irradiation of a photosensitizer. The sensitizer is excited from its singlet ground state to a first excited singlet state, which may undergo intersystem crossing to the first excited triplet state (^3S). In type I reactions (electron transfer), free radicals are involved. In type II reactions, triplet oxygen is converted to $^1O_2$, which reacts with substrate (see Gollnick, 1968; Foote and Clennan, 1995). Singlet oxygen can also be generated in dark reactions by chemiexcitation (Cilento, 1982).

hydro-8-oxodeoxyguanosine moieties (Piette, 1991). Upon replication, this leads to G:C to T:A transversions and is part of the mutagenic action of singlet oxygen (Decuyper-Debergh *et al.*, 1987; Kouchakdijan *et al.*, 1991; Piette, 1991; Epe *et al.*, 1996; Jeong *et al.*, 1998). It has been hypothesized that singlet oxygen produced endogenously in *Escherichia coli* stationary phase cells is responsible for DNA damage and the high mutation rates found under these conditions (Bridges and Timms, 1998). 7-Hydro-8-oxo-deoxyguanosine is also formed in mitochondrial DNA on treatment of human lung fibroblasts with methylene blue plus light. Interestingly, even in mitochondria, these lesions are repaired efficiently (Anson et al., 1998).

II. EFFECTS OF SINGLET OXYGEN ON GENE EXPRESSION

Singlet oxygen has been shown to mimic and mediate effects of UVA on gene expression. Induction of expression of heme oxygenase-1 (Basu-Modak and Tyrrell, 1993), interstitial collagenase (matrix metalloproteinase-1, MMP-1) (Scharffetter-Kochanek et al., 1993; Wlaschek et al., 1995), interleukin (IL)-1α/β, IL-6 (Wlaschek et al., 1997), intercellular adhesion molecule-1 (ICAM-1) (Grether-Beck et al., 1996), and Fas ligand (CD95-L; Morita et al., 1997) were reported.

Heme oxygenase-1 (HO-1) induction was the first system examined for testing whether singlet oxygen plays a key role in the induction of a protein in response of cells to UVA irradiation (Basu-Modak and Tyrrell, 1993). HO-1 is a microsomal enzyme catalyzing the rate-limiting step in heme catabolism, the oxygen and NADPH consuming breakdown of heme to form carbon monoxide, ferrous iron, and biliverdin, which, in turn, is reduced to bilirubin by biliverdin reductase. Its induction was proposed to be a beneficial adaptive response to oxidative stress (Keyse and Tyrrell, 1989; Applegate et al., 1991) as both biliverdin and bilirubin exhibit antioxidant properties (Stocker et al., 1987). The overall singlet oxygen quenching constant (chemical plus physical quenching) for biliverdin and bilirubin was determined as 2.3 and $3.2 \times 10^9 M^{-1} sec^{-1}$, respectively (Di Mascio et al., 1989). Moreover, ferritin synthesis is induced following UVA irradiation, depending on the induction of HO-1, most probably as a response to the iron released in the reaction catalyzed by HO-1 (Vile and Tyrrell, 1993).

With regard to signaling processes leading to HO-1 induction, the promoter region of the human HO-1 gene contains functional binding sites for transcription factors AP-1, AP-2, and NF-κB (Lavrovsky et al., 1994), all of which are known to be activated in response to oxidative stress. Transcription factor AP-1 is activated by the UVA irradiation of human skin cells (Djavaheri-Mergny et al., 1996). This may occur via the activation of upstream kinases (for review, see Karin and Hunter, 1995), such as the c-Jun-N-terminal kinases and p38-MAP kinases. Both were shown to be activated by UVA and $^1O_2$ generated photochemically by Rose Bengal plus light (Klotz et al., 1997, 1999). Further, AP-2 is responsive to UVA and to $^1O_2$ released from $NDPO_2$ (see Fig. 1) as was shown using ICAM-1 promoter constructs (Grether-Beck et al., 1996, 1997).

NF-κB has been shown to be induced on treatment with singlet oxygen generated photochemically by Rose Bengal or methylene blue plus light (Piret et al., 1995) or by UVA (Vile et al., 1995).

It is interesting to note that singlet oxygen generated outside the cells by NDPO$_2$ was not able to induce NF-κB (Schreck et al., 1992) or JNKs and p38 (Klotz et al., 1999).

In murine macrophages, HO-1 is inducible by hydrogen peroxide or lipopolysaccharide (LPS) each via AP-1 (Alam et al., 1995; Camhi et al., 1995, 1998); also, arsenite induces HO-1 in an AP-1-dependent fashion in a chicken hepatoma cell line. In the latter case, it has been shown that ERK and p38-MAPK activation by arsenite mediates this effect (Elbirt et al., 1998). There is, however, no explicit report as to which specific transcription factors are responsible for HO-1 induction in response to singlet oxygen (Ryter and Tyrrell, 1998).

The increased expression of MMPs by UVA has been proposed to contribute to photocarcinogenesis and photoaging; by degrading connective tissue components, MMPs would facilitate tumor cell invasion and impair proper organization and stability of connective tissue (for review, see Scharffetter-Kochanek et al., 1997). Collagenase (MMP-1) induction by singlet oxygen and UVA seems to be an indirect effect mediated by interleukins IL-1 α/β and IL-6. Singlet oxygen generated inside the cells by Rose Bengal plus light or outside by Rose Bengal–agarose plus light or by thermodecomposition of NDPO$_2$ induced expression of IL-1α/β and IL-6. Singlet oxygen is made responsible for a leakage of preformed cytosolic IL-1, which then binds to its receptor, thereby stimulating gene transcription of collagenase, its own gene, and of IL-6 (Wlaschek et al., 1997). IL-6, in turn, once synthesized and secreted, binds to the IL-6 receptor and stimulates collagenase synthesis (Wlaschek et al., 1994). IL-1 is known to exert its effects on gene expression via the activation of JNKs, eventually leading to an activation of transcription factor AP-1 (Muegge et al., 1993; Uciechowski et al., 1996) and by activating the NF-κB system (Stylianou et al., 1992). The promoter region of the MMP-1 gene contains functional binding sites for AP-1 (Angel et al., 1987).

Transcription factor AP-2 but neither AP-1 nor NF-κB has been proven to be responsible for singlet oxygen-mediated induction of ICAM-1 in human keratinocytes employing gel electrophoretic mobility-shift and reporter gene assays (Grether-Beck et al., 1996, 1997). The effects mentioned here are summarized in Fig. 2.

III. COMPARISON OF EFFECTS OF SINGLET OXYGEN AND OTHER OXIDANTS ON THE AP-1 PATHWAY

A variety of oxidative conditions are known to lead to the induction of *jun* and/or *fos* genes and to activate the DNA-binding activity and/or transactivat-

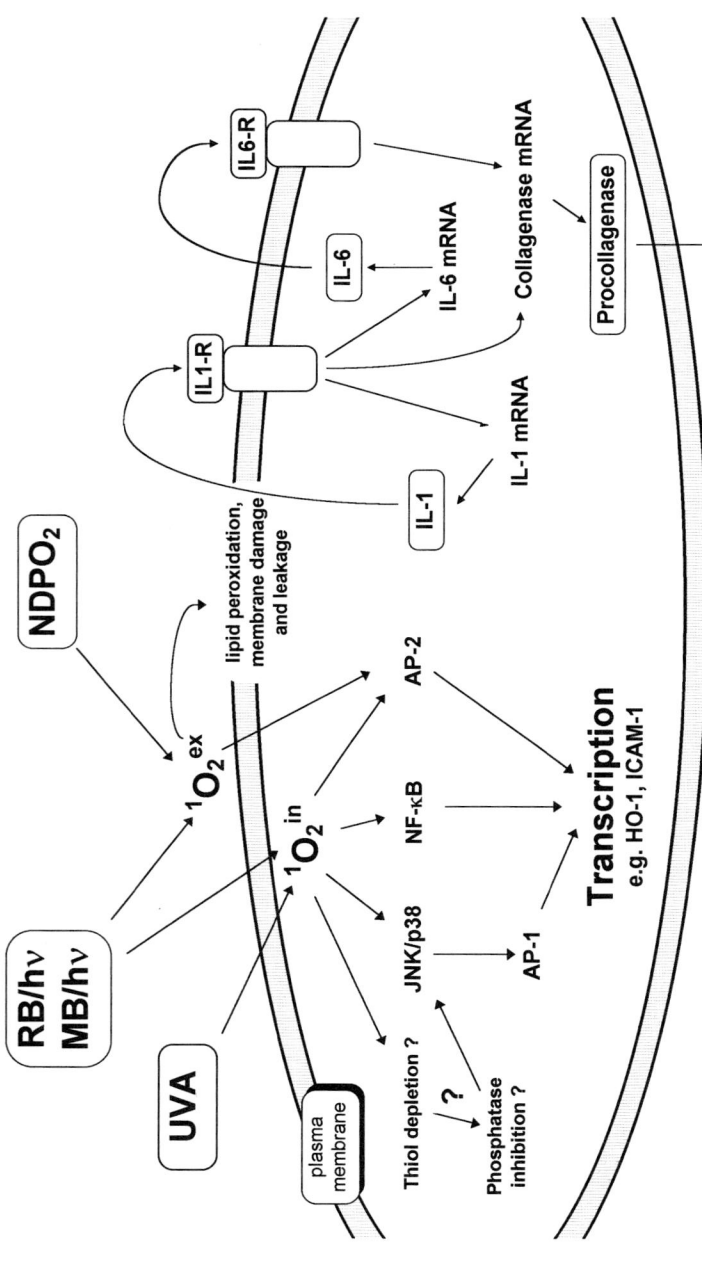

Figure 2 Scheme outlining effects on gene expression of singlet oxygen generated by UVA (320–400 nm), Rose Bengal plus light (RB/hν), methylene blue plus light (MB/hν), or NDPO$_2$. The scheme is based largely on the work of Basu-Modak and Tyrrell (1993), Scharffetter-Kochanek et al. (1993), Wlaschek et al. (1994, 1995, 1997), Piret et al. (1995), Vile et al. (1995), Grether-Back et al. (1996), and Klotz et al. (1997, 1999), IL1-R, interleukin-1 receptor; IL6-R, interleukin-6 receptor. See text.

ing capability of AP-1 transcription factors; this holds true for ultraviolet radiation in the UVC (Radler-Pohl *et al.*, 1993; Sachsenmaier *et al.*, 1994), UVB (Fisher *et al.*, 1996; Brenneisen *et al.*, 1998), and UVA (Djavaheri-Mergny *et al.*, 1996) regions, for treatment with hydrogen peroxide (Ishikawa *et al.*, 1997; Janssen *et al.*, 1997), for stressful stimuli such as hypoxia (Rupec and Baeuerle, 1995; Müller *et al.*, 1997), and for treatment with reactive nitrogen species such as nitric oxide (Janssen *et al.*, 1997) and peroxynitrite (Müller *et al.*, 1997). Hypoxia induces AP-1 but not NF-κB transactivation capacity in HeLa cells, with the latter but not the former being induced on reoxygenation (Rupec and Baeuerle, 1995). Pathways leading to an AP-1-dependent transcriptional activation do not necessarily have to be the same in all these cases. As AP-1 may be induced via the tracking of c-fos (e.g., ERK/p38 - Elk-1 - c-fos; or ERK/p38 - MAPKAP-K1/2-CREB -c-fos) or c-jun (e.g., JNK - c-Jun - c-jun; or JNK/p38 - ATF-2 - c-jun), more than one pattern of activation of upstream transcription factors or kinases can lead to the same effect, i.e., AP-1-dependent transcription. The ERK-, JNK-, and p38-MAP kinase activation patterns induced by a given reactive species may thus differ, but this does not necessarily mean the overall effects on gene expression are distinguishable. Table I presents the effects of some reactive oxygen or nitrogen species or treatments able to induce oxidative stress on MAP kinase activations; it shows that there are differences not only between different oxidative treatments but also between different sets of experiments with the same treatments using the same systems (e.g., the hypoxia/reperfusion experiments). What can also be seen from Table I is that there are both similarities between certain treatments and also clear differences, e.g., the ERK/JNK/p38 activation patterns of UVA and singlet oxygen derived from Rose Bengal plus light are similar, suggesting that the UVA and $^1O_2$ pathways coincide. It has, in fact, been shown that the UVA effect on JNK and p38 is mediated by $^1O_2$ (Klotz *et al.*, 1997, 1999). The same pattern of MAPK activation has also been found in human keratinocytes subjected to 5-aminolevulinate photodynamic treatment (Klotz *et al.*, 1998) and in murine keratinocytes treated with the benzoporphyrin derivative plus light (Tao *et al.*, 1996). Both treatments supposedly rely on the generation of $^1O_2$. However, the patterns induced by UVC/UVB and UVA are different, suggesting that there are probably different primary targets and different species mediating the irradiation effects, which is reasonable to propose, regarding the different absorption of the respective UV regions by biomolecules and tissues.

IV. INDUCTION OF TRANSCRIPTION FACTORS BY SINGLET OXYGEN: MECHANISM

Singlet oxygen generated photochemically by either UVA irradiation of cells or using photosensitizers that are membrane permeant (e.g., Rose Ben-

Table I Effect of Various Reactive Oxygen (ROS) or Nitrogen (RNS) Species or Oxidative Treatments on MAP Kinases[a]

Treatment	ERK	JNK	p38	Comment	Cell type/organ	Ref.
ROS						
Hydrogen peroxide (H_2O_2)	+	−	n.d.		NIH 3T3; perfused rat heart	Kyriakis et al. (1994); Knight and Buxton (1996)
	−	n.d.	n.d.	No activation of ERK, as opposed to superoxide in the same system	Rat aortic smooth muscle cells	Baas and Berk (1995)
	+	(+)	(+)	ERK2 was also strongly activated in HeLa, PC12, Rat-1, and primary aortic smooth muscle cells	NIH 3T3	Guyton et al. (1996)
	+	+	−	Note differential effect on JNK and p38 activity	Rat astrocytes; neonatal rat ventricular myocytes	Tournier et al. (1997); Clerk and Sugden (1997)
	(+)	+	+		Perfused rat heart	Sugden et al. (1997)
	+	+	+		HeLa cells	Wang et al. (1998)
Singlet oxygen ($^1O_2$)						
Rose bengal + light	−	+	+	Lipophilic Rose Bengal vs hydrophilic $NDPO_2$	Human skin fibroblasts	Klotz et al. (1997, 1999); Briviba et al. (1997)
$NDPO_2$	−	−	−			
Superoxide (O_2^-)	+	n.d.	n.d.	Superoxide was produced by incubating cells with the naphthoquinolinedione LY83583	Rat aortic smooth muscle cells	Baas and Berk (1995)

Agent				Cell type	Comments	References
RNS						
NO/NOx						
NO-saturated PBS	(+)	+	+	T cells (Jurkat)		Lander et al. (1996)
endogenous NO	+	n.d.	n.d.	Endothelial cells (human umbilical vein)	Pretreatment of cells overnight with LPS/IFN-γ to induce iNOS followed by a 15-min pulse of arginine	Lander et al. (1996)
SNP	+	+	+	T cells (Jurkat)	Activity peaks within 10 min of treatment	Lander et al. (1996)
SNAP	–	n.d.	n.d.	T cells (Jurkat)	ERKs are activated after pretreatment of cells with BSO	Lander et al. (1996)
	n.d.	+	n.d.	Primary bovine articular chondrocytes		Lo et al. (1996)
Peroxynitrite	+	+	+	WB-F344 rat liver epithelial cells		Schieke et al. (1999)
UV irradiation						
UVA (320–400 nm)	–	+	+	Human skin fibroblasts	Compare with singlet oxygen effects	Klotz et al. (1997, 1999)
UVB (280–320 nm)	(+)	(+)/–	n.d.	Mouse fibroblasts (3T3-4A); mouse keratinocytes (primary and C50 cells)	The (weak) activation of JNK after irradiation with a 270- to 330-nm lamp was abolished when wavelengths <300 nm were filtered out (Adler et al., 1995)	Adler et al. (1995); Dhanwada et al. (1995)
	(+)	+	n.d.	Human keratinocytes (primary and HaCaT, HeLa, HepG2, PC 12, A431, NIH 3T3	Good activator of JNK activity only weakly inducing ERK activation	Rosette and Karin (1996); Assefa et al. (1997); Brenneisen et al. (1998)

(*continues*)

Table I (*Continued*)

Treatment	ERK	JNK	p38	Comment	Cell type/organ	Ref.
UVC (< 280 nm)	+	+	+	"JNK inducibility not impaired by enucleation of cells" (Devary et al., 1993) vs "no more JNK inducibility after enucleation of cells" (Adler et al., 1995)	HeLa; mouse keratinocytes (primary and C50 cells); mouse fibroblasts (3T3-4A); COS1; CHO; Rat-1; many others	Sachsenmaier et al. (1994); Dérijard et al. (1994); Adler et al. (1995); Dhanwada et al. (1995); Raingeaud et al. (1995); Liu et al. (1996)
Others						
Hypoxia/reperfusion						
Hypoxia	−	−	n.d.		Rat kidney	Pombo et al. (1994)
Hypoxia/reperfusion	(+)	+	n.d.			
Hypoxia/reperfusion	+	+	n.d.		Perfused rat heart	Knight and Buxton (1996)
Hypoxia	+	−	n.d.		Human cervical carcinoma cells (HeLa)	Müller et al. (1997)
Hypoxia	−	−	n.d.		Perfused rat heart	Sugden et al. (1997)
Hypoxia/reperfusion	−	+	+			
Photodynamic therapy						
BPD-PDT	−	+	+		Murine keratinocytes (Pm212)	Tao et al. (1996)
ALA-PDT	−	+	+		Keratinocytes (HaCaT)	Klotz et al. (1998)
	n.d.	n.d.	+		Melanoma cells (Bro; SkMel-23)	Klotz et al. (1998)
CCl$_4$-induced metabolic oxidative stress	n.d.	+	↓	Activation of JNK and inactivation (!) of p38	Mouse liver	Mendelson et al. (1996)
4-Hydroxy-2-nonenal	(+)/−	+	+		RL 34 rat liver epithelial cells	Uchida et al. (1999)

[a] Abbreviations used: ALA, 5-aminolevulinic acid; BPD, benzoporphyrin derivative; BSO, buthionine sulfoximine; IFN, interferon; LPS, lipopolysaccharide; NMA, N$^\omega$-methyl-L-arginine; NOx, nitric oxide-related species; PDT, photodynamic treatment/therapy; SNAP, S-nitroso-N-acetylpenicillamine; SNP, sodium nitroprusside; +, activation by treatment; (+), weak activation (≤three-fold); −, no effect of treatment; n.d., not done.

gal) is known to activate signaling molecules such as JNK- and p38-MAPK (Klotz et al., 1997, 1999) or NF-κB (Piret et al., 1995; Vile et al., 1995). JNK and p38 are also activated in human skin fibroblasts by incubation with an intracellular generator of singlet oxygen, N,N'-di(2,3-dihydroxypropyl)-1,4-naphthalene dipropionamide endoperoxide (DHPNO$_2$; Dewilde et al., 1998). Neither activation of JNK and p38 nor of NF-κB is, however, achieved employing the hydrophilic NDPO$_2$ (Fig. 1) or Rose Bengal immobilized on agarose and kept extracellular as singlet oxygen source (Klotz et al., 1999).

Thus, it is reasonable to suggest that the activation of JNK/p38 and proposedly AP-1 as well as of the NF-κB pathway by singlet oxygen relies on signal-integrating mechanisms in the cytosol, as ^{1}O$_2$ generated outside cells by NDPO$_2$ or Rose Bengal–agarose is not sufficient for activation.

Cavigelli et al. (1996) and Knebel et al. (1996) have introduced a new aspect into the discussion as to which could be the cellular integrating point that leads to the activation of signaling pathways. Knebel et al. (1996) show that receptor tyrosine kinases (RTKs) are activated (phosphorylated) by inhibition of a phosphatase activity on treatment with UV (A, B, or C), hydrogen peroxide, or iodoacetamide. Phosphatase activity, i.e., dephosphorylation and inactivation of RTKs, was restored on the addition of thiol-regenerating agents, if not inhibited irreversibly by iodoacetamide. Cavigelli and co-workers (1996) find that the activation of JNK by sodium arsenite, potently inducing AP-1, which is reactive toward thiols (especially vicinal dithiols), is by inactivating a JNK phosphatase. These findings complement well results by Liu et al. (1996) that the activation of ERKs, JNKs, and p38 by arsenite in Rat-1 cells is abolished by the pretreatment of cells with N-acetylcysteine. The activation of JNK1 and p38 by arsenite is independent of Ras and is not inhibitable by the receptor poison suramin, quite different from ERK activation, which seems to depend on Ras and is inhibitable by suramin. Like arsenite, UVC activates all three types of MAP kinase; however, suramin (0.3 mM for 30 min prior to irradiation) inhibits the activation of all three MAP kinases to certain extents (that of ERKs is completely abolished). These data suggest a possible activation of MAP kinases independent of membraneous perturbations such as oxidant-induced receptor phosphorylation and clustering that were shown to be evoked by UVC, UVB, UVA, and hydrogen peroxide (Sachsenmaier et al., 1994; Gamou and Shimizu, 1995; Knebel et al., 1996; Rosette and Karin, 1996). H$_2$O$_2$, for instance, has been pointed out not only to inactivate membrane-bound phosphatases (Knebel et al., 1996) but also to diminish cytosolic general protein tyrosine phosphatase activity in mouse fibroblasts (Sullivan et al., 1994).

JNK activation by UVA in human skin fibroblasts is mimicked by simply incubating the cells with the tyrosine phosphatase inhibitor orthovanadate;

no further activation is found with UVA, pointing to the inactivation of a phosphatase by UVA treatment eventually leading to JNK activation (Klotz, Briviba, Sies, unpublished work). In fact, all protein tyrosine phosphatases known so far rely on the presence of a cysteine in their active site, which serves as a nucleophile accepting the phosphate moiety of the phosphatase substrate forming an intermediate phosphocysteine (Fauman and Saper, 1996). An oxidation of this cysteine residue, e.g., by singlet oxygen formed during UVA irradiation, would lead to the inactivation of the phosphatase concomitantly, allowing kinase activities to become predominant. $^1O_2$ efficiently depletes protein thiols; it does so even more efficiently when the source of singlet oxygen is hydrophobic and exhibits high protein affinity (Wagner *et al.*, 1993).

One of the signal integration points may be the depletion of cellular thiols leading to the inactivation or impaired reactivation of crucial enzymes resulting from a diminished cellular reduction capacity.

In summary, singlet oxygen is not only responsible for the oxidative damage of biomolecules, but also leads to the activation of signal transduction pathways. Both processes may also be regarded as connected: the cellular signaling effects may be a result of oxidative modifications of key target enzymes. Such a system may then be regarded as a feed-forward defense mechanism: Oxidative stress leads to the damage of biomolecules and, by way of some of these modifications, to the induction of a stress response.

ACKNOWLEDGMENTS

This study was supported by the Deutsche Forschungsgemeinschaft, SFB 503, Project B1, and the National Foundation for Cancer Research, Bethesda.

REFERENCES

Adler, V., Fuchs, S. Y., Kim, J., Kraft, A., King, M. P., Pelling, J., and Ronai, Z. (1995). Jun-NH₂-terminal kinase activation mediated by UV-induced DNA lesions in melanoma and fibroblast cells. *Cell Growth Differ.* **6**, 1437–1446.

Alam, J., Camhi, S., and Choi, A. M. (1995). Identification of a second region upstream of the mouse heme oxygenase-1 gene that functions as a basal level and inducer-dependent transcription enhancer. *J. Biol. Chem.* **270**, 11977–11984.

Angel, P., Imagawa, M., Chiu, R., Stein, B., Imbra, R.J., Rahmsdorf, H.J., Jonat, C., Herrlich, P., and Karin, M. (1987). Phorbol ester-inducible genes contain a common cis element recognized by a TPA-modulated trans-acting factor. *Cell (Cambridge, Mass.)* **49**, 729–739.

Anson, R. M., Croteau, D. L., Stierum, R. H., Filburn, C., Parsell, R., and Bohr, V. A. (1998). Homogenous repair of singlet oxygen-induced DNA damage in differentially transcribed regions and strands of human mitochondrial DNA. *Nucleic Acids Res.* **26**, 662–668.

Applegate, L. A., Lüscher, P., and Tyrrell, R. M. (1991). Induction of heme oxygenase: A general response to oxidant stress in cultured mammalian cells. *Cancer Res.* **51,** 974–978.

Assefa, Z., Garmyn, M., Bouillon, R., Merlevede, W., Vandenheede, J. R., and Agostinis, P. (1997). Differential stimulation of ERK and JNK activities by ultraviolet B irradiation and epidermal growth factor in human keratinocytes. *J. Invest. Dermatol.* **108,** 886–891.

Baas, A. S., and Berk, B. C. (1995). Differential activation of mitogen-activated protein kinases by H_2O_2 and O_2^- in vascular smooth muscle cells. *Circ. Res.* **77,** 29–36.

Basu-Modak, S., and Tyrrell, R. M. (1993). Singlet oxygen: A primary effector in the ultraviolet A/near-visible light induction of the human heme oxygenase gene. *Cancer Res.* **53,** 4505–4510.

Bonnett, R. (1995). Photosensitizers of the porphyrin and phthalocyanine series for photodynamic therapy. *Chem. Soc. Rev.* **21,** 19–33.

Brenneisen, P., Wenk, J., Klotz, L. O., Wlaschek, M., Briviba, K., Krieg, T., Sies, H., and Scharffetter-Kochanek, K. (1998). Central role of ferrous/ferric iron in the UVB-mediated signalling pathway leading to increased interstitial collagenase (MMP-1) and stromelysin-1 (MMP-3) mRNA levels in cultured human dermal fibroblasts. *J. Biol. Chem.* **273,** 5279–5287.

Bridges, B. A., and Timms, A. (1998). Effect of endogenous carotenoids and defective RpoS sigma factor on spontaneous mutation under starvation conditions in *Escherichia coli*: Evidence for the possible involvement of singlet oxygen. *Mutat. Res.* **403,** 21–28.

Briviba, K., Saha-Möller, C. R., Adam, W., and Sies, H. (1996). Formation of singlet oxygen in the thermal decomposition of 3-hydroxymethyl-3,4,4-trimethyl-1,2-dioxetane, a chemical source of triplet-excited ketones. *Biochem. Mol. Biol. Int.* **38,** 647–651.

Briviba, K., Klotz, L. O., and Sies, H. (1997). Toxic and signaling effects of chemically or photochemically generated singlet oxygen in biological systems. *Biol. Chem.* **378,** 1259–1265.

Camhi, S. L., Alam, J., Otterbein, L., Sylvester, S. L., and Choi, A. M. (1995). Induction of heme oxygenase-1 gene expression by lipopolysaccharide is mediated by AP-1 activation. *Am. J. Respir. Cell Mol. Biol.* **13,** 387–398.

Camhi, S. L., Alam, J., Wiegand, G. W., Chin, B. Y., and Choi, A. M. K. (1998). Transcriptional activation of the HO-1 gene by lipopolysaccharide is mediated by 5′distal enhancers: Role of reactive oxygen intermediates and AP-1. *Am. J. Respir. Cell Mol. Biol.* **18,** 226–234.

Cavigelli, M., Li, W. W., Lin, A., Su, B., Yoshioka, K., and Karin, M. (1996). The tumor promoter arsenite stimulates AP-1 activity by inhibiting a JNK phosphatase. *EMBO J.* **15,** 6269–6279.

Cilento, G. (1982). Electronic excitation in dark biological processes. *In* "Chemical and Biological Generation of Excited States" (W. Adam and G. Cilento, eds.), pp. 279–309. Academic Press, New York.

Clerk, A., and Sugden, P. H. (1997). Mitogen-activated protein kinases are activated by oxidative stress and cytokines in neonatal rat ventricular myocytes. *Biochem. Soc. Transac.* **25,** S566.

Decuyper-Debergh, D., Piette, J., and van de Vorst, A. (1987). Singlet oxygen-induced mutations in M13 lacZ phage DNA. *EMBO J.* **6,** 3155–3161.

Dérijard, B., Hibi, M., Wu, I., Barrett, T., Su, B., Deng, T., Karin, M., and Davis, R. J. (1994). JNK1: A protein kinase stimulated by UV light and ha-ras that binds and phosphorylates the c-Jun activation domain. *Cell (Cambridge, Mass.)* **76,** 1025–1037.

Devary, Y., Rosette, C., DiDonato, J. A., and Karin, M. (1993). NF-κB activation by ultraviolet light is not dependent on a nuclear signal. *Science* **261,** 1442–1445.

Dewilde, A., Pellieux, C., Pierlot, C., Wattre, P., and Aubry, J. M. (1998). Inactivation of intracellular and non-enveloped viruses by a non-ionic naphthalene. *Biol. Chem.* **379,** 1377–1379.

Dhanwada, K. R., Dickens, M., Neades, R., Davis, R. J., and Pelling, J. C. (1995). Differential effects of UV-B and UV-C components of solar radiation on MAP kinase signal transduction pathways in epidermal keratinocytes. *Oncogene* **11**, 1947–1953.

Di Mascio, P., and Sies, H. (1989). Quantification of singlet oxygen generated by thermolysis of 3,3′-(1,4-naphthylidene)dipropionate. Monomol and dimol photoemission and the effects of 1,4-diazabicyclo[2.2.2]octane. *J. Am. Chem. Soc.* **111**, 2909–2914.

Di Mascio, P., Kaiser, S., and Sies, H. (1989). Lycopene as the most efficient biological carotenoid singlet oxygen quencher. *Arch. Biochem. Biophys.* **274**, 532–538.

Di Mascio, P., Bechara, E. J. H., Medeiros, M. H. G., Briviba, K., and Sies, H. (1994). Singlet molecular oxygen production in the reaction of peroxynitrite with hydrogen peroxide. *FEBS Lett.* **355**, 287–289.

Di Mascio, P., Briviba, K., Sasaki, S. T., Catalani, L. H., Medeiros, M. H. G., Bechara, E. J. H., and Sies, H. (1997). The reaction of peroxynitrite with test-butyl hydroperoxide produces singlet molecular oxygen. *Biol. Chem.* **378**, 1071–1074.

Djavaheri-Mergny, M., Mergny, J. L., Bertrand, F., Santus, R., Mazière, C., Dubertret, L., and Mazière, J. C. (1996). Ultraviolet-A induces activation of AP-1 in cultured human keratinocytes. *FEBS Lett.* **384**, 92–96.

Elbirt, K. K., Whitmarsh, A. J., Davis, R. J., and Bonkovsky, H. L. (1998). Mechanism of sodium arsenite-mediated induction of heme oxygenase-1 in hepatoma cells. Role of mitogen-activated protein kinases. *J. Biol. Chem.* **273**, 8922–8931.

Epe, B., Ballmaier, D., Roussyn, I., Briviba, K., and Sies, H. (1996). DNA damage by peroxynitrite characterized with DNA repair enzymes. *Nucleic Acids Res.* **24**, 4105–4110.

Fauman, E. B., and Saper, M. A. (1996). Structure and function of the protein tyrosine phosphatases. *Trends Biochem. Sci.* **21**, 413–417.

Fisher, G. J., Datta, S. C., Talwar, H. S., Wang, Z. Q., Varani, J., Kang, S., and Voorhees, J. J. (1996). Molecular basis of sun-induced premature skin ageing and retinoid antagonism. *Nature (London)* **379**, 335–339.

Foote, C. S., and Clennan, E. L. (1995). Properties and reactions of singlet oxygen. *In* "Active Oxygen in Chemistry" (C. S. Foote, J. S. Valentine, A. Greenberg, and J. F. Liebman, eds.), pp. 105–141. Blackie Academic and Professional, London.

Fritsch, C., Abels, C, Goetz, A. E., Stahl, W., Bolsen, K., Ruzicka, T., Goerz, G., and Sies, H. (1997). Porphyrins preferentially accumulate in a melanoma following intravenous injection of 5-aminolevulinic acid. *Biol. Chem.* **378**, 51–57.

Gamou, S., and Shimizu, N. (1995). Hydrogen peroxide preferentially enhances the tyrosine phosphorylation of epidermal growth factor receptor. *FEBS Lett.* **357**, 161–164.

Gollnick, K. (1968). Type II photooxygenation reactions in solution. *Adv. Photochem.* **6**, 1–122.

Grether-Beck, S., Olaizola-Horn, S., Schmitt, H., Grewe, M., Jahnke, A., Johnson, J. P., Briviba, K., Sies, H., and Krutmann, J. (1996). Activation of transcription factor AP-2 mediates UVA radiation- and singlet oxygen-induced expression of the human intercellular adhesion molecule 1 gene. *Proc. Natl. Acad. Sci. U.S.A.* **93**, 14586–14591.

Grether-Beck, S., Buettner, R., and Krutmann, J. (1997). Ultraviolet A radiation-induced expression of human genes: Molecular and photobiological mechanisms, *Biol. Chem.* **378**, 1231–1236.

Guyton, K. Z., Liu, Y., Gorospe, M., Xu, Q., and Holbrook, N. J. (1996). Activation of mitogen-activated protein kinase by H_2O_2. *J. Biol. Chem.* **271**, 4138–4142.

Herrmann, G., Wlaschek, M., Bolsen, K., Prenzel, K., Goerz, G., and Scharffetter-Kochanek, K. (1996). Photosensitization of uroporphyrin augments the ultraviolet A-induced synthesis of matrix metalloproteinases in human dermal fibroblasts. *J. Invest. Dermatol.* **107**, 398–403.

Ishikawa, Y., Yokoo, T., and Kitamura, M. (1997). c-Jun/AP-1, but not NF-kappa B, is a mediator for oxidant-initiated apoptosis in glomerular mesangial cells. *Biochem. Biophys. Res. Commun.* **240**, 496–501.

Janssen, Y. M., Matalon, S., and Mossman, B. T. (1997). Differential induction of c-fos, c-jun, and apoptosis in lung epithelial cells exposed to ROS or RNS. *Am. J. Physiol.* **273**, L789–L796.

Jeong, J. K., Juedes, M. J., and Wogan, G. N. (1998). Mutations in the *supF* gene of pSP189 by hydroxyl radical and singlet oxygen: Relevance to peroxynitrite mutagenesis. *Chem. Res. Toxicol.* **11**, 550–556.

Kanofsky, J. R. (1989). Singlet oxygen production by biological systems. *Chem.-Biol. Interact.* **70**, 1–28.

Karin, M., and Hunter, T. (1995). Transcriptional control by protein phosphorylation: Signal transmission from the cell surface to the nucleus. *Curr. Biol.* **5**, 747–757.

Keyse, S. M., and Tyrrell, R. M. (1989). Heme oxygenase is the major 32kDa stress protein induced in human skin fibroblasts by UVA radiation, hydrogen peroxide, and sodium arsenite. *Proc. Natl. Acad. Sci. U.S.A.* **86**, 99–103.

Klotz, L. O., Briviba, K., and Sies, H. (1997). Singlet oxygen mediates the activation of JNK by UVA radiation in human skin fibroblasts. *FEBS Lett.* **408**, 289–291.

Klotz, L. O., Pellieux, C., Briviba, K., Pierlot, C., Aubry, J.-M., and Sies, H. (1999). Mitogen-activated protein kinase (p38-, JNK-, ERK-) activation pattern induced by extracellular and intracellular singlet oxygen and UVA. *Eur. J. Biochem.* **260**, 917–922.

Klotz, L. O., Fritsch, C., Briviba, K., Tsacmacidis, N., Schliess, F., and Sies, H. (1998). Activation of JNK and p38 but not ERK MAP kinases in human skin cells by 5-aminolevulinate-photodynamic therapy. *Cancer Res.* **58**, 4297–4300.

Knebel, A., Rahmsdorf, H. J., Ullrich, A., and Herrlich, P. (1996). Dephosphorylation of receptor tyrosine kinases as target of regulation by radiation, oxidants or alkylating agents. *EMBO J.* **15**, 5314–5325.

Knight, R. J., and Buxton, D. B. (1996). Stimulation of c-Jun kinase and mitogen-activated protein kinase by ischemia and reperfusion in the perfused rat heart. *Biochem. Biophys. Res. Commun.* **218**, 83–88.

Kouchakdijan, M., Bodepudi, V., Shibutani, S., Eisenberg, M., Johnson, F., Grollman, A. P., and Patel, D. J. (1991). NMR structural studies of the ionizing radiation adduct 7-hydro-8-oxodeoxyguanosine (8-oxo-7H-dG) opposite deoxyadenosine in a DNA duplex. 8-Oxo-7H-dG(syn)-dA(anti) alignment at lesion site. *Biochemistry* **30**, 1403–1412.

Kyriakis, J. M., Banerjee, P., Nikolakaki, E., Dai, T., Rubie, E. A., Ahmad, M. F., Avruch, J., and Woodgett, J. R. (1994). The stress-activated protein kinase subfamily of c-Jun kinases. *Nature (London)* **369**, 156–160.

Lander, H. M., Jacovina, A. T., Davis, R. J., and Tauras, J. M. (1996). Differential activation of mitogen-activated protein kinases by nitric oxide-related species. *J. Biol. Chem.* **271**, 19705–19709.

Lavrovsky, Y., Schwartzman, M. L., Levere, R. D., Kappas, A., and Abraham, N. G. (1994). Identification of binding sites for transcription factors NF-κB and AP-2 in the promoter region of the human heme oxygenase 1 gene. *Proc. Natl. Acad. Sci. U.S.A.* **91**, 5987–5991.

Liu, Y., Guyton, K. Z., Gorospe, M., Xu, Q., Lee, J. C., and Holbrook, N. J. (1996). Differential activation of ERK, JNK/SAPK and p38/CSBP/RK MAP kinase family members during the cellular response to arsenite. *Free Radical Biol. Med.* **21**, 771–781.

Lo, Y. Y. C., Wong, J. M. S., and Cruz, T. F. (1996). Reactive oxygen species mediate cytokine activation of c-Jun NH$_2$-terminal kinases. *J. Biol. Chem.* **271**, 15703–15707.

Mendelson, K. G., Contois, L. R., Tevosian, S. G., Davis, R. J., and Paulson, K. E. (1996). Independent regulation of JNK/p38 mitogen-activated protein kinases by metabolic oxidative stress in the liver. *Proc. Natl. Acad. Sci. U.S.A.* **93**, 12908–12913.

Morita, A., Werfel, T., Stege, H., Ahrens, C., Karmann, K., Grewe, M., Grether-Beck, S., Ruzicka, T., Kapp, A., Klotz, L. O., Sies, H., and Krutmann, J. (1997). Evidence that singlet oxygen-induced human T helper cell apoptosis is the basic mechanism of ultraviolet-A radiation phototherapy. *J. Exp. Med.* **186**, 1763–1768.

Muegge, K., Vila, M., Gusella, G. L., Musso, T., Herrlich, P., and Durum, S. K. (1993). Interleukin 1 induction of the c-jun promoter. *Proc. Natl. Acad. Sci. U.S.A.* **90**, 7054–7058.

Müller, J. M., Cahill, M. A., Rupec, R. A., Baeuerle, P. A., and Nordheim, A. (1997). Antioxidants as well as oxidants activate c-fos via Ras-dependent activation of extracellular-signal-regulated kinase 2 and Elk-1. *Eur. J. Biochem.* **244**, 45–52.

Müller, T., Haussmann, H.-J., and Schepers, G. (1997). Evidence for peroxynitrite as an oxidative stress-inducing compound of aqueous cigarette smoke fractions. *Carcinogenesis* **18**, 295–301.

Pierlot, C., Hajjam, S., Barthélémy, C., and Aubry, J. -M. (1996). Water-soluble naphthalene derivatives as singlet oxygen ($^1O_2$, $^1\Delta_g$) carriers for biological media. *J. Photochem. Photobiol. B: Biol.* **36**, 31–39.

Piette, J. (1991). Biological consequences associated with DNA oxidation mediated by singlet oxygen. *J. Photochem. Photobiol. B: Biol.* **11**, 241–260.

Piret, B., Legrand-Poels, S., Sappey, C., and Piette, J. (1995). NF-κB transcription factor and human immunodeficiency virus type 1 (HIV-1) activation by methylene blue photosensitisation. *Eur. J. Biochem.* **228**, 447–455.

Pombo, C. M., Bonventre, J. V., Avruch, J., Woodgett, J. R., Kyriakis, J. M., and Force, T. (1994). The stress-activated protein kinases are major c-Jun amino-terminal kinases activated by ischemia and reperfusion. *J. Biol. Chem.* **269**, 26546–26551.

Radler-Pohl, A., Sachsenmaier, C., Gebel, S., Auer, H. P., Bruder, J. T., Rapp, U., Angel, P., Rahmsdorf, H. J., and Herrlich, P. (1993). UV-induced activation of AP-1 involves obligatory extranuclear steps including Raf-1 kinase. *EMBO J.* **12**, 1005–1012.

Raingeaud, J., Gupta, S., Rogers, J. S., Dickens, M., Han, J., Ulevitch, R. J., and Davis, R. J. (1995). Pro-inflammatory cytokines and environmental stress cause p38 mitogen-activated protein kinase activation by dual phosphorylation on tyrosine and threonine. *J. Biol. Chem.* **270**, 7420–7426.

Rosette, C., and Karin, M. (1996). Ultraviolet light and osmotic stress: Activation of the JNK cascade through multiple growth factor and cytokine receptors. *Science* **274**, 1194–1197.

Rupec, R. A., and Baeuerle, P. A. (1995). The genomic response of tumor cells to hypoxia and reoxygenation. Differential activation of transcription factors AP-1 and NF-kappa B. *Eur. J. Biochem.* **234**, 632–640.

Ryter, S. W., and Tyrrell, R. M. (1998). Singlet molecular oxygen ($^1O_2$): A possible effector of eukaryotic gene expression. *Free Radical Biol. Med.* **24**, 1520–1534.

Sachsenmaier, C., Radler-Pohl, A., Zinck, R., Nordheim, A., Herrlich, P., and Rahmsdorf, H. J. (1994). Involvement of growth factor receptors in the mammalian UVC response. *Cell (Cambridge, Mass.)* **78**, 963–972.

Scharffetter-Kochanek, K., Wlaschek, M, Briviba, K., and Sies, H. (1993). Singlet oxygen induces collagenase expression in human skin fibroblasts. *FEBS Lett.* **331**, 304–306.

Scharffetter-Kochanek, K., Wlaschek, M., Brenneisen, P., Schauen, M., Blaudschun, R., and Wenk, J. (1997). UV-induced reactive oxygen species in photocarcinogenesis and photoaging. *Biol. Chem.* **378**, 1247–1257.

Schieke, S. M., Briviba, K., Klotz, L. O., and Sies, H. (1999). Activation pattern of mitogen-activated protein kinases elicited by peroxynitrite: attenuation by selenite supplementation. *FEBS Lett.* **448**, 301–303.

Schreck, R., Albermann, K., and Baeuerle, P. A. (1992). Nuclear factor κB: An oxidative stress-responsive transcription factor of eukaryotic cells (a review). *Free Radical Res. Commun.* **17**, 221–237.

Schulte-Herbrüggen, T., and Sies, H. (1989). The peroxidase/oxidase activity of soybean lipoxygenase. II. Triplet carbonyls and red photoemission during polyunsaturated fatty acid and glutathione oxidation. *Photochem. Photobiol.* **49**, 705–710.

Sies, H. (1986). Biochemistry of oxidative stress. *Angew. Chem., Int. Ed. Engl.* **25**, 1058–1071.

Sies, H., and Menck, C. F. (1992). Singlet oxygen induced DNA damage. *Mutat. Res.* **275**, 367–375.

Steinbeck, M. J., Khan, A. U., and Karnovsky, M. J. (1992). Intracellular singlet oxygen generation by phagocytosing neutrophils in response to particles coated with a chemical trap. *J. Biol. Chem.* **267**, 13425–13433.

Steinbeck, M. J., Khan, A. U., and Karnovsky, M. J. (1993). Extracellular production of singlet oxygen by stimulated macrophages quantified using 9,10-diphenylanthracene and perylene in a polystyrene film. *J. Biol. Chem.* **268**, 15649–15654.

Stocker, R., Yamamoto, Y., McDonagh, A. F., Glazer, A. N., and Ames, B. N. (1987). Bilirubin is an antioxidant of possible physiologic importance. *Science* **235**, 1043–1046.

Stylianou E., O'Neill, L. A., Rawlinson, L., Edbrooke, M. R., Woo, P., and Saklatvala, J. (1992). Interleukin 1 induces NF-κB through its type I but not its type II receptor in lymphocytes. *J. Biol. Chem.* **267**, 15836–15841.

Sugden, P. H., Fuller, S. J., Michael, A., and Clerk, A. (1997). Activation of mitogen-activated protein kinase subfamilies by oxidative stress in the perfused rat heart. *Biochem. Soc. Trans.* **25**, S565.

Sullivan, S. G., Chiu, D. T., Errasfa, M., Wang, J. M., Qi, J. S., and Stern, A. (1994). Effects of H_2O_2 on protein tyrosine phosphatase activity in HER14 cells. *Free Radical Biol. Med.* **16**, 399–403.

Tao, J., Sanghera, J. S., Pelech, S. L., Wong, G., and Levy, J. G. (1996). Stimulation of stress-activated protein kinase and p38 HOG1 kinase in murine keratinocytes following photodynamic therapy with benzoporphyrin derivative. *J. Biol. Chem.* **271**, 27107–27115.

Tournier, C., Thomas, G., Pierre, J., Jacquemin, C., Pierre, M., and Saunier, B. (1997). Mediation by arachidonic acid metabolites of the H_2O_2- induced stimulation of mitogen-activated protein kinases (extracellular-signal-regulated kinase and c-Jun NH_2- terminal kinase). *Eur J. Biochem.* **244**, 587–595.

Tyrrell, R. M. (1991). UVA (320-380 nm) radiation as an oxidative stress. In "Oxidative Stress: Oxidants and Antioxidants" (H. Sies, ed.), pp. 57–83. Academic Press, San Diego, CA.

Uchida, K., Shirahishi, M., Naito, Y., Torii, Y., Nakamura, Y., and Osawa, T. (1999). Activation of stress signaling pathways by the end product of lipid peroxydation. 4-Hydroxy-2-nonenal is a potent inducer of intracellular peroxide production. *J. Biol. Chem.* **274**, 2234–2242.

Uciechowski, P., Saklatvala, J., von der Ohe, J., Resch, K., Szamel, M., and Kracht, M. (1996). Interleukin 1 activates jun N-terminal kinases JNK1 and JNK2 but not extracellular regulated MAP kinase (ERK) in human glomerular mesangial cells. *FEBS Lett* **394**, 273–278.

Vile, G. F., and Tyrrell, R. M. (1993). Oxidative stress resulting from ultraviolet A irradiation of human skin fibroblasts leads to a heme oxygenase-dependent increase in ferritin. *J. Biol. Chem.* **268**, 14678–14681.

Vile, G. F., Tanew-Iliitschew, A., and Tyrrell, R. M. (1995). Activation of NF-κB in human skin fibroblasts by the oxidative stress generated by UVA radiation. *Photochem. Photobiol.* **62**, 463–468.

Wagner, J. R., Motchnik, P. A., Stocker, R., Sies, H., and Ames, B. N. (1993). The oxidation of blood plasma and low density lipoprotein components by chemically generated singlet oxygen. *J. Biol. Chem.* **268**, 18502–18506.

Wang, X., Martindale, J. L., Liu, Y., and Holbrook, N. J. (1998). The cellular response to oxidative stress: Influences of mitogen-activated protein kinase signalling pathways on cell survival. *Biochem. J.* **333**, 291–300.

Wlaschek, M., Heinen, G., Poswig, A., Schwarz, A., Krieg, T., and Scharffetter-Kochanek, K. (1994). UVA-induced autocrine stimulation of fibroblast-derived collagenase/MMP-1 by interrelated loops of interleukin-1 and interleukin-6. *Photochem. Photobiol.* **59**, 550–556.

Wlaschek, M., Briviba, K., Stricklin, G. P., Sies, H., and Scharffetter-Kochanek, K. (1995). Singlet oxygen may mediate the ultraviolet A-induced synthesis of interstitial collagenase. *J. Invest. Dermatol.* **104**, 194–198.

Wlaschek, M., Wenk, J., Brenneisen, P., Briviba, K., Schwarz, A., Sies, H., and Scharffetter-Kochanek, K. (1997). Singlet oxygen is an early intermediate in cytokine-dependent UV induction of interstitial collagenase in human dermal fibroblasts. *FEBS Lett.* **413**, 239–242.

2 Redox Regulation of Gene Expression

Dana R. Crawford,* Toshihide Suzuki,* and Kelvin J. A. Davies†

*Department of Biochemistry and Molecular Biology
Albany Medical College
Albany, New York 12208

†Ethel Percy Andrus Gerontology Center
University of Southern California
Los Angeles, California 90089

Oxidative stress is now known to modulate the expression of a number of genes, both in eukaryotic and in prokaryotic systems. These modulations have been observed in response to direct and indirect oxidative challenge and involve changes at transcriptional, mRNA stability, and signal transduction levels. A wide range of protein products to these modulated genes have been identified and include antioxidant enzyme, growth arrest, DNA repair, mitochondrial electron transport, cell adhesion, cytokine, and glucose-regulated proteins. This list will continue to expand as novel modulated gene species identified by new and sensitive experimental approaches are sequenced. Studies in basic experimental systems such as bacteria have led to invaluable insight into the identification and mechanism of action of oxidant-modulated genes and the subsequent elucidation of mammalian homologs. These latter genes now represent valuable potential diagnostic markers and therapeutic targets in the treatment of oxidative stress-related disease.

I. INTRODUCTION

The ability of cells to successfully mount and maintain a viable stress response following exposure to toxic stimuli is an important component of cellular and organismal homeostasis. Cells have evoked a plethora of protective mechanisms to cope with or prevent stress damage. Most notably, these include the heat shock proteins (HSP), glucose-response protein (GRP), and growth arrest and DNA damage (GADD) gene families. The modulation of expression of such essential stress-related genes is critical to protective responses. Their importance is underscored by their conserved sequences and modulations from organisms ranging from bacteria and yeast to human in response to a wide range of stress agents including heat shock, glucose deprivation, oxidants, radiation, and heavy metals.

It is now clear that cellular response to oxidative stress is a universal phenomena. Numerous genes have now been identified whose steady-state mRNA product levels are modulated by oxidant stress agents, including superoxide, hydrogen peroxide, nitric oxide, redox active quinones, hyperbaric oxygen, singlet oxygen, diethylmaleate, glutathione depletion, and others (Crawford *et al.*, 1995; Crawford, 1999). The earliest and most dramatic gene modulations to these agents were observed in bacteria, most notably those under the control of OxyR and SoxRS (Storz and Tartaglia, 1992; Greenberg *et al.*, 1990), following exposure to superoxide and hydrogen peroxide. Many of the modulated genes in the bacterial systems were identified as antioxidant enzymes and included superoxide dismutase (Hassan and Fridovich, 1977), catalase (Yashpe-Purer *et al.*, 1977), alkyl hydroperoxide reductase, glutathione reductase, and manganese superoxide dismutase.

Initial studies in mammalian systems were unsuccessful in identifying oxidant-modulated genes. Surprisingly, the same superoxide dismutase and catalase genes that were induced so strongly in bacteria showed little or no modulation in mammalian cells. For this reason, mammalian studies lagged behind those involving bacteria. In fact, for a short time, it was thought that oxidative stress might not be important at all in regulating gene expression in mammals. However, the subsequent discovery that mRNAs to the cellular protooncogenes c-*fos*, c-*myc*, and c-*jun* are induced by oxidative stress (Crawford and Cerutti, 1987; Nose *et al.*, 1991) indicated that redox regulation of gene expression must be important in mammals, especially since cellular protooncogenes play central roles in the most basic of biological processes: growth and differentiation. It was also shown that adaptation to oxidative stress in mammalian cells requires changes in the expression of several genes (Wiese *et al.*, 1995). We now know that a significant number of mammalian genes are modulated by oxidative stress in mammals.

II. METHODS OF ANALYSIS

A number of approaches now exist for studying the modulation of gene expression. Most operate at the level of mRNA, although protein electrophoresis is still a useful technique. In fact, polyacylamide gel electrophoresis led to two important discoveries in the field: the identification of 30–40 induced bacterial proteins by hydrogen peroxide as assessed by two-dimensional gel electrophoresis and the identification of induced heme oxygenase protein in cultured mammalian cells as assessed by one-dimensional gel electrophoresis (Christman *et al.*, 1985; Keyse and Tyrrell, 1989). Despite these important findings, these protein-based approaches have, for the most part, been supplanted by more powerful and sensitive nucleic acid-based techniques.

Nucleic-acid based techniques are primarily based on analysis at the level of cellular mRNA. Chronologically, they include differential hybridization and subtractive hybridization, followed by differential display, and then restriction fragment differential display, serial analysis of gene expression (SAGE), and gene chips.

Differential hybridization is a relatively simple technique. It consists of probing a cDNA library, successively, with reverse-transcribed, labeled cDNA obtained from the RNA of unstressed and stressed cells. In general, it lacks sensitivity, identifying only the more abundant mRNAs. However, it has been used successfully to identify CL100, a hydrogen peroxide-inducible mRNA that was originally identified in human fibroblasts (Keyse and Emslie, 1992). Subtractive hybridization is a much more challenging technique that also requires large amounts of sample, but is more sensitive. The basic approach here is to remove mRNAs common to both normal and stressed cells and then analyze for what remains ("subtracted" population). This technique was used successfully to detect the growth arrest and DNA damage genes that are induced by several types of stress, including oxidative (Fargnoli *et al.*, 1990). Differential display is a polymerase chain reaction (PCR)-based technique that requires only small amounts of sample. It utilizes different combinations of PCR primers to generate subpopulations of DNA species that can be analyzed on a sequencing gel (Liang and Pardee, 1992; Crawford *et al.*, 1994). It is relatively sensitive and is able to identify both induced and reduced species. This technique has led to the identification of hundreds of novel genes through their modulated mRNAs, some of which have been used as clinical targets. Its main drawbacks are a high number of false positives due to low template specificity during the annealing stage of PCR and products that are strongly biased for the 3′-untranslated region of the mRNA. Its basic approach has been supplanted by restriction fragment differential display, which gains the same advantages as conventional differential display but uses restriction enzyme-generated cDNA fragments as PCR templates. This modification allows for the use of much higher temperature and more specific annealing conditions as well as the ability to generate products that include the coding region and 5′-untranslated region. SAGE utilizes the generation of short sequence tags derived from cDNA populations (Velculescu et al., 1995). Subsequent ligation, cloning, and sequencing determine the abundance of a particular tag and therefore the level of expression of the corresponding gene. This technique has also successfully identified modulated mRNAs in a number of systems, although it is still plagued by several technical problems, such as unwanted linker ligation and its requirement for extensive sequencing. Finally, gene chips promise high sensitivity, ease of use, and, ultimately, comprehensiveness. At the moment, however, the analysis is costly, and most versions of this technique are limited only to those mRNAs that have been identified and partially sequenced. A comprehensive analysis thus awaits complete sequencing of the human genome.

III. MODULATION OF NUCLEAR GENE EXPRESSION

It is now known that the expression of a number of mammalian genes is modulated by oxidative stress. These modulations involve mostly inductions. A list of mRNAs induced by oxidative stress is presented in Table I. Surprisingly, few of these induced mRNAs involve the "classical" antioxidant enzymes superoxide dismutase, catalase, or glutathione peroxidase. This is surprising because a number of antioxidant enzymes are induced by hydrogen peroxide and superoxide in bacteria (Kullik and Storz, 1994; Storz and Tartaglia, 1992; Greenberg et al., 1990). In mammalian cells, modest inductions at best have been observed for antioxidant enzyme mRNAs and, in most cases, no modulation observed at all. This weak response led to a general impression that oxidative stress was probably unimportant in the modulation of mammalian gene expression. Several subsequent studies, however, changed this perception. Crawford and Cerutti (1987) found that mRNA to the protooncogenes c-*fos* and c-*myc* but not h-*ras* was induced in mouse epidermal JB6 cells by superoxide and hydrogen peroxide generated by xanthine/xanthine oxidase. Parallel analysis of antioxidant gene expression did not detect any modulation in antioxidant gene expression. Subsequent studies not only confirmed these inductions, but also demonstrated the induction of other protooncogene and immediate early gene mRNAs, including c-*jun*, *egr*, and JE (Shibanuma et al., 1988; Muehlematter et al., 1989; Nose et al., 1991). Other sources of oxidants also induced expression, including hydrogen peroxide and *t*-butyl hydroperoxide obtained and added to cell cultures from stock solutions (Nose et al., 1991; Muehlematter et al., 1989). These results revealed that not only were reactive oxygen species able to modulate the expression of mammalian gene expression, they also did so to a group of genes that are considered to be among the most important

Table I Oxidant-Induced mRNAs in Mammalian Cells

Gene	Oxidant stress	Reference
c-*fos*, c-*myc*	Xanthine/xanthine oxidase	Crawford and Cerutti (1987)
c-*fos*, c-*jun*, egr-1, JE	Hydrogen peroxide	Nose et al. (1991)
c-*fos*	t-Butyl hydroperoxide	Muehlematter et al. (1989)
c-*fos*, c-*jun*	Nitric oxide	Janssen et al. (1997)
Heme oxygenase	Multiple oxidants	Keyse and Tyrrell (1989)
gadd45 and *gadd153*	Hydrogen peroxide	Fornace et al. (1989)
CL100	Hydrogen peroxide	Keyse and Emslie (1992)
Interleukin-8	Hydrogen peroxide	DeForge et al. (1993)
γ-Glutamyl transpeptidase	Menadione	Kugelman et al. (1994)
Vimentin, cytochrome IV, RP-L4	Diethylmaleate	Ammendola et al. (1995)
c-*fos*, c-*jun*, c-*myc*, HSP	Xanthine/xanthine oxidase	Yamamoto et al. (1993)
Glucose-regulated proteins	Singlet oxygen, other oxidants	Gomer et al. (1991)
c-*fos* and zif/268	Nitric oxide	Morris (1995)
adapt15	Hydrogen peroxide	Crawford et al. (1996a,d)
adapt66(mafG)	Hydrogen peroxide	Crawford et al. (1996b)
adapt33	Hydrogen peroxide	Wang et al. (1996)
adapt73/PigHep3	Hydrogen peroxide	Crawford and Davies (1997)
adapt78	Hydrogen peroxide	Crawford et al. (1997)
Ref-1	Hydrogen peroxide, hypochlorite	Grosch et al. (1998)
Catalase, MnSOD, GPx	Hydrogen peroxide	Shull et al. (1991)
MnSOD	Xanthine/xanthine oxidase	Shull et al. (1991)
Mn-SOD	Hyperbaric oxygen	Stevens and Autor (1977)
MnSOD	Multiple oxidants	Stralin and Marklund (1994)
Numerous redox	p53 overexpression	Polyak et al. (1997)
NKEF-B	Hydrogen peroxide	Kim et al. (1997)

in the mammalian genome. Cellular protooncogenes are critical to cellular growth and differentiation, as evidenced by the observation that their aberrant expression leads to cancer (Anderson *et al.*, 1992). Therefore, reactive oxygen species must be important mediators of mammalian gene expression.

A second important study in the field was performed by Keyse and Tyrrell (1989). These investigators observed that the protein and mRNA to heme oxygenase is strongly induced by oxidative stress, also using a mammalian cell culture model system. The induction of heme oxygenase has now been observed in many model systems using a number of different inductants, so much so that heme oxygenase has become the prototype mRNA for assessing cellular oxidative stress at the mRNA level. In fact, it has been employed in a number of clinically related studies as a marker of oxidative stress-related disease due to its ready detection, its sustained induction (as compared with immediate early gene mRNAs such as c-*fos*), and its inducibility in many tissues and mammalian species.

A third study important to the oxidant stress field was carried out by Fornace *et al.* (1989). Using ultraviolet light as a stress agent, they discovered a number of genes that were induced by growth arrest and DNA damage. These so-called *"gadd"* genes were also found to be induced by oxidative stress. The two most studied of the *gadd* genes are *gadd45* and *gadd153*, and they have been implicated in growth arrest, cellular protection, apoptosis, and DNA repair (Zhan *et al.*, 1994; Smith *et al.*, 1994; Fornance *et al.*, 1992).

Since identification of these genes, a growing number of nuclear-encoded mRNAs have also been reported as modulated by reactive oxygen species. As indicated in Table I, these include CL100, interleukin-8, γ-glutamyl transpeptidase, vimentin, cytochrome IV, ribosomal protein L4, heat shock protein, glucose-regulated proteins, natural killer-enhancing factor B (NKEF-B), and apurinic endonuclease (Ref-1) (Kugelman *et al.*, 1994; Gomer *et al.*, 1991; Yamamoto *et al.*, 1993; DeForge *et al.*, 1993; Keyse and Emslie, 1992; Kim *et al.*, 1997; and Grosch *et al.*, 1998). In addition, 13 RNAs have been identified that are modulated following exposure to a pretreatment concentration of peroxide using the technique of differential display (Crawford *et al.*, 1996 a–d; Wang *et al.*, 1996). Seven of these are induced and six are reduced. The induced mRNAs have been designated *"adapts."* Interestingly, one of these *adapts, adapt15,* is expressed coordinately with *gadd45* and *gadd153* and appears to be a member of the same *gadd* family. Other *adapts* studied include *adapt33*, a novel and apparently untranslated RNA (Wang *et al.*, 1996); *adapt66*, a homolog of the transcriptional regulator *mafG* (Crawford *et al.*, 1996b); *adapt73*, a cardiogenic shock-inducible homolog (Crawford and Davies, 1997); and *adapt78*, whose gene maps to a region of human chromosome 21 thought to be important in dementia (Crawford *et al.*, 1997). It is interesting to note that we also see the induction of *gadd153, gadd45,* and heme oxygenase mRNAs during adaptive response.

Oxidative damage by reactive nitrogen species is also a growing field of interest. Nitric oxide is the prototype reactive nitrogen species, and it has been found to induce the levels of several mRNAs. These include mRNAs to the immediate early genes c-*fos* and zif/268 in rat striatal neurons (Morris, 1995), c-*fos* and c-*jun* in rat lung epithelial cells (Janssen *et al.*, 1997), and c-*fos* in retinal pigment epithelium following injection of the nitric oxide donor sodium nitroprusside into the vitreous cavity of rat eye (Ohki *et al.*,

1995). Nitric oxide also induced heme oxygenase mRNA and protein in vascular smooth muscle cells (Hartsfield *et al.*, 1997). A biphasic regulation of calmodulin-dependent protein kinase II (CaMKII α) by the nitric oxide donor spermine NONOate was observed in hippocampal granule cells consisting of a strong induction of the mRNA followed by a later dramatic reduction in levels (Johnston and Morris, 1995). Downregulation of certain other mRNAs by nitric oxide was also observed, including angiotensin II type 1 receptor in vascular smooth muscle cells (Ichiki *et al.*, 1998) and soluble guanylate cyclase mRNA in rat pulmonary artery smooth muscle cells (Filippov *et al.*, 1997).

Although early studies reported a lack of antioxidant enzyme modulation by oxidative stress in mammalian cells, subsequent studies have revealed moderate modulation of these genes. Of these antioxidant enzymes genes, MnSOD has clearly exhibited the greatest overall modulation. In tracheobronchial epithelia, catalase and, to a lesser extent, MnSOD and glutathione peroxidase, but not copper,zinc-superoxide dismutase (Cu,Zn-SOD) mRNA levels are elevated in response to hydrogen peroxide. However, only MnSOD mRNA levels are elevated in these same cells in response to xanthine/xanthine oxidase (Shull *et al.*, 1991). MnSOD mRNA is also elevated in neonatal rat lung following hyperbaric lung treatment (Stevens and Autor, 1977). No modulation of Cu,Zn-SOD mRNA was observed in this study. In fact, Cu,Zn-SOD is so rarely modulated by oxidative stress that it often used as an unmodulated internal marker to normalize the levels of other genes whose levels are modulated by oxidative stress. Although MnSOD is the most frequently modulated antioxidant gene mRNA, its modulation by reactive oxygen species is significantly less than that reported for other agents not considered to be classic oxidants, most notably interleukin-1 (IL-1), endotoxin, and tumor necrosis factor (TNF). The observations of Stralin and Marklund (1994) are typical. They observed no induction of Cu,Zn-SOD and modest (2- to 3-fold) induction of MnSOD in human fibroblasts using a variety of oxidants. This MnSOD induction is well below the sometimes 20- to 100-fold induction reported for MnSOD mRNA induction by IL-1, endotoxin, and TNF (Gwinner *et al.*, 1995; Suzuki *et al.*, 1993).

It should be noted that the modulation of gene expression by oxidative stress is not limited to exogenous oxidant sources, even though most such studies on cells, tissue, or even whole organisms are usually performed by exposure to exogenous sources of reactive oxygen species. Significant oxidative stress also occurs via the intracellular generation of a prooxidant state and to a wide variety of agents not usually considered to be oxidants in and of themselves. As examples, a role for reactive oxygen species has been demonstrated in the induction of vascular cell adhesion molecule-1 by interleukin-1β (Marui *et al.*, 1993); the induction of cyclooxygenase-2 by lipopolysaccharide, interleukin-1, and tumor necrosis factor (Feng et al., 1995); and in tyrosine phosphorylation and MAP kinase stimulation by platelet-derived growth factor (PDGF), the latter of which has been found to be dependent on PDGF-generated hydrogen peroxide in rat vascular smooth muscle cells (Sundaresan et al., 1995).

One of the more impressive demonstrations of the modulation of gene expression via an intracellular-generated oxidative stress, and one with important physiological ramifications, was reported by Polyak et al. (1997). In this study, the genes that mediate the apoptogenic effects of p53 overexpression were examined. p53 is inactivated in a large percentage of human cancers, and its overexpression mediates cellular growth arrest and apoptosis. Using a model system in which overexpression of p53 led to apoptosis, 14 genes were found to be strongly induced as analyzed using the technique of SAGE. Surprisingly, sequence analysis revealed that many of these genes were redox related. They included a novel glutathione S-transferase; a gene also induced by the apoptogen- and reactive oxygen generating quinone-etoposide; a proline oxidoreductase; p21, a growth arrest protein known to be induced by hydrogen peroxide; a serum amyloid protein inducible by oxidative stress; a galectin family member capable of generating superoxide; an mRNA inducible by the reactive oxygen generating TNF α; and an NADH oxidoreductase homolog. The authors propose a mechanism of action for p53 that involves the p53 transcriptional induction of a number of genes that increase the levels of intracellular reactive oxygen species, mitochondrial damage by these oxidants, and activation of apoptotic caspases by released calcium. These redox-related genes that mediate the action of p53 are also potential clinical targets against p53-related cancer.

IV. MODULATION OF MITOCHONDRIAL GENE EXPRESSION

Mitochondria are a major target of oxidative damage. This damage includes the loss of membrane potential, release of calcium, depletion of ATP, lipid peroxidation, protein oxidation, DNA damage, loss of electron transport capacity, and more (Zhang et al., 1990; Farber et al., 1990; Bindoli, 1988; Shay and Werbin, 1987). It is now known that, like nuclear genes, the expression of mitochondrial genes is modulated by reactive oxygen species. These modulations usually involve the downregulation of mitochondrial RNA levels. Decreases in overall mitochondrial transcription have been reported in several different experimental systems in response to xanthine/xanthine oxidase (Vincent et al., 1994) and peroxyl radicals (Kristal et al., 1994 a, b). Kristal et al. (1994a) identified two specific mitochondrial genes that were involved in this downregulation: 12S ribosomal RNA and NADH dehydrogenase subunit 4 mRNA. We have observed a dramatic downregulation in the steady-state levels of at least six specific mitochondrial RNAs in HA-1 hamster cells exposed to hydrogen peroxide (Crawford *et al.,* 1996c), including 16S rRNA, 12S rRNA, and mRNAs to NADH dehydrogenase subunit 6, ATPase subunit 6, and cytochrome oxidase subunits I and III. Importantly, these downregulations were preferential for mitochondrial RNA as compared with cytoplasmic RNAs. This latter observation further underscores the hypersensitivity of mitochondria to oxidative stress. Although downregulation of mitochondrial RNA levels by oxidative stress has been reported, these RNA species can also be induced, as reported in response to transformation (Coral *et al.,* 1989) and hypoxemia (Coral-Debrinski *et al.,* 1991).

V. STABLE VERSUS TRANSIENT GENE EXPRESSION

One of the more intriguing observations of aberrant gene expression is its comparative effect as a function of time. Modulations that occur transiently may have different effects than those that do long term. A number of stable model systems exhibiting resistance to oxidative stress have now been described. Although long-term compensatory and secondary modulations can be drawbacks of this approach, they do possess advantages over the transient models described earlier. These include less experiment-to-experiment variability, a clearly documented resistant phenotype, and valuable models for pathologies where chronic prooxidant states either cause or exacerbate cell damage.

There are several popular models of stable cell lines resistant to chronic oxidative stress. Most were developed as hydrogen peroxide-resistant lines, but others include hyperbaric oxygen-, redox quinone-, and excitotoxin-resistant lines as well. These lines were developed by long-term exposure to the oxidative stresses just mentioned followed by selection of cells exhibiting stress resistance. Because the stable lines developed resistance to oxidative stress, the logical explanation for their resistance was a change in the expression of antioxidant enzymes and related proteins. These lines have therefore been subsequently characterized with respect to the expression of these proteins and their corresponding mRNAs.

Spitz *et al.* (1990) observed a dramatic induction of catalase (20-fold) accompanied by modest elevations in Cu,Zn superoxide dismutase and selenium-dependent glutathione peroxidase in hydrogen peroxide-resistant HA-1 Chinese hamster fibroblasts, designated OC14. Cross-resistance to hyperbaric oxygen damage was also observed in these cells, although resistance was not as great as compared to peroxide. Subsequent analysis of these cells revealed elevated heme oxygenase, c-*jun*, and *gadd153* mRNA levels as well (Guyton et al., 1996). Hydrogen peroxide-resistant human myeloid HL-60 cells were reported with 16-fold higher levels of catalase mRNA in the most resistant cell line (Yamada et al., 1991). Once again, amplification of the catalase gene was observed and localized near or at the original catalase gene locus on chromosome 11. Chinese hamster cell line R-8, which exhibits a 10-fold resistance to hydrogen peroxide as compared to control cells, also exhibited a corresponding 10-fold increase in catalase activity but not superoxide dismutase, glutathione peroxidase, or glutathione reductase, suggesting that resistance was associated with the enhanced catalase activity (Sawada et al., 1988). In the latter case, however, resistance was lost with time in culture in the absence of peroxide, suggesting a mode of resistance distinct from catalase gene amplification. In neonatal cardiac myocytes, exposure to continuously generated hydrogen peroxide resulted in catalase mRNA induction but not superoxide dismutase or selenium glutathione peroxidase (Lai et al., 1996). It is interesting to note that resistance to peroxide in bacteria also correlates with elevated catalase levels (Winquist et al., 1984).

Hyperbaric oxygen has also been used as an oxidant stress to generate stable-resistant cell lines. O_2R95 cells, derived from Chinese hamster HA-1 fibroblasts, exhibit resistance to 95% O_2 (Sullivan et al., 1992). These cells also exhibit increased levels of catalase, superoxide dismutase, glutathione peroxidase, heme oxygenase, c-*jun*, and *gadd153* mRNAs. These elevations were very similar to those observed for the OC14 hydrogen peroxide-resistant cells mentioned earlier and may help explain the cross-resistance of these cells lines to oxidative stress. Menadione-resistant Chinese hamster ovary K1 cells exhibited increased levels of glutathione S-transferase, glutathione peroxidase, and heme oxygenase (Vallis and Wolf, 1996). Catalase was not measured. Another redox quinone-resistant cell line, M1HQ, exhibited resistance to hydroquinone and cross-resistance to several other but not all stresses (Colinas et al., 1996). These cells exhibited elevated level of glutathione peroxidase, glutathione reductase, quinone reductase, and γ-glutamyl transpeptidase. Catalase was not measured in this study either. rC18 cells are a PC12 pheochromocytoma neuronal-like cell line derivative developed by long-term exposure to the Alzheimer's disease-derived amyloid β protein (Sagara et al., 1996). This protein generates hydrogen peroxide in culture and, expectedly, the rC18 cells are also resistant to hydrogen peroxide and *t*-butyl hydroperoxide killing. These cells exhibit highly elevated levels of catalase and glutathione peroxidase, but not the superoxide dismutases and further suggest the importance of hydrogen peroxide in amyloid β protein toxicity of brain neurons. The same cell line also exhibited increased transcriptional activity of the redox-sensitive transcription factor

NF-κB, suggesting that this factor helps mediate the resistance of neuronal cells to oxidative stress (Lezoualc'h *et al.*, 1998). Another neuronal model of oxidative stress resistance was developed by exposing mouse nerve cell line HT-22 to the excitotoxin glutamate, a neurotransmitter known to generate reactive oxygen species in target cells (Sagara *et al.*, 1998). Once again, catalase levels were elevated in the stress-resistant cells but not glutathione peroxidase nor superoxide dismutase.

Overall, as expected, elevations in at least some antioxidant enzymes levels were observed in these stable oxidant stress cell lines. With the exception of catalase and heme oxygenase, however, the modulations were not consistent and varied from cell line to cell line. Somewhat surprising, however, is the central importance that catalase appears to play, suggesting that this enzyme may play a much greater role in the cell than the simple protection against occasional high intracellular levels of hydrogen peroxide. It also suggests that catalase is a potential valuable clinical target for oxidative stress-related pathologies. Heme oxygenase, where evaluated, was also elevated consistently in these model systems and further underscores the potential of this gene as a clinical marker of oxidative stress.

Finally, although transient versus stable gene expression can have quite different effects on cells, there are also similarities between the two. For example, similar changes in gene expression were observed in control cells exposed transiently to menadione and long-term stable menadione-resistant cells, suggesting that, at least in some cases, stable lines may be the result of immortalization of a normally transient adaptive stress response (Vallis and Wolf, 1996).

VI. MODES OF REGULATION

Cells modulate their levels of mRNA by three major mechanisms: transcription, mRNA stability, and precursor processing. Little is known about which of these mechanisms are involved in oxidant gene regulation, mainly because the identification of oxidant-modulated genes is a relatively recent observation. However, there is evidence for all three mechanisms of mRNA level regulation by oxidant-regulated genes.

Most analyses in this area have assessed the contribution of transcription. The induction of both c-*fos* and c-*jun* mRNA levels is known to be mediated, at least in part, by increased gene transcription (Devary *et al.*, 1991; Crawford *et al.*, 1988). Heme oxygenase induction by direct oxidant exposure or by depletion of intracellular glutathione in human fibroblasts, and induction of both glutathione *S*-transferase Ya subunit and NAD(P)H:quinone oxidoreductase by hydrogen peroxide, also involves increased transcription (Tyrrell *et al.*, 1993; Rushmore *et al.*, 1991; Rushmore and Pickett, 1990). Thus, oxidant-responsive elements would be expected to exist somewhere in these genes. In fact, these genes have served as valuable models for this important area of oxidant research. Oxidant-responsive elements have now been identified for several genes whose inductions depend, at least in part, on transcription. These include c-*fos*, c-*jun*, heme oxygenase, glutathione S-transferase Ya subunit, and NAD(P)H:quinone oxidoreductase. Much of the work in this area has been done on the glutathione S-transferase Ya subunit gene by Pickett and co-workers (Rushmore *et al.*, 1991; Rushmore and Pickett, 1990). Ironically, the identified element in this gene has been designated the antioxidant-responsive element, as it is bound and activated by a series of compounds classically considered to be antioxidants. However, it has been determined that their ability to stimulate transcription is based on their ability to produce a certain amount of reactive oxygen species. The antioxidant-responsive element has been found to be present in the promoter regions of both glutathione S-transferase Ya subunit and NAD(P)H:quinone oxidoreductase genes (Pinkus *et al.*, 1995; Bergelson *et al.*, 1994; Rushmore and Pickett, 1990). Both genes are stimulated transcriptionally at this site by hydrogen peroxide (Rushmore and Pickett, 1990). The c-*fos* gene is also induced transcriptionally by hydrogen peroxide, and this induction depends on the serum response element (Amstad *et al.*, 1992). An AP-1 response element appears to be important in the stimulation of c-*jun* transcription (Devary *et al.*, 1991). For the heme oxygenase gene, several oxidant-responsive elements reside close to the TATA box in the promoter. However, a critical enhancer region with an AP-1-like sequence is also present in this gene more than 4 kb upstream of the TATA box (Alam and Den, 1992). The presence of this AP-1 site suggests that the induction of c-*fos* and c-*jun* mRNA and their subsequent protein products during oxidative stress contributes to the transcriptional induction of heme oxygenase.

Several important transcriptional factors have been identified that are mediators of oxidative stress. These factors are induced or activated by reactive oxygen species and then bind and activate nuclear genes involved in overall cellular response to oxidative stress. Most notably, they include AP-1 and NF-κB. Consistent with this, oxidative stress is known to induce the mRNA levels of both c-*jun* and c-*fos* and activates NF-κB (Crawford and Cerutti, 1987; Nose *et al.*, 1991; Schreck *et al.*, 1992). Another level of redox regulation of transcription factors can occur during DNA binding. The binding and subsequent activity of a number of transcription factors, including c-Fos, c-Jun, Sp1, v-Ets, v-Rel, v-Myb, and p53, are affected by their redox state (Burdon, 1995; Ammendola *et al.*, 1994). Usually reduction favors binding, with NF-κB being the exception. Cysteine residues are especially important in the reduction and subsequent activation of most of these transcription factors. This introduces yet another level of regulation, that of the reductase molecule. Extensive studies on AP-1 have determined that Ref-1, a DNA repair enzyme, stimulates the DNA-binding activity of Fos–Jun dimers by reducing a critical cysteine residue (Xanthoudakis et al., 1994).

We have found that *adapt66*, a *mafG*-homolog, is induced by hydrogen peroxide (Crawford et al., 1996b). *mafG* is a member of the *maf* family of genes that encode nuclear proteins that recognize AP-1-like response elements. Members of this family include MafB, MafK, MafF, and MafG and are all able to heterodimerize with each other, c-Fos, and erythroid cell-specific transcription factor NF-E2 to affect the transcription of target genes. The modulation of *mafG* homolog/*adapt66* by oxidative stress, and the known interaction of the protein product of this gene with oxidant-inducible c-*fos*, may represent important mechanisms by which oxidative stress can modulate gene expression.

There have been very few reports on the effect of oxidative stress on mRNA stabilization. In vascular smooth muscle cells, heme oxygenase mRNA levels are increased by a combination of decreased mRNA turnover and increased gene transcription (Hartsfield et al., 1997). We have assessed the contribution of mRNA stability to the induction of *adapt15*, *adapt33*, c-*fos*, c-*jun*, and *gadd45* mRNA following exposure of HA-1 cells to hydrogen peroxide. A significant increase in the stability of *adapt33* and *gadd45* but not *adapt15*, c-*fos*, nor c-*jun* mRNAs was observed (Wang et al., 1996). Intracellular chelation of calcium dramatically inhibited the induction of *adapt33* mRNA by hydrogen peroxide, indicating a role for calcium in this stabilization. We also observe a significant acceleration of the processing of *adapt15* precursor RNA following the exposure of HA-1 cells to hydrogen peroxide. This apparently accounts for some of the observed increase in the steady-state levels of mature *adapt15* mRNA that we see in these cells following oxidative stress. Exposure of cells to peroxide also leads to an apparent degradation of mitochondrial precusor RNAs, which may also involve precursor processing (Crawford et al., 1996c).

VII. OXIDANT STRESS-RELATED PATHOLOGIES

Oxidative stress has been linked, to varying extents, to a number of diseases and disorders (Cerutti, 1985; Sies, 1991). In fact, it is probable that most pathologies involve oxidative stress, at least to some extent. Identification of genes involved in the etiology of oxidant-mediated pathology has already led to important insights into the cellular response to stress and mechanisms of oxidant damage and will continue to do so as more and more involved genes are uncovered. These genes also represent potential clinical targets for oxidant-related disease and disorders. In fact, some are already being utilized or targeted in clinical studies. Examples of pathologies for which the aberrant expression of oxidant-modulated genes have been implicated are shown in Table II.

One of the most important targets for oxidant-related damage is the brain. For one, oxidative stress is associated with a wide range of brain pathologies, many of which are mediated by excitotoxicity. Excitotoxicity is thought to play an important role in neurodegeneration and is associated with the generation of reactive oxygen species (Bondy and LeBel, 1993). Neuronal loss through neurodegeneration underlies many neural disorders and diseases, including Alzheimer's disease, stroke and ischemia, seizure activity, Huntington's disease, aging, Parkinson's disease, and more. Second, the high lipid content of the brain makes lipid peroxides and aldehyde byproducts likely triggers for cellular damage and gene modulation. Finally, brain damage affects other parts of the body, e.g., muscular function. The oxidant-inducible genes heme oxygenase and c-*fos* are associated with brain damage and dysfunction as is the antioxidant gene Cu,Zn-superoxide dismu-

Table II Examples of Human Pathologies for which the Aberrant Expression of Oxidant-Modulated Genes Has Been Implicated

Pathology	Gene	Expression	Reference
Neurodegeneration	Heme oxygenase-1	Increased	Castellani *et al.* (1995)
Alzheimer's disease	Heme oxygenase-1	Increased	Schipper *et al.* (1995)
Parkinson's disease	Heme oxygenase-1	Increased	Schipper *et al.* (1998)
Brain trauma	c-*fos*	Increased	Sharp and Sagar (1994)
Down's syndrome	Cu,Zn-SOD	Increased	Schickler *et al.* (1989)
Down's syndrome	*adapt78*	Increased	Crawford *et al.* (1997)
Cancer	Mn-SOD	Decreased	St. Clair and Oberley (1991)
			Li *et al.* (1995)
			Yan *et al.* (1996)
Inflammation	Cyclooxygenase-2, IL-8	Increased	DeForge *et al.* (1993)
			Feng *et al.* (1995)
Cataractogenesis	c-*fos*	Increased	Spector (1995)
Atheroscherosis	VCAM-1	Increased	Marui *et al.* (1993)
HIV infection	Mn-SOD	Decreased	Flores *et al.* (1993)
Aging protection	Cu,Zn-SOD, catalase	Increased	Orr and Sohal (1994)

tase. Specifically, increased levels of heme oxygenase-1 protein are associated with the neurofibrillary pathology of Alzheimer's disease as well as progressive supranuclear palsy, subacute sclerosing panencephalitis, Pick disease, and corticobasal degeneration (Castellani et al., 1995). High levels of heme oxygenase-1 are also observed in the hippocampal astrocytes of Alzheimer's disease obtained postmortem (Schipper et al., 1995) and in the astroglia cells of the substantia nigra of postmortem idiopathic Parkinson's disease tissue (Schipper et al., 1998). c-fos induction is associated with a number of brain pathological stimuli, including ischemia, hypoxia, seizures, and cortical injury (Sharp and Sagar, 1994). Overexpression of Cu,Zn-superoxide dismutase has been implicated in Down's syndrome. It is located on human chromosome 21, the chromosome responsible for Down's where there is an extra copy of the region of this chromosome containing Cu,Zn-superoxide dismutase and other genes. Transgenic mice overexpressing the Cu,Zn-superoxide dismutase gene produce Down's-like phenotypes as well as increased lipid peroxidation (Schickler et al., 1989; Avraham et al., 1988). Paradoxically, brain nigral cells, the target of neuron degeneration in patients with Parkinson's disease, survive better when Cu,Zn-superoxide dismutase is overexpressed in transgenic mouse (Nakao et al., 1995). Another recently discovered gene, *adapt78*, also appears to be overexpressed in Down's syndrome patients (Crawford et al., 1997).

Cancer is also associated with oxidative stress. In laboratory model systems, chronic oxidant treatment has been shown to generate tumors in rodents, and reactive oxygen species generated from activated neutrophils transform target fibroblasts in cell culture (Slaga et al., 1981; Weitzman et al., 1985; Cerutti, 1985). Although the genes involved in these transfromations were not identified, subsequent studies clearly demonstrated that reactive oxygen species are capable of inducing the levels of cellular protooncogenes (Crawford and Cerutti, 1987; Shibanuma et al., 1988; Muehlematter et al., 1989; Nose et al., 1991). Classic tumor promotion involves the modulation of gene expression (Cerutti, 1985). Therefore, the observation that oxidant tumor promoters such as hydrogen peroxide and t-butyl hydroperoxide modulate cellular protooncogenes, which are mediators of tumor promotion, was consistent with a role for oxidative stress in tumor development (Sun et al., 1993). Tumor-suppressor genes are also involved in carcinogenesis. It has been demonstrated that a decreased level of Mn-SOD is associated with a number of tumor cell lines (St. Clair and Oberley, 1991). Thus, MnSOD may act as a cellular tumor suppressor gene, especially since overexpression of this gene leads to growth suppression (Li et al., 1995; Yan et al., 1996).

Another pathology associated with oxidative stress in diabetes. Several studies on antioxidant levels revealed modulations in expression in streptozotocin-induced diabetic rats. Kakkar *et al.*, (1995) found increased catalase levels in liver, heart, and pancreas but decreased levels in kidney; increased glutathione peroxidase in pancreas and kidney; and increased superoxide dismutase activity in liver, heart, and pancreas. In diabetic rats, increased mRNA expression for Cu,Zn superoxide dismutase and glutathione peroxidase and decreased catalase were observed (Reddi and Bollineni, 1997). Wohaieb and Godin (1987) observed increased catalase, glutathione reductase, and Cu,Zn superoxide dismutase in the pancreas and increased catalase in the heart. However, livers exhibited decreases in the levels of these enzymes. Analysis of human blood determined significantly higher glutathione peroxidase but lower superoxide dismutase activities in diabetic patients (Matkovics *et al.*, 1982). These results indicate a varied and complicated pattern of modulation of these enzymes, not only from model to model but also from tissue to tissue. Nonetheless, the modulation of antioxidant gene expression in diabetic mice is further evidence for the importance of oxidative stress in the etiology of the disease and its complications.

Ethanol consumption produces oxidative stress and therefore is one of the concerns in alcohol disease. The liver is the major target of ethanol toxicity. Chronic ethanol exposure of rats leads to decreased antioxidant factors, including glutathione peroxidase and glutathione levels in liver (Rouach *et al.*, 1997) and periferal rat nerve (Bosch-Morell *et al.*, 1998). However, the levels of catalase, glutathione peroxidase, glutathione reductase, and reduced glutathione were elevated in the livers of rats (Oh *et al.*, 1998); the level of manganese superoxide dismutase elevated in rat liver (Koch *et al.*, 1994); and the mRNA levels of catalase elevated, superoxide dismutase reduced, and glutathione peroxidase unaffected in rat brain, all on alcohol-supplemented diets (Omodeo-Sale *et al.*, 1997). Thus, although ethanol ingestion leads to oxidative stress and the modulation of expression of oxidant-related genes, the pattern of modulation appears to be quite varied, as also observed earlier for diabetes.

In addition, a number of other genes have been identified whose expressions are modulated by reactive oxygen species in association with a disease state. For example, reactive oxygen species are strongly involved in the induction of vascular cell adhesion molecule-1 (VCAM-1) by interleukin 1β (Marui *et al.*, 1993). Both oxidative stress and VCAM-1 expression are early features in the pathogenesis of atheroschlerosis and other inflammatory diseases. Cyclooxygenase-2 is also induced by reactive oxygen species and this induction may contribute to the deleterious amplification of prostanoids during inflammation (Feng *et al.*, 1995). Interleukin-8 is induced by reactive oxygen species, perhaps acting as an early signal to recruitment neutrophils during inflammation (DeForge *et al.*, 1993). During cataractogenesis, c-*fos* mRNA levels are elevated (Spector, 1995). Suppressed MnSOD levels are observed in cells infected with the human immunodeficiency virus (Flores *et al.*, 1993), possibly leading to increased oxidative stress, which appears to promote viral replication. Finally, the levels of antioxidant gene mRNAs appear to be important in aging (Orr and Sohal, 1994). Clearly, a number of diseases and disorders are associated with oxidative stress, and undoubtedly, more oxidant-modulated genes will be found to play critical roles.

Finally, the observation that mitochondrial RNA levels are modulated by reactive oxygen species may prove to have important clinical significance. In resting cells, the mitochondrion is the main site for the intracellular production of oxidants. A number of human diseases have now been linked to aberrations in the mitochondrial genome (Wallace, 1992; Taylor *et al.*, 1994; Bandy and Davison, 1990). Reactive oxygen species have been implicated in these alterations. This is not surprising since, as indicated previously, reactive oxygen species are generated continuously in the mitochondria, and this generation occurs proximal to its genome. Furthermore, mitochondrial DNA is not protected by histones. Some of these mitochondrial genome alterations may occur in genes whose RNA products are modulated by oxidative stress and may lead to an improper modulation during oxidative and other stress. We speculate that the degradation of mitochondrial RNAs in HA-1 fibroblasts, represents an early stage "shut down" of mitochondria, which may be important in combating the damaging effects of oxidative stress.

VIII. CONCLUDING REMARKS

The last decade or so has produced a dramatic increase in our knowledge and understanding of the modulation of gene expression by reactive oxygen species. This is a remarkable progression, given that it was not so long ago when it was uncertain as to whether reactive oxygen species were even important in the modulation of gene expression in mammals.

One obvious question is what is the function of mammalian oxidant-inducible genes? Predictably, the answer is many. The list of inducible mRNAs presented in Table I includes genes involved in transcriptional regulation, DNA damage response, DNA repair, cytoplasmic stress response, signal transduction, inflammation response, antioxidant enzymes, and others. Because oxidants affect gene expression, damage DNA, trigger intracellular signal transduction pathways, and induce a prooxidant state among other actions, modulation of the expression of these genes make intuitive sense. However, the induction of expression of some genes is not so obvious at first, e.g., heme oxygenase. In this case, subsequent studies revealed that induction of this enzyme is a protective response against oxidative stress, detoxifying the prooxidant heme, leading to the production of the antooxidants biliverdin and bilirubin, and releasing iron that induces a second protective protein, ferritin. Understanding the actions of the products of certain oxidant-modulated genes and why they might be involved in cellular oxidative stress response will be one challenge in the field in future years.

Clearly, much more study will be needed to elucidate the exact details of oxidant-modulated gene expression and the genes involved. Important areas of future research include the identification and characterization of the binding proteins to mRNAs modulated at the mRNA stability level; the continued identification and characterization of redox-sensitive transcriptional factors; the sequence identification of oxidant-response elements; the further elucidation of signal transduction pathways triggered by reactive oxygen species; identification of the cellular reductases and oxidases that act on gene regulatory factors; and the grouping of mediators into those involved in protective stress response, proliferation, differentiation, and apoptosis. The continued identification of oxidant-modulated genes is also important, including those modulated by reactive nitrogen species.

In performing these analyses, one cannot underestimate the importance of improving technological methodology. Many of the oxidant-modulated genes identified to date were detected using improved technology for the time, such as subtractive hybridization and differential display, and this will continue to be the case. Current techniques, such as restriction fragment differential display, SAGE, and gene chip hybridization combined with gene family analysis and bioinformatics, will identify more of these important genes, even those expressing at very low levels in the cells. The same holds true at the protein level, where protein chips and protein–protein interaction approaches will identify and define the roles of protein products in oxidant protection and signal transduction.

The studies described in this chapter have clearly contributed to the study and understanding of how mammalian cells respond to oxidative stress. Gradually, more and more of this knowledge will be used clinically. In fact, this approach is already being used from knowledge gained from so-called "adaptive reponse" studies. An adaptive response is a phenomena whereby exposure of biological cells, tissues, or even organisms to sublethal concentrations of a toxic agent elicits a protective cellular response that protects cells against the damaging effects of lethal concentrations. It has been described in several systems in response to oxidative stress (Crawford and Davies, 1994). These include bovine vascular endothelial cells following hydrogen peroxide exposure (Lu et al., 1993); HA-1 hamster cells in response to hydrogen peroxide (Crawford et al., 1996a,d; Wiese et al., 1995); and myocardial cells in response to several oxidant-based stresses (Das et al., 1995). Preconditioning studies based on adaptive response studies are now being used clinically and will eventually be used to treat oxidative-related pathologies. Oxidant-modulated genes identified in these and other systems represent potential pharmaceutical targets and will ultimately lead to new directions in the areas of detection and therapy of oxidative stress-related diseases and disorders.

REFERENCES

Alam, J., and Den, Z. (1992). Distal AP-1 binding sites mediate basal level enhancement and TPA induction of the mouse heme oxygenase-1 gene. *J. Biol. Chem.* **267,** 21894–21900.
Ammendola, R., Mesuraca, M., Russo, T., and Cimino, F. (1994). The DNA-binding efficiency of Sp1 is affected by redox changes. *Eur. J. Biochem.* **225,** 483–489.
Ammendola, R., Fiore, F., Esposito, F., Caserta, G., Mesuraca, M., Russo, T., and Cimino, F. (1995). Differentially expressed mRNAs as a consequence of oxidative stress in intact cells. *FEBS Lett.* **371,** 209–213.
Amstad, P. A., Krupitza, G., and Cerutti, P. A. (1992). Mechanism of c-fos induction by active oxygen. *Cancer Res.* **52,** 3952–3960.
Anderson, M. W., Reynolds, S. H., You, M., and Maronpot, R. M. (1992). Role of proto-oncogene activation in carcinogenesis. *Environ. Health Perspect.* **98,** 13–24.
Avraham, K. B., Schickler, M., Sapoznikov, D., Yarom, R., and Groner, Y. (1988). Down's syndrome: Abnormal neuromuscular junction in tongue of transgenic mice with elevated levels of human Cu/Zn-superoxide dismutase. *Cell (Cambridge, Mass.)* **54,** 823–829.
Bandy, B., and Davison, A. J. (1990). Mitochondrial mutations may increase oxidative stress: Implications for carcinogenesis and aging? *Free Radical Biol. Med.* **8,** 523–539.
Bergelson, S., Pinkus, R., and Daniel, V. (1994). Intracellular glutathione levels regulate Fos/Jun induction and activation of glutathione S-transferase gene expression. *Cancer Res.* **54,** 36–40.
Bindoli, A. (1988). Lipid peroxidation in mitochondria. *Free Radical Biol. Med.* **5,** 247–261.
Bondy, S. C., and LeBel, C. P. (1993). The relationship between excitotoxicity and oxidative stress in the central nervous system. *Free Radical Biol. Med.* **14,** 633–642.
Bosch-Morell, F., Martinez-Soriano, F., Colell, A., Fernandez-Checa, J. C., and Romero, F. J. (1998), Chronic ethanol feeding induces cellular antioxidants decrease and oxidative stress in rat periferal nerves. *Free Radical Biol. Med.* **25,** 365–368.
Burdon, R. H. (1995). Superoxide and hydrogen peroxide in relation to mammalian cell proliferation. *Free Radical Biol. Med.* **18,** 775–794.
Castellani, R., Smith, M. A., Richey, P. L., Kalaria, R., Gambetti, P., and Perry, G. (1995). Evidence for oxidative stress in Pick disease and corticobasal degeneration. *Brain Res.* **696,** 268–271.
Cerutti, P. A. (1985). Prooxidant states and tumor promotion. *Science* **227,** 375–381.
Christman, M. F., Morgan, R. W., Jacobson, F. S., and Ames, B. N. (1985). Positive control of a regulon for defenses against oxidative stress and some heat-shock proteins in Salmonella typhimurium. *Cell (Cambridge, Mass.)* **41,** 753–762.
Colinas, R. J., Hunt, D. H., Walsh, A. C., and Lawrence, D. A. (1996). Hydroquinone resistance in a murine myeloblastic leukemia cell line. *Biochem. Pharmacol.* **52,** 945–956.
Corral, M., Paris, B., Baffet, G., Tichonicky, L., Guguen-Guillouzo, C., Kruh, J., and Defer, N. (1989). Increased levels of the mitochondrial ND5 transcript in chemically induced rat hepatomas. *Exp. Cell Res.* **184,** 158–166.

Corral-Debrinski, M., Stepien, G., Shoffner, J. M., Lott, M. T., Kanter, K., and Wallace, D. C. (1991). Hypoxemia is associated with mitochondrial DNA damage and gene induction. Implications for cardiac disease. *JAMA, J. Am. Med. Assoc.* **266**, 1812–1816.

Crawford, D. R. (1999). Regulation of mammalian gene expression by reactive oxygen species. In "Reactive Oxygen Species in Biological Systems" (D. Gilbert and C. Colton, eds.), Plenum, New York pp. 155–171.

Crawford, D. R., and Cerutti, P. A. (1987). Expression of oxidant stress-related genes in tumor promotion of mouse epidermal cells JB6. In "Proceedings of the Second Conference on Anticarcinogens and Radioprotectors" (O. F. Nygaard, ed.), pp. 183–190. Plenum, New York.

Crawford, D. R., and Davies, K. J. A. (1994). Adaptive response and oxidative stress. *Environ. Health Perspect.* **102**, 25–28.

Crawford, D. R., and Davies, K. J. A. (1997). Modulation of a cardiogenic shock inducible RNA by chemical stress: Adapt73/PigHep3. *Surgery* **121**, 581–587.

Crawford, D. R., Zbinden, I., Amstad, P. and Cerutti, P. A. (1988). Oxidant stress induces the proto-oncogenes c-fos and c-myc in mouse epidermal cells. *Oncogene* **3**, 27–32.

Crawford, D. R., Edbauer-Nechamen, C. A., Lowry, C. V., Salmon, S. L., Kim, Y. K., Davies, J. M. S., and Davies, K. J. A. (1994). Assessing gene-expression during oxidative stress. In "Methods in Enzymology" (L. Packer, ed.), Vol. 234, pp. 175–217. Academic Press, San Diego, CA.

Crawford, D. R., Edbauer-Nechamen, C. A., Schools, G. P., Salmon, S. L., Davies, J. M. S., and Davies, K. J. A. (1995), Oxidant-modulated gene expression. In "The Oxygen Paradox" (K. J. A. Davies, ed.), pp. 327–336, Cooperativa Libraria Editrice Università di Padova, Padova, Italy Press.

Crawford, D. R., Schools, G. P., Salmon, S. L., and Davies, K. J. A. (1996a). Hydrogen peroxide induces the expression of *adapt15*, a novel RNA associated with polysomes in hamster HA-1 cells. *Arch. Biochem. Biophys.* **325**, 256–264.

Crawford, D. R., Leahy, K. P., Wang, Y., Schools, G. P., Kochheiser, J. C., and Davies, K. J. A. (1996b). Oxidative stress induces the levels of a mafG homolog in hamster HA-1 cells. *Free Radical Biol. Med.* **21**, 521–525.

Crawford, D. R., Wang, Y., Schools, G. P., Kochheiser, J., and Davies, K. J. A. (1996c). Downregulation of mammalian mitochondrial RNAs during oxidative stress. *Free Radical Biol. Med.* **22**, 551–559.

Crawford, D. R., Schools, G. P., and Davies, K. J. A. (1996d). Oxidant-inducible *adapt15* is associated with growth arrest and DNA damage-inducible *gadd153* and *gadd45*. *Arch. Biochem. Biophys.* **329**, 137–144.

Crawford, D. R., Leahy, K. P., Abramova, N., Lan L., Wang, Y., and Davies, K. J. A. (1997). Hamster *adapt78* mRNA is a Down syndrome critical region homologue that is inducible by oxidative stress. *Arch. Biochem. Biophys.* **342**, 6–12.

Das, D. K., Maulik, N., and Moraru, I. I. (1995). Gene expression in acute myocardial stress. Induction by hypoxia, ischemia, reperfusion, hyperthermia and oxidative stress. *J. Mol. Cell. Cardiol.* **27**, 181–193.

DeForge, L. E., Preston, A. M., Takeuchi, E., Kenney, J., Boxer, L. A., and Remick, D. G. (1993). Regulation of interleukin 8 gene expression by oxidant stress. *J. Biol. Chem.* **268**, 25568–25576.

Devary, Y., Gottlieb, R. A., Lau, L. F., and Karin, M. (1991). Rapid and preferential activation of the c-jun gene during the mammalian UV response. *Mol. Cell. Biol.* **11**, 2804–2811.

Farber, J. L., Kyle, M. E., and Coleman, J. B. (1990). Mechanisms of cell injury by activated oxygen species. *Lab. Invest.* **62**, 670–679.

Fargnoli, J., Holbrook, N. J., and Fornace, A. J., Jr., (1990), Low-ratio hybridization subtraction. *Anal. Biochem.* **187**, 364–373.

Feng, L., Xia, Y., Garcia, G. E., Hwang, D., and Wilson, C. B. (1995). Involvement of reactive oxygen intermediates in cyclooxygenase-2 expression induced by interleukin-1, tumor necrosis factor-alpha, and lipopolysaccharide. *J. Clin. Invest.* **95**, 1669–1675.

Filippov, G., Bloch, D. B., and Bloch, K. D. (1997). Nitric oxide decreases stability of mRNAs encoding soluble guanylate cyclase subunits in rat pulmonary artery smooth muscle cells. *J. Clin. Invest.* **100**, 942–948.

Flores, S. C., Marecki, J. C., Harper, K. P., Bose, S. K., Nelson, S. K., and McCord, J. M. (1993). Tat protein of human immunodeficiency virus type 1 represses expression of manganese superoxide dismutase in HeLa cells. *Proc. Natl. Acad. Sci. U. S. A.* **90**, 7632–7636.

Fornace, A. J., Jr., Nebert, D. W., Hollander, M. C., Luethy, J. D., Papathanasiou, M., Fargnoli, J., and Holbrook, N. J. (1989). Mammalian genes coordinately regulated by growth arrest signals and DNA-damaging agents. *Mol. Cell. Biol.* **9**, 4196–4203.

Fornace, A. J., Jr., Jackman, J., Hollander, M. C., Hoffman-Liebermann, B., and Liebermann, D. A. (1992). Genotoxic-stress-response genes and growth-arrest genes. gadd, MyD, and other genes induced by treatments eliciting growth arrest. *Ann. N. Y. Acad. Sci.* **663**, 139–153.

Gomer, C. J., Ferrario, A., Rucker, N., Wong, S., and Lee, A. S. (1991). Glucose regulated protein induction and cellular resistance to oxidative stress mediated by porphyrin photosensitization. *Cancer Res.* **51**, 6574–6579.

Greenberg, J. T., Monach, P., Chou, J. H., Josephy, P. D., and Demple, B. (1990). Positive control of a global antioxidant defense regulon activated by superoxide-generating agents in *Escherichia coli*. *Proc. Natl. Acad. Sci. U.S.A* **87**, 6181–6185.

Grosch, S., Fritz, G., and Kaina, B. (1998). Apurinic endonuclease (Ref-1) is induced in mammalian cells by oxidative stress and involved in clastogenic adaptation. *Cancer Res.* **58**, 4410–4416.

Guyton, K. Z., Liu, Y., Gorospe, M., Xu, Q., and Holbrook, N. J. (1996). Activation of mitogen-activated protein kinase by H2O2. Role in cell survival following oxidant injury. *J. Biol. Chem.* **271**, 4138–4142.

Gwinner, W., Tisher, C. C., and Nick, H. S. (1995). Regulation of manganese superoxide dismutase in glomerular epithelial cells: Mechanisms for interleukin 1 induction. *Kidney Int.* **48**, 354–362.

Hartsfield, C. L., Alam, J., Cook, J. L., and Choi, A. M. (1997). Regulation of heme oxygenase-1 gene expression in vascular smooth muscle cells by nitric oxide. *Am. J. Physiol.* **273**, L980–L988.

Hassan, H. M., and Fridovich, I. (1977). Regulation of the synthesis of superoxide dismutase in Escherichia coli. *J. Biol. Chem.* **252**, 7667–7672.

Ichiki, T., Usui, M., Kato, M., Funakoshi, Y., Ito, K., Egashira, K., and Takeshita, A. (1998). Downregulation of angiotension II type 1 receptor gene transcription by nitric oxide. *Hypertension* **31**, 342–438.

Janssen, Y. M., Matalon, S., and Mossman, B. T. (1997). Differential induction of c-fos, c-jun, and apoptosis in lung epithelial cells exposed to ROS or RNS. *Am. J. Physiol.* **273**, L789–L796.

Johnston, H. M., and Morris, B. J. (1995). N-methyl-D-asparate and nitric oxide regulate the expression of calcium/calmodulin-dependent kinase II in the hippocampal dentate gyrus. *Brain Res.* **31**, 141–150.

Kakkar, R., Kalra, J., Mantha, S. V., and Prasad, K. (1995). Lipid peroxidation and activity of antioxidant enzymes in diabetic rats. *Mol. Cell. Biochem.* **151**, 113–119.

Keyse, S. M., and Emslie, E. A. (1992). Oxidative stress and heat shock induce a human gene encoding a protein-tyrosine phosphatase. *Nature (London)* **359**, 644–647.

Keyse, S. M., and Tyrrell, R. M. (1989). Heme oxygenase is the major 32-kDa stress protein induced in human skin fibroblasts by UVA radiation, hydrogen peroxide, and sodium arsenite. *Proc. Natl. Acad. Sci. U.S.A.* **86,** 99–103.

Kim, A. T., Sarafian, T. A., and Shau, H. (1997). Characterization of antioxidant properties of natural killer-enhancing factor-B and induction of its expressionby hydrogen peroxide. *Toxicol. Appl. Pharmacol.* **147,** 135–142.

Koch, O. R., De Leo, M. E., Borrello, S., Palombibi, G., and Galeotti, T. (1994). Ethanol treatment up-regulates the expression of mitochondrial manganese superoxide dismutase in rat liver. *Biochem. Biophys. Res. Commun.* **201,** 1356–1365.

Kristal, B. S., Chen, J., and Yu, B. P. (1994a). Sensitivity of mitochondrial transcription to different free radical species. *Free Radical, Biol. Med.* **16,** 323–329.

Kristal, B. S., Kim, J. D., and Yu, B. P. (1994b). Tissue-specific susceptibility to peroxyl radical-mediated inhibition of mitochondrial transcription. *Redox Rep.* **1,** 51–55.

Kugelman, A., Choy, H. A., Liu, R., Shi, M. M., Gozal, E., and Forman, H. J. (1994), gamma-Glutamyl transpeptidase is increased by oxidative stress in rat alveolar L2 epithelial cells. *Am. J. Respir. Cell Mol. Biol.* **11,** 586–592.

Kullik, I., and Storz, G. (1994). Transcriptional regulators of the oxidative stress response in prokaryotes and eukaryotes. *Redox Rep.* **1,** 23–29.

Lai, C.-C., Peng, M., Huang, L., Huang, W.-H., and Chiu, T. H. (1996). Chronic exposure of neonatal cardiac myocytes to hydrogen peroxide enhances the expression of catalase. *J. Mol. Cell. Cardiol.* **28,** 1157–1163.

Lezoualc'h, F., Sagara, Y., Holsboer, F., and Behl, C. (1998). High constitutive NF-κB activity mediates resistance to oxidative stress in neuronal cells. *J. Neurosci.* **18,** 3224–3232.

Li, J. J., Oberley, L. W., St. Clair, D. K., Ridnour, L. A., and Oberley, T. D. (1995). Phenotypic changes induced in human breast cancer cells by overexpression of manganese-containing superoxide dismutase. *Oncogene* **10,** 1989–2000.

Liang, P., and Pardee, A. B. (1992). Differential display of eukaryotic messenger RNA by means of the polymerase chain reaction. *Science* **257,** 967–971.

Lu, D., Maulik, N., Moraru, I. I., Kreutzer, D. L., and Das, D. K. (1993). Molecular adaptation of vascular endothelial cells to oxidative stress. *Am. J. Physiol.* **264,** C715–C722.

Marui, N., Offermann, M. K., Swerlick, R., Kunsch, C., Rosen, C. A., Ahmad, M., Alexander, R. W., and Medford, R. M. (1993). Vascular cell adhesion molecule-1 (VCAM-1) gene transcription and expression are regulated through an antioxidant-sensitive mechanism in human vascular endothelial cells. *J. Clin. Invest.* **92,** 1866–1874.

Matkovics, B., Varga, S. I., Szabo, L., and Witas, H. (1982). The effect of diabetes on the activities of the peroxide metabolism enzymes. *Horm. Metab. Res.* **14,** 77–79.

Morris, B. J. (1995). Stimulation of immediate early gene expression in striatal neurons by nitric oxide. *J. Biol. Chem.* **270,** 24740–24744.

Muehlematter, D., Ochi, T., and Cerutti, P. (1989). Effects of tert-butyl hydroperoxide on promotable and non-promotable JB6 mouse epidermal cells. *Chem.-Biol. Interact.* **71,** 339–352.

Nakao, N., Frodl, E. M., Widner, H., Carlson, E., Eggerding, F. A., Epstein, C. J., and Brundin, P. (1995). Overexpressing Cu/Zn superoxide dismutase enhances survival of transplanted neurons in a rat model of Parkinson's disease. *Nat. Med.* **1,** 226–231.

Nose, K., Shibanuma, M., Kikuchi, K., Kageyama, H., Sakiyama, S., and Kuroki, T. (1991). Transcriptional activation of early-response genes by hydrogen peroxide in a mouse osteoblastic cell line. *Eur. J. Biochem.* **201,** 99–106.

Oh, S. I., Kim, C.-I., Chun, H. J., and Park, S. C. (1998). Chronic ethanol consumption affects glutathione status in rat liver. *J. Nutr.* **128,** 758–763.

Ohki, K., Yoshida, K., Hagiwara, M., Harada, T., Takamura, M., Ohashi, T., Matsuda, H., and Imaki, J. (1995). Nitric oxide induces c-fos gene expression via cyclic AMP response element binding protein (CREB) phosphorylation in rat retinal pigment epithelium. *Brain Res.* **696,** 140–144.

Omodeo-Sale, F., Gramigna, D., and Campaniello, R. (1997). Lipid peroxidation and antioxidant systems in rat brain: Effect of chronic alcohol consumption. *Neurochem. Res.* **22,** 577–582.

Orr, W. C., and Sohal, R. S. (1994). Extension of life-span by overexpression of superoxide dismutase and catalase in Drosophila melanogaster. *Science* **263,** 1128–1130.

Pinkus, R., Weiner, L. M., and Daniel, V. (1995). Role of quinone-mediated generation of hydroxyl radicals in the induction of glutathione S-transferase gene expression. *Biochemistry* **34,** 81–88.

Polyak, K., Xia, Y., Zweier, J. L., Kinzler, K. W., and Vogelstein, B., (1997). A model for p53-induced apoptosis. *Nature (London)* **389,** 300–305.

Reddi, A. S., and Bollineni, J. S. (1997). Renal cortical expression of mRNAs for antioxidant enzymes in normal and diabetic rats. *Biochem. Biophys. Res. Commun.* **235,** 598–601.

Rouach, H., Fataccioli, V., Gentil, M., French, S. W., Morimoto, M., and Nordmann, R. (1997). Effect of chronic ethanol feeding on lipid peroxidation and protein oxidation in relation to liver pathology. *Hepatology* **25,** 351–355.

Rushmore, T. H., and Pickett, C. B. (1990). Transcriptional regulation of the rat glutathione S-transferase Ya subunit gene. Characterization of a xenobiotic-responsive element controlling inducible expression by phenolic antioxidants. *J. Biol. Chem.* **265,** 14648–14653.

Rushmore, T. H., Morton, M. R., and Pickett, C. B. (1991). The antioxidant responsive element. Activation by oxidative stress and identification of the DNA consensus sequence required for functional activity. *J. Biol. Chem.* **266,** 11632–11639.

Sagara, Y., Dargusch, R., Klier, F. G., Schubert, D., and Behl, C. (1996). Increased antioxidant enzyme activity in amyloid β protein-resistant cells. *J. Neurosci.* **16,** 497–505.

Sagara, Y., Dargusch, R., Chambers, D., Davis, J., Schubert, D., and Maher, P. (1998). Cellular mechanisms of resistance to chronic oxidative stress *Free Radical Biol. Med.* **24,** 1375–1389.

Sawada, M., Sofuni, T., and Ishidate, M. (1988). Induction of chromosomal aberrations in active oxygen-generating systems. *Muta. Res.* **197,** 133–140.

Schickler, M., Knobler, H., Avraham, K. B., Elroy-Stein, O., and Groner, Y. (1989). Diminished serotonin uptake in platelets of transgenic mice with increased Cu/Zn-superoxide dismutase activity. *EMBO J.* **8,** 1385–1392.

Schipper, H. M., Cisse, S., and Stopa, E. G. (1995). Expression of heme oxygenase-1 in the senescent and Alzheimer-diseased brain. *Ann. Neurol.* **37,** 758–768.

Schipper, H. M., Liberman, A., and Stopa, E. G. (1998). Neural heme oxygenase-1 expression in idiopathic Parkinson's disease. *Exp. Neurol.* **150,** 60–68.

Schreck, R., Albermann, K., and Baeuerle, P. A. (1992). Nuclear factor kappa B: An oxidative stress-responsive transcription factor of eukaryotic cells. *Free Radical Res. Commun.* **17,** 221–237.

Sharp, F. R., and Sagar, S. M. (1994). Alterations in gene expression as an index of neuronal injury: Heat shock and the immediate early gene response. *Neurotoxicology* **15,** 51–59.

Shay, J. W., and Werbin, H. (1987). Are mitochondrial DNA mutations involved in the carcinogenic process? *Mutat. Res.* **186,** 149–160.

Shibanuma, M., Kuroki, T., and Nose, K. (1988). Induction of DNA replication and expression of proto-oncogene c-myc and c-fos in quiescent Balb/3T3 cells by xanthine/xanthine oxidase. *Oncogene* **3,** 17–21.

Shull, S., Heintz, N. H., Periasamy, M., Manohar, M., Janssen, Y. M., Marsh, J. P., and Mossman, B. T. (1991). Differential regulation of antioxidant enzymes in response to oxidants. *J. Biol. Chem.* 266, 24398–24403.

Sies, H. (1991). Oxidative stress: From basic research to clinical application. *Am. J. Med.* 91, 31S–38S.

Slaga, T. J., Klein-Szanto, A. J., Triplett, L. L., Yotti, L. P., and Trosko, K. E. (1981). Skin tumor-promoting activity of benzoyl peroxide, a widely used free radical-generating compound. *Science* 213, 1023–1025.

Smith, M. L., Chen, I. T., Zhan, Q., Bae, I., Chen, C. Y., Gilmer, T. M., Kastan, M. B., O'Connor, P. M., and Fornace, A. J., Jr. (1994). Interaction of the p53-regulated protein Gadd45 with proliferating cell nuclear antigen. *Science* 266, 1376–1380.

Spector, A. (1995). Oxidative stress-induced cataract: Mechanism of action. *FASEB J.* 9, 1173–1182.

Spitz, D. R., Elwell, J. H., Sun, Y., Oberley, L. W., Oberley, T. D., Sullivan, S. J., and Roberts, R. J. (1990). Oxygen toxicity in control and hydrogen H_2O_2-resistant Chinese hamster fibroblast cell lines. *Arch. Biochem. Biophys.* 279, 249–260.

St. Clair, D. K., and Oberley, L. W. (1991). Manganese superoxide dismutase expression in human cancer cells: A possible role of mRNA processing. *Free Radical Res. Commun.* 12-13 (Pt. 2), 771–778.

Stevens, J. B., and Autor, A. P. (1977). Induction of superoxide dismutase by oxygen in neonatal rat lung. *J. Biol. Chem.* 252, 3509–3514.

Storz, G., and Tartaglia, L. A. (1992). OxyR: A regulator of antioxidant genes. *J. Nutr.* 122, 627–630.

Stralin, P., and Marklund, S. L. (1994). Effects of oxidative stress on expression of extracellular superoxide dismutase, CuZn-superoxide dismutase and Mn-superoxide dismutase in human dermal fibroblasts. *Biochem. J.* 298, 347–352.

Sullivan, S. J., Oberley, T. D., Roberts, R. J., and Spitz, D. R. (1992). A stable O_2 resistant cell line: Role of lipid peroxidation byproducts in O_2 mediated injury. *Am. J. Physiol.* 262, L748–L756.

Sun, Y., Colburn, N. H., and Oberley, L. W. (1993). Decreased expression of manganese superoxide dismutase mRNA and protein after immortalization and transformation of mouse liver cells. *Oncol. Res.* 5, 127–132.

Sundaresan, M., Yu, Z. X., Ferrans, V. J., Irani, K., and Finkel, T. (1995). Requirement for generation of H_2O_2 for platelet-derived growth factor signal transduction. *Science* 270, 296–299.

Suzuki, K., Tatsumi, H., Satoh, S., Senda, T., Nakata, T., Fujii, J., and Taniguchi, N. (1993). Manganese-superoxide dismutase in endothelial cells: Localization and mechanism of induction. *Am. J. Physiol.* 265, H1173–H1178.

Taylor, R. W., Birch-Machin, M. A., Bartlett, K., Lowerson, S. A., and Turnbull, D. M. (1994). The control of mitochondrial oxidations by complex III in rat muscle and liver mitochondria. Implications for our understanding of mitochondrial cytopathies in man. *J. Biol. Chem.* 269, 3523–3528.

Tyrrell, R. M., Applegate, L. A., and Tromvoukis, Y. (1993). The proximal promoter region of the human heme oxygenase gene contains elements involved in stimulation of transcriptional activity by a variety of agents including oxidants. *Carcinogenesis (London)* 14, 761–765.

Vallis, K. A., and Wolf, C. R. (1996). Relationship between the adaptive response to oxidants and stable menadione-resistance in Chinese hamster ovary cell lines. *Carcinogenesis (London)* 17, 649–654.

Velculescu, V. E., Zhang, L., Vogelstein, B., and Kinzler, K. W. (1995). Serial analysis of gene expression. *Science* **270**, 484–487.

Vincent, F., Corral-Debrinski, M., and Adolphe, M. (1994). Transient mitochondrial transcript level decay in oxidative stressed chondrocytes. *J. Cell. Phys.* **158**, 128–132.

Wallace, D. C. (1992). Mitochondrial genetics: A paradigm for aging and degenerative diseases? *Science* **256**, 628–632.

Wang, Y., Crawford, D. R., and Davies, K. J. A. (1996). adapt33, a novel oxidant-inducible RNA from hamster HA-1 cells. *Arch. Biochem. Biophys.* **332**, 255–260.

Weitzman, S. A., Weitberg, A. B., Clark, E. P., and Stossel, T. P. (1985). Phagocytes as carcinogens: Malignant transformation produced by human neutrophils. *Science* **227**, 1231–1233.

Wiese, A. G., Pacifici, R. E., and Davies, K. J. A. (1995). Transient adaptation to oxidative stress in mammalian cells. *Arch. Biochem. Biophys.* **318**, 231–240.

Winquist, L., Rannug, U., Rannug, A., and Ramel, C. (1984). Protection from toxic and mutagenic effects of H_2O_2 by catalase induction in *Salmonella typhimurium*. *Mutat. Res.* **141**, 145–147.

Wohaieb, S. A., and Godin, D. V. (1987). Alterations in free radical tissue-defense mechanisms in streptozocin-induced diabetes in rat. *Diabetes* **36**, 1014–1018.

Xanthoudakis, S., Miao, G. G., and Curran, T. (1994). The redox and DNA-repair activities of Ref-1 are encoded by nonoverlapping domains. *Proc. Natl. Acad. Sci. U.S.A.* **91**, 23–27.

Yamada, M., Hashinaka, K., Inazawa, J., and Abe, T. (1991). Expression of catalase and myeloperoxidase genes in hydrogen peroxide-resistant HL-60 cells. *DNA Cell Biol.* **10**, 735–742.

Yamamoto, N., Maki, A., Swann, J. D., Berezesky, I. K., and Trump, B. F. (1993). Induction of immediate early and stress genes in rat proximal tubule epithelium following injury: The significance of cytosolic ionized calcium. *Renal Failure* **15**, 163–171.

Yan, T., Oberley, L. W., Zhong, W., and St Clair, D. K. (1996). Manganese-containing superoxide dismutase overexpression causes phenotypic reversion in SV40-transformed human lung fibroblasts. *Cancer Res.* **56**, 2864–2871.

Yashpe-Purer, Y., Henis, Y., and Yashpe, J. (1977). Regulation of catalase level in *Escherichia coli* K12. *Can. J. Microbiol.* **23**, 84–91.

Zhan, Q., Lord, K. A., Alamo, I., Jr., Hollander, M. C., Carrier, F., Ron, D., Kohn, K. W., Hoffman, B., Liebermann, D. A., and Fornace, A. J., Jr. (1994). The gadd and MyD genes define a novel set of mammalian genes encoding acidic proteins that synergistically suppress cell growth. *Mol. Cell. Biol.* **14**, 2361–2371.

Zhang, Y., Marcillat, O., Giulivi, C., Ernster, L., and Davies, K. J. (1990). The oxidative inactivation of mitochondrial electron transport chain components and ATPase. *J. Biol. Chem.* **265**, 16330–16336.

3 Rac, Superoxide, and Signal Transduction

Hamdy H. Hassanain and
Pascal J. Goldschmidt-Clermont
Heart and Lung Institute
Ohio State University
Columbus, Ohio 43210

GTP-binding proteins of the Ras superfamily regulate a wide variety of cellular activities. The Rac protein is a member of the Rho branch of this extended family. Rac has been implicated in the processes involved in assembly of the actin cytoskeleton, stimulating the formation of lamellipodia and membrane ruffles. In phagocytic cells, Rac induces the activation of the NADPH oxidase, leading to free radical production, a key step for the successful elimination of infectious agents. The intracellular pathways leading to the generation of reactive oxygen species (ROS) are best characterized in these specialized cells. In nonphagocytic cells, Rac also regulates the production of intracellular superoxide and other reactive oxygen species, shown to act as essential intracellular second messengers for several cytokines and growth factors. For example, Rac functions as part of a redox-dependent signal transduction pathway leading to NF-κB activation. In plants, resistance to pathogens such as fungi, bacteria, and viruses depends on the release of ROS, which is termed the oxidative burst. Rac molecules play an important role in this defense mechanism against plant pathogens by regulating the production of the intracellular ROS. The Rac gene has been conserved throughout evolution, such that its protein would regulate the production of ROS in most cells and can transduce its signal pathway in mammals as well as in plants. The amazing conservation of the effector domain of Rac proteins, as well as of other proteins responsible for the oxidative burst, over a time span of 1 billion years of evolution, may indicate that Rac regulatory functions and interaction with proteins are essential.

I. INTRODUCTION

GTP-binding proteins of the Ras superfamily regulate a wide variety of cellular activities. These proteins function as molecular switches that cycle between the active GTP-bound state, which interacts with downstream targets, and the inactive GDP-bound state (Hall, 1994; Van Aelst and D'Souza-Schorey, 1997; Boguski and McCormick, 1993; Irani and Goldschmidt-Clermont, 1998). The level of the active form depends on the balance between the rate of activation-induced GDP/GTP exchange and the rate of inactivation, i.e., hydrolysis of GTP (Fig. 1). Both of these reactions are very slow and are controlled by the GTPase-activating protein (GAP) and guanine nucleotide-exchange factors (GEF), either accelerator or inhibitor of the exchange process (Parrini *et al.*, 1997). Mammalian cells contain an estimated 50 to 100 different GTP-binding proteins involved in regulating

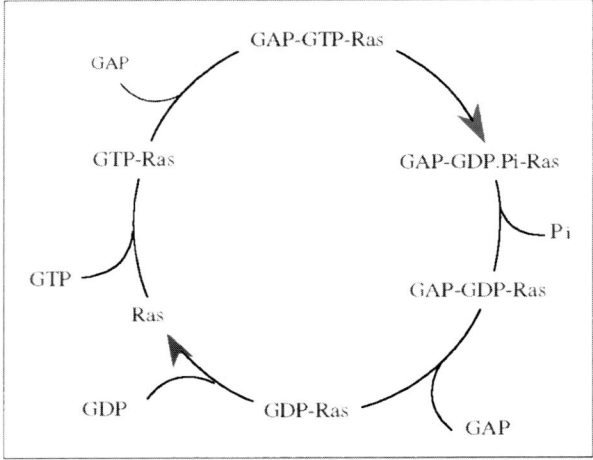

Figure 1 Ras nucleotide triphosphatase cycle. The top arrow indicates the irreversible step in the cycle. The bottom arrow corresponds to the reaction that is accelerated by exchanger molecules. Ras bound to GTP represents the activated conformation of Ras. P_i, inorganic phosphate; GAP, GTPase-activating protein. (**See color reproduction in color plate section.**)

many different biological processes, ranging from protein biosynthesis and membrane trafficking to communicating signals from the outside of the cell to the inside (Geyer and Wittinghofer, 1997). GTP-binding proteins contain a set of five conserved sequence elements by which they can be identified easily as being a member of this class of proteins. These elements are necessary for guanine nucleotide and Mg^{2+} binding for GTPase activity and a conformational switch site (Geyer and Wittinghofer, 1997). Furthermore, almost all Ras-related proteins contain a C-terminal cysteine-containing motif that serves as attachment points for one or two prenyl groups such as farnesyl or geranyl (de Vos *et al.*, 1988; Lowy *et al.*, 1993). Enzymes required for posttranslational modifications, such as farnesyl transferase, are therefore important regulators of functions. Such posttranslational modifications are necessary for membrane binding and biological activity (Omer and Gibbs, 1994).

The Ras superfamily can be subdivided further based on sequence similarities into Ras, Rho/Rac, Rab, Ran, Rad, and Arf subfamilies. These subfamilies can be associated with specific different biological functions of their members (Boguski and McCormick, 1993). Rho/Rac GTPases form a subgroup of the Ras superfamily of 20- to 30-kDa GTP-binding proteins that regulate a wide spectrum of cellular functions. These proteins are expressed ubiquitously across species, from yeast to human. Mammalian Rho-like GTPases comprise at least 10 distinct proteins: RhoA, B, C, D, and E; Rac1 and 2; RacB and E; and Cdc42 and TC10 (Van Aelst and D'Souza-Schorey, 1997). Rho proteins from various species are conserved, as they share 50–55% homology in their amino acid sequences. Research on Ras and Rho proteins has been facilitated considerably by the fact that dominant negative inhibitory forms of each protein can be obtained by mutating the highly conserved amino acid 17 (Ras numbering) from serine/threonine to asparagine. This mutation in Ras, N17Ras, has long used as a dominant inhibitor of Ras function, but the discovery that the same mutation in Rac generated a dominant inhibitor (Ridley *et al.*, 1992) has led to the introduction of similar mutations in many GTPases for the study of their function in cells. Similarly, a number of mutations have been used to activate Ras constitutively. Most mutations that lock Ras in the activated state result in the loss of ability of Ras to hydrolyze GTP. Alternatively, mutations that accelerate the exchange of the bound GDP, which in cells containing a large excess of GTP over GDP increase Ras interaction with GTP, can also induce the constitutive activation of Ras. Valine 12 and its mutation have been used to study the effect of Ras-related proteins in cells. Using a combination of dominant negative and constitutively activated mutants of Rac, it has been shown that Rac can regulate many processes, including actin organization, cytokinesis, transcription, secretion, the activity of NADPH oxidase of phagocytes, and endocytosis (Ridley *et al.*, 1992). A substantial challenge for the field has been to sort out the signaling pathways involved in each of these responses, signaling reactions that are likely to provide specificity to the system.

In phagocytic cells, Rac1 induces activation of the NADPH oxidase, leading to free radical production. Intracellular pathways leading to the generation of ROS are best characterized in these specialized cells. The superoxide-free radical ($O_2^{\cdot-}$) is generated by the multimolecular β-nicotinamide adenine dinucleotide phosphate (NADPH)–oxidase complex, which includes Rac, as a regulator for the stability of the enzymatic complex (Abo *et al.*, 1991; Diekmann *et al.*, 1994; Heyworth *et al.*, 1993; Knaus *et al.*, 1991). In nonphagocytic cells, several reports have suggested that Rac1 regulates the production of intracellular reactive oxygen species that were shown to act as essential intracellular second messengers for several cytokines and growth factors (Baeuerle and Henkel, 1994; Siebenlist *et al.*, 1994). For example, Rac1 functions as part of a redox-dependent signal transduction pathway leading to NF-κB activation (Sulciner *et al.*, 1996).

II. ACTIVATION OF RAC PROTEINS

Rho, Rac, and Cdc42 are bound predominantly in the cytoplasm to proteins known as GDP dissociation inhibitors (GDIs). Two GDIs have been characterized so far and bind tightly to all Rho family proteins (Ridley, 1995). This interaction is dependent on the posttranslational modification (by prenylation) of Rho family proteins at the C terminus (Boguski and McCormick, 1993). Binding to GDIs prevents the interaction of these proteins with the plasma membrane. The first step in the activation of Rho family proteins is believed to require dissociation of the small GTP-binding proteins from GDIs. In fact, when cells are activated, an increase in the membrane localization of Rac and Cdc42 has been observed (Aepfelbacher *et al.*, 1994; Quinn *et al.*, 1993). Next, to convert Rac to its active form after dissociating from GDIs, exchange of GDP for GTP is required (Ridley, 1996). Like all other members of the Ras superfamily, the activity of Rac is determined by the ratio of their GTP/GDP-bound forms in the cell. The ratio of these two forms is regulated by the opposing effects of guanine nucleotide exchange factors (GEFs), which enhance the exchange of bound GDP for GTP as well as the GTpase-activating proteins (GAPs), which increase the intrinsic rate of hydrolysis of bound GTP (Van Aelst and D'Souza-Schorey, 1997). In case of Rho GTPases, it is believed that Dbl homology (DH)-containing proteins represent cellular Rho GEFs; indeed, a number of DH-containing proteins have been shown to act as GEFs *in vitro*

(Cerione and Zheng, 1996). The DH family comprises a number of proteins that share two conserved domains: a DH domain of approximately 250 amino acids for which the Dbl oncogene product is the prototype and a pleckstrin homology (PH) domain of approximately 100 amino acids (Cerione and Zheng, 1996). The Dbl oncogene was originally discovered for its ability to induce tumorigenicity when expressed in NIH 3T3 cells (Eva and Aaronson, 1985). Biochemical analysis has shown that Dbl is able to release GDP from the human homolog of Cdc42 *in vitro*. Furthermore, structure and function analysis of Dbl demonstrated that the DH domain was essential and sufficient for this activity and that the Dbl domain was also necessary to induce oncogenicity (Hart *et al.*, 1991; Hart and Roberts, 1994). Since the discovery of Dbl, a growing number of mammalian proteins containing both a DH and a PH domain have been reported (Cerione and Zheng, 1996). GEF proteins such as Tiam 1, Vav, Trio, Ost, Bcr, Abr, and Ect2 showed specificity for Rac as well as Rho and Cdc42 *in vitro* (Chuang *et al.*, 1995; Crespo *et al.*, 1997; Debant *et al.*, 1996; Habets *et al.*, 1994). However, Tiam l is now known to act as Rac-specific GEF. The Tiam l gene was originally identified as being capable of conferring an invasive phenotype when introduced into a noninvasive lymphoma cell line, and indeed Rac itself is capable of inducing an invasive phenotype in these cells (Van Aelst and D'Souza-Schorey, 1997). Furthermore, overexpression of Tiam 1 in fibroblasts elicited the formation of membrane ruffles and activation of JNK in a Rac-dependent manner and that the intact amino-terminal PH was essential for these activities (Michiels *et al.*, 1995, 1997).

The first GAP protein specific for the Rho subfamily was purified by biochemical analysis and was named p50Rho-GAP (Van Aelst and D'Souza-Schorey, 1997). This protein was shown to have GAP activity toward Rho, Cdc42, and Rac *in vitro* (Hall, 1990; Lancaster *et al.*, 1994). Additional proteins that exhibit GAP activity for Rho GTPases have been identified in mammalian, *Sacramycese cerevisiae, Drosophila* and *C. elegans* (Van Aelst and D'Souza-Schorey, 1997). These proteins shared a related GAP domain that spans 140 amino acids, but bears no significant resemblance to Ras GAP (Van Aelst and D'Souza-Schorey, 1997). Specificities of GAP proteins to members of the Rho subfamily can vary between *in vitro* and *in vivo* assays. Although GAP proteins such as p50Rho-GAP, Bcr, Abr, p190GAP, 3BP-1, RalBP1, N chimerin, and β chimerin exhibit GAP activity for Rac in cell-free assays, interactions *in vivo* appear to be more restricted. For example, targets of p50Rho-GAP *in vitro* include Cdc42, Rac, and Rho; however, *in vivo*, it appears to be restricted specifically to Rho only (Ridley *et al.*, 1993). N and β chimerins have been demonstrated to exhibit GAP activity toward Rac GTPases, and indeed microinjection of the chimerin domain into fibroblasts prevented Rac- and Cdc42-induced cytoskeleton rearrangements (Diekmann *et al.*, 1991; Kozma *et al.*, 1996; Leung *et al.*, 1993; Manser *et al.*, 1995).

III. RAC AND RESPIRATORY BURST IN PHAGOCYTES

In phagocytic cells, Rac induces activation of the nicotinamide adenine dinucleotide phosphate (NADPH) oxidase leading to free radical production. Intracellular pathways leading to the generation of ROS are best characterized in these specialized cells. The superoxide-free radical ($O_2^{\cdot-}$) is generated by the multimolecular NADPH oxidase complex, which includes Rac (Abo *et al.*, 1991; Bokoch, 1994; Jones, 1994; Knaus *et al.*, 1991). Rac was shown to stabilize the assembly of several proteins, to form the protein structure responsible for NADPH oxidase activity (Dagher *et al.*, 1995; Knaus *et al.*, 1991). Thus on exchanging its GDP for GTP, Rac triggers the clustering of a multiprotein complex, which in the presence of the cofactor NADPH catalyzes the generation of superoxide (DeLeo and Quinn, 1996; Diekmann *et al.*, 1994; Patriarca *et al.*, 1971). The catalytic activity is provided by a flavocytochrome, cytochrome b558, an integral membrane protein composed of two subunits: glycoprotein (gp) 91phox and p22phox flavoprotein (Fig. 2). The enzyme also requires FAD as a cofactor and catalyzes the following reaction: $NADPH + 2O_2 \rightarrow 2\ O_2^{\cdot-} + NADP + H^+$. The activity of the b558 subunit is dependent on its interaction with additional components of the complex: p67phox, p47phox, p40phox, and Rac. These subunits are cytoplasmic in resting phagocytes, but join b558 at the b558 at the membrane on activation of the respiratory burst (Dusi *et al.*, 1996). They assemble through the interaction of src homology domain-3 (SH3) modular elements with domains rich in poly-L-proline (de Mendez *et al.*, 1996). The function of these components of the NADPH oxidase complex is not fully elucidated. Perhaps they might help mediation the active conformation of the flavocytochrome, when the enzyme produces large amounts of superoxide.

Two isoforms of Rac, Rac1 and Rac2, once bound to GTP, promote the assembly of the NADPH oxidase and time the stability of this multimolecular complex through the hydrolysis rate of the bound GTP (Dorseuil *et al.*, 1996; Kwong *et al.*, 1995; Leusen *et al.*, 1996b). Rac2 has a higher affinity for the NADPH oxidase than Rac1 and seems to be associated constitutively with membranes, whereas Rac1 shifts from the cytosol to the membrane together with the other b558 ligands on stimulation of the respiratory burst. It has been hypothesized that Rac2 represents a specialized Rac isoform designed to induce the production of bactericidal concentrations of superoxide anion, whereas Rac1 could be involved in the generation of smaller amounts of superoxide. In support of this concept, Rac2 represents 95%

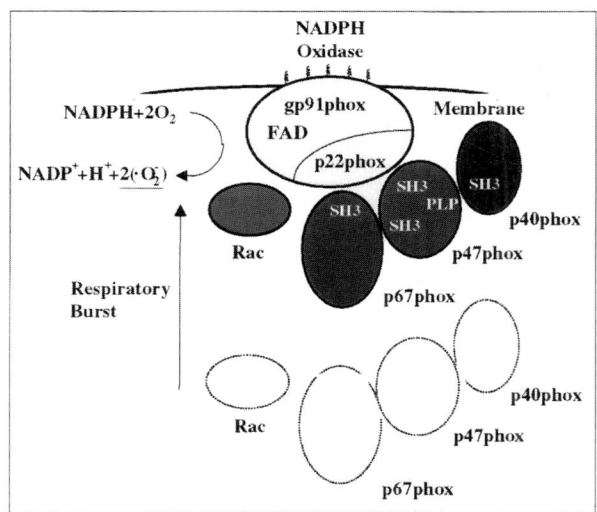

Figure 2 NADPH oxidase and respiratory burst. NADPH oxidase is a multimolecular enzymatic complex that produces superoxide anion. A membrane glycoprotein, gp91phox, and its associated subunit, p22phox, compose the flavocytochrome b_{558}. The oxidase catalyzes the formation of superoxide from molecular oxygen and NADPH in an FAD-dependent fashion. On activation of the respiratory burst in neutrophils, the NADPH oxidase is switched on by its interaction with several cytoplasmic proteins: p67phox, p47phox, and p40phox. Clustering of proteins in this complex is mediated by the interaction of modular units, SH3, with domains rich in proline residues. The stability of this complex is controlled and timed by GTP-Rac and the hydrolysis of the bound GTP. (See color reproduction in color plate section.)

of total Rac in neutrophils, the major superoxide producing phagocyte, and depends on b558 for interaction with membranes. In contrast, Rac1 depends on its interaction with p67phox to activate the NADPH oxidase, and mutations of p67phox that suppress this interaction mediate rare forms of chronic granulomatous disease (CGD) (Leusen et al., 1996a; Tauber et al., 1983).

CGD is a group of inherited disorders characterized by an extreme susceptibility to pyogenic infections, potentially fatal, believed to result from single mutations of the genes coding for the components of the phagocytic NADPH oxidase (Roos, 1994, 1996; Wientjes and Segal, 1995). The patients, usually infants or children with cutaneous abscesses and other bacterial or fungal pyogenic infections, are lacking phagocytes with efficient bactericidal activity due to the absence of respiratory burst. For two-thirds of the patients, the defect is in the X-linked gene encoding gp91phox. About 30% of CGD cases are induced by autosomally inherited defects in the p47phox subunit of the NADPH oxidase. The remaining patients have mutations in p22phox and the other components of the complex. A CGD-like transgenic model was reproduced in the mouse by targeted disruption of the gene for gp91phox or for p47phox (Jackson *et al.,* 1995; Pollock *et al.,* 1995). Much effort has been committed to the development of gene therapy strategies for the management of patients with this disorder. Virus-mediated gene transfer has been used to restore NADPH activity in phagocytes from patients with CGD, and CGD is a likely major gene disease to be treated successfully with gene transfer in the foreseeable future (Bjorgvinsdottir *et al.,* 1997; Ding *et al.,* 1996; Foster *et al.,* 1993; Kume and Dinauer, 1994; Thrasher *et al.,* 1995).

IV. RAC IN NONPHAGOCYTIC CELLS

In fibroblasts, studies have shown that Cdc42 appears to activate Rac, which in turn activates Rho (Nobes and Hall, 1995). Moreover, Ras has been shown to activate Rac (Ridley *et al.,* 1992). However, the molecular links between these proteins in mammalian cells are still uncertain. A similar cascade controls bud formation and morphogenesis in yeasts *S. cerevisiae* and *Schizosaccharomyces pombe* (Matsui *et al.,* 1996). The Rho subfamily links cell surface receptors to the assembly and organization of the actin cytoskeleton. Extracellular signals can activate Rho GTPases through multiple receptors. The addition of lysophosphatidic acid (LPA) to quiescent fibroblasts induces the formation of actin stress fibers, a response that can be blocked with the expression of the bacterial enzyme C3 transferase that ribosylates and inactivates Rho proteins (Ridley and Hall, 1992). Furthermore, growth factors such as PDGF, insulin, and bombesin induce actin polymerization at the plasma membrane and consequently induce lamellipodia formation and surface membrane ruffling in many cell types (Ridley *et al.,* 1992). This ruffling response can be inhibited by dominant-negative mutant of Rac, RacN17, thereby establishing a Rac-regulated signaling pathway linking growth factor receptors to the polymerization of actin at the plasma membrane. The activation of Cdc42 by bradykinin results in the formation of filopodia and the subsequent formation of a lamellipodia. Filopodia formation can be inhibited by a dominant-negative mutant of Cdc42, Cdc42N17, whereas RacN17 inhibited only membrane-ruffling formation (Kozma *et al.,* 1995; Nobes and Hall, 1995). Because bradykinin, LPA, and bombesin receptors all belong to the seven transmembrane domain heterotrimeric G protein-coupled receptor family, trimeric G proteins are likely to play a role in the activation of their respective small GTP-binding proteins (Van Aelst and D'Souza-Schorey, 1997). It been shown that an activated mutant form of subunit of the heterotrimeric G protein G12 stimulates JNK activity in a Ras- and Rac-dependent manner (Collins et al., 1996). Another group has shown that the heteromeric G proteins, β–γ subunits, are involved in the signaling from m1 and m2 muscaric receptors (mACHRs) to JNK kinase (Coso et al., 1996). Furthermore, studies in S. cerevisiae showed that pheromone signaling is mediated by subunits of heterotrimeric G proteins (Whiteway et al., 1989). Taken together, these results suggest the involvement of G proteins in the activation of at least some small GTP-binding proteins.

Several studies have pointed toward the involvement of phosphotidylinositide 3-kinase (PI3 kinase) in PDGF and insulin-induced cytoskeletal rearrangements. Treatment of fibroblasts with the PI3 kinase inhibitor wortmannin inhibits membrane ruffling induced by PDGF, epidermal growth factor (EGF), and insulin, although not by microinjected Rac protein (Kotani et al., 1994; Nobes et al., 1995; Wennström et al., 1994). Furthermore, PDGF could stimulate the level of Rac GTP by increasing the Rac interaction with GTP in a PI3 kinase-dependent manner (Hawkins et al., 1995). Hence, PI3 kinase appears to function upstream of Rac for the induction of membrane ruffling in response to extracellular growth factors. Morever, a constitutively active PI3 kinase mutant has shown to trigger membrane ruffles and stress fibers in a Rac- and Rho-dependent manner (Reif et al., 1996). Interestingly, this active mutant failed to induce Rac/Rho signaling pathways that regulate gene transcription (Reif et al., 1996). A possible explanation for this obervation is that this specific cellular effect of Rho proteins is linked to upstream regulatory proteins, which may determine the interaction with various effector pathways leading to diverse biological responses. This may explain how Rho proteins can regulate such a wide variety of biological functions. The mechanism by which PI3 activates Rac is unknown but could be mediated by phosphoinositides that are phosphorylated on the third carbon of the inositol ring (Kandzari, 1996).

Furthermore, accumulating data indicate that Rac is involved in regulating nuclear proteins. It has been shown that Rac and Cdc42 can regulate the activation of JNK and reactivating p38PK (Seger and Krebs, 1995). The expression of constitutively active mutants of Rac and Cdc42 in NIH 3T3, HeLa, and Cos cells resulted in the stimulation of JNK and p38 activity (Coso et al., 1995; Minden et al., 1995). Both JNK and p38 can be activated strongly by inflammatory cytokines, TNF-α and IL-1β, as well as by cellular stress factors such as heat shock, UV, and ionizing radiation (Kyriakis and Avruch, 1996). Activated JNKs translocate to the nucleus and phosphorylate transcription factors such as c-Jun, ATF2, and Elk (Derijard et al., 1994; Gille et al., 1995; Gupta et al., 1995; Pulverer et al., 1991). However, activated p38 phosphorylates ATF3, ELK, MAX, and cAMP response element-binding protein homologous protein/growth arrest DNA damage 153 (CHOP/GADD153), MAPKAP-K2, and, more recently, 3pK (Ludwig et al., 1996; Rouse et al., 1994; Wang and Ron, 1996; Zervos et al., 1995). Although Rac has been shown to stimulate the transcription activity of PEAS, a member of the Ets family, in a JNK-dependent manner, the identity of the molecules that link Rac and Cdc42 to JNK and p38 is not yet clear (Van Aelst and D'Souza-Schorey, 1997). Both JNK and p38 can be activated directly by the dual specificity kinase SEK1 (also called MEKK4 or JNKK), whereas MKK3 and MKK6 activate p38 specifically (Derijard et al., 1995; Jiang et al., 1996; Raingeaud et al., 1996; Sanchez et al., 1994). It has been reported that the family of serine/threonine kinases known as PAKs may connect Rac and Cdc42 to JNK and p38 because they bind *in vitro* to the activated form of Rac and Cdc42 and become activated (Bagrodia et al., 1995; Knaus et al., 1995; Manser et al., 1995; Martin et al., 1995). By analogy with the Ste20 kinase cascade in *S. cerevisiae,* it was suggested that PAK regulates the activity of MEKK (Van Aelst and D'Souza-Schorey, 1997). However, some groups did not observe an increase in JNK on coexpression of PAK1 with the activated form of Cdc42 or Rac (Teramoto et al., 1996b), and a mutant of Rac that fails to bind to PAK remains a potent activator for JNK (Westwick et al., 1997). This suggests the possibility that other kinases are connecting Rho proteins to JNK. In support of this possibility, MLK3 (also called SPRK) and MEKK4 have become suitable candidates as effectors of Rac/Cdc42-induced JNK activation. Studies by different groups have shown that MLK3 and MEKK4 are regulated by Cdc42 and Rac and selectively activate the JNK pathway (Gallo et al., 1994; Gerwins et al., 1997; Teramoto et al., 1996b; Tibbles et al., 1996). Furthermore, Cdc42 and Rac bind to MLK3 both *in vitro* and *in vivo,* and the coexpression of activated Cdc42 and Rac mutants increased the activity of MLK3 in Cos-7 cells (Gallo et al., 1994; Teramoto et al., 1996a). A kinase-inactive mutant of MEKK4 blocks Cdc42/Rac stimulation of the JNK pathway and both MLK3 and MKK4 have been shown to activate SEK1. Taken together, these data support the possibility that both MLK3 and MEKK4 are effectors of Rac/Cdc42-induced JNK activation. However, it remains to be shown that direct binding of MLK3 to activated small GTP-binding proteins results in its activation and whether MEKK4 binds directly to the activated forms of Rac and Cdc42. It is possible that different kinases are involved in mediating the signal from Cdc42 and Rac to JNK (or p38), depending on the cell type and the extracellular stimulus.

It has been shown that the injection of Rac, Cdc42, and Rho into Swiss 3T3 fibroblasts stimulated cell cycle progression through the G_I phase and subsequent DNA synthesis, whereas injection of dominant-negative forms of these GTPases blocked the stimulation of DNA synthesis in response to serum (Olson *et al.*, 1995). It was also shown that Rac is required for v-Abl to activate a mitogenic program in NIH 3T3 cells (Renshaw *et al.*, 1996). Furthermore, Rac, Rho, and Cdc42 have been shown to have transforming activity in some cell lines. Cells expressing constitutively active mutants of Rac and Rho displayed enhanced growth in low serum, anchorage independence and caused tumor formation when injected into nude mice (Khosravi-Far *et al.*, 1995). Moreover, the GEF for Rac, Tiam 1, can transform NIH 3T3, which strengthens the role of Rac in transformation (van *et al.*, 1995). Truncation or amplification of the normal Tiam 1 protein has been reported to cause invasiveness in T lymphoma cell variants (Habets *et al.*, 1994), and cell clones that are invasive *in vitro* produced experimental metastasis in nude mice. These observations suggest a possible role of Rac in carcinogenesis. Indeed it was shown that the expression of activated Rac in lymphoma cells results in a transformed phenotype with metastatic potential for cells that could not otherwise penetrate into surrounding fibroblasts (Michiels *et al.*, 1995).

V. RAC AND ROS PRODUCTION

Activation of the transcription factor NF-κB is stimulated by a variety of agents, including cytokines, viruses, phorbol esters, and UV light (Baeuerle and Henkel, 1994; Siebenlist *et al.*, 1994). In unstimulated cells, NF-κB consists of a cytoplasmic complex of homodimeric or heterodimeric Rel-related proteins bound to a member of the IκB family of inhibitor proteins. Agents that stimulate NF-κB result in the phosphoylation and degradation of the IκB inhibitory subunit.

Previous studies have suggested that the activation of NF-κB may be redox dependent (Israel *et al.*, 1992; Meyer *et al.*, 1993; Schenk *et al.*, 1994; Schreck *et al.*, 1991; Staal *et al.*, 1990; Westendorp *et al.*, 1995). In particular, micromolar doses of exogenous hydrogen peroxide (H_2O_2) stimulate NF-κB activation, whereas chemical antioxidants inhibit such activation (Schreck *et al.*, 1991). Furthermore, agents that stimulate NF-κB, such as cytokines, UV light, and phorbol ester, appear capable of increasing intracellular reactive oxygen species (Lo and Cruz, 1995; Meier *et al.*, 1989; Sundaresan *et al.*, 1996). Several lines of evidences suggest that Rac1 regulates the production of intracellular ROS that were shown to act as essential intracellular second messengers for several cytokines and growth factors in nonphagocytic cells (Baeuerle and Henkel, 1994; Siebenlist *et al.*, 1994). Thus, Rac1 may function as part of a redox-dependent signal transduction pathway leading to NF-κB activation (Beg and Baltimore, 1996; Cooper *et al.*, 1996; Sulciner *et al.*, 1996; Wang *et al.*, 1996; Wu *et al.*, 1996). Our studies and studies from other laboratories have shown that transient expression of a constitutively active mutant of Rac1, Rac1V12, leads to increased NF-κB activity, whereas the expression of a dominant-negative Rac1 mutant, Rac1N17, inhibits basal and interleukin 1-stimulated NF-κB activity (Sulciner *et al.*, 1996). We have also demonstrated that Rac proteins function downstream of Ras in the activation of NF-κB (Fig. 3). In addition, Rac1V12 stimulation of NF-κB activity is shown to be independent of the ability of Rac1 proteins to activate the family of c-jun amino-terminal kinases (Sulciner *et al.*, 1996). Another line of evidence has shown that exogenous H_2O_2 stimulated serum response element (SRE)-dependent reporter gene activity in a dose- and time-dependent manner (Kim *et al.*, 1997).

RAS, SUPEROXIDE AND SIGNAL TRANSDUCTION

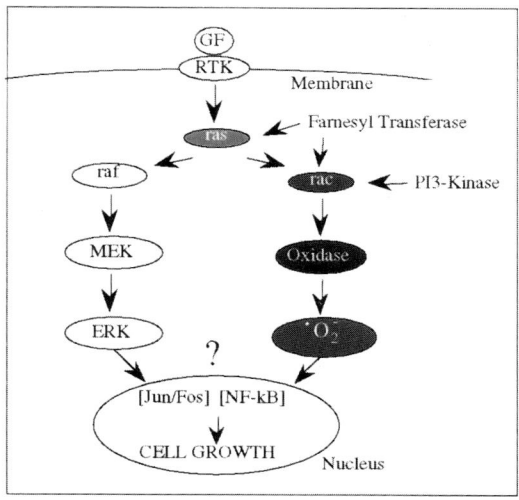

Figure 3 An alternative effector pathway for Ras. The pathway includes Rac, NADPH oxidase, and superoxide and shows specifically how Ras, Rac, and superoxide regulate mitogenesis and how cell growth remains essentially uncharacterized (question mark). MEK, mitogen-activated protein kinase-kinase; ERK, MAP-kinase; NF-κB, nuclear factor; RTK, receptor tyrosine kinase; PI, phosphatidylinositol. (**See color reproduction in color plate section.**)

In addition, Rac1 and the subsequent phospholipase A_2 activation is essential for the H_2O_2 signaling pathway that is independent of the JNK signaling cascade known to be activated by Rac1 as well (Kim and Kim, 1997; Peppelenbosch et al., 1992). Thus, Rac regulates cytokine-stimulated NF-κB activation via a redox-dependent pathway and is independent of the JNK signaling cascade. Another example for redox signaling is the activation of the OxyR transcription factor in response to elevated H_2O_2 in *Escherichia coli*, which in turn activates the antioxidant genes coding for antioxidant pathways (Zheng et al., 1998). Such a direct effect of oxidants on gene expression has not yet been reported in eukaryotes.

Previous studies from our laboratory showed that NIH 3T3 cells transformed stably with a constitutively active isoform of p21Ras (H-RasV12) produced large amounts of reactive oxygen species detected by electron paramagnetic resonance (EPR) using 5,5-dimethyl-1-pyrroline-*N*-oxide (DMPO) as the spin trap (Irani et al., 1997). Superoxide dismutase (SOD) quenched the observed signals, whereas catalase had no effect. These results suggested that the observed signals was attributable to $O_2^{\cdot-}$ trapping rather than to ·OH derived from H_2O_2. Production of $O_2^{\cdot-}$ by NIH 3T3 transformed stably with H-RasV12 was confirmed by a Lucigenin-enhanced chemiluminescence assay, which has specificity for $O_2^{\cdot-}$ (Gyllenhammar, 1987). The $O_2^{\cdot-}$ production was suppressed by the expression of a dominant-negative isoform of Rac1 as well as by treatment with a farnesyl protein transferase (FPTase) inhibitor, which inhibits Ras-dependent transformation and results in morphological reversion of Ras-transformed cells (Kohl et al., 1993). Ras-transformed cells have the ability to progress through the cell cycle even under conditions of confluence and growth factors deprivation, as these cells displayed a greater rate of DNA synthesis than control cells in such conditions (Irani et al., 1997). The Ras-induced mitogenic response of Ras-transformed cells was inhibited by treating cells with the antioxidant *N*-acetyl-L-cysteine (NAC), which reduced DNA synthesis. Thus, proliferation of these cells was directly dependent on the concentration of superoxide. Moreover, overexpression of a dominant-negative isoform of Rac1 (RacN17) inhibited both superoxide generation and unchecked proliferation. In contrast, NIH 3T3 fibroblasts transformed with a constitutively activated isoform of the serine and threonine kinase, Raf, which is known to mediate Ras activation of MAP-kinase, neither produced detectable amounts of superoxide nor were inhibited significantly in their proliferation by antioxidants. In addition, the mitogenic-activated protein kinase (MAPK) activity was decreased and c-Jun N-terminal kinase (JNK) was not activated in chronically Ras-transformed cells. In conclusion, these results indicate that H-RasV12-induced transformation can lead to the production of $O_2^{\cdot-}$

through one or more pathways involving Racl (Irani et al., 1997) independently of Ras activation of the MAP-kinase pathway (Fig. 4).

The implication of a reactive oxygen species, probably $O_2^{\cdot-}$, as a mediator of Ras-induced cell cycle progression independent of MAPK and JNK suggests a possible mechanism for the effects of antioxidants against Ras-induced cellular transformation beyond oxidant-mediated DNA damage. Although there is no evidence in mammalian cells, for proteins capable of sensing superoxide or H_2O_2 as in bacteria (Bunn and Poyton, 1996), there is growing evidence supporting the concept that many signal-transducing proteins and transcription factors are highly sensitive to the redox state of the cells (Hidalgo et al., 1997; Irani et al., 1997; Sundaresan et al., 1995). Characterization of the molecular reactions responsible for such a "sensing" function will require further work (Fig. 5).

Furthermore, it has been demonstrated that the transient expression of a constitutively active mutant of Rac1 (Rac1V12) in NIH 3T3 cells leads

SMALL GTP BINDING PROTEINS AND THEIR KINASE CASCADE

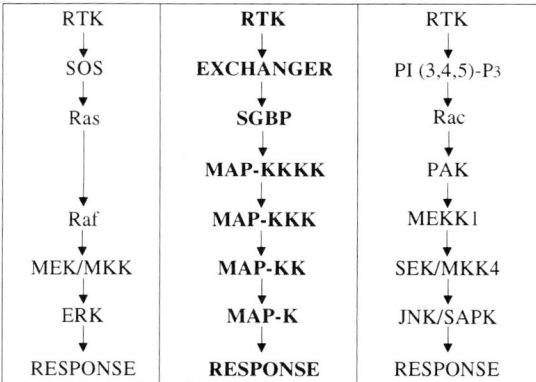

Figure 4 MAP-K cascade as a effector pathway for activated Ras. A generic cascade (in bold) showing mitogen-activating protein kinases linked in series and transducing signals mediated by small GTP-binding proteins (SGBP) under the control of a receptor tyrosine kinase (RTK), which induces the exchange of the nucleotide bound to SGBP. The successive Ks following MAP-K indicate the activity of kinases upstream from MAP-K. Such a cascade has been characterized in the case of Ras and also Rac or Cdc42 (not shown). SOS is an exchanger protein discovered in *Drosophila* (Son of Sevenless). MEK (or MKK for MAP-KK) is the mitogen- and extracellular-regulated kinase; ERK corresponds to extracellular-regulated kinase. PI (3,4,5)-P_3 is a putative exchanger molecule for Rac and is produced by phosphatidylinositol 3-kinase (PI3-K). PAK is a p21-activated, serine and threonine kinase, PAK65. SEK (or MKK4) is the stress-activated protein kinase (SAPK) activator, SEK-1. JNK is the c-Jun N-terminal kinase.

SUPEROXIDE, CHAPERONES AND PROTEIN CONFORMATION

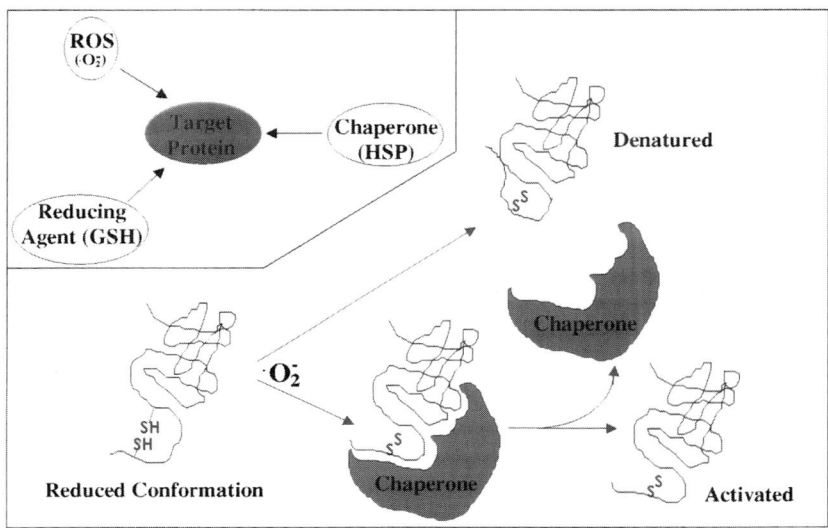

Figure 5 Cellular components are never exclusively exposed to oxidants. Instead, they are exposed to oxidants in a context where a large excess of reducing agents is also present, such as GSH, which functions as a true sulfhydryl buffer. Because of the very strong activity of the enzyme glutathione reductase, the oxidized conformation of glutathione (GSSG) is converted rapidly to GSH, such that the ration of GSH to GSSG in most cells is maintained ≥ 500. Therefore, the oxidized conformation of targeted proteins is likely to be reduced rapidly by the abundant cellular antioxidants. Moreover, intracellular oxidants are promptly inactivated by dismutases, catalases, and cellular antioxidants. Hence, their oxidizing effects are expected to be limited to a site directly surrounding their producing units. In addition, cells also contain large amounts of chaperone protein, such as heat shock proteins (HSP). These protein chaperones are known to protect their ligands against the damaging action of various stresses, including heat shock (Huot *et al.*, 1996). Thus, in the absence of chaperones, superoxide (and other ROS) might oxidize proteins to result in their denaturation, which would be followed rapidly by their degradation (Stadtman *et al.*, 1992). Instead, in the presence of chaperones, the titrated oxidation of targeted proteins such as actin or other protein/enzymes might result in conformational changes that could contribute to their activation (DalleDonne *et al.*, 1995; Heyworth *et al.*, 1997). Alternatively, the relative sensitivity of proteins to oxidants in a given system might result in the disruption of the steady state for this system, resulting in substantial reorganization of the affected system (Burdon *et al.*, 1990; Mirabelli *et al.*, 1989; Kamata *et al.*, 1996). (**See color reproduction in color plate section.**)

to a significant increase in ROS as detected by EPR spectroscopy using the spin trap DEMPO [5-(diethoxyphosphory)-5-methyl-1-pyrroline-*N*-oxide] (Frejaville *et al.*, 1995). In contrast, the expression of a dominant-negative Rac1 mutant (Rac1N17) inhibits the production of ROS in NIH 3T3 cells induced to produce ROS.

As mentioned earlier, Rac1 interacts with p67phox to activate the NADPH oxidase and mutation of p67phox that suppress this interaction could cause chronic granulomatous disease. Interestingly, fibroblasts from patients with established CGD are capable of producing superoxide through a flavoprotein-dependent enzymatic system (an NADPH oxidase complex) (Emmendorffer *et al.*, 1993). Such data support the concept that more than one NADPH oxidase system contributes to superoxide production in the body: one system is responsible for the bactericidal activity of phagocytes, whereas at least one other NADPH oxidase complex is responsible for the production of superoxide in nonphagocytic cells. Several groups have reported that Rac activation of NADPH oxidase is not limited to phagocytes (Jones *et al.*, 1994; 1996; Rajagopalan *et al.*, 1996; Zweier *et al.*, 1994). Although the specific structure of the putative nonphagocytic NADPH oxidase(s) has yet to be characterized and although the key protein subunit of this enzymatic activity has yet to be cloned (the flavocytochrome itself), it seems at least functionally similar to the enzymatic complex found in phagocytes (Rajagopalan *et al.*, 1996). Like Ras, Rac seems to have effects that depend on the activity of downstream kinases, particularly PAK65 and c-JUN N-terminal kinase (Minden *et al.*, 1995; Knaus *et al.*, 1995). However, some effects of Rac are independent of these kinases, and the effectors for the several essential nonkinase-mediated activities of Rac have remained uncharacterized (Joneson *et al.*, 1996; Lamarche *et al.*, 1996).

The role ascribed to superoxide and derived oxidants in biology is clearly expanding. By analogy with nitric oxide (NO), whose activity at low concentration is to transduce signals within vessels and neurons (Lowenstein *et al.*, 1994), high concentrations of NO produce damage to cells and microorganisms (Dugas *et al.*, 1995). Superoxide and probably other oxidants function as messengers at a low concentration, whereas larger amounts are required for cidal activity. In addition to cancer and infections, oxidants have been implicated in the genesis of many disease entities. For example, ROS production has been detected during tissue reperfusion after a period of ischemia and has been shown to contribute to injury to the heart (Korthuis and Granger, 1993), the brain (Chan, 1996), and the gastrointestinal tract (Bulkley, 1994) following acute myocardial infarction, stroke, or mesenteric ischemic insult. Through metal-catalyzed oxidation reactions, proteins can be denatured or cleaved (Stadtman, 1992). In the pathogenesis of systemic sclerosis, oxidants have been shown to interact with specific protein-bound metals to generate peptide fragments with antigenic properties that are able to break self-tolerance (Casciola-Rosen *et al.*, 1997).

Perhaps more interesting will be the definition of the role of superoxide and other ROS as signal-transducing molecules involved in the remodeling of tissues, either spontaneously as a result of genetic alteration of the key regulatory proteins involved in free radical generating, degrading, and sensing pathways or following specific injuries. In a transgenic mouse model, overexpression of H-Ras in the cardiac tissue results in a hypertrophic phenotype, with increased myofibrillar disarray, ventricular wall thickness, and juvenile mortality (Gottshall *et al.*, 1997). Overexpression of an intracellular enzyme that catabolizes superoxide, Cu,Zn-superoxide dismutase protects tissue from ischemic injuries (Yang *et al.*, 1994). While the production of oxidants by phagocytes has been traditionally implicated as the main source of superoxide and other ROS in injured tissues, the discovery of the widespread use by cells of oxidants as signaling molecules is likely to improve our understanding of the contribution of such molecules to the biology of normal and injured tissues. The precise orchestration of the targeted production of oxygen radicals at strategic sites of cells is likely to be timed by the hydrolysis of GTP, bound to the triphosphatase member of the Ras superfamily.

VI. RAC AND THE CYTOSKELETON

The ability of a eukaryotic cell to maintain or change its shape and its degree of attachment to the substratum in response to extracellular signals is largely dependent on rearrangement of the actin cytoskeleton. Cytoskeletal rearrangements play a crucial role in cell motility, cytokinesis, and phagocytosis. The actin cytoskeleton of animal cells is composed of actin filaments and many specialized actin-binding proteins (Small, 1994; Stossel, 1993; Zigmond, 1996). Profilin is an important actin-binding protein that stimulates the recharging of ADP actin with ATP, perhaps to accommodate the dynamic changes to the actin cytoskeleton that follow cell stimulation (Goldschmidt-Clermont *et al.*, 1992). Filamentous actin is generally organized into a number of discrete structures: (1) filopodia, the finger-like protrusions that contain a tight bundle of long actin filaments in the direction of the protrusion; (2) lamellipodia, the thin protrusive actin sheets that extend the edges of cultured fibroblasts and many motile cells (membrane ruffles observed at the leading edge of cells result from lamellipodia that lift up away from the substrate and fold backward); and (3) actin stress fibers, the bundles of actin filaments that traverse the cell and are linked to the extracellular matrix through focal adhesions. It is important, therefore, for the polymerization of cortical actin to be tightly regulated. This regulation of actin polymerization, for the most part, is orchestrated by small GTP-binding proteins of the Rho family (Van Aelst and D'Souza-Schorey, 1997).

In fibroblasts, Rac has been shown to be a key control element in the reorganization of the actin cytoskeleton induced by growth factors and RasV12 (Ridley and Hall, 1992). Injection of RacV12 is sufficient to induce lamellipodia and membrane ruffles and subsequent stress fiber formation, whereas microinjection of RacN17, a dominant-negative mutant of Rac, prior to the addition of growth factor or together with RasV12 abolishes these effects (Ridley *et al.*, 1992; Van Aelst and D'Souza-Schorey, 1997). Furthermore, both Rac and Cdc42 have been shown to induce the assembly of multimolecular focal complexes (MFC) at the plasma membrane of fibroblasts (Nobes and Hall, 1995). These complexes, which are most apparent as staining positively for vinculin and phosphotyrosine around the leading edge of the lamellipodia and at the tips of filopodia, are morphologically distinct from Rho regulated focal adhesions, yet the protein components of MFC appear similar to those in Rho-triggered focal adhesions. In addition, it has been demonstrated that Rac regulates hepatocyte growth factor (HGF) and RasV12-induced motility in MDCK cells (Ridley *et al.*, 1995). In mammals, three serine/threonine kinases known as PAK isoforms have been identified that bind Rac and Cdc42 in a GTP-dependent manner and stimulate the activity of the kinase (as described earlier) (Bagrodia *et al.*, 1995; Brown *et al.*, 1996; Knaus *et al.*, 1995; Manser *et al.*, 1995; Martin *et al.*, 1995). PAK homologs have been identified in *S. cerevisiae, C. elegans,* and *Drosophila* and have been shown to be involved in mediating the effects of Rho-like GTPases on the cytoskeleton (Chen *et al.*, 1996; Cvrckova *et al.*, 1995; Harden *et al.*, 1996; Van Aelst and D'Souza-Schorey, 1997). However, in mammalian cells the role of PAK isoforms remains unclear.

In addition to PAK, several other proteins have been isolated that bind to Rac and/or Cdc42, among them a 34-kDa protein, POR1 (partner of Rac). This protein was isolated in a two-hybrid system and plays a role in Rac-mediated membrane ruffling (Van Aelst *et al.*, 1996). POR1 interacts specifically with Rac in a GTP-dependent manner. Deletion mutants of POR1 inhibited the induction of membrane ruffles by RacV12, whereas a synergistic effect of wild-type POR1 with RasV12 was observed for the induction of membrane ruffling. Moreover, a mutant Rac that failed to bind POR1 also failed to induce membrane ruffling (Joneson *et al.*, 1996). Interestingly, POR1 also interacts with the GTPase ARF6 (D'Souza-Schorey *et al.*, 1997). ARF6 is the least conserved member of the ARF family of GTPases; in addition to its role in regulating peripheral membrane trafficking, ARF6 and its activated mutant ARF6 (Q67L) have been shown to elicit cytoskeletal rearrangements at cell surface. Cytoskeletal rearrangements induced by ARF (Q67L) could be inhibited by coexpression of the deletion mutants of POR1 but not with the dominant-negative Rac mutant, Rac(S17N) (D'Souza-Schorey et al., 1997). These findings indicate that ARF6 and Rac function on distinct signaling pathways to mediate cytoskeletal reorganization and suggest a role for POR1 as an important regulatory element in orchestrating cytoskeletal rearrangements at the cell periphery induced by ARF6 and Rac. Depending on the nature of extracellular stimuli, it is possible that POR1 could interact with ARF6, Rac, or both to establish highly specific patterns of cytoskeletal rearrangements at the plasma membrane. Another protein with a potential role in cytoskeletal organization is IQGAPA (Bagrodia et al., 1995; Brill et al., 1996; Kuroda et al., 1996). This protein was found to interact with both Rac and Cdc42 and localizes to membrane ruffles; however, its function remains to be established (Van Aelst and D'Souza-Schorey, 1997).

As mentioned earlier, constitutively activated isoforms of Ras and Rac not only alter the mitogenic activity of cells, but they are also known for their strong impact on the organization of the actin cytoskeleton (Nobes and Hall, 1995; Ridley et al., 1992; Ridley and Hall, 1992). It is tempting to speculate that superoxide (and perhaps other ROS) generation, resulting from the activation of Ras and/or Rac, also contributes to the effects of small GTP-binding protein on the actin cytoskeleton (Crawford et al., 1996; Kamata et al., 1996; Mirabelli et al., 1989; Nobes and Hall, 1995; Ridley et al., 1992; Ridley and Hall, 1992). Such contributions could be mediated, either directly or indirectly, by the interaction of ROS with the proteins of the actin cytoskeleton: (i) superoxide could affect the activity of molecules belonging to signal transduction pathways regulating the actin cytoskeleton; consequently, interactions of regulatory proteins for actin with inositol phospholipids, calcium, and perhaps other signaling molecules known to be involved in actin regulation (Janmey, 1994) might be altered; and (ii) alternatively, ROS could oxidize actin and/or binding proteins directly (Chai et al., 1994; Mirabelli et al., 1989). Whatever the specific mechanism involved in the reorganization of the actin cytoskeleton by ROS might be, the net result of such interactions is likely to require the contribution of other, concurrently activated, signaling pathways (Hartwig et al., 1995; Peppelenbosch et al., 1995).

Interestingly, polyphosphoinositides such as phosphatidylinositol 4,5-bisphosphate [PI(4,5)P2] and phosphatidylinositol 3,4,5-triphosphate [PI(3,4,5)P3] have been proposed as regulatory molecules for actin-binding proteins and for small GTP-binding proteins (Hawkins et al., 1995; Tsai et al., 1989). Hence superoxide (and other ROS) production resulting from the activation of Rac might function to amplify signals generated by the turnover of inositol phospholipids and targeted at the actin cytoskeleton

(Hartwig et al., 1995; Hawkins et al., 1995; Janmey, 1994). Future research aimed at investigating these issues should further our understanding on the redox control of the superstructure of cells.

VII. RAC GENE IN DEVELOPMENT

Current understanding of the role of the Rho GTPase subfamily in development has been shaped mainly from studies using *Drosophila* and *C. elegans* as a model system (Van Aelst and D'Souza-Schorey, 1997). Many lines of evidences have indicated that Rac, Rho, and Cdc42 function in the receptor-mediated signaling cascades that regulate actin dynamics in such a way that provides the cell with polarity and force required for motility (Hall, 1998). Zipkin *et al.*, (1997) has characterized a widely expressed Rho family member, mig-2, in *C. elegans*. They have shown that mutation in the mig-2 gene inhibits cell migration and axon guidance. Another study has demonstrated that the unc-73 gene product from *C. elegans* activates Rac *in vitro* and can stimulate actin polymerization at the plasma membrane when expressed in Rat2 cells growing in culture (Steven *et al.*, 1998).

In *Drosophila melanogaster,* the homologs of Rac1, Rac2, and Cdc42 have been named Drac1, Drac2, and DCdc42, respectively (Van Aelst and D'Souza-Schorey, 1997). These genes are highly expressed in the nervous system and in the mesoderm during neuronal and muscle development (Luo *et al.*, 1996). In developing neurons, studies have shown that the expression of constitutively activated or dominant-negative mutants of DRac1 perturbed the initiation and elongation of axonal outgrowth. These perturbations appear to result from defects in the actin cytoskeleton and that the oscillation of DRac1 between its GTP and GDP states is critical for the developing axon (Luo *et al.*, 1996). In the presence of active DRac1, the accumulation of F-actin in the dorsal neuronal clusters persisted through later stages of development, whereas no F-actin accumulation was observed in the presence of a dominant-negative Drac1. Furthermore, DRac1 and DCdc42 regulate muscle development most likely by their effects on membrane fusion and actin cytoskeleton, respectively (Luo *et al.*, 1996). Expression of the activated mutant of DRac1 inhibited myoblasts fusion, whereas the dominant-negative mutant generates excessively fused muscle fibers in later developmental stages (Luo *et al.*, 1996). DRac1 has also been shown to play a role during dorsal closure of the epidermis during fly development, a process in which the lateral epidermal cells migrate over the amnioserosa of the developing embryo and join at the dorsal midline (Martinez-Arias, 1993). Interestingly, the *Drosophila* homolog of c-Jun kinase, DJNK, has been shown to play a role in dorsal closure: embryos lacking DJNK were defective in dorsal closure (Riesgo-Escovar *et al.*, 1996). It has been demonstrated that dorsal closure is also blocked by expression of the dominant-negative mutant of DCdc42 (Riesgo-Escovar *et al.*, 1996). Therefore, it is possible that both DCdc42 and DRac1 function in the DJNK pathway of *Drosophila*. In addition, the *Drosophila* MAPK kinase HEP, which phosphorylates DJNK, is required for the expression of the puckered gene product that is needed for dorsal closure to occur (Nusslein-Volhard, 1984). The fly PAK homolog was isolated and shown to bind the GTP-bound form of DRac1 and DCdc42 during embryogenesis (Harden *et al.*, 1996). Furthermore, DPAK levels were elevated in focal complexes along the leading edges of epidermal cells. Thus, it has been proposed that DPAK may regulate the turnover of DRac1-dependent focal complex at the leading edge. Taken together, it appears that an analogous cascade to those described for mammals and yeast, which includes DRac1, DPAK, HEP, and DJNK, exists in *Drosophila* and may regulate the dorsal closure during fly development.

A role for DRac1 and DCdc42 in the development of fly wing disc epithelium and wing hairs has also been demonstrated (Poodry, 1980; Van Aelst and D'Souza-Schorey, 1997). Expression of dominant-negative DCdc42 in a subset of disc epithelial cells indicates that DCdc42 controls epithelial cell shape changes by modulating the basic actin cytoskeleton during pupal and larval development (Eaton *et al.*, 1995). In contrast, the expression of dominant-negative DRac1 during disc development did not appear to affect cytoskeletal elements that regulate cell shape at that level. However, DRac1 activity appears to be required for actin assembly at the adherens junction as the latter is disrupted on DRacN17 expression (Eaton *et al.*, 1995). However, cells expressing DRac1N17 have multiple, but structurally normal, wing hairs. The failure to restrict hair outgrowth to a single site suggests that DRac1 may function to select the site for hair formation. Whether this function is related to its role in organizing junctional actin is unclear. However, because cells expressing DRac1N17 exhibited a disorganized microtubule network in the apical region, the site selection for hair outgrowth may require an intact microtubule network. It has been proposed that the regulation of adherens junction assembly and microtubule organization in response to Rac1 activity may be coupled directly to the suppression of inappropriate hair growth (Eaton *et al.*, 1995). Finally, dominant-negative DRac1, DRac1N17, inhibits border cell migration during *Drosophila* oogenesis (Murphy and Montell, 1996).

VIII. OXIDATIVE BURST IN PLANTS

It is becoming increasingly clear that plants and animals share a multitude of common signal transduction elements. One of the clearest analogies to date has been the finding that plants possess genes that show high homology to animal and yeast genes coding for small signal-transducing, GTP-binding proteins of the Ras superfamily. In plants, resistance to pathogens such as fungi, bacteria, and viruses depends on the release of ROS (oxidative burst) (Mehdy, 1994). As in phagocytes, during the oxidative burst of plants, ROS generates toxic intermediates that result from successive one-electron steps in the reduction of molecular O_2. The predominant species detected in plant–pathogen interactions are superoxide anion ($O_2^{\cdot-}$), hydrogen peroxide (H_2O_2), and hydroxyl radical (OH^-). Studies have shown that H_2O_2 generated by the oxidative burst functions as a selective signal for the induction of cellular defense genes in surrounding cells (Levine *et al.*, 1994). This response is also observed in plants when they are infected by nonpathogen or an avirulent strain of a pathogen, including the hypersensitivity response (Levine *et al.*, 1994; Tenhaken *et al.*, 1995). The rapid production of reactive oxygen intermediates triggers the programmed cell death of cells in the affected area, thereby killing pathogens within the apoptotic-infected cells. In addition, the produced superoxide drives the cross-linking of cell wall structural proteins, creating a stronger barrier against the invading pathogen (Bradley *et al.*, 1992).

The oxidative burst and production of H_2O_2 is an effective bactericidal mechanism that could prove to be an ancient response, developed to protect cells against microorganisms. In *Arabidopsis thaliana,* no true homologs of Ras, Cdc42, and Rho were found, but a large number of Rac-like genes were detected (Winge *et al.*, 1997). A protein described as PsRop1 was found in *Pisum sativum* (Yang and Watson, 1993) and has been implicated in tip growth and movement of the generative cell in pollen tubes (Lin *et al.*, 1996). In *Gossypium hirsutum,* two similar Rac-like proteins, Rac 9 and Rac 13, have been characterized and found to be expressed preferentially in developing cotton fibers (Delmer *et al.*, 1995; Trainin, 1996). Expression of Rac 13 was shown to be high during cotton fiber development at a time when the fiber cells displayed changes in cortical microtubule and cell wall microfibril alignments, changes thought to be regulated by actin cytoskeleton (Delmer *et al.*, 1995). In *A. thalinana,* the Rac gene family consists of at least 10 genes. Expression analysis of 5 of *A. thaliana* Rac genes showed that 4 of them—ARac1,3,4, and 5—are expressed ubiquitously in all tissues examined, whereas ARac2 is only expressed in the root, hypocotyl, and stem (Winge *et al.*, 1997). Rac proteins in *A. thaliana* are all highly conserved, with 80–95% identity, and the N-terminal part, including the effector domain, shares considerable homology to animal Rac proteins. Similarities between the oxidative burst observed in animal phagocytes and in plant cells exposed to microbial elicitors suggest that plant Rac proteins may have similar functions as their animal counterparts (Abo *et al.*, 1991). The phagocyte NADPH oxidase system is composed of five essential components: two cytosolic proteins, p47 and p67 phox, the Rac protein, and the large and the small subunit of cytochrome b558, gp91-phox and p22-phox. Immunological analysis of plant cell extracts suggests that components of the NADPH oxidase system in plants are related immunologically to human counterparts (Tenhaken et al., 1995), and the recently found homologs of gp91-phox in *O. sativa* (X93301 and D46291) and *A. thaliana* show that the gp91-phox subunit is highly conserved between animals and plants. The amazing conservation of the effector domain in Rac proteins, as well as the proteins responsible for the oxidative burst, over a time span of 1 billion years or more, may indicate that plant Rac protein regulatory functions and interaction with proteins have been highly conserved throughout evolution.

REFERENCES

Abo, A., Pick, E., Hall, A., Totty, N., Teahan, C. G., and Segal, A. W. (1991). Activation of the NADPH oxidase involves the small GTP-binding protein p21rac1. *Nature (London)* **353**, 668–670.

Aepfelbacher, M., Vauti, F., Weber, P. C., and Glomset, J. A. (1994). Spreading of differentiating human monocytes is associated with a major increase in membrane-bound CDC42. *Proc. Natl. Acad. Sci. U.S.A.* **91**, 4263–4267.

Baeuerle, P. A., and Henkel, T. (1994). Function and activation of NF-kappa B in the immune system. *Annu. Rev. Immunol.* **12**, 141–179.

Bagrodia, S., Taylor, S. J., Creasy, C. L., Chernoff, J., and Cerione, R. A. (1995). Identification of a mouse p21Cdc42/Rac activated kinase *J. Biol. Chem.* **270**, 22731–22737.

Beg, A. A., and Baltimore, D. (1996). An essential role for NF-kappa B in preventing TNF-alpha-induced cell death. *Science* **274**, 782–784.

Bjorgvinsdottir, H., Ding, C., Pech, N., Gifford, M. A., Li, L. L., and Dinauer, M. C. (1997). Retroviral-mediated gene transfer of gp91phox into bone marrow cells rescues defect in host defense against *Aspergillus fumigatus* in murine X-linked chronic granulomatous disease. *Blood* **89**, 41–48.

Boguski, M. S., and McCormick, F. (1993). Proteins regulating Ras and its relatives. *Nature (London)* **366**, 643–654.

Bokoch, G. M. (1994). Regulation of the human neutrophil NADPH oxidase by the Rac GTP-binding proteins. *Curr. Opin. Cell Biol.* **6**, 212–218.

Bradley, D. J., Kjellbom, P., and Lamb, C. J. (1992). Elicitor- and wound-induced oxidative cross-linking of a proline-rich plant cell wall protein: A novel, rapid defense response. *Cell (Cambridge, Mass.)* **70**, 21–30.

Brill, S., Li, S., Lyman, C. W., Church, D. M., Wasmuth, J. J., Weissbach, L., Bernards, A., and Snijders, A. J. (1996). The Ras GTPase-activating-protein-related human protein IQGAP2 harbors a potential actin binding domain and interacts with calmodulin and Rho family GTPases. *Mol. Cell. Biol.* **16**, 4869–4878.

Brown, J. L., Stowers, L., Baer, M., Trejo, J., Coughlin, S., and Chant, J. (1996). Human Ste20 homologue hPAK1 links GTPases to the JNK MAP kinase pathway. *Curr. Biol.* **6**, 598–605.

Bulkley, G. B. (1994). Reactive oxygen metabolites and reperfusion injury: Aberrant triggering of reticuloendothelial function. *Lancet* **344**, 934–936.

Bunn, H. F., and Poyton, R. O. (1996). Oxygen sensing and molecular adaptation to hypoxia. *Physiol. Rev.* **76**, 839–885.

Burdon, R. H., Gill, V., and Rice-Evans, C. (1990). Oxidative stress and tumour cell proliferation. *Free Radical Res. Commun.* **11**, 65–76.

Casciola-Rosen, L., Wigley, F., and Rosen, A. (1997). Scleroderma autoantigens are uniquely fragmented by metal-catalyzed oxidation reactions: Implications for pathogenesis. *J. Exp. Med.* **185**, 71–79.

Cerione, R. A., and Zheng, Y. (1996). The Dbl family of oncogenes. *Curr. Opin. Cell Biol.* **8**, 216–222.

Chai, Y. C., Ashraf, S. S., Rokutan, K., Johnston, R. B. J., and Thomas, J. A. (1994). S-thiolation of individual human neutrophil proteins including actin by stimulation of the respiratory burst: Evidence against a role for glutathione disulfide. *Arch. Biochem. Biophys.* **310**, 273–281.

Chan, P. H. (1996). Role of oxidants in ischemic brain damage. *Stroke* **27**, 1124–1129.

Chen, W., Chen, S., Yap, S. F., and Lim, L. (1996). The *Caenorhabditis elegans* p21-activated kinase (CePAK) colocalizes with CeRac1 and CDC42Ce at hypodermal cell boundaries during embryo elongation. *J. Biol. Chem.* **271**, 26362–26368.

Chuang, T. H., Xu, X., Kaartinen, V., Heisterkamp, N., Groffen, J., and Bokoch, G. M. (1995). Abr and Bcr are multifunctional regulators of the Rho GTP-binding protein family. *Proc. Natl. Acad. Sci. U.S.A.* **92**, 10282–10286.

Collins, L. R., Minden, A., Karin, M., and Brown, J. H. (1996). Galpha12 stimulates c-Jun NH2-terminal kinase through the small G proteins Ras and Rac. *J. Biol. Chem.* **271**, 17349–17353.

Cooper, J. T., Stroka, D. M., Brostjan, C., Palmetshofer, A., Bach, F. H., and Ferran, C. (1996). A20 blocks endothelial cell activation through a NF-kappaB-dependent mechanism. *J. Biol. Chem.* **271**, 18068–18073.

Coso, O. A., Chiariello, M., Yu, J. C., Teramoto, H., Crespo, P., Xu, N., Miki, T., and Gutkind, J. S. (1995). The small GTP-binding protein Rac1 and Cdc42 regulate the activity of the JNK/SAPK signaling pathway. *Cell (Cambridge, Mass.)* **81**, 1137–1146.

Coso, O. A., Teramoto, H., Simonds, W. F., and Gutkind, J. S. (1996). Signaling from G protein-coupled receptors to c-Jun kinase involves beta gamma subunits of heterotrimeric G proteins acting on a Ras and Rac1-dependent pathway. *J. Biol. Chem.* **271**, 3963–3966.

Crawford, L. E., Milliken, E. E., Irani, K., Zweier, J. L., Becker, L. C., Johnson, T. M., Eissa, N. T., Crystal, R. G., Finkel, T., and Goldschmidt-Clermont, P. J. (1996). Superoxide-mediated actin response in post-hypoxic endothelial cells. *J. Biol. Chem.* **271**, 26863–26867.

Crespo, P., Schuebel, K. E., Ostrom, A. A., Gutkind, J. S., and Bustelo, X. R. (1997). Phosphotyrosine-dependent activation of Rac-1 GDP/GTP exchange by the vav proto-oncogene product. *Nature (London)* **385**, 169–172.

Cvrckova, F., De Virgilio, C., Manser, E., Pringle, J. R., and Nasmyth, K. (1995). Ste20-like protein kinases are required for normal localization of cell growth and for cytokinesis in budding yeast. *Genes Dev.* **9**, 1817–1830.

Dagher, M. C., Fuchs, A., Bourmeyster, N., Jouan, A., and Vignais, P. V. (1995). Small G proteins and the neutrophil NADPH oxidase. *Biochimie* **77**, 651–660.

DalleDonne, I., Milzani, A., and Colombo, R. (1995). H_2O_2-treated actin: Assembly and polymer interactions with cross-linking proteins. *Biophys. J.* **69**, 2710–2719.

Debant, A., Serra-Pages, C., Seipel, K., O'Brien, S., Tang, M., and Park, S. H. (1996). The multidomain protein Trio binds the LAR transmembrane tyrosine phosphatase, contains a protein kinase domain, and has separate rac- specific and rho-specific guanine nucleotide exchange factor domains. *Proc. Natl. Acad. Sci. U.S.A.* **93**, 5466–5471.

DeLeo, F. R., and Quinn, M. T. (1996). Assembly of the phagocyte NADPH oxidase: Molecular interaction of oxidase proteins. *J. Leukocyte Biol.* **60**, 677–691.

Delmer, D. P., Pear, J. R., Andrawis, A., and Stalker, D. M. (1995). Genes encoding small GTP-binding proteins analogous to mammalian rac are preferentially expressed in developing cotton fibers. *Mol. Gen. Genet.* **248**, 43–51.

de Mendez, I., Adams, A. G., Sokolic, R. A., Malech, H. L., and Leto, T. L. (1996). Multiple SH3 domain interactions regulate NADPH oxidase assembly in whole cells. *EMBO J.* **15**, 1211–1220.

Derijard, B., Hibi, M., Wu, I. H., Barrett, T., Su, B., Deng, T., Karin, M., and Davis, R. J. (1994). JNK1: A protein kinase stimulated by UV light and Ha-Ras that binds and phosphorylates the c-Jun activation domain. *Cell (Cambridge, Mass.)* **76**, 1025–1037.

Derijard, B., Raingeaud, J., Barrett, T., Wu, I. H., Han, J., Ulevitch, R. J., and Davis, R. J. (1995). Independent human MAP-kinase signal transduction pathways defined by MEK and MKK isoforms. *Science* **267**, 682–685.

de Vos, A. M., Tong, L., Milburn, M. V., Matias, P. M., Jancarik, J., Noguchi, S., Nishimura, S., Miura, K., Ohtsuka, E., and Kim, S. H. (1988). Three-dimensional structure of an oncogene protein: Catalytic domain of human c-H-ras p21. *Science* **239**, 888–893.

Diekmann, D., Brill, S., Garrett, M. D., Totty, N., Hsuan, J., Monfries, C., Hall, C., Lim, L., and Hall, A. (1991). Bcr encodes a GTPase-activating protein for p21rac. *Nature (London)* **351**, 400–402.

Diekmann, D., Abo, A., Johnston, C., Segal, A. W., and Hall, A. (1994). Interaction of Rac with p67 phox and regulation of phagocytic NADPH oxidase activity. *Science* **265**, 531–533.

Ding, C., Kume, A., Bjorgvinsdottir, H., Hawley, R. G., Pech, N., and Dinauer, M. C. (1996). High-level reconstitution of respiratory burst activity in a human X-linked chronic granulomatous disease (X-CGD) cell line and correction of murine X-CGD bone marrow cells by retroviral-mediated gene transfer of human gp91phox. *Blood* **88**, 1834–1840.

Dorseuil, O., Reibel, L., Bokoch, G. M., Camonis, J., and Gacon, G. (1996). The Rac target NADPH oxidase p67phox interacts preferentially with Rac2 rather than Rac1. *J. Biol. Chem.* **271**, 83–88.

D'Souza-Schorey, C., Boshans, R. L., McDonough, M., Stahl, P. D., and Van Aelst, L. (1997). A role for POR1, a Rac1-interacting protein, in ARF6-mediated cytoskeletal rearrangements. *EMBO J.* **16**, 5445–5454.

Dugas, B., Debre, P., and Moncada, S. (1995). Nitric oxide, a vital poison inside the immune and inflammatory network. *Res. Immunol.* **146**, 664–670.

Dusi, S., Della, B. V., Donini, M., Nadalini, K. A., and Rossi, F. (1996). Mechanisms of stimulation of the respiratory burst by TNF in nonadherent neutrophils: Its independence of lipidic transmembrane signaling and dependence on protein tyrosine phosphorylation and cytoskeleton. *J. Immunol.* **157**, 4615–4623.

Eaton, S., Auvinen, P., Luo, L., Jan, Y. N., and Simons, K. (1995). CDC42 and Rac1 control different actin-dependent processes in the Drosophila wing disc epithelium. *J. Cell Biol.* **131**, 151–164.

Emmendorffer, A., Roesler, J., Elsner, J., Raeder, E., Lohmann-Matthes, M. L., and Meier, B. (1993). Production of oxygen radicals by fibroblasts and neutrophils from a patient with x-linked chronic granulomatous disease. *Eur. J. Haematol.* **51**, 223–227.

Eva, A., and Aaronson, S. A. (1985). Isolation of a new human oncogene from a diffuse B-cell lymphoma. *Nature (London)* **316**, 273–275.

Foster, W. R., Ungar, L. H., and Schwaber, J. S. (1993). Significance of conductances in Hodgkin-Huxley models. *J. Neurophysiol.* **70**, 2502–2518.

Frejaville, C., Karoui, H., Tuccio, B., Le Moigne, F., Culcasi, M., Pietri, S., Lauricella, R., and Tordo, P. (1995). 5-(Diethoxyphosphoryl)-5-methyl-1-pyrroline N-oxide: A new efficient phosphorylated nitrone for the in vitro and in vivo spin trapping of oxygen-centered radicals. *J. Med. Chem.* **38**, 258–265.

Gallo, K. A., Mark, M. R., Scadden, D. T., Wang, Z., Gu, Q., and Godowski, P. J. (1994). Identification and characterization of SPRK, a novel src-homology 3 domain-containing proline-rich kinase with serine/threonine kinase activity. *J. Biol. Chem.* **269**, 15092–15100.

Gerwins, P., Blank, J. L., and Johnson, G. L. (1997). Cloning of a novel mitogen-activated protein kinase kinase kinase, MEKK4, that selectively regulates the c-Jun amino terminal kinase pathway. *J. Biol. Chem.* **272**, 8288–8295.

Geyer, M. and Wittinghofer, A. (1997). GEFs, GAPs, GDIs and effectors: Taking a closer (3D) look at the regulation of Ras-related GTP-binding proteins. *Curr. Opin. Struct. Biol.* **7**, 786–792.

Gille, H., Strahl, T., and Shaw, P. E. (1995). Activation of ternary complex factor Elk-1 by stress-activated protein kinases. *Curr. Biol.* **5**, 1191–1200.

Goldschmidt-Clermont, P. J., Furman, M. I., Wachsstock, D., Safer, D., Nachmias, V. T., and Pollard, T. D. (1992). The control of actin nucleotide exchange by thymosin beta 4 and profilin. A potential regulatory mechanism for actin polymerization in cells. *Mol. Biol. Cell* **3**, 1015–1024.

Gottshall, K. R., Hunter, J. J., Tanaka, N., Dalton, N., Becker, K. D., Ross, J. J., and Chien, K. R. (1997). Ras-dependent pathways induce obstructive hypertrophy in echo-selected transgenic mice. *Proc. Natl. Acad. Sci. U.S.A.* **94**, 4710–4715.

Gupta, S., Campbell, D., Derijard, B., and Davis, R. J. (1995). Transcription factor ATF2 regulation by the JNK signal transduction pathway. *Science* **267**, 389–393.

Gyllenhammar, H. (1987). Lucigenin chemiluminescence in the assessment of neutrophil superoxide production. *J. Immunol. Methods* **97**, 209–213.

Habets, G. G., Scholtes, E. H., Zuydgeest, D., van der Kammen, R. A., Stam, J. C., Berns, A., and Collard, J. G. (1994). Identification of an invasion-inducing gene, Tiam-1, that encodes a protein with homology to GDP-GTP exchangers for Rho-like proteins. *Cell (Cambridge, Mass.)* **77**, 537–549.

Hall, A. (1990). ras and GAP—who's controlling whom? *Cell (Cambridge, Mass.)* **61**, 921–923.

Hall, A. (1994). Small GTP-binding proteins and the regulation of the actin cytoskeleton. *Annu. Rev. Cell Biol.* **10**, 31–54.

Hall, A. (1998). Rho GTPases and the actin cytoskeleton. *Science* **279**, 509–514.

Harden, N., Lee, J., Loh, H. Y., Ong, Y. M., Tan, I., Leung, T., Manser, E., and Lim, L. (1996). A Drosophila homolog of the Rac- and Cdc42-activated serine/threonine kinase PAK is a potential focal adhesion and focal complex protein that colocalizes with dynamic actin structures. *Mol. Cell Biol.* **16**, 1896–1908.

Hart, C. M., and Roberts, J. W. (1994). Deletion analysis of the lambda tR1 termination region. Effect of sequences near the transcript release sites, and the minimum length of rho-dependent transcripts. *J. Mol. Biol.* **237**, 255–265.

Hart, M. J., Eva, A., Evans, T., Aaronson, S. A., and Cerione, R. A. (1991). Catalysis of guanine nucleotide exchange on the CDC42Hs protein by the dbl oncogene product. *Nature (London)* **354**, 311–314.

Hartwig, J. H., Bokoch, G. M., Carpenter, C. L., Janmey, P. A., Taylor, L. A., Toker, A., and Stossel, T. P. (1995). Thrombin receptor ligation and activated Rac uncap actin filament barbed ends through phosphoinositide synthesis in permeabilized human platelets. *Cell (Cambridge, Mass.)* **82**, 643–653.

Hawkins, P. T., Eguinoa, A., Qiu, R. G., Stokoe, D., Cooke, F. T., Walters, R., Wennstrom, S., Claesson-Welsh, L., Evans, T., and Symons, M. (1995). PDGF stimulates an increase in GTP-Rac via activation of phosphoinositide 3-kinase. *Curr. Biol.* 5, 393–403.

Heyworth, P. G., Knaus, U. G., Settleman, J., Curnuttle, J. T., and Bokoch, G. M. (1993). Regulation of NADPH oxidase by Rac GTPase activity protein(s). *Mol. Biol. Cell* 4, 1217–1221.

Heyworth, P. G., Robinson, J. M., Ding, J., Ellis, B. A., and Badwey, J. A. (1997). Cofilin undergoes rapid dephosphorylation in stimulated neutrophils and translocates to ruffled membranes enriched in products of the NADPH oxidase complex. Evidence for a novel cycle of phosphorylation and dephosphorylation. *Histochem. Cell. Biol.* 108, 221–233.

Hidalgo, E., Ding, H., and Demple, B. (1997). Redox signal transduction: Mutations shifting [2Fe-2S] centers of the SoxR sensor-regulator to the oxidized form. *Cell (Cambridge, Mass.)* 88, 121–129.

Huot, J., Houle, F., Spitz, D. R., and Landry, J. (1996). HSP27 phosphorylation-mediated resistance against actin fragmentation and cell death induced by oxidative stress. *Cancer Res.* 56, 273–279.

Irani, K. and Goldschmidt-Clermont J. (1998). Ras, superoxide and signal transduction. *Biochem. Pharmacol.* 55, 1339–1346.

Irani, K., Xia, Y., Zweier, J. L., Sollott, S. J., Der, C. J., Fearon, E. R., Sundaresan, M., Finkel, T., and Goldschmidt-Clermont, P. J. (1997). Mitogenic signaling mediated by oxidants in Ras-transformed fibroblasts. *Science* 275, 1649–1652.

Israel, N., Gougerot-Pocidalo, M. A., Aillet, F., and Virelizier, J. L. (1992). Redox status of cells influences constitutive or induced NF-kappa B translocation and HIV long terminal repeat activity in human T and monocytic cell lines. *J. Immunol.* 149, 3386–3393.

Jackson, S. H., Gallin, J. I., and Holland, S. M. (1995). The p47phox mouse knock-out model of chronic granulomatous disease. *J. Exp. Med.* 182, 751–758.

Janmey, P. A. (1994). Phosphoinositides and calcium as regulators of cellular actin assembly and disassembly. *Annu. Rev. Physiol.* 56, 169–191.

Jiang, Y., Chen, C., Li, Z., Guo, W., Gegner, J. A., Lin, S., and Han, J. (1996). Characterization of the structure and function of a new mitogen-activated protein kinase (p38beta). *J. Biol. Chem.* 271, 17920–17926.

Jones, O. T. (1994). The regulation of superoxide production by the NADPH oxidase of neutrophils and other mammalian cells. *BioEssays* 16, 919–923.

Jones, S. A., Wood, J. D., Coffey, M. J., and Jones, O. T. (1994). The functional expression of p47-phox and p67-phox may contribute to the generation of superoxide by an NADPH oxidase-like system in human fibroblasts. *FEBS Lett.* 355, 178–182.

Jones, S. A., O'Donnell, V. B., Wood, J. D., Broughton, J. P., Hughes, E. J., and Jones, O. T. (1996). Expression of phagocyte NADPH oxidase components in human endothelial cells. *Am. J. Physiol.* 271, 626–634.

Joneson, T., McDonough, M., Bar-Sagi, D., and Van Aelst, L. (1996). RAC regulation of actin polymerization and proliferation by a pathway distinct from Jun kinase. *Science* 274, 1374–1376; *erratum*: 276, 185 (1997).

Kamata, H., Tanaka, C., Yagisawa, H., Matsuda, S., Gotoh, Y., Nishida, E., and Hirata, H. (1996). Suppression of nerve growth factor-induced neuronal differentiation of PC12 cells. N-acetylcysteine uncouples the signal transduction from ras to the mitogen-activated protein kinase cascade. *J. Biol. Chem.* 271, 33018–33025.

Kandzari, D. E. (1996). Regulation of the actin cytoskeleton by inositol phospholipid pathways. *Subcell. Biochem.* 5(3), 210–208.

Khosravi-Far, R., Solski, P. A., Clark, G. J., Kinch, M. S., and Der, C. J. (1995). Activation of Rac1, RhoA, and mitogen-activated protein kinases is required for Ras transformation. *Mol. Cell. Biol.* 15, 6443–6453.

Kim, B. C., and Kim, J. H. (1997). Role of Rac GTPase in the nuclear signaling by EGF. *FEBS Lett.* **407**, 7–12.

Kim, J. H., Kwack, H. J., Choi, S. E., Kim, B. C., Kim, Y. S., Kang, I. J., and Kumar, C. C. (1997). Essential role of Rac GTPase in hydrogen peroxide-induced activation of c-fos serum response element. *FEBS Lett.* **406**, 93–96.

Knaus, U. G., Heyworth, P. G., Evans, T., Curnutte, J. T., and Bokoch, G. M. (1991). Regulation of phagocyte oxygen radical production by the GTP-binding protein Rac 2. *Science* **254**, 1512–1515.

Knaus, U. G., Morris, S., Dong, H. J., Chernoff, J., and Bokoch, G. M. (1995). Regulation of human leukocyte p21-activated kinases through G protein-coupled receptors. *Science* **269**, 221–223.

Kohl, N. E., Mosser, S. D., deSolms, S. J., Giuliani, E. A., Pompliano, D. L., Graham, S. L., Smith, R. L., Scolnick, E. M., Oliff, A., and Gibbs, J. B. (1993). Selective inhibition of ras-dependent transformation by a farnesyltransferase inhibitor. *Science* **260**, 1934–1937.

Korthuis, R. J., and Granger, D. N. (1993). Reactive oxygen metabolites, neutrophils, and the pathogenesis of ischemic-tissue/reperfusion. *Clin. Cardiol.* **16**, 119–126.

Kotani, K., Yonezawa, K., Hara, K., Ueda, H., Kitamura, Y., Sakaue, H., Ando, A., Chavanieu, A., Calas, B., and Grigorescu, F. (1994). Involvement of phosphoinositide 3-kinase in insulin- or IGF-1-induced membrane ruffling. *EMBO J.* **13**, 2313–2321.

Kozma, R., Ahmed, S., Best, A., and Lim, L. (1995). The Ras-related protein Cdc42Hs and bradykinin promote formation of peripheral actin microspikes and filopodia in Swiss 3T3 fibroblasts. *Mol. Cell. Biol.* **15**, 1942–1952.

Kozma, R., Ahmed, S., Best, A., and Lim, L. (1996). The GTPase-activating protein n-chimaerin cooperates with Rac1 and Cdc42Hs to induce the formation of lamellipodia and filopodia. *Mol. Cell. Biol.* **16**, 5069–5080.

Kume, A., and Dinauer, M. C. (1994). Retrovirus-mediated reconstitution of respiratory burst activity in X-linked chronic granulomatous disease cells. *Blood* **84**, 3311–3316.

Kuroda, S., Fukata, M., Kobayashi, K., Nakafuku, M., Nomura, N., Iwamatsu, A., and Kaibuchi, K. (1996). Identification of IQGAP as a putative target for the small GTPases, Cdc42 and Rac1. *J. Biol. Chem.* **271**, 23363–23367.

Kwong, C. H., Adams, A. G., and Leto, T. L. (1995). Characterization of the effector-specifying domain of Rac involved in NADPH oxidase activation. *J. Biol. Chem.* **270**, 19868–19872.

Kyriakis, J. M., and Avruch, J. (1996). Protein kinase cascades activated by stress and inflammatory cytokines. *BioEssays* **18**, 567–577.

Lamarche, N., Tapon, N., Stowers, L., Burbelo, P. D., Aspenstrom, P., Bridges, T., Chant, J., and Hall, A. (1996). Rac and Cdc42 induce actin polymerization and G1 cell cycle progression independently of p65PAK and the JNK/SAPK MAP kinase cascade. *Cell (Cambridge, Mass.)* **87**, 519–529.

Lancaster, C. A., Taylor-Harris, P. M., Self, A. J., Brill, S., van, E. H., and Hall, A. (1994). Characterization of rhoGAP. A GTPase-activating protein for rho-related small GTPases. *J. Biol. Chem.* **269**, 1137–1142.

Leung, T., How, B. E., Manser, E., and Lim, L. (1993). Germ cell beta-chimaerin, a new GTPase-activating protein for p21rac, is specifically expressed during the acrosomal assembly stage in rat testis. *J. Biol. Chem.* **268**, 3813–3816.

Leusen, J. H. W., de, K. A., Hilarius, P. M., Ahlin, A., Palmblad, J., Smith, C. I., Diekmann, D., Hall, A., Verhoeven, A. J., and Roos, D. (1996a). Disturbed interaction of p21-rac with mutated p67-phox causes chronic granulomatous disease. *J. Exp. Med.* **184**, 1243–1249.

Leusen, J. H. W., Verhoeven, A. J., and Roos, D. (1996b). Interactions between the components of the human NADPH oxidase: A review about the intrigues in the phox family. *Front. Biosci.* **1**, 72–90.

Levine, A., Tenhaken, R., Dixon, R., and Lamb, C. (1994). H2O2 from the oxidative burst orchestrates the plant hypersensitive disease resistance response. *Cell (Cambridge, Mass.)* **79**, 583–593.

Lin, Y., Wang, Y., Zhu, J., and Yang, Z. (1996). Localization of a Rho GTPase implies a role in tip growth and movement of the generative cell in pollen tubes. *Plant Cell* **8**, 293–303.

Lo, Y. Y., and Cruz, T. F. (1995). Involvement of reactive oxygen species in cytokine and growth factor induction of c-fos expression in chondrocytes. *J. Biol. Chem.* **270**, 11727–11730.

Lowenstein, C. J., Dinerman, J. L., and Snyder, S. H. (1994). Nitric oxide: A physiologic messenger. *Ann. Intern. Med.* **120**, 227–237.

Lowy, D. R., Johnson, M. R., DeClue, J. E., Cen, H., Zhang, K., Papageorge, A. G., Vass, W. C., Willumsen, B. M., Valentine, M. B., and Look, A. T. (1993). Cell transformation by ras and regulation of its protein product. *Ciba Found. Symp.* **176**, 67–80; discussion: pp. 80–84.

Ludwig, S., Engel, K., Hoffmeyer, A., Sithanandam, G., Neufeld, B., Palm, D., Gaestel, M., and Rapp, U. R. (1996). 3pK, a novel mitogen-activated protein (MAP) kinase-activated protein kinase, is targeted by three MAP kinase pathways. *Mol. Cell. Biol.* **16**, 6687–6697.

Luo, L., Hensch, T. K., Ackerman, L., Barbel, S., Jan, L. Y., and Jan, Y. N. (1996). Differential effects of the Rac GTPase on Purkinje cell axons and dendritic trunks and spines. *Nature (London)* **379**, 837–840.

Manser, E., Chong, C., Zhao, Z. S., Leung, T., Michael, G., Hall, C., and Lim, L. (1995). Molecular cloning of a new member of the p21-Cdc42/Rac-activated kinase (PAK) family. *J. Biol. Chem.* **270**, 25070–25078.

Martin, G. A., Bollag, G., McCormick, F., and Abo, A. (1995). A novel serine kinase activated by rac1/CDC42Hs-dependent autophosphorylation is related to PAK65 and STE20. *EMBO J.* **14**, 1970–1978.

Martinez-Arias, A. (1993). Development and patterning of the larval epidermis of Drosophila. *In* "Development of Drosophila," pp. 517–608. Cold Spring Harbor Lab. Press, Cold Spring Harbor, NY.

Matsui, Y., Matsui, R., Akada, R., and Toh (1996). Yeast src homology region 3 domain-binding proteins involved in bud formation. *J. Cell Biol.* **133**, 865–878.

Mehdy, M. (1994). Active oxygen species in plant defense against pathogens. *Plant Physiol.* **105**, 467–472.

Meier, B., Radeke, H. H., Selle, S., Younes, M., Sies, H., Resch, K., and Habermehl, G. G. (1989). Human fibroblasts release reactive oxygen species in response to interleukin-1 or tumour necrosis factor-alpha. *Biochem. J.* **263**, 539–545.

Meyer, M., Schreck, R., and Baeuerle, P. A. (1993). H2O2 and antioxidants have opposite effects on activation of NF-kappa B and AP-1 in intact cells: AP-1 as secondary antioxidant-responsive factor. *EMBO J.* **12**, 2005–2015.

Michiels, F., Habets, G. G., Stam, J. C., van der Kammen, R. A., and Collard, J. G. (1995). A role for Rac in Tiam1-induced membrane ruffling and invasion. *Nature (London)* **375**, 338–340.

Michiels, F., Stam, J. C., Hordijk, P. L., van der Kammen, R. A., Ruuls-Van, S. L., Feltkamp, C. A., and Collard, J. G. (1997). Regulated membrane localization of Tiam1, mediated by the NH2-terminal pleckstrin homology domain, is required for Rac-dependent membrane ruffling and C-Jun NH2-terminal kinase activation. *J. Cell Biol.* **137**, 387–398.

Minden, A., Lin, A., Claret, F. X., Abo, A., and Karin, M. (1995). Selective activation of the JNK signaling cascade and c-Jun transcriptional activity by the small GTPases Rac and Cdc42Hs. *Cell (Cambridge, Mass.)* **81**, 1147–1157.

Mirabelli, F., Salis, A., Vairetti, M., Bellomo, G., Thor, H., and Orrenius, S. (1989). Cytoskeletal alterations in human platelets exposed to oxidative stress are mediated by oxidative and Ca2+-dependent mechanisms. *Arch. Biochem. Biophys.* **270**, 478–488.

Murphy, A. M., and Montell, D. J. (1996). Cell type-specific roles for Cdc42, Rac, and RhoL in Drosophila oogenesis. *J. Cell Biol.* **133**, 617–630.

Nobes, C. D., and Hall, A. (1995). Rho, rac, and cdc42 GTPases regulate the assembly of multimolecular focal complexes associated with actin stress fibers, lamellipodia, and filopodia. *Cell (Cambridge, Mass.)* **81**, 53–62.

Nobes, C. D., Hawkins, P., Stephens, L., and Hall, A. (1995). Activation of the small GTP-binding proteins rho and rac by growth factor receptors. *J. Cell Sci.* **108**, 225–233.

Nusslein-Volhard, C. E. (1984). Mutations affecting the pattern of the larval cuticle in *Drosophila melanogaster*. I. Zygotic loci on the second chromosome. *Arch. Dev. Biol.* **193**, 267–282.

Olson, M. F., Ashworth, A., and Hall, A. (1995). An essential role for Rho, Rac, and Cdc42 GTPases in cell cycle progression through G1. *Science* **269**, 1270–1272.

Omer, C. A., and Gibbs, J. B. (1994). Protein prenylation in eukaryotic microorganisms: Genetics, biology and biochemistry. *Mol. Microbiol.* **11**, 219–225.

Parrini, M. C., Giglione, C., and Parmeggiani, A. (1997). Co-ordination and specificity of the action of GTPase-activating proteins and GDP/GTP-exchange factors on Ras. *Biochem. Soc. Trans.* **25**, 997–1001.

Patriarca, P., Cramer, R., Moncalvo, S., Rossi, F., and Romeo, D. (1971). Enzymatic basis of metabolic stimulation in leucocytes during phagocytosis: The role of activated NADPH oxidase. *Arch. Biochem. Biophys.* **145**, 255–262.

Peppelenbosch, M. P., Tertoolen, L. G., den Hertog, J., and de Laat, S. W. (1992). Epidermal growth factor activates calcium channels by phospholipase A2/5-lipoxygenase-mediated leukotriene C4 production. *Cell (Cambridge, Mass.)* **69**, 295–303.

Peppelenbosch, M. P., Qiu, R. G., de Vries-Smits, A. M., Tertoolen, L. G., de Laat, S. W., McCormick, F., Hall, A., Symons, M. H., and Bos, J. L. (1995). Rac mediates growth factor-induced arachidonic acid release. *Cell (Cambridge, Mass.)* **81**, 849–856.

Pollock, J. D., Williams, D. A., Gifford, M. A., Li, L. L., Du, X., Fisherman, J., Orkin, S. H., Doerschuk, C. M., and Dinauer, M. C. (1995). Mouse model of X-linked chronic granulomatous disease, an inherited defect in phagocyte superoxide production. *Nat. Genet.* **9**, 202–209.

Poodry, C. (1980). Imaginal discs: Morphology and development. *Genet. Biol. Drosophila.* **2**, 407–441.

Pulverer, B. J., Kyriakis, J. M., Avruch, J., Nikolakaki, E., and Woodgett, J. R. (1991). Phosphorylation of c-jun mediated by MAP kinases. *Nature (London).* **353**, 670–674.

Quinn, M. T., Evans, T., Loetterle, L. R., Jesaitis, A. J., and Bokoch, G. M. (1993). Translocation of Rac correlates with NADPH oxidase activation. Evidence for equimolar translocation of oxidase components. *J. Biol. Chem.* **268**, 20983–20987.

Raingeaud, J., Whitmarsh, A. J., Barrett, T., Derijard, B., and Davis, R. J. (1996). MKK3- and MKK6-regulated gene expression is mediated by the p38 mitogen-activated protein kinase signal transduction pathway. *Mol. Cell. Biol.* **16**, 1247–1255.

Rajagopalan, S., Kurz, S., Munzel, T., Tarpey, M., Freeman, B. A., Griendling, K. K., and Harrison, D. G. (1996). Angiotensin II-mediated hypertension in the rat increases vascular superoxide production via membrane NADH/NADPH oxidase activation. Contribution to alterations of vasomotor tone. *J. Clin. Invest.* **97**, 1916–1923.

Reif, K., Nobes, C. D., Thomas, G., Hall, A., and Cantrell, D. A. (1996). Phosphatidylinositol 3-kinase signals activate a selective subset of Rac/Rho-dependent effector pathways. *Curr. Biol.* **6**, 1445–1455.

Renshaw, M. W., Lea-Chou, E., and Wang, J. Y. (1996). Rac is required for v-Abl tyrosine kinase to activate mitogenesis. *Curr. Biol.* **6**, 76–83.

Ridley, A. J. (1995). Rho-related proteins: Actin cytoskeleton and cell cycle. *Curr. Opin. Genet. Dev.* **5**, 24–30.
Ridley, A. J. (1996). Rho: Theme and variations. *Curr. Biol.* **6**, 1256–1264.
Ridley, A. J., and Hall, A. (1992). The small GTP-binding protein rho regulates the assembly of focal adhesions and actin stress fibers in response to growth factors. *Cell (Cambridge, Mass.)* **70**, 389–399.
Ridley, A. J., Paterson, H. F., Johnston, C. L., Diekmann, D., and Hall, A. (1992). The small GTP-binding protein rac regulates growth factor-induced membrane ruffling. *Cell (Cambridge, Mass.)* **70**, 401–410.
Ridley, A. J., Self, A. J., Kasmi, F., Paterson, H. F., Hall, A., Marshall, C. J., and Ellis, C. (1993). rho family GTPase activating proteins p190, bcr and rhoGAP show distinct specificities in vitro and in vivo. *EMBO J.* **12**, 5151–5160.
Ridley, A. J., Comoglio, P. M., and Hall, A. (1995). Regulation of scatter factor/hepatocyte growth factor responses by Ras, Rac, and Rho in MDCK cells. *Mol. Cell. Biol.* **15**, 1110–1122.
Riesgo-Escovar, J. R., Jenni, M., Fritz, A., and Hafen, E. (1996). The Drosophila Jun-N-terminal kinase is required for cell morphogenesis but not for DJun-dependent cell fate specification in the eye. *Genes Dev.* **10**, 2759–2768.
Roos, D. (1994). The genetic basis of chronic granulomatous disease. *Immunol. Rev.* **138**, 121–157.
Roos, D. (1996). X-CGDbase: A database of X-CGD-causing mutations. *Immunol. Today.* **17**, 517–521.
Rouse, J., Cohen, P., Trigon, S., Morange, M., Alonso-Llamazares, A., Zamanillo, D., Hunt, T., and Nebreda, A. R. (1994). A novel kinase cascade triggered by stress and heat shock that stimulates MAPKAP kinase-2 and phosphorylation of the small heat shock proteins. *Cell (Cambridge, Mass.)* **78**, 1027–1037.
Sanchez, I., Hughes, R. T., Mayer, B. J., Yee, K., Woodgett, J. R., Avruch, J., Kyriakis, J. M., and Zon, L. I. (1994). Role of SAPK/ERK kinase-1 in the stress-activated pathway regulating transcription factor c-Jun. *Nature (London)* **372**, 794–798.
Schenk, H., Klein, M., Erdbrugger, W., Droge, W., and Schulze-Osthoff, K. (1994). Distinct effects of thioredoxin and antioxidants on the activation of transcription factors NF-kappa B and AP-1. *Proc. Natl. Acad. Sci. U.S.A.* **91**, 1672–1676.
Schreck, R., Rieber, P., and Baeuerle, P. A. (1991). Reactive oxygen intermediates as apparently widely used messengers in the activation of the NF-kappa B transcription factor and HIV-1. *EMBO J.* **10**, 2247–2258.
Seger, R., and Krebs, E. G. (1995). The MAPK signaling cascade. *FASEB J.* **9**, 726–735.
Siebenlist, U., Franzoso, G., and Brown, K. (1994). Structure, regulation and function of NF-kappa B. *Annu. Rev. Cell Biol.* **10**, 405–455.
Small, J. V. (1994). Lamellipodia architecture: Actin filament turnover and the lateral flow of actin filaments during motility. *Semin. Cell Biol.* **5**, 157–163.
Staal, F. J., Roederer, M., and Herzenberg, L. A. (1990). Intracellular thiols regulate activation of nuclear factor kappa B and transcription of human immunodeficiency virus. *Proc. Natl. Acad. Sci. U.S.A.* **87**, 9943–9947.
Stadtman, E. R. (1992). Protein oxidation and aging. *Science* **257**, 1220–1224.
Steven, R., Kubiseski, T. J., Zheng, H., Kulkarni, S., Mancillas, J., Ruiz, M. A., Hogue, C. W., Pawson, T., and Culotti, J. (1998). UNC-73 activates the Rac GTPase and is required for cell and growth cone migrations in C. elegans. *Cell (Cambridge, Mass.)* **92**, 785–795.
Stossel, T. P. (1993). On the crawling of animal cells. *Science* **260**, 1086–1094.
Sulciner, D. J., Irani, K., Yu, Z. X., Ferrans, V. J., Goldschmidt-Clermont, P., and Finkel, T. (1996). rac1 regulates a cytokine-stimulated, redox-dependent pathway necessary for NF-kappa B activation. *Mol. Cell. Biol.* **16**, 7115–7121.

Sundaresan, M., Yu, Z. X., Ferrans, V. J., Irani, K., and Finkel, T. (1995). Requirement for generation of H2O2 for platelet-derived growth factor signal transduction. *Science* **270**, 296–299.

Sundaresan, M., Yu, Z. X., Ferrans, V. J., Sulciner, D. J., Gutkind, J. S., Irani, K., Goldschmidt-Clermont, P. J., and Finkel, T. (1996). Regulation of reactive-oxygen-species generation in fibroblasts by Racl. *Biochem J.* **318**, 379–382.

Tauber, A. I., Borregaard, N., Simons, E., and Wright, J. (1983). Chronic granulomatous disease: A syndrome of phagocyte oxidase deficiencies. *Medicine (Baltimore)* **62**, 286–309.

Tenhaken R., Levine, A., Brisson, L. F., Dixon, R. A., and Lamb, C. (1995). Function of the oxidative burst in hypersensitive disease resistance. *Proc. Natl. Acad. Sci. U.S.A.* **92**, 4158–4163.

Teramoto, H., Crespo, P., Coso, O. A., Igishi, T., Xu, N., and Gutkind, J. S. (1996a). The small GTP-binding protein rho activates c-Jun N-terminal kinases/stress-activated protein kinases in human kidney 293T cells. Evidence for a Pak-independent signaling pathway. *J. Biol. Chem.* **271**, 25731–25734.

Teramoto, H., Coso, O. A., Miyata, H., Igishi, T., Miki, T., and Gutkind, J. S. (1996b). Signaling from the small GTP-binding proteins Rac1 and Cdc42 to the c-Jun N-terminal kinase/stress-activated protein kinase pathway. A role for mixed lineage kinase 3/protein-tyrosine kinase 1, a novel member of the mixed lineage kinase family. *J. Biol. Chem.* **271**, 27225–27228.

Thrasher, A. J., de Alwis, M., Casimir, C. M., Kinnon, C., Page, K., Lebkowski, J., Segal, A. W., and Levinsky, R. J. (1995). Generation of recombinant adeno-associated virus (rAAV) from an adenoviral vector and functional reconstitution of the NADPH-oxidase. *Gene. Ther.* **2**, 481–485.

Tibbles, L. A., Ing, Y. L., Kiefer, F., Chan, J., Iscove, N., Woodgett, J. R., and Lassam, N. J. (1996). MLK-3 activates the SAPK/JNK and p38/RK pathways via SEK1 and MKK3/6. *EMBO J.* **15**, 7026–7035.

Trainin, T. (1996). In vitro prenylation of the small GTPase Rac13 of cotton. *Plant Physiol.* **112**, 1491–1497.

Tsai, M. H., Hall, A., and Stacey, D. W. (1989). Inhibition by phospholipids of the interaction between R-ras, rho, and their GTPase-activating proteins. *Mol. Cell. Biol.* **9**, 5260–5264.

Van Aelst, L., and D'Souza-Schorey, C. (1997). Rho GTPases and signaling networks. *Genes Dev.* **11**, 2295–2322.

Van Aelst, L., Joneson, T., and Bar-Sagi, D. (1996). Identification of a novel Rac1-interacting protein involved in membrane ruffling. *EMBO J.* **15**, 3778–3786.

van, Leeuwen, F. N., van der Kammen RA, Habets, G. G., and Collard, J. G. (1995). Oncogenic activity of Tiam1 and Rac1 in NIH3T3 cells. *Oncogene* **11**, 2215–2221.

Wang, C. Y., Mayo, M. W., and Baldwin, A. S. J. (1996). TNF- and cancer therapy-induced apoptosis: Potentiation by inhibition of NF-kappa B. *Science* **274**, 784–787.

Wang, X. Z., and Ron, D. (1996). Stress-induced phosphorylation and activation of the transcription factor CHOP (GADD153) by p38 MAP Kinase. *Science* **272**, 1347–1349.

Wennström, S., Hawkins, P., Cooke, F., Hara, K., Yonezawa, K., Kasuga, M., Jackson, T., Claesson-Welsh, L., and Stephens, L. (1994). Activation of phosphoinositide 3-kinase is required for PDGF-stimulated membrane ruffling. *Curr. Biol.* **4**, 385–393.

Westendorp, M. O., Shatrov, V. A., Schulze-Osthoff, K., Frank, R., Kraft, M., Los, M., Krammer, P. H., Droge, W., and Lehmann, V. (1995). HIV-1 Tat potentiates TNF-induced NF-kappa B activation and cytotoxicity by altering the cellular redox state. *EMBO J.* **14**, 546–554.

Westwick, J. K., Lambert, Q. T., Clark, G. J., Symons, M., Van, Aelst, L., Pestell, R. G., and Der, C. J. (1997). Rac regulation of transformation, gene expression, and actin organization by multiple, PAK-independent pathways. *Mol. Cell. Biol.* **17**, 1324–1335.

Whiteway, M., Hougan, L., Dignard, D., Thomas, D. Y., Bell, L., Saari, G. C., Grant, F. J., O'Hara, P., and MacKay, V. L. (1989). The STE4 and STE18 genes of yeast encode potential beta and gamma subunits of the mating factor receptor-coupled G protein. *Cell (Cambridge, Mass.)* **56**, 467–477.

Wientjes, F. B., and Segal, A. W. (1995). NADPH oxidase and the respiratory burst. *Semin. Cell. Biol.* **6**, 357–365.

Winge, P., Brembu, T., and Bones, A. M. (1997). Cloning and characterization of rac-like cDNAs from *Arabidopsis thaliana*. *Plant Mol. Biol.* **35**, 483–495.

Wu, M., Lee, H., Bellas, R. E., Schauer, S. L., Arsura, M., Katz, D., FitzGerald, M. J., Rothstein, T. L., Sherr, D. H., and Sonenshein, G. E. (1996). Inhibition of NF-kappa B/Re1 induces apoptosis of murine B cells. *EMBO J.* **15**, 4682–4690.

Yang, G., Chan, P. H., Chen, J., Carlson, E., Chen, S. F., Weinstein, P., Epstein, C. J., and Kamii, H. (1994). Human copper-zinc superoxide dismutase transgenic mice are highly resistant to reperfusion injury after focal cerebral ischemia. *Stroke* **25**, 165–170.

Yang, Z., and Watson, J. C. (1993). Molecular cloning and characterization of rho, a rasrelated small GTP- binding protein from the garden pea. *Proc. Natl. Acad. Sci. U.S.A.* **90**, 8732–8736.

Zervos, A. S., Faccio, L., Gatto, J. P., Kyriakis, J. M., and Brent, R. (1995). Mxi2, a mitogen-activated protein kinase that recognizes and phosphorylates Max protein. *Proc. Natl. Acad. Sci. U.S A.* **92**, 10531–10534.

Zheng, M., Slund, F., and Storz, G. (1998). Activation of the OxyR transcription factor by reversible disulfide bond formation. *Science* **279**, 1718–1721.

Zigmond, S. H. (1996). Signal transduction and actin filament organization. *Curr. Opin. Cell Biol.* **8**, 66–73.

Zipkin, I. D., Kindt, R. M., and Kenyon, C. J. (1997). Role of a new Rho family member in cell migration and axon guidance in *C. elegans*. *Cell (Cambridge, Mass.)* **90**, 883–894.

Zweier, J. L., Broderick, R., Kuppusamy, P., Thompson-Gorman, S., and Lutty, G. A. (1994). Determination of the mechanism of free radical generation in human aortic endothelial cells exposed to anoxia and reoxygenation. *J. Biol. Chem.* **269**, 24156–24162.

4 Redox Regulation of Ion Channels

Suneil K. Koliwad,* Anna K. Brzezinska,† and Stephen J. Elliott‡

*Department of Pediatrics
Baylor College of Medicine
Houston, Texas 77030

†Department of Physiology
Medical College of Wisconsin
Milwaukee, Wisconsin 53226

‡Doctors Medical Center
Modesto, California 95350

I. ION CHANNELS

No field in physiology, it seems, is more confusing than that of electrophysiology, which has a language all to itself. Yet the past two decades have witnessed a massive increase in our understanding of ion channels and in our ability to identify their molecular makeup and their biophysical behavior. As the role of ion channels in signal transduction and cell physiology has been progressively elucidated, so too has the modulation of channel function by oxidation/reduction reactions. The following section covers some basic electrophysiological concepts, after which the role of redox regulation in channel function will be discussed.

Ion channels are protein macromolecules that form pores in the membranes of mammalian cells. It is not known when ion channels evolved, but their presence in primitive organisms suggests that they are an integral component of the cell membrane. The lipid membrane that encloses living cells prevents intracellular contents from dissipating into the extracellular environment and protects the cell from this environment. However, no cell could exist if the cell membrane was totally impermeable. Cell viability requires not only the influx of substrates for energy production and anabolism, but also the efflux of waste molecules and products of metabolism. Channels form one mechanism by which ions traverse the cell membrane.

Channels allow for fluxes as great as 10^7 ions per second. Channels are generally classified and named according to their functional characteristics. These characteristics include (1) ionic selectivity, (2) the electrical conductance generated by the passage of ions, (3) the relationship between current amplitude through the channel and the membrane potential, (4) the molecular and biophysical factors that cause the channel to open and close, and (5) the kinetics of activation and inactivation. More recently, the application of molecular biology to electrophysiology has led to new nomenclatures based on the amino acid sequences of wild-type and mutant channel proteins.

Ion channels obey Ohm's law (V = IR, where V is voltage, I is current, and R is resistance). The electrical conductance, G, of a single channel reflects the ease with which the channel allows a flow of ions and is given by the ratio of current amplitude to electrical driving force (i.e., the voltage across the membrane). An ion channel that has 20 pS conductance (1 picoSiemen = 10^{-12} S) will conduct 1 pA of current when subjected to a driving force of 50 mV. The molecular structure of a channel determines its conductance for any one type of ion. In this way, two different K^+ channels each might be highly selective for K^+, yet have very different conductances.

In electrophysiology, the current–voltage relationship is typically shown as an x–y plot in which the amplitude and direction of current is plotted on the ordinate and the difference in potential across the membrane is plotted on the abscissa. Rectification refers to that property of a channel wherein current flows preferentially in one direction. The inward-rectifier K^+ channel of vascular endothelial cells conducts current preferentially in the inward direction.

The relative distribution of open channels in any one membrane determines the inherent permeability of that membrane to each ionic species. Ions move through open channels in accordance with their electrical and chemical gradients. The membrane potential at which inward diffusion and outward diffusion are equal for any one ionic species is the Nernst equilibrium potential for that species. Under physiologic conditions, the Nernst equilibrium potentials for Na^+, Ca^{2+}, and K^+ are approximately 70, 150, and −90 mV, respectively. Thus, at a membrane potential of −90 mV, no net flux of K^+ will occur. Conversely, if a membrane is impermeable to all ions but K^+, the membrane potential will come to the Nernst potential for K^+. Mammalian cell membranes are highly permeable to K^+ because constitutive channels that are selective for K^+ are open in the resting state. For this reason, the resting membrane potential (approximately −60 mV) is close to the Nernst equilibrium potential for K^+ (Fig. 1). In living cells, the membrane potential is determined by (1) the electrical charge of each ionic species, (2) the concentrations of the ions on each side of the membrane,

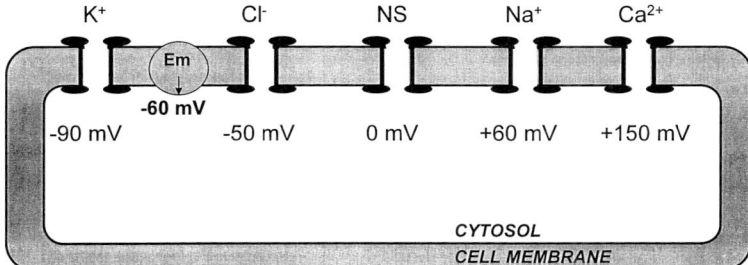

Figure 1 Schematic depiction of Nernst equilibrium potentials for cations as they pertain to the plasma membrane of mammalian cells. Compared with other cations, membrane permeability is greatest for K^+. For this reason, the membrane potential is close to the Nernst potential for K^+ under physiologic conditions. NS, nonselective; Em, membrane potential.

and (3) the permeability of the membrane to each type of ion. The opening and closing of channels alter the permeability of the membrane to specific ions dramatically and hence can alter the membrane potential dramatically.

Because the cell membrane is an insulating layer that separates two ionic solutions, it forms an electrical capacitor. Capacitance (C) is a measure of the charge (Q) that is transferred in order to create a potential difference (V). In this way, $C = Q/V$. Cell membranes have an electrical capacitance of 1 $\mu F/cm^2$. This high capacitance defines the lower limit to the quantity of ions (i.e., the charge, Q) that must move in order to generate an electrical signal. Capacitance retards the change in voltage resulting from a flow of current and hence determines the rapidity with which the charge must move to generate the signal. The product of membrane resistance (R) and membrane capacitance (C) defines the membrane time constant (τ_M). Depending on the particular cell membrane, R ranges from 10 to 10^6 Ohms/cm^{-2}, and τ_M from 10 μs to 1 s. In other words, the number and type of ion channels that are open under resting conditions vary immensely from one cell type to another. This wide diversity of channels provides a way for particular channels to open and close in response to the myriad of chemical and biophysical signals with which the cell membrane is presented. Moreover, the diversity of channels provides a basis for selectivity in matching incoming signals with downstream effector pathways that exist within the cell.

In the past several decades, the signals that cause ion channels to open and close have been investigated intensively. In vascular smooth muscle, for example, numerous molecules that either open or close specific types of K^+ channels have been identified. Some of these molecules are present endogenously *in vivo*, whereas others are biological toxins and/or pharmacologic agents. The sensitivity of channels to such molecules represents a characteristic by which channels are identified and categorized. Thus, the Ca^{2+}-activated

K⁺ channel is opened by increases in cytosolic [Ca^{2+}]. It is also sensitive to the snake venom, iberiotoxin, and is known synonymously as the iberiotoxin-sensitive K⁺ channel. Application of iberiotoxin to the external (i.e., extracellular) aspect of the channel causes the channel to close. Because it has a relatively large conductance (180–260 pS), it has yet another name, the maxi-K⁺ channel. Some channels, such as the voltage-gated, L-type Ca^{2+} channel, open and close in response not to second messengers but to the biophysical event of depolarization. "Gating" is a term that refers to the opening and closing of the channel pore.

When analyzing the activity of single channels, electrophysiologists calculate the so called open probability (P_o). A channel that is open 100% of the time has a P_o of 1, whereas a channel that is open 50% of the time has P_o of 0.5. Many channels are not limited to being either simply closed or simply open. Various substates can exist, such that a single channel protein can generate multiple levels of current amplitude. Alternatively, multiple levels of current amplitude that are recorded in a path of membrane might be due to the presence of multiple channels that are opening and closing in a nonsynchronized fashion. In analyzing channel activity from such recordings, a quantitative measure of channel activity (NP_o) is given by the product of the P_o and the number of open channels (or open states), N. The complexity of electrical recordings is in part the reason why electrophysiologists refer to channel "activity" when referring to the function of a channel in the open state.

The application of molecular biology to the field of ion channels has led to an explosion of information with respect to the relationship between structure and function of channel proteins. In this molecular age, it is possible that all channel proteins will ultimately be classified by their genetic makeup rather than by their functional phenotype. Not only have channel proteins been sequenced and cloned, but site-directed mutagenesis has helped identify residues that are critically important to channel function.

The voltage-gated, L-type Ca^{2+} channel of cardiac myocytes and vascular smooth muscle is an example of a channel that has been subjected to numerous molecular manipulations using the *Xenopus laevis* oocyte expression system. Injection of cDNA clones with base substitutions results in the expression of channels with substituted residues. The *Xenopus* system is favored by electrophysiologists because the oocyte is largely void of channels that conduct cations and because the currents that are generated by expressed channels have relatively large amplitudes. As a result, the *Xenopus* system allows study of channel function without the confounding influence of other cation channels that are constitutively expressed and endogenously active in mammalian cells.

The genes encoding many different types of ion channels and their chromosomal location have been identified. In numerous instances, specific channels have been associated with specific clinical diseases. For example, the ΔF508 mutation in the epithelial chloride channel gene located on chromosome 7q is responsible for approximately 70% of the cases of cystic fibrosis.

Redox regulation of ion channels has long been investigated at the level of the channel protein. Our understanding of how thiol redox status directly affects the transcription and translation of channel genes is much less developed. Subsequent sections discuss several key examples of how thiol oxidation and reduction reversibly alter channel activity and cell function.

II. THE PATCH CLAMP

The patch-clamp technique derives its name from the fact that the potential across a patch of cell membrane can be fixed (or "clamped") and the resultant ionic currents recorded. Less frequently, current across the membrane is clamped and the resultant membrane potential is measured. Central to the technique is the formation of a seal between a glass pipette and the membrane (Hamill *et al.*, 1981). The seal must be electrically tight, such that the electrical resistance is in the order of 1–10 gΩ (1 gigaohm = $10^{12} \Omega$). Absence of "leak" current around the edge of the seal is critically important to achieving successful and reliable recordings. When the seal is tight and resistance is high, leak current becomes small. Patch-clamp configurations commonly used to study ion channels in mammalian cells are (1) whole cell, (2) cell attached, (3) excised inside-out, and (4) excised outside-out (Fig. 2). Together, these configurations enable detailed investigation of the structure–function relationships that characterize each ion channel. Moreover, the patch-clamp technique stands alone as a method by which a single protein molecule, albeit commonly multimeric in structure, can be functionally assayed in living cells.

In whole-cell configuration, the pipette first makes a seal on the plasma membrane. A pulse of suction applied through the pipette ruptures the membrane and creates direct continuity between the pipette solution and the interior of the cell. In whole-cell configuration, the entire plasma membrane is voltage clamped, and all ion transport systems in the plasma membrane, including active pumps, transporters, exchangers, and channels, contribute to the overall measured current. Although the whole-cell configuration is expedient, it has several disadvantages. The configuration does not allow for the resolution of electrical activity at the single-channel

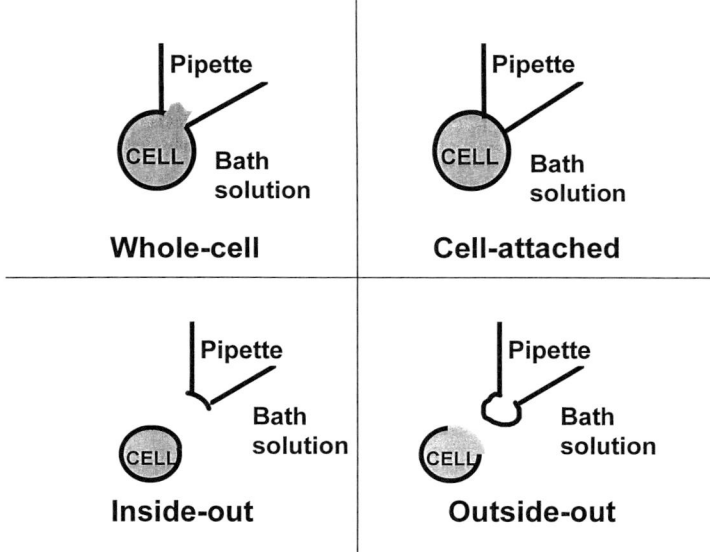

Figure 2 Patch-clamp configurations used most commonly to study ion channel activity in mammalian cells. In each configuration, the pipette contains the recording electrode, whereas the ground electrode is in continuity with the bath solution.

level. Moreover, channel function may be altered inadvertently by the inclusion or omission of substances in the pipette solution.

In the cell-attached configuration, the pipette tip forms a tight seal on the external aspect of the plasma membrane, leaving the membrane essentially intact. The current that is recorded is generated by channels contained in that portion of membrane within the circumference of the pipette tip. It may be that there are no channels, one channel, or several channels within the area of membrane under the pipette. The cell-attached configuration affords the ability to measure the electrical activity of single channels without disrupting the cytoplasmic contents of the cell. Thus, channels that require receptor ligands or intracellular messengers can still be recorded and analyzed. However, there are disadvantages: the intracellular milieu is largely inaccessible and the ionic characteristics of this milieu are not manipulated readily. Moreover, the opening of one channel in the membrane patch can cause a change in the membrane potential of the cell, which in turn alters the activity of other channels.

In the inside-out configuration, the pipette first forms a seal on the external aspect of the plasma membrane. The pipette is then pulled away from the cell, resulting in excision of a membrane patch. The cytosolic (i.e., the inside) aspect of the membrane patch faces outward toward the bath solution; hence the name, inside-out. The external aspect of the membrane patch faces the solution contained within the pipette. The inside-out configuration allows for rigorous control of the ionic conditions on both sides of the membrane patch. However, some channels regulated by intracellular messengers may close and become electrically silent on separation from the cytosolic milieu. This phenomenon can be turned to the advantage of the investigator if the addition of a messenger molecule to the bath restores channel activity.

The outside-out configuration is created by first achieving a whole-cell configuration followed by withdrawal of the pipette from the cell. As the pipette is extracted from the cell, remnants of the plasma membrane adhere to the pipette. The free edges of this adherent membrane seal to each other, forming a membrane "sack" on the end of the pipette. The extracellular aspect of the membrane sack faces outward to the bath solution.

III. REDOX REGULATION OF ION CHANNELS

Within cells of the arterial wall, membrane potential and cell-to-cell signaling are governed by an array of ion channel types, many of which are functionally modulated by diffusible factors such as secondary messengers, pharmacologic agents, and biological toxins. Furthermore, many ion channels possess thiol-containing residues that are susceptible to oxidation. By virtue of their strategic location in a channel protein, such residues may influence the function of a channel greatly, including its ionic conductance and its gating characteristics. Since the development of patch-clamp technology, sulfhydryl reagents have been used experimentally in the ongoing effort to understand channel structure and function. Many of these studies have yielded important mechanistic information. A comprehensive review of all patch-clamp studies that have utilized redox-active agents would be voluminous. A 1998 review of the effect of reactive oxygen species on ion transport physiology has been published (Kourie, 1998). For the purposes of this chapter, we focus on several representative channels whose activity is altered directly by oxidation and reduction of thiol groups constitutively present in the channel protein. First, the newly discovered glutathione-operated cation channel of vascular endothelial cells will be discussed. Second, the evidence that thiol oxidation/reduction reversibly alters K^+ channel activity will be reviewed. Finally, the role of thiol redox status in Ca^{2+} channel activity will be covered.

A. The Glutathione-Operated Cation Channel of Vascular Endothelial Cells

The effects of sulfhydryl reagents on ion flux have been described for many types of ion channels (Chiamvimonvat et al., 1995; Koivisto et al., 1993; Lei et al., 1992). In large part, these studies have documented changes in either the magnitude of current (ionic conductance) flowing through open channels or, more commonly, changes in the gating characteristics of the channel. Gating of a channel to the open position either more frequently or for longer durations will increase the open probability (P_o) for that channel. Conversely, gating to the closed position more frequently or for longer durations will decrease the open probability.

Historically, most studies in this field have used nonphysiological agents such as thimerosal, mersalyl, Ag^+, Hg^+, 2,2'-dithiodipyridine, N-ethylmaleimide, diethyl maleate, or 5,5'-dithiobis-2-nitrobenzoic acid (DTNB) to test the effects of thiol oxidation or alkylation on channel function. Dithiothreitol is the compound employed most commonly to investigate the effects of thiol reduction. To physiologists, such compounds carry several disadvantages in that the compounds are nonphysiological, are often required in relatively high concentrations, and have no inherent selectivity or specificity for the thiols on channel proteins. Nevertheless, sulfhydryl reagents have been useful tools in the armamentarium to investigate structure–function relationships. Reciprocal effects exerted by oxidants and reductants often constitute convincing evidence that channel activity is indeed altered by the redox status of the cell. Moreover, sulfhydryl reagents have been used in combination with molecular techniques such as site-directed mutagenesis to identify strategic cysteines responsible not only for the secondary and tertiary structures of the channel but also for its function.

Studies of pharmacological sulfhydryl reagents have clearly shown that the redox state of channel thiols can influence the gating of selected channels. However, it has been discovered that oxidized glutathione (GSSG) activates a previously unidentified channel in vascular endothelial cells (Koliwad et al., 1996b). The glutathione-operated cation channel is opened by concentrations of GSSG (low micromolar) that are within the levels estimated to exist *in vivo*. The importance of the glutathione-operated channel lies in its effect on membrane potential. Depolarization of the membrane lowers the driving force for the entry of external Ca^{2+}. Ca^{2+} entry activates Ca^{2+}-dependent nitric oxide synthase by endothelial-dependent agonists.

It has been long recognized that oxidized glutathione can oxidize protein thiols (Kosower and Kosower, 1978). Oxidation of a target protein can result in the formation of an intramolecular disulfide bond between adjacent sulfhydryls. Alternatively, GSSG may form a mixed disulfide, characterized by a disulfide bond between the protein and the glutathione. In either case, the formation of a disulfide link where none formerly existed often alters protein structure and function. Reversible thiol oxidation, and hence thiol-disulfide balance, was emphasized by Gilbert (1982, 1984), who proposed that this might represent a mechanism by which intracellular enzyme activity is dynamically modulated. Studies in the field of electrophysiology suggest that the hypothesis forwarded by Gilbert might be extended to channel proteins and their activity.

A key clue to the existence of the GSSG-operated cation channel in vascular endothelial cells lay in the observation that hydroperoxide oxidants cause membrane depolarization (Koliwad et al., 1996a). In calf pulmonary artery cells, sublethal doses of t-butyl hydroperoxide (TB) activate a current that has a conductance of 4.6 nS (nanoSiemens). Figure 3 displays the electrical currents obtained using the whole-cell configuration when the cell is maintained at a holding potential of -60 mV and step pulses are applied from -120 to 110 mV. The currents observed in control cells are characteristic of inward-rectifier K^+ (K_{ir}) channel activity. K_{ir} channels are constitutively present in vascular endothelial cells and are open under physiological conditions. In contrast, currents observed in TB-treated cells are of much greater amplitude than those in control cells and demonstrate a linear current–voltage relationship. A concomitant shift in the reversal potential from -61 to -5 mV is indicative of cell membrane depolarization.

The effect of TB on membrane currents and membrane potential can be replicated in control cells if whole-cell patches are formed using a pipette solution that contains GSSG. GSSG diffuses into the cytosol, where it activates membrane currents identical to those seen in treated cells. Channels, pumps, and transporters can all contribute to the currents recorded using the whole-cell configuration. However, patch-clamp experiments at the level of the single channel have confirmed that ion channels are indeed responsible for the currents recorded at the whole-cell level. Channels are activated by applying GSSG, but not TB or reduced glutathione (GSH), to the intracellular aspect of the membrane. Channels activated by GSSG are equal in ionic conductance (28 pS) and ionic selectivity to those channels activated by the incubation of intact cells with hydroperoxides. Ionic selectivity of the GSSG-operated channel is nonselective in that the selectivity profile for $Na^+:K^+:Ca^{2+}$ is 1:1:0.5. Because current flow is inward at negative membrane potentials and because the concentration of extracellular Na^+ greatly exceeds that of K^+ or Ca^{2+}, channel opening promotes Na^+ entry. The increase in membrane permeability to Na^+ causes the membrane potential to depolarize toward the Nernst equilibrium potential for Na^+. Ca^{2+} entry via this channel may contribute to the increase in resting cytosolic $[Ca^{2+}]$ during prolonged oxidant stress (Elliott et al., 1989).

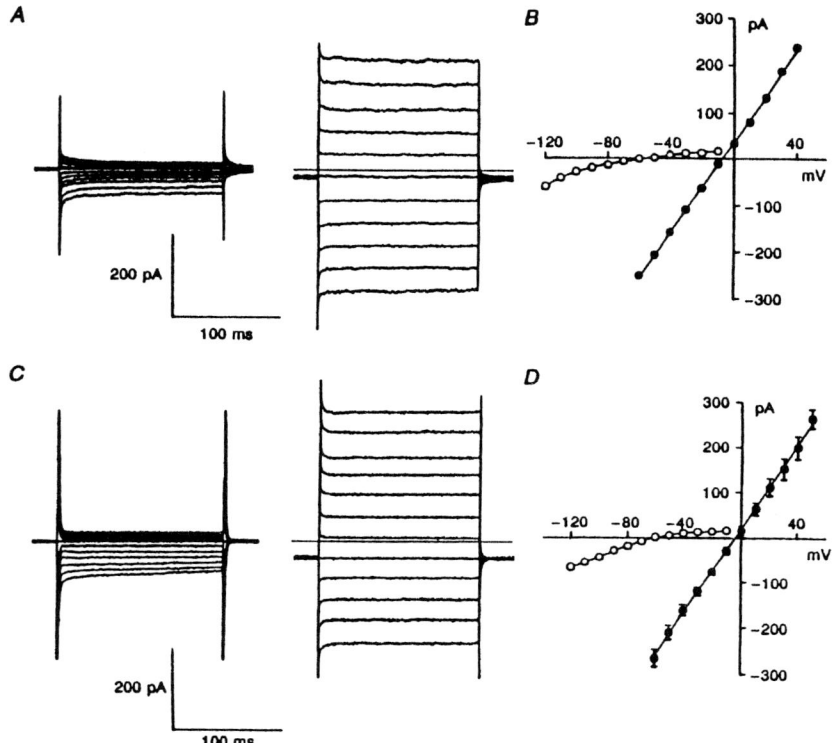

Figure 3 Activation of whole-cell currents in endothelial cells by TB and GSSG. (A) Currents recorded in a control cell (left) and a cell exposed to TB (400 μM, 1 hr (right). (B) Current–voltage relationships for control (open symbols) and treated cells (closed symbols) are shown. Note that the control cell currents reflect inward-rectifier K$^+$ channel activity. In contrast, treated cells demonstrate a depolarized reversal potential and a large-conductance, bidirectional current. (C, left) Currents recorded in a control cell immediately after pipette break in. In this case, the pipette contained 2 mM GSSG. (Right) The same cell 20 min later, after GSSG has had time to diffuse into the cell interior. (D) Current–voltage relationships are shown for currents obtained immediately after break in (open symbols) and 20 min after break in (closed symbols). In all cases, the pipette contained a high K$^+$ solution and the bath contained a physiological solution. Note the striking similarity between currents obtained by exposing cells to TB (A and B) and currents obtained by introducing GSSG into the cell interior (C and D). Reproduced with permission of the Physiological Society (London) from Koliwad *et al.*, (1996b).

A concentration as low as 20 μM GSSG is sufficient to maximally activate channels. However, the time course of channel activation is concentration dependent between 20 μM and 2 mM GSSG applied to the internal aspect of the membrane patch. Figure 4 shows that a characteristic effect of GSSG is to trigger multiple levels of channel activity in any one patch of plasma membrane. Application of 2 mM GSSG to a membrane already exposed to 20 μM GSSG decreases the time required to reach maximal channel activity (quantified by NP_o). GSH and dithiothreitol each close the GSSG-operated channel, providing strong evidence that the chemical mechanism of channel gating is redox based. Importantly, channel activation that is triggered by the incubation of intact cells with TB is mediated by intracellular GSSG. Channel activity is inhibited rapidly by the addition of GSH to the internal aspect of a membrane patch excised from a TB-treated cell (Fig. 5).

Activation of GSSG-operated channels and membrane depolarization together profoundly influence endothelial cell Ca^{2+} signaling. The basis for this is as follows. During receptor stimulation of vascular endothelial cells, extracellular Ca^{2+} enters the cell, leading to activation of nitric oxide synthase and other Ca^{2+}-dependent enzymes. Ca^{2+} enters the cell in response to declining levels of releasable Ca^{2+} within microsomal stores (for review, see Berridge, 1995). For this reason, receptor-stimulated Ca^{2+} entry has been variously termed capacitative, store dependent, or store operated. Depolarization decreases the electrical gradient for Ca^{2+} entry, thereby inhibiting the nitric oxide signaling pathway. During the early stages of endothelial cell oxidative stress, the effects on Ca^{2+} signaling are selective for capacitative Ca^{2+} entry and are potentiated either by increases in GSSG or by shifts in the intracellular glutathione redox balance toward GSSG (Henschke and Elliott, 1995; Elliott *et al.*, 1989, 1995; Elliott and Doan, 1993). This finding has been confirmed independently in Jurkat T lymphocytes and human T cells (Staal *et al.*, 1994), cell types that are vastly different to vascular endothelial cells. The glutathione-operated cation channel, therefore, might be responsible for membrane depolarization and impaired signal transduction in many different mammalian cell types.

B. Redox Modulation of K_{Ca} Channels

In vascular smooth muscle cells, plasma membrane permeability to K^+ is a major determinant of membrane potential. K^+ channel activation hyperpolarizes the cell, which in turn decreases the activity of voltage-operated Ca^{2+} channels, leading to muscle relaxation. Conversely, inhibition of K^+ channel activity depolarizes the cell, thereby activating voltage-operated Ca^{2+} channels, leading to muscle contraction. The vascular smooth muscle cell contains four main types of K^+ channel: the large-conductance Ca^{2+}-activated K^+ (K_{Ca}) channel, the voltage-dependent K^+ (K_v) channel, the ATP-dependent K^+ (K_{ATP}) channel, and the inwardly rectifying K^+ (K_{IR}) channel

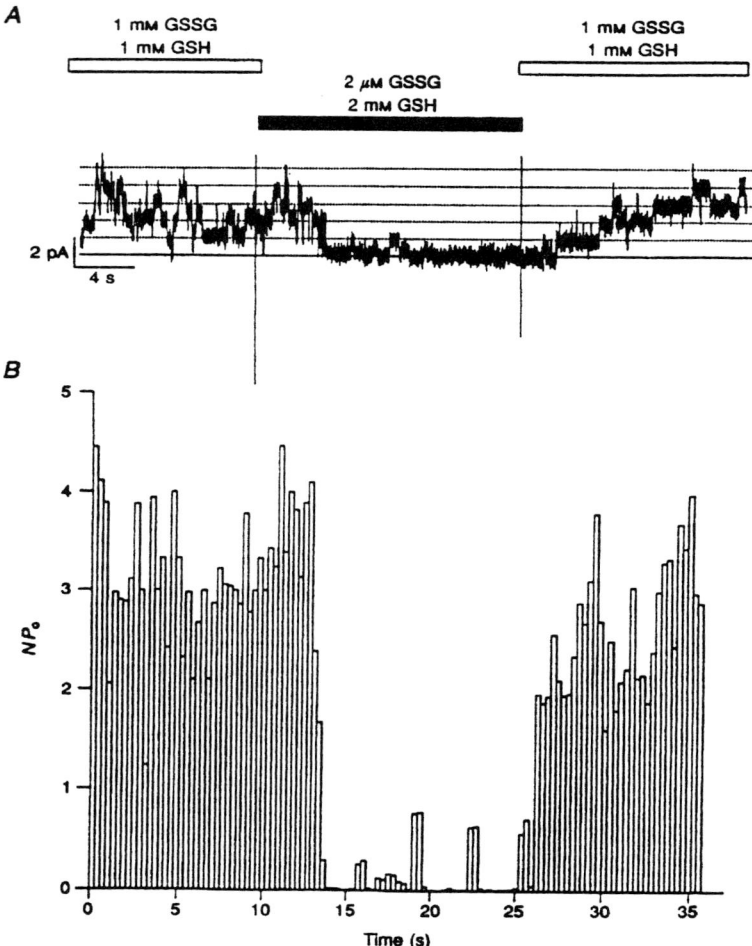

Figure 4 The glutathione-operated cation channel is activated by GSSG. (A) A continuous record from an inside-out membrane patch excised from a control cell. On excision, the internal aspect of the membrane was exposed to the bath solution, which contained GSSG and GSH in the concentrations shown. The solution was first changed to one containing an excess of GSH over GSSG, and then changed back to the original concentrations. The lowest horizontal line represents the closed state. Vertical noise is due to bath exchange. Note that GSSG stimulated channel activity, whereas GSH returned the channel to its closed state. (B) The histogram of NP_o versus time is shown. For additional details, the reader is referred to the original publication. Reproduced with permission of the Physiological Society (London) from Koliwad et al., (1996b).

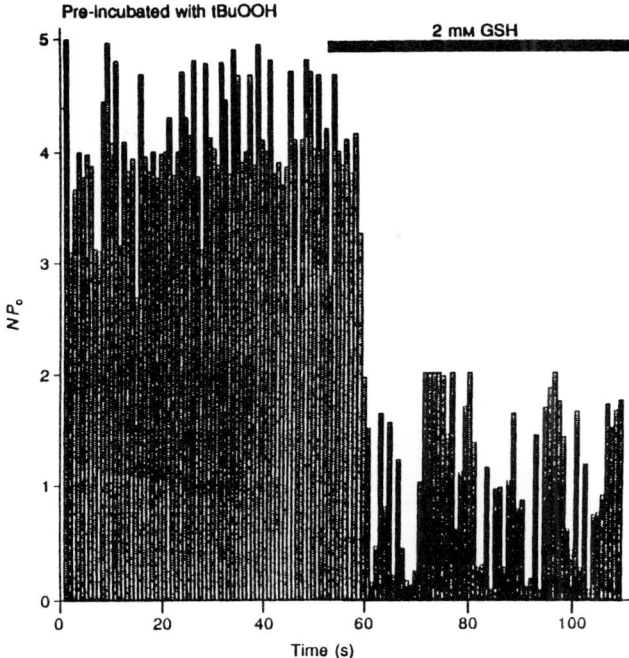

Figure 5 GSH closes channels activated during oxidant stress. A histogram of NP_o versus time from an inside-out membrane patch excised from a cell treated with TB (indicated as tBuOOH). The record began immediately after patch excision. GSH (2 mM) was then added to the solution, bathing the interior aspect of the membrane. Reproduced with permission of the Physiological Society (London) from Koliwad et al., (1996b).

(for review, see Nelson and Quayle, 1995). A hallmark of K_{Ca} channels is their reliance on cytosolic Ca^{2+} for activation. Because the unitary conductance of K_{Ca} channels is so large, the opening and closing of relatively few channels can result in significant changes in membrane potential. The crosstalk among K_{Ca} channel activity, membrane potential, and intracellular Ca^{2+} means that K_{Ca} channels are key modulators of muscle cell contraction. K_{Ca} channels are involved in setting the resting tone of numerous arteries, including those in the mesenteric, coronary, and cerebral circulatory beds.

K_{Ca} channels are expressed widely and are believed to exist in a broad range of functional phenotypes (Reinhart et al., 1989; Farley and Rudy, 1988). A human K_{Ca} channel gene product, *hSlo*, has been cloned and expressed (McCobb et al., 1995; Tseng-Crank et al., 1994). The *Slo* gene derives its name from the original analysis of a *Drosophila* mutant, known as *Slowpoke*, in which the absence of K_{Ca} currents impairs repolarization in flight muscles. At the time of this writing, only one mammalian gene for K_{Ca} channels has been identified (Adelman *et al.*, 1992), which in the human has been mapped to chromosome 10q23.1 (McCobb *et al.*, 1995). Alternative splicing and/or numerous posttranslational chemical mechanisms, including redox modification, might contribute to phenotypic diversity within the family of K_{Ca} channels. However, thiol oxidation and reduction are more likely to dynamically and reversibly modulate K_{Ca} channel activity. The deduced amino acid sequence of the putative pore region for *hSlo* cDNA *hSlo*1.1 contains a single cysteine residue (C342). A 191 amino acid *hSlo* β subunit, when coexpressed with *hSlo* (now also referred to as the *hSlo* α subunit), increases the sensitivity of the expressed channel both to Ca^{2+} and to voltage (Tseng-Crank *et al.*, 1996). The deduced amino acid sequence of the β subunit includes six cysteine residues. As will be discussed later, thiol reduction and oxidation alter the gating characteristics of the channel, suggesting that one or more cysteines are strategically located in or near the channel pore or near the Ca^{2+}-sensing region of the channel. To date, the Ca^{2+} sensor has not been identified. The proposed structure and membrane topology of the human K_{Ca} channel are shown schematically in Fig. 6.

The *hSlo* channel protein contains some 27 cysteine residues, of which 5 are located in the putatively cytoplasmic tail located between the S10 hydrophobic region and the carboxy terminus. DiChiara and Reinhart (1997) examined the activity of *hSlo* expressed in human embryonic kidney 293 cells and in *Xenopus laevis* oocytes. They found that intracellular dithiothreitol increases the voltage sensitivity, accelerates the kinetics of channel activation by one-third, and increases single-channel P_o. These changes occurred in the absence of any increase in single-channel conductance. In contrast, they found that H_2O_2 decreases voltage sensitivity and P_o. Similarly, Cai and Sauvé (1997) demonstrated that the thiol oxidant DTNB inhibited K_{Ca} activity in bovine aortic endothelial cells, an effect that could be reversed by GSH (Fig. 7).

Rundown of channel activity is often a problematic aspect of electrophysiological recording. The term refers to the progressive attenuation of electrical current after a membrane patch is excised from a cell. The phenomenon can provide a critical clue that an intracellular messenger is required for channel activity. Oxidation of a channel protein might contribute to rundown. In the work of *hSlo* channels by DiChiara and Reinhart (1997), H_2O_2 (300 μM) accelerated (whereas dithiothreitol slowed) a time-dependent decrease in voltage sensitivity in excised membrane patches.

In equine tracheal smooth muscle K_{Ca} channels, Kotlikoff and co-workers showed that K_{Ca} channel activity could be modulated by thiol-reactive compounds applied to the cytosolic aspect of the plasma membrane (Wang *et al.*, 1997). Thiol oxidizing agents such as diamide decreased chan-

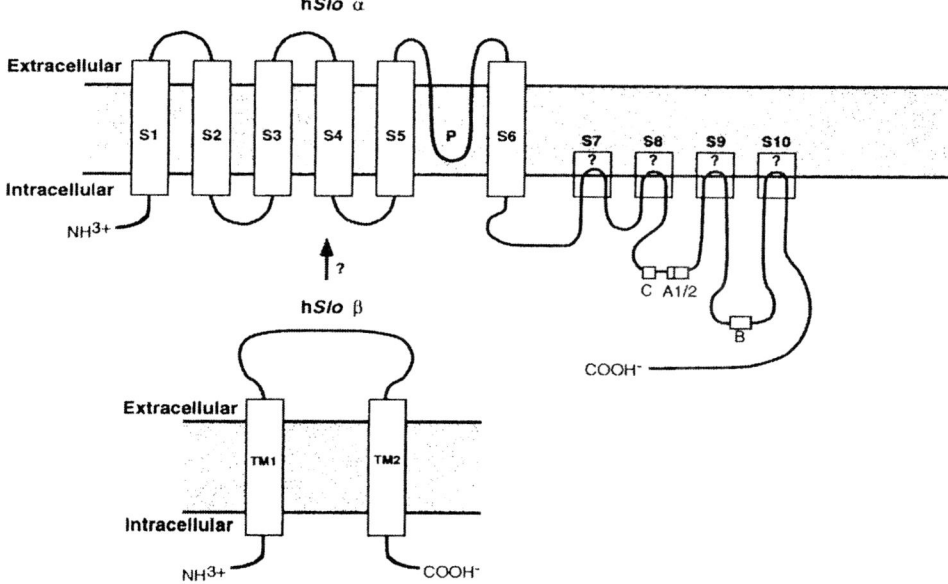

Figure 6 Proposed structure and membrane topology of the human maxi-K^+ channel hSlo (also designated as hSlo α) and hSlo β, a modulatory subunit. The channel protein contains 27 cysteine residues, of which 5 are located between the S10 hydrophobic region and the carboxy terminus. Reproduced with permission from Gribkoff et al., (1997).

nel activity, evidenced by a decrease in NP_o by as much as 90%. Conversely, GSH (170 μM) increased K_{Ca} NP_o nearly 10-fold and reversed the inhibition induced by diamide. Analysis of channel kinetics and current–voltage profiles indicated that GSH alters channel gating, specifically by decreasing the mean closed time. Thus, the increase in NP_o induced by GSH is a result of an increase in P_o. As would be expected, the alkylating agent, N-ethylmaleimide, inhibited channel activity and prevented the inhibitory effect of diamide. The effect of N-ethylmaleimide was replicated using (2-aminoethyl) methane thiosulfonate, a compound that covalently forms a mixed disulfide with cysteine (Stauffer and Karlin, 1994).

Pharmacologic agents that open or block channels are useful tools in the study of channel structure–function relationships and channel distribution. Iberiotoxin is a 37 amino acid scorpion venom peptide that blocks the extracellular vestibule of K_{Ca} channels. The selectivity of iberiotoxin for K_{Ca} channels is conferred in large part by its tertiary structure resulting from its three disulfide (Cys–Cys) bonds. Although it might be fundamentally self-evident, the redox state of compounds that gate channels open and closed

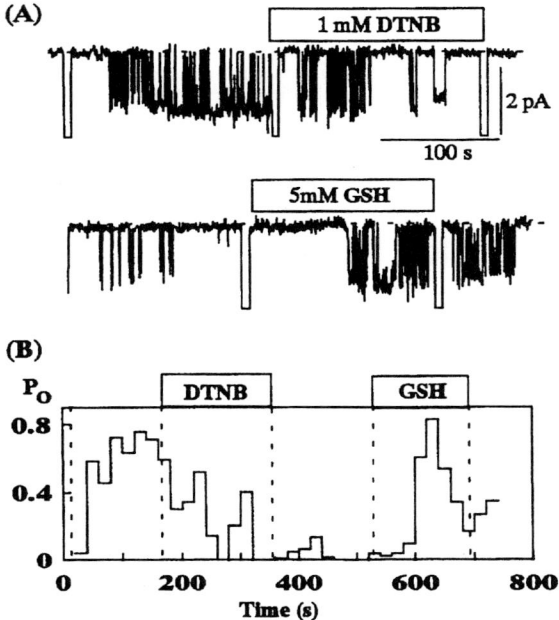

Figure 7 Thiol oxidation inhibits endothelial cell K_{Ca} channels. Inside-out membrane patches were excised and channel activity recorded. Channel activity was inhibited by DTNB (1 mM), an effect that could be reversed by the addition of GSH (5 mM). (B) The accompanying changes in NP_o. The reader is referred to the original publication for additional details. Reproduced with permission from Cai and Sauvé (1997).

is just as important as the redox state of the channel protein itself. A case in point is that of the endothelial glutathione-operated channel discussed earlier. Oxidation of the messenger molecule, glutathione, results in channel opening, whereas reduction leads to channel closure. In other words, a thiol redox reaction requires two substrates, and in electrophysiology, one of those substrates is the channel protein and the other is typically a soluble compound.

In a study of K_{Ca} channels in rabbit pulmonary artery myocytes, Lee and colleagues (1994) likewise found NP_o to be altered by glutathione. However, these investigators showed that channel activity was decreased, rather than increased, by GSH. These results might be related to the vascular bed of origin. In several key functions, the vascular responses of pulmonary arteries are opposite to those of systemic arteries. For example, hypoxia triggers vasoconstriction of pulmonary arteries and vasodilation in systemic arteries. At the ion channel level, differences in K_{Ca} channel amino acid sequence and tertiary structure might account for at least some of the differences in vascular responses observed in pulmonary versus systemic circulations.

The role of thiol oxidation in pulmonary artery K^+ channels has received attention because K^+ channels appear to be involved in hypoxic pulmonary vasoconstriction and because thiol oxidants reverse the effect of hypoxia (Weir et al., 1983). Intracellular GSSG increases, whereas GSH decreases, K^+ currents recorded in rat pulmonary artery smooth muscle cells (Archer et al., 1993). Extracellular GSH appears to have a similar effect (Yuan et al., 1994). Whole-cell recording results from the contribution of multiple types of K^+ channels that contribute variably to the overall recorded current. It is not known whether the amino acid sequence, tertiary folding, and function of K_{Ca} channels expressed in pulmonary artery smooth muscle cells are identical to those properties of systemic artery K_{Ca} channels. It is therefore possible that the basis for the inhibitory effects of GSH on K^+ currents observed in pulmonary artery myocytes is related to its action on one or more K^+ channels that are unique to the pulmonary circulation.

Maximal channel activity might depend on specific cysteine residues existing in a reduced or an oxidized state. For example, maximal channel activity might occur when a readily accessible cysteine located in the cytosolic tail of a channel protein is oxidized and a less accessible residue in a hydrophobic region of the channel is reduced. In this scenario, the redox state of the cysteines under the conditions of study would determine whether reduction by GSH or oxidation by GSSG increases or decreases channel activity. Precise identification of the target cysteine residues will depend on patch-clamping studies carried out in expression systems using K_{Ca} mutants in which specific cysteine residues have been exchanged for nonthiol residues. Gene homology and identity between amino acid sequences will become more pertinent as those channels that appear unaffected by glutathione (Thuringer and Findlay, 1997) are compared with those that are.

The modulation of K_{Ca} channels by glutathione and other thiol-reactive molecules is likely to represent a key mechanism by which blood flow and tissue perfusion are modulated under conditions of oxidative stress. In the cerebral circulation, for example, we have found that the thiol oxidant peroxynitrite is a contractile agonist of both single myocytes and intact vessels (Elliott et al., 1998). At the same time, peroxynitrite inhibits K_{Ca} channel activity, an effect that might explain its macrophysiologic actions (Brzezinska et al., 1998a,b).

C. Redox Regulation of the Ryanodine Receptor Ca^{2+} Channel

Like K^+ channels, Ca^{2+} channel proteins also contain thiol residues that are targets for oxidation. This section focuses on the ryanodine receptor as an example of how Ca^{2+} channel activity is affected by oxidation and reduction.

In terminology that is somewhat confusing, the ryanodine receptor (RyR) *is* a Ca^{2+}-selective ion channel. Ryanodine receptors are located within the membrane of sarcoplasmic reticulum, which serves as the major intracellular Ca^{2+} store of skeletal muscle cells. Ryanodine receptors are responsible for conducting Ca^{2+} from the sarcoplasmic reticulum lumen to the cytosol. The subsequent increase in cytosolic $[Ca^{2+}]$ activates the troponin–myosin–actin complex, leading to contraction. Ryanodine receptor activity is modulated by both physiological and pharmacological agents (Meissner, 1994; Coronado *et al.*, 1994; Franzini-Armstrong and Protasi, 1997). The channel is activated by micromolar concentrations of cytosolic Ca^{2+}, as well as by adenine nucleotides and caffeine. Conversely, the channel is inhibited or blocked by Mg^{2+}, ruthenium red, and doxorubicin. Calmodulin modulates the response of the channel to Ca^{2+}; at high $[Ca^{2+}]$, calmodulin decreases channel P_o, whereas at low $[Ca^{2+}]$, calmodulin increases P_o.

The ryanodine receptor is a key component of the mechanism that couples muscle cell depolarization with cell shortening, a mechanism referred to as excitation–contraction coupling. The myocyte plasma membrane possesses invaginating blind pouches, termed T-tubules. Sarcoplasmic reticulum abuts each side of a T-tubule to form a "triad" (Fig. 8). Located within the plasma membrane, including that portion of the membrane that forms the T-tubule, are dihydropyridine receptors. Dihydropyridine receptors are voltage-gated, L-type Ca^{2+} channels. In response to the arrival of a depolarizing action potential, L-type channels open, thereby allowing external Ca^{2+} to flux into the cytosolic cleft between the T-tubule plasma membrane and the sarcoplasmic reticulum membrane. Ryanodine receptor Ca^{2+} channels, located in the sarcoplasmic reticulum membrane, are activated by the increase in $[Ca^{2+}]$ occurring within the cleft. This phenomenon, known as Ca^{2+}-induced Ca^{2+} release (CICR), results in a large release of Ca^{2+}, presumably by way of "recruitment" of additional ryanodine receptors responding to Ca^{2+} released by other ryanodine receptors.

The ryanodine receptor molecule is exceptionally large, consisting of some 5000 residues and totaling about 560 kDa (Takeshima *et al.*, 1989). There are three known isoforms; isoform 1 is found mainly in skeletal muscle cells, whereas isoform 2 is the predominant form expressed in cardiac myocytes. Isoforms 1 and 2 are 67% homologous (Hakamata *et al.*, 1992). The molecule appears to possess a large, hydrophilic amino-terminal that forms a cytoplasmic "foot."

The ryanodine receptor interacts with several constitutively expressed membrane proteins. FKBP12 is a 12-kDa molecule that is associated with the cytoplasmic foot and modulates channel gating (Brillantes *et al.*, 1994; Collins, 1991). Triadin is a membrane protein that forms oligomers via disulfide bonds. It is thought that triadin may modulate ryanodine receptor

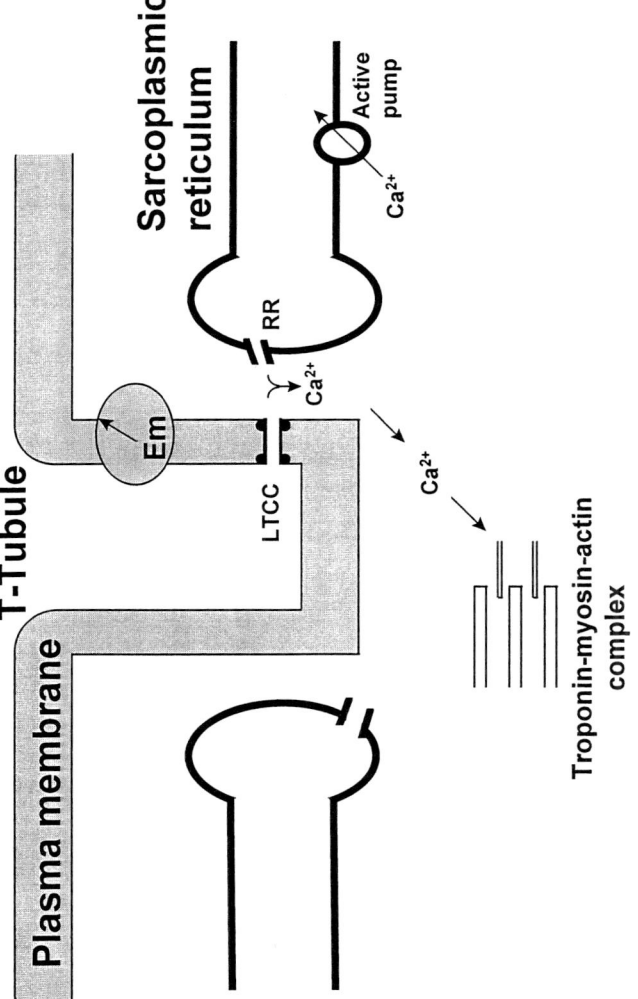

Figure 8 Ryanodine receptor Ca^{2+} channels and their role in excitation–contraction coupling. The plasma membrane of skeletal muscle cells forms invaginations known as T-tubules. In response to depolarization of the membrane potential (Em), voltage-operated, L-type Ca^{2+} channels (LTCC) open, causing an increase in Ca^{2+} in the cleft between the T-tubule and the sarcoplasmic reticulum. In response, ryanodine receptor Ca^{2+} channels (RR) open. The subsequent increase in cytosolic Ca^{2+} leads to activation of the troponin–myosin–actin complex and cell contraction. Ca^{2+} is actively pumped into the lumen of the sarcoplasmic reticulum.

function via its binding to dihydropyridine receptors or its binding to calsequestrin on the luminal aspect of the sarcoplasmic reticulum membrane (Guo et al., 1994; Fan et al., 1995).

The stability of the ryanodine receptor/triadin complex is linked to the oxidation state of thiol groups that react readily with the fluorogenic maleimide compound, 7-diethylamino-3-(4'-amleimidylphenyl)-4-methylcoumarin (CPM) (Liu and Pessah, 1994; Liu et al., 1994). CPM has been used to show that thiol groups in the ryanodine receptor/triadin complex are regulated allosterically by the binding of channel ligands. Nanomolar ryanodine, millimolar caffeine, and micromolar Ca^{2+}, each of which are activators of the ryanodine receptor Ca^{2+} channel, stabilize the ryanodine receptor/triadin complex. Likewise, the thiol oxidant, 1,4-naphthoquinone, oxidizes sulfhydryls on ryanodine receptors and triadin, stabilizes the ryanodine receptor/triadin complex, and triggers Ca^{2+} efflux from sarcoplasmic reticulum. Conversely, nanomolar CPM binds to triadin and the ryanodine receptor only in the presence of channel inhibitors, such as neomycin or ruthenium red.

These data are consistent with a series of reports by Salama and colleagues showing that a variety of pharmacological thiol oxidants trigger the opening of ryanodine receptors and the release of Ca^{2+} from sarcoplasmic reticulum (Abramson et al., 1988, 1995; Stoyanovsky et al., 1994; Salama et al., 1992; Prabhu and Salama, 1990; Abramson and Salama, 1988; Trimm et al., 1986). In 1997, Abramson and co-workers reported on the ability of glutathione to modulate the ryanodine receptor (Zable et al., 1997). They found that dithiothreitol, β-mercaptoethanol, and GSH each decreased the gating activity of channels that had been reconstituted in lipid bilayers (Zable et al., 1997).

Several groups have investigated the effects of sulfhydryl oxidation on the Ca^{2+} dependence of ryanodine receptor Ca^{2+} channels (Stuart et al., 1992; Favero et al., 1995; Marengo et al., 1998). Ryanodine receptors exhibit at least three different types of activity profiles in response to cytosolic $[Ca^{2+}]$. In sarcoplasmic reticulum vesicles isolated from mammalian cardiac tissue, channels exhibit sigmoidal activation by micromolar cytosolic Ca^{2+}. Although it would be reasonable to predict that each of the three ryanodine receptor isoforms displayed a single Ca^{2+} dependency, Marengo et al. (1998) found that such is not the case. These investigators found that sulfhydryl oxidants modified the type of response to cytosolic $[Ca^{2+}]$. Regardless of whether the ryanodine receptor channels were isolated from skeletal muscle, brain cortex, or cardiac muscle, extensive oxidation produced a Ca^{2+} dependence typical for channels isolated from cardiac tissue. Alternatively, channel closure ensued. Thiol groups important in these responses appear to be located on the cytosolic aspect of the membrane, suggesting that the redox state of the cysteine residues in the large cytoplasmic domain modulates the Ca^{2+} dependency of ryanodine receptor Ca^{2+} channels.

The overall effect, therefore, of oxidant stress on ryanodine receptor activity is to activate the release of Ca^{2+} from sarcoplasmic reticulum. In this regard, oxidant stress would be expected to promote contraction and increase tissue requirements for metabolic substrates.

In conclusion, ion channels are membrane proteins that are functionally responsive to oxidation and reduction of constitutive thiol residues. Dynamic alteration of channel structure–function relationships as a result of oxidative modification represents a major mechanism by which oxidants modulate many cell functions, potentially resulting in altered tissue perfusion and altered tissue demands for oxygen and energy substrate. The scientific horizon for the redox regulation of ion channels includes study of the mechanisms by which oxidative stress and the redox environment might alter the transcription and translation of those genes encoding for ion channel proteins.

ACKNOWLEDGMENTS

This work was supported by Grant 96013570 from the American Heart Association, National Office (SJE), and by Grant R01-NS38133 from the National Institutes of Health (SJE).

REFERENCES

Abramson, J. J., and Salama, G. (1988). Sulfhydryl oxidation and Ca^{2+} release from sarcoplasmic reticulum. *Mol. Cell. Biochem.* **82**, 81–84.

Abramson, J. J., Cronin, J. R., and Salama, G. (1988). Oxidation induced by phthalocyanine dyes causes rapid calcium release from sarcoplasmic reticulum vesicles. *Arch. Biochem. Biophys.* **263**, 245–255.

Abramson, J. J., Zable, A. C., Favero, T. G., and Salama, G. (1995). Thimerosal interacts with the Ca^{2+} release channel ryanodine receptor from skeletal muscle sarcoplasmic reticulum. *J. Biol. Chem.* **270**, 29644–29647.

Adelman, J. P., Shen, K. Z., Kavanaugh, M. P., Warren, R. A., Wu, Y. N., Lagrutta, A., Bond, C. T., and North, R. A. (1992). Calcium-activated potassium channels expressed from cloned complementary DNAs. *Neuron* **9**, 209–216.

Archer, S. L., Huang, J., Post, J. M., Hume, J. R., and Weir, E. K. (1993). t-Butyl hydroperoxide and glutathione modulate an outward K^+ current in rat pulmonary vascular smooth muscle cells. *Circulation* **88**, 139–143.

Berridge, M. J. (1995). Capacitative calcium entry. *Biochem. J.* **312**, 1–11.

Brillantes, A. B., Ondrias, K., Scott, A., Kobrinsky, E., Ondriasova, E., Moschella, M. C., Jayaraman, T., Landers, M., Ehrlich, B. E., and Marks, A. R. (1994). Stabilization of calcium release channel (ryanodine receptor) function by FK506-binding protein. *Cell (Cambridge, Mass.)* **77**, 513–523.

Brzezinska, A. K., Chilian, W. M., and Elliott, S. J. (1998a). Peroxynitrite inhibits K^+ currents and causes freshly isolated cerebral artery smooth muscle cells to contract. *FASEB J.* **12**, A1001.

Brzezinska, A. K., Chilian, W. M., and Elliott, S. J. (1998b). Peroxynitrite inhibits maxi-K^+ channels and is a contractile agonist of cerebral artery smooth muscle cells. *Circulation* **98**, I-803.

Cai, S., and Sauvé, R. (1997). Effects of thiol-modifying agents on a K(Ca2+) channel of intermediate conductance in bovine aortic endothelial cells. *J. Membr. Biol.* **158**, 147–158.

Chiamvimonvat, N., O'Rourke, B., Kamp, T. J., Kallen, R. G., Hofmann, F., Flockerzi, V., and Marban, E. (1995). Functional consequences of sulfhydryl modification in the pore-forming subunits of cardiovascular Ca^{2+} and Na^+ channels. *Circ. Res.* **76**, 325–334.

Collins, J. H. (1991). Sequence analysis of the ryanodine receptor: Possible association with a 12 kD, FK506-binding immunophilin/protein kinase C inhibitor. *Biochem. Biophys. Res. Commun.* **178**, 1288–1290.

Coronado, R., Morrissette, J., Sukhareva, M., and Vaughan, D. M. (1994). Structure and function of ryanodine receptors. *Am. J. Physiol.* **266**, C1485–C1504.

DiChiara, T. J., and Reinhart, P. H. (1997). Redox modulation of *hslo* Ca^{2+}-activated K^+ channels. *J. Neurosci.* **17**, 4942–4955.

Elliott, S. J., and Doan, T. N. (1993). Oxidant stress inhibits the store-dependent Ca^{2+} influx pathway of vascular endothelial cells. *Biochem. J.* **292**, 385–393.

Elliott, S. J., Eskin, S. G., and Schilling, W. P. (1989). Effect of t-butyl-hydroperoxide on bradykinin-stimulated changes in cytosolic calcium in vascular endothelial cells. *J. Biol. Chem.* **264**, 3806–3810.

Elliott, S. J., Doan, T. N., and Henschke, P. N. (1995). Reductant substrate for glutathione peroxidase modulates oxidant-inhibition of Ca^{2+} signaling in endothelial cells. *Am. J. Physiol.* **268**, H278-H287.

Elliott, S. J., Lacey, D. J., Chilian, W. M., and Brzezinska, A. K. (1998). Peroxynitrite is a contractile agonist of cerebral artery smooth muscle cells. *Am. J. Physiol.* **275**, H1585–H1591.

Fan, H., Brandt, N. R., Peng, M., Schwartz, A., and Caswell, A. H. (1995). Binding sites of monoclonal antibodies and dihydropyridine receptor alpha 1 subunit cytoplasmic II-III loop on skeletal muscle triadin fusion peptides. *Biochemistry* **34**, 14893–14901.

Farley, J., and Rudy, B. (1988). Multiple types of voltage-dependent Ca^{2+}-activated K^+ channels of large conductance in rat brain synaptosomal membranes. *Biophys. J.* **53**, 919–934.

Favero, T. G., Zable, A. C., and Abramson, J. J. (1995). Hydrogen peroxide stimulates the Ca^{2+} release channels from skeletal sarcoplasmic reticulum vesicles. *J. Biol. Chem.* **270**, 2557–2563.

Franzini-Armstrong, C., and Protasi, F. (1997). Ryanodine receptors of striated muscles: A complex channel capable of multiple interactions. *Physio. Rev.* **77**, 699–729.

Gilbert, H. F. (1982). Biological disulfides: The third messenger? *J. Biol. Chem.* **257**, 12086–12091.

Gilbert, H. F. (1984). Redox control of enzyme activities by thiol/disulfide exchange. *In* "Methods, Enzymology" (F. Wold and K. Moldave, eds.), Vol. 107, pp. 330–351. Academic Press, New York.

Gribkoff, V. K., Starrett, J. E., Jr., and Dworetzky, S. I. (1997). The pharmacology and molecular biology of large-conductance calcium-activated potassium channels. *Adv. Pharmacol.* **37**, 319–348.

Guo, W., Jorgensen, A. O., Jones, L. R., and Campbell, K. P. (1994). Characterization and ultrastructural localization of a novel 90-kDa protein unique to skeletal muscle junctional sarcoplasmic reticulum. *J. Biol. Chem.* **269**, 28359–28365.

Hakamata, Y., Nakai, J., Takeshima, H., and Imoto, K. (1992). Primary structure and distribution of a novel ryanodine receptor/calcium release channel from rabbit brain. *FEBS Lett.* **312**, 229–235.

Hamill, D. P., Marty, A., Neher, E., Sakmann, B., and Sigworth, F. J. (1981). Improved patch-clamp techniques for high-resolution current recording from cells and cell-free membrane patches. *Pfluegers Arch.* **391**, 85–100.

Henschke, P. N., and Elliott, S. J. (1995). Oxidized glutathione decreases luminal Ca^{2+} content of the endothelial cell $Ins(1,4,5)P_3$-sensitive Ca^{2+} store. *Biochem. J.* **312**, 485–489.

Koivisto, A., Siemen, D., and Nedergaard, J. (1993). Reversible blockade of the calcium-activated nonselective cation channel in brown fat cells by the sulfhydryl reagents mercury and thimerosal. *Pfluegers Arch.* **425**, 549–551.

Koliwad, S. K., Kunze, D. L., and Elliott, S. J. (1996a). Oxidant stress activates a non-selective cation channel responsible for membrane depolarization in calf vascular endothelial cells. *J. Physiol. (London)* **491**, 1–12.

Koliwad, S. K., Elliott, S. J., and Kunze, D. L. (1996b). Oxidized glutathione mediates channel activation and membrane depolarization in vascular endothelial cells during oxidant stress. *J. Physiol. (London)* **495**, 37–49.

Kosower, N. S., and Kosower, E. M. (1978). The glutathione status of cells. *Int. Rev. Cytol.* **54**, 109–153.

Kourie, J. I. (1998). Interaction of reactive oxygen species with ion transport mechanisms. *Am. J. Physiol.* **275**, C1–C24.

Lee, S., Park, M., So, I., and Earm, Y. E. (1994). NADH and NAD modulates Ca^{2+}-activated K^+ channels in small pulmonary arterial smooth muscle cells of the rabbit. *Pfluegers Arch.* **427**, 378–380.

Lei, S. Z., Pan, Z.-H., Aggarwal, S. K., Chen, H. V., Hartman, J., Sucher, N. J., Lipton, S. A., Pan, Z. H., Chen, H. S., and Lipton, S. A. (1992). Effect of nitric oxide production on the redox modulatory site of the NMDA receptor-channel complex. *Neuron* **8**, 1087–1099.

Liu, G., and Pessah, I. N. (1994). Molecular interaction between ryanodine receptor and glycoprotein triadin involves redox cycling of functionally important hyper-reactive sulfhydryls. *J. Biol. Chem.* **269**, 33028–33034.

Liu, G., Abramson, J. J., Zable, A. C., and Pessah, I. N. (1994). Direct evidence for the existence and functional role of hyper-reactive sulfhydryls on the ryanodine receptor-triadin complex selectively labeled by the coumarin maleimide 7-diethylamino-3-(4′maleimidylphenyl)-4-methylcoumarin. *Mol. Pharmacol.*, **45**, 189–200.

Marengo, J. J., Hidalgo, C., and Bull, R. (1998). Sulfhydryl oxidation modifies the calcium dependence of ryanodine-sensitive calcium channels of excitable cells. *Biophys. J.* **74**, 1263–1277.

McCobb, D. P., Flowler, N. L., Featherstone, T., Lingle, C. L., Saito, M., Krause, K.-H., and Salkoff, L. (1995). A human calcium-activated potassium channel gene expressed in vascular smooth muscle. *Am. J. Physiol.* **269**, H767–H777.

Meissner, G. (1994). Ryanodine receptor/Ca^{2+} release channels and their regulation by endogenous effectors. *Annu. Rev. Physiol.* **56**, 485–508.

Nelson, M. T., and Quayle, J. M. (1995). Physiological roles and properties of potassium channels in arterial smooth muscle. *Am. J. Physiol.* **268**, C799–C822.

Prabhu, S. D., and Salama, G. (1990). The heavy metal ions Ag^+ and Hg^{2+} trigger calcium release from cardiac sarcoplasmic reticulum. *Arch. Biochem. Biophys.* **277**, 47–55.

Reinhart, P. H., Chung, S., and Levitan, I. B. (1989). A family of calcium-dependent potassium channels from rat brain. *Neuron* **2**, 1031–1041.

Salama, G., Abramson, J. J., and Pike, G. K. (1992). Sulphydryl reagents trigger Ca^{2+} release from the sarcoplasmic reticulum of skinned rabbit psoas fibres. *J. Physiol. (London)* **454**, 389–420.

Staal, F. J. T., Anderson, M. T., Staal, G. E. J., Herzenberg, L. A., and Gitler, C. (1994). Redox regulation of signal transduction: Tyrosine phosphorylation and calcium influx. *Proc. Natl. Acad. Sci. U.S.A.* **91**, 3619–3622.

Stauffer, D. A., and Karlin, A. (1994). Electrostatic potential of the acetylcholine binding sites in the nicotinic receptor probed by reactions of binding-site cysteines with charged methanethiosulfonates. *Biochemistry* **33**, 6840–6849.

Stoyanovsky, D. A., Salama, G., and Kagan, V. E. (1994). Ascorbate/iron activates Ca^{2+}-release channels of skeletal sarcoplasmic reticulum vesicles reconstituted in lipid bilayers. *Arch. Biochem. Biophys.* **308**, 214–221.

Stuart, J., Pessah, I. N., Favero, T. G., and Abramson, J. J. (1992). Photooxidation of skeletal muscle sarcoplasmic reticulum induces rapid calcium release. *Arch. Biochem. Biophys.* **292**, 512–521.

Takeshima, H., Nishimura, S., Matsumoto, T., Ishida, H., Kangawa, K., Minamino, N., Matsuo, H., Ueda, M., Hanaoka, M., and Hirose, T. (1989). Primary structure and expression from complementary DNA of skeletal muscle ryanodine receptor. *Nature (London)* **339**, 439–445.

Thuringer, D., and Findlay, I. (1997). Contrasting effects of intracellular redox couples on the regulation of maxi-K channels in isolated myocytes from rabbit pulmonary artery. *J. Physiol. (London)* **500**, 583–592.

Trimm, J. L., Salama, G., and Abramson, J. J. (1986). Sulfhydryl oxidation induces rapid calcium release from sarcoplasmic reticulum vesicles. *J. Biol. Chem.* **261**, 16092–16098.

Tseng-Crank, J., Foster, C. D., Krause, J. D., Mertz, R., Godinot, N., DiChiara, T. J., and Reinhart, P. H. (1994). Cloning, expression, and distribution of functionally distinct Ca^{2+}-activated K^+ channel isoforms from human brain. *Neuron* **13**, 1315–1330.

Tseng-Crank, J., Godinot, N., Johansen, T. E., Ahring, P. K., Strobaek, D., Mertz, R., Foster, C. D., Olesen, S. P., and Reinhart, P. H. (1996). Cloning, expression, and distribution of a Ca^{2+}-activated K^+ channel β-subunit from human brain. *Proc. Natl. Acad. Sci. U.S.A.* **93**, 9200–9205.

Wang, Z. W., Nara, M., Wang, Y. X., and Kotlikoff, M. I. (1997). Redox regulation of large conductance Ca^{2+}-activated K^+ channels in smooth muscle cells. *J. Gen. Physiol.* **110**, 35–44.

Weir, E., Will, J., Lundquist, L., Eaton, J., and Chesler, E. (1983). Diamide inhibits pulmonary vasoconstriction induced by hypoxia or prostaglandin F2α. *Proc. Soc. Exp. Biol. Med.* **173**, 96–103.

Yuan, X.-J., Tod, M. L., Rubin, L. J., and Blaustein, M. P. (1994). Deoxyglucose and reduced glutathione mimic effects of hypoxia on K^+ and Ca^{2+} conductances in pulmonary artery cells. *Am. J. Physiol.* **267**, L52–L63.

Zable, A. C., Favero, T. G., and Abramson, J. J. (1997). Glutathione modulates ryanodine receptor from skeletal muscle sarcoplasmic reticulum. Evidence for redox regulation of the Ca^{2+} release mechanism. *J. Biol. Chem.* **272**, 7069–7077.

5 Cell Ca^{2+} in Signal Transduction: Modulation in Oxidative Stress

Julio Girón-Calle and Henry Jay Forman
Department of Environmental Health Sciences
School of Public Health
University of Alabama at Birmingham
Birmingham, Alabama 35294

The cytosolic free calcium concentration is tightly regulated, and changes in it are part of many cellular signaling pathways. In these pathways, binding of agonists to receptors in the plasma membrane trigger a release of calcium from the endoplasmic reticulum and/or uptake from the extracellular space through the plasma membrane. Exposure to reactive oxygen species and lipid peroxides, as well as xenobiotics that generate these reactive species, may cause increases in the cytosolic calcium concentration as well. Calcium homeostasis may be disrupted by these agents to the point of causing necrotic or apoptotic cell death. Accumulating evidence shows that nonlethal oxidative stress causes reversible increases in the cytosolic calcium concentration in many cell types, including hepatocytes, myocytes, endothelial cells, smooth muscle cells, and phagocytes. In the majority of cases described so far, release from the endoplasmic reticulum and/or entry from the extracellular medium is responsible for these increases. These calcium movements may be due to the activation of signaling pathways shared by classic agonists, but may also be due to oxidative modification of calcium channels, pumps, and exchangers located in the plasma and endoplasmic reticulum membranes. Reversible oxidation of protein sulfhydryl groups is a mechanism by which the function of these channels and transporters may be modified by oxidative stress. In this manner, oxidative stress may modulate the responses to classic agonists that depend on increases in the cytosolic calcium concentration. Ca^{2+} participates in a multitude of signal transduction pathways. Thus, it is likely that a large number of subtle and physiologically relevant interactions between reversible changes in the intracellular calcium concentration and the generation of reactive oxygen species will be discovered as this field advances.

I. Ca^{2+} IN SIGNAL TRANSDUCTION AND CELL DEATH

Changes in the intracellular concentration of Ca^{2+} play a very important role in regulating cell function in prokaryotic and eukaryotic cells. The cytosolic free Ca^{2+} concentration ($[Ca^{2+}]_c$) is about $10^{-7} M$, four orders of magnitude lower than in the extracellular medium. This means that much energy is spent to keep this steep gradient between intracellular and extracellular concentrations. It has been proposed that the reason cells evolved the capacity to maintain such a low intracellular concentration is that Ca^{2+} phosphate precipitates easily in water (Clapham, 1995). Regulatory mechanisms in which Ca^{2+} plays a role take advantage of this situation, benefiting from the tight regulation of Ca^{2+} concentration imposed by the widespread presence of phosphate as the main intracellular buffer. This section of the chapter offers an overview of the basic mechanisms that mammalian cells use to maintain a specific Ca^{2+} concentration and compartmentalization and how regulatory pathways make use of changes in Ca^{2+} concentration to transmit signals. Many of the references that are cited are reviews that describe these mechanisms in more detail. Later, we will refer to this information to consider how these mechanisms and information pathways may be affected by changes in Ca^{2+} concentration induced by oxidative stress. The possible involvement of disruptions in Ca^{2+} homeostasis in pathological conditions and cell death will also be reviewed briefly.

A. Maintenance of Intracellular Ca^{2+} Levels

$[Ca^{2+}]_c$ is maintained at a low steady-state ($\sim 10^{-7} M$) by Ca^{2+}-ATPases in the plasma membrane and endoplasmic reticulum and a Na^+/Ca^{2+} antiport located in the plasma membrane (Fig. 1). Ca^{2+} ATPases present in plasma membrane and endoplasmic reticulum are encoded by two different groups of tissue-specific genes and do not show a high degree of homology between them. They function as Ca^{2+} pumps using energy in the form of ATP to channel Ca^{2+} out of the cell or into the endoplasmic reticulum (Pozzan *et al.*, 1994). Thapsigargin is a tumor promoter that inhibits Ca^{2+} uptake into the endoplasmic reticulum by binding to the Ca^{2+}-ATPase pump in this organelle. As the intraluminal Ca^{2+} concentration is in a steady state, with

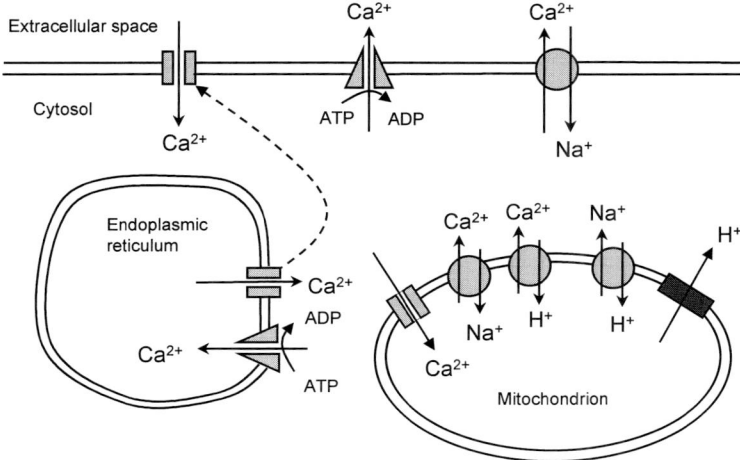

Figure 1 Cellular Ca^{2+} pools and the mechanisms that maintain a steady-state concentration of Ca^{2+}. Channels or uniporters (rectangles), ATPase pumps (triangles), and exchangers (circles) maintain cellular Ca^{2+} homeostasis. Ca^{2+}-ATPases in the plasma and the endoplasmic reticulum membranes spend energy in the form of ATP in order to pump Ca^{2+} out of the cytosol so that a low basal concentration of about $0.1 \times 10^{-6} M$ is maintained in the cytosol. Mitochondria drive energy from the electrochemical gradient generated by the respiratory chain (darker rectangle) for the uptake of Ca^{2+} and use exchangers to get Ca^{2+} out against this gradient. Many signal transduction pathways depend on an increase in $[Ca^{2+}]_c$, which is achieved by the opening of channels in the endoplasmic reticulum and the plasma membranes. Channels in the endoplasmic reticulum are receptor operated (IP_3, cyclic ADP ribose). Channels in the plasma membrane may be voltage or receptor operated and may also be activated by the capacitative mechanism in response to Ca^{2+} release from the endoplasmic reticulum (arrow). The mitochondrial Ca^{2+} pool does not seem to play an important role in nonlethal signal transduction.

continuous exit and entry of Ca^{2+} out of and into the organelle, inhibition of the Ca^{2+}-ATPase empties the endoplasmic reticulum Ca^{2+} pool passively (Inesi and Sagara, 1992; Lytton et al., 1991). Thapsigargin has been widely used as an experimental tool to study the origin of the Ca^{2+} responsible for increases in the cytosolic concentration. The movement of Ca^{2+} out of the cell by the Na^+/Ca^{2+} exchanger depends on high extracellular and low intracellular Na^+ concentrations. Three genes that encode this protein and are expressed in a tissue-specific manner have been cloned so far (Nicoll et al., 1996). The low intracellular Na^+ concentration is achieved by Na^+/K^+-ATPase, which consumes 30 to 50% of the ATP produced in cells.

Ca^{2+} channels facilitate the movement of Ca^{2+} from the extracellular medium and from the lumen of the endoplasmic reticulum into the cytosol. On opening of a channel, Ca^{2+} moves in the direction favored by the difference of concentration across the plasma and the endoplasmic reticulum membranes. Excitable cells, such as neurons and muscle cells, have additional voltage-dependent channels in the plasma membrane. These are controlled by membrane potential and/or changes in specific ion concentration, often K^+ and Ca^{2+} itself. In nonexcitable cells the principal channels in the plasma membrane are voltage independent. These are often regulated by G proteins coupled to specific receptors, but others respond to a signal generated by the emptying of the endoplasmic reticulum (see later). Channels in the endoplasmic reticulum are characterized by being activated by increases in inositol (1,4,5)-trisphosphate (IP_3) (Clapham, 1995; Pozzan et al., 1994). As is the case with Ca^{2+}-ATPases present in plasma membrane and endoplasmic reticulum, Ca^{2+} channels of the plasma membrane and the IP_3 receptor are not similar proteins. The IP_3 receptor in the endoplasmic reticulum is a tetrameric complex that is autoregulated in the sense that its channel activity is coinhibited by IP_3 and cytosolic Ca^{2+} (Hajnoczky and Thomas, 1994). More recently, another Ca^{2+} channel in addition to the IP_3 receptor has been described in several cell types. It is also a tetrameric protein called the ryanodine receptor that releases a separate Ca^{2+} pool from the endoplasmic reticulum in response to cyclic ADP-ribose (Galione, 1992; Guse et al., 1995). Although different estimates of the free Ca^{2+} concentration inside the endoplasmic reticulum have been given depending on the technique used, it is clear that it is much higher than in the cytosol (Meldolesi and Pozzan, 1998).

The mitochondria matrix contains a pool of Ca^{2+}, although release of Ca^{2+} from this organelle does not seem to play an important role in physiological signal transduction. Nevertheless, changes in the mitochondrial Ca^{2+} concentration that vary as a consequence of changes in the cytosolic concentration regulate mitochondrial dehydrogenases (Fisher et al., 1973). In this way, activation of energy-consuming metabolic reactions in the cytosol are matched by oxidative phosphorylation in mitochondria (Denton and McCormack, 1993). Uptake of Ca^{2+} in this organelle is driven by the mitochondrial electrochemical gradient through a uniporter in the inner membrane that is inhibited by ruthenium red and Mg^{2+} and is mostly regulated by the mitochondrial inner membrane potential and the $[Ca^{2+}]_c$. Release of Ca^{2+} depends on Na^+/Ca^{2+} and H^+/Ca^{2+} antiports that work against the electrochemical gradient and could be carriers or gated pores. In any case, and through the intervention of a Na^+/H^+ antiport if necessary, the net balance of Ca^{2+} efflux is the exchange of two molecules of Na^+ for one of Ca^{2+} (Gunter and Pfeiffer, 1990; Hallestrap et al., 1993).

A fourth mechanism for movement of Ca^{2+} across the inner mitochondrial membrane is probably only relevant in situations that threaten cell survival and consists of what has been called the permeability transition (Bernardi et al., 1994; Gunter and Pfeiffer, 1990). This is a phenomenon stimulated by Ca^{2+} together with prooxidants or ATP and is inhibited by cyclosporin, Mg^{2+}, and ADP, among other agents. It consists of a nonspecific loss of the permeability barrier formed by the inner membrane after accumulation or overload of Ca^{2+}. Not only Ca^{2+}, but also other small molecules and even small matrix proteins, leak out of the mitochondria as well (Igbavboa *et al.*, 1989). The nature of the so-called megapore that is responsible for the permeability transition is a controversial matter. Evidence indicates that it may be formed by a disruption of one or several of the carrier proteins located in the inner mitochondrial membrane, most likely with the nonexclusive involvement of the ADP/ATP carrier (Brustovetsky and Klingenberg, 1996; Hallestrap *et al.*, 1997; Lohret *et al.*, 1996). The participation of the permeability transition in normal physiology is not clear. Although reversal of the transition has been achieved *in vitro* on Ca^{2+} chelation (Crompton and Costi, 1988; Gunter and Pfeiffer, 1990), it is hard to imagine how it may have a role in normal, nonpathological Ca^{2+} homeostasis because of the nonspecificity of the leakage from the mitochondrial matrix. However, an irreversible change in mitochondrial permeability may be part of the triggering mechanism for apoptosis (see later).

B. Changes in the Intracellular Ca^{2+} Concentration in Signal Transduction Events

1. IP_3 Activation of Ca^{2+} Release from the Endoplasmic Reticulum

Increases in $[Ca^{2+}]_c$ over the levels that are kept by the machinery described earlier are responsible for the activation of a variety of processes inside cells. In the most classic pathway, a receptor protein exposed to the outer side of the plasma membrane is activated by binding of a ligand. This activation signal is carried by GTP-binding proteins (G proteins) to phospholipase C, which catalyzes the hydrolysis of phosphatidylinositol (4,5)-bisphosphate to diacylglycerol and IP_3. Finally, IP_3 acting as a second messenger binds to the IP_3 receptor activating Ca^{2+} release from the endoplasmic reticulum (Berridge and Irvine, 1989). An alternative pathway for signaling Ca^{2+} release from the endoplasmic reticulum relies on binding of an extracellular ligand to a protein tyrosine kinase receptor protein in the plasma membrane, which then directly activates phospholipase C by phosphorylation, leading to the production of IP_3 (Fantl *et al.*, 1993). G proteins activate phospholipase C β, whereas protein tyrosine kinase receptors activate the γ isoform of phospholipase C. Examples of receptors activating Ca^{2+} release through one or the other pathway have been reviewed (see, for instance, Clapham, 1995) and include adrenergic, muscarinic, and purinergic receptors; receptors for serotonin, glucagon, thromboxanes, platelet-activating factor, and f-Met-Leu-Phe (through the receptor/G protein/phospholipase C β/IP$_3$ pathway); receptors for epidermal growth factor; and platelet and fibroblast-derived growth factors (through the tyrosine kinase receptor/phospholipase Cγ/IP$_3$ pathway). The average cytosolic concentration of Ca^{2+} may increase up to 10^{-6} M from the basal level of about 0.1×10^{-6} M on stimulation of Ca^{2+} release through IP$_3$ production. In areas proximal to the endoplasmic reticulum Ca^{2+} concentration can be significantly higher for a very short period following IP$_3$ binding (Clapham, 1995).

2. Uptake of Extracellular Ca^{2+}

Increases of cytosolic Ca^{2+} can also be achieved by opening the channels in the plasma membrane. Opening of these channels has been proposed as occurring through a mechanism of capacitative Ca^{2+} entry in response to depletion of the intracellular Ca^{2+} stores that follow IP$_3$ stimulation. In this model, both the release of Ca^{2+} from the endoplasmic reticulum and the opening of plasma membrane Ca^{2+} channels are dependent on IP$_3$ production, although the latter is an indirect response to IP$_3$ (Bennett *et al.*, 1998). Different second messengers have been proposed to play the role of carrying the signal between the Ca^{2+}-depleted endoplasmic reticulum and the plasma membrane for the opening of Ca^{2+} channels (Clapham, 1995), including a molecule with an unknown structure that has been called the Ca^{2+} influx factor (Randriamampita and Tsien, 1993).

3. Mitochondrial Ca^{2+}

As mentioned earlier, mitochondria do not seem to play an important role in physiological signal transduction. Because the mitochondrial Ca^{2+} uniporters responsible for the uptake of Ca^{2+} have a lower affinity for Ca^{2+} than the endoplasmic reticulum ATPase, uptake occurs only when the [Ca^{2+}]$_c$ is above approximately 0.5×10^{-6} M. Such a condition occurs transiently when the endoplasmic reticulum releases Ca^{2+} in response to IP$_3$, particularly for mitochondria located next to the endoplasmic reticulum (Pozzan *et al.*, 1994). At rest, however, mitochondria seem to contain little Ca^{2+} as compared with endoplasmic reticulum (Carafoli, 1987; Somlyo, 1985). Thus mitochondria are not a source of Ca^{2+} for receptor-mediated signaling pathways. Nevertheless, the capacity of mitochondria to pick up Ca^{2+} at higher than normal intracellular concentrations, and the eventual release of this accumulated Ca^{2+}, may be of relevance in pathological situations in which severe damage is inflicted on cells. In this context, the release of Ca^{2+} from mitochondria or failure of mitochondria to function as a buffer through Ca^{2+} uptake in cells subjected to severe oxidative stress will be discussed later in this chapter.

4. Targets of the Increase in $[Ca^{2+}]_c$

Cellular responses to increases in $[Ca^{2+}]_c$ are given by activation of a great variety of Ca^{2+}-binding proteins, including phospholipases C, A_2, and D, Ca^{2+}-dependent forms of protein kinase C, calmodulin, gelsolin, α-actinin, and annexins. These proteins are modulators of a multitude of cellular reactions, cytoskeletal dynamics, and signal transduction pathways (Dennis, 1997; Divecha and Irvine, 1995; Klee, 1988; Maki *et al.*, 1992; Moss, 1997; Newton, 1995). In this manner, changes in the $[Ca^{2+}]_c$ control metabolic responses to changing external conditions, gene expression, and cell cycle progression. Most of these proteins remain mostly in an unbound inactive state until there is sufficient elevation of the $[Ca^{2+}]_c$ on stimulation of the cell. In contrast, there is a subset of Ca^{2+}-binding proteins that have a higher binding capacity, so that even in the absence of stimulation, they may be considered another potential pool of releasable Ca^{2+}. This is the case of annexins, which are a family of proteins that bind to phospholipids in a Ca^{2+}-dependent manner (Moss, 1997; Raynal and Pollard, 1994).

II. LETHAL ALTERATIONS OF THE INTRACELLULAR Ca^{2+} CONCENTRATION: NECROSIS, APOPTOSIS, AND OXIDATIVE STRESS

Oscillations in the steady-state intracellular Ca^{2+} concentration have profound effects on signal transduction pathways. Long-lasting increases in the intracellular Ca^{2+} concentration above the range in which these oscillations take place are characteristic of pathological situations such as ischemia–reperfusion injury and chemical poisoning, in which lethal oxidative stress is involved. At the cellular level these increases have been associated with cell death by necrosis or apoptosis. At the subcellular level, mitochondria play a major role that may explain the dynamics and deleterious effects of large increases in $[Ca^{2+}]_c$. A main issue in the study of major disruptions of Ca^{2+} homeostasis is whether large increases in the intracellular Ca^{2+} concentration are a causative factor or just one consequence of massive cell damage.

A. Severe Oxidative Stress, Ca^{2+} Homeostasis, and Cell Death

Oxidative stress seems to be responsible for cellular damage during the reperfusion phase of ischemia–reperfusion injury in organs, such as heart, liver, and intestine, and during exposure to many xenobiotics. It has been proposed that increases in $[Ca^{2+}]_c$ are the result of oxidative stress and are the cause of cellular damage and death (Farber *et al.*, 1990; Hallestrap *et al.*, 1993; Stone *et al.*, 1989). *In vitro* studies using diverse cell types, but most often hepatocytes, led to a proposed series of metabolic events (Albano et al., 1991; Bellomo et al., 1982; 1984; Carini et al., 1995; Nieminen et al., 1997). Hydrogen peroxide and lipid peroxides, whether added to cells or generated within, as in the metabolism of xenobiotics, are reduced by glutathione in a reaction catalyzed by glutathione peroxidase. This reaction generates glutathione disulfide, which is then reduced by glutathione reductase consuming NADPH. If the exposure to peroxides is great enough, the balance of the redox intermediates goes toward the oxidized species, glutathione disulfide, and $NADP^+$ with formation of protein–glutathione mixed disulfides. Then, the Ca^{2+} homeostasis is disturbed by the oxidation of NADPH and sulfhydryl groups and the $[Ca^{2+}]_c$ increases, leading to an uncontrolled activation of Ca^{2+}-dependent proteins, such as proteases, lipases, and nucleases that cause cell damage and eventually cell death (Orrenius and Bellomo, 1991; Orrenius et al., 1992).

Mitochondria seem to play a very important role in this process. As described previously, mitochondria do not store much Ca^{2+} and do not intervene in receptor-mediated signal transduction. Nevertheless, when the $[Ca^{2+}]_c$ rises above normal levels, enough to allow substantial uptake of Ca^{2+} through the relatively low-affinity mitochondrial Ca^{2+} uniporter, this organelle begins taking up Ca^{2+}. The capacity of mitochondria for the storage of Ca^{2+} is very high. It has been estimated that over 3 μmol Ca^{2+}/mg protein, equivalent to 3 mM, may be stored by mitochondria (Scarpa and Azzone, 1970). Ca^{2+} overload eventually leads to loss of the inner membrane permeability barrier, with massive release of Ca^{2+} during which Ca^{2+} homeostasis is completely lost (Kass et al., 1992; Lötscher et al., 1980; Richter and Kass, 1991). In addition, the uptake of Ca^{2+} is energy dependent and therefore linked to NADH, which drives electron transport. When mitochondrial NADPH is oxidized, NADH becomes oxidized through the transhydrogenase-catalyzed reaction (Livingston et al., 1992). Thus, oxidation of mitochondrial NADPH also impairs the mitochondrial uptake of Ca^{2+}. Liu and co-workers (1997) found that chaperon proteins of the grp family in the endoplasmic reticulum were able to prevent a rise in the $[Ca^{2+}]_c$ concentration and oxidative stress after treatment with iodoacetamide, an alkylating toxicant. Nevertheless, Ca^{2+} overloading of mitochondria seemed to be the critical factor leading to cell death.

The enzymatic reduction of hydrogen peroxide and lipid peroxides by glutathione peroxidases prevents formation of the much more reactive hydroxyl and alkoxyl radicals by the metal-catalyzed decomposition of peroxides; however, when the oxidative load overwhelms this system the redox balance is lost, sulfhydryl groups are oxidized, and Ca^{2+} homeostasis is lost. This general hypothesis is also supported by studies *in vivo* showing that alterations in Ca^{2+} homeostasis and mitochondria are an early event in tissue damage and by *in vitro* experiments using isolated mitochondria (Vercesi *et al.*, 1997). Nevertheless, as discussed by Farber, *et al.*, (1990), cell death due to exposure to oxidants can occur independently of alternations in Ca^{2+} homeostasis. Indeed, preventing the alteration of Ca^{2+} homeostasis caused by oxidants does not always avoid cell death. Thus, loss of Ca^{2+} homeostasis may be only one possible mode of oxidative damage-induced cell death.

B. Oxidative Stress, Ca^{2+} Homeostasis, and Apoptosis

Increases in the intracellular Ca^{2+} concentration and exposure to oxidants may cause programmed cell death (Kerr *et al.*, 1972). Although necrosis is frequently a manifestation of pathological conditions such as those referred to earlier, apoptosis may be triggered by the pathological dysregulation of Ca^{2+} homeostasis but also by physiological Ca^{2+}-dependent processes as well. Necrosis typically follows severe disruptions in the Ca^{2+} homeostasis, characterized by large increases in the intracellular Ca^{2+} concentration. Lower increases in the intracellular Ca^{2+} concentration may lead to apoptosis.

Many agents that cause oxidative stress may cause apoptosis as well. In some cases, such as radiation, various anticancer drugs, tumor necrosis factor α, nitric oxide, hydrogen peroxide, glutathione depletion, and lipid peroxides, there appears to be a cause and effect relationship between oxidative stress and apoptosis. Furthermore, several antioxidants, including *N*-acetylcysteine, thioredoxin, and superoxide dismutase, have been shown to have antiapoptotic activity (Buttke and Sandström, 1994; Higuchi *et al.*, 1996; Hockenbery *et al.*, 1993; Jacobson, 1996; Slater *et al.*, 1995b). The antiapoptotic activity of the product of the prooncogene bcl-2 has been proposed to be due to a supposed antioxidant activity of the Bcl-2 protein (Hockenbery *et al.*, 1993; Jacobson, 1996). Nevertheless, no enzymatic antioxidant activity has been ascribed to Bcl-2. Indeed, it has even been suggested that Bcl-2 is a prooxidant, to which cells respond by increasing their antioxidant capacity (Steinman, 1995).

A mechanism by which oxidative stress may cause apoptosis is by increasing the cytosolic concentration of Ca^{2+} (Slater *et al.*, 1995a). Interestingly, agents such as thapsigargin and Ca^{2+} ionophores may cause apoptosis through elevation of the $[Ca^{2+}]_c$, whereas buffering of increases in $[Ca^{2+}]_c$ by intracellular Ca^{2+} chelators inhibits apoptosis (Jiang *et al.*, 1994; Kruman *et al.*, 1998; Lam *et al.*, 1994). It has now been shown that Bcl-2 can prevent apoptosis by controlling Ca^{2+} efflux from the endoplasmic reticulum (Lam *et al.*, 1994) and perhaps from mitochondria as well (Petit *et al.*, 1996).

Cytochrome c has been identified as a proapoptotic factor in cytosolic extracts, and it has also been observed that its concentration in cytoplasm increases in apoptotic cells. Thus, release of cytochrome c from mitochondria has been proposed to be a common mediator of apoptosis (Liu *et al.*, 1996). Although it has been reported that Bcl-2 prevents the release of cytochrome c, it also may prevent apoptosis by exerting an unspecified function, perhaps related to Ca^{2+} regulation, downstream of this event (Bossy-Wetzel *et al.*, 1998; Rosse *et al.*, 1998). The release of cytochrome c may be related to the Ca^{2+}-driven permeability transition (Scarlett and Murphy, 1997). Nonetheless, cytochrome c is located in the intermembrane space, whereas the permeability transition is primarily involved with the opening of pores in the inner membrane so that the relationship is unclear. Furthermore, it has been reported that cytochrome c is released from mitochondria in HeLa cells before any disruption in the mitochondrial inner membrane potential occurs (Bossy-Wetzel *et al.*, 1998).

III. MODULATION OF Ca^{2+} SIGNALING BY OXIDATIVE STRESS

Nonlethal, oxidative stress-induced changes in the intracellular concentration of Ca^{2+} and its effects on classic signal transduction pathways are examined in this section. This is a more recently developed area of research than that concerned with cell death by oxidative stress and the subsequent loss of Ca^{2+} homeostasis. Oxidative stress-caused changes in the intracellular concentration of Ca^{2+} do not always represent a threat to cells, but may modulate "normal" or "physiological" signal transduction pathways (Suzuki *et al.*, 1997). The intervention of reactive oxygen species and changes in the intracellular concentration of Ca^{2+} in apoptosis, as discussed earlier, also represent an example of this, as apoptosis occurs during normal tissue differentiation and growth. The effect of reactive oxygen species on Ca^{2+} signaling may be due to the mobilization of Ca^{2+}, which might disturb or modulate subsequent Ca^{2+} mobilization by classic agonists. It may also be given by modifications of Ca^{2+} channels, exchangers, or ATPase pumps that may not be apparent until activation by classic agonists occurs.

Hydrogen peroxide, *t*-butyl hydroperoxide, lipid peroxidation primary and secondary products, and the reactive oxygen species generating systems hypoxanthine/xanthine oxidase, glucose/glucose oxidase, and quinones are the agents used most frequently to model cellular oxidative stress *in vitro*. In numerous, but not all, cell types, they cause an increase in $[Ca^{2+}]_c$, which in some cases has been correlated with alterations in the response of the cells to different physiological stimuli. Experiments illustrating this kind of situation are reviewed in this second section of the chapter. The Ca^{2+} pools and the Ca^{2+} transporters that are the target of reactive oxygen species and how cell functions may be modulated by the oxidative stress-induced changes in the intracellular Ca^{2+} concentration are described. At this point in time, however, it is not clear what factors cause differences in response among cell types, Ca^{2+} channels, and Ca^{2+} transporters.

A. Technical Note: A Brief Introduction to Methods Used for the Study of Intracellular Ca^{2+} Concentration and Fluxes

Determination of $[Ca^{2+}]_c$ in living cells relies mostly on the use of fluorescent indicators that are added to the cells as plasma membrane-permeable derivatives, such as INDO 1-AM and FURA 2-AM. Once inside cells, ester bonds are hydrolyzed in these molecules and the indicator becomes trapped in a nonpermeable form (Kao, 1994). Ca^{2+}-bound indicators have specific fluorescent properties that can be followed by fluorescent spectrophotometry or fluorescent microscopy.

The Ca^{2+} homeostasis and signaling machinery can be disrupted experimentally in different ways of interest to investigators. Intracellular Ca^{2+} buffers such as BAPTA eliminate or attenuate changes in $[Ca^{2+}]_c$. This is useful to check whether cell responses depend on these changes. To assess entry from the extracellular medium, Ca^{2+} chelators and Ca^{2+}-free buffers are used. Also, the Ca^{2+} ionophore A23187, which allows free entry through the plasma membrane, is used frequently to mimic physiological Ca^{2+} entry (Livingston et al., 1992; Hoyal et al., 1996a,b). A different approach is to measure fluxes of Ca^{2+}, which can be achieved by using the isotope $^{45}Ca^{2+}$ (Elliott and Schilling, 1991). Treatment of cells with digitonin or saponin is used to allow the direct access of agonists or other agents to intracellular targets such as the IP_3-sensitive channels in the endoplasmic reticulum, this way bypassing the plasma membrane permeability barrier (Renard et al., 1992).

Release of Ca^{2+} from the endoplasmic reticulum can be eliminated by allowing the passive emptying of this organelle using the Ca^{2+} ATPase pump inhibitors thapsigargin or 2,5-di(t-butyl)-1,4-benzohydroquinone prior to the addition of other agents (Lytton et al., 1991; Kass et al., 1989). The study of mitochondrial Ca^{2+} frequently requires the use of uncoupling agents and metabolic inhibitors in order to check for the effect of disrupting oxidative phosphorylation and the respiratory chain on Ca^{2+} movement across the inner membrane. Ruthenium red, an inhibitor of Ca^{2+} uptake by mitochondria, is also used frequently (Gunter and Pfeiffer, 1990).

B. Release from the Endoplasmic Reticulum and Entry from the Extracellular Medium

In classical Ca^{2+} signal transduction, Ca^{2+} release from the endoplasmic reticulum is often followed by entry from the extracellular space according to the capacitative mechanism explained earlier (Section I,B 2). After treatment with reactive oxygen species, particularly hydroperoxides, several cell types have shown a response consisting of a transient increase in the $[Ca^{2+}]_c$ that may be followed by a more permanent increase when the oxidative treatment reaches a threshold. In most cases, release from the endoplasmic reticulum and entry from the extracellular space have been found to be responsible for the transient and permanent increases, respectively. It has been shown that release from the endoplasmic reticulum may be due to decreased Ca^{2+} ATPase activity or to increased outflow through Ca^{2+} channels. Entry from the extracellular space may be due to capacitative entry in response to the depletion of the endoplasmic reticulum Ca^{2+} pool, but it also may be due to inhibition of plasma membrane Ca^{2+} ATPases or the Na^+/Ca^{2+} exchanger.

A transient increase followed by a permanent elevation in the $[Ca^{2+}]_c$ has been found in several cell types treated with reactive oxygen species and lipid peroxidation products (Fig. 2). The transient increase may occur in a matter of seconds and is due to release of Ca^{2+} from the IP_3-sensitive Ca^{2+} pool. On the other side, the permanent increase proceeds gradually and may go on for many minutes, and is due to entry from the extracellular medium. The intracellular Ca^{2+} concentration rises from the basal level of $0.1-0.2 \times 10^{-6} M$ to up to no more than $1 \times 10^{-6} M$, thus creating changes in the Ca^{2+} concentration in the same order of magnitude, or at least not necessarily much higher, than those due to stimulation by classic agonists. This pattern of changes in the $[Ca^{2+}]_c$ with initial release from the IP_3-sensitive Ca^{2+} pool has been described in endothelial cells (Doan *et al.*, 1994; Dreher *et al.*, 1995; Volk *et al.*, 1997), ventricular myocytes (Goldhaber and Liu, 1994), and smooth muscle cells (Roveri *et al.*, 1992) treated with hydrogen peroxide or hydrogen peroxide generating systems, hepatocytes treated with 4-hydroxynonenal (Carini *et al.*, 1996), and chicken B cells treated with hydrogen peroxide (Qin *et al.* 1996).

In some of the cases just mentioned, Ca^{2+} mobilization was found to be dependent on the activation of certain components of signal transduction pathways. Thus, Qin *et al.*, (1996) found that the hydrogen peroxide-induced Ca^{2+} mobilization occurred after $p53/p56^{lyn}$-dependent $p72^{syk}$ activation. $p53/p56^{lyn}$ and $p72^{syk}$ are nonreceptor protein kinases of the Src and Syk families that intervene in lymphocyte activation.

Carini *et al.*, (1996) found that incubation of isolated hepatocytes with low micromolar concentrations of 4-hydroxynonenal caused first a transient increase in $[Ca^{2+}]_c$ that peaked at $0.4 \times 10^{-6} M$ 5 min after addition of the aldehyde. After that, a second rise went on until the end of the incubation (30 min) when $[Ca^{2+}]_c$ was $0.9 \times 10^{-6} M$. The first, transient increase, but not the second, permanent increase, was inhibited by U73122, an inhibitor

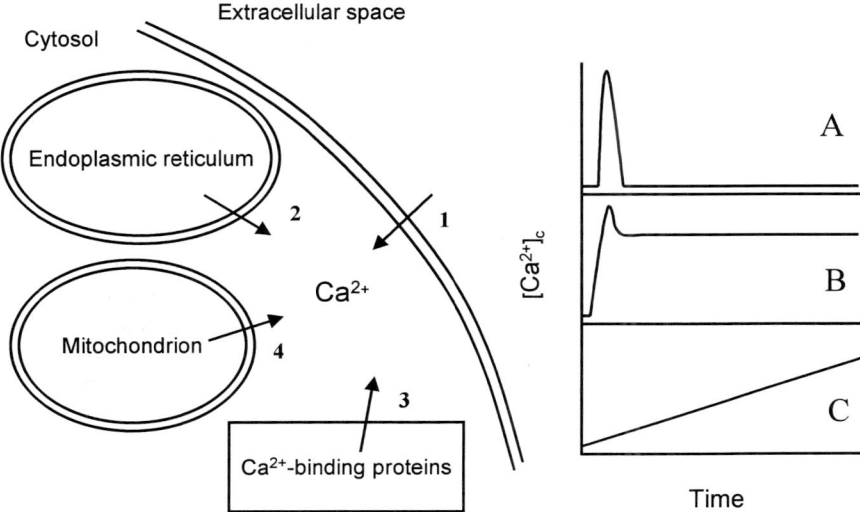

Figure 2 Changes in $[Ca^{2+}]_c$ due to oxidative stress. Arrows represent net Ca^{2+} movement from different Ca^{2+} pools into the cytosol. The right-hand side of the figure is an attempt to summarize the results of the studies discussed in the text, in which $[Ca^{2+}]_c$ was determined in a variety of cell types exposed to reactive oxygen species or lipid peroxidation products. These graphs do not represent any particular actual measurements. Transient Ca^{2+} increases like the one shown in graph A may be due to release of Ca^{2+} from the endoplasmic reticulum (2), occur in a matter of seconds or very few minutes after exposure, and are characteristic of low-level oxidative stress. Under relatively stronger oxidative stress, they are typically followed (graph B) or substituted (graph C) by a more sustained and gradual elevation of $[Ca^{2+}]_c$ due to the opening of channels in the plasma membrane (1). This opening may be due to, or independent of, the capacitative mechanism in response to release from the endoplasmic reticulum, when the latter occurs. Release of Ca^{2+} from Ca^{2+}-binding proteins (3) might explain increases in $[Ca^{2+}]_c$ in alveolar macrophages treated with peroxides and may prove to be a novel mechanisms for Ca^{2+} signaling. All these changes in $[Ca^{2+}]_c$ are of a magnitude similar to those caused by classic agonist stimulation, from basal levels of about $0.1-0.2 \times 10^{-6}$ M to not more than 10^{-6} M. They mimic the Ca^{2+} release by classic agonists and this way modulate cell responses to the agonists.

of phospholipase C, and also by thapsigargin. It was concluded that the first increase follows the phospholipase C/IP_3 pathway for release of Ca^{2+} from the endoplasmic reticulum. The second increase could not be explained by capacitative influx in response to the depletion of the IP_3-sensitive store, but had to be due to alteration of the signaling for the capacitative mechanism itself or of the influx channels.

Phospholipase C activation was also found to be involved in the increase of $[Ca^{2+}]_c$ caused by the exposure of endothelial cells to hydrogen peroxide

(Volk et al., 1997). These cells were treated with hydrogen peroxide and superoxide generated by hypoxanthine/xanthine oxidase or with hydrogen peroxide generated by glucose/glucose oxidase. Relatively high rates of generation of the reactive oxygen species by these enzymatic systems led to a gradual increase of $[Ca^{2+}]_c$ that went on for 40 min and peaked at 0.6×10^{-6} M. This kind of treatment was responsible for drastic decreases in viability, and the gradual increase in $[Ca^{2+}]_c$ was found to be due to increased uptake from the extracellular medium. At lower concentrations of xanthine oxidase and glucose oxidase, no toxicity was observed and the elevation of $[Ca^{2+}]_c$ occurred very rapidly and was transient. These transients lasted for not more than 15 sec and peaked at a similar concentration as before ($\sim 0.6 \times 10^{-6}$ M). In the case of the hydrogen peroxide and superoxide generating system hypoxanthine/xanthine oxidase, it was determined that hydrogen peroxide was responsible for the increase in $[Ca^{2+}]_c$. Interestingly, bolus addition of the peroxide had to be of a concentration much higher than the concentration achieved by the enzymatic systems in order to get the same effect. Transients were diminished by the phospholipase C inhibitor U73122 and were due to release from the endoplasmic reticulum. The effect of Ca^{2+} mobilization by these low levels of hydrogen peroxide on the Ca^{2+} signaling that occurs after stimulation with the agonist ATP and histamine was studied, but no alteration of the Ca^{2+} mobilization characteristic of these agonists was observed.

Elliott, Schilling, and co-workers described the effect of incubating vascular endothelial cells with 400×10^{-6} M t-butyl hydroperoxide on the subsequent Ca^{2+} mobilization caused by the physiological agonist bradykinin (Elliott et al., 1992; Schilling and Elliott, 1992). Treatment for 30 min caused inhibition of the Ca^{2+} influx provoked by bradykinin. Longer treatments caused inhibition of release from the IP_3-sensitive intracellular pool and eventually a loss of responsiveness and a progressive increase in the basal Ca^{2+} level. This corresponds with initial inhibition of Ca^{2+} influx, followed in time by inhibition of efflux, as determined using $^{45}Ca^{2+}$ to measure Ca^{2+} movement (Elliott and Schilling, 1991).

1. Ca^{2+} ATPase Pumps and IP_3-Sensitive Channels in the Endoplasmic Reticulum

Reactive oxygen species are believed to be involved in ischemia–reperfusion damage to cardiac muscle. Because Ca^{2+}-ATPase pumps and IP_3-sensitive channels might be targets of this oxidative damage, preparations of sarcoplasmic reticulum have been used to study the susceptibility of these proteins to inactivation by reactive oxygen intermediates. Permeabilized hepatocytes and endothelial and smooth muscle cells have also been used for these studies.

Treatment of preparations of sarcoplasmic reticulum with hydrogen peroxide (Boraso and Williams, 1994) and with a singlet oxygen plus superoxide anion generating system (Holmberg et al., 1991; Xiong et al., 1992) casued an increase in the probability of the Ca^{2+} release channels being open. The same result was found when the channel protein reconstituted in phospholipid bilayers was treated with the ascorbate/iron couple (Stoyanovsky et al., 1994). In preparations treated with hypoxanthine/xanthine oxidase, both stimulation of IP_3 release and inhibition of Ca^{2+} ATPases were found. Superoxide anion, not hydrogen peroxide, was responsible for this effect, and cysteine protected the Ca^{2+}-ATPase from inhibition, suggesting the possibility of modification of sulfhydryl groups (Suzuki and Ford, 1991, 1992). Treatment with the hydroxyl radical generating system hydrogen peroxide/Fe^{3+}/nitriloacetic acid caused inhibition of the Ca^{2+}-ATPase as well. ATP exhibited a protective effect, suggesting that the target of the modification is localized in the ATP-binding site of the pump (Xu et al., 1997).

Treatment of permeabilized hepatocytes with t-butyl hydroperoxide caused both inhibition of uptake and sensitization of release of Ca^{2+} from the endoplasmic reticulum. Interestingly, these effects were also caused by glutathione disulfide and were reversed by the sulfhydryl-reducing agent dithiothreitol (Rooney et al., 1991). It was described later, also in permeabilized hepatocytes, that not only glutathione disulfide but also cystine increases the sensitivity of the IP_3-dependent channels. Dithiothreitol, cysteine, and glutathione reversed the sensitization of the IP_3-dependent channels, whereas sulfhydryl group-modifying agents did not have any effect on IP_3-sensitive channels but inhibited Ca^{2+}-ATPase pumps (Renard et al., 1992). Thus, the release of Ca^{2+} from the IP_3-sensitive pool may be regulated by the formation of mixed disulfides. In the paper mentioned earlier, Rooney and co-workers (1991), using fluorescent microscopy, showed that the Ca^{2+} oscillatory waves caused by t-butyl hydroperoxide exhibited the same kinetic characteristics as those caused by stimulation with the receptor-mediated agonist phenylephrine. The latter, unlike t-butyl hydroperoxide, activates phospholipase C.

Grover and Samson (1997) have compared the inhibition by hydrogen peroxide of the Ca^{2+}-ATPase pump in permeabilized endothelial and smooth muscle cells. They found that the pump is inhibited much more easily in smooth muscle than in endothelial cells, which may be of physiological relevance as the endothelium constitutes the first barrier of defense against attack by reactive oxygen species in the vasculature. Accordingly, in arterial rings the inhibition by hydrogen peroxide of the smooth muscle-dependent contraction induced by cyclopiazonic acid and angiotensin was much greater than in the endothelium-dependent, cyclopiazonic acid- and bradykinin-induced relaxation.

2. Ca^{2+} ATPase Pumps, Na^+/Ca^{2+} Exchanger, and Ca^{2+} Channels in the Plasma Membrane

Influx of Ca^{2+} into isolated hepatocytes and a few other cell types due to alterations of Ca^{2+}-ATPase pumps and the Na^+/Ca^{2+} exchanger of the plasma membrane has been described. Lipid antioxidants and/or sulfhydryl-reducing agents have proved to be effective in preventing this inhibition in many cases.

Treatment of isolated hepatocytes with Fe^{3+} caused a vigorous lipid peroxidation that was associated with entry of Ca^{2+} into the cell. Lipid antioxidants prevented this influx, which otherwise went on for the duration of the incubations (180 min), and led to an increase in the concentration of Ca^{2+} in endoplasmic reticulum and mitochondria (Albano et al., 1991). In the same experimental system, Carini et al., (1995) described that the opening of Ca^{2+} channels, rather than the Na^+/Ca^{2+} exchanger functioning in a reverse mode, was responsible for the increase in the intracellular Ca^{2+} concentration. Some reports had previously shown that treatment with the redox cycling quinone menadione (Carini et al., 1994) or hypoxia (Haigney et al., 1992) was able to convert the Na^+/Ca^{2+} exchanger to a reverse mode that was the cause of net Ca^{2+} entry into the cells. In isolated hepatocytes, inactivation of the Ca^{2+}-ATPase pump of the plasma membrane by treatment with cytotoxic doses of menadione was also reversed by the addition of dithiothreitol, showing that this activity may be regulated by the oxidation/reduction of sulfhydryl groups (Nicotera et al., 1985). Inhibition of the Ca^{2+}-ATPase was also found in red blood cells after treatment with Fe^{2+}; this inhibition was prevented by lipid antioxidants (Rohn et al., 1996). In rat alveolar macrophages, a sustained elevation of the intracellular Ca^{2+} concentration due to treatment with t-butyl hydroperoxide could also be reversed by the addition of dithiothreitol (Hoyal et al., 1996b).

Treatment of cardiac myocytes with hydrogen peroxide caused entry of Ca^{2+} into the cells due to the opening of channels, which made the plasma membrane leaky (Wang et al., 1995). In these cells, treatment with rose bengal, a generator of singlet oxygen and superoxide anion, also induced an electrogenic Na^+/Ca^{2+} exchange that was proposed as the cause of oxidative stress-induced arrhythmias (Matsuura and Shattock, 1991). A situation similar to the one described earlier in hepatocytes was found in smooth muscle cells treated with hydrogen peroxide (Roveri et al., 1992). After a rapid increase, $[Ca^{2+}]_c$ went down but to a new steady-state level higher than before treatment with the peroxide. The first increase was due to release from the IP_3-sensitive intracellular store, whereas the subsequently sustained increased steady-state level was due to uptake through plasma membrane channels. The sustained increase was inhibited by lipid antioxidants and sulfhydryl group reducing agents.

C. Release from Other Ca^{2+} Pools

A pattern of changes in the cytosolic concentration of Ca^{2+} in response to challenge with reactive oxygen species has been observed numerous times. It consists of an early and transient increase due to the release of Ca^{2+} from the endoplasmic reticulum and a subsequent gradual increase or elevated basal level due to entry from the extracellular space. Primary isolates of rat alveolar macrophages show a similar pattern, but in this case it was found that the primary source of Ca^{2+} was not the endoplasmic reticulum, extracellular Ca^{2+}, or even mitochondria (Hoyal et al., 1998). Treatment of alveolar macrophages with a concentration of t-butyl hydroperoxide or hydrogen peroxide equal or below $50 \times 10^{-6} M$ causes an increase of up to twice the basal level, after which the intracellular Ca^{2+} concentration goes back to the baseline concentration. Treatment with concentrations of peroxides higher than $50 \times 10^{-6} M$ causes an indefinitely sustained elevation of Ca^{2+} concentration of a similar magnitude (Murphy et al., 1995). Activation of the respiratory burst with ADP or a phorbol ester after these kinds of treatments showed that peroxide concentrations causing a transient elevation of the $[Ca^{2+}]_c$ caused an enhancement of the respiratory burst, whereas peroxide concentrations causing a sustained elevation of the $[Ca^{2+}]_c$ had an inhibitory effect (Hoyal et al., 1996a). The ADP-stimulated, but not the phorbol ester-stimulated respiratory burst depends largely on the release of Ca^{2+} from the endoplasmic reticulum, but the effect on the respiratory burst activated by both agents of treatment with hydroperoxides was similar. Confocal microscopy of macrophages loaded with a Ca^{2+} fluorescent indicator showed that the Ca^{2+} increase started close to the inner side of the plasma membrane seconds after the addition of t-butyl hydroperoxide and increased inward subsequently. However, the increase in the $[Ca^{2+}]_c$ initiated by the receptor-mediated agonist ADP started from deep inside the cell, consistent with release from the endoplasmic reticulum. Once the classic Ca^{2+} stores were ruled out as possible sources for the peroxide-induced release of Ca^{2+}, it was proposed that annexin VI is the source of this Ca^{2+} (Hoyal et al., 1996b). Annexin VI is a member of the annexin family of proteins, which are characterized by binding to phospholipids in a Ca^{2+}-dependent manner, and are present in cells in amounts large enough to constitute a sizable, potentially releasable, pool of Ca^{2+} (Barwise and Walker, 1996; Gerke and Moss, 1997).

Mitochondria might constitute a source for Ca^{2+} after exposure to oxidative stress due to the fact that its capacity to retain Ca^{2+} depends on the concentration of NADPH. Nevertheless, it is not clear whether this property may be of relevance only for lethal oxidative stress, as discussed earlier in Section II, or also in nonlethal situations.

IV. CONCLUDING REMARKS

Oxidants are recognized not only for their potential to create random havoc in cells but also for participating in signal transduction. As Ca^{2+} transients play a major role in signal transduction, oxidative dysregulation or modulation of Ca^{2+} mobilization is a major underlying mechanism for pathologic and physiologic effects of oxidants, respectively. In the near future, it is likely that the intensive research of this area of investigation now underway will lead to a greater understanding and recognition of this fundamental aspect of biological regulation. In addition, the generation of reactive oxygen species occurs in response to increases in $[Ca^{2+}]_c$ (Goldman et al., 1998). This is separate from the participation of increased $[Ca^{2+}]_c$ on the assembly of the respiratory burst oxidase of phagocytes. Thus, instead of just a consequence of oxidative stress, Ca^{2+} transients and the generation or exposure to reactive oxygen species should be considered as interacting phenomena in normal cell physiology.

REFERENCES

Albano, E., Bellomo, G., Parola, M., Carini, R., and Dianzani, M. V. (1991). Stimulation of lipid peroxidation increases the intracellular calcium content of isolated hepatocytes. *Biochim. Biophys. Acta* **1091**, 310–314.

Barwise, J. L., and Walker, J. H. (1996). Annexins II, IV, V, and VI relocate in response to rises in intracellular Ca^{2+} in human foreskin fibroblasts. *J. Cell Sci.* **109**, 247–255.

Bellomo, G., Jewell, S. A., Thor, H., and Orrenius, S. (1982). Regulation of intracellular Ca^{2+} compartmentation: Studies with isolated hepatocytes and t-butyl hydroperoxide. *Proc. Natl. Acad. Sci. U.S.A.* **79**, 6842–6846.

Bellomo, G., Thor, H., and Orrenius, S. (1984). Increase in cytosolic Ca^{2+} concentration during t-butyl hydroperoxide metabolism by isolated hepatocytes involves NADPH oxidation and mobilization of intracellular Ca^{2+} stores. *FEBS Lett.* **168**, 38–42.

Bennett, D. L., Bootman, M. D., Berridge, M. J., and Cheek, T. R. (1998). Ca^{2+} entry into PC12 cells initiated by ryanodine receptors or inositol 1,4,5-trisphosphate receptors. *Biochim J.* **329**, 349–357.

Bernardi, P., Broekmeier, K. M., and Pfeiffer, D. R. (1994). Recent progress on regulation of the permeability transition pore; a cyclosporin-sensitive pore in the inner mitochondrial membrane. *J. Bioenerg. Biomembr.* **26**, 509–517.

Berridge, M. J., and Irvine, R. F. (1989). Inositol phosphates and cell signaling. *Nature (London)* **341**, 197–205.

Boraso, A., and Williams, A. J. (1994). Modification of the gating of the cardiac sarcoplasmic reticulum Ca^{2+}-release channel by H_2O_2 and dithiothreitol. *Am. J. Physiol.* **267**, H1010–H1016.

Bossy-Wetzel, E. Newmeyer, D.D., and Green, D. R. (1998). Mitochondrial cytochrome c release in apoptosis occurs upstream of DEVD-specific caspase activation and independently of mitochondrial transmembrane depolarization. *EMBO J.* **17**, 37–49.

Brustovetsky, N., and Klingenberg, M. (1996). Mitochondrial ADP/ATP carrier can be reversibly converted into a large channel by Ca^{2+}. *Biochemistry* **35**, 8483–8488.

Buttke, T. M., and Sandström, P. A. (1994). Oxidative stress as a mediator of apoptosis. *Immunol. Today* **15**, 7–10.

Carafoli, E. (1987). Intracellular Ca^{2+} homeostasis. *Annu. Rev. Biochem.* **56**, 395–433.

Carini, R., Bellomo, G., Dianzani, M. U., and Albano, E. (1994). Evidence for a sodium-dependent calcium influx in isolated rat hepatocytes undergoing ATP depletion. *Biochem. Biophys. Res. Commun.* **202**, 360–366.

Carini, R., Bellomo, G., Dianzani, M. U., and Albano, E. (1995). The operation of Na^+/Ca^{2+} exchanger prevents intracellular Ca^{2+} overload and hepatocyte killing following iron-induced lipid peroxidation. *Biochem. Biophys. Res. Commun.* **208**, 813–818.

Carini, R., Bellomo, G., Paradisi, L., Dianzani, M. U., and Albano, E. (1996). 4-Hydroxynonenal triggers Ca^{2+} influx in isolated rat hepatocytes. *Biochem. Biophys. Res. Commun.* **218**, 772–776.

Clapham, D. E. (1995). Ca^{2+} signaling. *Cell (Cambridge, Mass.)* **80**, 259–268.

Crompton, M., and Costi, A. (1988). Kinetic evidence for a heart mitochondrial pore activated by Ca^{2+}, inorganic phosphate and oxidative stress. A potential mechanism for mitochondrial dysfunction during cellular Ca^{2+} overload. *Eur. J. Biochem.* **178**, 489–501.

Dennis, E. A. (1997). The growing phospholipase A_2 superfamily of signal transduction enzymes. *Trends Biochem. Sci.* **22**, 1–2.

Denton, R. M., and McCormack, J. G. (1993). Calcium and the regulation of intramitochondrial dehydrogenases. *Methods Toxicol.* **2**, 390–403.

Divecha, N., and Irvine, R. F. (1995). Phospholipid signaling. *Cell (Cambridge, Mass.)* **80**, 269–278.

Doan, T. N., Gentry, D. L., Taylor, A. A., and Elliott, S. J. (1994). Hydrogen peroxide activates agonist-sensitive Ca^{2+}-flux pathways in canine venous endothelial cells. *Biochem. J.* **297**, 209–215.

Dreher, D., Jornot, L., and Junod, A. F. (1995). Effects of hypoxanthine-xanthine oxidase on Ca^{2+}-stores and protein synthesis in human endothelial cells. *Circ. Res.* **76**, 388–395.

Elliott, S. J., and Schilling, W. P. (1991). Oxidative stress inhibits bradykinin-stimulated $45Ca^{2+}$ flux in pulmonary vascular endothelial cells. *Am. J. Physiol.* **260**, H549–H556.

Elliott, S. J., Meszaros, J. G., and Schilling, W. P. (1992). Effect of oxidant stress on calcium signaling in vascular endothelial cells. *Free Radical Biol. Med.* **13**, 635–650.

Fantl, W. J., Johnson, D. E., and Williams, L. T. (1993). Signaling by receptor tyrosine kinases. *Annu. Rev. Biochem.* **62**, 453–481.

Farber, J. L., Kyle, M. E., and Coleman, J. B. (1990). Biology of disease. Mechanisms of cell injury by activated oxygen species. *Lab. Invest.* **62**, 670–679.

Fisher, A. B., Scarpa, A., LaNoue, K. F., Bassett, D., and Williamson, J. R. (1973). Respiration of rat lung mitochondria and the influence of Ca^{2+} on substrate utilization. *Biochemistry* **12**, 1438–1445.

Galione, A. (1992). Ca^{2+}-induced Ca^{2+} release and its modulation by cyclic ADP-ribose. *Trends Pharmacol. Sci.* **13**, 304–306.

Gerke, V., and Moss, S. E. (1997). Annexins and membrane dynamics. *Biochem. Biophys. Acta* **1357**, 129–154.

Goldhaber, J. I., and Liu, E. (1994). Excitation-contraction coupling in single guinea-pig ventricular myocytes exposed to hydrogen peroxide. *J. Physiol (London)* **477**, 135–147.

Goldman, R., Moshonov, S., and Zor, U. (1998). Generation of reactive oxygen species in a human keratinocyte cell line: Role of calcium. *Arch. Biochem. Biophys.* **350**, 10–18.

Grover, A. K., and Samson, S. E. (1997). Peroxide resistance of ER Ca^{2+} pump in endothelium: Implications to coronary artery function. *Am. J. Physiol.* **273**, C1250–C1258.

Gunter, T. E., and Pfeiffer, D. R. (1990). Mechanisms by which mitochondria transport calcium. *Am. J. Physiol.* **258**, C755–C786.

Guse, A. H., da Silva, C. P., Emmrich, F., Ashamu, G. A., Potter, B. V., and Mayr, G. W. (1995). Characterization of cyclic adenosine diphosphate-ribose-induced calcium release in T lymphocyte cell lines. *J. Immunol.* **155**, 3353–3359.

Haigney, M. C., Miyata, H., Lakatta, E. G., Stern, M. D., and Silverman, H. S. (1992). Dependence of hypoxic cellular calcium loading on Na^+-Ca^{2+} exchange. *Circ. Res.* **71**, 547–557.

Hajnoczky, G., and Thomas, A. P. (1994). The inositol trisphosphate calcium channel is inactivated by inositol trisphosphate. *Nature (London)* **370**, 474–477.

Hallestrap, A. P., Griffiths, E. J., and Connern, C. P. (1993). Mitochondrial calcium handling and oxidative stress. *Biochem. Soc. Trans.* **21**, 353–358.

Hallestrap, A. P., Woodfield, K. Y., and Connern, C. P. (1997). Oxidative stress, thiol reagents, and membrane potential modulate the mitochondrial permeability transition by affecting nucleotide binding to the adenine nucleotide translocase. *J. Biol. Chem.* **272**, 3346–3354.

Higuchi, H., Kurose, I., Kato, S., Miura, S., and Ishii, H. (1996). Ethanol-induced apoptosis and oxidative stress in hepatocytes. *Alcohol: Clin. Exp. Res.* **20**, 340A–346A.

Hockenbery, D. M., Oltvai, Z. N., Yin, X. M., Milliman, C. L., and Korsmeyer, S. J. (1993). Bcl-2 functions in an antioxidant pathway to prevent apoptosis. *Cell (Cambridge Mass.)* **75**, 241–251.

Holmberg, S. R., Cumming, D. V., Kusama, Y., Hearse, D. J., Poole-Wilson, P. A., Shattock, M. J., and Williams, A. J. (1991). Reactive oxygen species modify the structure and function of the cardiac sarcoplasmic reticulum calcium-release channel. *Cardioscience* **2**, 19–25.

Hoyal, C. R., Gozal, E., Zhou, H., Foldenauer, K., and Forman, H. J. (1996a). Modulation of the rat alveolar macrophage respiratory burst by hydroperoxides is calcium dependent. *Arch. Biochem. Biophys.* **326**, 166–171.

Hoyal, C. R., Thomas, A. P., and Forman, H. J. (1996b). Hydroperoxide-induced increases in intracellular calcium due to annexin VI translocation and inactivation of plasma membrane Ca^{2+}-ATPase. *J. Biol. Chem.* **271**, 29205–29210.

Hoyal, C. R., Girón-Calle, J., and Forman, H. J. (1998). The alveolar macrophage as a model of calcium signaling in oxidative stress. *J. Toxicol. Environ. Health B Crit. Rev.* **1**, 117–134.

Igbavboa, U., Zwizinski, C. W., and Pfeiffer, D. R. (1989). Release of mitochondrial matrix proteins through a Ca^{2+}-requiring, cyclosporin-sensitive pathway. *Biochem. Biophys. Res. Commun.* **161**, 619–625.

Inesi, G., and Sagara, Y. (1992). Thapsigargin, a high affinity and global inhibitor of intracellular Ca^{2+} transport ATPases. *Arch. Biochem. Biophys.* **298**, 313–317.

Jacobson, M. D. (1996). Reactive oxygen species and programmed cell death. *Trends Biochem. Sci.* **21**, 83–86.

Jiang, S., Chow, S. C., Nicotera, P., and Orrenius, S. (1994). Intracellular Ca^{2+} signals activate apoptosis in thymocytes: Studies using the Ca^{2+}-ATPase inhibitor thapsigargin. *Exp. Cell Res.* **212**, 84–92.

Kao, J.P. (1994). Practical aspects of measuring [Ca^{2+}] with fluorescent indicators. *Methods Cell Biol.* **40**, 155–181.

Kass, G. E., Duddy, S. K., Moore, G. A., and Orrenius, S. (1989). 2,5-Di-(tert-butyl)-1,4-benzohydroquinone rapidly elevates cytosolic Ca^{2+} concentration by mobilizing the inositol 1,4,5-trisphosphate-sensitive Ca^{2+} pool. *J. Biol. Chem.* **264**, 15192–15198.

Kass, G. E., Juedes, M. J., and Orrenius, S. (1992). Cyclosporin A protects hepatocytes against prooxidant-induced cell killing. A study on the role of mitochondrial Ca^{2+} cycling in cytotoxicity. *Biochem. Pharmacol.* **44**, 1995–2003.

Kerr, J. F., Wyllie, A. H., and Currie, A. R. (1972). Apoptosis: A basic biological phenomenon with wide-ranging implications in tissue kinetics. *Br. J. Cancer* **26**, 239–257.

Klee, C. B. (1988). Ca^{2+}-dependent phospholipid- (and membrane-) binding proteins. *Biochemistry* **27**, 6645–6653.

Kruman, I., Guo, Q., and Mattson, M. P. (1998). Calcium and reactive oxygen species mediate staurosporine-induced mitochondrial dysfunction and apoptosis in PC12 cells. *J. Neurosci. Res.* **51**, 293–308.

Lam, M., Dubyak, G., Chen, L., Nunez, G., Miesfeld, R. L., and Distelhorst, C. W. (1994). Evidence that BCL-2 represses apoptosis by regulating endoplasmic reticulum-associated Ca^{2+} fluxes. *Proc. Natl. Acad. Sci. U.S.A.* **91**, 6569–6573.

Liu, X., Kim, C.N., Yang, J., Jemmerson, R., and Wang, X. (1996). Induction of apoptotic program in cell-free extracts: requirement for dATP and cytochrome c. *Cell (Cambridge, Mass.)* **86**, 147–157.

Liu, H., Bowes, R. C., 3rd, van de Water B., Sillence, C., Nagelkerke, J. F., and Stevens, J. L. (1997). Endoplasmic reticulum chaperones GRP78 and calreticulin prevent oxidative stress, Ca^{2+} disturbances, and cell death in renal epithelial cells. *J. Biol. Chem.* **272**, 21751–21759.

Livingston, F. R., Lui, E. M., Loeb, G. A., and Forman, H. J. (1992). Sublethal oxidant stress induces a reversible increase in intracellular calcium dependent on NAD(P)H oxidation in rat alveolar macrophages. *Arch. Biochem. Biophys.* **299**, 83–91.

Lohret, T. A., Murphy, R. C., Drgoñ, T., and Kinnally, K. W. (1996). Activity of the mitochondrial multiple conductance channel is independent of the adenine nucleotide translocase. *J. Biol. Chem.* **271**, 4846–4849.

Lötscher, H. R., Winterhalter, K. H., Carafoli, E., and Richter, C. (1980). Hydroperoxide-induced loss of pyridine nucleotides and release of calcium from rat liver mitochondria. *J. Biol. Chem.* **255**, 9325–9330.

Lytton, J., Westlin, M., and Hanley, M. R. (1991). Thapsigargin inhibits the sarcoplasmic or endoplasmic reticulum Ca-ATPase family of calcium pumps. *J. Biol. Chem.* **266**, 17067–17071.

Maki, A., Berezesky, I. K., Fargnoli, J., Holbrook, N. J., and Trump, B. F. (1992). Role of $[Ca^{2+}]_i$ in induction of c-fos, c-jun, and c-myc mRNA in rat PTE after oxidative stress. *FASEB J.* **6**, 919–924.

Matsuura, H., and Shattock, M. J. (1991). Membrane potential fluctuations and transient inward currents induced by reactive oxygen intermediates in isolated rabbit ventricular cells. *Circ. Res.* **68**, 319–329.

Meldolesi, J., and Pozzan, T. (1998). The endoplasmic reticulum Ca^{2+} store: A view from the lumen. *Trends Biochem. Sci.* **23**, 10–14.

Moss, S. E. (1997). Annexins. *Trends Cell Biol.* **7**, 87–89.

Murphy, J. K., Hoyal, C. R., Livingston, F. R., and Forman, H. J. (1995). Modulation of the alveolar macrophage respiratory burst by hydroperoxides. *Free Radical Biol. Med.* **18**, 37–45.

Newton, A. C. (1995). Protein kinase C: Structure, function, and regulation. *J. Biol. Chem.* **270**, 28495–28498.

Nicoll, D. A., Quednau, B. D., Qui, Z., Xia, Y. R., Lusis, A. J., and Philipson, K. D. (1996). Cloning of a third mammalian Na^+-calcium exchanger, NCX3. *J. Biol. Chem.* **271**, 24914–24921.

Nicotera, P., Moore, M., Mirabelli, F., Bellomo, G., and Orrenius, S. (1985). Inhibition of hepatocyte plasma membrane Ca^{2+}-ATPase activity by menadione metabolism and its restoration by thiols. *FEBS Lett.* **181**, 149–153.

Nieminen, A. L., Byrne, A. M., Herman, B., and Lemasters, J. J. (1997). Mitochondrial permeability transition in hepatocytes induced by t-BuOOH: NAD(P)H and reactive oxygen species. *Am. J. Physiol.* **272**, C1286–C1294.

Orrenius, S., and Bellomo, G. (1991). Metabolic regulation in oxidative stress. In "Oxidative Damage and Repair" (K. J. A. Davies, ed.), pp. 449–457. Pergamon, Pasadena, CA.

Orrenius, S., Burkitt, M. J., Kass, G. E., Dypbukt, J. M., and Nicotera, P. (1992). Calcium ions and oxidative cell injury. *Ann. Neurol.* **32** S33–S42.

Petit, P. X., Susin, S. A., Zamzami, N., Mignotte, B., and Kroemer, G. (1996). Mitochondria and programmed cell death: Back to the future. *FEBS Lett.* **396**, 7–13.

Pozzan, T., Rizzuto, R., Volpe, P., and Meldolesi, J. (1994). Molecular and cellular physiology of intracellular calcium stores. *Physiol. Rev.* **74**, 595–636.

Qin, S., Inazu, T., Takata, M., Kurosaki, T., Homma, Y., and Yamamura, H. (1996). Cooperation of tyrosine kinases p72^{syk} and p53/56^{lyn} regulates calcium mobilization in chicken B cell oxidant stress signaling. *Eur. J. Biochem.* **236**, 443–449.

Randriamampita, C., and Tsien, R. Y. (1993). Emptying of intracellular Ca^{2+} stores releases a novel small messenger that stimulates Ca^{2+} influx. *Nature (London)* **364**, 809–814.

Raynal, P., and Pollard, H. B. (1994). Annexins: The problem of assessing the biological role for a gene family of multifunctional calcium- and phospholipid-binding proteins. *Biochim. Biophys. Acta* **1197**, 63–93.

Renard, D. C., Seitz, M. B., and Thomas, A. P. (1992). Oxidized glutathione causes sensitization of calcium release to inositol 1,4,5-trisphosphate in permeabilized hepatocytes. *Biochem. J.* **284**, 507–512.

Richter, C., and Kass, G. E. (1991). Oxidative stress in mitochondria: Its relationship to cellular Ca^{2+} homeostasis, cell death, proliferation, and differentiation. *Chem. Biol. Interact.* **77**, 1–23.

Rohn, T. T., Hinds, T. R., and Vincenzi, F. F. (1996). Inhibition of Ca^{2+}-pump ATPase and the Na^+/K^+-pump ATPase by iron-generated free radicals. Protection by 6,7-dimethyl-2,4-DI-l-pyrrolidinyl-7H-pyrrolo[2,3-d] pyrimidine sulfate (U-89843D), a potent, novel, antioxidant/free radical scavenger. *Biochem. Pharmacol.* **51**, 471–476.

Rooney, T. A., Renard, D. C., Sass, E. J., and Thomas, A. P. (1991). Oscillatory cytosolic calcium waves independent of stimulated inositol 1,4,5-trisphosphate formation in hepatocytes. *J. Biol. Chem.* **266**, 12272–12282.

Rosse, T., Olivier, R., Monney, L., Rager, M., Conus, S., Fellay, I., Jansen, B., and Borner, C. (1998). Bcl-2 prolongs cell survival after Bax-induced release of cytochrome c. *Nature (London)* **391**, 496–499.

Roveri, A., Coassin, M., Maiorino, M., Zamburlini, A., van Amsterdam, F. T., Ratti, E., and Ursini, F. (1992). Effect of hydrogen peroxide on calcium homeostasis in smooth muscle cells. *Arch. Biochem. Biophys.* **297**, 265–270.

Scarlett, J. L., and Murphy, M. P. (1997). Release of apoptogenic proteins from the mitochondrial intermembrane space during the mitochondrial permeability transition. *FEBS Lett.* **418**, 282–286.

Scarpa, A., and Azzone, G. F. (1970). The mechanism of ion translocation in mitochondria. 4. Coupling of K^+ efflux with Ca^{2+} uptake. *Eur. J. Biochem.* **12**, 328–335.

Schilling, W. P., and Elliott, S. J. (1992). Ca^{2+} signaling mechanisms of vascular endothelial cells and their role in oxidant-induced endothelial cell dysfunction. *Am. J. Physiol.* **262** (6 Pt. 2), H1617–H1630.

Slater, A. F., Nobel, C. S., and Orrenius, S. (1995a). The role of intracellular oxidants in apoptosis. *Biochim. Biophys. Acta* **1271**, 59–62.

Slater, A. F., Stefan, C., Nobel, I., van den Dobbelsteen, D. J., and Orrenius, S. (1995b). Signaling mechanisms and oxidative stress in apoptosis. *Toxicol. Lett.* **82–83**, 149–153.

Somlyo, A. P. (1985). The messenger across the gap. *Nature (London)* **316**, 298–299.
Steinman, H. M. (1995). The Bcl-2 oncoprotein functions as a pro-oxidant. *J. Biol. Chem.* **270**, 3487–3490.
Stone, D., Darley-Usmar, V., Smith, D. R., and O'Leary, V. (1989). Hypoxia-reoxygenation induced increase in cellular Ca^{2+} in myocytes and perfused hearts: The role of mitochondria. *J. Mol. Cell. Cardiol.* **21**, 963–973.
Stoyanovsky, D. A., Salama, G., and Kagan, V. E. (1994). Ascorbate/iron activates Ca^{2+}-release channels of skeletal sarcoplasmic reticulum vesicles reconstituted in lipid bilayers. *Arch. Biochem. Biophys.* **308**, 214–221.
Suzuki, Y. J., and Ford, G. D. (1991). Inhibition of Ca^{2+}-ATPase of vascular smooth muscle sarcoplasmic reticulum by reactive oxygen intermediates. *Am. J. Physiol.* **261**, H568–H574.
Suzuki, Y. J., and Ford, G. D. (1992). Superoxide stimulates IP_3-induced Ca^{2+} release from vascular smooth muscle sarcoplasmic reticulum. *Am. J. Physiol.* **262**, H114–H116.
Suzuki, Y. J., Forman, H. J., and Sevanian, A. (1997). Oxidants as stimulators of signal transduction. *Free Radical Biol. Med.* **22**, 269–285.
Vercesi, A. E., Kowaltowski, A. J., Grijalba, M. T., Meinicke, A. R., and Castilho, R. F. (1997). The role of reactive oxygen species in mitochondrial permeability transition. *Biosci. Rep.* **17**, 43–52.
Volk, T., Hensel, M., and Kox, W. J. (1997). Transient Ca^{2+} changes in endothelial cells induced by low doses of reactive oxygen species: Role of hydrogen peroxide. *Mol. Cell. Biochem.* **171**, 11–21.
Wang, S. Y., Clague, J. R., and Langer, G. A. (1995). Increase in calcium leak channel activity by metabolic inhibition or hydrogen peroxide in rat ventricular myocytes and its inhibition by polycation. *J. Mol. Cell. Cardiol.* **27**, 211–222.
Xiong, H., Buck, E., Stuart, J., Pessah, I. N., Salama, G., and Abramson, J. J. (1992). Rose bengal activates the Ca^{2+} release channel from skeletal muscle sarcoplasmic reticulum. *Arch. Biochem. Biophys.* **292**, 522–528.
Xu, K. Y., Zweier, J. L., and Becker, L. C. (1997). Hydroxyl radical inhibits sarcoplasmic reticulum Ca^{2+}-ATPase function by direct attack on the ATP binding site. *Circ. Res.* **80**, 76–81.

6 Induction of Protein Tyrosine Phosphorylation by Oxidative Stress and Its Implications

Gary L. Schieven
Bristol-Myers Squibb Pharmaceutical Research Institute
Princeton, New Jersey 08543

Oxidative stress has been shown to activate tyrosine phosphorylation signal transduction pathways in a wide variety of cell types. These pathways are normally under the control of ligand binding to receptors. Oxidative stress is able to bypass this normal receptor control to activate pathways that are often the central regulatory pathways for cellular proliferation and functional responses. Ionizing radiation, ultraviolet (UV) radiation, and chemical agents can activate these tyrosine phosphorylation pathways. In the case of exposure of B cells to ionizing radiation, activation of the pathways leads to the induction of programmed cell death. In other cases, activation of signal pathways may induce a pattern of gene expression that would be protective. Phosphotyrosine phosphatases (PTP) are highly sensitive to inhibition by oxidative stress, and inhibition of these enzymes leads to the accumulation of phosphotyrosine signaling. A second potential mechanism by which oxidative stress may activate the pathways is by the induction of receptor dimerization or aggregation of signaling proteins.

I. INTRODUCTION

Oxidative stress has been defined as the oxidative damage inflicted on biological systems by reactive oxygen species (ROS) (Sies, 1986). The damaging effect of oxidative stress on cellular components leads cells to either make defensive responses to protect themselves or, in the most extreme cases, undergo apoptosis. Although DNA damage is a well-known consequence of exposure to oxidative stress induced by radiation or chemical agents, studies indicate that tyrosine phosphorylation signal pathways mediate many important cellular responses to oxidative stress. Tyrosine phosphorylation is a central regulatory pathway for many cell types. Activation of these pathways by oxidative stress outside of normal receptor control can lead to productive responses to repair damage or, alternatively, may lead to uncoordinated activation of signal pathways in a chaotic fashion leading to cell death. This chapter reviews the types of signaling that can be activated by various forms of oxidative stress and the examines underlying molecular mechanisms by which oxidative stress may induce signaling.

Although any cell types might conceivably be exposed to oxidative stress, much research has focused on endothelial cells and on lymphocytes. These two cell types are among those most likely to be exposed to oxidative stress arising from environmental agents, ischemia–reperfusion injury, inflammation, or radiation. Cell surface receptors that regulate the growth and function of these cells transmit signals to the nucleus via signal transduction pathways that frequently have tyrosine phosphorylation as their central theme. In each case of tyrosine phosphorylation-mediated signal transduction, tyrosine kinases are the proximal steps in a signal cascade. Receptor tyrosine kinases, such as the receptors for epidermal growth factor (EGF), platelet derived growth factor (PDGF), and insulin, have intrinsic tyrosine kinase activity in intracellular domains. Other receptors, such as T-cell and B-cell antigen receptors, or cytokine receptors, such as interleukin (IL)-2 or erythropoietin receptors, lack intrinsic tyrosine kinase activity and instead associate with nonreceptor tyrosine kinases (Weiss and Littman, 1994). Numerous studies have shown that oxidative stress can activate signal pathways regulated by both types of receptors, resulting in ligand-independent signaling outside of normal receptor control.

II. TYROSINE PHOSPHORYLATION SIGNAL PATHWAYS

Both the growth factor receptors with intrinsic tyrosine kinase activity such as EGF or PDGF receptor and the receptors that need to use nonreceptor tyrosine kinases have several features in common. First, the receptors are activated by dimerization. Binding of ligands induces dimerization, either by ligands that themselves are dimers, such as the case with PDGF, or dimerization induced by conformation change following ligand binding (Ullrich and Schlessinger, 1990). Dimerization induces activation of the tyrosine kinases. In the case of growth factor receptors such as EGF receptor, the kinase domains phosphorylate each other in *trans* (Ullrich and Schlessinger, 1990).

However, many receptors lack intrinsic tyrosine kinase activity but nonetheless activate tyrosine phosphorylation signal pathways. For these receptors, nonreceptor tyrosine kinases are activated by the aggregation of receptors into dimers or multimers and phosphorylate receptor subunits. The T-cell receptor is a particularly well-characterized example of this type of receptor, and T-cell receptor signaling is illustrated in Fig. 1. The first kinases to be activated by the T-cell receptor in complex with the accessory molecule CD4 are the Src-family kinases Lck and Fyn. Src-family kinases are frequently the signaling kinases proximal to a receptor. The activation of these kinases by the T-cell receptor leads to the phosphorylation of pairs of tyrosines in immunoreceptor tyrosine activation motifs (ITAM) sequences present in invariant chains of the T-cell receptor (Weiss and Littman, 1994). Lck is essential for T-cell activation, and Fyn, while not essential, also contributes to activation. These ITAM tyrosine phosphorylations lead to the recruitment of a second class of nonreceptor tyrosine kinases: Syk and ZAP-70 kinases. In T cells, ZAP-70 plays the major role, whereas Syk is important in B cells and myeloid cells. The ZAP-70 and Syk kinases bind to the pairs of phosphotyrosine residues in the ITAM sequences by tandem

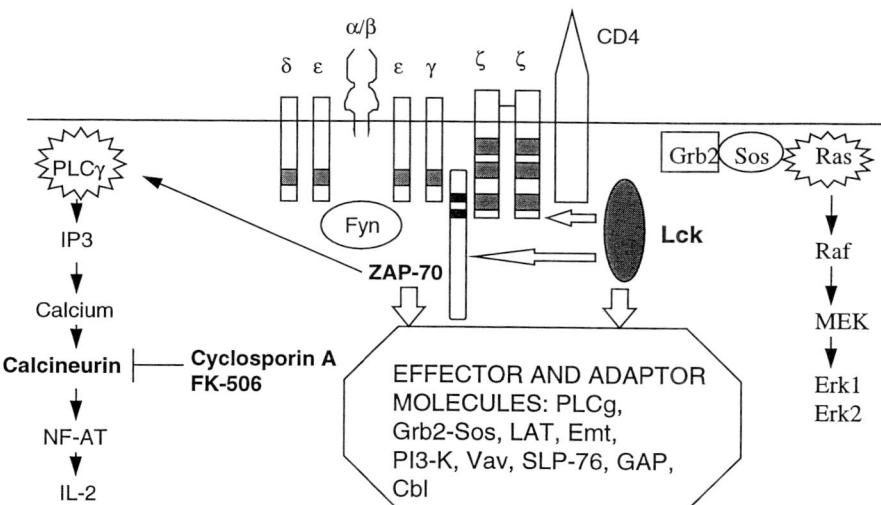

Figure 1 Signaling by the T-cell receptor in CD4$^+$ T cells. The gray boxes in the invariant chains (ε, δ, γ, and ζ) of the T-cell receptor represent ITAM sequences. The darker shaded boxes in ZAP-70 represent SH2 domains.

N-terminal SH2 domains. SH2 domains bind phosphotyrosine in the context of specific amino acid sequences. Lck then phosphorylates ZAP-70, leading to ZAP-70 activation. Autophosphorylation of ZAP-70 leads to the generation of multiple phosphotyrosine sites that can interact with the SH2 domains of a variety of downstream signaling molecules. These phosphotyrosine–SH2 interactions lead to the generation of signaling complexes that contain downstream signaling molecules such as phospholipase Cγ (PLCγ), -phosphatidylinositol 3-kinase (PI-3 kinase), and adapter molecules such as Grb2. The interaction of SH3 domains of signaling proteins with proline-rich regions also aids in the formation of signaling complexes (Pawson, 1995).

The combined action of Lck, Fyn, and ZAP-70 results in the phosphorylation and activation of these signal molecules. One important pathway is initiated by Grb2, which leads to the activation of the Ras-Raf-MAP kinase signal cascade via recruitment of Sos, the GTP exchange protein for Ras. This cascade leads to the activation of the mitogen-activated protein (MAP) kinases Erk1 and Erk2. Another key signal cascade in T cells is the phosphorylation and activation of PLCγ, resulting in the generation of inositol-trisphosphate that results in intracellular calcium mobilization. The increased calcium results in the activation of the calcium-sensitive phosphatase calcineurin, which dephosphorylates the nuclear factor of activated T cells (NF-AT). NF-AT translocates to the nucleus to induce transcription of IL-2, the major autocrine growth factor for T cells. This pathway is essential for T-cell activation, and the calcineurin step is the target of the important immunosuppressive drugs cyclosporin A and FK-506.

Activation of these multiple tyrosine kinase-dependent signaling pathways leads to the activation of transcription factors such as AP-1, NF-κB, NF-AT, and TCF/Elk-1. In the case of cytokine receptors such as IL-2 and interferon-γ receptors, another class of nonreceptor tyrosine kinases known as the JAK family kinases are activated following receptor dimerization (Ihle, 1995). The JAK kinases phosphorylate STAT (signal transducers and activators of transcription) proteins. STAT proteins each contain an SH2 domain that recognizes phosphotyrosine on STAT proteins of the same or other types, leading to the formation of homo- or heterodimers of STAT proteins. These dimers migrate to the nucleus and activate transcription directly.

Signal complex formation by the receptor tyrosine kinases is quite similar to that of the receptors that lack intrinsic tyrosine kinase activity. Receptors in a dimer phosphorylate each other in *trans* at multiple sites, leading to multiple SH2-docking sites to recruit many of the downstream signaling molecules described earlier for the T-cell receptor (Ullrich and Schlessinger, 1990). Among the molecules recruited are nonreceptor tyrosine kinases such as Src, which is essential for PDGF receptor-induced mitogenesis (Twamley-

Sein *et al.*, 1993). Similarly, JAK kinases can also be recruited to receptor tyrosine kinases to aid in downstream signaling.

Interestingly, receptors that make use of tyrosine phosphorylation signal pathways can also induce the generation of reactive oxygen species. PDGF receptor activation by PDGF induces the production of hydrogen peroxide (Sundarensan *et al.*, 1995). This hydrogen peroxide is required for productive PDGR receptor signaling, as overexpression of catalase can block cell responses (Sundarensan *et al.*, 1995). Similarly, EGF binding to the EGF receptor induces production of hydrogen peroxide by a mechanism requiring the tyrosine kinase activity of the receptor (Bae *et al.*, 1997). The elimination of hydrogen peroxide production by catalase also inhibited the extent of cellular tyrosine phosphorylation, including the phosphorylation of key substrates such as PLCγ. In another example where the generation of hydrogen peroxide contributes to signaling, changes in integrin-mediated adhesion in synovial fibroblasts have been found to lead to the generation of ROS via a Rac1-dependent pathway (Kheradmand *et al.*, 1998), resulting in activation of NF-κB, a transcription factor whose activation by many stimuli is dependent on ROS (Schreck *et al.*, 1991).

What are the implications of the action of oxidative stress on these signal pathways? Oxidative stress can cause damage to cellular components, and thus one effect that would be expected is damage to parts of the signal apparatus, blocking signaling and impairing cell responses to normal ligands. One important example of this is the effect of oxidative stress on JAK family kinases. These kinases are inhibited by oxidative stress (Duhe *et al.*, 1998), leading to faulty signaling through key cytokine receptors important in the immune response such as IL-2 and IL-4. This may be one mechanism by which oxidative stress suppresses immune responses.

Surprisingly, oxidative stress can also activate tyrosine kinase signal pathways. This has been observed for a variety of different agents that induce oxidative stress, including ionizing radiation, ultraviolet radiation, and chemical agents such as hydrogen peroxide. Because each of these agents can act differently on cells, they will be considered separately. The action of each of these agents provides clues as to the molecular pathways that oxidative stress employs to activate signal transduction by tyrosine phosphorylation pathways in vertebrate cells.

III. IONIZING RADIATION

Ionizing radiation is well known to kill cells via the induction of DNA damage. In animals, certain cell types are much more sensitive than others, and this difference in sensitivity is taken advantage of for radiation therapy of malignancies. For example, rapidly dividing cells are known to be sensitive. One of the most successful uses of radiation therapy is for the treatment of B-cell leukemia, followed by bone marrow transplantation. Studies have shown that ionizing radiation induces cellular tyrosine phosphorylation in leukemic B-cell precursors, leading to the induction of programmed cell death. Treatment with tyrosine kinase inhibitors such as genistein and herbimycin A inhibited the induction of cellular tyrosine phosphorylation and prevented cell death in multiple leukemic B-cell precursor cell lines (Uckun *et al.*, 1992), demonstrating that the activation of tyrosine phosphorylation signal pathways plays an important role in radiation-induced death in these cells. In contrast, treatment with the phosphotyrosine phosphatase inhibitor orthovanadate greatly enhanced radiation-induced tyrosine phosphorylation and the induction of apoptosis in the cells (Uckun *et al.*, 1992). Radiation-induced tyrosine phosphorylation was found to induce the activation of multiple serine/threonine kinases, including protein kinase C (PKC) (Uckun *et al.*, 1993). The activation of PKC by radiation increases the expression of immediate early genes such as c-*jun* and *EGRI* (Sherman *et al.*, 1990; Hallahan *et al.*, 1991). Tyrosine kinase inhibitors such as herbimycin and genistein can block the radiation-induced activation of PKC and other serine kinases (Uckun *et al.*, 1992), as well as the radiation-induced activation of c-*jun*, c-*fos*, c-*myc*, and c-*Ha-ras* (Chae *et al.*, 1993; Prasad *et al.*, 1995). Subsequent studies have shown that the Src-family kinases Lck and Lyn can be activated by ionizing radiation. Activation of Src-family kinases is a key step in the activation of downstream signaling protein tyrosine kinases (PTK) such as Btk (Saouaf *et al.*, 1994; Li *et al.*, 1997), which is essential in B-cell development (Desiderio, 1997). Btk is activated by ionizing radiation, and chicken DT-40 B cells deficient in BTK are resistant to radiation-induced death (Uckun *et al.*, 1996), indicating a key role for this PTK in radiation-induced apoptosis in B cells.

The Abl tyrosine kinase is also activated by ionizing radiation, leading to the activation of the stress-activated protein kinase JNK (Kharbanda *et al.*, 1995). Although the Abl tyrosine kinase is known to phosphorylate nuclear proteins, studies indicate that a key substrate of Abl in irradiated cells is PI-3 kinase (Yuan *et al.*, 1997). PI-3 kinase activation is an important element of growth factor receptor signaling and is important for cell survival. Following irradiation, Abl has been reported to phosphorylate the p85 subunit of PI-3 kinase and inhibit its function both *in vitro* and in irradiated cells (Yuan *et al.*, 1997). Interestingly, a stress-activated protein with antioxidant properties, known as PAG, has been identified as an Abl SH3-binding protein and a physiological inhibitor of Abl tyrosine kinase activity (Wen and Van Etten, 1998). Inhibition of Abl may thus be a defensive measure taken by cells in response to oxidative stress.

Taken together, these results indicate that radiation activates tyrosine kinases in leukemic B-cell precursors, leading to a signal cascade that culminates in the induction of programmed cell death. PTP appear to be able to limit this response. An important role for PTP for suppressing pro-apoptotic signals, even in the absence of radiation, is suggested by the finding that the PTP inhibitor bis (maltolato) oxovanadium(IV) (BMOV), which is highly active against B cells, activates signaling pathways involving Syk and PLCγ that induce apoptosis in leukemic immature B cells (Schieven et al., 1995).

IV. ULTRAVIOLET RADIATION

Ultraviolet radiation induces the expression of a variety of immediate early genes in what is known as the mammalian UV response (Ronai et al., 1990; Holbrook and Fornace, 1991; Karin and Herrlich, 1989). The UV response includes the activation of AP-1, NF-κB, and TCF/Elk-1 plus the transcriptional activation of *fos* and *jun* (Sachsenmaier et al., 1994). Although ultraviolet radiation is well known to induce DNA damage by the production of thymidine dimers, studies have shown that UV activates tyrosine kinase signal pathways and that this activation is responsible for much of the UV-induced gene expression observed in the mammalian UV response.

Ultraviolet radiation can activate NF-κB in enucleated cells (Devary et al., 1993), demonstrating that DNA damage is not required for this event. The mammalian UV response in HeLa cells has been found to require active Src-family kinases plus active Ras and Raf (Devary et al., 1992). UV radiation activates the EGF receptor in A431 cells directly (Schieven et al., 1994), and expression of EGF receptor and other receptor tyrosine kinases is essential for the response in cells such as HeLa cells (Sachsenmaier et al., 1994). Because Src and related PTK can bind to and are activated by receptor tyrosine kinases, it is likely that Src activation is downstream of receptor tyrosine kinase activation.

The induction of signals by UV is not limited to cells expressing receptor tyrosine kinases. Lymphocytes are highly sensitive to specific wavelengths of UV, particularly UVC (252 nm) and UVB (302 nm), but the longer wavelength UVA radiation gives little biological response in these cells (Kripke, 1984). UV is immunosuppressive *in vivo,* giving improved graft survival in a variety of transplantation models (Pamphilon et al., 1991). However, UV activates signal pathways in lymphocytes that are normally under the control of the T-cell receptor. UVB and UVC both induce cellular tyrosine phosphorylation in T and B cells in a pattern similar to that observed for antigen receptor stimulation (Schieven et al., 1993b, 1994; Schieven and Ledbetter, 1993). Normal human peripheral blood T lymphocytes as well as T cell lines respond to UV irradiation with strong calcium signaling, a key step in T-cell activation. However, B cells and other leukocytes isolated from blood do not give a calcium response to UV irradiation (Schieven *et al.*, 1993b; Schieven and Ledbetter, 1993). UV induces the tyrosine phosphorylation of PLCγ-1 in T cells, activating calcium signaling by the same pathway as in normal antigen receptor signaling, as in both cases calcium signals can be blocked by tyrosine kinase inhibitors (Schieven *et al.*, 1993b). UV induces activation of the key tyrosine kinases Syk in B cells (Schieven *et al.*, 1993b) and ZAP-70 in T cells (Schieven *et al.*, 1994). These kinases are essential for normal antigen receptor signaling. Thus UV activates the precise signal pathways normally under receptor control.

Studies demonstrate that UV activates these pathways by acting on the receptors themselves. In T cells, the UV induction of tyrosine kinase activation and downstream cellular responses requires the surface expression of antigen receptor and CD45 (Schieven *et al.*, 1994). This has been demonstrated using cell lines deficient in T-cell receptor expression and through the use of internalizing antibodies that remove the T-cell receptor from the cell surface. In either case, lack of T-cell receptor surface expression prevents UV induction of cellular tyrosine phosphorylation, ZAP-70 tyrosine kinase activation, or calcium signaling. Thus UV activates signal pathways normally under receptor control by acting on the receptors themselves, either directly or indirectly.

Confocal microscopy studies of cells irradiated with UV have revealed that UV irradiation induces clustering of receptors for EGF, IL-1, and tumor necrosis factor (TNF) (Rosette and Karin, 1996). These receptors are known to be activated by dimerization. The UV irradiation appeared to cause higher order aggregation than dimerization, as the aggregation was visible by light microscopy. The mechanism by which UV might induce such aggregation is unknown, but given the disparate nature of the various cell types and receptors that can be activated, from T-cell receptors to EGF receptors, the common element must rely on physical properties of transmembrane proteins in a lipid bilayer. Possible mechanisms include lipid peroxidation, inducing protein aggregation and direct absorption of UV light by the proteins.

V. CHEMICAL OXIDANTS

Chemical agents can also induce the activation of tyrosine phosphorylation signal pathways. One of the most well-studied agents is hydrogen peroxide. Treatment of lymphocytes with 1–10 mM hydrogen peroxide induces cellular tyrosine phosphorylation in a pattern quite similar to antigen receptor signaling (Schieven *et al.*, 1993c). The underlying cause of this similarity was revealed by the finding that hydrogen peroxide treatment leads to the activation of the Syk-family kinases ZAP-70 and Syk in T cells and B cells, respectively (Schieven *et al.*, 1993a, 1994). The effect of hydrogen peroxide on Syk-family kinases is indirect, as direct treatment of the immunoprecipitated Syk does not activate it (Schieven *et al.*, 1993a). Additional studies of various Syk mutants expressed in Syk-deficient cells indicate that Syk activation induced by oxidative stress is due to Syk autophosphorylation as opposed to phosphorylation of Syk by other tyrosine kinases (Qin *et al.*, 1998). Activation of the Syk-family kinases leads to the activation of downstream signaling pathways such as calcium mobilization via the tyrosine phosphorylation of PLCγ (Schieven *et al.*, 1993a), leading to increased inositol trisphosphate production (Schieven *et al.*, 1993c).

Tyrosine kinase activation may be observed not only by the experimental addition of exogenous hydrogen peroxide to cells, but also as a result of endogenous production of hydrogen peroxide. Activation of the neutrophil respiratory burst has been reported to result in increased cellular tyrosine phosphorylation (Fialkow *et al.*, 1993). Induction of hydrogen peroxide production by nonhydrolyzable GTP analogs to bypass cell surface receptors (and thus avoid ligand-dependent tyrosine kinase activation) has been reported to induce the activation of multiple tyrosine kinases, including the Src-family kinase Hck, Syk, and Btk (Brumell *et al.*, 1996). The activation of Hck was indirect, as direct exposure of immunoprecipitated Hck to oxidizing agents did not activate the kinase. Human neutrophils produce hydrogen peroxide in response to adhesion, which is accompanied by the activation of the Fgr and Lyn members of the Src-family of tyrosine kinases in these cells (Yan and Berton, 1996). Elimination of hydrogen peroxide by catalase or inhibition of its production by NADPH oxidase inhibitors reduced Fgr and Lyn activation, and cells from patients with chronic granulomatous disease that are deficient in hydrogen peroxide production showed reduced Fgr and Lyn activation.

Detailed studies of the activation of the Src-family kinase Lck by hydrogen peroxide suggest that oxidative stress may be able to activate this key T-cell kinase by methods outside of previously described pathways under normal receptor control. Like all other Src-family kinases, Lck contains two major sites of tyrosine phosphorylation: Tyr-394, whose phosphorylation is activating for the kinase, and the C-terminal Tyr-505, whose phosphorylation is inhibitory when the kinase's own SH2 domain binds the phosphorylated Tyr-505 site. Treatment of cells containing Lck results in the activation of Lck kinase activity and phosphorylation of the main autophosphorylation site, Tyr-394 (Hardwick and Sefton, 1995). Mutations at this site abrogated the activation of the kinase by hydrogen peroxide. Interestingly, hydrogen peroxide treatment of cells also induced phosphorylation of Tyr-394 in a kinase inactive mutant of Lck, demonstrating that hydrogen peroxide can induce other tyrosine kinases to phosphorylate Lck. In addition, hydrogen peroxide treatment also induced activation of Lck that was phosphorylated at the negative regulatory site, Tyr-505 (Hardwick and Sefton, 1997). Src family kinases phosphorylated at this site can be activated if their SH2 or SH3 domains bind to a high-affinity ligand that competes effectively with the intracellular interactions of these domains that hold the kinase in an inactive conformation, as has been revealed by X-ray crystallography (Moarefi et al., 1997; Xu et al., 1997). This result suggests that hydrogen peroxide can induce the formation of such ligands that recruit Lck into a signaling complex with the activating ligand on another protein, which must be either a phosphotyrosine sequence or an exposed proline-rich binding sequence for the SH3 domain. Surprisingly, mutant forms of Lck defective in the sites of fatty acylation that permit association with the inner leaf of the plasma membrane could still be activated by the treatment of cells with hydrogen peroxide, even though the mutant forms of Lck could no longer respond to receptor stimulation (Yurchak et al., 1996).

Activation of downstream signaling kinases by hydrogen peroxide can occur by several mechanisms. Hydrogen peroxide activates protein kinase C via direct tyrosine phosphorylation of PKC by the hydrogen peroxide-induced tyrosine phosphorylation of multiple PKC isoforms (Konishi et al. 1997). Activation of the mitogen-activated protein kinases such as extracellular signal-regulated protein kinases ERK1 and ERK2 by hydrogen peroxide or other oxidative stress can be accomplished by activation of a signal cascade leading from receptor or nonreceptor tyrosine kinases. In vascular smooth muscle cells, hydrogen peroxide induces complex formation of Shc-Grb2-Sos with receptor tyrosine kinase, causing activation of Ras by the GTP exchange protein SOS (Rao, 1996). This process leads to Raf activation and subsequent activation of MEK kinases that in turn induce Erk1 and Erk2 activation. These effects occur in many cell types. In Jurkat T cells, ZAP-70 is required for hydrogen peroxide-induced activation of the MAP kinases Erk1 and Erk2 (Griffith et al., 1998). In cardiac myocytes, expression of dominant-negative Ras or Raf or overexpression of the Csk negative regulatory kinase of Src inhibits Erk1 activation (Aikawa et al., 1997). Interestingly, inhibition of hydrogen peroxide-induced Erk activation by a selective kinase inhibitor increased hydrogen peroxide-induced apoptosis, suggesting that Erk activation may protect cells from the consequences of oxidative stress (Aikawa et al., 1997).

Activation of such signal pathways results in downstream activation of transcription factors. A particularly interesting example is NF-κB. Hydrogen peroxide can activate NF-κB through tyrosine kinase-dependent mechanisms in lymphocytes (Schieven *et al.*, 1993c), and NF-κB is well known to require oxidative stress for its activation by many stimuli (Schreck *et al.*, 1991). However, studies suggest that tyrosine phosphorylation pathways can activate NF-κB in a manner bypassing the need for oxidative stress. Pervanadate and ischemia–reperfusion have been found to activate NF-κB by a pathway involving the tyrosine phosphorylation of IκB, which normally holds NF-κB in an inactive complex in the cytosol (Imbert *et al.*, 1996). The PTP inhibitors BMOV and pV(phen) [sodium oxodiperoxo(1,10-phenanthroline)vanadate(V)] can both activate NF-κB (Krejsa *et al.*, 1997). Strikingly, the activation of NF-κB by the PTP inhibitors BMOV or pV(phen) was not inhibited by pyrrolidine dithiocarbamate (PDTC) (Krejsa *et al.*, 1997). Similarly, the NF-κB activation by BMOV was not inhibited by a second antioxidant, N-acetylcysteine (NAC). Indeed, the activation was increased. These examples suggest that one important role of oxidative stress in NF-κB activation may be to activate tyrosine phosphorylation signal pathways by means that PTP inhibitors mimic.

Hydrogen peroxide-induced tyrosine phosphorylation signaling can lead to changes in gene expression, such as has been demonstrated for the expression of Fas (Suhara *et al.* 1998). Fas is a member of the TNF receptor family and can induce programmed cell death. Cells expressing Fas-L (Fas ligand) (Nagata and Golstein, 1998) induce activation of Fas, also known as APO-1 or CD95. Exposure of endothelial cells to hydrogen peroxide was found to induce Fas expression in a tyrosine kinase-dependent manner, as the expression could be blocked by a tyrosine kinase inhibitor and was augmented by the presence of the PTP inhibitor vanadate (Suhara *et al.*, 1998). Treatment of the hydrogen peroxide-treated cells with the anti-Fas antibody led to the induction of apoptosis (Suhara *et al.*, 1998). Thus hydrogen peroxide appears to be able to induce cell death both directly through damage to cells and indirectly by the induction of Fas expression.

How might hydrogen peroxide induce the activation of the tyrosine kinases that lead to the changes in gene expression? One important finding is that unlike UV, hydrogen peroxide does not require the expression of cell surface receptors to activate these pathways in lymphocytes. T cells deficient in CD3, and thus the T-cell receptor, could not signal in response to UV, but were still responsive to hydrogen peroxide treatment (Schieven *et al.*, 1994). ZAP-70 was still activated and calcium signaling still occurred, albeit with altered kinetics. The signaling was not identical to wild-type cells, indicating that the surface receptors contributed to the response, but they were not essential. Thus hydrogen peroxide must be able to act on targets other than cell surface receptors.

The phosphotyrosine phosphatases act in partnership with the tyrosine kinases to regulate signaling. PTP can play dual roles in both the positive and the negative regulation of signaling. The transmembrane PTP CD45 regulates signaling positively by dephosphorylating Src-family kinases in lymphocytes, permitting them to be activated (Neel, 1997). However, PTP also regulates signaling negatively, dephosphorylating kinase substrates and terminating signaling.

PTP contain a highly reactive cysteine residue in their active site that acts as the phosphoacceptor during the phosphatase reaction, forming a phosphocysteine intermediate that is hydrolyzed to release free phosphate. PTPs are extremely sensitive to oxidative stress because the active site cysteine must be in its reduced form for the enzyme to be active (Guan and Dixon, 1993). For this reason, oxidizing agents such as hydrogen peroxide are potent inhibitors of PTP (Hecht and Zick, 1992).

Inhibition of PTP is thus a likely mechanism by which oxidizing agents can activate signal pathways. Inhibition of PTP would permit the accumulation of cellular tyrosine phosphorylation. Because many tyrosine kinases, such as receptor tyrosine kinases as well as nonreceptor tyrosine kinases, are activated by tyrosine phosphorylation, this process could lead to increased kinase activation. This phenomenon has been observed with the nonoxidizing PTP inhibitor BMOV (Schieven *et al.*, 1995). Treatment of B lymphocytes with this inhibitor results in the activation of the Syk tyrosine kinase and phosphorylation of downstream substrates of Syk such as PLCγ (Schieven *et al.*, 1995).

The combination of kinase activation and phosphatase inhibition has the potential to drive total phosphorylation to high levels. One of the most striking examples of this phenomenon is the action of peroxovanadium compounds such as pervanadate. Pervanadate can induce extremely high levels of tyrosine phosphorylation in lymphocytes and other cell types (Schieven *et al.*, 1993c; Secrist *et al.*, 1993; O'Shea *et al.*, 1992). In addition to being a potent inhibitor of PTP, pervanadate activates Src-family and other tyrosine kinases in lymphocytes and other cell types (Schieven *et al.*, 1993c; Fantus *et al.*, 1989; Secrist *et al.*, 1993). Studies with the stable peroxovanadium compound pV(phen) have revealed that oxidative stress plays an important role in the activation of kinases by these compounds. The antioxidant PDTC was found to strongly inhibit the intracellular oxidation and protein tyrosine phosphorylation induced by pV(phen), but it did not affect the ability of pV(phen) to inhibit PTP such as PTP1B or CD45 (Krejsa *et al.*, 1997). This approach provided a means to separate the effects of oxidative stress and PTP inhibition for the mechanism of action of the peroxovanadium compound. In the presence of PDTC, the activation of kinases such as Syk was inhibited (Krejsa *et al.*, 1997). Syk could respond normally to antigen receptor stimulation in the presence of PDTC, demonstrating that the antioxidant did not interfere with normal functioning of the enzyme or the signal pathways regulating its function (Krejsa *et al.*, 1997). These results indicate that peroxovanadium compounds do not necessarily achieve all their effects simply as a consequence of PTP inhibition, but rather the effects of oxidative stress can extend beyond PTP inhibition to the activation of kinases by other means. The activation is not direct, as it has been shown that immunoprecipitated tyrosine kinases are not activated directly by peroxovanadium compounds or by hydrogen peroxide (Krejsa *et al.*, 1997; Schieven *et al.*, 1993c).

VI. DIMERIZATION HYPOTHESIS

Taken together, these findings suggest that there are at least two pathways by which oxidative stress can activate signal pathways. The first is inhibition of PTP. However, the examples cited earlier demonstrate that PTP inhibition can account for only some of the effects of oxidative stress on signal pathways. A second potential pathway is the induction of protein dimerization or multimerization by oxidative stress. Clustering or aggregation of signal proteins is an essential component of intracellular signal transduction by both receptor and nonreceptor tyrosine kinases (Klemm *et al.*, 1998). Oxidative stress might induce receptor aggregation, as has been observed previously for the UV-induced activation of receptor pathways (Rosette and Karin, 1996). For example, the oxidizing agent diamide, which activates Lck in T cells (Nakamura *et al.*, 1993), has been reported to strongly induce the aggregation of membrane proteins (Kosower *et al.*, 1981). However, dimerization as a means of activation is not limited to transmembrane receptors. Induction of such protein aggregates by oxidative stress would be expected to activate the signal pathways by mimicking the aggregation of intracellular proteins into signal complexes that normally occurs on stimulation by ligands.

This process would not be independent of PTP inhibition, but might actually contribute to inhibition of the transmembrane PTP. While dimerization of receptor tyrosine kinases leads to their activation, dimerization of transmembrane phosphotyrosine phosphatases leads to their inhibition. The crystal structure of RPTPα has shown that the PTP can form dimers in which one member of the dimer obstructs the active site of the other with an amino-terminal wedge (Bilwes *et al.*, 1996). Studies of chimeric molecules consisting of the extracellular domain of the EGF receptor in combination with the intracellular PTP domains of CD45 have shown that the addition of EGF to induce the dimerization of the chimeric molecules led to inhibition of CD45 enzymatic activity in cells (Majeti *et al.*, 1998).

Thus oxidative stress might activate signal pathways by inducing receptor aggregation, promoting activation of signal pathways normally under ligand control. Even if the effect were small, the concomitant inhibition of PTP by oxidative stress would serve to prolong and extend the signals initiated by receptor aggregation. Receptor aggregation would be expected to have a second important effect in addition to kinase activation, the enhancement of the formation of signal complexes. The combination of kinase activation, signal complex formation, and inhibition of PTP would be expected to be synergistic, initiating and propagating signals that would be amplified more than usual in kinase signal cascades due to PTP inhibition. This hypothesis can provide a starting point for further experimental approaches to elucidate the pathways by which oxidative stress results in the activation of tyrosine phosphorylation signal pathways.

REFERENCES

Aikawa, R., Komuro, I, Yamazaki, T., Zou, Y., Kudoh, S., Tanaka, M., Shiojima, I., Hiroi, Y., and Yazaki, Y. (1997). Oxidative stress activates extracellular signal-regulated kinases through Src and Ras in cultured cardiac myocytes of neonatal rats. *J. Clin. Invest.* **100**, 1813–1821.

Bae, Y. S., Kang, S. W., Seo, M. S., Baines, I. C., Tekle, E., Chock, P. B., and Rhee, S. G. (1997). Epidermal growth factor (EGF)-induced generation of hydrogen peroxide. Role in EGF receptor-mediated tyrosine phosphorylation. *J. Biol. Chem.* **272**, 217–221.

Bilwes, A. M., den Hertog, J., Hunter, T., and Noel, J. P. (1996). Structural basis for inhibition of receptor protein-tyrosine phosphatase-alpha by dimerization. *Nature (London)* **382**, 555–559.

Brumell, J. H., Burkhardt, A. L., Bolen, J. B., and Grinstein, S. (1996). Endogenous reactive oxygen intermediates activate tyrosine kinases in human neutrophils. *J. Biol. Chem.* **271**, 1455–1461.

Chae, H. P., Jarvis, L. J., and Uckun, F. M. (1993). Role of tyrosine phosphorylation in radiation-induced activation of c-jun protooncogene in human lymphohematopoietic precursor cells. *Cancer Res.* **53**, 447–451.

Desiderio, S. (1997). Role of Btk in B cell development and signaling. *Curr. Opin. Immunol.* **9**, 534–540.

Devary, Y., Gottlieb, R. A., Smeal, T., and Karin, M. (1992). The mammalian ultraviolet response is triggered by activation of src tyrosine kinases. *Cell (Cambridge, Mass.)* **71**, 1081–1091.

Devary, Y., Rosette, C., Didonato, J. A., and Karin, M. (1993). NF-κB activation by ultraviolet light not dependent on a nuclear signal. *Science* **261**, 1442–1445.

Duhe, R. J., Evans, G. A., Erwin, R. A., Kirken, R. A., Cox, G. W., and Farrar, W. L. (1998). Nitric oxide and thiol redox regulation of Janus kinase activity. *Proc. Natl. Acad. Sci. U.S.A.* **95**, 126–131.

Fantus, I. G., Kadota, S., Deragon, G., Foster, B., and Posner, B. (1989). Pervanadate [peroxide(s) of vanadate] mimics insulin action in rat adipocytes via activation of the insulin receptor tyrosine kinase. *Biochemistry* **28**, 8864.

Fialkow, L., Chan, C. K., Grinstein, S., and Downey, G. P. (1993). Regulation of tyrosine phosphorylation in neutrophils by the NADPH oxidase. Role of reactive oxygen intermediates. *J. Biol. Chem.* **268**, 17131–17137.

Griffith, C. E., Zhang, W., and Wange, R. L. (1998). ZAP-70-dependent and independent activation of Erk in Jurkat T cells. Differences in signaling induced by H_2O_2 and CD3 cross-linking. *J. Biol. Chem.* **273**, 10771–10776.

Guan, K. L., and Dixon, J. E. (1993). Evidence of protein-tyrosine-phosphatase catalysis proceeding via a cysteine-phosphate intermediate. *J. Biol. Chem.* **266**, 17026–17030.

Hallahan, D. E., Sukhatme, V. P., Sherman, J. L., Virudachalam, S., Kufe, D., and Weichselbaum, R. R. (1991). Protein kinase C mediates x-ray inducibility of nuclear signal transducers EGR1 and JUN. *Proc. Natl. Acad. Sci. U.S.A.* **88**, 2156–2160.

Hardwick, J. S., and Sefton, B. M. (1995). Activation of the Lck tyrosine protein kinase by hydrogen peroxide requires the phosphorylation of Tyr-394. *Proc. Natl. Acad. Sci. U.S.A.* **92**, 4527–4531.

Hardwick, J. S., and Sefton, B. M. (1997). The activated form of the Lck tyrosine protein kinase in cells exposed to hydrogen peroxide is phosphorylated at both Tyr-394 and Tyr-505. *J. Biol. Chem.* **272**, 25429–25432.

Hecht, D., and Zick, Y. (1992). Selective inhibition of protein tyrosine phosphatase activities by H_2O_2 and vanadate in vitro. *Biochem. Biophys. Res. Commun.* **188**, 773–779.

Holbrook, N. J., and Fornace, A. J., Jr. (1991). Response to adversity: Molecular control of gene activation following genotoxic stress. *New Biol.* **3**, 825–833.

Ihle, J. N. (1995). The *Janus* protein tyrosine kinases in hematopoetic cytokine signaling. *Semin. Immunol.* **7**, 247–254.

Imbert, V., Peyron, J. F., Rupec, R. A., Livolsi, A., Pahl, H. L., Traenckner, E. B., Mueller-Dieckmann, C., Farahifar, D., Rossi, B., Auberger, P., Baeuerle, P. A., and Rupec, R. A., L. A., and Pahl, H. L. (1996). Tyrosine phosphorylation of IκB-α activates NF-κB without proteolytic degradation of IκB-α. *Cell (Cambridge, Mass.)* **86**, 787–798.

Karin, M., and Herrlich, P. (1989). Cis and transacting genetic elements responsible for induction of specific genes by tumor promoters, serum factors and stress. *In* "Genes and Signal Transduction in Multistage Carcinogenesis" (N. H. Colburn, ed.), pp. 415–440. Dekker, New York.

Kharbanda, S., Ren, R. B., Pandey, P., Shafman, T. D., Feller, S. M., Weichselbaum, R. R., and Kufe, D. W. (1995). Activation of the c-Abl tryosine kinase in the stress response to DNA-damaging agents. *Nature (London)* **376**, 785–788.

Kheradmand, F., Werner, E., Tremble, P., Symons, M., and Werb, Z. (1998). Role of Rac1 and oxygen radicals in collagenase-1 expression induced by cell shape change. *Science* **280**, 898–902.

Klemm, J. D., Schreiber, S. L., and Crabtree, G. R. (1998). Dimerization as a regulatory mechanism in signal transduction. *Annu. Rev. Immunol.* **16**, 569–592.

Konishi, H., Tanaka, M., Takemura, Y., Matsuzaki, H., Ono, Y., Kikkawa, U., and Nishizuka, Y. (1997). Activation of protein kinase C by tyrosine phosphorylation in response to H_2O_2. *Proc. Natl. Acad. Sci. U.S.A.* **94**, 11233–11237.

Kosower, N. S., Kosower, E. M., Zipser, Y., Faltin, Z., and Shomrat, R. (1981). Dynamic changes of red cell membrane thiol groups followed by bimane fluorescent labeling. *Biochim. Biophys. Acta* **640**, 748–759.

Krejsa, C. M., Nadler, S. G., Esselstyn, J. M., Kavanagh, T. J., Ledbetter, J. A., and Schieven, G. L. (1997). Role of oxidative stress in the action of vanadium phosphotyrosine phosphatase inhibitors. *J. Biol. Chem.* **272**, 11541–11549.

Kripke, M. L. (1984). Immunological unresponsiveness induced by ultraviolet radiation. *Immunol. Rev.* **80**, 87–102.

Li, Z., Wahl, M. I., Eguinoa, A., Stephens, L. R., Hawkins, P. T., and Witte, O. N. (1997). Phosphatidylinositol 3-kinase-gamma activates Bruton's tyrosine kinase in concert with Src family kinases. *Proc. Natl. Acad. Sci. U.S.A.* **94**, 13820–13825.

Majeti, R., Bilwes, A. M., Noel, J. P., Hunter, T., and Weiss, A. (1998). Dimerization-induced inhibition of receptor protein tyrosine phosphatase function through an inhibitory wedge. *Science* **279**, 88–91.
Moarefi, I., LaFevre-Bernt, M., Sicheri, F., Huse, M., Lee, C. H., Kuriyan, J., and Miller, W. T. (1997). Activation of the src-family tyrosine kinase Hck by SH3 domain displacement. *Nature (London)* **385**, 650–653.
Nagata, S., and Golstein, P. (1998). The Fas death factor. *Science* **267**, 1449–1456.
Nakamura, K., Hori, T., Sato, N., Sugie, K., Kawakami, T., and Yodoi, J. (1993). Redox regulation of a src family protein tyrosine kinase p56 lck in T cells. *Oncogene* **8**, 3133–3139.
Neel, B. G. (1997). Role of phosphatases in lymphocyte activation. *Curr. Opin. Immunol.* **9**, 405–420.
O'Shea, J. J., McVicar, D. W., Bailey, T. L., Burns, C., and Smyth, M. J. (1992). Activation of human peripheral blood T lymphocytes by pharmacological induction of proteintyrosine phosphorylation. *Proc. Natl. Acad. Sci. U.S.A.* **89**, 10306–10310.
Pamphilon, D. H., Alnaqdy, A. A., and Wallington, T. B. (1991). Immunomodulation by ultraviolet light: Clinical studies and biological effects. *Immunol. Today* **12**, 119–123.
Pawson, T. (1995). Protein modules and signalling networks. *Nature (London)* **373**, 573–579.
Prasad, A. V., Mohan, N., Chandrasekar, B., and Meltz, M. L. (1995). Induction of transcription of "immediate early genes" by low-dose ionizing radiation. *Radiat. Res.* **143**, 263–272.
Qin, S., Kurosaki, T., and Yamamura, H. (1998). Differential regulation of oxidative and osmotic stress induced Syk activation by both autophsophorylation and SH2 domains. *Biochemistry* **37**, 5481–5486.
Rao, G. N. (1996). Hydrogen peroxide induces complex formation of SHC-Grb2-SOS with receptor tyrosine kinase and activates Ras and extracellular signal regulated protein kinases goup of mitogen-activated kinases. *Oncogene* **13**, 713–719.
Ronai, Z. A., Lambert, E., and Weinstein, I. B. (1990). Inducible cellular responses to ultraviolet light irradiation and other mediators of DNA damage in mammalian cells. *Cell Biol. Toxicol.* **6**, 105–126.
Rosette, C., and Karin, M. (1996). Ultraviolet light and osmotic stress: Activation of the JNK cascade through multiple growth factor and cytokine receptors. *Science* **274**, 1194–1197.
Sachsenmaier, C., Radler-Pohl, A., Zinck, R., Nordheim, A., Herrlich, P., and Rahmsdorf, H. J. (1994). Involvement of growth factor receptors in the mammalian UV response. *Cell (Cambridge, Mass.)* **78**, 963–972.
Saouaf, S. J., Mahajan, S., Rowley, R. B., Kut, S. A., Fargnoli, J., Burkhardt, A. J., Tsukada, S., Witte, O. N., and Bolen, J. B. (1994). Temporal differences in the activation of three classes of non-transmembrane protein tyrosine kinases following B-cell antigen receptor surface engagement. *Proc. Natl. Acad. Sci. U.S.A.* **91**, 9524–9528.
Schieven, G. L., and Ledbetter, J. A. (1993). Ultraviolet radiation induces differential calcium signals in human peripheral blood lymphocyte subsets. *J. Immunother.* **14**, 221–225.
Schieven, G. L., Kirihara, J. M., Burg, D. L., Geahlen, R. L., and Ledbetter, J. A. (1993a). p72^{syk} tyrosine kinase is activated by oxidizing conditions which induce lymphocyte tyrosine phosphorylation and Ca^{2+} signals. *J. Biol. Chem.* **268**, 16688-16692.
Schieven, G. L., Kirihara, J. M., Gilliland, L. K., Uckun, F. M., and Ledbetter, J. A. (1993b). Ultraviolet radiation rapidly induces tyrosine phosphorylation and calcium signaling in lymphocytes. *Mol. Biol. Cell* **4**, 523–530.
Schieven, G. L., Kirihara, J. M., Myers, D. E., Ledbetter, J. A., and Uckun, F. M. (1993c). Reactive oxygen intermediates activate NF-KB in a tyrosine kinase dependent mechanism and in combination with vanadate activate the p56lck and p59fyn tyrosine kinases in human lymphocytes. *Blood* **82**, 1212–1220.

Schieven, G. L., Mittler, R. S., Nadler, S. G., Kirihara, J. M., Bolen, J. B., Kanner, S. B., and Ledbetter, J. A. (1994). ZAP-70 tyrosine kinase, CD45 and T cell receptor involvement in UV and H_2O_2 induced T cell signal transduction. *J. Biol. Chem.* **269**, 20718–20726.

Schieven, G. L., Wahl, A. F., Myrdal, S., Grosmaire, L., and Ledbetter, J. A. (1995). Lineage-specific induction of B cell apoptosis and altered signal transduction by the phosphotyrosine phosphatase inhibitor bis(maltolato)oxovanadium(IV). *J. Biol. Chem.* **270**, 20824–20831.

Schreck, R., Rieber, P., and Baeuerle, P. A. (1991). Reactive oxygen intermediates as apparently widely used messengers in the activation of the NF-KB transcription factor and HIV-1. *EMBO J.* **10**, 2247.

Secrist, J. P., Burns, L. A., Karnitz, L., Koretzky, G. A., and Abrahams, R. T. (1993). Stimulatory effects of the protein tyrosine phosphatase inhibitor, pervanadate, on T-cell activation events. *J. Biol. Chem.* **268**, 5886–5893.

Sherman, M. L., Datta, R., Hallahan, D. E., Weichselbaum, R. R., and Kufe, D. W. (1990). Ionizing radiation regulates expression of the c-jun protooncogene. *Proc. Natl. Acad. Sci. U.S.A.* **87**, 5663–5666.

Sies, H. (1986). Biochemistry of oxidative stress. *Angew. Chem., Int. Ed. Engl.* **25**, 1058–1071.

Suhara, T., Fukuo, K., Sugimoto, T., Morimoto, S., Nakahashi, T., Hata, S., Shimizu, M., and Ogihara, T. (1998). Hydrogen peroxide induces up-regulation of Fas in human endothelial cells. *J. Immunol.* **160**, 4042–4047.

Sundarensan, M., Yu, Z. X., Ferrans, V. J., Irani, K., and Finkel, T. (1995). Requirement for generation of H_2O_2 for platelet-derived growth factor signal transduction. *Science* **270**, 296–299.

Twamley-Sein, G. M., Pepperkok, R., Ansorge, W., and Courtneidge, S. A. (1993). The Src family tyrosine kinases are required for platelet-derived growth factor-mediated signal transduction in NIH 3T3 cells. *Proc. Natl. Acad. Sci. U.S.A.* **90**, 7696–7700.

Uckun, F. M., Tuel-Ahlgren, L., Song, C. W., Waddick, K., Myers, D. E., Kirihara, J., Ledbetter, J. A., and Schieven, G. L. (1992). Ionizing radiation stimulates unidentified tyrosine-specific protein kinases in human B-lymphocyte precursors triggering apoptosis and clonogenic cell death. *Proc. Natl. Acad. Sci. U. S. A.* **89**, 9005–9009.

Uckun, F. M., Schieven, G. L., Tuel-Ahlgren, L. M., Dibirdik, I., Myers, D. E., Ledbetter, J. A., and Song, C. W. (1993). Tyrosine phosphorylation is a mandatory proximal step in radiation-induced activation of the protein kinase C signaling pathway in human B-lymphocyte precursors. *Proc. Natl. Acad. Sci. U.S.A.* **90**, 252–256.

Uckun, F. M., Waddick, K. G., Mahajan, S., Jun, X., Takata, M., Bolen, J., and Kurosaki, T. (1996). BTK as a mediator of radiation-induced apoptosis in DT-40 lymphoma B cells. *Science* **273**, 1096–1100.

Ullrich, A., and Schlessinger, J. (1990). Signal transduction by receptors with tyrosine kinase activity. *Cell (Cambridge, Mass.)* **61**, 203–212.

Weiss, A., and Littman, D. R. (1994). Signal transduction by lymphocyte antigen receptors. *Cell (Cambridge, Mass.)* **76**, 263–274.

Wen, S. T., and Van Etten, R. A. (1998). The PAG gene product, a stress-induced protein with antioxidant properties, is an Abl SH3-binding protein and a physiological inhibitor of c-Able tyrosine kinase activity. *Genes Dev.* **11**, 2456–2467.

Xu, W., Harrison, S. C., and Eck, M. J. (1997). Three-dimensional structure of the tyrosine kinase c-src. *Nature (London)* **385**, 595–602.

Yan, S. R., and Berton, G. (1996). Regulation of Src family tyrosine kinase activities in adherent human neutrophils. Evidence that reactive oxygen intermediates produced by adherent neutrophils increase the activity of the p58$^{c\text{-}fgr}$ and p53/56^{lyn} tyrosine kinases. *J. Biol. Chem.* **271**, 23464–23471.

Yuan, Z. M., Utsugisawa, T., Huang, Y., Ishiko, T., Nakada, S., Kharbanda, S., Weichselbaum, R., and Kufe, D. (1997). Inhibition of phosphatidylinositol 3-kinase by c-Abl in the genotoxic stress response. *J. Biol. Chem.* **272**, 23485–23488.

Yurchak, L. K., Hardwick, J. S., Amrein K., Pierno K., and Sefton, B. M. (1996). Stimulation of phosphorylation of Tyr394 by hydrogen peroxide reactivates biologically inactive, non-membrane-bound forms of Lck. *J. Biol. Chem.* **271**, 12549–12554.

7 Regulation of Signal Transduction and Gene Expression by Reactive Nitrogen Species

Ami A. Deora and Harry M. Lander
Department of Biochemistry
Cornell University Medical College
New York, New York 10021

It is now clear that nitric oxide and related chemical species are ubiquitous molecules that play crucial roles in human physiology. Nitric oxide (NO), which is synthesized in most tissues, is diffusible and can take on several chemical forms, each of which has its own reactive specificity. The major chemical modification produced by these species is nitrosylation of the target, generally a protein iron or thiol. The direct interaction of nitric oxide with the protein target may result in activation, inactivation, or switching of protein function and subsequent modulation of gene expression. Some physiological events regulated by this type of signaling include vasodilation, cytotoxicity, inflammation, and synaptic plasticity. The mechanistic understanding of nitric oxide-triggered signaling will likely have far-reaching clinical implications, especially in understanding and combating artherosclerosis and inflammation. This chapter provides an overview of the interaction of various redox forms of NO with cellular targets, the mechanisms involved, and how this signaling results in the regulation of gene expression.

I. INTRODUCTION

Nitric oxide was the first gas known to act as a messenger in mammals. Research in the field of nitric oxide biology has revolutionized our understanding of cellular behavior and has made a biological impact on an unforeseen scale.

Nitric oxide was first implicated in vascular smooth muscle relaxation by Gruetter *et al.* (1979). Furchgott and Zawadzki (1980) made a landmark discovery when they reported that the intact endothelium was essential for vascular smooth muscle relaxation by acetylcholine and called this substance endothelium-derived growth factor. In 1986, Furchgott and Ignarro independently proposed that NO was endothelium-derived relaxing factor because of their similar biological properties. Experimental evidence came from seminal studies by Ignarro *et al.* (1987) and Palmer *et al.* (1987). They identified the chemical nature of endothelium-derived relaxing factor as NO, which precipitated the identification of the pivotal role of NO in various cellular functions, including vascular relaxation, immune defense, and neuronal plasticity (Nathan, 1992). In 1992, NO was *Science* magazine's "Molecule of the Year."

Nitric oxide, one of the 10 smallest molecules found in nature, is a paramagnetic radical, thus making it very reactive. It is derived through the oxidation of one of the terminal guanidino-nitrogen atoms of L-arginine (Palmer *et al.*, 1988) by the enzyme nitric oxide synthase (NOS), which exists in three isoforms. nNOS (type I) (Bredt *et al.*, 1991) and eNOS (type III) (Lamas *et al.*, 1992) were initially cloned from neuronal and endothelial cells, respectively, are calcium-calmodulin dependent, and are expressed constitutively under most conditions. iNOS (type II) (Xie *et al.*, 1992), which was first identified in macrophages, is calcium independent and inducible. Many tissues have now been shown to express these isoforms. The regulation of these enzymes is complex and requires five cofactors—FAD, FMN, heme, calmodulin, and tetrahydrobiopterin—and three cosubstrates—L-arginine, NADPH, and O_2 (Nathan and Xie, 1994).

Redox reactions are essential events regulating signal transduction. Various redox mediators such as NO, superoxide, and hydroxyl radical are known to trigger redox signaling, which leads to gene expression and the concomitant regulation of biological processes. This chapter attempts to marshall the studies involving NO-mediated signal transduction and gene expression and to explore the physiological relevance and potential future directions.

A. Chemistry of Reactive Nitrogen Species and Its Interaction with Cellular Targets

Nitric Oxide has an unpaired electron on the nitrogen atom, making it highly reactive. Its reaction with redox modulators yields many reactive species. Hence, the term reactive nitrogen species (RNS) is used in this chapter and refers to those species whose origin is the free radical · NO, but whose final chemical nature depends on its interaction with local redox modulators and the redox milieu of the cell (Lander, 1997).

Chemically, NO can exist in three redox forms: nitrosonium cation (NO^+), nitric oxide (·NO), and nitroxyl anion (NO^-). In the presence of superoxide anion (O_2^-), NO combines to form peroxynitrite ($ONOO^-$), a strong prooxidant species (Stamler et al., 1992a) (Fig. 1). In aqueous aerobic solutions, NO predominantly forms nitrite (NO_2^-). In the presence of oxyhemoglobin and oxymyoglobin, NO is completely oxidized to nitrate (NO_3^-) (Ignarro et al., 1993).

Covalent interactions of RNS with cellular macromolecules are responsible for various physiological and pathological consequences. Proteins containing iron and thiol groups are the major cellular targets of RNS. Extensive studies have been performed characterizing the RNS–iron interaction. The iron, as Fe^{+2} or Fe^{+3}, can be targeted when either in a heme group or in an iron–sulfur cluster (Ignarro, 1991; Bredt and Synder, 1994). The physiological significance of S-nitrosothiol adduction is now widely accepted. S-Nitrosothiols at critical active site thiol residues are reported to regulate the function of several proteins (Stamler, 1994; Lander et al., 1996a; Xu et al., 1998). Under more extreme conditions, such as severe oxidative and nitrosative stress, RNS reacts at other targets such as amino groups on DNA and tyrosine residues on proteins. RNS modification of target proteins results in modulation of their functional properties and can propagate downstream signals (Fig. 1).

Many of the studies referenced herein use RNS donors as an exogenous source of RNS. These donors are structurally different compounds and are metabolized by enzymes or undergo spontaneous chemical hydrolysis to release RNS. These compounds have aided pharmacological, biochemical, and molecular studies, greatly. Commonly used RNS donors are listed according to their chemical group in Table I along with their generic structures (Bauer et al., 1995).

II. SIGNAL TRANSDUCTION BY REACTIVE NITROGEN SPECIES

Reactive nitrogen species serve as an extraordinarily widespread effector of cellular functions. Generally, RNS-responsive targets serve sensory and regulatory roles in signal transduction. The target recognizes RNS and transduces the chemical signal into a functional response.

In principle, NO does not differ from other hormonal second messengers. Once released it can act on many cells and at a distance. However, unlike hormones, it is an economical signal because it is produced from

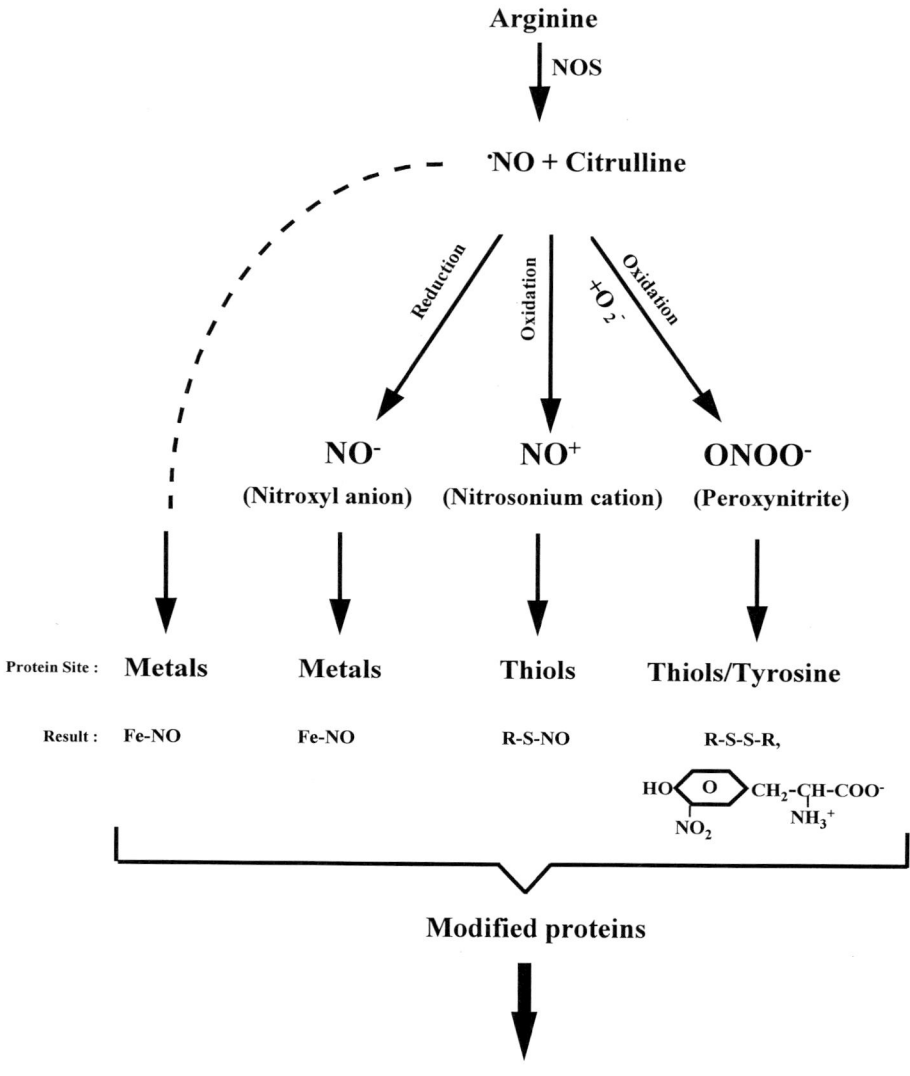

Figure 1 Different congeners of nitric oxide and their site of interaction on proteins.

Table I Commonly Used RNS Donors[a]

RNS donors	Exemplary structures	Notes
1. Organic nitrates and nitrites		
Nitroglycerin	R_3C-ONO_2	Enzymatic metabolism to yield NO
Isosorbide dinitrate		
Isoamyl nitrite	R_3C-ONO	Enzymatic metabolism and chemical hydrolysis
Isobutyl nitrite		
2. Ferrous nitro complexes		
Sodium nitroprusside	$[(CN)_5Fe^{2+}]NO$	Chemical hydrolysis
3. Sydnonimines		
Molsidomine	$R_2-N=\overset{\pm}{\underset{N\diagdown O\diagup}{\mid}}=N-R_1$	Metabolism followed by chemical hydrolysis
3-Morpholinylsydnoneimine		
Pirsidomine		
4. S-Nitrosothiols		
S-Nitrosoglutathione		
S-Nitrosocysteine	$R-S-NO$	Chemical hydrolysis
S-Nitroso-N-acetylpenicillamine		
5. Nucleophile adducts		
Diethylamine-NO	$R_2N-[N(O)NO]^-$	Chemical hydrolysis
Spermine-NO		
NONOates		

[a] After Bauer et al. (1995).

the abundant and recyclable substrate arginine and does not use vesicular machinery to be released. Moreover it uses receptor systems that are already found in the general regulatory machinery of cells and can pass through and between cells with ease, making it an ideal inter- and intracellular messenger. RNS utilize means of communication used by both protein kinases, which control function by covalent modification, i.e., phosphorylation, and reactive oxygen species that signal through redox events and coordinative interactions with metals. Some of the signaling pathway involving RNS are discussed in the following section.

A. Interaction with Heme Iron

Redox-sensitive heme groups of metalloproteins located at catalytic or allosteric sites of proteins are natural sensors of redox reactive species such as RNS. A variety of heme-containing proteins exhibit RNS-responsive control.

1. Guanylyl Cyclase

Guanosine 3′,5′-monophosphate (cGMP) is utilized as an intracellular amplifier and second messenger by a wide spectrum of ligands to elicit diverse physiological responses. cGMP synthesis is catalyzed by multiple types of soluble and particulate guanylyl cyclase (Schmidt et al., 1993).

Soluble guanylyl cyclase is a family of heterodimeric heme proteins exhibiting a pyridine hemochrome visual absorption spectrum typical for ferroprotoporphyrin IX and contains copper and iron as transition metals (Gerzer et al., 1981). The activation of soluble guanylyl cyclase by RNS is the principal mechanism of action in various RNS-induced cellular events such as smooth muscle relaxation (Ignarro and Kadowitz, 1985) and inhibition of platelet adhesion (Radomski et al., 1987). Many cellular consequences of RNS-induced guanylyl cyclase activation are listed in Table II. RNS bind to the heme moiety of guanylyl cyclase and, by disrupting the plane of the heme-iron, induce a conformational change that activates the enzyme allosterically (Ignarro et al., 1984). The cGMP synthesized further modulates an array of mediators including ion channels, phosphodiesterases, and protein kinases (Schmidt et al., 1993). Activation of guanylyl cyclase was the first RNS-induced signaling mechanism reported. To date, it remains the most important physiological pathway of RNS signaling.

2. Cytochrome P450 and Related Enzymes

Reactive nitrogen species can inhibit the enzyme activity of cytochrome P450 and related enzymes, including NOS itself and indoleamine 2,3-dioxygenase. Unlike guanylyl cyclase, these proteins utilize the heme prosthetic group in catalysis. The binding of RNS to either Fe(II) or Fe(III) heme may interfere with the conversion of ferric to ferrous iron required for catalysis, resulting in attenuation of enzyme activity.

The enzyme NOS is regulated negatively and positively (Nathan and Xie, 1994). Positive regulation is achieved through allosteric activation by Ca^{2+}-calmodulin (for eNOS and nNOS) or through enzyme levels (for iNOS). Negative regulation is achieved via interaction of RNS with the heme group of NOS. This prevents catalysis and generation of RNS by NOS, thus representing a negative feedback mechanism (Buga et al., 1993; Griscavage et al., 1994; Abu-Soud et al., 1995; Hurshman and Marletta, 1995).

The in vivo relevance of inhibiting indoleamine 2,3-dioxygenase in interferon-γ-primed mononuclear phagocytes is to prevent the total depletion of tryptophan and/or other indoleamines important for cellular function and to prevent the accumulation of a high concentration of potentially toxic metabolites of the kynurenine pathway (Thomas et al., 1994).

In summary, when heme-Fe is involved in catalysis directly, its modification by RNS typically yields an inhibition of enzymatic activity. It is important to note, however, that not all heme groups of heme-containing enzymes are modified by RNS. For example, cyclooxygenase-1 is a heme-containing enzyme whose heme group is protected from RNS (Hajjar et al., 1995).

Table II Cellular Events Regulated by the RNS/cGMP Signaling Pathway

Cell type	Cellular event	Reference
Smooth muscle cells	Vasodilation	Ignarro and Kadowitz (1985)
Platelets	Antiaggregation	Radomski et al. (1987)
Sheep mitral cells	Olfactory memory formation by potentiation of glutamate release	Kendrick et al. (1997)
Human neutrophils	Attenuates platelet-activating factor-induced release of elastase	Patrick et al. (1997)
Human myocardium	Negative inotropic effect	Flesch et al. (1997)
Human hepatocellular carcinoma cells	Hypoxic regulation of erythropoietin production	Yoshioka et al. (1997)
Human γδ T lymphocytes	CD95-induced apoptosis	Sciorati et al. (1997)
Bovine chondrocytes	Disruption of focal adhesion signaling complex	Clancy et al. (1997)
Rat smooth muscle cells, myocytes, pinealocytes, alveolar epithelial cells	Inhibits spontaneous depolarization of L-type Ca^{2+} channels	Schobersgerber et al. (1997)
Rat thyroid follicular, rat-2 fibroblasts cells	Activation of transcription from AP-1 responsive promoters	Pilz et al. (1995)
Rat aorta	Inhibition of α_1-adrenergic receptor-induced c-fos and c-jun mRNA	Okazaki et al. (1996)
Mice splenic B cells	Inhibition of apoptosis by increasing expression of bcl-2	Genaro et al. (1995)
Dendrites of hippocampal granule cells of rat	Downregulation of prodynorphin mRNA, upregulation of proenkephalin mRNA	Johnston and Morris (1994a)
Rat pheochromocytoma PC12 cells	Activation of transcription factor AP-1	Haby et al. (1994)
Mice hypothalamic GT1 cell line	Repression of hypothalamic gondotrophin-releasing hormone gene expression	Belsham et al. (1996)

B. Interaction with Fe–S Proteins

Iron–sulfur proteins are polymetallic structures whose iron atoms are linked to inorganic sulfides and are usually liganded to proteins by cysteine thiolates. These proteins are sensitive to oxido reduction and have long been known as targets of O_2^- and H_2O_2. In *in vitro* models, RNS were shown to yield complexes with [Fe–S] clusters (Butler *et al.*, 1988). It is thought that peroxynitrite may react with iron–sulfur clusters and, in contrast to a reversible reaction of RNS with heme, its reaction with iron–sulfur clusters results in the dissolution of the cluster (Castro *et al.*, 1994; Henry *et al.*, 1993).

1. Aconitase

In addition to its participation in the citric acid cycle, the mammalian [4Fe–4S] aconitase is also a regulator of iron homeostasis. RNS inhibit aconitase activity by disrupting the [Fe–S] clusters. This disruption exposes its RNA-binding site, permitting binding of the protein, now called the iron-regulatory protein 1, to the iron-responsive element on mRNA of transferrin receptor and ferritin. When bound to an iron-responsive element at the 3' end of transferrin receptor mRNA, it stabilizes the mRNA. However, when bound to an iron-responsive element at the 5' end of ferritin mRNA, it inhibits translation (Drapier *et al.*, 1993; Weiss *et al.*, 1993; Jaffrey *et al.*, 1994). Thus RNS signal through the proteins involved in iron metabolism and regulate iron homeostasis (Fig. 2). A second iron-regulatory protein has been identified and found to bind the iron-responsive element. Like-iron-regulatory protein 1, it was found to be sensitive to iron concentrations and RNS. It is known to regulate iron hepatic homeostasis during acute liver inflammation (Cairo and Pietrangelo, 1995).

2. SoxR

In response to microbial attack, activated macrophages produce various free radical species, including superoxide anion, hydrogen peroxide, and NO. However, bacteria possess several protective mechanisms against macrophage-induced oxidative and nitrosative stress. The SoxR protein is a bacterial transcriptional activator requiring an iron–sulfur center for its activity. In solution it is a homodimer and each monomer possesses a [2Fe–2S] cluster (Hidalgo *et al.*, 1995). Oxidation or nitrosylation reversibly activates the FeS centers of SoxR protein, which triggers the soxS gene. The protein product of soxS further activates transcription of approximately 12 regulon genes, which include various defense proteins such as manganese superoxide dismutase, the DNA repair enzyme endonuclease IV, and glucose-6-phosphate dehydrogenase. Activation of soxRS regulon genes confers resistance to attack by activated macrophages (Hidalgo and Demple, 1994;

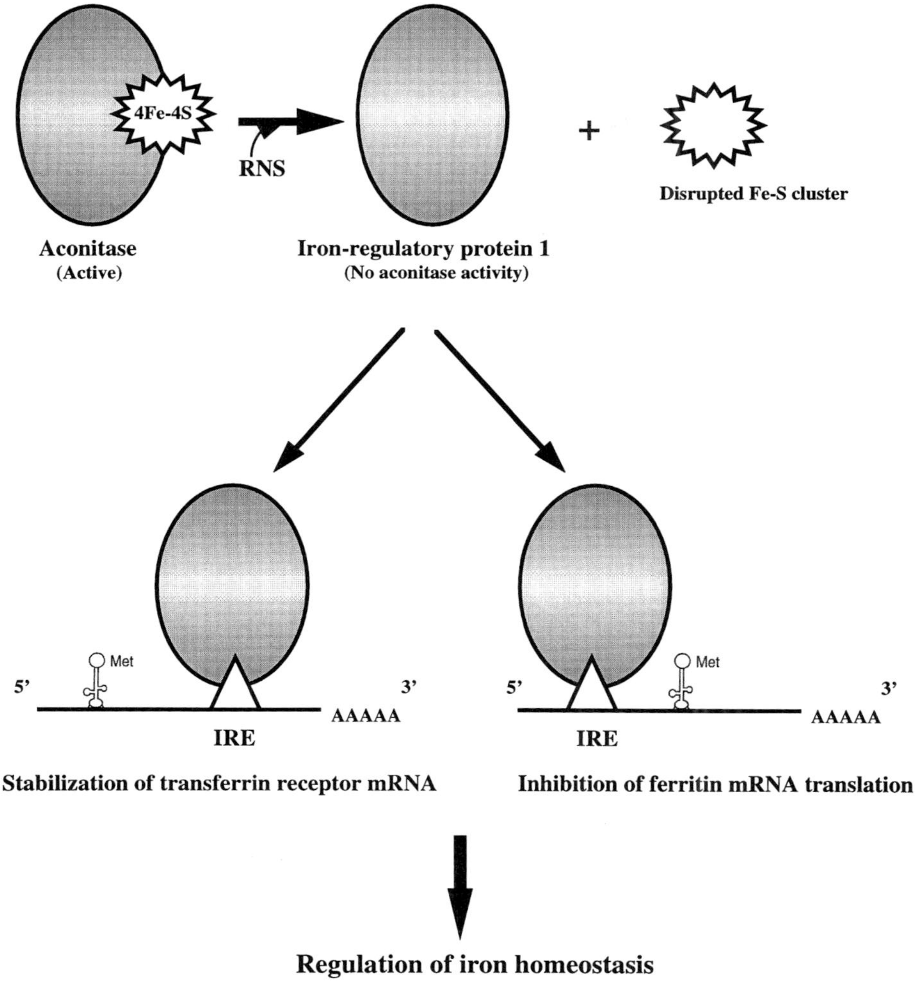

Regulation of iron homeostasis

Figure 2 RNS-induced conversion of acontiase to iron-regulatory protein 1. IRE, iron-responsive element.

Nunoshiba *et al.*, 1995). Thus, the redox-sensing FeS centers detect oxidative and nitrosative stress and help bacteria survive by triggering the coordinated expression of various defense genes.

C. Interaction with Thiols

Cysteine residues are known to be important for maintaining the native conformation of proteins, are critical residues at the active sites of enzymes, and are the most reactive residues to RNS at physiological pH. Furthermore, cysteine residues are sites for the covalent attachment of other regulatory molecules, e.g., lipid and ADP-ribose. Hence, any modification at this site may have implications extending to other signaling pathways. RNS react with thiols to form a variety of oxidized thiol species, including sulfenic acids (DeMaster et al., 1995), disulfides (Lei et al., 1992), mixed disulfides (Luperchio et al., 1996), nitrosothiol (Williams, 1985), and covalent NAD–thiol linkages (McDonald and Moss, 1993). As discussed later, modification of protein thiols by RNS alters the function and activity of various proteins, ion channels, receptors, and transcription factors.

Under physiological aerobic conditions, oxidation of NO yields NO^+-like species that have a high propensity for nucleophilic centers such as thiols and forms nitrosothiol (RSNO) (Williams, 1985; Stamler et al., 1992b; Stamler, 1994). Thiol nitrosylation may also be mediated by the oxidative activation of NO through binding to transition metals. Nitrosothiol formation by the metal ion-mediated formation of NO^+ is likely to be faster than via reactions of NO and O_2 (Kharitonov et al., 1995). Also, a peroxynitrite anion is capable of nitrosylating thiols (Radi et al., 1991; Wu et al., 1994). Nitrosothiols have a longer half-life than free NO and are an important pool from which redox signals can be generated. The S-nitrosation of protein thiols is a form of posttranslational modification and may either activate or inactivate protein function.

1. Proteins Activated by S-Nitrosothiol Formation

a. Ras The monomeric G protein, Ras, is a key element of various signaling pathways. It is implicated in the regulation of proliferation and differentiation by tyrosine kinase and G protein-coupled receptors. Activation of Ras involves guanine nucleotide exchange factor-mediated exchange of GDP for GTP and subsequent interaction with effector proteins. Effector proteins transduce signals via several pathways and trigger cellular responses (Khosravi-Far and Der, 1994; Overbeck et al., 1995; Denhardt, 1996).

RNS were found to activate Ras in human T cells, rat pheochromocytoma cells, and human endothelial cells. Nitrosothiol formation at a single cysteine residue, Cys^{-118} of Ras, was found to trigger GDP/GTP exchange (Lander et al., 1995a, 1996a, 1997). Thus, RNS activate Ras by a mechanism akin to that of growth factors. However, it differs in that it bypasses the requirement for guanine nucleotide exchange factors by directly modifying Ras and triggering GDP/GTP exchange.

Mitogen-activated protein kinases are important components of the Ras-dependent signal transduction pathway. S-nitrosylated Ras was found to trigger activation of all the three mitogen-activated protein kinases. They include extracellular signal-regulated kinase, c-Jun NH_2-terminal kinase, and p38 MAP kinase (Lander et al., 1996b). Interestingly, c-Jun NH_2-terminal kinase and p38 MAP kinase are also activated by proinflammatory cytokines and environmental stress (Cano and Mahadevan, 1995). Further downstream, RNS were found to activate the transcription factor, NF-κB. This activation was observed in human peripheral blood mononuclear cells, T cells, and rat pheochromocytoma cells and depended on S-nitrosothiol formation at the Cys^{-118} residue of Ras (Lander et al., 1993, 1995a,b). Identifying the signals immediate to RNS-induced Ras activation (which includes Ras effectors such as phosphoinositide 3-kinase, Raf-1, and protein kinase C-zeta) and downstream intermediate signals leading to NF-κB activation will help in deciphering the RNS-induced Ras pathway. Because Ras is also activated by various other redox modulators such as hemin, mercuric chloride, and hydrogen peroxide (Lander et al., 1995b), it may serve as a sensor of cellular redox stress and enable the cell to respond appropriately to the external milieu.

b. Calcium Release Channel (Ryanodine Receptor) Ryanodine receptors belong to a multigene family of channel proteins. They are localized at the junctional sarcoplasmic reticulum in muscle and the endoplasmic reticulum in epithelial and neuronal cells. These receptors are sensitive to the muscle-paralyzing alkaloid ryanodine and are responsible for the release of Ca^{2+} ions from intracellular stores, which activates contraction (Coronado et al., 1994). Ion channels are reported to be redox regulated via their sulfhydryl groups (Abramson and Salama, 1988; Oba et al., 1992). Stoyanovsky et al. (1997) and Xu et al. (1998) reported RNS-induced ryanodine receptor channel opening and Ca^{2+} release from skeletal and cardiac sarcoplasmic reticulum into the cytoplasm. They observed that in response to a muscle action potential, polynitrosylation of up to 12 free thiols of the cardiac calcium release channel reversibly activated the channel, which then released Ca^{2+} from the sarcoplasmic reticulum. Channel activity was monitored by incorporating calcium release channels into proteoliposomes. Interestingly, in contrast to S-nitrosylation, the oxidation of thiol groups has no effect on channel function, suggesting a specificity of nitrosothiol-induced activation. Thus, direct interaction of RNS with thiols of the cardiac calcium release channel regulates force in the contracting muscle and controls excitation–contraction coupling.

c. Tissue Plasminogen Activator The normal endothelium secretes cardioprotective mediators such as RNS and tissue plasminogen activator (t-PA). t-PA is involved in the activation of the fibrinolytic system (Lijnen and Collen, 1997). S-nitrosylation of tissue-type plasminogen activator at Cys^{-83} endows the enzyme with vasodilatory and antiplatelet properties and enhances the catalytic efficiency of plasminogen activation in the presence of fibrin (Stamler et al., 1992c). S-nitrosylated t-PA was also able to attenuate cardiac necrosis after myocardial ischemia–reperfusion and inhibited a neutrophil–endothelium interaction. This latter effect may involve a decrease in expression of the adhesion molecule, P-selectin (Delyani et al., 1996). Hence, S-nitrosylation of t-PA converts a simple protease into a pleiotropic antithrombotic agent.

d. Calcium-Dependent Potassium Channels The endothelium controls vascular smooth muscle tone by secreting relaxing and contracting factors. Nitric oxide, also called endothelium-derived relaxing factor, activates calcium-dependent potassium channels leading to hyperpolarization of the vascular smooth muscle cell and is a major mechanism of vasodilation. The mechanism of activation involves S-nitrosylation of thiols and the eventual disulfide formation of vicinal thiols accompanied by the release of NO$^-$ (Bolotina et al., 1994). Hence RNS regulates vascular tone by activation of the calcium-dependent potassium channels in addition to activation of guanylyl cyclase.

2. Proteins Inactivated by Nitrosothiol Formation

a. N-Methyl-D-aspartic Acid Receptor In the central nervous system, glutamate is the main excitatory neurotransmitter. Activation of the N-methyl-D-aspartic acid (NMDA) subtype of excitatory amino acid receptors by glutamate results in increased activity of NOS in the brain. The NO produced acts as a neurotransmitter and is implicated in a variety of functions, including synaptic plasticity, cerebral circulation, regulation of circadian rhythm, and production of cerebrospinal fluid. However, excessive NMDA receptor activation leads to overproduction of NO, resulting in neuronal damage due to cell death (Dawson et al., 1993). Paradoxically, the NMDA receptor is inhibited by RNS at a redox-modulatory site. Inactivation of the NMDA receptor is thought to be due to S-nitrosylation at thiol residues and eventual disulfide bond formation (Lipton et al., 1993). RNS provide a neuroprotective signal by inhibiting the NMDA receptor and preventing neural injury caused by high levels of glutamate and RNS.

b. Long Chain Fatty Acylation of Neuronal Proteins RNS play a very important role in neuronal signaling. Hess et al. (1993) observed that RNS triggered the immediate collapse of neuronal growth cones in rat dorsal root ganglion neurons *in vitro*. One of the possible mechanisms may be RNS-induced cysteine modification and subsequent inhibition of thioester-linked long chain fatty acylation of proteins involved in growth cone maintenance. Long chain fatty acylation of neuronal proteins is very important for growth cone motility and neurite processes. These growth cone proteins include GAP-43, a growth cone constituent, and SNAP-25, a synaptic protein important in late stages of axon growth and in synaptogenesis. RNS were found to inhibit ongoing palmitoylation of these two proteins, which may be the mechanism utilized in mediating growth cone collapse (Hess et al., 1993). Thus, by regulating neurite outgrowth, RNS may play an important role in remodeling during neuronal development.

c. Glyceraldehyde-3-phosphate Dehydrogenase Glyceraldehyde-3-phosphate dehydrogenase is a key enzyme of the glycolytic pathway. RNS react with an active site cysteine of glyceraldehyde-3-phosphate dehydrogenase (Cys^{-149}), which promotes subsequent direct binding of NADH to Cys^{-149} and inhibition of glyceraldehyde-3-phosphate dehydrogenase catalytic activity. This results in depression of glycolysis and is one of the cytotoxic mechanisms of RNS. The first modification step, S-nitrosylation is reversible, unlike the irreversible inactivation by subsequent NADH modification (McDonald and Moss, 1993). Hence, S-nitrosylation of glyceraldehyde-3-phosphate dehydrogenase may also be a means to protect it from irreversible inhibition by oxidative damage triggered by inflammatory cytokines (Mohr et al., 1996). Berata and Berata (1995) reported an increase in glyceraldehyde-3-phosphate dehydrogenase mRNA levels in murine microvascular endothelial cells stimulated with tumor necrosis factor-α and IFN-γ. This increase in mRNA levels depended on NO synthesis. This may be an RNS-induced adaptive mechanism to compensate for the inhibition of glyceraldehyde-3-phosphate dehydrogenase enzyme activity.

d. Epidermal Growth Factor Receptor Epidermal growth factor is an important mitogenic factor. The epidermal growth factor receptor is a member of the tyrosine kinase receptor superfamily and possesses two prominent cysteine-rich domains in its extracellular region (Ullrich and Schlessinger, 1990). Estrada et al. (1997) reported that RNS act directly on the epidermal growth factor receptor, resulting in a reversible reaction with sensitive thiol groups. This abrogates its transphosphorylation and tyrosine kinase activities and may be another mechanism by which RNS exert antimitotic effects.

e. Zinc-Finger Proteins Various proteins involved in gene expression are zinc-finger proteins. In these proteins, zinc is either bound to two cysteine and two histidine residues or zinc atoms are bound to cysteine ligands only. Thus, these proteins contain thiol-rich environments (Klug and Rhodes, 1987). Zinc is not affected by redox chemistry, however, it is proposed that the interaction of RNS with cysteine displaces zinc from the protein, resulting in abrogation of activity. Some of these proteins inhibited by RNS include

Fpg, a zinc containing repair enzyme (Wink and Laval, 1994), and LAC9, a yeast transcription factor (Kroncke *et al.*, 1994). The zinc-finger proteins postulated to be targets of RNS include GAL-4, HTLV-1 Tax protein, and proteins with the cysteine-rich LIM motif (Drapier and Bouton, 1996). These examples demonstrate how RNS can specifically target proteins that have evolved the use of metal-thiol chemistry to perform various functions.

f. Caspases Caspases are a family of at least 10 proteases that cleave after an aspartic acid residue in a consensus sequence. They are important effectors in apoptotic signaling. All caspases contain a conserved cysteine residue in the active site (Alnemri *et al.*, 1996). Mohr *et al.* (1997) demonstrated that caspase-3 activity was attenuated in the actinomycin D-induced leukemic cell line U937 by S-nitrosylation and oxidation of a critical thiol group. A similar observation was made by Li *et al.* (1997). They reported reversible inhibition of seven members of the caspase family *in vitro* by interacting human recombinant caspases with RNS donors. Hence RNS may play a role in rescuing cells from apoptotic machinery. Moreover, inhibition of caspase-1 by RNS will also affect processing and maturation of pro-inflammatory cytokines, IL-1β and IL-18, and thus RNS may also control secretion of these cytokines.

g. Transcription Factors Transcription factors are proteins that transduce signals to the transcriptional apparatus by binding to specific DNA sequences. Studies suggest that mammalian and bacterial transcription factors can be regulated by S-nitrosylation.

i. NF-κB. The classic, inactive cytoplasmic form of NF-κB exists in trimer of three subunits: p65, p50, and IκBα. On activation with a stimuli, the inhibitory subunit, IκBα, dissociates and the dimer of p50/p65 translocates to the nucleus. In addition to NF-κB-binding sites in the immunoglobulin κ chain gene, the site has been identified in many other genes, including cytokines, cytokine receptors, cell adhesion molecules, genes of the HIV provirus (Baldwin, 1996), and, interestingly, in the iNOS gene (Xie *et al.*, 1994; Chartrain *et al.*, 1994).

Reactive nitrogen species were implicated to play an indirect role in the activation of NF-κB in human peripheral mononuclear cells (Lander *et al.*, 1993). Subsequently, NF-κB activation in human T cells and rat pheochromocytoma cells was found to be dependent on RNS-induced S-nitrosylation of Ras at the Cys^{-118} residue and subsequent Ras activation (Lander *et al.*, 1995a,b). These studies were the first to demonstrate that RNS can trigger an NF-κB response and may provide a mechanistic basis by which RNS trigger NF-κB-dependent gene expression.

In contrast, in nonlymphoid astroglial cells, Park *et al.* (1997) observed that RNS derived from a spermine NONOate inhibits formation of an NF-κB–DNA complex. Matthews *et al.* (1996) studied the direct interaction of RNS with recombinant p50 and p65 subunits. RNS donors inactivated the DNA-binding activity of the recombinant subunits. The p50 subunit was S-nitrosylated at Cys^{-62} residue and this modification seemed to be responsible for inhibition in DNA binding. The Cys^{-62} residue of the p50 subunit is conserved in the NF-κB transcription factor family. Moreover, it is located in the peptide loop, which makes specific contacts with the DNA consensus sequence. This residue is redox sensitive, as its oxidation and subsequent intersubunit disulfide linkage also abrogate its DNA-binding activity (Ghosh *et al.*, 1995; Muller *et al.*, 1995). Because the promoter of the iNOS gene contains an NF-κB-binding motif, RNS inhibit iNOS gene expression by inhibiting NF-κB activation, thus regulating the enzyme by feedback inhibition (Togashi *et al.*, 1997).

Hence, like oxidative stress, nitrosative stress activates NF-κB and may evoke an inflammatory response in some settings. In other settings, RNS can downregulate its own synthesis, likely through a direct modification of the NF-κB subunits. This dual functionality of RNS highlights the complex signaling behavior of NO-derived free radical species.

ii. AP-1. The AP-1 transcription factor mainly controls the expression of genes involved in cell proliferation. It is a heterodimer of Fos and Jun proteins, products of the protooncogenes, c-*fos* and c-*jun*. It interacts with the DNA regulatory element known as the AP-1-binding site.

Like NF-κB, both Jun and Fos possess a conserved cysteine residue in their DNA-binding domain that is sensitive to redox changes (Abate *et al.*, 1990; Xanthoudakis and Curran, 1992). Nikitovic *et al.* (1998) reported inhibition of AP-1 DNA binding by RNS *in vitro*. The inhibition was mediated by RNS-induced modification of Fos–Cys^{-154} and Jun–Cys^{-272}, and the effect was abolished by dithiothreitol. Tabuchi *et al.* (1996) observed that RNS-induced inhibition of AP-1 DNA binding in mice cerebellar granule cells. In contrast to inhibition, RNS were found to activate AP-1 DNA binding via a cGMP-dependent mechanism (Pilz *et al.*, 1995). Thus, similar to NF-κB, AP-1 can also be activated or inhibited depending on the cellular setting. Clearly, *in vivo* studies are badly needed to determine the physiological relevance of the cell culture studies.

iii. Oxy R. In a cell, intramolecular disulfide bond formation often results in the inactivation of protein activity. However, bacteria utilize this mechanism to activate the transcription factor OxyR, which, like SoxR, rescues bacteria from oxidative and nitrosative stress. It induces the expression of protective genes such as hydrogen peroxidase I and glutathione reductase. The conserved residues, Cys^{-199} and Cys^{-208}, are critical for its activation by oxidative stress. Activation is mediated by intramolecular disulfide bond-induced conformational changes between these two cysteine residues. Nitrosative stress is also known to activate OxyR by reacting with cysteine residues (Hausladen et al., 1996; Zheng et al., 1998). It is noteworthy how bacterial transcription factors have evolved sophisticated mechanisms to respond to and survive stressful redox environments.

D. Interaction with Heme and Thiol

1. Hemoglobin

When administered systemically, cell-free hemoglobin leads to hypertension. This effect is thought to be due to scavenging of RNS by hemoglobin via its heme-iron. Hence, many experimental studies utilize hemoglobin to determine RNS specific effects. Nevertheless, direct interaction of hemoglobin with RNS was not found to modulate any functional properties of hemoglobin. Jia et al. (1996) and Gow and Stamler (1998) demonstrated that hemoglobin can be S-nitrosylated. They elucidated a new reaction highlighting the importance of RNS in the respiratory cycle and dynamic properties of hemoglobin in vasoregulation. According to their observation, in the microcirculation and venous system, RNS reside predominantly on the T-state (deoxy) α-chain heme-iron of hemoglobin. The β chain of hemoglobin possesses a highly reactive thiol group at cysteine-93, which is conserved among mammalian species. When venous blood enters the lungs, oxygen favors an allosteric transition with RNS group exchange from α-chain heme to the β chain (Cys^{-93}) thiol and O_2 attaches to the heme. This R state (oxy) structure of hemoglobin then reenters the circulation and, when faced with an O_2 gradient in resistance vessels, releases both O_2 and RNS. The RNS released may bind to the abundant glutathione present in erythrocytes and the S-nitrosoglutathione formed dilate the blood vessels. This augments O_2 delivery to the peripheral tissues. This study highlights how redox-regulated residues are conserved at a strategic site and how allosteric changes in protein conformation control dynamic functions of the protein.

2. Cyclooxygenase

Cyclooxygenase catalyzes the oxidation of arachidonic acid to prostaglandin H2, the precursor for prostacyclin and thromboxane A_2. The effect of RNS on cyclooxygenase is controversial. It has been reported to inhibit (Mathews et al., 1995; Habib et al., 1997), activate (Salvemini et al., 1993; Hajjar et al., 1995; Salvemini, 1997), or have no effect (Tsai et al., 1994) on enzymatic activity. Like hemoglobin, both heme-iron and cysteine residues are potential targets. Heme-iron is present in the active site of the enzyme and is essential for catalytic function, whereas three cysteine residues are present in the catalytic domain. Hajjar *et al.* (1995) reported an increase in enzyme activity due to allosteric changes induced by S-nitrosothiol formation. However, Tsai *et al.* (1994) and Karthein *et al.* (1987) reported interaction of RNS at the heme moiety of the enzyme.

The variable effect of RNS on cyclooxygenase activity may be due to the different cell types and experimental settings utilized. More studies need to be done to gain a mechanistic understanding of this interaction.

E. Cytotoxic Mechanisms

At high concentrations, RNS are cytotoxic. This action of RNS is utilized as a defense mechanism against pathogens and tumor cells. Some of the cytotoxic signals are discussed.

RNS inactivate several mitochondrial iron-sulfur enzymes involved in ATP synthesis. These include NADH:ubiquinone oxidoreductase, NADH:succinate oxidoreductase, and cis-acotinase (Nathan, 1992; Lowenstein *et al.*, 1994). Inactivation of GAPDH by S-nitrosylation inhibits glycolysis (see Section II,C,2,b). RNS bind the non-heme iron of ribonucleotide reductase, attenuating its activity and inhibiting DNA synthesis. Quenching of the tyrosyl radical by RNS may be another mechanism involved in inactivation of this important enzyme. Iron of the iron-storage protein, ferritin, is also a target of RNS. This interaction leads to the release of free iron, which may cause lipid peroxidation (Nathan, 1992). Like other free radicals, RNS can damage DNA by base deamination, resulting in neurotoxicity (Bredt and Synder, 1994).

The hallmark discovery of the existence of endogenous NO provided a mechanistic understanding of various physiological processes. Its myriad roles extend from bacterial transcription factors to the mammalian nervous system. Extensive studies in the field of RNS-induced signaling have helped us to appreciate the sophistication as well as simplicity of this protean molecule. In response to environmental cues, RNS can posttranscriptionally modify proteins, leading to allosteric or catalytic modulation and specific signaling. It is apparent that thiol groups are major targets of RNS, and the extensive studies discussed earlier suggest that S-nitrosylation of cysteine residues is as crucial to signal transduction as is phosphorylation of tyrosine, serine, and threonine residues.

III. GENE EXPRESSION BY REACTIVE NITROGEN SPECIES

Cell signaling involves moving information from the extracellular milieu to the nucleus. As discussed in the previous section, RNS propagate many signaling pathways that convey information about environmental changes. The posttranscriptional modification of proteins by RNS leads to an alteration in gene expression. This section explores genes whose expression is known to be regulated by RNS.

A. Vascular System

RNS, via cGMP, mediate the activity to endothelium-derived relaxing factor. In addition to its vasodilatory effects, RNS also exhibit various antiatherogenic properties. These include platelet antiaggregating effects, inhibition of vascular smooth muscle cell and endothelial cell proliferation, and prevention of leukocyte adhesion.

1. Inhibition of Cell Proliferation

a. Vascular Smooth Muscle The abnormal proliferation of vascular smooth muscle cells is implicated in various pathological situations, including atherosclerotic lesions, restenosis after balloon angioplasty, and vascular wall thickening in hypertension (Ross, 1993; Schwartz et al., 1995). RNS may play a protective role by inhibiting the proliferation of vascular smooth muscle cells. Mechanisms underlying this reversible inhibition involve regulation at the genetic level and have been elucidated by Ishida et al. (1997). When treated with the RNS donor, S-nitroso-N-acetylpenicillamine, vascular smooth muscle cells enhanced mRNA levels of p21 protein, a cyclin-dependent kinase (Cdk) inhibitor, and maintained it for 30 hr, resulting in increased p21 protein levels. This led to increased association with one of its substrates, Cdk2, thus attenuating its activity. Retinoblastoma protein (pRb) is hyperphosphorylated by Cdk2, facilitating G_1–S phase transition. Hence, the RNS-induced increase of p21 protein results in the inhibition of Cdk2 activity and ultimately decreases the phosphorylation of pRb protein (Fig. 3). This hinders cell cycle progression and cell division. Whether cGMP is involved in this antiproliferative pathway is not yet known.

Okazaki et al. (1996) observed that RNS released from endothelial cells inhibited the expression of c-*fos* and c-*jun* mRNA induced by α_1-adrenergic receptors in rat aorta. This inhibition was cGMP dependent. These receptors play important roles in the regulation of blood vessel contraction and smooth muscle proliferation. Hence, the RNS-induced inhibition of vascular smooth muscle cell proliferation may, in part, involve inhibition of the protooncogenes c-*fos* and c-*jun* (Fig. 3). One way RNS may protect against atherogenesis is by regulating VSMC proliferation.

b. Endothelial Cells Vascular endothelial growth factor (VEGF) is synthesized by smooth muscle cells and acts as a mitogen for endothelial

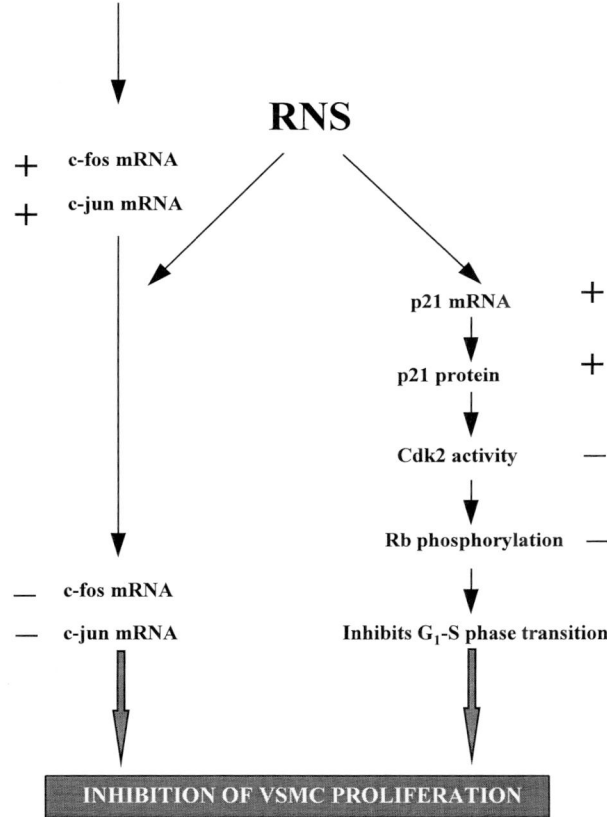

Figure 3 RNS-induced inhibiton of vascular smooth muscle cell proliferation. +, increased; −, decreased; VSMC, vascular smooth muscle cell.

cells. Arterial injury was found to upregulate VEGF gene expression. This increased expression is due to protein kinase C-induced binding of AP-1 to the VEGF promoter. van der Zee *et al.* (1997) reported that VEGF increases NO synthesis in endothelial cells. However, RNS released from regenerating endothelial cells were found to downregulate VEGF expression to basal levels. This attenuation was due to RNS-induced inhibition of AP-1 DNA binding (Tsurumi *et al.*, 1997). Thus, the regenerated endothelium regulates proliferation by negative feedback inhibition and utilizes RNS-induced signaling.

Chronic hypoxia leads to pulmonary hypertension and remodeling of pulmonary arteries concomitant with vascular cell proliferation. Hypoxia increases gene expression of vascular endothelial cell growth factor (Tuder *et al.*, 1995) and platelet-derived growth factor (Kourembanas *et al.*, 1993). Interestingly, RNS prevent this increase and therefore inhibit vascular cell proliferation in this setting.

2. Inflammatory Response

In the process of atherogenesis, the endothelial-leukocyte adhesion molecules, vascular cell adhesion molecule-1 (VCAM-1), intracellular adhesion molecule-1 (ICAM-1), and endothelial-leukocyte adhesion molecule-1 (E-selectin) play pivotal roles in the recruitment of leukocytes to the site of lesion formation (Ross, 1993). In human saphenous vein endothelial cells, RNS donors inhibit VCAM-1 gene expression induced by the cytokines IL-1α, IL-1β, IL-4, and tumor necrosis factor-α. In an *in vitro* assay, reduced VCAM-1 levels were associated with reduced monocyte adhesion. RNS were also found to regulate gene expression of E-selectin, ICAM-1, and the inflammatory cytokines IL-6 and IL-8 negatively (De Caterina *et al.*, 1995). Each of these genes has NF-κB-specific DNA-binding motifs in their promoters and the mechanism of RNS-induced gene repression seems to be via inhibition of NF-κB. Peng *et al.* (1995) observed that RNS prevents NF-κB activation by stabilizing its inhibitor, IκB, and augmenting its mRNA levels. These studies provide insight into the antiatherogenic properties of RNS. In apparent conflict to these reports, RNS have been implicated in endothelial cell activation. For example, Villarete and Remick (1995) found that RNS increased cytokine-induced IL-8 expression in endothelial cells.

Polyinosinic-polycytidylic acid [poly(I-C)] is a double-stranded RNA and mimics a virally infected state in cells. It induces an inflammatory response by enhancing the adherence of leukocytes to activated endothelial cells. This is achieved by activating the transcription factor NF-κB and gene expression of the adhesion molecules, VCAM-1 and E-selectin (Sedmak *et al.*, 1994). Faruqi *et al.* (1997) observed that NO synthesized by eNOS was essential for a poly(I-C)-induced inflammatory response as the NOS inhibitor, N^G-methyl-L-arginine, abrogated the inflammatory response.

The apparent conflict between these studies may be due to the experimental settings used. Because RNS can take many forms, depending on the redox environment, its ultimate action is equally variable. *In vivo* studies are awaited to determine the actual role of RNS in the progression of artherogenesis.

3. Regulatory Response

Carbon monoxide is another gaseous molecule reported to share many properties similar to NO, including stimulation of guanylyl cyclase, vasodilation, and neurotransmission (Morita *et al.*, 1995; Morita and Kourembanas, 1995; Verma *et al.*, 1993). It is released during heme oxygenase-catalyzed oxidative degradation of heme to biliverdin. Heme oxygenase-1 is an inducible isoform of this enzyme (Choi and Alam, 1996). Hartsfield *et al.* (1997) observed a potent increase in heme oxygenase-1 mRNA levels by RNS in rat aortic smooth muscle cells. The activation was independent of cGMP and suggests a coordinated regulation between the two gaseous cellular messengers.

Because heme oxygenase-1 activity is not regulated by exogenous signals rapidly, more studies are needed to determine the physiological importance of carbon monoxide as a signaling molecule.

B. Immune System

RNS is a cytotoxic agent that plays a crucial role in immune defense mechanisms. However, a role in immune regulation is emerging and reports indicate a modulatory role in some of the developmental stages of immune cells.

1. Differentiation

Early growth response-1, a DNA-binding protein belonging to the zinc-finger family, plays a crucial role in the differentiation process of macrophages and primes macrophages for activation and an inflammatory response. On stimulation of a rat pulmonary alveolar macrophage cell line with either IFN-γ and LPS or an RNS donor, early growth response-1 gene expression was found to be inhibited, suggestive of a regulatory role for RNS in macrophage-induced inflammation (Henderson *et al.*, 1994).

Rothe *et al.* (1996) reported the regulation of interleukin-12 gene expression by RNS in a murine macrophage cell line. IL-12 recruits T-helper subsets to differentiate along the T-helper 1 subset pathway. These T-helper 1 cells are implicated in delayed hypersensitivity responses. RNS induced the gene expression of one of the IL-12 subunits, p40. Excess production of the p40 subunit is antagonistic for IL-12 synthesis and is detrimental for Th1 subset differentiation. Hence, RNS modulate IL-12 synthesis by an autoregulatory feedback mechanism and hinder the differentiation of T-helper 1 cells.

2. Apoptosis

In the development of the immune system, negative selection of B- and T-cell lineages is very important and is mediated by the process of apoptosis (Cohen, 1991). Many genes, including bcl-2, bcl-x, and bax, are key factors controlling the apoptotic fate of the cells. RNS were found to inhibit apoptosis in *ex vivo*-cultured mice splenic B cells in both unstimulated and MHC-I-activated B cells. Increased expression of the protooncogene bcl-2, at both mRNA and protein levels, was implicated in RNS-induced inhibition of apoptosis and was reported to be mediated by the activation of guanylyl cyclase. This study helps in identifying an RNS-dependent pathway controlling B-cell apoptosis (Genaro *et al.*, 1995).

Mannick *et al.* (1994) observed that endogenous NO synthesized by a human B-cell line also inhibited apoptosis and maintained Epstein–Barr virus latency. Zta is an Epstein–Barr virus transactivator and is a member of the bZIP family of transcription factors. It mediates Epstein–Barr virus reactivation by switching from latent to lytic infection (Rooney *et al.*, 1988). Zta expression is regulated positively by Zta protein and negatively by NF-κB. RNS were found to downregulate Zta expression either by inactivating Zta function or by activating NF-κB. This inhibitory effect of RNS was independent of cGMP.

These two studies indicate an important physiological role for RNS in the regulation of apoptosis in B-cell biology.

In the human myeloid leukemia cell line HL-60, RNS inhibited cell proliferation and increased expression of the monocyte-specific enzyme naphthylbutyrate esterase and the monocytic marker CD14. Thus RNS help the myeloid lineage to differentiate and commit to a monocytic lineage (Magrinat *et al.*, 1992). This effect of RNS could be exploited in the treatment of human myeloid leukemia.

3. Inflammatory Response

RNS were found to increase the gene expression of inflammatory cytokines TNF-α and IL-1β in the myeloid leukemic cell line (Magrinat *et al.*, 1992). Lander *et al.* (1993) also reported on RNS-induced increase in TNF-α production in human peripheral blood mononuclear cells. Interestingly, in combination with IFNγ, TNF-α induces NOS gene expression. Therefore, this could be a positive feedback mechanism critical for an inflammatory response.

IL-8 is induced by various inflammatory stimuli, including lipopolysaccharide, tumor necrosis factor-α, and IL-1. It is a potent neutrophil chemoattractant (Baggiolini *et al.*, 1989). RNS were found to be required for IL-8 gene expression in human whole blood when stimulated with lipopolysaccharide. Also, adding an RNS donor alone stimulated IL-8 production (Villarete and Remick, 1997). RNS-induced regulation of gene expression of this inflammatory cytokine may be mediated by the activation of NF-κB (Lander *et al.*, 1993).

Taken together, the bulk of the studies that examined the effect of RNS on gene expression on immune cells suggest a pro-inflammatory effect of

RNS. It is likely that therapeutic intervention during chronic inflammation will be better achieved with specific inhibitors of NOS gene expression or enzymatic activity.

C. Nervous System

RNS are unique messengers in central and peripheral nervous systems. In the central nervous system, they are thought to play a significant role in synaptic plasticity and are known to influence neurophysiological phenomena underlying memory. This includes long-term potentiation in the hippocampus and long-term depression in the cerebellum. RNS are also known to be involved in electrocortical activation and modulation of pain perception. In the peripheral nervous system, RNS released from neurons are involved in gastric relaxation in response to a food bolus and in relaxation of the ileocolonic junction and anal sphincter (Synder, 1992; Yun et al., 1997). In the region of the brain containing NOS activity, NO has been reported to alter the pattern of gene expression in both cGMP-dependent and -independent manners. Some examples are discussed.

In dendrites of hippocampal granule cells, an RNS donor, a cGMP analog, or NMDA triggered a switch in opioid gene expression by downregulating prodynorphin mRNA and upregulating proenkaphalin mRNA, thus aiding in the long-term regulation of hippocampal excitability (Johnston and Morris, 1994a).

The immediate early genes, c-fos, c-jun, junB, junD, and zif/268, are stimulated by several second messengers and they regulate alterations in patterns of late response gene expression. RNS donors were found to stimulate c-fos and junB gene expression in rat pheochromocytoma PC12 cells by activating the transcription factor AP-1 in a cGMP-dependent manner (Haby et al., 1994). In response to an RNS donor, rat striatal neurons upregulated c-fos and zif/268 gene expression in a cGMP-independent manner (Morris, 1995).

In a hypothalamic cell line, GT1, an RNS donor, a cGMP analog, or NMDA effectively repressed hypothalamic gonadotrophin-releasing hormone gene expression and this may be a significant mechanism in RNS-induced long-term synaptic changes, especially in memory formation (Belsham et al., 1996). Long-term potentiation of synaptic transmission involves the modulation of synaptic architecture and involves the neuronal cytoskeleton. In hippocampal granule cells, RNS donors and NMDA significantly augment the gene expression of microtubule-associated protein 2, a component of the neuronal cytoskeleton. This effect was cGMP dependent. Hence, RNS facilitate dendritic remodeling crucial for long-term potentiation (Johnston and Morris, 1994b).

NO synthesized by NOS is essential for nerve growth factor-induced differentiation in PC12 cells (Penunova and Enikolpov, 1995). Poluha et al., (1997) found that RNS potently activated gene expression of the cyclin-dependent kinase inhibitor p21^{WAF1} and the pathway was partially dependent on p53. The RNS-induced pathway was found to be mandatory for neurite extension and may regulate cell proliferation of PC12 cells during nerve growth factor-induced differentiation.

Information regarding detailed signaling events leading to changes in gene expression is not mechanistic. Thus, in addition to discovering more genes whose expression is controlled by RNS, future studies must focus on deciphering the signaling pathways.

In conclusion, elegant experiments have identified RNS as second messengers. Unlike classical messengers, RNS chemically and reversibly modify targets and thus trigger signals that regulate gene expression. Diverse and paradoxic events in every organ of the body are attributed directly or indirectly to RNS, and many more significant pathways and cellular outcomes are yet to be deciphered. The field of redox signaling is in its infancy and future in-depth studies in the field of RNS-induced signaling and gene expression will provide insight into physiological and pathophysiological conditions and will yield targets for drug development.

REFERENCES

Abate, C., Patel, L., Rauscher, F. J., III, and Curran, T. (1990). Redox regulation of fos and jun DNA-binding activity *in vitro*. *Science* **249,** 1157–1161.

Abramson, J. J., and Salama, G. (1988). Sulfhydryl oxidation and Ca^{+2} release from sarcoplasmic reticulum. *Mol. Cell. Biochem.* **82,** 81–84.

Abu-Soud, H. M., Wang, J., Rousseau, D. L., Fukuto, J. M., Ignarro, L. J., and Stuehr, D. J. (1995). Neuronal nitric oxide synthase self-inactivates by forming a ferrous-nitrosyl complex during aerobic catalysis. *J. Biol. Chem.* **270,** 22997–23006.

Alnemri, E. S., Livingston, D. J., Nicholson, D. W., Salvesen, G., Thornberry, N. A., Wong, W. W., and Yuan, J. (1996). Human ICE/CED-3 protease nomenclature. *Cell (Cambridge, Mass.)* **87,** 171.

Baggiolini, M., Waltz, A., and Kunkel, S. L. (1989). Neutrophil-activating peptide-1/interleukin 8, a novel cytokine that activates neutrophils. *J. Clin. Invest.* **84,** 1045–1049.

Baldwin, A. S. (1996). The NF-κB and IκB proteins: New discoveries and insights. *Annu. Rev. Immunol.* **14,** 649–681.

Bauer, J. A., Booth, B. P., and Fung, H. L. (1995). Nitric oxide donors: Biochemical pharmacology and therapeutics. *Adv. Pharmacol.* **34,** 361–381.

Belsham, D. D., Wetsel, W. C., and Mellon, P. L. (1996). NMDA and nitric oxide act through the cGMP signal transduction pathway to repress hypothalamic gonadotropin-releasing hormone gene expression. *EMBO J.* **15,** 538–547.

Berata, J., and Berata, M. (1995). Stimulation of glyceraldehyde-3-phosphate dehydrogenase mRNA levels by endogenous nitric oxide in cytokine-activated endothelium. *Biochem. Biophys. Res. Commun.* **217,** 363–369.

Bolotina, V. M., Najibi, S., Palacino, J. J., Pagano, P. J., and Cohen, R. A. (1994). Nitric oxide directly activates calcium-dependent potassium channels in vascular smooth muscle. *Nature (London)* **368**, 850–853.

Bredt, D. S., and Synder, S. H. (1994). Nitric oxide: A physiologic messenger molecule. *Annu. Rev. Biochem.* **63**, 175–195.

Bredt, D. S., Hwang, P. M., Glatt, C. E., Lowenstein, C., Reed, R. R., and Synder, S. H. (1991). Cloned and expressed nitric oxide synthase structurally resembles cytochrome P-450 reductase. *Nature (London)* **351**, 714–718.

Buga, G. M., Griscavage, J. M., Rogers, N. E., and Ignarro, L. J. (1993). Negative feedback regulation of endothelial cell function by nitric oxide. *Circ. Res.* **73**, 808–812.

Butler, A. R., Glidewell, C., and Li, M.S. (1988). Nitrosyl complexes of iron-sulfur cluster. *Adv. Inorg. Chem.* **32**, 335–392.

Cairo, G., and Pietrangelo, A. (1995). Nitric oxide-mediated activation of iron-regulatory protein controls hepatic iron metabolism during acute inflammation. *Eur. J. Biochem.* **232**, 358–363.

Cano, E., and Mahadevan, L. C. (1995). Parallel signal processing among mammalian MAPKs. *Trends Biochem. Sci.* **20**, 117–122.

Castro, L., Rodriguez, M., and Radi, R. (1994). Aconitase is readily inactivated by peroxynitrite, but not by its precursor, nitric oxide. *J. Biol. Chem.* **269**, 29409–29415.

Chartrain, N. A., Geller, D. A., Koty, P. P., Sitrin, N. F., Nussler, A. K., Hoffman, E. P., Billiar, T. R., Hutchinson, N. I., and Mudgett, J. S. (1994). Molecular cloning, structure, and chromosomal localization of the human inducible nitric oxide synthase gene. *J. Biol. Chem.* **269**, 6765–6772.

Choi, A. M. K., and Alam, J. (1996). Heme oxygenase-1: Function, regulation, and implication of a novel stress-inducible protein in oxidant-induced lung injury. *Am. J. Respir. Cell Mol. Biol.* **15**, 9–19.

Clancy, R. M., Rediske, J., Tang, X., Nijher, N., Frenkel, S., Philips, M., and Abramson, S. B. (1997). Outside-in signaling in the chondrocyte. Nitric oxide disrupts fibronectin induced assembly of a subplasmalemmal actin/rhoA/focal adhesion kinase signaling complex. *J. Clin. Invest.* **100**, 1789–1796.

Cohen, J. J. (1991). Programmed cell death in immune system. *Adv. Immunol.* **50**, 55–85.

Coronado, R., Morrissette, J., Sukhareva, M., and Vaughan, D. M. (1994). Structure and function of ryanodine receptors. *Am. J. Physiol.* **266**, C1485–C1504.

Dawson, V. L., Dawson, T. M., Bartley, D. A., Uhl, G. R., and Snyder, S. H. (1993). Mechanisms of nitric oxide mediated neurotoxicity in primary brain cultures. *J. Neurosci.* **13**, 2651–2661.

De Caterina, R. D., Libby, P., Peng, H. B., Thannickal, V. J., Rajavashisth, T. B., Gimbrone, M. A. Jr., Shin, W. S., and Liao, J. K. (1995). Nitric oxide decreases cytokine-induced endothelial activation. *J. Clin. Invest.* **96**, 60–68.

Delyani, J. A., Nossuli, T. O., Scalia, R., Thomas, G., Garvey, J. S., and Lefer, A. M. (1996). S-nitrosylated tissue-type plasminogen activator protects against myocardial ischemia/reperfusion injury in cats: Role of endothelium. *J. Pharmacol. Exp. Ther.* **279**, 1174–1180.

DeMaster, E. G., Quast, B. J., Redfern, B., and Nagasawa, H. T. (1995). Reaction of nitric oxide with the free sulfhydryl group of human serum albumin yields a sulfenic acid and nitrous oxide. *Biochemistry* **34**, 11494–11499.

Denhardt, D. T. (1996). Signal transducing protein phosphorylation cascades mediated by Ras/Rho proteins in the mammalian cell: The potential for multiplex signalling. *Biochem. J.* **318**, 729–747.

Drapier, J. C., and Bouton, C. (1996). Modulation by nitric oxide of metalloprotein regulatory activities. *BioEssays* **18**, 549–556.

Drapier, J. C., Hirling, H., Wietzerbin, J., Kaldy, P., and Kuhn, L. C. (1993). Biosynthesis of nitric oxide activates iron regulatory factor in macrophages. *EMBO J.* **12**, 3643–3649.

Estrada, C., Gomez, C., Martin-Nieto, J., De Frutos, T., Jimenez, A., and Villalobo, A. (1997). Nitric oxide reversibly inhibits the epidermal growth factor receptor tyrosine kinase. *Biochem. J.* **326**, 369–376.

Faruqi, T. R., Erzurum, S. C., Kaneko, F. T., and Dicarleto, P. E. (1997). Role of nitric oxide in poly(I-C)-induced endothelial cell expression of leukocyte adhesion molecules. *Am. J. Physiol.* **273**, 2490–2497.

Flesch, M., Kilter, H., Cremers, B., Lenz, O., Sudkamp, M., Kuhn-Regnier, F., and Bohm, F. (1997). Acute effects of nitric oxide and cyclic GMP on human myocardial contractility. *J. Pharmacol. Exp. Ther.* **281**, 1340–1349.

Furchgott, R. F., and Zawadzki J. V. (1980). The obligatory role of endothelial cells in the relaxation of arterial smooth muscle by acetylcholine. *Nature (London)* **288**, 373–376.

Genaro, A. M., Hortelano, S., Alvarez, A., Martinez-A, C., and Bosca, L. (1995). Splenic B lymphocytes programmed cell death is prevented by nitric oxide release through mechanisms involving sustained Bcl-2 levels. *J. Clin. Invest.* **95**, 1884–1890.

Gerzer, R., Bohme, E., Hofmann, F., and Schultz, G. (1981). Soluble guanylate cyclase purified from bovine lung contains heme and copper. *FEBS Lett.* **132**, 71–74.

Ghosh, G., van Duyne, G., Ghosh, S., and Sigler, P. B. (1995). Structure of NF-kappaB p50 homodimer bound to a kappa B site. *Nature (London)* **373**, 303–310.

Gow, A. J., and Stamler, J. S. (1998). Reactions between nitric oxide and haemoglobin under physiological conditions. *Nature (London)* **391**, 169–173.

Griscavage, J. M., Fukuto, J. M., Komori, Y., and Ignarro, L. J. (1994). Nitric oxide inhibits neuronal nitric oxide synthase by interacting with their heme prosthetic group: Role of tetrahydrobiopterin in modulating inhibitory action of nitric oxide. *J. Biol. Chem.* **269**, 21644–21649.

Gruetter, C. A., Barry, B. K., McNamar D. B., Gruetter D. Y., Kadowitz P. J., and Ignarro L. J. (1979). Relaxation of bovine coronary artery and activation of coronary arterial guanylate cyclase by nitric oxide, nitroprusside and a carcinogenic nitrosoamine. *J. Cyclic Nucleotide Protein Phosphorylation Res.* **5**, 211–224.

Habib, A., Bernard, C., Lebret, M., Creminon, C., Esposito, B., Tedgui, A., and Maclouf, J. (1997). Regulation of the expression of cyclooxygenase-2 by nitric oxide in rat peritoneal macrophages. *J. Immunol.* **158**, 3845–3851.

Haby, C., Lisovoski, F., Aunis, D., and Zwiller, J. (1994). Stimulation of the cyclic GMP pathway by No induces expression of the immediate early genes c-fos and jun B in PC12 cells. *J. Neurochem.* **62**, 496–501.

Hajjar, D. P., Lander, H. M., Pearse, S. F. A., Upmacis, R. K., and Pomerantz, K. B. (1995). Nitric oxide enhances prostaglandin-H synthase-1 activity by a heme-independent mechanism: Evidence implicating nitrosothiols. *J. Am. Chem. Soc.* **117**, 3340-3346.

Hartsfield, C. L., Alam, J., Cook, J. L., and Choi, A. M. K. (1997). Regulation of heme oxygenase-1 gene expression in vascular smooth muscle cells by niric oxide. *Am. J. Physiol.* **273**, L980–L988.

Hausladen, A., Privaille, C. T., Keng, T., De Angelo, J., and Stamler, J. S. (1996). Nitrosative stress: Activation of the transcription factor OxyR. *Cell (Cambridge, Mass.)* **86**, 719–729.

Henderson, S. A, Lee, P. H., Aeberhard, E. E., Adams, J. W., Ignarro, L. J., Murphy, W. J., and Sherman, M. P. (1994). Nitric oxide reduces early growthy response-1 gene expression in rat lung macrophages treated with interferon-γ and lipopolysaccharide. *J. Biol. Chem.* **41**, 25239–25242.

Henry, Y., Lepoivre, M., Drapier, J. C., Ducrocq, C., Boucher, J. L., and Guissani, A. (1993). EPR characterization of molecular targets for NO in mammalian cells and organelles. *FASEB J.* **7**, 1124–1134.

Hess, D. T., Patterson, S. I., Smith, D. S., and Pate Skene, J. H., (1993). Neuronal growth cone collapse and inhibition of protein fatty acylation by nitric oxide. *Nature (London)* **366**, 562–565.

Hidalgo, E., and Demple, B. (1994). An iron-sulfur center essential for transcriptional activation by the redox-sensing Sox-R protein. *EMBO J.* **13**, 138–146.

Hidalgo, E., Bollinger, J. M., Jr., Bradley, T. M., Walsh, C. T., and Demple, B. (1995). Binuclear [2Fe-2S] clusters in the *Escherichia coli* SoxR protein and role of the metal centers in transcription. *J. Biol. Chem.* **270**, 20908–20914.

Hurshman, A. R., and Marletta, M. A. (1995). Nitric oxide complexes of inducible nitric oxide synthase: Spectral characterization and effect on catalytic activity. *Biochemistry* **34**, 56278–5634.

Ignarro, L. J. (1991). Signal transduction mechanisms involving nitric oxide. *Biochem. Pharmacol.* **41**, 485–490.

Ignarro, L. J., Wood, K. S., and Wolin, M. S. (1984). Regulation of purified soluble guanylate cyclase by porphyrins and metalloporphrins: A unifying concept. *Adv. Cyclic Nucleotide Protein Phosphorylation. Res.* **17**, 267–274.

Ignarro, L. J., and Kadowitz, P. J. (1985). The pharmacological and physiological role of cyclic GMP in vascular smooth muscle relaxation. *Ann. Pharmacol. Toxicol.* **25**, 171–191.

Ignarro, L. J., Buga, G. M., Wood, K. S., Byrns, R. E., and Chaudhuri, G. (1987). Endothelium-derived relaxing factor produced and released from artery and vein is nitric oxide. *Proc. Natl. Acad. Sci. U.S.A.* **84**, 9265–9268.

Ignarro, L. J., Fukuto, J. M., Griscavage, J. M., Rogers, N. E., and Byrns, R. E. (1993). Oxidation of NO in aqueous solution to nitrite but not nitrate: Comparison with enzymatically formed nitric oxide. *Proc. Natl. Acad. Sci. U.S.A.* **90**, 8103–8107.

Ishida, A., Sasaguri, T., Kosaka, C., Nojima, H., and Ogata, Y. (1997). Induction of the cyclin-dependent kinase inhibitor p21sdi1/cip1/Waf1 by nitric oxide-generating vasodilator in vascular smooth muscle cells. *J. Biol. Chem.* **272**, 10050–10057.

Jaffrey, S. R., Cohen, N. A., Rouault, T. A., Klausner, R. D., and Snyder, S. H. (1994). The iron-responsive element binding protein: A target for synaptic actions of nitric oxide. *Proc. Natl. Acad. Sci. U.S.A.* **91**, 12994–12998.

Jia, L., Bonaventura, C., Bonaventura, J., and Stamler, J. (1996). S-nitrosohaemoglobin: A dynamic activity of blood involved in vascular control. *Nature (London)* **380**, 221–226.

Johnston, H. M., and Morris, B. J. (1994a). Nitric oxide alters proenkephalin and prodynorphin gene expression in hippocampal granule cells. *Neuroscience* **61**, 435–439.

Johnston, H. M., and Morris, B. J. (1994b). NMDA and nitric oxide increase microtubute-associated protein 2 gene expression in hippocampal granule cells. *Ann. Neurochem.* **63**, 379–382.

Karthein, R., Nastainczyk, W., and Ruf, H. (1987). EPR study of ferric native prostaglandin H synthase and its ferrous NO derivative. *Eur. J. Biochem.* **166**, 173–180.

Kendrick, K. M., Guevara-Guzman, R., Zorilla, J., Hinton, H. R., Broad, K. D., Himmack, M., and Ohkura, S. (1997). Formation of olfactory memories mediated by nitric oxide. *Nature (London)* **388**, 670–674.

Kharitonov, V. G., Sundquist, A. R., and Sharma, V. S. (1995). Kinetics of nitrosation of thiols by nitric oxide in the presence of oxygen. *J. Biol. Chem.* **270**, 158–164.

Khosravi-Far, R., and Der, C. J. (1994). The Ras signal transduction pathway. *Cancer Metastasis Rev.* **13**, 67–89.

Klug, A., and Rhodes, D. (1987). Zinc fingers: A novel protein motif for nucleic acid recognition. *Trends Biochem. Sci.* **12**, 464–469.

Kourembanas, S., McQuillan, L. P., Leung, G. K., and Faller, D. V. (1993). Nitric oxide regulates the expression of vasoconstrictors and growth factors by vascular endothelium under both normoxia and hypoxia. *J. Clin. Invest.* **92**, 99–104.

Kroncke, K. D., Feshel, K., Schmidt, T., Zenke, F. T., Dasting, I., Wesener, J. R., Bettermann, H., Breunig, K. D., and Kolb-Bachofen, V. (1994). Nitric oxide destroys zinc-sulfur clusters inducing zinc release from metallothionein and inhibition of the zinc finger-type yeast transcription activator Lac9. *Biochem. Biophys. Res. Commun.* **200**, 1105–1110.

Lamas, S., Marsden, P. A., Li, G. K., Tempst, P., and Michel, T. (1992). Endothelial nitric oxide synthase: Molecular cloning and characterization of a distinct constitutive enzyme isoform. *Proc. Natl. Acad. Sci. U.S.A.* **89**, 6348–6352.

Lander, H. M. (1997). An essential role for free radicals and derived species in signal transduction. *FASEB J.* **11**, 118–124.

Lander, H. M., Sehajpal, P., Levine, D. M., and Novogrodsky, A. (1993). Activation of human peripheral blood mononuclear cells by nitric-oxide generating compounds. *J. Immunol.* **150**, 1509–1516.

Lander, H. M., Ogiste, J. S., Pearce, S. F. A., Levi, R., and Novogrodsky, A. (1995a). Nitric oxide-stimulated guanine nucleotide exchange on p21^{ras}. *J. Biol. Chem.* **270**, 7017–7020.

Lander, H. M., Ogiste, J. S., Teng, K. K., and Novogrodsky, A. (1995b). p21^{ras} as a common signaling target of reactive free radicals and cellular redox stress. *J. Biol. Chem.* **270**, 21195–21198.

Lander, H. M., Milbank, A. J., Tauras, J. M., Hajjar, D. P., Hempstead, B. L., Schwartz, G. D., Kraemer, R. T., Mirza, U. A., Chait, B. T., Campbell-Burk, S., and Quilliam, L. A. (1996a). Redox regulation of cell signalling. *Nature (London)* **381**, 380–381.

Lander, H. M., Jacovina, A. T., Davis, R. J., and Tauras, J. M. (1996b). Differential activation of mitogen-activated protein kinases by nitric oxide-related species. *J. Biol. Chem.* **271**, 19705–19709.

Lander, H. M., Hajjar, D. P., Hempstead, B. L., Mirza, U. A., Chait, B. T., Campbell, S., and Quilliam, L. A. (1997). A molecular redox switch on p21^{ras}. *J. Biol. Chem.* **272**, 4323–4326.

Lei, S. Z., Pan, Z. H., Aggarwal, S. K., Chen, H. S., Hartman, J., Sucher, N. J., and Lipton, S.A. (1992). Effect of nitric oxide production on the redox modulatory site of the NMDA receptor-channel complex. *Neuron* **8**, 1087–1099.

Li, J., Billiar, T. R. Talalnian, R. V., and Kim, Y. M. (1997). Nitric oxide reversibly inhibits seven members of the caspase family via S-nitrosylation. *Biochem. Biophys, Res. Commun.* **240**, 419–424.

Lijnen, H. R., and Collen, D. (1997). Endothelium in hemostasis and thrombosis. *Prog. Cardiovasc. Dis.* **39**, 343–350.

Lipton, S. A., Choi, Y. B., Pan, Z. H., Lei, S. Z., Chen, H.S. V., Sucher, N. J., Loscalzo, J., Singel, D.J., and Stamler, J. S. (1993). A redox-based mechanism for the neuroprotective and neurodestructive effects of nitric oxide and related nitroso-compounds. *Nature (London)* **364**, 626–632.

Lowenstein, C. J., Dinerman, J. L., and Snyder, S. H. (1994). Nitric oxide: A physiologic messenger. *Ann. Intern. Med.* **120**, 227–237.

Luperchio, S., Tamir, S., and Tannenbaum, S. R. (1996). NO-induced oxidative stress and gluthathione metabolism in rodent and human cells. *Free Radical Biol. Med.* **21**, 513–519.

Magrinat, G., Mason, S. N., Shami, P. J., and Weinberg, J. B. (1992). Nitric oxide modulation of human leukemia cell differentiation and gene expression. *Blood* **80**, 1880–1884.

Mannick, J. B., Asano, K., Izumi, K., Kieff, E., and Stamler, J. S. (1994). Nitric oxide produced by human B lymphocytes inhibits apoptosis and Epstein-Barr virus reactivation. *Cell (Cambridge, Mass.)* **79**, 1137–1146.

Mathews, J. S., McWilliams, P. J., Key, B. J., and Keen, M. (1995). Inhibition of prostacyclin release from cultured endothelial cells by nitrovasodilator durgs. *Biochim. Biophys. Acta* **1269**, 237–242.

Matthews, J. R., Botting, C. H., Panico, M., Morris, H. R., and Hay, R. T. (1996). Inhibition of NF-κB DNA binding by nitric oxide. *Nucleic Acids Res.* **24**, 2236–2242.

McDonald, L. J., and Moss, J. (1993). Stimulation by nitric oxide of an NAD linkage to glyceraldehyde-3-phosphate dehydrogenase. *Proc. Natl. Acad. Sci. U.S.A.* **90**, 6238–6241.

Mohr, S., Stamler, J. S., and Brune, B. (1996). Posttranslational modification of glyceraldehyde-3-phosphate dehydrogenase by S-nitrosylation and subsequent NADH attachment. *J. Biol. Chem.* **271**, 4209–4214.

Mohr, S., Zech, B., Lapetina, E. G., and Brune, B. (1997). Inhibition of caspase-3 by S-nitrosation and oxidation caused by nitric oxide. *Biochem. Biophys. Res. Commun.* **238**, 387–391.

Moncada, S. (1997). The biology of nitric oxide. *Funct. Neurol.* **12**, 135–140.

Morita, T., and Kourembanas, S. (1995). Endothelial cell expression of vasoconstrictors and growth factors is regulated by smooth muscle cell-derived carbon monoxide. *J. Clin. Invest.* **96**, 2676–2682.

Morita, T., Perrella, M. A., Lee, M.-E., and Kourembanas, S. (1995). Smooth muscle cell-derived carbon monoxide is a regulator of vascular cGMP. *Proc. Natl. Acad. Sci. U.S.A.* **92**, 1475–1479.

Morris, B. J. (1995). Stimulation of immediate early gene expression in striatal neurons by nitric oxide. *J. Biol. Chem.* **270**, 24740–24744.

Muller, C. W., Rey, F. A., Sodeka, M., Verdine, G. L., and Harrison, S. C. (1995). Structure of the NF-kappa B p50 homodimer bound to DNA. *Nature (London)* **373**, 311–317.

Nathan, C. (1992). Nitric oxide as a secretory product of mammalian cells. *FASEB J.* **6**, 3051–3064.

Nathan, C., and Xie, Q. W. (1994). Nitric oxide synthases: Roles, tolls and controls. *Cell (Cambridge, Mass.)* **78**, 915–918.

Nikitovic, D., Holmgren, A., and Spyrou, G. (1998). Inhibition of AP-1 binding by nitric oxide involving cysteine residues in Jun and Fos. *Biochem. Biophys. Res. Commun.* **242**, 109–112.

Nunoshiba, T., Derojas-Walker, T., Tannenbaum, S. R., and Demple, B. (1995). Roles of nitric oxide in inducible resistance of *Escherichia coli* to activated murine macrophages. *Infect. Immun.* **63**, 794–798.

Oba, T., Yamaguchi, M., Wand, S., and Johnson, J. D. (1992). Modulation of the Ca^{2+} channel voltage sensor and excitation-contraction coupling by silver. *Biophys. J.* **63**, 1416–1420.

Okazaki, M., Hu, Z., Fujinaga, M., and Hoffman, B. B. (1996). Adrenergic receptor activation of proto-oncogene expression in arterial smooth muscle. *Recept. Signal Transduct.* **6**, 165–178.

Overbeck, A. F., Brtva, T. R., Cox, A. D., Graham, S. M., Huff, S. Y., Khosravi-Far, R., Quilliam, L. A., Solski, P. A., and Der, C. J. (1995). Guanine nucleotide exchange factors: Activators of Ras superfamily protein. *Mol. Reprod. Dev.* **42**, 468–476.

Palmer, R. M. J., Ferrige, A. G., and Moncada, S. (1987). Nitric oxide release accounts for the biological activity of endothelium-derived relaxing factor. *Nature (London)* **327**, 524–526.

Palmer, R. M. J., Rees, D. D., Ashton, D. S., and Moncada, S. (1988). L-arginine is the physiological precursor for the formation of nitric oxide in endothelium dependent relaxation. *Biochem. Biophys. Res. Commun.* **153**, 1251–1256.

Park, S. K., Lin, H. L., and Murphy, S. (1997). Nitric oxide regulates nitric oxide synthase-2 gene expression by inhibiting NF-kappaB binding to DNA. *Biochem. J.* **322**, 609–613.

Patrick, D. A., Moore, E. E., Offner, P. J., Barnett, C. C., Barkin, M., and Silliman, C. C. (1997). Nitric oxide attenuates platelet-activating factor priming for elastase release in human neutrophils via a cyclic guanosine monophosphate-dependent pathway. *Surgery* **122**, 196–202.

Peng, H., Libby, P., and Liao, J. K. (1995). Induction and stabilization of IκBα by nitric oxide mediates inhibition of NFκB. *J. Biol. Chem.* **270**, 14214–14219.

Penunova, N., and Enikolopov, G. (1995). Nitric oxide triggers a switch to growth arrest during differentiation of neuronal cells. *Nature (London)* **375**, 68–73.

Pilz, R. B., Suhasini, M., Idriss, S., Meinkoth, J. L., and Boss, G. R. (1995). Nitric oxide and cGMP analogs activate transcription from AP-1-responsive promoters in mammalian cells. *FASEB J.* **9**, 552–558.

Poluha, W., Schonhoff, C. M., Harrington, K. S., Lachyankar, M. B., Crosbie, N. E., Bulseco, D. A., and Ross, A. H. (1997). A novel, nerve growth factor-activated pathway involving nitric oxide, p53, p21^{WAF1} regulates neuronal differentiation of PC12 cells. *J. Biol. Chem.* **272**, 24002–224007.

Radi, R., Beckman, J. S., Bush, K. M., and Freeman, B.A. (1991). Peroxynitrite oxidation of sulfhydryls. The cytotoxic potential of superoxide and nitric oxide. *J. Biol. Chem.* **266**, 4244–4250.

Radomski, M. W., Palmer, R. M. J., and Moncada, S. (1987). The anti-aggregating properties of vascular endothelium: Interactions between prostacyclin and nitric oxide. *Br. J. Pharmacol.* **92**, 639–646.

Rooney, C., Taylor, N., Countryman, J., Jenson, H., Kolman, J., and Miller, G. (1988). Genome rearrangements actvate the Epstein-Barr virus gene whose product disrupts latency. *Proc. Natl. Acad. Sci. U.S.A.* **85**, 9801–9805.

Ross, R. (1993). The pathogenesis of artherosclerosis: A perspective of the 1990s. *Nature (London)* **362**, 801–809.

Rothe, H., Hartmann, B., Geerlings, P., and Kolb, H. (1996). Interleukin-12 gene expression of macrophages is regulated by nitric oxide. *Biochem. Biophys. Res. Commun.* **224**, 159–163.

Salvemini, D.(1997). Regulation of cyclooxygenase enzyme by nitric oxide. *Cell. Mol. Life Sci.* **53**, 576–582.

Salvemini, D., Misko, T. P., Masferrer, J. L., Seibert, K., Currie, M. G., and Needleman, P. (1993). Nitric oxide activates cyclooxygenase enzymes. *Proc. Natl. Acad. Sci. U.S.A.* **90**, 7240–7244.

Schmidt, H. H. H. W., Lohmann, S. M., and Walter, U. (1993). The nitric oxide and cGMP signal transduction system: Regulation and mechanism of action. *Biochim. Biophys. Acta* **1178**, 153–175.

Schobersberger, W., Friedrich, F., Hoffmann, G., Volkl, H., and Dietl, P. (1997). Nitric oxide donors inhibit spontaneous depolarizations by L-type Ca2+ currents in alveolar epithelial cells. *Am. J. Physiol.* **272**, L1092–L1097.

Schwartz, S. M., deBlois, D., and O'Brien, E. R. M. (1995). The intima: Soil for artherosclerosis and restenosis. *Circ. Res.* **77**, 445–465.

Sciorati, C., Rovere, P., Ferrarini, M., Heltai, S., Manfredi, A. A., and Clementi, E. (1997). Autocrine nitric oxide modulates CD95-induced apoptosis in $\gamma\delta$ T lymphocytes. *J. Biol. Chem.* **272**, 23211–23215.

Sedmark, D. D., Knight, D. A., Vook, N. C., and Waldman, J. W. (1994). Divergent patterns of ELAM-1, ICAM-1 and VCAM-1 expression in cytomegalovirus infected endothelial cells. *Transplantation* 1379–1385.

Stamler, J. S. (1994). Redox signaling: Nitrosylation and related target interactions of nitric oxide. *Cell, (Cambridge, Mass.)* **78**, 931–936.

Stamler, J. S., Single, D., and Loscalzo, J. (1992a). Biochemistry of nitric oxide and its redox-activated forms. *Science* **258**, 1898–1902.

Stamler, J. S., Simon, D.I., Osborne, J. A., Mullins, M. E., Jaraki, O., Michel, T., Singel, D. J., and Loscalzo, J. (1992b). S-nitrosylation of proteins with nitric oxide: Synthesis and characterization of biologically active compounds. *Proc. Natl. Acad. Sci. U.S.A.* **89**, 444-448.

Stamler, J. S., Simon, D.I., Jaraki, O., Osborne, J. A., Francis, S., Mullins, M., Single, D., and Loscalzo, J. (1992c). S-nitrosylation of tissue-type plasminogen activator confers vasodilatory and antiplatelet properties on enzyme. *Proc. Natl. Acad. Sci. U.S.A.* **89**, 8087-8089.

Stoyanovsky, D., Murphy, T., Anno, P. R., Kim, Y. M., and Salama, G. (1997). Nitric oxide activates skeletal and cardiac ryanodine receptors. *Cell Calcium* **21**, 19-29.

Synder, S. H. (1992). Nitric oxide: First in a new class of neurotransmitters. *Science* **257**, 494-496.

Tabuchi, A., Oh, E., Taoka, A., Sakuri, H., Tsuchiya, T., and Tsuda, M. (1996). Rapid attenuation of AP-1 transcriptional factors associated with nitric oxide (NO)-mediated neuronal death. *J. Biol. Chem.* **271**, 31061-31067.

Thomas, S. R., Mohr, D., and Stocker, R. (1994). Nitric oxide inhibits indoleamine 2,3-dioxygenase activity in interferon-γ primed mononuclear phagocytes. *J. Biol. Chem.* **269**, 14457-14464.

Togashi, H., Sasaki, M., Frohman, E., Taira, E., Ratan, R. R., Dawson, T. M., and Dawson, V. L. (1997). Neuronal (type I) nitric oxide synthase regulates nuclear factor-κB activity and immunologic (type II) nitric oxide synthase expression. *Proc. Natl. Acad. Sci. U.S.A.* **94**, 2676-2680.

Tsai, A. L., Wei, C. H., and Kulmacz, R. J. (1994). Interaction between nitric oxide and prostaglandin H synthase. *Arch. Biochem. Biophys.* **313**, 367-372.

Tsurumi, Y., Murohara, T., Krasinski, K., Chen, D., Witzenbichler, B., Kearney, M., Couffinhal, T., and Isner, J. M. (1997). Reciprocal relation between VEGF and NO in the regulation of endothelial integrity. *Nat. Med.* **3**, 879-886.

Tuder, R. M., Flook, B. E., and Voelkel, N. F. (1995). Increased gene expression for VEGF and the VEGF receptors KDR/FIK and Flt in lungs exposed to acute or to chronic hypoxia. *J. Clin. Invest.* **95**, 1798-1807.

Ullrich, A., and Schlessinger, J. (1990). Signal transduction by receptors with tyrosine kinase activity. *Cell (Cambridge, Mass.)* **61**, 203-212.

van der Zee, R., Murohara, T., Luo, Z., Zollman, F., Passeri, J., Lekutat, C., and Isner, J. M. (1997). Vascular endothelial growth factor (VEGF)/vascular permeability factor (VPF) augments nitric oxide release from quiescent rabbit and human vascular endothelium. *Circulation* **95**, 1030-1037.

Verma, A., Hirsch, D. J., Glatt, C. E., Ronnett, G. V., and Snyder, S. H. (1993). Carbon monoxide: A putative neural messenger. *Science* **259**, 381-384.

Villarete, L. H., and Remick, D. G. (1995). Nitric oxide regulation of IL-8 expression in human endothelial cells. *Biochem. Biophys. Res. Commun.* **211**, 671-676.

Villarete, L. H., and Remick, D. G. (1997). Nitric oxide regulation of interleukin-8 gene expression. *Shock* **7**, 29-35.

Weiss, G., Goossen, B., Doppler, W., Fuchs, D., Pantopoulos, K., Werner-Felmayer, G., Watcher, H., and Hentze, M. W. (1993). Translational regulation via iron-responsive elements by the nitric oxide/NO-synthase pathway. *EMBO J.* **12**, 3651-3657.

Williams, D. L. H. (1985). S-nitrosation and the reactions of S-nitrosocompounds. *Chem. Soc. Rev.* **14**, 171-196.

Wink, D. A., and Laval, J. (1994). The Fpg protein, a DNA repair enzyme, is inhibited by the biomediator nitric oxide *in vitro* and *in vivo*. *Carcinogenesis (London)* **15**, 2125-2129.

Wu, M., Pritchard, K. A., Kaminiski, P. M., Fayngersh, R. P., Hintze, T. H., and Wolin, M. S. (1994). Involvement of nitric oxide and nitrosothiols in relaxation of pulmonary arteries to peroxynitrite. *Am. J. Physiol.* **266,** H2108–H2113.

Xanthoudakis, S., and Curran T. (1992). Identification and characterization of Ref-1, a nuclear protein that facilitates AP-1 DNA-binding activity. *EMBO J* **11,** 653–665.

Xie, Q-W., Cho, H. J., Calaycay, J., Mumford, R. A., Swiderek, K. M., Lee, T. D., Ding, A., Troso, T., and Nathan, C. (1992). Cloning and characterization of inducible nitric oxide synthase from mouse macrophages. *Science* **256,** 225–228.

Xie, Q-W., Kashiwabara, Y., and Nathan, C. (1994). Role of transcription factor NF-kappa B/Rel in induction of nitric oxide synthase. *J. Biol. Chem.* **269,** 4705–4708.

Xu, L., Eu, J. P., Meissner, G., and Stamler, J. S. (1998). Activation of cardiac calcium release channel (ryanodine receptor) by poly-S-nitrosylation. *Science* **279,** 234–237.

Yoshioka, K., Thompson, J., Miller, M. J., and Fisher, J. W. (1997). Inducible nitric oxide synthase expression and erythropoietin production in human hepatocellular carcinoma cells. *Biochem. Biophys. Res. Commun.* **232,** 702–706.

Yun, H. Y., Dawson, V. L., and Dawson, T. M. (1997). Nitric oxide in health and disease of the nervous system. *Mol. Psychiatry* **2,** 300–310.

Zheng, M., Aslund, F., and Storz, G. (1998). Activation of the OxyR transcription factor by reversible disulfide bond formation. *Science* **279,** 1718–1721.

II Redox-Sensitive Molecular Processes and Cellular Responses

8 Reactive Oxygen Species as Costimulatory Signals of Cytokine-Induced NF-κB Activation Pathways

Patrick A. Baeuerle
Micromet GmbH
D-82152 Martinsried, Germany

Inflammation relies on the coordinated expression of numerous genes in response to primary and secondary inflammatory stimuli. Transcription factor NF-κB has been recognized as a central regulator of inflammatory genes that is activated by the proinflammatory cytokines tumor necrosis factor-α (TNF) and interleukin-1β (IL-1) but also by many other pathogenic conditions. Among the numerous activators of NF-κB are oxidants and many conditions causing a prooxidant state in cells, such as radiation, transition metals, and toxins. In many instances, administration of antioxidants or expression of antioxidative enzymes was found to prevent NF-κB activation. With a detailed understanding of the signaling events and the discovery of most, if not all, protein components leading to NF-κB activation in response to TNF and IL-1, the stage is now set to investigate at a molecular level how oxidants and antioxidants affect NF-κB activation and at which level they act on cytokine-induced pathways. This chapter reviews the most recent insights into NF-κB activation by cytokine-induced pathways and discusses a model in which reactive oxygen species (ROS) act as important costimulatory signals rather than second messengers of these pathways.

I. FUNCTIONS OF TRANSCRIPTION FACTOR NF-κB

In inflamed tissue, many genes are turned on that are not normally needed by the organism. Only upon challenge of cells and tissues with bacteria, viruses, or physical and chemical stressors, are proinflammatory genes activated and their encoded proteins expressed. The newly synthesized proteins serve to defend the organism and to put it into a stage of alert. Many are signaling proteins, such as cytokines, growth factors, or chemokines, which activate the immune system; others are enzymes that synthesize inflammatory mediators or are receptors that, depending on the cell type, make cells respond better to inflammatory, growth-stimulatory, or apoptotic signals.

Because the often tiny amounts of pathogens are not able to stimulate proinflammatory gene expression in many cells, the organism employs a potent amplification system by producing the proinflammatory cytokines TNF and IL-1. These proteins can elicit more or less the same genomic response as the primary pathogenic stimulus. Because they can be produced in vast amounts and are diffusible, they can reach many cells in the organism. In the worst case, overproduced cytokines cause a deadly systemic syndrome, called septic shock. In the optimal case, they make sure that the defense systems of the body are rapidly, coordinately, and widely activated in response to infection and trauma. If TNF and IL-1 are not produced in an acute and transient fashion but continuously, they can play a causative role in chronic proinflammatory diseases, such as rheumatoid arthritis, asthma, or inflammatory bowel disease (Feldman et al., 1996). Their role in such diseases is best documented by the beneficial therapeutic effect of TNF-neutralizing antibodies and receptor domains (reviewed by Eigler et al., 1997).

The proinflammatory cytokines TNF and IL-1 and many primary pathogenic stimuli both activate the same transcription regulating protein, called NF-κB, which is responsible for a coordinate induction of proinflammatory genes (for reviews, see Baeuerle and Henkel, 1994; Verma et al., 1995; Siebenlist et al., 1995; Baeuerle and Baltimore, 1996; Baldwin, 1996; Barnes and Karin, 1997). NF-κB was shown to bind the regulatory DNA elements of more than 50 proinflammatory genes (for a list, see Baeuerle and Baichwal, 1997). In most cases, the deletion of the NF-κB DNA-binding motif in promoters was found to ablate the transcriptional response of the genes to TNF or IL-1. NF-κB binds to DNA as a dimer, which is frequently composed of the DNA-binding subunits p50 and RelA/p65. Three other NF-κB DNA-binding subunits are known, c-Rel, RelB, and p52, which dimerize among each other. NF-κB proteins occur in most cell types. In some, such as mature B cells or certain neurons, NF-κB is found in an active nuclear form, whereas in most cell types it is inactivated. Null mutant mice for all five DNA-binding NF-κB subunits have been produced. Their phenotypes are consistent with an important role of NF-κB proteins in regulating immune and inflammatory responses (reviewed by Sha, 1998).

A hallmark of NF-κB is that it does not need to be newly synthesized but can be rapidly activated from a preexisting cytoplasmic form, which makes NF-κB particularly suitable as an immediate-early regulator of inflammation. In a rather unique fashion, NF-κB is kept inactive by association with IκB, an inhibitory subunit (Baeuerle and Baltimore, 1988). There are a number of IκB subunits known, from α to ε, which allow for a differential control of NF-κB proteins (reviewed by Baldwin, 1996). In response to a large number of proinflammatory stimuli, various IκB forms are irreversibly inactivated by proteasome-mediated degradation. The released NF-κB is rapidly translocated into the nucleus where it will initiate transcription binding to regulatory DNA elements of genes. The degradation of IκB proteins in response to TNF and IL-1 is controlled by phosphorylation of two serine residues in their N-terminal domains. The newly phosphorylated IκB is selectively recognized by the ubiquitin-conjugating enzymes and is subject to polyubiquitination, the prerequisite for a rapid degradation by the proteasome. The cytokine-inducible kinases responsible for IκB-α phosphorylation on serines 32 and 36 were identified only recently (see later).

Some stimuli such as hypoxia, reoxygenation, and the tyrosine phosphatase inhibitor pervanadate activate NF-κB without causing IκB degradation (Imbert *et al.*, 1996). These stimuli trigger phosphorylation of IκB-α on tyrosine 42. The tyrosine-phosphorylated IκB associates with the SH2 domains of the p85-α subunit of phosphatidylinositol 3-kinase and NF-κB is activated, which also relies on the kinase activity of the catalytic subunit (Béraud *et al.*, 1999). For the majority of the more than 100 stimuli reported to activate NF-κB (for a list, see Baeuerle and Baichwal, 1997) it is not yet known by which mechanism IκB is inactivated.

Mice lacking a functional RelA/p65 gene die early *in utero* from apoptosis of liver cells, suggesting a role of NF-κB in protecting cells from apoptosis (reviewed by Sha, 1998). These results were corroborated by the finding that overexpression of dominant-negative IκB mutant proteins sensitize cells for proapoptotic stimuli, such as TNF and certain anticancer compounds (reviewed in Baichwal and Baeuerle, 1997). NF-κB is apparently inducing genes whose products protect cells from a simultaneous proapoptotic effect of the stimulus. It is well established that TNF-receptor 1 (TNFR1) will not only activate NF-κB but at the same time induces the activation of caspases leading to apoptotic cell death. As long as the NF-κB pathway is functional and activated at the same time as the proapoptotic pathway, cells may be protected. However, if cells are not capable of activating NF-κB or are blocked in subsequent steps of new protein synthesis, e.g., as a consequence of a viral infection or cell damage, the apoptosis pathway prevails and cells are killed. This regulatory system allows for a selective killing by TNF of dysfunctional, overly stressed, or infected cells.

II. HOW TNF AND IL-1 ACTIVATE NF-κB

The key trigger in activation of the NF-κB–IκB complex is the *de novo* phosphorylation of IκB. Most of the components that control this event in response to binding of TNF and IL-1 to their respective cell surface receptors are now identified (Fig. 1). The reader will find most of the relevant references in reviews by Maniatis (1997), Stancovski and Baltimore (1997), Baeuerle (1998), and Baichwal and Baeuerle (1998).

TNF receptor-1 is the subtype that plays a predominant role in the proinflammatory actions of TNF. Binding of the trimeric TNF ligand to the extracellular domain of TNFR1 induces aggregation of the cytoplasmic death domains (DDs), which also serve in many other receptors as protein–protein interaction motifs. Upon trimerization, the DDs of TNFR1 assemble a signaling complex containing at least three different protein components. First, the TNF receptor-associated death domain protein (TRADD) is recruited. Subsequently, the TNF receptor-associated factor 2 (TRAF2) and the receptor-interacting protein (RIP) associate with the adapter TRADD. Of the three proteins, TRADD, TRAF2, and RIP, only RIP has a known intrinsic enzymatic activity in its N-terminal serine/threonine-specific kinase domain. Although RIP is essential for NF-κB induction by TNF, its kinase activity appears not to be required for this particular function. Apart from activating inflammatory genes through NF-κB, the TRADD–RIP–TRAF2 signaling complex activates at least two additional pathways. TRADD can activate apoptosis through activation of a caspase cascade, and TRAF2 can indirectly activate Jun N-terminal kinase (JNK), thereby activating AP-1-dependent gene expression. While initial experiments with transdominant negative mutants suggested a role of TRAF2 in NF-κB activation, results from knockout mice show that TRAF2 may be solely responsible for JNK/AP-1 activation (Lee *et al.*, 1997).

TRAF2 is a member of a new family of signaling proteins that is characterized by the presence of a conserved TRAF domain of approximately 230 amino acids in the carboxy-terminal portion and multiple Zn-finger motifs within the amino-terminal half. The TRAF domain mediates homotypic and heterotypic interactions and is also involved in binding to other signaling molecules and TNF receptor superfamily members. A total of six TRAF proteins have been identified to date that are involved in signaling by various TNF receptor superfamily members and in other cellular functions.

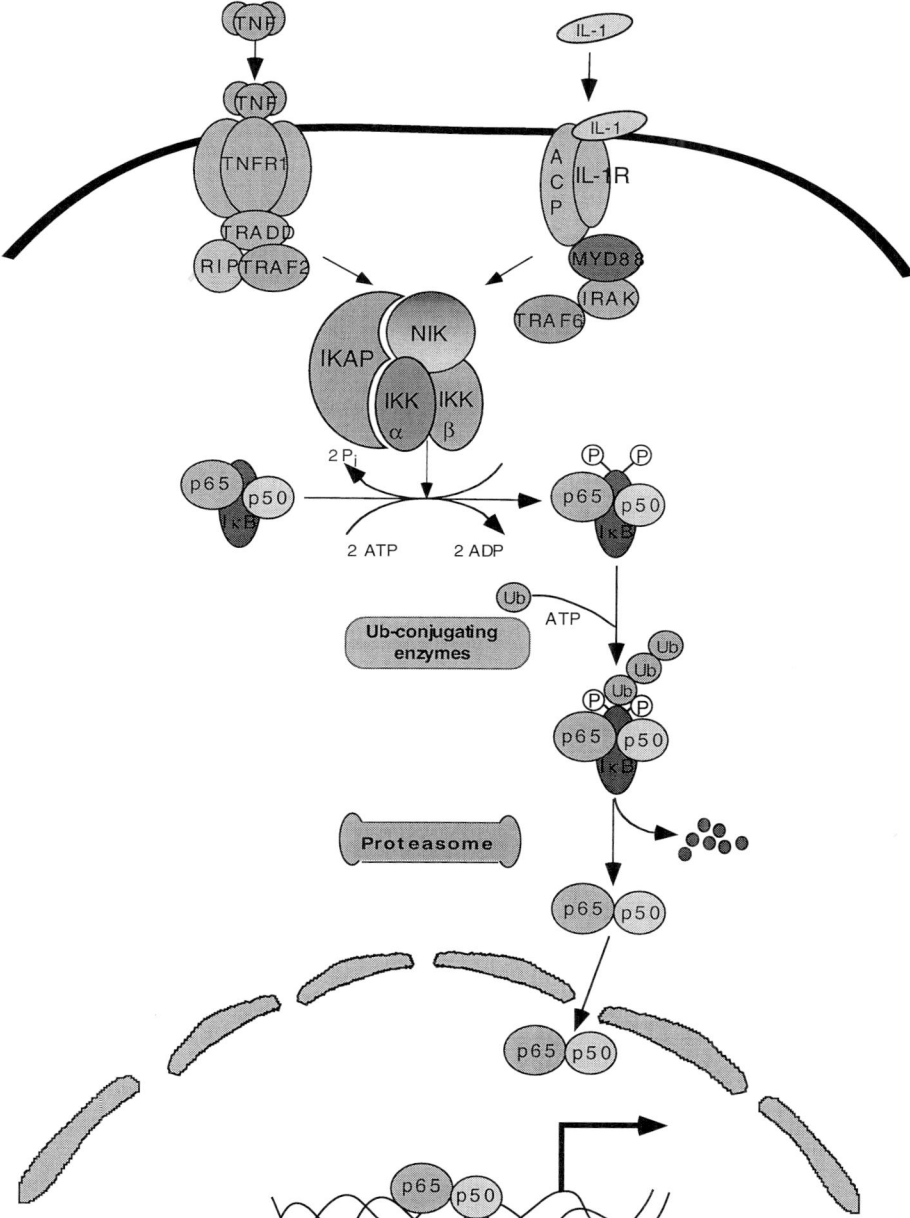

Figure 1 Known and newly discovered proteins in TNF- and IL-1-induced pathways of NF-κB activation. For abbreviations, see text.

Although the IL-1 receptor is entirely unrelated in sequence to members of the TNF receptor superfamily, NF-κB activation by IL-1 involves a TRAF protein as well. Activation of IL-1R by IL-1 induces dimerization with an accessory membrane-spanning protein, called AcP, which also belongs to the IL-1 receptor family. Like the trimerized TNF receptor-1, the activated IL-1R/AcP complex first recruits an adapter protein and, subsequently, a serine/threonine kinase and a TRAF protein. The TRADD-equivalent adapter protein is called MyD88 (Wesche et al., 1997), the kinase is IL-1 receptor-activated kinase (IRAK), and the TRAF protein is TRAF6. Neither RIP nor IRAK kinases can phosphorylate IκB directly. How then do two unrelated receptors using their own specific signaling complexes—TRADD–RIP–TRAF2 for TNFR-1 and MyD88–IRAK–TRAF6 for IL-1R/AcP—both trigger phosphorylation of IκB on the same serine residues?

The first shared signaling protein in TNF and IL-1 pathways emerged as a novel MAP3 kinase-related protein called NF-κB-inducing kinase (NIK), which was identified in a yeast two-hybrid screen in search for TRAF2-interacting proteins. NIK, which can also interact with TRAF6, activates NF-κB when overexpressed and a kinase-inactive NIK mutant protein blocks NF-κB activation by both TNF and IL-1, as well as in response to TRAF2 and TRAF6 overexpression. These and other observations place NIK downstream from the TRAFs and at a point where TNF and IL-1 pathways converge.

The link between NIK, which like IRAK and RIP cannot phosphorylate IκB by itself, and the signal-induced phosphorylation of IκB was finally established by the discovery of a NIK-binding protein kinase, called IκB kinase α (IKK-α). The protein was discovered independently in a yeast two-hybrid screen using NIK as bait and as a component of a purified IκB kinase complex. IKK-α, which has a conserved helix–loop–helix motif, was cloned previously but its function remained unknown. An EST data base search and biochemical purification revealed the existence of a related kinase, called IKK-β, which shares more than 50% overall identity with IKK-α. The two IKKs can phosphorylate various IκB proteins at their critical serine residues and apparently act as heterodimers in a large complex with NIK to phosphorylate and ultimately trigger the degradation of IκB. Consistent with a critical role of IKKs, their overexpression activates NF-κB; the expression of kinase-inactive forms specifically inhibit TNF and IL-1 signaling; their activity is stimulated by cytokines; and the purified recombinant proteins show the proper substrate specificity.

Dimerization between IKKs is mediated by their leucine zipper motifs and appears necessary for kinase activity. Despite its affinity for IKKs, NIK has not been observed so far in highly purified IκB kinase complexes. Only in one case using IκB substrate affinity purification (Cohen et al., 1998) was NIK observed as a component of a large 700- to 800-kDa IκB kinase complex that also contained NF-κB, IκB, and a novel 150-kDa protein. The 150-kDa protein can independently bind NIK and IKKs via distinct domains and was dubbed IKAP (for IκB kinase complex-associated protein). IKAP can assemble an active complex of NIK, IKK-α, and IKK-β and hence may function as a scaffold protein with the potential to control activation and inactivation of the IKK complex via assembly and disassembly of its kinase components.

A major focus of research is now to understand how NIK is activated in response to the assembly of upstream signaling complexes and how their signal is relayed through a complex of three distinct kinases. With the identification of IKKs and NIK, a contiguous pathway is outlined that links TNF and IL-1 pathways from the cell surface to the cytoplasmic NF-κB–IκB complex. Do we now know all the essential components required for activation of NF-κB in response to TNF (and IL-1)? In support of this idea is the fact that all presently known protein components can physically interact with each other in a consecutive fashion, e.g., TNF with TNFR1 with TRADD with RIP with TRAF2 with NIK with IKKs with IκB with NF-κB. Perhaps there are more signaling components to be discovered that associate peripherally with the known ones and allow branching into other signaling pathways or modulate the amplitude of signaling. In view of these most recent revelations, a role of ROS as messengers in cytokine-induced NF-κB activation pathways needs to be revisited.

III. NF-κB ACTIVATION BY PROOXIDANT CONDITIONS

A number of reviews have covered in depth the involvement of ROS and their inhibitors in NF-κB activation (Schreck *et al.*, 1992; Schreck and Baeuerle, 1994; Baeuerle *et al.*, 1996; Sen and Packer, 1996; Flohé *et al.*, 1997; Piette *et al.*, 1997; Schulze-Osthoff and Baeuerle, 1998; Schulze-Osthoff *et al.*, 1995). This chapter therefore focuses on those stimuli and inhibitors that may be most relevant and reliable to study the involvement of ROS in what now is the best understood NF-κB activation pathway.

An increasing number of oxidants has been found to activate NF-κB but it is presently unknown whether these stimuli feed into the cytokine-induced signaling pathways of NF-κB activation or whether they take alternative routes involving unrelated kinases and distinct mechanisms of IκB inactivation. The addition of hydrogen peroxide to the cell culture medium was found to activate NF-κB at micromolar concentrations in a number of cell types, such as HeLa, endothelial cells, several T-cell lines, and some primary cells (e.g., Schreck *et al.*, 1991; Los *et al.*, 1995). Likewise, hydrogen peroxide induces the new expression of proteins encoded by NF-κB-regulated genes. Examples include ICAM-1 (Bradley et al., 1993), interleukin-8 (DeForge et al., 1993), and inducible nitric oxide synthase (Adcock et al., 1994). Hydrogen peroxide may be a physiological stimulus of the transcription factor as monocytes and neutrophils produce and release large amounts of hydrogen peroxide during inflammation. Smaller amounts of the membrane-permeable oxidant can be produced by almost every cell type as a by-product of mitochondrial respiration as well as by many enzymatic reactions using dioxygen as the electron acceptor. The hydrogen peroxide that may activate NF-κB *in situ* could therefore arise from extracellular sources, such as stimulated immune cells, or be produced intracellularly. In the case of an intracellular production, a particular stimulus could change the normal cellular turnover of hydrogen peroxide, leading to either its increased production or its decreased elimination by enzymes. Other prooxidant conditions reported to activate NF-κB in intact cells are hypochlorite (Schoonbroodt et al., 1997), singlet oxygen (Legrand-Poels et al., 1995), hyperoxia (Li et al., 1997), UV light (Vile et al., 1995), and reoxygenation (Rupec and Baeuerle 1995; Imbert et al., 1996).

In trying to understand how hydrogen peroxide and other oxidants activate NF-κB, one must keep in mind that these are toxic compounds that can lead to covalent modification of many proteins in the cell. Hydrogen peroxide, for instance, can cause cysteine residues to undergo oxidation, resulting in the formation of disulfide bonds or the (reversible) addition of glutathione to cysteine residues, a reaction called thiolation. The most likely mode of ROS action may thus involve the oxidative damage and subsequent inactivation of a protein that functions to keep NF-κB in its inactive state. Currently there are two classes of such proteins known: IκB proteins and yet to be defined phosphatases that counteract NF-κB-activating kinases. Selective oxidative damage of any of these proteins may be able to activate NF-κB but has not been observed to date.

It is also possible that ROS-mediated damage can activate rather than inhibit components of the known signaling pathway. Key to the activation of TNF and IL-1 pathways is an aggregation of signaling proteins (such as TRAFs, TRADD, RIP, IRAK, etc). One can imagine that oxidants promote a cross-linking or aggregation of activating signaling components. A third possibility is that a separate signaling pathway has evolved that is specialized in responding to increased levels of ROS. Such a pathway could converge into known NF-κB activation pathways or activate NF-κB by a separate mechanism. Hydrogen peroxide is known to be an activator of MAP kinases (e.g., Mueller et al., 1997a,b) and a potent costimulus of protein tyrosine phosphorylation *in vivo* (Heffetz and Zick, 1989). Oxidation of a crucial cysteine residue in the active site of protein tyrosine phosphatases may account for the effect of hydrogen peroxide on tyrosine phosphorylation. Overexpression of kinase-inactive mutant proteins will allow to investigate whether hydrogen peroxide uses a pathway for NF-κB activation containing any of the newly discovered kinases. Key to the elucidation of a NF-κB-activating pathway is also the biochemical analysis of how IκB is inactivated, particularly what kind of covalent modifications the inhibitor undergoes in the course of NF-κB activation.

Given the toxic nature of hydrogen peroxide it is not surprising to find that NF-κB activation is only seen within a small concentration range of the oxidant (Meyer et al., 1993). Unlike cytokine ligands for surface receptors, high concentrations of hydrogen peroxide do not saturate the response but will inhibit NF-κB activation. The concentration window at which a ROS species induces NF-κB may vary greatly with the cell type. It has been reported that certain cell types will not show any NF-κB activation by hydrogen peroxide. This could be related to the efficiency of the ROS-metabolizing enzyme status of the cell. It may be so efficient that the effective ROS concentration is shifted toward very high concentrations. Alternatively, hydrogen peroxide elimination may be so inefficient that cells get severely damaged before they can activate NF-κB. It is equally possible that not every cell type expresses a particular protein that functions as a sensor/receptor for the ROS. A rather trivial explanation for an apparent insensitivity of cultured cells for hydrogen peroxide is that the cell culture medium may contain high levels of catalase activity that will rapidly decompose the oxidant.

The observation that ROS take considerably longer than cytokines to activate NF-κB (Schreck et al., 1991) may indicate that they act in a very indirect fashion. In the simplest case, protein synthesis may be transiently impaired by ROS, resulting in loss of IκB due to its short half-life. However, the observation that hydrogen peroxide induces IκB phosphorylation and degradation argues for an effect of this particular ROS on IκB kinase regulation (Kretz-Remy et al., 1998). Because the protein synthesis inhibitor cycloheximide does not prevent NF-κB activation by hydrogen peroxide in a T-cell line (Schreck et al., 1991), an involvement of newly synthesized cytokines as autocrine stimuli is unlikely.

There are more than 100 conditions known to date to activate NF-κB in various cell types (for an updated list, see Baeuerle and Baichwal, 1997). A number of these stimuli were reported to not only activate NF-κB but also to induce a transient and rapid production of ROS (see Schreck et al., 1992). The prooxidant intracellular state induced by such stimuli is characterized by a depletion of the reduced form of glutathione and a concomitant increase of its oxidized form, glutathione disulfide, and an increased cellular production of ROS, such as superoxide and hydrogen peroxide. Using various techniques to detect ROS production, a wide variety of conditions known to activate NF-κB were found to cause a transient and rapid increase in the cellular production of hydrogen peroxide and superoxide (reviewed by Schreck et al., 1992; Schreck and Baeuerle, 1994). It was thus tempting to speculate that ROS may serve as a common second messenger for NF-κB activation (Schreck and Baeuerle, 1991). As described earlier, discoveries suggest that TNF and IL-1 signaling pathways function as a contiguous signaling system involving protein–protein interactions. To date, no ROS-generating enzyme has been found as a component in TNF and IL-1 pathways. It is also not known whether any of their known signaling components are subject to a physiologically relevant, intracellular redox regulation. As discussed in the following chapter, we are nevertheless faced with the inhibitory effect of antioxidants on TNF and IL-1 signaling and, even more compelling, the inhibitory effect of overexpressed antioxidative proteins in this seemingly well-understood pathway of NF-κB activation.

IV. INHIBITORY EFFECT OF ANTIOXIDANTS AND REDOX-REGULATING PROTEINS ON NF-κB ACTIVATION BY CYTOKINES

Numerous studies showed that structurally unrelated chemicals at a rather high micro- or millimolar concentration can prevent NF-κB activation in response to many stimuli, including TNF and IL-1 (reviewed in Schreck et al., 1992; Schreck and Baeuerle, 1994; Sen and Packer, 1996; Flohé et al., 1997; Piette et al., 1997; Schulze-Osthoff and Baeuerle, 1998; Schulze-Osthoff et al., 1995). Among these chemicals are glutathione precursors (N-acetylcysteine), thiols (dithothreitol, cysteine and derivatives, lipoic acid), dithiocarbamates, metal chelators, vitamin E and derivatives, salicylates, and flavonoids. A common denominator of such compounds is that they have the potential—among other activities—to neutralize ROS within the cell. This is reflected by their stabilizing effect in intact cells on reduced glutahione levels, their prevention of ROS production, or their protective effect on ROS-induced cell death. In each case it is hard to discern by what chemical activity such compounds prevent NF-κB activation: ROS scavenging, transition metal chelation, glutathione stabilization, kinase or protease inhibition, combinations thereof, or other compound-specific effects. There is now a vast literature on the effects of antioxidative chemicals on NF-κB activation that gives a rather inconsistent picture. Considerable cell type-specific effects are observed and occasionally an antioxidant is found to inhibit one pathway but not another one in the same cell type (e.g., Brennan and O'Neill, 1995; Bonizzi et al., 1996).

In conclusion, although antioxidant chemicals can be helpful reagents in some cases, they do not appear as a very reliable tool in studying a role of ROS in NF-κB activation. Their effects vary with cell types; they need to be used at rather high concentrations; they may act pleiotropically; and, depending on their concentration, they can even act as prooxidants. A preferred, less artifact-prone approach in studying the intracellular role of ROS is the genetic manipulation of the level of cytoplasmic enzymes and other proteins that the cell employs to control its intracellular ROS levels. Stable overexpression of mitochondrial Mn-dependent superoxide dismutase was found early on to protect cells from the cytotoxic effect of TNF (Wong *et al.*, 1988). Interestingly, the TNF-inducible Mn-SOD gene has been discovered to be regulated by NF-κB (Jones *et al.*, 1997).

The use of stably transfected cell lines allows for the selection of lines that have a defined and moderate but still effective levels of overexpressed antioxidative proteins. Transient overexpression usually is more problematic because it may lead to variable and aberrantly high enzyme levels. Transfection efficiencies are distinct for every cell (and cell type), which may yield inconclusive results because some cells will remain untransfected whereas others have low to very high levels of the transfected gene and consequently enzyme levels.

A number of studies now show that the stable or transient overexpression of cellular proteins involved in ROS metabolism has an inhibitory effect on NF-κB activation by cytokines. Overexpression of catalase (Schmidt *et al.*, 1995), glutathione peroxidase (Kretz-Remy *et al.*, 1996; Renard *et al.*, 1997), thioredoxin (Schenk *et al.*, 1994), and a novel thioredoxin peroxidase (Jin *et al.*, 1997) was found to prevent NF-κB activation in response to TNF, IL-1, and/or phorbol ester. Likewise, overexpression of the small heat shock protein Hsp27 was found to exert an antioxidative effect in cells and to prevent TNF-induced NF-κB activation (Mehlen *et al.*, 1996, 1997). Hsp27 can increase intracellular glutathione levels, thereby exerting an antioxidative function. Overexpression of Cu/Zn-dependent superoxide dismutase did not inhibit NF-κB activation by TNF (Schmidt *et al.*, 1995). Rather, a potentiation of NF-κB activation by TNF was observed on overexpression of this enzyme, which was perhaps due to an increased production of hydrogen peroxide from superoxide. Taken together, these results provide the best indication to date that prooxidant conditions play a role in signaling and NF-κB activation by the proinflammatory cytokines TNF and IL-1. It is very hard to dismiss the evidence provided by the overexpression of five very distinct proteins that only have their antioxidative activity in common.

V. HOW CAN ANTIOXIDANTS AND ENZYMES AFFECT NF-κB ACTIVATION BY CYTOKINES?

Although TNF-inducible ROS production has been observed in a number of cell lines, an obligatory role in TNF signaling is far from being proven.

It is also not understood how TNF can increase the intracellular levels of superoxide and hydrogen peroxide. TNF and other apoptotic stimuli are thought to impair mitchondrial function, resulting in cytochrome c release and the production of large amounts of ROS (Cai and Jones, 1998). The protective effect of Mn-SOD overexpression on TNF cytotoxicity may be related to this mitochondrial ROS production. It was, however, observed that low levels of ROS can be produced very rapidly in response to TNF, which may not necessarily involve mitochondria (Meier et al., 1989).

When activated by cytokine binding, the TNFR-1 is known to produce a host of distinct signals inside the cell (Fig. 2). Not only does TNF activate NF-κB via the TRADD–RIP–NIK–IKK pathway, but also a proapoptotic pathway via TRADD (or FADD) and caspase-8, the JNK pathway via

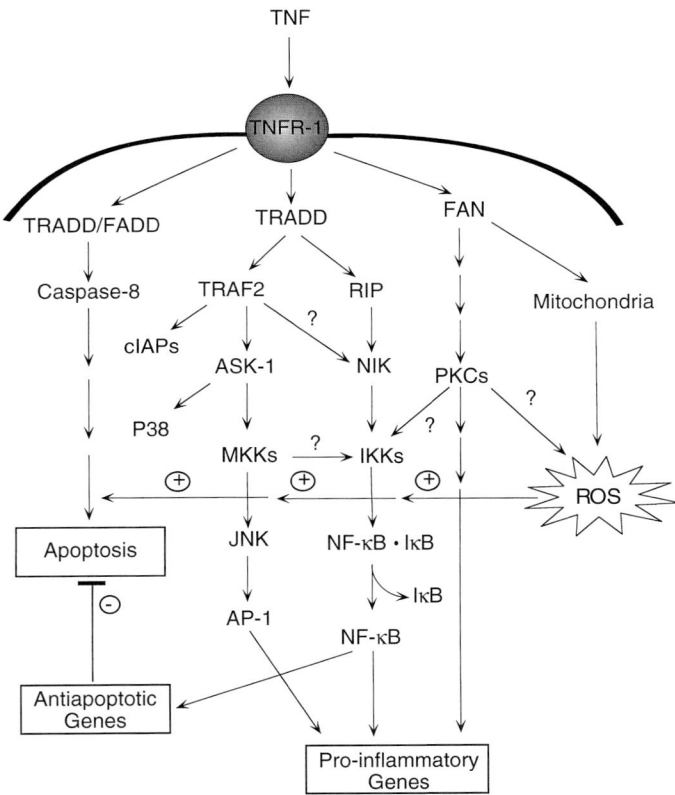

Figure 2 The TNF signaling network. The scheme illustrates the various signaling pathways activated by TNFR-1 and the potential cross-talk between components of the branches.

TRAF2 (Liu et al., 1996; Lee et al., 1997), an antiapoptotic pathway via cIAPs (Rothe et al., 1995), a tyrosine kinase pathway via Pyk2 (Tokiwa et al., 1996), and a pathway activating atypical protein kinase C (PKC) (Diaz-Meco et al., 1993), presumably via the TNFR-binding WD repeat protein FAN (Adam-Klages et al., 1996). It is conceivable that one or more of these pathways produce ROS transiently. Although this may not occur in the branch leading to NF-κB activation, inducibly generated ROS may act as a costimulus on the simultaneously activated NF-κB pathway and on other TNF-induced pathways. This idea is expressed in Figs. 2 and 3. The NF-κB pathway may still function without a parallel production of ROS but may be less effective if ROS are eliminated by antioxidant chemicals or overexpressed enzymes. In support of a modulatory role of ROS is that hydrogen peroxide has a strong costimulatory effect on TNF-induced NF-κB activation (Ginn-Pease and Whisler, 1996; Janssen et al., 1998).

Future research needs to investigate whether individual signaling components in the NF-κB pathway of TNF are redox sensitive in that they signal better if oxidized. An example of a bacterial signaling protein that is activated on oxidation (i.e., by disulfide bonding) is OxyR, a transcription factor turning on genes specifically in response to hydrogen peroxide (Zheng and Storz, 1998). A systematic mutation analysis may identify reactive cysteine residues in TNF signaling components. This approach could produce mutant proteins that abolish the costimulatory effect of ROS as well as the sensitivity of the pathway to inhibition by antioxidants or overexpressed antioxidative enzymes. It is, however, also conceivable that modulatory redox-sensitive proteins are not within the NF-κB pathway but control its signaling components laterally.

Future research also needs to identify the enzymatic system that generates the small burst of ROS production observed in response to inflammatory cytokines. From macrophages it is known that PKC is a key regulator of a massive oxidative burst that involves activation of a membrane-bound flavoenzyme, called NADPH oxidase (reviewed in Halliwell and Gutteridge, 1989). An atypical protein kinase C species is thought to be involved in NF-κB activation by TNF (Diaz-Meco et al., 1993) and there is evidence for the existence of less active forms of NADPH oxidases in many other cell types (Meier et al., 1989). Specific oxidase inhibitors and genetic studies are suitable to investigate the role of atypical PKCs and NADPH oxidases in the TNF-induced miniature oxidative burst. It has also been suggested that mitochondria are an important source of ROS for both cytotoxicity and NF-κB activation by TNF (Schulze-Osthoff et al., 1993).

A redox-sensitive element in the TNF pathway could be identified by genetic experiments. Overexpression of individual signaling components will activate NF-κB, as shown for TNFR-1, TRADD, TRAFs, IKKs, and

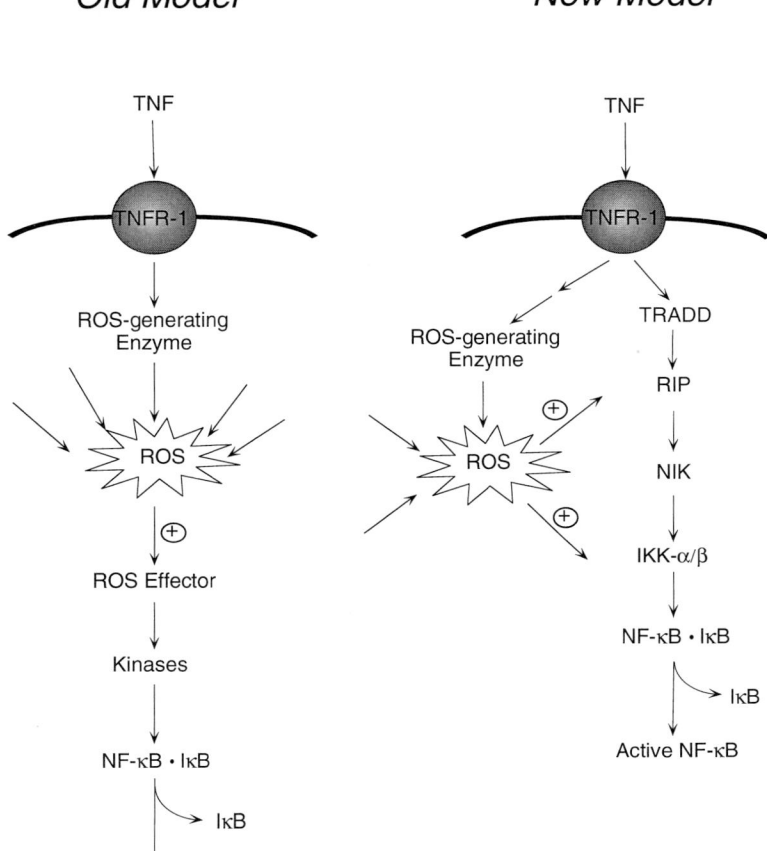

Figure 3 Comparison between pathways in which ROS function either as messengers (old model) or as costimulatory signals (new model).

NIK. In such an experiment, the activating signal (the overexpressed protein) will enter at different points along the pathway. Measuring the effect of a cooverexpressed ROS-scavenging enzyme on NF-κB activation may allow to pinpoint a ROS-dependent signaling step (provided overexpression does not overcome the requirement for a costimulation by ROS). In another experiment, cells that overexpress the various signaling components would be treated with hydrogen peroxide. Expression levels must be titrated for each component such that the costimulatory effect of the ROS is still detectable.

VI. WHY SHOULD ROS MODULATE PROINFLAMMATORY CYTOKINE SIGNALING?

A complex enzymatic machinery and millimolar levels of the cell's antioxidant glutathione are continuously engaged in maintaining a reduced intracellular milieu that protects the biological activity of cysteine residues in proteins and minimizes the damage introduced by toxic waste products of dioxygen to DNA and lipids (reviewed by Sies, 1995). Likewise, there is a continuous repair of oxidatively damaged lipids, proteins, and DNA. Any functional perturbation of the cell as a result of an encounter with toxins, microbes, or stress conditions may perturb the intracellular redox balance and shift it toward a prooxidant condition (and increased cell damage). Hence a prooxidant condition may provide a universal *intracellular* signal for cellular distress. The kind of signal provided by proinflammatory cytokines to cells is also a distress signal; however, an *extracellular* one which, for instance, is produced by remote immune cells that have encountered a pathogen. Extracellular cytokine distress signals are able to diffuse through the entire organism and can bind to a large number of cells and cell types. It seems very meaningful if cells that have already experienced some internal distress, i.e., as a result of infection, intoxication, or damage, react most vigorously to the external cytokine stimulus, whereas virtually unaffected healthy cells remain only weakly reponsive. An enhanced intracellular level of ROS would lend itself to increase the susceptibility of a particular cell to proinflammatory cytokine signals and mark it for either a stronger or a different kind of cytokine response.

Depending on the level of oxidative stress a cell is exposed to, the inflammatory cytokines may cause very distinct effects. In the absence of internal oxidative stress, both NF-κB activation and the apoptotic response to TNF may be weak (Fig. 4A). In a healthy cell, the small amounts of ROS produced in response to TNF are scavenged rapidly and efficiently and cannot exert a strong costimulatory effect. At mild oxidative stress (Fig. 4B), the preexisting ROS will add to the TNF-induced ROS and produce a strong costimulatory signal for both NF-κB and apoptosis. As long as there is a potent NF-κB response, the proapoptotic response may be kept in check by the antiapoptotic effect of NF-κB (Fig. 4B). In this alerted stage, the slightest impairment of NF-κB function may, however, cause a rapid and vigorous cell death. Finally, with higher levels of oxidative stress or any other condition affecting NF-κB function, the NF-κB response is increasingly

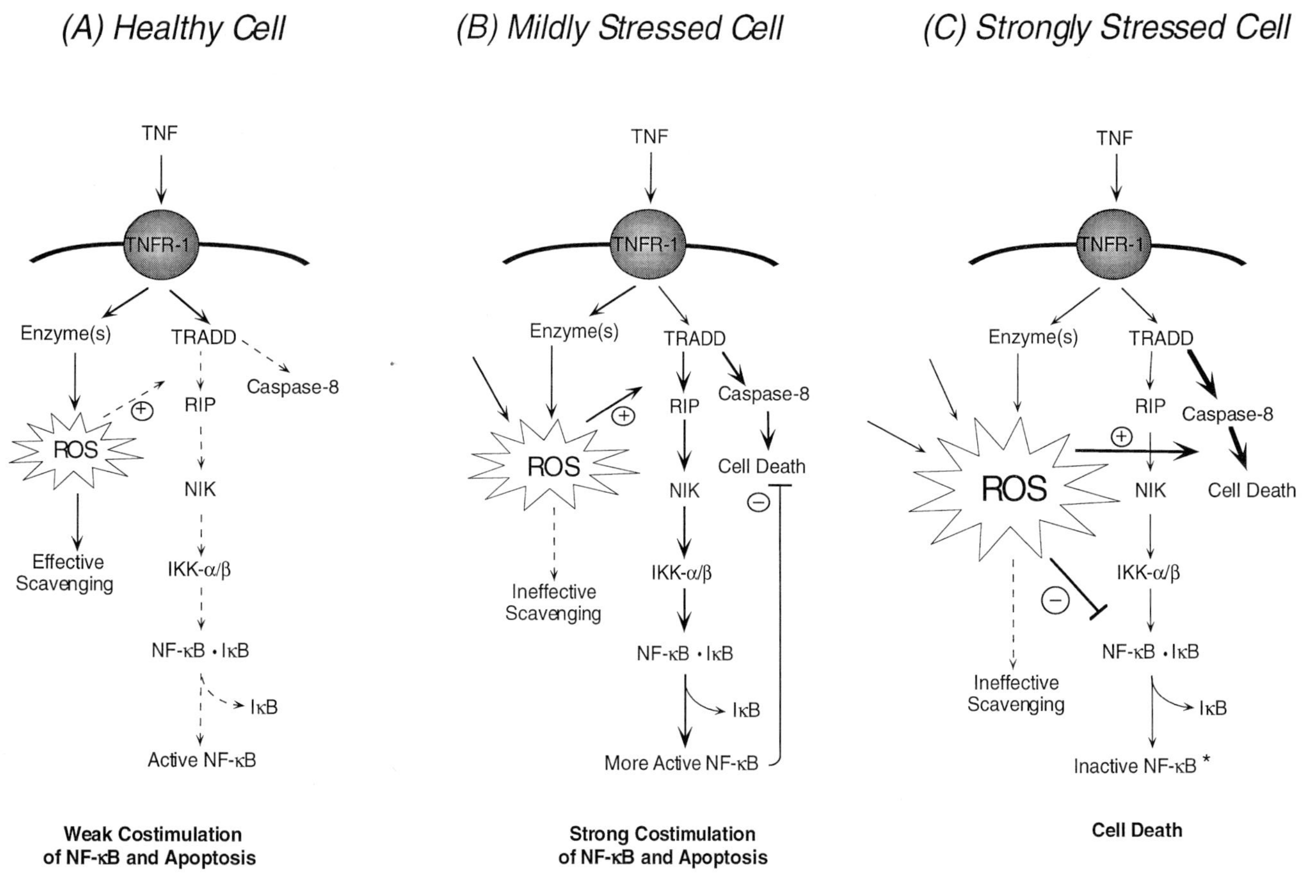

Figure 4 The effect of increasing intracellular ROS levels on TNF-induced NF-κB activation and apoptosis. For details, see text.

impaired. The proapoptotic pathway initiated by TNF will prevail and selectively kill those cells in response to TNF that have their redox balance shifted strongest toward a prooxidant condition (Fig. 4C).

Why does TNF need to induce ROS production itself? A healthy cell will be able to scavenge the small amounts of ROS produced in response to TNF quickly. In a distressed cell, however, TNF-produced ROS will add to the preexisting oxidative stress and cause a different response to the cytokine. Hence ROS production in response to TNF may have evolved as a parallel, costimulatory signal for NF-κB activation. The preexisting redox conditions within the cell will decide greatly whether this TNF-induced ROS signal is eliminated quickly or can add up with the existing ROS levels to elicit anything from a most productive NF-κB activation all the way to programmed cell death.

VII. CONCLUSION: ROS MAY ACT AS COSTIMULATORY SIGNALS

Some years ago, we have proposed that ROS may serve a second messenger role in NF-κB activation by a variety of stimuli (Schreck and Baeuerle, 1991). With new insights into the pathways activating NF-κB in response to TNF and IL-1, such a requirement has not been supported in the sense that this pathway itself generates a ROS signal and employs a ROS receptor as a downstream signaling component. TNF and IL-1 NF-κB-activating pathways rather appear as a series of protein–protein interaction and serine/threonine kinase phosphorylation steps. Nevertheless, independent data obtained from many laboratories do support an inhibitory effect of antioxidative compounds and enzymes on cytokine-induced NF-κB activation and an activation of NF-κB through prooxidant conditions. An appealing way to reconcile all presently known data is to assume that ROS act as modulators and costimulatory signals rather than as second messengers of proinflammatory cytokine signaling.

The TNFR-1 has emerged as a highly pleiotropic signaling molecule that can send many distinct signals into the cytoplasm and nucleus. Instead of assuming that the ROS signal and the NF-κB signal are produced by the same branch, they may in fact be produced in parallel reactions. As discussed earlier, the ROS signal would then act on the parallel NF-κB pathway and modulate its amplitude. This modulation would be further dependent on the preexisting cellular redox status. The identification of the enzymatic mechanism by which TNF and IL-1 produce ROS and of the redox-activated signaling component in the NF-κB pathway would help greatly in validating this hypothesis.

ACKNOWLEDGMENTS

I am grateful to Erika Treuenfels for preparing the figures. I want to apologize to many of my colleagues for not directly citing all of their relevant and stimulating work.

REFERENCES

Adam-Klages, S., Adam, D., Wiegmann, K., Struve, S., Kolanus, W., Schneider-Mergener, J., and Krönke, M. (1996). FAN, a novel WD-repeat protein, couples the p55 TNF-receptor to neutral sphingomyelinase. *Cell (Cambridge, Mass.)* **86,** 937–947.

Adcock, I. M., Brown, C. R., Kwon, O., and Barnes, P. J. (1994). Oxidative stress-induced NF-κB binding and inducible NOS mRNA in human epithelial cells. *Biochem. Biophys. Res. Commun.* **199,** 1518–1524.

Baeuerle, P. A. (1998). Pro-inflammatory signaling: Last pieces in the NF-κB puzzle? *Curr. Biol.* **8,** R19–R22.

Baeuerle, P. A., and Baichwal, V. R. (1997). NF-κB as frequent target for immunosuppressive and anti-inflammatory molecules. *Adv. Immunol.* **6,** 111–137.

Baeuerle, P. A., and Baltimore, D. (1988). IκB: A specific inhibitor of the NF-κB transcription factor. *Science* **242,** 540–546.

Baeuerle, P. A., and Baltimore, D. (1996). NF-κB: Ten years after. *Cell (Cambridge, Mass.)* **87,** 13–20.

Baeuerle, P. A., and Henkel, T. (1994). Function and activation of NF-κB in the immune system. *Annu. Rev. Immunol.* **12,** 141–179.

Baeuerle, P. A., Rupec, R. A., and Pahl, H. L. (1996). Reactive oxygen intermediates as second messengers of a general pathogen response. *Pathol. Biol.* **44,** 29–35.

Baichwal, V. R., and Baeuerle, P. A. (1997). Apoptosis: Activate NF-κB or die? *Curr. Biol.* **7,** 94–96.

Baichwal, V. R., and Baeuerle, P. A. (1998). Kinases in pro-inflammatory signal transduction pathways: New opportunities for drug discovery. *Annu. Rev. Med. Chem.* **33,** 233–242.

Baldwin, A. S. (1996). The NF-κB and I-κB proteins: New discoveries and insights. *Annu. Rev. Immunol.* **14,** 649–681.

Barnes, P. J., and Karin, M. (1997). Nuclear factor-κB—a pivotal transcription factor in chronic inflammatory diseases. *N. Engl. J. Med.* **336,** 1066–1071.

Béraud, C., Henzel, W. J., and Baeuerle, P. A. (1999). Involvement of regulatory and catalytic subunits of phosphoinositide 3-kinase in NF-kappa B activation. *Proc. Natl. Acad. Sci. USA* **96,** 429–434.

Bonizzi, G., Dejardin, E., Biret, B., Piette, J., Merville, M. P., and Bours, V. (1996). IL-1β induces NF-κB in epithelial cells independently of the production of reactive oxygen intermediates. *Eur. J. Biochem.* **142,** 544–549.

Bradley, J. R., Johnson, D. R., and Pober, J. S. (1993). Endothelial activation by hydrogen peroxide. Selective increases of intracellular adhesion molecule-1 and major histocompatibility complex I. *Am. J. Pathol.* **142,** 1598–1609.

Brennan, P., and O'Neill, L. A. (1995). Effects of oxidants and antioxidants and nuclear factor kappa B activation in three different cell lines: Evidence against a universal hypothesis involving oxygen radicals. *Biochim. Biophys. Acta* **1260,** 167–175.

Cai, J., and Jones, D. P. (1998). Superoxide in Apoptosis. *J. Biol. Chem.* **273,** 11401–11404.

Cohen, L., Henzel, W. J., and Baeuerle, P. A. (1998). IKAP is a scoffold protein of the IκB kinase complex. *Nature* **395,** 292–296.

DeForge, L. E., Preston, A. M., Takeuchi, E., Kenney, J., Boxer, L. A., and Remick, D. G. (1993). Regulation of interleukin 8 gene expression by oxidant stress. *J. Biol. Chem.* **268**, 25568–25576.

Diaz-Meco, M. T., Berra, E., Municio, M. M., Sanz, L., Lozano, J., Dominguez, I., DiazGolpe, V., Lein de Lera, M. T., Alcami, J., Paya, C. V., Arenzana-Seisdedos, F., Virelizier, J. L., and Moscat, J. (1993). A dominant negative protein kinase C-ζ subspecies blocks NF-κB activation. *Mol. Cell. Biol.* **13**, 4770–4775.

Eigler, A., Sinha, B., Hartmann, G., and Endres, S. (1997). Taming TNF: Strategies to restrain this proinflammatory cytokine. *Immunol. Today* **18**, 487–492.

Feldman, M., Brennan, F. M., and Maini, R. N. (1996). Rheumatoid arthritis. *Cell (Cambridge, Mass.)* **85**, 307–310.

Flohé, L., Brigelius-Flohe, R., Saliou, C., Traber, M. G., and Packer, L. (1997). Redox regulation of NF-κB activation. *Free Radical Biol. Med.* **22**, 1115–1126.

Ginn-Pease, M. E., and Whisler, R. L. (1996). Optimal NF-κB-mediated transcriptional responses in Jurkat T cells exposed to oxidative stress are dependent on intracellular glutathione and costimulatory signals. *Biochem. Biophys. Res. Commun.* **226**, 695–702.

Halliwell, B., and Gutteridge, J. M. C. (1989). "Free Radicals in Biology and Medicine," 2nd ed. Clarendon Press, Oxford.

Heffetz, D., and Zick, Y. (1989). Hydrogen peroxide potentiates phosphorylation of novel putative substrates for then insulin receptor kinase in intact Fao cells. *J. Biol. Chem.* **264**, 10126–10132.

Imbert, V., Rupec, R. A., Auberger, P., Pahl, H. L., Traenckner, E. B.-M., Müller-Dieckmann, C., Farahifar, D., Rossi, B., Baeuerle, P. A., and Peyron, J.-F. (1996). Tyrosine phosphorylation activates NF-κB without proteolytic degradation of IκB-α. *Cell (Cambridge, Mass.)* **86**, 787–798.

Janssen, Y. M. W., Macara, I., and Mossman, B. T. (1999). Cooperativity between reactive oxygen/nitrogen species and tumor necrosis factor in the activation of NF-κB in lung epithelial cells. Submitted for publication.

Jin, D.-Y., Chae, H. Z., Rhee, S. G., and Jeang, K.-T. (1997). Regulatory role for a novel human thioredoxin peroxidase in NF-κB activation. *J. Biol. Chem.* **272**, 30952–30961.

Jones, P. L., Ping, D., and Boss, J. M. (1997). Tumor necrosis factor alpha and interleukin-1β regulate the murine managnaese superoxide dismutase gene through a complex intronic enhancer involving C/EBP-β and NF-κB. *Mol. Cell. Biol.* **17**, 6970–6981.

Kretz-Remy, C., Mehle, P., Mirault, M.-E., and Arrigo, A.-P. (1996). Inhibition of IκB-alpha phosphorylation and degardation and subsequent NF-κB activation by glutathione peroxidase overexpression. *J. Cell Biol.* **133**, 1083–1093.

Kretz-Remy, C., Bates, E. E., and Arrigo, A.-P. (1998). Amino acid analogs activate NF-kappaB through redox-dependent IkappaB-alpha degradation by the proteasome without apparent IkappaB-alpha phosphorylation. Consequence on HIV-1 long terminal repeat activation. *J. Biol. Chem.* **273**, 3180–3191.

Lee, S. Y., Reichlin, A., Santana, A., Sokol, K. A., Nussenzweig, M. C., and Choi, Y. (1997). TRAF-2 is essential for JNK but not NF-kappaB activation and regulates lymphocyte proliferation and survival. *Immunity* **5**, 703–713.

Legrand-Poels, S., Bours, V., Piret, B., Pflaum, M., Epe, B., Rentier, B., and Piette, J. (1995). Transcription factor NF-kappa B is activated by photosensitization generating oxidative damages. *J. Biol. Chem.* **270**, 6925–6934.

Li, Y., Zhang, W., Mantell, L. L., Kazzaz, J. A., Fein, A. M., and Horowitz, S. (1997). Nuclear factor-κB is activated by hyperoxia but does not protect from cell death. *J. Biol. Chem.* **272**, 20646–20649.

Liu, Z.-G., Hu, H., Goeddel, D. V., and Karin, M. (1996). Dissection of TNF receptor 1 effector functions: JNK activation is not linked to apoptosis, while NF-κB activation prevents cell death. *Cell (Cambridge, Mass.)* **87**, 565–576.

Los, M., Dröge, W., Stricker, K., Baeuerle, P. A., and Schulze-Osthoff, K. (1995). Hydrogen peroxide as a potent activator of T lymphocyte functions. *Eur. J. Immunol.* **25**, 159–165.

Maniatis, T. (1997). Catalysis by a multiprotein IκB kinase complex. *Science* **278**, 818–819.

Mehlen, P., Kretz-Remy, C., Preville, X., and Arrigo, A.-P. (1996). Human hsp27, Drosophila hsp27 and human alphaB-crystallin expression-mediated increase in glutathione is essential for the protective activity of these proteins against TNF-alpha-induced cell death. *EMBO J.* **15**, 2695–2706.

Mehlen, P., Hickey, E., Weber, L. A., and Arrigo, A.-P. (1997). Large unphosphorylated aggregates as the active form of hsp27 which controls intracellular reactive oxygen species and glutathione levels and generates a protection against TNF-alpha in NIH-3T3-ras cells. *Biochem. Biophys. Res. Commun.* **241**, 187–192.

Meier, B., Radeke, H. H., Selle, S., Sies, H., Resch, K., and Habermehl, G. G. (1989). Human fibroblasts release reactive oxygen species in response to interleukin-1 and tumor necrosis factor alpha. *Biochem. J.* **263**, 539–545.

Meyer, M., Schreck, R., and Baeuerle, P. A. (1993). H_2O_2 and antioxidants have opposite effects on activation of NF-κB and AP-1 in intact cells: AP-1 as secondary antioxidant-responsive factor. *EMBO J.* **12**, 2005–2015.

Mueller, J. M., Cahill, M., Baeuerle, P. A., and Nordheim, A. (1997a). Antioxidants as well as oxidants activate *c-fos* via Ras-dependent activation of extracellular-signal-regulated kinase 2 and Elk-1. *Eur. J. Biochem.* **244**, 45–52.

Mueller, J. M., Krauss, B., Kaltschmidt, C., Baeuerle, P. A. and Rupec, R. (1997b). Hypoxia induces *c-fos* transcription via a MAPK-dependent pathway. *J. Biol. Chem.* **272**, 23435–23339.

Piette, J., Piret, B., Bonizzi, G., Schoonbroodt, S., Merville, M.-P., Legrand-Poels, S., and Bours, V. (1997). Multiple redox regulation in NF-κB transcription factor activation. *Biol. Chem.* **378**, 1237–1245.

Renard, P., Zachari, M.-D., Bougelet, C., Mirault, M.-E., Haegeman, G., Remacle, J., and Raes, M. (1997). Effects of antioxidant enzymes modulation on interleukin-1-induced NF-κB activation. *Biochem. Pharmacol.* **53**, 149–160.

Rothe, M., Pan, M.-G., Henzel, W. J., Ayres, T. M., and Goeddel, D. V. (1995). The TNF receptor 2-TRAF signaling complex contains two novel proteins related to baculoviral inhibitor of apopotosis proteins. *Cell (Cambridge, Mass.)* **83**, 1243–1252.

Rupec, R. A., and Baeuerle, P. A. (1995). The genomic response of tumor cells to hypoxia and reoxygenation: Differential activation of transcription factors AP-1 and NF-κB. *Eur. J. Biochem.* **234**, 632–640.

Schenk, H., Klein, M., Erdbuerger, W., Droege, W., and Schulze-Osthoff, K. (1994). Distinct effects of thioredoxin and other antioxidants on the activation of NF-κB and AP-1. *Proc. Natl. Acad. Sci. U.S.A.* **91**, 1672–1676.

Schmidt, K. N., Amstad, P., Cerutti, P., and Baeuerle, P. A. (1995). The roles of hydrogen peroxide and superoxide as messengers in the activation of transcription factor NF-κB. *Chem. Biol.* **2**, 13–21.

Schoonbroodt, S., Legrand-Poels, S., Best-Belpomme, M., and Piette, J. (1997). Hypochlorous acid activates NF-κB transcription factor in T lymphocytes. *Biochem. J.* **321**, 777–785.

Schreck, R., and Baeuerle, P. A. (1991). A role of oxygen radicals as second messengers. *Trends Cell Biol.* **1**, 39–42.

Schreck, R., and Baeuerle, P. A. (1994). Methods of assessing oxygen radicals as mediators in the activation of inducible eukaryotic transcription factor NF-κB. *In* "Methods in Enzymology" (L. Packer, ed.), Vol. 234, pp. 151–163. Academic Press, San Diego, CA.

Schreck, R., Rieber, P., and Baeuerle, P. A. (1991). Reactive oxygen species as apparently widely used messengers in the activation of the NF-κB transcription factor and HIV-1. *EMBO J.* **10**, 2247–2258.

Schreck, R., Albermann, K., and Baeuerle, P. A. (1992). NF-κB: An oxidative stress-responsive transcription factor of eukaryotic cells (a review). *Free Radical Res. Commun.* **17**, 221–237.

Schulze-Osthoff, K., and Baeuerle, P. A. (1998). Regulation of gene expression by oxidative stress. *Adv. Mol. Cell Biol.* **25**, 15–44.

Schulze-Osthoff, K., Beyaert, R., Vandervoorde, V., Haegeman, G., and Fiers, W. (1993). Depletion of the mitochondrial electron transport abrogates the cytotoxic and gene-inductive effects of TNF. *EMBO J.* **12**, 3095–3104.

Schulze-Osthoff, K., Los, M., and Baeuerle, P. A. (1995). Redox signaling by transcription factors NF-κB and AP-1 in lymphocytes. *Biochem. Pharmacol.* **50**, 735–741.

Sen, C. K., and Packer, L. (1996). Redox regulation of transcriptional activators. *Free Radical Biol. Med.* **21**, 335–348.

Sha, W. C. (1998). Regulation of immune responses by NF-kappaB/Rel transcription factors. *J. Exp. Med.* **187**, 143–146.

Siebenlist, U., Brown, K., and Franzoso, G. (1995). NF-κB: A mediator of pathogen and stress responses. *In* "Inducible Gene Expression" (P. A. Baeuerle, ed.), Vol. 1, pp. 93–141. Birkhaeuser, Boston.

Sies, H. (1995). Strategies of antioxidant defense: relations to oxidative stress. *NATO ASI Ser., Ser. H.* **H92**, 165–186.

Stancovski, I., and Baltimore, D. (1997). NF-κB activation: The IκB kinase revealed? *Cell (Cambridge, Mass.)* **91**, 299–302.

Tokiwa, G., Dikic, I., Lev, S., and Schlessiger, J. (1996). Activation of Pyk2 by stress isgnals and coupling with JNK signaling pathways. *Science* **273**, 792–794.

Verma, I. M., Stevenson, J. K., Schwarz, E. M., Van Antwerp, D., and Miyamoto, S. (1995). Rel/NF-κB/IκB family: Intimate tales of association and dissociation. *Genes Dev.* **9**, 2723–2735.

Vile, G. T., Tanew-Iliitschew, A., and Tyrrell, R. M. (1995). Activation of NF-κB in human skin fibroblasts by the oxidative stress generated by UVA radiation. *Photochem. Photobiol.* **62**, 436–468.

Wesche, H., Henzel, W. J., Shillinglaw, W., Li, S., and Cao, Z. (1997). MyD88: an adapter that recruits IRAK to the IL-1R complex. *Immunity* **7**, 837–847.

Wong, G. H. W., Elwell, J. H., Oberley, L. W., and Goeddel, D. V. (1988). Manganeous superoxide dismutase is essential for cellular resistance to cytotoxicity of tumor necrosis factor: Possible protective mechanism. *Cell (Cambridge, Mass.)* **58**, 923–931.

Zheng, M., and Storz, G. (1998). Activation of the OxyR transcription factor by reversible disulfide bond formation. *Science* **279**, 1718–1721.

9 Redox Regulation of NF-κB

Takashi Okamoto, Toshifumi Tetsuka, Sinichi Yoshida, and Takumi Kawabe
Department of Molecular Genetics
Nagoya City University Medical School
Nagoya 467, Japan

NF-κB is an inducible cellular transcription factor that activates various cellular and viral genes. In resting cells, NF-κB exists as a molecular complex with an inhibitory molecule IκB and is located in the cytosol. On stimulation of the cells by various agents, such as proinflammatory cytokines interleukin (IL)-1 and tumor necrosis factor (TNF), IκB is dissociated and NF-κB is translocated to the nucleus and activates the expression of target genes. We found that the redox control mechanism is involved in the DNA-binding activity of NF-κB and that the cellular-reducing catalyst thioredoxin (Trx), together with kinases, is primarily involved as an effector molecule in this signaling pathway. Trx has been found to associate with the redox-sensitive cysteine within the DNA-binding loop of NF-κB. Effects of antioxidants in blocking NF-κB activation can be explained by the involvement of radical oxygen intermediates (ROI) in this pathway. These findings support the idea that redox regulation involving ROI and Trx plays a crucial role in the signal transduction pathway leading to NF-κB activation and thus contributes a great deal to the understanding of the pathogenetic processes of various diseases, such as acquired immunodeficiency syndrome (AIDS), hematogenic cancer cell metastasis, and rheumatoid arthritis (RA), in which NF-κB plays a major role.

I. INTRODUCTION: TRANSCRIPTION FACTOR NF-κB AND ITS BIOLOGICAL FUNCTIONS

NF-κB regulates expression of a wide variety of cellular and viral genes (Gilmore, 1990; Baeuerle and Baichwal, 1997; Baldwin, 1996; Thanos and

Maniatis, 1995; Okamoto et al., 1997). These genes include cytokines such as IL-2, IL-6, IL-8, GM-CSF, and TNF, cell adhesion molecules such as ICAM-1 and E-selectin, inducible nitric oxidase synthase (iNOS), and viruses such as human immunodeficiency virus (HIV) and cytomegalovirus. (Schindler and Baichwal, 1994; Okamoto et al., 1989; Stade et al., 1990; Mukaida et al., 1990; Roebuck et al., 1995; Donnelly et al., 1993; Schreck and Baeuerle, 1990; Staynov et al., 1995; Xie et al., 1994; Sen and Baltimore, 1986; Nabel and Baltimore, 1987). Through the causal relationship with these genes, NF-κB is considered to be involved in the currently intractable diseases such as AIDS, hematogenic cancer cell metastasis, and RA as shown in Fig. 1. Although the genes induced by NF-κB are variable according to the context of cell lineage and are also under the control of other transcription factors, NF-κB is considered to play a major role in the regulation of expression of these genes and thus contributes a great deal to the pathogenesis.

Additionally, another biological action of NF-κB as an inhibitor of apoptosis has been demonstrated, although the molecular mechanism has yet to be elucidated (Beg and Baltimore, 1996; Wu et al., 1996). It is thus hoped that disruption of this antiapoptotic mechanism would make cells more vulnerable to killing. Because proapoptotic signals elicited by TNF

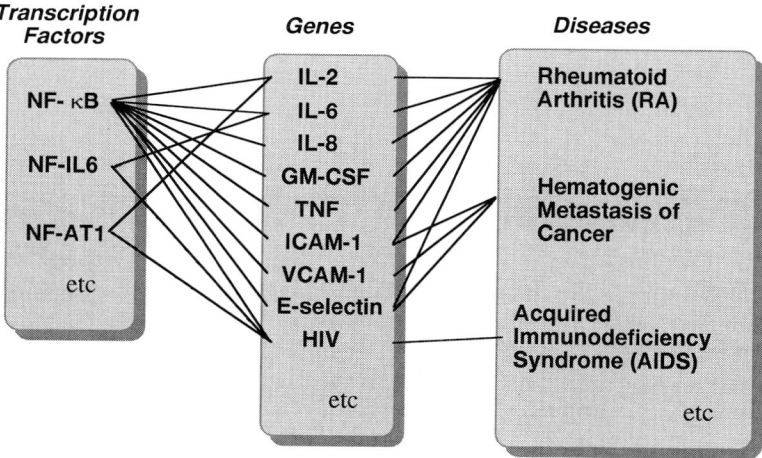

Figure 1 Relationships between transcription factors and diseases: A pathogenetic paradigm regarding NF-κB. Transcription factors are causally associated with various genes known to be involved in pathogenetic processes of the relevant diseases. This diagram shows cross-correlations between transcription factors and diseases with attention to transcription factor NF-κB. This diagram indicates that some disease processes may be controlled by modifying the actions of transcription factors.

and radiation are also known to stimulate the NF-κB activation pathway, selective inhibition of the antiapoptotic action of NF-κB should augment the antitumor effects of these agents. It has not yet been clarified whether this antiapoptotic action of NF-κB is mediated by the induction of a novel gene(s) as some NF-κB target genes, such as p53, Fas ligand, and c-Myc, have been implicated in apoptosis (Klefstrom et al., 1997; Lane, 1992; Wu and Lozano, 1994). An apparently unexpected action of anti-NF-κB reagents in the protection of neuronal cell death (Grilli et al., 1996) indicates that either the pro- or the antiapoptotic action of NF-κB may be determined by the context of the cellular signaling cascade. It is possible that the antiapoptotic action of NF-κB is mediated through direct protein–protein interaction, although the target protein has not been identified. Therefore, the biochemical intervention of NF-κB should conceivably interfere with the pathogenic process and would be effective for treatment.

A hypothesis has been proposed whereby divergent agents activate NF-κB by increasing ROI. Previous studies have revealed that these reducing enzymes, together with ROI, are involved in cell signaling (Holmgren, 1985, 1989; Ziegler, 1985; Allen, 1993). The term "redox regulation" has thus been proposed, indicating the active role of oxido-reductive modifications of proteins in regulating their activities. Oxidation and reduction of biomolecules are now considered to be "signals" in certain instances and are utilized for the maintenance of cellular homeostasis. In particular, H_2O_2 has been implicated as a common second messenger in the various pathways leading to NF-κB activation. This hypothesis is based on several lines of evidence. First, most of the agents activating NF-κB tend to trigger the formation of ROI (Schreck et al., 1991; Schmidt et al., 1995; Los et al., 1995). Second, H_2O_2 or organic hydroperoxide has been shown to activate NF-κB in some cell lines in the abscence of any physiological stimulus. Third, antioxidants were shown to be effective in blocking NF-κB activation in response to diverse stimuli (Schreck et al., 1991; Roederer et al., 1990; Suzuki et al., 1992, 1994; Merin et al., 1996; Meyer et al., 1993; Biswas et al., 1993; Packer et al., 1995). However, direct evidence is still lacking to indicate where and how ROI are generated and involved in signaling. This chapter focuses on the nature of redox regulation of NF-κB and its possible therapeutic implications.

II. INVOLVEMENT OF NF-κB IN VARIOUS PATHOLOGIES

A. AIDS

The pivotal role of NF-κB in the HIV life cycle, especially in the virus reactivation process within latently infected cells, has been widely accepted.

After activation through intracellular signaling pathways, such as those elicited by the T-cell receptor antigen complex or by receptors for IL-1 or TNF, NF-κB initiates HIV gene expression by binding to the target DNA element within the promoter region of HIV LTR (Okamoto et al., 1989, 1990; Nabel and Baltimore, 1987; Bohnlein et al., 1988). The virus-encoded transactivator Tat is produced and triggers explosive viral replication (Arya et al., 1985; Sodroski et al., 1985; Okamoto and Wong-Staal, 1986).

Because activation of HIV gene expression by cellular transcription factor NF-κB conceptually precedes the production of Tat, NF-κB can be ascribed to be a critical determinant between maintenance and breakdown of the viral latency. Various attempts have been carried out to control the signaling pathways to NF-κB activation in view of the therapeutic intervention of AIDS pathology. For example, antioxidant compounds known to block NF-κB activation cascade have been suggested to be effective in preventing the clinical development of AIDS by blocking HIV replication (Roederer et al., 1990; Suzuki et al., 1992; Merin et al., 1996). Because viral transcription is the only step for the amplification of viral genetic information, the clinically relevant anti-NF-κB compound should be effective in blocking viral replication from the latent status and in substantially decreasing the viral load in virus-infected individuals (Ho, 1998; Perelson et al., 1997; Okamoto, 1995; Tozawa et al., 1995).

B. Cancer Metastasis

Another situation in which NF-κB plays a role is hematogenic cancer cell metastasis (Tozawa et al., 1995). NF-κB induces E-selectin on the surface of vascular endothelial cells (Schindler and Baichwal, 1994; Montgomery et al., 1991; Whelan et al., 1991). Because some cancer cells constitutively express a ligand for E-selectin, called sialyl-Lewis^X and sialyl-Lewis^A antigens, on their cell surface, the induction of E-selectin is considered to be a rate-determining step of cancer cell–endothelial cell interaction (Dejana et al., 1988; Takada et al., 1993). Based on these findings, we examined this phenomenon with regard to the role of NF-κB in E-selectin induction (Tozawa et al., 1995). When human umbilical venous endothelial cells (HUVEC) were treated with IL-1 or TNF, nuclear translocation of NF-κB was observed, followed by the augmented expression of E-selectin. We examined the cell-to-cell interaction between HUVEC and the QG90 cell, a tumor cell line derived from human small cell carcinoma of the lung expressing sialyl-Lewis^X antigen, and found that IL-1 was able to induce the attachment of cancer cells to HUVEC. However, pretreatment of HUVEC with N-acetylcysteine, aspirin, or pentoxyphillin efficiently blocked the cell-to-cell attachment in a dose-dependent manner. Therefore, use of antioxidants in the interference of the NF-κB signaling cascade is considered to be feasible for the prevention of tumor metastasis.

C. Rheumatoid Arthritis

Evidence has indicated the involvement of NF-κB in the pathogenesis of rheumatoid arthritis (Alvaro-Garcia *et al.*, 1991; Ulfgren *et al.*, 1995; Sakurada *et al.*, 1996). Handel *et al.* (1996) have demonstrated the presence of NF-κB subunit proteins, p65 and p50, in the nuclei of synovial lining cells of fresh synovial tissue obtained from patients with RA, indicating activation of NF-κB *in situ*. Because of its regulatory role in the gene expression of IL-1, TNF, IL-2, IL-6, IL-8, GM-CSF, chemokines such as RANTES and MIP-1 α, ICAM-1, E-selectin, and iNOS that are known to be overexpressed in the rheumatoid synovium, NF-κB is considered to be a major transcriptional regulator in the expansion and maintenance of the chronic inflammatory response in the affected joints. For example, sustained NF-κB activation would induce the production of cytokines and thus activate the maturation of B lymphocytes to produce antibodies, whereas GM-CSF and chemokine production, together with the overexpression of cell adhesion molecules, would support the recruitment of leukocytes from the bloodstream, thus augmenting the local inflammatory response. Additionally, it has been noted that NF-κB blocks the cellular apoptotic response (Beg and Baltimore, 1996; Wu *et al.*, 1996), which is probably involved in promoting synovial cell proliferation of the affected joints.

Some of the effective antirheumatic drugs, including corticosteroids, aspirin, and gold compounds, are now known to block the NF-κB cascade. Because features of inflammatory responses observed in RA are also found in other chronic inflammatory processes irrespective of the affected organs or tissues, NF-κB is probably universally involved. Therefore, the clinical applicability of antirheumatic drugs should be evaluated in other pathologies in which NF-κB plays a role.

III. SIGNAL TRANSDUCTION PATHWAY TO ACTIVATE NF-κB

NF-κB consists of two subunit molecules, p65 and p50, and usually exists as a molecular complex with an inhibitory molecule, IκBα, in the cytosol, (Gilmore, 1990; Baeuerle and Baichwal, 1997; Baldwin, 1996; Thanos and Maniatis, 1995; Okamoto *et al.*, 1997; Sen and Baltimore, 1986; Nabel and Baltimore, 1987). On stimulation of cells, such as by proinflammatory cytokines, IL-1 and TNF, IκBα undergoes phosphorylation and rapid degradation via the 26S proteasome. These processes allow the nuclear translocation of NF-κB and the activation of gene expression (Fig. 2).

The activity of NF-κB is regulated by an upstream regulatory mechanism. TNF and IL-1 initiate their signaling through receptor-associated signal transducers, TRAF2 and TRAF6, respectively (Rothe et al., 1995; Cao et al., 1996). These distinct TNF and IL-1 pathways merge at the level of the protein kinase NIK, NF-κB-inducing kinase, (Malinin et al., 1997). Using a yeast two-hybrid screen, a 85-kDa protein kinase, called CHUK (conserved helix–loop–helix ubiquitous kinase), was found to be associated with NIK and has been redesignated as IκB kinase α (IKK-α) (Zandi et al., 1997). A second IKK, IKK-β has also been identified (Woronicz et al., 1997). IKK-α and IKK-β exist in a heterocomplex that is able to interact with NIK.

Moreover, the involvement of ROI is suggested in at least one of the upstream steps of the NF-κB activation pathway, as the signaling was blocked efficiently by pretreatment of the cells with antioxidants such as N-acetyl-L-cysteine (NAC) or α-lipoic acid (Schreck et al., 1991; Roederer et al., 1990; Suzuki et al., 1992, 1994; Merin et al., 1996; Biswas et al., 1993; Packer et al., 1995). Therefore, antioxidants are now considered to be effective NF-κB inhibitors. Moreover, we found that NAC could also block the induction of Trx (Sachi et al., 1995). Therefore, anti-NF-κB actions of antioxidants are considered to be twofold: (1) blocking the signaling immediately downstream of the signal elicitation and (2) suppressing induction of the redox effector Trx.

A. Origin of Redox Signaling: Involvement of Radical Oxygen Intermediates

Figure 3 illustrates the intracellular redox cascade involved in the successive reduction of oxygen by the addition of four electrons and a cellular antioxidant defense system. Among ROI, hydrogen peroxide has the longest half-life and is considered to be a major mediator of the oxidative signal.

Figure 2 Signal transduction pathways for NF-κB activation. The first step involves kinase pathways such as by NF-κB and IκB kinases. The second step involves "redox regulation" by thioredoxin (Trx). After stimulation of the cells by TNF or IL-1, for example, radical oxygen intermediates (ROI) are produced. ROI not only induce the activation of kinase cascade but also the production of Trx. TRAF is known to be associated with the TNF receptor and is known to stimulate NF-κB activation. Similarly, IRAK (IL-1 receptor-associated kinase) is considered to participate in NF-κB activation through its ability to phosphorylate IKK. Phosphorylation of NF-κB or IκB will release NF-κB. However, NF-κB must go through the Trx-mediated reduction of the redox-sensitive cysteine to recognize the target DNA sequence (κB site). The phosphorylated IκB will be ubiquitinated and then degraded by proteasome or other proteases.

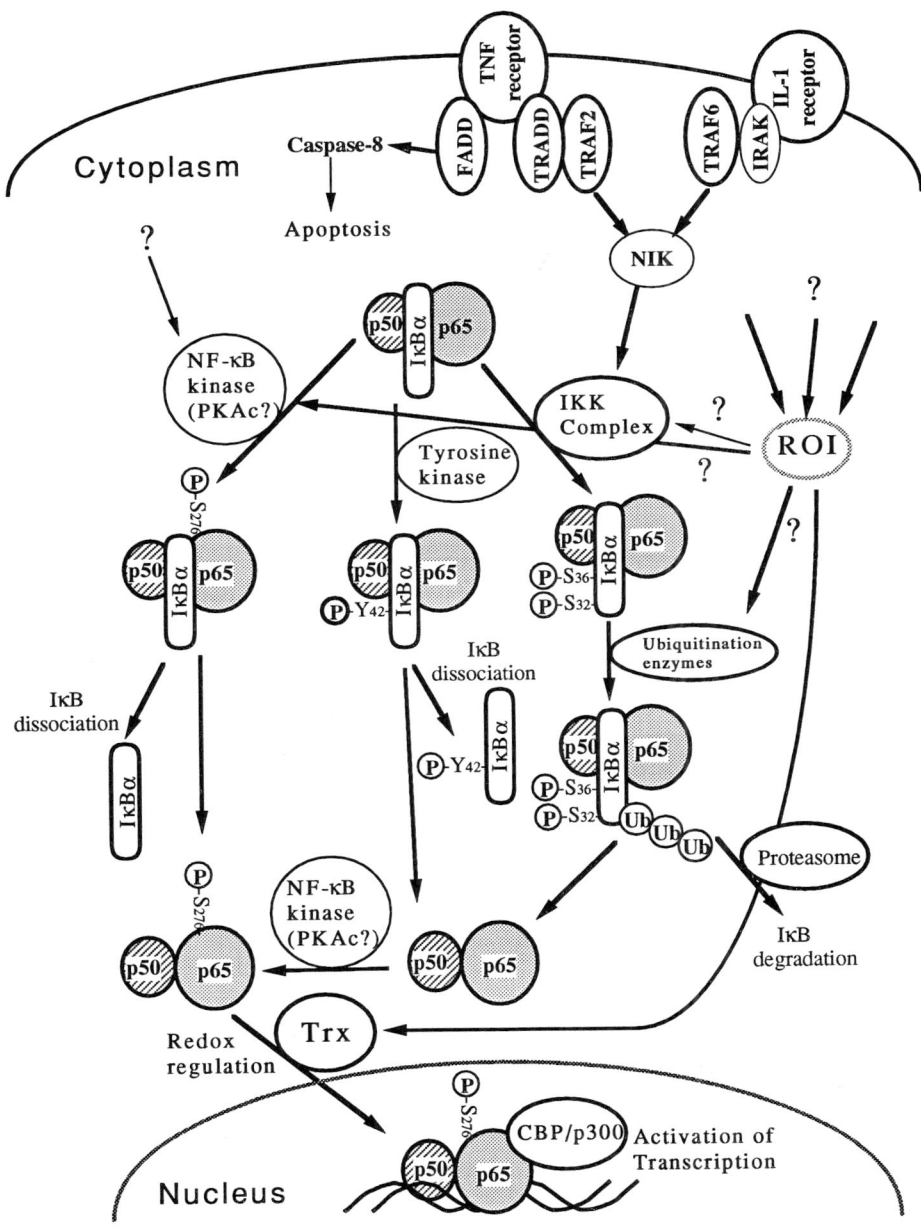

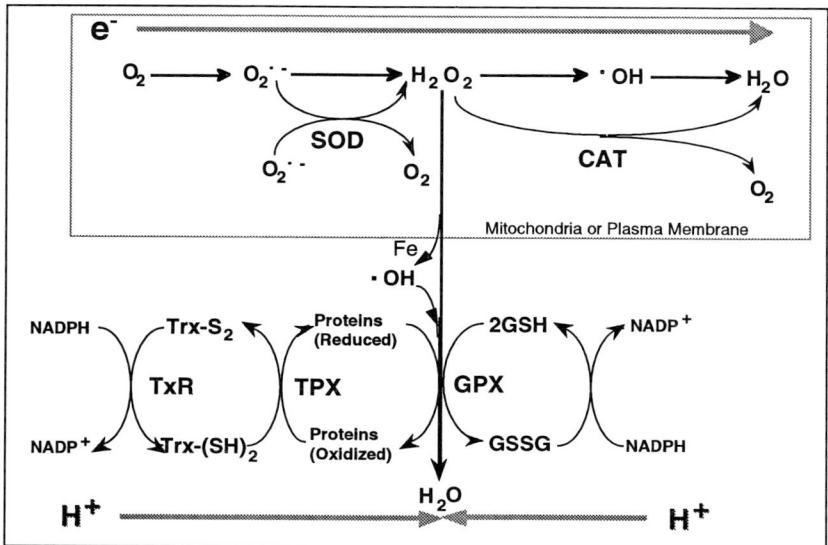

Figure 3 Cellular redox system. Successive reduction of oxygen by the addition of four electrons generates reactive oxygen species called ROI. Among these, ROI hydrogen peroxide has the longest half-life and is considered to be a mediator of the oxidative signal. In order to maintain redox homeostasis, there are multiplicated antioxidant defense mechanisms within the cells. These include superoxide dismutase (SOD), catalase (CAT), glutathione (GSH), glutathione peroxidase (GPX), thioredoxin (Trx), and Trx peroxidase (TPX). Unlike other antioxidant enzymes, the Trx system may be involved more specifically in the redox repairment of the oxidized protein molecules. The oxidized or reduced status of the protein confers biological information by regulating the activity of the relevant protein molecule.

However, cellular-reducing system such as Trx and GSH counteract the action of hydrogen peroxide (Holmgren, 1985, 1989; Ziegler, 1985; Allen, 1993). Whereas GSH is directly involved in scavenging ROI, Trx appears to participate in this cascade by repairing the oxidized proteins through its reducing activity. This reversible oxidation and reduction of a functional protein determine its activity. Therefore, GSH itself may not be directly involved in the redox signaling but the intensity of the oxidative signal may be determined by the intracellular GSH level. Thus, total GSH/GSSG content could be a useful indicator for the responsiveness of the cellular redox signaling. This redox regulatory pathway is involved in the NF-κB activation pathway.

B. Activation of NF-κB by the Kinase Cascade and Trx-Mediated Redox Regulation

There are at least two independent steps in the NF-κB activation cascade: kinase pathways and redox-signaling pathways. These two distinct pathways are involved in the NF-κB activation cascade in a coordinate fashion, which may contribute to fine tune, as well as fail-safe, regulation of NF-κB activity. In most of the cases, NF-κB dissociation by the kinase cascade is a primary step of NF-κB activation. At least two distinct types of kinase pathways are known to be involved in NF-κB activation: NF-κB kinase and IκB kinase (Fig. 2). The IKK complex is considered to be responsible for IκB phosphorylation in a wide variety of signal transduction pathway (Zandi et al., 1997; Woronicz et al., 1997). However, NF-κB was shown to be phosphorylated in some cell lines in response to stimulation with TNF or IL-1 followed by NF-κB-dependent gene expression (Ostrowski et al., 1991; Naumann and Scheidereit, 1994; Li et al., 1994). We found a 43-kD serine kinase, called NF-κB kinase, that is associated with NF-κB (Hayashi et al., 1993b). This kinase phosphorylates the p65 subunit of NF-κB and dissociates it from IκB. The catalytic subunit of protein kinase A (PKAc) was found to be associated with the NF-κB/IκB complex in the cytoplasm (Zhong et al., 1997). However, when overexpressed in cultured cells, PKAc alone did not stimulate or augment NF-κB-dependent gene expression (unpublished observation; Feuillard et al., 1991; Neumann et al., 1995; Ollivier et al., 1996).

After dissociation from IκB, however, NF-κB must go through redox regulation by the cellular-reducing catalyst, thioredoxin (Trx) (Okamoto et al., 1992, 1997; Hayashi et al., 1993a), in order to recognize the target DNA sequence and induce transcription. Trx is a cellular-reducing catalyst and is known to participate in redox reactions through the reversible oxidation of its active center dithiol to a disulfide (Figs. 2 and 3). Interestingly, human Trx had been identified initially as a factor responsible for the induction of the α subunit of the interleukin-2 receptor, which has been revealed to be under the control of NF-κB (Tagaya et al., 1989). It has been demonstrated *in vitro* that NF-κB cannot bind to the κB DNA sequence of the target genes until it is reduced (Okamoto et al., 1992; Hayashi et al., 1993a; Molitor et al., 1991; Toledano and Leonard, 1991). Based on the estimation of the high local pI value near one of the conserved cysteine residues, we have assigned the cysteine residue at the 62nd amino acid position of the p50 subunit as a target of the redox regulation (Hayashi et al., 1993a) which was confirmed by a site-directed mutagenesis study (Matthews et al., 1992), in which the cysteine 62 substitution abolished DNA-binding activity.

C. Structural Basis for Trx-Mediated Redox Regulation of NF-κB

Structural biological approaches have provided the molecular mechanism of the redox regulation of NF-κB by Trx. Two independent groups have demonstrated the three-dimensional (3D) structure of the NF-κB subunit p50 homodimer cocrystallized with the target DNA (Ghosh *et al.*, 1995; Müller *et al.*, 1995). NF-κB appears to have a novel DNA-binding structure called β barrel, a group of β sheets stretching toward the target DNA. There is a loop in the tip of the β barrel structure that intercalates with the nucleotide bases and is considered to make a direct contact with DNA. This DNA-binding loop contains the cysteine 62 that we predicted to be the target of redox regulation as a hydrogen acceptor from Trx. Although this cysteine was replaced with alanine in both studies, presumably because of technical reasons, for crystallization, these observations confirm our earlier speculations (Okamoto *et al.*, 1992; Hayashi *et al.*, 1993a). Additionally, Qin *et al.*, (1995) have solved the 3D structure of the Trx, molecule that is associated with the DNA-binding loop of p50 by using nuclear magnetic resonance. A boot-shaped hollow on the surface of Trx containing the redox-active cysteines could stably recognize the DNA-binding loop of p50 (Ghosh *et al.*, 1995; Müller *et al.*, 1995) and is likely to reduce the oxidized cysteine by donating protons in a structure-dependent fashion. Therefore, the reduction of NF-κB by Trx is considered to be specific and dependent on the structural compatibility between the target protein and Trx. However, the disulfide bridge formation between Trx and NF-κB might be transient, as the binding of Trx to the NF-κB DNA-binding loop would block the recognition of DNA because of the apparent competition of the same cysteine residue (Fig. 4). In favor of this model, we have demonstrated that NF-κB and Trx concomitantly migrated to the nucleus in the rheumatoid synovial cells during the first phase of the NF-κB activation process and that Trx was relocated in the cytoplasm after 30 min of stimulation whereas the NF-κB nuclear predominance was still observed for several hours (Fig. 5) (Sakurada *et al.*, 1996).

In addition to these findings, our *in vitro* binding study has demonstrated that the zinc ion is required for the DNA-binding activity of NF-κB as well as the reduction of the redox-sensitive cysteine (Hayashi *et al.*, 1993a; Qin *et al.*, 1995). We found that even the fully activated NF-κB could still be blocked by monovalent gold ion by a redox mechanism (Yang *et al.*, 1995). We found that the zinc ion is a necessary component of the active NF-κB and that the addition of monovalent gold ion could efficiently block its activity by oxidizing the redox-active cysteines on NF-κB. Because gold did not appear to replace zinc, it was suggested that gold ion oxidizes these thiolate anions on NF-κB into disulfides and thus abrogates the DNA-

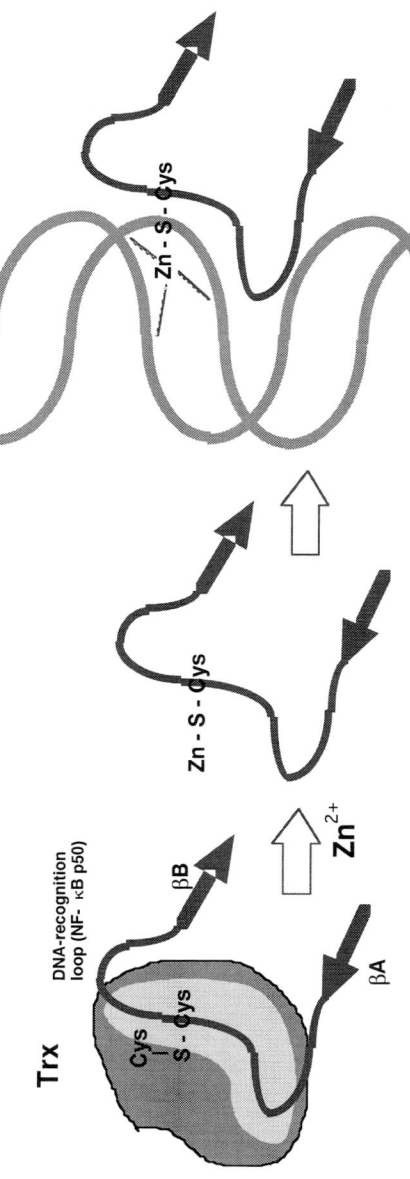

Figure 4 Reduction of the redox-sensitive cysteine on NF-κB by Trx. A boot-shaped hollow on the surface of Trx, also containing the redox-active cysteines, could stably recognize the DNA-binding loop of p50 and reduce the oxidized cysteine by donating protons in a structure-dependent way [based on the 3D structure of the Trx molecule with the DNA-binding loop peptide of p50 (Ghosh et al., 1995; Müller et al., 1995; Qin et al., 1995)]. Zinc is required to make NF-κB competent for DNA binding (Yang et al., 1995). We thus assume that the active intermediate of NF-κB is associated with zinc.

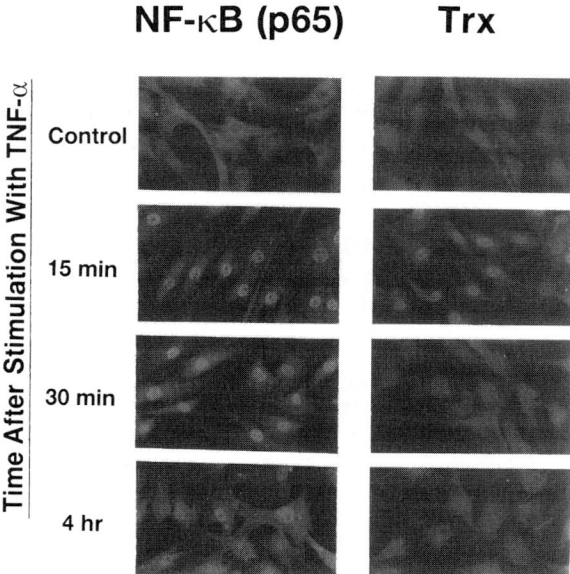

Figure 5 Nuclear translocation of NF-κB and Trx in rheumatoid synovial fibroblasts (RSF). RSF were treated with 10 ng/ml of IL-1β for various time periods (0, 15 min, 30 min, and 4 hr). The cells were then reacted with rabbit antibodies against p65 of the NF-κB subunit or Trx and subsequently stained with FITC-conjugated goat antirabbit IgG. Note the concomitant nuclear translocation of p65 and Trx only at the stage after 15 min of TNF stimulation. This colocalization of NF-κB (p65) and Trx suggests the interaction between these two proteins in the cultured cells during the NF-κB activation cascade.

binding activity because of its higher oxidation potential over the zinc ion (Yang *et al.*, 1995). It is interesting to note that gold compounds have been used successfully for the treatment of RA (Skosey, 1993; Insel, 1996). Our findings could explain why gold is effective in RA. It may be that the gold compound is potentially effective in other diseases where NF-κB plays a pathological role.

IV. CONCLUSION: REDOX REGULATION OF NF-κB

With regard to the redox regulation of NF-κB, we would like to propose the following model: (1) Generation of ROI in response to the extracellular stimuli (proinflammatory cytokines such as TNF and IL-1, irradiation, and other oxidative stress, etc.) would activate one of the kinase cascades that lead to IκB dissociation (such as NF-κB kinase and IκB kinases). (2) After the dissociation of IκB, NF-κB will move to the nucleus. During this nuclear translocation, NF-κB is associated with Trx, and the redox active cysteine (e.g., the cysteine 62 of p50) is reduced in a structure-dependent manner. A disulfide bridge between NF-κB and Trx might be transiently formed. (3) Association of the zinc ion with the redox-sensitive cysteine of NF-κB will eventually dissociate Trx and Trx will be relocated to the cytoplasm. Finally, the zinc-associated NF-κB will bind to the target DNA.

The findings of the redox control mechanism in the NF-κB activation cascade as described in this chapter should provide a basis for the development of a novel therapeutic strategy against various pathologies in which NF-κB plays a substantial role. A transcription factor-directed pathological paradigm regarding NF-κB, such as depicted in Fig. 1, should provide useful insights for the development of a novel therapeutic strategy as signal transduction pathways, transcription factors, target genes, and their biological activities are causally associated.

ACKNOWLEDGMENTS

This work was supported by grants-in-aid from the Ministry of Health and Welfare, from the Ministry of Education, Science, Culture, and Sports of Japan, and from the Human Science Foundation of Japan.

REFERENCES

Allen, J. F. (1993). Redox control of transcription: Sensors, response regulators, activators and repressors. *FEBS Lett.* **332**, 203–207.

Alvaro-Garcia, J. M., Zvaifler, N. J., Brown, C. B., Kaushansky, K., and Firestein, G. S. (1991). Cytokines in chronic arthritis. VI. Analysis of the synovial cells involved in granulocyte macrophage colony-stimulating factor production and gene expression in rheumatoid arthritis and its regulation by IL-1 and TNF-α. *J. Immunol* **146**, 3365–3372.

Arya, S. K., Guo, C., Josephs, S. F., and Wong-Staal, F. (1985). Trans-activator gene of human T-lymphotropic virus type III (HTLV-III). *Science* **229**, 69–73.

Baeuerle, P. A., and Baichwal, V. R. (1997), NF-κB as a frequent target for immunosuppressive and anti-inflammatory molecules. *Adv. Immunol.* **65**, 111–137.

Baldwin, A. S., Jr. (1996), The NF-κB and IκB proteins: New discoveries and insights. *Annu. Rev. Immunol.* **14**, 649–683.

Beg, A. A., and Baltimore, D. (1996). An essential role for NF-κB in preventing TNF-α-induced cell death. *Science* **274**, 782–784.

Biswas, D. K., Dezube, B. J., Ahlers, C. M., and Pardee, A. B. (1993). Pentoxifylline inhibits HIV-1 LTR-driven gene expression by blocking NF-κB action. *J. AIDS* **6**, 778–786.

Bohnlein, E., Lowenthal, J. W., Siekevitz, M., Ballard, D. W., Franza, B. R., and Greene, W. C. (1988). The same inducible nuclear proteins regulates mitogen activation of both the interleukin-2 receptor-alpha gene and type 1 HIV. *Cell (Cambridge, Mass.)* **53**, 827–836.

Cao, Z., Xiong, J., Takeuchi, M., Kurama, T., and Goeddel, D. V. (1996). TRAF6 is a signal transducer for interleukin-1. *Nature (London)* **383**, 443–446.
Dejana, E., Bertocci, F., Bortolami, M. C., Regonesi, A., Tonta, A., Breviario, F., and Giavazzi, R. (1988). Interleukin 1 promotes tumor cell adhesion to cultured human endothelial cells. *J. Clin. Invest.* **82**, 1466–1470.
Donnelly, R. P., Crofford, L. J., Freeman, S. L., Buras, J., Remmers, E., Wilder, R. L., and Fenton, M. J. (1993). Tissue-specific regulation of IL-6 production by IL-4. Differential effects of IL-4 on nuclear factor-kappa B activity in monocytes and fibroblasts. *J. Immunol.* **151**, 5603–5612.
Feuillard, J., Gouy, H., Bismuth, G., Lee, L. M., Debré, P., and Korner, M. (1991). NF-kappa B activation by tumor necrosis factor alpha in the Jurkat T cell line is independent of protein kinase A, protein kinase C, and Ca (2+)-regulated kinase. *Cytokine* **3**, 257–265.
Ghosh, G., Van Duyne, G., Ghosh, S., and Sigler, P. B. (1995). Structure of NF-kappa B p50 homodimer bound to a kappa B site. *Nature (London)* **373**, 303–310.
Gilmore, T D. (1990). NF-κB, KBF-1, dorsal and related matters. *Cell (Cambridge, Mass.)* **62**, 841–843.
Grilli, M., Pizzi, M., Memo, M., and Spano, P. (1996). Neuroprotection by aspirin and sodium salicylate through blockade of NF-κB activation. *Science* **274**, 1383–1393.
Handel, M. L., McMorrow, L. B., and Gravallese, E. M. (1996). Nuclear factor-κB in rheumatoid synovium. Localization of p50 and p60. *Arthritis Rheum* **38**, 1762–1770.
Hayashi, T., Ueno, Y., and Okamoto, T. (1993a). Oxidoreductive regulation of nuclear factor kappa B. Involvement of a cellular reducing catalyst thioredoxin. *J. Biol. Chem.* **268**, 11380–11388.
Hayashi, T., Sekine, T., and Okamoto, T. (1993b). Identification of a new serine kinase that activates NF-κB by direct phosphorylation. *J. Biol. Chem.* **268**, 26790–26795.
Ho, D. D. (1998). Toward HIV eradication or remission: The tasks ahead. *Science* **280**, 1866–1867.
Holmgrem, A. (1985). Thioredoxin. *Annu. Rev. Biochem.* **54**, 237–271.
Holmgren, A. (1989). Thioredoxin and glutaredoxin systems. *J. Biol. Chem.* **264**, 13963–13966.
Insel, P. A. (1996). Analgesic-antipyretic and anti-inflammatory agents and drugs employed in the treatment of gout. In "The Pharmacological Basis of Therapeutics" (J. G. Hardman et al., eds.), pp. 670–681. Macmillan, New York.
Klefstrom, J., Arghi, E., Littlewood, T., Jaattela, M., Saksela, E., Evan, G. I. and Alitalo, K. (1997). Induction of TNF-sensitive cellular phenotype by c-Myc involves p53 and impaired NF-κB activation. *Cell (Cambridge, Mass.)* **16**, 7382–7392.
Lane, D. P. (1992). p53, guardian of the genome. *Nature (London)* **358**, 15–16.
Li, C. - C. H., Dai, R. -M., Chen, E., and Longo, D. L. (1994). Phosphorylation of NF-KB1-p50 is involved in NF-κB activation and stable DNA binding. *J. Biol. Chem.* **269**, 30089–30092.
Los, M., Schenk, H., Hexel, K., Baeuerle, P. A., Droge, W., and Schulze-Osthoff, K. (1995). IL-2 gene expression and NF-κB activation through CD28 requires reactive oxygen production by 5-lipoxygenase. *EMBO J.* **14**, 3731–3740.
Malinin, N. L., Boldin, M. P., Kovalenko, A. V., and Wallach, D. (1997), MAP3K-related kinase involved in NF-κB induction by TNF and IL-1. *Nature (London)* **385**, 540–544.
Matthews, J. R., Wakasugi, N., Virelizier, J. -L., Yodoi, J., and Hay, R. T. (1992). Thioredoxin regulates the DNA binding activity of NF-kappa B by reduction of a disulfide bond involving cystein 62. *Nucleic Acids Res.* **20**, 3821–3830.
Merin, J. P., Matsuyama, M., Kira, T., Baba, M., and Okamoto, T. (1996). α-Lipoic acid blocks HIV-1 LTR-dependent expression of hygromycin resistance in THP-1 stable transformants. *FEBS Lett.* **394**, 9–13.

Meyer, M., Schreck, R., and Baeuerle, P. A. (1993). H$_2$O$_2$ and antioxidants have opposite effects on activation of NF-κB and AP-1 in intact cells: AP-1 as secondary antioxidant-responsive factor. *EMBO J.* **12**, 2005–2015.

Molitor, J. A., Ballard, D. W., and Greene, W. C. (1991). Kappa-B-specific DNA binding proteins are differentially inhibited by enhancer mutations and biological oxidation. *New Biol.* **3**, 987–996.

Montgomery, K. F., Osborn, L., Hession, C., Tizard, R., Goff, D., Vassallo, C., Tarr, P. I., Bomsztyk, K., Lobb, R., Harlan, J. M., and Pohlman, T. H. (1991). Activation of endothelial-leukocyte adhesion molecule 1 (ELAM-1) gene transcription. *Proc. Natl. Acad. Sci. U.S.A.* **88**, 6523–6527.

Mukaida, N., Mahe, Y., and Matsushima, K. (1990). Cooperative interaction of nuclear factor-15. appa B- and cis-regulatory enhancer binding protein-like factor binding elements in activating the interleukin-8 gene by pro-inflammatory cytokines. *J. Biol. Chem.* **265**, 21128–21133.

Müller, C. W., Rey, F. A., Sodeoka, M., Verdine, G. L., and Harrison, S. C. (1995). Structure of the NF-kappa B p50 homodimer bound to DNA. *Nature (London)* **373**, 311–317.

Nabel, G., and Baltimore, D. (1987). An inducible transcription factor activates expression of human immunodeficiency virus in T cells. *Nature (London)* **326**, 711–713.

Naumann, M., and Scheidereit, C. (1994). Activation of NF-κB in vivo is regulated by mutiple phosphorylations. *EMBO J.* **13**, 4597–4607.

Neumann, M., Grieshammer, T., Chuvpilo, S., Kneitz, B., Lohoff, M., Schimpl, A., Franza, B. R., Jr., and Serfling, E. (1995). Re1A/p65 is a molecular target for the immunosuppressive action of protein kinase A. *EMBO J.* **14**, 1991–2004.

Okamoto, T. (1995). Regulatory proteins of human immunodeficiency virus and therapy. In "Anti-AIDS Drug Development: Challenges Strategies and Prospects" (P. Mohan and M. Baba, eds.), pp. 117–128. Harwood Chur, Academic Publishers, Switzerland.

Okamoto, T., and Wong-Staal, F. (1986). Demonstration of virus-specific transcriptional activator(s) in cells infected with HTLV-III by an in vitro cell-free system. *Cell (Cambridge, Mass.)* **47**, 29–35.

Okamoto, T., Matsuyama, T., Mori, S., Hamamoto, Y., Kobayashi, N., Yamamoto, N., Josephs, S. F., Wong-Staal, F., and Shimotohno, K. (1989). Augmentation of human immunodeficiency virus type 1 gene expression by tumor necrosis factor alpha. *AIDS Res. Hum Retroviruses* **5**, 131–138.

Okamoto, T., Benter, T., Josephs, S. F., Sadaie, M. R., and Wong-Staal, F. (1990). Transcriptional activation from the long-terminal repeat of human immunodeficiency virus in vitro. *Virology* **177**, 606–614.

Okamoto, T., Ogiwara, H., Hayashi, T., Mitsui, A., Kawabe, T., and Yodoi, J. (1992). Human thioredoxin/adult T cell leukemia-derived factor activates the enhancer binding protein of human immunodeficiency virus type 1 bt thiol redox control mechanism. *Int. Immunol.* **4**, 811–819.

Okamoto, T., Sakurada, S., Yang, J. -P., and Merin, J. P. (1997). Regulation of NF-κB and Disease control: Identification of a novel serine kinase and thioredoxin as effectors for signal transduction pathway for NF-κB activation. *Curre. Top. Cell. Regul.* **35**, 149–161.

Ollivier, V., Parry, G. C. N., Cobb, R. R., de Prost, D., and Mackman, N. (1996). Elevated cyclic AMP inhibits NF-κB-mediated transcription in human monocytic cells and endothelial cells. *J. Biol. Chem.* **271**, 20828–20835.

Ostrowski, J., Sims, J. E., Sibley, C. H., Valentine, M. A., Dower, S. K., Meier, K. E., and Bomsztyk, K. (1991). A serine/threonine kinase activity is closely associated with a 65-kDa phosphoprotein specifically recognized by the kappa B enhancer element. *J. Biol. Chem.* **266**, 12722–12733.

Packer, L., Witt, E. H., and Tritschler, H. J. (1995). α-lipoic acid as a biological antioxidant. *Free Radical Biol. Med.* **19**, 227–250.

Perelson, A. S., Essunger, P., Cao, Y., Vesanen, M., Hurley, A., Saksela, K., Markowitz, M., and Ho, D. D. (1997). Decay characteristics of HIV-1-infected compartments during combination therapy. *Nature (London)* **387**, 188–191.

Qin, J., Clore, G. M., Kennedy, W. M. P., Huth, J. R., and Gronenborn, A. M. (1995). Solution structure of human thioredoxin in a mixed disulfide intermediate complex with its target peptide from the transcription factor NF-kappa B. *Structure* **3**, 289–297.

Roebuck, K. A., Rahman, A., Lakshminarayanan, V., Janakidevi, K., and Malik, A. B. (1995). H_2O_2 and tumor necrosis factor-alpha activate intercellular adhesion molecule 1 (ICAM-1) gene transcription through distinct cis-regulatory elements within the ICAM-1 promoter. *J. Biol. Chem.* **270**, 18966–18974.

Roederer, M., Staal, F. J. T., Raju, P. A., Ela, S. W., Herzenberg, L. A., and Herzenberg, L. A. (1990). Cytokine-stimulated human immunodeficiency virus replication is inhibited by N-acetyl-Lcysteine. *Proc. Natl. Acad. Sci. U.S.A.* **87**, 4884–4888.

Rothe, M., Sarma, V., Dixit, V. M., and Goeddel, D. V. (1995). TRAF2-mediated activation of NF-κB by TNF receptor 2 and CD40. *Science* **269**, 1424–1427.

Sachi, Y., Hirota, K., Masutani, H., Toda, K., Okamoto, T., Takigawa, M., and Yodoi, J. (1995). Three NF-kappa B binding sites in the human E-selectin gene required for maximal tumor necrosis factor alpha-induced expression. *Immunol. Lett.* **44**, 189–193.

Sakurada, S., Kato, T., and Okamoto, T. (1996). Induction of cytokines and ICAM-1 by proinflammatory cytokines in primary rheumatoid synovial fibroblats and infibition by N-acetyl-L-cysteine and aspirin. *Int. Immunol.* **8**, 1483–1493.

Schindler, U., and Baichwal, V. R. (1994). Three NF-kappa B binding sites in the human E-selectin gene required for maximal tumor necrosis factor alpha-induced expression. *Mol. Cell. Biol.* **14**, 5820–5831.

Schmidt, K. N., Traenckner, E. B., Meier, B., and Baeuerle, P. A. (1995). Induction of oxidative stress by okadaic acid is required for activation of transcription factor NF-κB. *J. Biol. Chem.* **270**, 27136–27142.

Schreck, R., and Baeuerle, P. A. (1990). NF-kappa B as inducible transcriptional activator of the granulocyte-macrophage colony-stimulating factor gene. *Mol. Cell. Biol.* **10**, 1281–1286.

Schreck, R., Rieber, P., and Baeuerle, P. A. (1991). Reactive oxygen intermediates as apparently widely used messengers in the activation of the NF-κB transcription factor and HIV-1. *EMBO J.* **10**, 2247–2258.

Sen, R., and Baltimore, D. (1986). Inducibility of kappa immunoglobulin enhancer-binding protein NF-kappa B by a posttranslational mechanism. *Cell (Cambridge, Mass.)* **46**, 705–716.

Skosey, J. L. (1993). Treatment of rheumatoid arthritis. In "Arthritis and Allied Conditions" (D. J. McCarty and W. J. Koopman, eds.), pp. 603–614. Lea & Febiger, Philadelphia.

Sodroski, J., Patarca, R., and Rosen, C. (1985). Location of the trans-activating region on the genome of human T-cell lymphotropic virus type III. *Science* **229**, 74–77.

Stade, B. G., Messer, G., Riethmuller, G., and Johnson, J. P. (1990). Structural characteristics of the 5' region of the human ICAM-1 gene. *Immunobiology* **182**, 79–87.

Staynov, D. Z., Cousins, D. J., and Lee, T. H. (1995). A regulatory element in the promoter of the human granulocyte-macrophage colony-stimulating factor gene that has related sequences in other T-cell-expressed cytokine genes. *Proc. Natl. Acad. Sci. U.S.A.* **92**, 3606–3610.

Suzuki, Y. J., Aggarwal, B. B., and Packer, L. (1992). Alpha-lipoic acid is a potent inhibitor if NF-kappa B activation in human T cells. *Biochem. Biophys. Res. Commun.* **189**, 1709–1715.

Suzuki, Y. J., Mizuno, M., and Packer, L. (1994). Signal transduction for nuclear factor-kappa B activation. Proposed location of antioxidant-inhibitable step. *J. Immunol.* **153,** 5008–5015.
Tagaya, Y., Maeda, Y., Mitsui, A., Kondo, N., Matsui, H., Hamuro, J., Brown, J., Arai, K. I., Yokota, T., Wakasugi, H., and Yodoi, J. (1989). ATL-derived factor (ADF), an IL-2 receptor/Tac inducer homologous to thioredoxin; possible involvement of dithiol-reduction in the IL-2 receptor induction. *EMBO J.* **8,** 757–764.
Takada, A., Ohmori, K., Yoneda, T., Tsuyuoka, K., Hasegawa, A., Kiso, M., and Kannagi, R. (1993). Contribution of carbohydrate antigens sialyl Lewis A and sialyl Lewis X to adhesion of human cancer cells to vascular endothelium. *Cancer Res.* **53,** 354–361.
Thanos, D., and Maniatis, T. (1995). NF-κB: A lesson in family values. *Cell (Cambridge, Mass.)* **80,** 529–532.
Toledano, M. B., and Leonard, W. J. (1991). Modulation of transcription factor NF-kappa B binding activity by oxidation-reduction in vitro. *Proc. Natl. Acad. Sci. U.S.A.* **88,** 4328–4332.
Tozawa, K., Sakurada, S., Kohri, K., and Okamoto, T. (1995). Effects of anti-nuclear factor kappa B reagents in blocking adhesion of human cancer cells to vascular endothelial cells. *Cancer Res.* **55,** 4162–4167.
Ulfgren, A. K., Lindblad, S., Klareskog, L., Andersson, J., and Andersson, U. (1995). Detection of cytokine producing cells in the synovial membrane from patients with rheumatoid arthritis. *Ann. Rheum. Dis.* **54,** 654–659.
Whelan, J., Ghersa, P., Huijsduijnen, R. H., Gray, J., Chandra, G., Talabot, F., and DeLamarter, J. F. (1991). An NF kappa B-like factor is essential but not sufficient for cytokine induction of endothelial leukocyte adhesion molecule 1 (ELAM-1) gene transcription. *Nucleic Acids Res.* **19,** 2645–2653.
Woronicz, J. D., Gao, X., Cao, Z., Rothe, M., and Goeddel, D. V. (1997). IκB kinase-β: NF-κB activation and complex formation with IκB kinase-α and NIK. *Science* **278,** 866–869.
Wu, H., and Lozano, G. (1994). NF-κB activation of p53: A potential mechanism for suppressing cell growth in response to stress. *J. Biol. Chem.* **269,** 20067–20074.
Wu, M., Lee, H., Bellas, R. E., Schauer, S. L., Arsura, M., Katz, D., FitzGerald, M. J., Rothstein, T. L., Sherr, D. H., and Sonenshein, G. E. (1996). Inhibition of NF-κB/Rel induces apoptosis of murine B cells. *EMBO J* **15,** 4682–4690.
Xie, Q. -W., Kashiwabara, Y., and Nathan, C. (1994). Role of transcription factor NF-kappa B/Rel in induction of nitric oxide synthase. *J. Biol. Chem.* **269,** 4705–4708.
Yang, J. P., Merin, J. P., Nakano, T., Kato, T., Kitade, Y., and Okamoto, T. (1995). Inhibition of the DNA-binding activity of NF-kappa B by gold compounds in vitro. *FEBS Lett.* **361,** 89–96.
Zandi, E., Rothwarf, D. M., Delhase, M., Hayakawa, M., and Karin, M. (1997). The IκB kinase complex (IKK) contains two kinase subunits, IKKα and IKKβ, necessary for IκB phosphorylation and NF-κB activation. *Cell (Cambridge, Mass.)* **91,** 243–252.
Zhong, H., Su Yang, H., Erdjument-Bromage, H., Tempst, P., and Ghosh, S. (1997). The transcriptional activity of NF-κB is regulated by the IκB-associated PKAc subunit through a cyclic-AMP independent mechanism. *Cell (Cambridge, Mass.)* **89,** 413–424.
Ziegler, D. M. (1985). Role of reversible oxidation-reduction of enzyme thios-disulfides in metabolic regulation. *Annu. Rev. Biochem.* **54,** 305–329.

10 Oxidants and Antioxidants in Apoptosis: Role of Bcl-2

Chandan K. Sen

Lawrence Berkeley National Laboratory
University of California
Berkeley, California 94720
and Department of Physiology
University of Kuopio
Finland

I. INTRODUCTION

Programmed cell death or apoptosis represents a highly controlled form of cell death in which single cells are selectively eliminated without the release of cellular debris and perturbation of neighboring tissues (Ashkenazi and Dixit, 1998; Granville *et al.*, 1998; Green, 1998; King and Cidlowski, 1998; Schulze-Osthoff *et al.*, 1998). Apoptosis regulates several key physiological functions. For example, developing lymphocytes undergo extensive cell death during selection of the immune repertoire. Redox-dependent mechanisms have been shown to be involved in the regulation of a wide variety of molecular processes (Schulze-Osthoff *et al.*, 1995; 1997; Baeuerle *et al.*, 1996; Sen and Packer, 1996; Sen, 1998), including apoptosis. Reactive oxygen species (ROS), antioxidants, and other determinants of the intracellular redox state have been shown to regulate apoptosis in several types of cell (Stoian *et al.*, 1996). It has been observed in various experimental models that the activation of apoptosis is associated with the generation of ROS. Perhaps one of the primary functions of apoptosis evolutionarily is to purify tissues from ROS-overproducing cells without affecting adjacent cells. Although ROS are thought to be implicated in a wide variety of apoptosis processes, it has been shown that apoptosis may also take place in the absence or decreased presence of ROS (Jacobson, 1996). In certain cases such as apoptosis caused via activation of the Fas receptor, ROS has been shown to have an inhibitory effect (Clement and Stamenkovic, 1996).

Understanding the fundamental mechanism of apoptosis is crucial to developing therapeutic strategies for controlling apoptosis in diseased tissues (Holzman, 1997). Such information may be used to protect healthy cells against apoptosis and to selectively kill diseased cells. For example, inhibitors of apoptosis may be utilized to induce resistance to chemotherapeutic drugs and irradiation, and inducers of apoptosis may be used to control neoplastic events that result from uncontrolled cell proliferation. The aim of this chapter is to provide a brief overview of our current knowledge of the role of reactive species and antioxidants in regulating apoptosis. Specifically, the role of Bcl-2 family proteins in the redox regulation of apoptosis is discussed.

II. INDUCTION OF APOPTOSIS BY REACTIVE OXYGEN SPECIES

Measurements of intracellular ROS in cells treated with oxidants or antioxidants have shown a correlation between levels of reactive oxygen and the induction or suppression of apoptosis (Packham *et al.*, 1996). Two primary lines of evidence suggest that ROS are common mediators in several apoptosis signaling pathways: (i) that an elevated level of oxidants and/or oxidative damage markers is detected in activated cells *en route* to apoptosis and (ii) that antioxidants suppress apoptosis induced by diverse stimuli in many cells (Chau *et al.*, 1998). In several cases, the activation of lymphocytes is followed by the accumulation of intracellular ROS, which may serve as intracellular signaling molecules. The ultimate outcome of this increased ROS formation, i.e., lymphocyte proliferation versus programmed cell death, is thought to be dictated by macrophage-derived costimulatory molecules that bolster or diminish lymphocyte antioxidant defenses (Buttke and Sandstrom, 1995). The importance of reactive oxygen species in the cytokine-mediated degradation of sphingomyelin to ceramide, a proapoptotic intermediate, has been highlighted (Singh *et al.*, 1998). Similar to cytokines, such as TGF-β 1, ROS such as hydrogen peroxide increase the expression of c-*fos* and c-*jun* genes rapidly and may induce cell death by apoptosis (Lafon *et al.*, 1996). A role of p53 to regulate the intracellular redox state and to induce apoptosis by a pathway that is dependent on ROS production has been proposed (Johnson *et al.*, 1996). In thymocyte apoptosis, intracellular oxidation has been claimed to be an obligate early component (Bustamante *et al.*, 1995; Offen *et al.*, 1995). Superoxide is produced by mitochondria isolated from apoptotic cells due to a switch from the normal four-electron reduction of molecular oxygen to a one-electron reduction when cytochrome c is released from mitochondria (Cai and Jones, 1998). In several experimental situations, evidence shows that apoptosis is preceded by the accumulation of intracellular ROS or elevated levels of oxidative damage markers, which are discussed in the following section.

Signals generated by T-cell receptor cross-linking, phorbol 12-myristate-13-acetate + Ca^{2+} ionophore, glucocorticoids, or ionizing radiation all stimulate apoptotic cell death in thymocytes by signals that are initially distinct from each other. However, when these stimuli were administered to thymocyte cultures that were maintained under an atmosphere containing less than 20 ppm oxygen as opposed to one that contained 18.5% molecular oxygen, cell death was inhibited or abrogated, suggesting that the induction of death by all three different stimuli depends on the presence of molecular oxygen. N-Acetyl-L-cysteine inhibited the induction of death, suggesting a possible involvement of ROS in causing death (McLaughlin *et al.*, 1996).

Apoptosis is associated with a sequential dysregulation of mitochondrial function that precedes cell shrinkage and nuclear fragmentation. Cytofluorometrically purified cells with a reduced mitochondrial transmembrane potential are initially incapable of oxidizing hydroethidine into ethidium. On short-term *in vitro* culture, such cells acquire the capacity of hydroethidine oxidation, thus revealing a second step of apoptosis marked by the mitochondrial generation of ROS. Finally, cells reduce their volume, a step that is delayed by radical scavengers, indicating the implication of ROS in the apoptotic process (Zamzami *et al.*, 1995). 6-Carboxy-2′,7′-dichlorodihydrofluorescein diacetoxymethyl ester is often used as another fluorometric probe to measure intracellular ROS. Using this measurement approach, it has been shown that serum deprivation-induced apoptosis is preceded by a significant increase of ROS in PC12 cells and rat cortical neurons. N,N′-Diphenyl-*p*-phenylenediamine, an antioxidant, reduced ROS production induced by serum deprivation and recovered cell survival (Satoh *et al.*, 1996). Serum deprivation of cerebellar granule neurons has also been shown to be followed by an elevation of intracellular peroxide level preceding apoptosis (Atabay *et al.*, 1996). These observations suggest that ROS play a causative role in serum deprivation-induced apoptosis. Neuronally differentiated PC12 cells undergo synchronous apoptosis when deprived of nerve growth factor. Similar to the effect of serum deprivation, the withdrawal of nerve growth factor has been shown to be followed by an accumulation of intracellular ROS, which in turn is followed by apoptosis. In this case, ROS appeared to be required for apoptosis because cell death was prevented by the free radical spin trap, N-*tert*-butyl-α-phenylnitrone, and the antioxidant, N-acetyl-L-cysteine. ROS production was blocked by actinomycin D, cycloheximide, and caspase protease inhibitors, suggesting that ROS generation occurred downstream of new mRNA and protein synthesis and activation of caspases (Schulz *et al.*, 1997). In neurons, apoptosis has also been shown to be caused by ROS generated by neurotransmitter-linked mechanisms (Manev *et al.*, 1995).

Oxidative stress appears to be a common requirement for apoptosis that occurs during mouse embryonic development. Using an *in vitro* culture system in which digit individualization of developing limbs normally occurs, the effect of different antioxidants on the cell death that takes place at interdigits has been examined. The addition of phenol, dimethyl sulfoxide, or 2′,7′-dichlorodihydrofluorescein diacetate to murine developing limbs in culture prevented digit individualization as well as the typical interdigital cell death. Two ROS-sensitive dyes, 3-(4,5-dimethylthiazol)-2,5-diphenyl tetrazolium bromide and 2′,7′-dichlorodihydrofluorescein diacetate, showed that the cells at the interdigits contain elevated levels of ROS (Salas-Vidal *et al.*, 1998).

Oxygen radicals generated by NADPH oxidase may contribute directly or indirectly to apoptosis induced by camptothecin, a topoisomerase I inhibitor, in human leukemia and in neutrophilic-differentiated cells (Hiraoka *et al.*, 1998). Also, treatment of cells with certain topoisomerase inhibitors, e.g., β-lapachone, is accompanied by a remarkable increase in intracellular hydrogen peroxide in human leukemia HL-60 cells (Chau *et al.*, 1998). Erbstatin, a tyrosine kinase inhibitor, induces apoptosis in human small cell lung carcinoma cells. Erbstatin treatment resulted in elevated levels of intracellular hydrogen peroxide generation. Erbstatin-induced apoptosis was inhibited by antioxidants, whereas erbstatin-inhibited tyrosine phosphorylation was not affected by them. Erbstatin-induced hydrogen peroxide production and DNA fragmentation were partially suppressed by the inhibition of protein synthesis, suggesting that erbstatin-induced apoptosis is likely due to hydrogen peroxide generation via a newly synthesized protein (Simizu *et al.*, 1996). Polyunsaturated fatty acids such as arachidonic acid have been shown to kill Hep G2 cells by a cytochrome P4502E1-dependent mechanism (Fig. 1). The cytotoxicity of arachidonic acid was suggested to involve a lipid peroxidation type of mechanism because the toxicity was associated with elevated levels of malondialdehyde and 4-hydroxy-2-nonenal (Chen *et al.*, 1997). Snake venom-induced apoptosis of endothelial cells has also been shown to be ROS mediated (Suzuki *et al.*, 1997).

The inhibitory effects of N-(4-hydroxyphenyl)retinamide on tumorigenesis and tumor growth have been shown to result from its ability to induce apoptosis via a ROS-dependent mechanism (Oridate *et al.*, 1997). Arsenite represents another inducer of apoptosis that may function via a ROS-dependent mechanism. An increase of intracellular peroxide levels was accompanied with arsenite-induced apoptosis (Chen *et al.*, 1998). Activation of caspase-3 activity, poly(ADP-ribose)polymerase (a DNA repair enzyme) degradation, and release of cytochrome c from mitochondria to the cytosol are involved in arsenite-induced apoptosis, and Bcl-2 antagonizes arsenite-

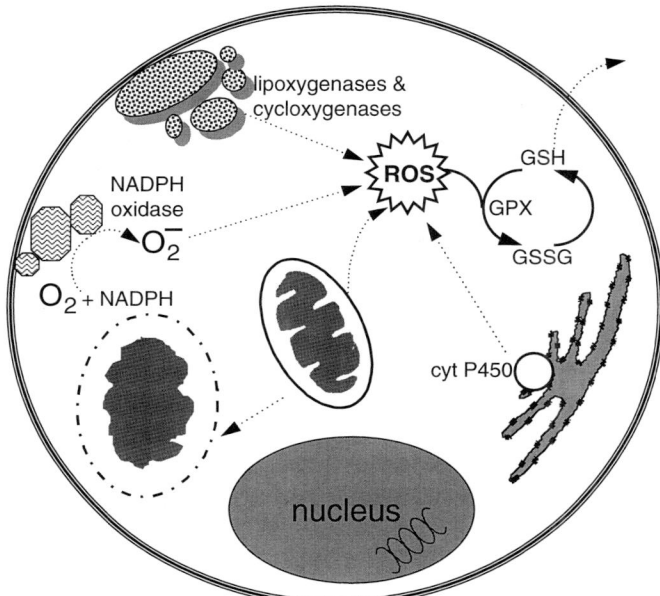

Figure 1 Possible sources of reactive oxygen species (ROS) in a cell undergoing apoptosis. Dotted arrows represent events that may be induced in response to an apoptosis-triggering signal. Most forms of apoptosis are associated with a rapid loss of intracellular GSH. In such GSH-deficient cells, the activity of glutathione peroxidase (GPX), a ROS-decomposing enzyme, may be limited, resulting in increased levels of intracellular ROS. In several cases of apoptosis, such as that induced by tumor necrosis factor-α (see Chapter 11 this volume), mitochondria may serve as a considerable source of ROS. In some cases, mitochondrial swelling, disintegration of the outer membrane, and distension of the inner membrane are suspected to precede apoptosis. Depending on the stimulus inducing apoptosis, NADPH oxidase, cytochrome P450, and lipoxygenase/cycloxygenase may also contribute to elevated levels of intracellular ROS formation.

induced apoptosis by a mechanism that interferes with the activity of caspase-3. Results of this work led to a working hypothesis that arsenite-induced apoptosis is triggered by the generation of hydrogen peroxide through the activation of flavoprotein-dependent superoxide-producing enzymes (such as NADPH oxidase) and that hydrogen peroxide might play a role as a mediator to induce apoptosis through the release of cytochrome c to cytosol (Chen *et al.*, 1998).

III. ANTIAPOPTOTIC PROPERTIES OF ANTIOXIDANTS

Numerous studies have reported the antiapoptotic property of endogenous as well as exogenous antioxidants. Some of these studies showing that antioxidants may protect from apoptosis in a variety of systems have been discussed. Mutations in human Cu/Zn superoxide dismutase-1 cause approximately 20% of cases of familial amyotrophic lateral sclerosis. Expression of two familial amyotrophic lateral sclerosis-related mutant superoxide dismutases (A4V and V148G) caused apoptosis of differentiated PC12 cells, superior cervical ganglion neurons, and hippocampal pyramidal neurons, suggesting that superoxide dismutases may have an antiapoptotic function in neuronal cells (Ghadge et al., 1997). The antiapoptotic property of Mn-superoxide dismutase, a mitochondrial antioxidant, has been also reported (Briehl et al., 1997). Thioredoxin peroxidase is a member of a newly discovered family of proteins that are conserved from yeast to mammals and to which the natural killer enhancing factor belongs. These proteins are antioxidants that function as peroxidases only when coupled to a sulfhydryl-reducing system. Overexpression of human thioredoxin peroxidase II has been shown to protect against apoptosis induced by serum deprivation, ceramide, or etoposide. Thioredoxin peroxidase, like Bcl-2, was able to inhibit the release of cytochrome c from mitochondria to cytosol and inhibited lipid peroxidation in cells. Thioredoxin peroxidase II, unlike Bcl-2, could prevent hydrogen peroxide accumulation in cells, suggesting that it functions upstream of Bcl-2 in protecting from apoptosis and may be implicated as an endogenous regulator of apoptosis (Zhang et al., 1997). Apoptosis is associated with myocardial ischemic reperfusion injury. The presence of apoptotic cells and DNA fragmentation in the myocardium were abolished by preperfusing the hearts in the presence of ebselen, a glutathione peroxidase mimic (Maulik et al., 1998).

The intracellular thiol status of human polymorphonuclear leukocytes has been shown to influence the path of apoptosis. Sulfhydryl oxidation by diethylmaleate alone was shown to induce apoptosis, suggesting the presence of a redox-sensitive, thiol-mediated pathway of apoptosis (Watson et al., 1996). Cancer cells contain a high level of intracellular GSH and are relatively more resistant to inducible apoptosis (Chau et al., 1998). Most forms of apoptosis are preceded by a severe loss of intracellular reduced glutathione (GSH), a response often associated with oxidative stress. The ability of cellular GSH to regulate the ceramide path to apoptosis has been demonstrated. Treatment of rat primary astrocytes with tumor necrosis factor-α or interleukin-1β led to a decrease in intracellular GSH and a rapid degradation of sphingomyelin to ceramide. Pretreatment of astrocytes with N-acetyl-L-cysteine prevented a cytokine-induced decrease in GSH and a degradation of sphingomyelin to ceramide, whereas treatment of astrocytes with diamide, a thiol-oxidizing agent, alone caused a degradation of sphingomyelin to ceramide. Besides astrocytes, N-acetyl-L-cysteine also blocked cytokine-mediated ceramide production in rat primary oligodendrocytes, microglia, and C6 glial cells. These results suggest that the intracellular level of GSH may play a critical role in the regulation of cytokine-induced generation of ceramide, leading to apoptosis of brain cells (Singh et al., 1998). 2-Deoxy-D-ribose, the most reducing sugar, has been shown to induce apoptosis in normal human fibroblasts by a mechanism involving glutathione metabolism. 2-Deoxy-D-ribose was found to provoke disruption of the actin filament network and detachment from the substratum, while at the same time increasing the expression of several integrins and cell adhesion molecules. N-Acetyl-L-cysteine fully blocked 2-deoxy-D-ribose-induced apoptosis by preventing GSH depletion, while it also inhibited actin filament network disruption and mitochondrial depolarization (Kletsas et al., 1998).

A functional role of ROS both for the TGF-β 1-induced signal pathway in normal cells and for the induction of apoptosis in transformed cells has been suggested mostly based on antioxidant-inhibition results. TGF-β-treated normal fibroblasts can induce apoptosis of transformed cells. The overall process was inhibited by antioxidants and radical scavengers (Langer et al., 1996). N-Acetyl-L-cysteine inhibited TGF-β 1-induced apoptosis of adenocarcinoma cells consistently, but did not influence cell cycle arrest caused by the cytokine (Lafon et al., 1996). The antiapoptotic property of phenolic antioxidants has been also evident. Cultured A431 epidermoid cells exposed to UVB (120–2400 J/m^2) develop numerous blebs on their surface, detach from the plastic dish, and undergo injury and death. The lipophilic antioxidant vitamin E inhibited UVB-induced apoptosis. Moreover, vitamin E was observed to be effective in stimulating cell recovery when it was added after the end of UVB irradiation (Straface et al., 1995). Antioxidants such as α-tocopherol phosphate, 6-hydroxy-2,5,7,8-tetramethylchroman-2-carboxylic acid (trolox), propylgallate, ascorbate, and diphenylphenylenediamine and the iron chelator desferrioxamine protected against arachidonic acid-induced lipid peroxidation and apoptosis of Hep G2 cells (Chen et al., 1997). The cancer chemopreventive retinoid N-(4-hydroxyphenyl)-*all-trans* retinamide causes apoptosis of human leukemic cell lines. Antioxidants such as N-acetyl-L-cysteine, ascorbic acid, α-tocopherol, and deferroxamine suppress such apoptosis (Delia et al., 1995). Arsenite treatment results in the accumulation of intracellular peroxides followed by apoptosis N-acetyl-L-cysteine, diphenylene iodonium (an inhibitor of NADPH oxidase), 4,5-dihydro-1,3-benzenedisulfonic acid (a selective scavenger of superoxide anion), and catalase significantly inhibited arsenite-induced apoptosis and elevation of intracellular ROS (Chen et al., 1998).

Antioxidants have been shown to play a critical role in the regulation of programmed cell death, even when death is induced by apparently nonoxidative stimuli. During spermatogenesis, most of the testicular germ cells degenerate by an apoptotic process that is under hormonal control. N-

Acetyl-L-cysteine suppressed apoptosis in human testicular germ cells *in vitro* (Erkkila *et al.*, 1998). Antioxidants have also been shown to suppress ROS-induced apoptosis in human spermatozoa (Lopes *et al.*, 1998).

IV. PROAPOPTOTIC PROPERTIES OF ANTIOXIDANTS

Apart from ROS-detoxifying effects, antioxidant molecules may function as regulators of signal transduction pathways. For example, the possibility that N-acetylcysteine, unlike other antioxidants, suppresses neuronal apoptosis by regulating cell cycle progression has been proposed (Ferrari *et al.*, 1995). In several other cases, however, potentiation of apoptosis by antioxidants has been reported to occur by pathways that are apparently independent of ROS detoxification. N-Acetylcysteine has been shown to elevate p53 expression posttranscriptionally by increasing the rate of p53 mRNA translation. In this way, N-acetylcysteine induced apoptosis in several transformed cell lines and transformed primary cultures but not in normal cells (Liu *et al.*, 1998). NF-κB induction has been shown to play a role in protecting cells from programmed cell death (see Chapter 9, this volume). HIV-1 infection of primary monocytic cells and myeloid cell lines results in sustained NF-κB activation. Inhibition of TNFα-induced NF-κB activation using the antioxidant N-acetyl-L-cysteine resulted in increased apoptosis in both U937 and U9-IIIB cells (DeLuca *et al.*, 1998). Pyrrolidinedithiocarbamate and N-acetyl-L-cysteine have been used as antioxidants to prevent apoptosis in lymphocytes, neurons, and vascular endothelial cells. However, in rat and human smooth muscle cells, both pyrrolidinedithiocarbamate and N-acetyl-L-cysteine induce apoptosis. These observations suggest that certain antioxidants may be useful to induce apoptosis in vascular smooth muscle cells because they may help prevent their proliferation in arteriosclerotic lesions (Tsai *et al.*, 1996).

Curcumin, widely used as a spice and coloring agent in food, possesses potent antioxidant, anti-inflammatory, and antitumor promoting activities. Paradoxically, curcumin has been found to be capable of inducing apoptosis in human cancer cells by a ROS-dependent mechanism (Kuo *et al.*, 1996). Similar to the effect of N-acetyl-L-cysteine, curcumin may also stimulate the death of cancer cells by enhancing p53 expression. Treatment of cells with p53 antisense oligonucleotide effectively prevented curcumin-induced intracellular p53 protein increase and apoptosis, but sense p53 oligonucleotide could not (Jee *et al.*, 1998). Garlic is thought to have a role in the prevention and treatment of cancer. Ajoene, a major compound of garlic, has been shown to induce ROS formation and cause apoptosis in human leukemic cells, but not in peripheral mononuclear blood cells of healthy donors (Dirsch *et al.*, 1998). Gallic acid, a naturally occurring plant phenol with antioxidative activity, has also been shown to induce cell death in promyelocytic leukemia HL-60RG cells by a ROS-dependent mechanism (Inoue *et al.*, 1994). Carotenoids such as canthaxanthin inhibit growth and trigger apoptosis in cancer cells (Palozza *et al.*, 1998).

Antioxidants such as melatonin and retinoic acid have oncostatic properties. They have been shown to suppress the growth of hormone-responsive breast cancer. Using the estrogen receptor-positive MCF-7 human breast tumor cell line, it has been shown that a sequential treatment regimen of melatonin followed by *all-trans* retinoic acid arrests breast tumor cell proliferation *in vitro* by enhancing apoptosis (Eck *et al.*, 1998). Antioxidant vitamins in combination with direct-acting apoptotic agents (X-rays, chemotherapeutic agents, and hyperthermia) or in combination with indirect-acting apoptotic agents (adenosine $3',5'$-cyclic monophosphate, butyric acid, and interferon) produce a greater extent of apoptotic death in cancer cells in culture. In contrast to vitamin-induced apoptosis in cancer cells, normal cells never undergo apoptotic death after treatment with vitamins, not including retinoids (Cole and Prasad, 1997).

V. Bcl-2

The first discovered regulator of apoptosis in mammalian cells was Bcl-2. Bcl-2 was initially identified as a mitochondrial- and perinuclear-associated protein that prolongs the life span of a variety of cell types by interfering with apoptosis. Now, several members of the evolutionarily conserved Bcl-2 family are known (Table I). These proteins share homology, clustered within four conserved regions, namely Bcl-2 homology (BH1-4) domains, which control the ability of these proteins to dimerize and function as regulators of apoptosis (Fig. 2). To a varying extent, all Bcl-2 proteins regulate apoptosis: some are proapoptotic whereas others are antiapoptotic (Table I). These proteins serve as checkpoints in the apoptosis cascade upstream of caspases (Chao and Korsmeyer, 1998). Hydrogen peroxide-induced apoptosis of PC12 pheochomocytoma cells has been shown to be associated with an increased expression of bax and bak and an inhibition of bad expression. In the case of apoptosis induced by serum deprivation, however, a decreased expression of bad, bax, and bak has been observed. When the same cells were incubated in a nerve growth factor-enriched apoptosis-sparing medium, the expression of both bax and bak increased whereas bad expression remained inhibited. Bcl-xL expression was increased by hydrogen peroxide but was unaffected by serum deprivation or long-term nerve growth factor treatment. These results indicate that the expression of

Table I Bcl-2-Related Proteins

Antiapoptotic[a]	Proapoptotic
Bcl-2 (Hockenbery et al., 1990)	Bax (Oltvai et al., 1993)
Bcl-xL (Boise et al., 1993)	Bak (Chittenden et al., 1995)
Mcl-1 (Zhou et al., 1997)	Mtd/Bok (Hsu et al., 1997; Inohara et al., 1998)
Bcl-w (Gibson et al., 1996)	Bcl-xS (Minn et al., 1996)
A-1 (Lin et al., 1993)	Bik (Boyd et al., 1995)
E1B 19K* (Han et al., 1996)	Harakiri (Inohara et al., 1997)
BHRF1* (Henderson et al., 1993)	Bim (O'Connor et al., 1998)
CED-9* (Hengartner and Horvitz, 1994)	Blk (Hegde et al., 1998)
	Bad (Yang et al., 1995)
	Bid (Wang et al., 1996)

[a] Bcl-2 homologs are marked with an asterisk. Others are Bcl-2 family proteins.

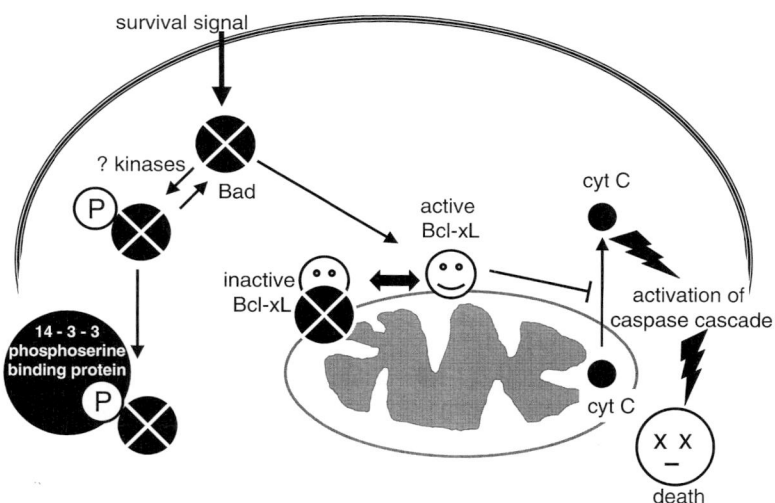

Figure 2 Role of Bad in the survival signaling pathway. In response to an appropriate survival signal, proapoptotic Bad is phosphorylated by as yet unidentified kinase(s). Phosphorylated Bad is sequestered, for example, by the phosphoserine-binding protein 14-3-3. In the absence of a survival signal, Bad is hypophosphorylated or dephosphorylated and is available to bind to the antiapoptotic protein Bcl-xL. Active Bcl-xL prevents the release of cytochrome c from mitochondria to the cytosol; Bcl-xL bound to Bad is unable to do so, resulting in cytochrome c release to the cytosol, which is followed by the activation of caspases that leads to apoptosis (for details, see Adams and Cory, 1998; Brady and Gil-Gomez, 1998; Chao and Korsmeyer, 1998; Kelekar and Thompson, 1998; Minn et al., 1998; Newton and Strasser, 1998; Wang and Reed, 1998).

bax, bak, bad, and Bcl-xL is altered in a stimulus-dependent manner but cannot be used to define whether a cell will undergo or survive apoptosis (Maroto and Perez-Polo, 1997).

Bcl-2 is an integral membrane protein that functions as a suppressor of programmed cell death. It contains a COOH-terminal signal anchor sequence that is selective for import and insertion of Bcl-2 into the mitochondrial outer membrane and, by a different mechanism, can also direct the protein to other membrane sites. Bcl-2$-/-$ mice complete embryonic development, but display growth retardation and early mortality postnatally. Hematopoiesis, including lymphocyte differentiation, is initially normal, but thymus and spleen undergo massive apoptotic involution (Veis *et al.*, 1993). The mechanism by which the Bcl-2 protein acts to prevent cell death remains elusive. One mechanism by which Bcl-2 has been proposed to act is by decreasing the net cellular generation of reactive oxygen species (Hockenbery *et al.*, 1993). Although elevated levels of Bcl-2 may protect against cell death, it has been shown that such surviving cells may suffer from genomic instability and have a high chance of mutagenesis. It has been observed that Bcl-2 overexpression may impair nucleotide excision repair in UV-irradiated cells (Liu *et al.*, 1997). In the next section, several lines of evidence have been discussed suggesting that the antiapoptotic property of Bcl-2 may act via an antioxidant pathway. These evidences, however, do not rule out other mechanisms of action of Bcl-2. Data from cultured cells in a near-anaerobic atmosphere, where the generation of oxidants would be expected not to occur or at least be reduced greatly, show that Bcl-2 can protect cells from apoptosis. These results suggest that the antiapoptotic property of Bcl-2 proteins may be independent of antioxidant action (Jacobson and Raff, 1995). It has been shown that Bcl-2 family proteins may regulate apoptosis by regulating cytochrome c release, which has been identified to be a proapoptotic factor (Fig. 2). Cytochrome c is released from the outer surface of the inner mitochondrial membrane at early steps of apoptosis and, combining with some cytosolic proteins, activates conversion of the latent apoptosis-promoting protease pro-caspase-9 to its active form. Cytochrome c release can be initiated by the proapoptotic protein Bax. This process is blocked by the antiapoptotic proteins Bcl-2 and Bcl-xL (Skulachev, 1998). Findings that Bcl-2 can also function as an ion channel and as an adapter or docking protein are beginning to provide insights into the molecular mechanisms through which these proteins regulate the programmed cell death pathway in physiological and disease conditions (Wang and Reed, 1998).

VI. Bcl-2 FUNCTIONS VIA AN ANTIOXIDANT MECHANISM?

Bcl-2 is localized to intracellular sites of oxygen-free radical generation, including mitochondria, endoplasmic reticula, and nuclear membranes.

Hockenberry et al. (1993) observed that antioxidants that scavenge peroxides, N-acetylcysteine and glutathione peroxidase, countered apoptotic death, whereas manganese superoxide dismutase did not. Bcl-2 protected cells from hydrogen peroxide- and menadione-induced oxidative deaths. Following an apoptotic signal, cells sustained progressive lipid peroxidation. Overexpression of Bcl-2 functioned to suppress lipid peroxidation completely. Based on these results, it was proposed that Bcl-2 regulates an antioxidant pathway at sites of free radical generation (Hockenbery et al., 1993). This proposal, however, was soon challenged radically by work claiming that Bcl-2 is actually a prooxidant that poses oxidant insult to cells and thus results in the upregulation of endogenous antioxidant defenses. Expression of Bcl-2 in superoxide dismutase-deficient *Escherichia coli* resulted in increased transcription of the KatG catalase-peroxidase, a 13-fold increase in KatG activity, and a 100-fold increase in resistance to hydrogen peroxide. In addition, the mutation rate was increased 3-fold, and katG and oxyR, a transcriptional regulator of katG induction, were required for aerobic survival. In support of a prooxidant mechanism in eukaryotic cells, a 73% increase in superoxide dismutase activity in a murine B-cell line overexpressing Bcl-2 was shown (Steinman, 1995). Although interesting, this line of evidence has remained overshadowed by the remarkable amount of information describing the possible antioxidant property of Bcl-2.

Several studies claim that the antiapoptotic property of Bcl-2 proteins is dependent on their antioxidant properties. Although it is not clear whether the ability of some Bcl-2 family proteins to prevent apoptosis is indeed mediated by an antioxidant mechanism, some evidence shows that proteins such as P26-Bcl-2 have antioxidant properties (Desoize, 1994). Reactive oxygen species are known to perturb intracellular calcium homeostasis (Sen and Packer, 1996, Sen et al., 1996). The endoplasmic reticulum is the major intracellular reservoir of Ca^{2+} in nonmuscle cells, sequestering Ca^{2+} for use in intracellular signaling, and is a prime target of oxidative damage. The Bcl-2 protein is localized to intracellular membranes, including the endoplasmic reticulum. Elevated levels of intracellular Bcl-2 function like an antioxidant (Sen and Packer, 1996; Sen et al., 1996) because it protects the endoplasmic reticulum Ca^{2+} pool in cells exposed to reactive oxygen species (Distelhorst et al., 1996). Most autosomal dominant-inherited forms of early onset Alzheimer's disease are caused by mutations in the presenilin-1 gene on chromosome 14. Presenilin-1 is an integral membrane protein with six to nine membrane-spanning domains and is expressed in neurons throughout the brain wherein it is localized mainly in the endoplasmic reticulum. The mechanism or mechanisms whereby presenilin-1 mutations promote neuron degeneration in Alzheimer's disease are unclear. Alzheimer's disease is associated with neuronal apoptosis that is preceded by the deposition of amyloid β peptide, oxidative stress, and disruption of $[Ca^{2+}]_i$ homeostasis. Bcl-2 has been shown to prevent apoptosis after the withdrawal of nerve growth factor from differentiated PC12 cells expressing the mutant presenilin-1 gene. Elevations of $[Ca^{2+}]_i$ in response to thapsigargin, an inhibitor of the endoplasmic reticulum Ca^{2+}-ATPase, were increased in cells expressing mutant presenilin-1, and this adverse effect was abolished in cells expressing Bcl-2. Similar to Bcl-2, other antioxidants also protected cells against the adverse consequences of the presenilin-1 mutation (Guo et al., 1997).

The contention that Bcl-2 may have strong antioxidant properties was firmly supported by a study reporting the redox state of various redox-active intracellular agents such as glutathione and NAD^+/NADH in Bcl-2 overexpressing cells. Activities of antioxidant enzymes as well as levels of glutathione and pyridine nucleotides in control and Bcl-2 transfectants in two different neural cell lines, rat pheochromocytoma PC12 and the hypothalamic GnRH cell line GT1-7, were determined. Both neural cell lines overexpressing Bcl-2 had elevated total glutathione levels when compared with control transfectants. The ratios of oxidized glutathione to total glutathione in PC12 and GT1-7 cells overexpressing Bcl-2 were reduced significantly. In addition, the NAD^+/NADH ratio of Bcl-2 expressing PC12 and GT1-7 cells was two- to threefold less than that of control cell lines. Results of this study showed that overexpression of Bcl-2 shifts the cellular redox potential to a more reduced state, without affecting the major cellular antioxidant enzymes consistently (Ellerby et al., 1996). In further support of the notion that Bcl-2 may prevent apoptosis by an antioxidant-dependent mechanism, it has been shown that Bcl-2 protects against hyperoxia-induced apoptosis of pheochomocytoma (PC) cells. PC12 cells grown under a high oxygen atmosphere showed chromatin condensation and apoptosis. Such hyperoxia-induced death was blocked by the antioxidant vitamin E, suggesting that oxygen toxicity may have played a causative role in triggering the death program. Although a large number of PC12 cells transfected with the control vector died in a 50% O_2 atmosphere within 6 days, Bcl-2-transfected PC12 cells survived and proliferated (Kubo et al., 1996). Ectopic overexpression of Bcl-2 in HL-60 cells also attenuated the accumulation of intracellular hydrogen peroxide induced by the treatment of cells with a topoisomerase inhibitor, β-lapachone. Higher levels of Bcl-2 conferred resistance to apoptosis induced by β-lapachone, which is known to kill cells by a ROS-dependent mechanism (Chau et al., 1998). Overexpression of Bcl-2 also provided complete protection against cytochrome P4502E1-dependent arachidonic acid-induced oxidative damage and apoptosis (Chen et al., 1997). In HL-60, Bcl-2 delayed curcumin-induced apoptosis, a process that has been shown to be antioxidant inhibitable and thus ROS dependent (Kuo et al., 1996). Bcl-2 has been also shown to protect against apoptosis by blocking cytochrome c release and preventing superoxide production. The switch in electron transfer provides a mechanism for redox signaling that is concomitant with the cytochrome c-dependent activation of caspases. The block of cytochrome c release provides a mechanism for the apparent antioxidant function of Bcl-2 (Cai and Jones, 1998).

Dopamine produces a time- and dose-dependent increase in cell death in clonal catecholaminergic cells derived from the central nervous system. Apoptosis seems to be produced by dopamine autoxidation, as intracellular peroxides increase after dopamine treatment and cell death can be inhibited by antioxidants such as catalase and N-acetylcysteine. N-Acetylcysteine produced a dose-dependent decrease in dopamine-induced cell death; this correlated with a decrease in peroxide formation (Masserano *et al.*, 1996). Oxidation in dopamine-treated cells has been shown to activate stress-activated protein kinase and Jun N-terminal kinase pathways that signal for death in these cells (Luo *et al.*, 1998). Bcl-2 overexpression has been observed to protect against dopamine-induced apoptosis via a antioxidant-dependent mechanism (Offen *et al.*, 1997). Bcl-2 has been also observed to protect against other ROS-dependent apoptosis processes. UVA radiation results in "immediate" apoptosis, a phenomenon where DNA laddering is noted after only 4 hr of radiation. Overexpression of Bcl-2 in fibroblasts inhibited the UVA-induced immediate apoptosis. Following UVA radiation, one of the general oxidative stress responses is the expression of the heme oxygenase 1 gene. Bcl-2 overexpression also blocked UVA-induced heme oxygenase 1 expression. These results indicate that the protective effect of Bcl-2 against UVA-induced apoptosis was mediated by an antioxidant-dependent mechanism (Pourzand *et al.*, 1997).

The cancer chemopreventive retinoid N-(4-hydroxyphenyl)-*all-trans* retinamide causes apoptosis of human leukemic cell lines by a mechanism that is antioxidant inhibitable. In response to N-(-4-hydroxyphenyl)-*all-trans* retinamide treatment, the levels of Bcl-2 mRNA were markedly diminished. The protective role of Bcl-2 on cell death by N-(4-hydroxyphenyl)-*all-trans* retinamide was evident in HL-60 as well as 697 pre-B leukemia and Jurkat T-acute lymphocytic leukemia (T-ALL) cells constitutively expressing high levels of Bcl-2 proteins due to gene transfer manipulation (Delia *et al.*, 1995). Although these results show that Bcl-2 and antioxidants may both inhibit apoptosis induced by N-(4-hydroxyphenyl)-*all-trans* retinamide, there was no evidence clarifying whether the antiapoptotic role of Bcl-2 was mediated via an antioxidant-dependent mechanism (Delia *et al.*, 1995). A possible antioxidant property of Bcl-2 has been observed in an experimental system using exogenous ROS insult of cells. Mock-transfected PC12 rat pheochomocytoma cells and PC12 cells transfected with the Bcl-2 gene were subjected to oxidative stress by incubation in the presence of the azo-initiator of lipid peroxyl radicals, 2,2′-azobis(2,4-dimethylvaleronitrile). Analysis of the pattern of changes in parinaric acid-labeled phospholipids after exposure to 2,2′-azobis(2,4-dimethylvaleronitrile) showed significant oxidation of phosphatidylcholine, phosphatidylethanolamine, phosphatidylserine, phosphatidylinositol, and sphingomyelin. This study showed that antioxidant protection by Bcl-2-related product(s) is phospholipid specific and that aminophospholipids are relatively less protected than other phospholipids. The vitamin E analog, 2,2,5,7,8-pentamethyl-6-hydrochromane, acted as an effective antioxidant in preventing the oxidation of parinaric acid-labeled membrane phospholipids in the presence of 2,2′-azobis(2,4-dimethylvaleronitrile), and the extent of protection was approximately the same in both cell lines. Unlike the agents used to generate oxidative stress in other studies, the temperature-driven generation of peroxyl radicals by 2,2′-azobis(2,4-dimethylvaleronitrile) is not dependent on intracellular metabolism. Thus, these results may be taken as proof for antioxidant protection rather than abrogation of radical generation afforded by the Bcl-2 transfection of PC12 cells (Tyurina *et al.*, 1997).

Another line of evidence claims that certain antioxidants inhibit apoptosis by a Bcl protein-dependent pathway. Lipopolysaccharide-induced apoptosis of human umbilical venous endothelial cells has been correlated with a reduction in Bcl-2 levels. Furthermore, lipopolysaccharide treatment upregulated Bax, which heterodimerizes with and thereby inhibits Bcl-2 activity (Fig. 2). Antioxidants such as N-acetylcysteine and the combination of vitamins C and E completely inhibited lipopolysaccharide-induced apoptosis. The reduction of lipopolysaccharide-induced apoptosis by vitamins C and E was paralleled by an increase in Bcl-2 and a decreases in Bax protein levels (Haendeler *et al.*, 1996). Bcl-2, in turn, appears to be dependent on intracellular glutathione for its ability to prevent apoptosis. This was observed while studying the mechanism by which Bcl-2 expression inhibits radiation-induced apoptosis in two mouse lymphoma cell lines: line LY-as expressed low levels of Bcl-2 and was radiation sensitive and line LY-ar expressed a 30-fold higher amount of Bcl-2 protein compared to LY-as and was radiation resistant. LY-ar could be made radiation sensitive simply by depleting intracellular thiol pools, suggesting that the antiapoptotic activity of Bcl-2 was impaired under conditions of reduced thiol depletion in the cell (Mirkovic *et al.*, 1997). Also, LY-as could be made relatively radiation resistant by enhancing the intracellular GSH pool, showing that indeed cellular GSH status is a key determinant of radiation-induced apoptosis (Vlachaki and Meyn, 1998).

Although the notion that the antiapoptotic properties of Bcl-2 are mediated via an antioxidant pathway is well supported, the question whether Bcl-2 proteins function upstream or downstream of oxidant generation appears to be dependent on the experimental model in question. Experiments using oxidant-generating systems as inducers of apoptosis show that proteins belonging to the Bcl-2 family function downstream of oxidant production. Bcl-xL-transfected cells are resistant to apoptosis following incubation in low serum medium or exposure to γ-irradiation, the sphingomyelin ceramide, or compounds that increase intracellular levels of oxidants. Anti-IgM cross-linking, ceramide, and γ-irradiation treatments elevated intracellular peroxide production, which was prevented by treatment with known antioxidants. Cells overexpressing Bcl-xL had a similar rise in intracellular oxidants as control cells, indicating that Bcl-xL modified the response of the cell to oxidants while having no detectable influence on the endogenous production of oxidants following apoptotic stimuli. These data implicate Bcl-xL as a potent death repressor in B lymphocytes and support the hypothesis that Bcl-xL regulates survival decisions within susceptible cells by functioning downstream of oxidant production (Fang et al., 1995).

Evidence discussed in this section lends firm credence to the contention that one of the mechanisms by which Bcl-2 may exert their antiapoptotic property is likely to be antioxidant related. A related question that follows is whether Bcl-2 is a general antioxidant that may protect cells from any type of oxidant insult or if its ability is limited to the regulation of pathways involved in apoptosis. To address this question the possible correlation between the expression of high levels of Bcl-2 and the susceptibility of human Burkitt's lymhoma cell lines to hydrogen peroxide-induced killing has been examined (Lee and Shacter, 1997). Nonapoptotic death induced by hydrogen peroxide was not prevented by Bcl-2, suggesting that Bcl-2 does not function simply by detoxifying ROS but perhaps enhances the resistance of a specific component(s) of the apoptosis cascade that is sensitive to ROS. This contention is supported by the observation that the inability of the Bcl-2 protein to appropriately localize is associated with the loss of its antiapoptotic activity. Deletion of the signal anchor sequence in the protein has been shown to render Bcl-2 cytosolic and impair its ability to prevent apoptotic death. When the predicted transmembrane domain of the Bcl-2 signal anchor was replaced with that of the signal anchor of the yeast outer mitochondrial membrane protein, Mas70p, the Bcl-2/Mas70p hybrid was found to be very similar to Bcl-2 in its distribution within transfected cells, in its ability to heterodimerize with Bax, and in its ability to suppress apoptosis. These results are consistent with a model in which the transmembrane segment contributes to the function of Bcl-2 by targeting and anchoring the protein to strategic membrane locations in the cell (Nguyen et al., 1994).

VII. SUMMARY

In some cases of apoptosis, oxidation-reduction or redox-dependent mechanisms play a central role. This opens up the opportunity to modulate apoptosis by agents that may alter the intracellular redox status, including antioxidants. At present, the exact molecular mechanisms involved in the redox regulation of apoptosis remain elusive. Observational studies show that antioxidants may have antiapoptotic or proapoptotic properties, depending on the type of cells and the apoptosis-inducing signal involved. A trend showing that antioxidants may prevent apoptosis of healthy cells and may promote the apoptosis of diseased cells appears to be emerging. Such an observation, however, is not yet supported by reasonable mechanism-based details. It should be kept in mind that antioxidants may have signal transduction regulatory properties that are independent of their ability to detoxify ROS or to repair oxidative damage. As such, a single line of evidence showing sensitivity to antioxidant treatment should not be interpreted as an indication for the involvement of ROS. A major function of Bcl-2 family proteins is to regulate apoptosis. Although a considerable bulk of evidence suggests that the Bcl-2 protein may have antioxidant properties, it is clear that the antioxidant-independent mechanism of action of Bcl-2 exists as well. A strategic localization of Bcl-2 protein inside the cell seems to be necessary to render its antiapoptotic property functional.

ACKNOWLEDGMENT

Supported by NIH GM27345, Finnish Ministry of Education, and Juho Vainio Foundation, Helsinki.

REFERENCES

Adams, J. M., and Cory, S. (1998). The Bcl-2 protein family: Arbiters of cell survival. *Science* **281,** 1322–1326.
Ashkenazi, A., and Dixit, V. M. (1998). Death receptors: Signaling and modulation. *Science* **281,** 1305–1308.
Atabay, C., Cagnoli, C. M., Kharlamov, E., Ikonomovic, M. D., and Manev, H. (1996). Removal of serum from primary cultures of cerebellar granule neurons induces oxidative stress and DNA fragmentation: Protection with antioxidants and glutamate receptor antagonists. *J. Neurosci. Res.* **43,** 465–475.
Baeuerle, P. A., Rupec, R. A., and Pahl, H. L. (1996). Reactive oxygen intermediates as second messengers of a general pathogen response. *Pathol. Biol.* **44,** 29–35.
Boise, L. H., Gonzalez-Garcia, M., Postema, C. E., Ding, L., Lindsten, T., Turka, L. A., Mao, X., Nunez, G., and Thompson, C. B. (1993). bcl-x, a bcl-2-related gene that functions as a dominant regulator of apoptotic cell death. *Cell (Cambridge, Mass.)* **74,** 597–608.
Boyd, J. M., Gallo, G. J., Elangovan, B., Houghton, A. B., Malstrom, S., Avery, B. J., Ebb, R. G., Subramanian, T., Chittenden, T., Lutz, R. J., *et al.* (1995). Bik, a novel death-inducing protein shares a distinct sequence motif with Bcl-2 family proteins and interacts with viral and cellular survival-promoting proteins. *Oncogene* **11,** 1921–1928.

Brady, H. J., and Gil-Gomez, G. (1998). Bax. The pro-apoptotic Bcl-2 family member, Bax. *Int. J. Biochem. Cell Biol.* **30**, 647–650.

Briehl, M. M., Baker, A. F., Siemankowski, L. M., and Morreale, J. (1997). Modulation of antioxidant defenses during apoptosis. *Oncol. Res.* **9**, 281–285.

Bustamante, J., Slater, A. F., and Orrenius, S. (1995). Antioxidant inhibition of thymocyte apoptosis by dihydrolipoic acid. *Free Radical Biol. Med.* **19**, 339–347.

Buttke, T. M., and Sandstrom, P. A. (1995). Redox regulation of programmed cell death in lymphocytes. *Free Radical Res.* **22**, 389–397.

Cai, J., and Jones, D. P. (1998). Superoxide in apoptosis. Mitochondrial generation triggered by cytochrome c loss. *J. Biol. Chem.* **273**, 11401–11404.

Chao, D. T., and Korsmeyer, S. J. (1998). BCL-2 family: Regulators of cell death. *Annu. Rev. Immunol.* **16**, 395–419.

Chau, Y. P., Shiah, S. G., Don, M. J., and Kuo, M. L. (1998). Involvement of hydrogen peroxide in topoisomerase inhibitor beta-lapachone-induced apoptosis and differentiation in human leukemia cells. *Free Radical Biol. Med.* **24**, 660–670.

Chen, Q., Galleano, M., and Cederbaum, A. I. (1997). Cytotoxicity and apoptosis produced by arachidonic acid in Hep G2 cells overexpressing human cytochrome P4502E1. *J. Biol. Chem.* **272**, 14532–14541.

Chen, Y. C., Lin-Shiau, S. Y., and Lin, J. K. (1998). Involvement of reactive oxygen species and caspase 3 activation in arsenite-induced apoptosis. *J. Cell. Physiol.* **177**, 324–333.

Chittenden, T., Harrington, E. A., O'Connor, R., Flemington, C., Lutz, R. J., Evan, G. I., and Guild, B. C. (1995). Induction of apoptosis by the Bcl-2 homologue Bak. *Nature (London)* **374**, 733–736.

Clement, M. V., and Stamenkovic, I. (1996). Superoxide anion is a natural inhibitor of FAS-mediated cell death. *EMBO J.* **15**, 216–225.

Cole, W. C., and Prasad, K. N. (1997). Contrasting effects of vitamins as modulators of apoptosis in cancer cells and normal cells: A review. *Nutr. Cancer* **29**, 97–103.

Delia, D., Aiello, A., Formelli, F., Fontanella, E., Costa, A., Miyashita, T., Reed, J. C., and Pierotti, M. A. (1995). Regulation of apoptosis induced by the retinoid N-(4-hydroxyphenyl) retinamide and effect of deregulated bcl-2. *Blood* **85**, 359–367.

DeLuca, C., Kwon, H., Pelletier, N., Wainberg, M. A., and Hiscott, J. (1998). NF-kappaB protects HIV-1-infected myeloid cells from apoptosis. *Virology* **244**, 27–38.

Desoize, B. (1994). Anticancer drug resistance and inhibition of apoptosis. *Anticancer Res.* **14**, 2291–2294.

Dirsch, V. M., Gerbes, A. L., and Vollmar, A. M. (1998). Ajoene, a compound of garlic, induces apoptosis in human promyeloleukemic cells, accompanied by generation of reactive oxygen species and activation of nuclear factor kappaB. *Mol. Pharmacol.* **53**, 402–407.

Distelhorst, C. W., Lam, M., and McCormick, T. S. (1996). Bcl-2 inhibits hydrogen peroxide-induced ER Ca2+ pool depletion. *Oncogene* **12**, 2051–2055.

Eck, K. M., Yuan, L., Duffy, L., Ram, P. T., Ayettey, S., Chen, I., Cohn, C. S., Reed, J. C., and Hill, S. M. (1998). A sequential treatment regimen with melatonin and all-trans retinoic acid induces apoptosis in MCF-7 tumour cells. *Br. J. Cancer* **77**, 2129–2137.

Ellerby, L. M., Ellerby, H. M., Park, S. M., Holleran, A. L., Murphy, A. N., Fiskum, G., Kane, D. J., Testa, M. P., Kayalar, C., and Bredesen, D. E. (1996). Shift of the cellular oxidation-reduction potential in neural cells expressing Bcl-2. *J. Neurochem.* **67**, 1259–1267.

Erkkila, K., Hirvonen, V., Wuokko, E., Parvinen, M., and Dunkel, L. (1998). N-acetyl-L-cysteine inhibits apoptosis in human male germ cells in vitro. *J. Clin. Endocrinol. Metab.* **83**, 2523–2531.

Fang, W., Rivard, J. J., Ganser, J. A., LeBien, T. W., Nath, K. A., Mueller, D. L., and Behrens, T. W. (1995). Bcl-xL rescues WEHI 231 B lymphocytes from oxidant-mediated death following diverse apoptotic stimuli. *J. Immunol.* **155**, 66–75.

Ferrari, G., Yan, C. Y., and Greene, L. A. (1995). N-acetylcysteine (D- and L-stereoisomers) prevents apoptotic death of neuronal cells. *J. Neurosci.* **15,** 2857–2866.

Ghadge, G. D., Lee, J. P., Bindokas, V. P., Jordan, J., Ma, L., Miller, R. J., and Roos, R. P. (1997). Mutant superoxide dismutase-1-linked familial amyotrophic lateral sclerosis: Molecular mechanisms of neuronal death and protection. *J. Neurosci.* **17,** 8756–8766.

Gibson, L., Holmgreen, S. P., Huang, D. C., Bernard, O., Copeland, N. G., Jenkins, N. A., Sutherland, G. R., Baker, E., Adams, J. M., and Cory, S. (1996). bcl-w, a novel member of the bcl-2 family, promotes cell survival. *Oncogene* **13,** 665–675.

Granville, D. J., Carthy, C. M., Hunt, D. W., and McManus, B. M. (1998). Apoptosis: molecular aspects of cell death and disease. *Lab. Invest.* **78,** 893–913.

Green, D. R. (1998). Apoptotic pathways: the roads to ruin. *Cell (Cambridge, Mass.)* **94,** 695–698.

Guo, Q., Sopher, B. L., Furukawa, K., Pham, D. G., Robinson, N., Martin, G. M., and Mattson, M. P. (1997). Alzheimer's presenilin mutation sensitizes neural cells to apoptosis induced by trophic factor withdrawal and amyloid beta-peptide: Involvement of calcium and oxyradicals. *J. Neurosci.* **17,** 4212–4222.

Haendeler, J., Zeiher, A. M., and Dimmeler, S. (1996). Vitamin C and E prevent lipopolysaccharide-induced apoptosis in human endothelial cells by modulation of Bcl-2 and Bax. *Eur. J. Pharmacol.* **317,** 407–411.

Han, J., Sabbatini, P., Perez, D., Rao, L., Modha, D., and White, E. (1996). The E1B 19K protein blocks apoptosis by interacting with and inhibiting the p53-inducible and death-promoting Bax protein. *Genes Dev.* **10,** 461–477.

Hegde, R., Srinivasula, S. M., Ahmad, M., Fernandes-Alnemri, T., and Alnemri, E. S. (1998). Blk, a BH3-containing mouse protein that interacts with Bcl-2 and Bcl- xL, is a potent death agonist. *J. Biol. Chem.* **273,** 7783–7786.

Henderson, S., Huen, D., Rowe, M., Dawson, C., Johnson, G., and Rickinson, A. (1993). Epstein-Barr virus-coded BHRF1 protein, a viral homologue of Bcl-2, protects human B cells from programmed cell death. *Proc. Natl. Acad. Sci. U.S.A.* **90,** 8479–8483.

Hengartner, M. O., and Horvitz, H. R. (1994). *C. elegans* cell survival gene ced-9 encodes a functional homolog of the mammalian proto-oncogene bcl-2. *Cell (Cambridge, Mass.)* **76,** 665–676.

Hiraoka, W., Vazquez, N., Nieves-Neira, W., Chanock, S. J., and Pommier, Y. (1998). Role of oxygen radicals generated by NADPH oxidase in apoptosis induced in human leukemia cells. *J. Clin. Invest.* **102,** 1961–1968.

Hockenbery, D., Nunez, G., Milliman, C., Schreiber, R. D., and Korsmeyer, S. J. (1990). Bcl-2 is an inner mitochondrial membrane protein that blocks programmed cell death. *Nature (London)* **348,** 334–336.

Hockenbery, D. M., Oltvai, Z. N., Yin, X. M., Milliman, C. L., and Korsmeyer, S. J. (1993). Bcl-2 functions in an antioxidant pathway to prevent apoptosis. *Cell (Cambridge, Mass.)* **75,** 241–251.

Holzman, D. (1997). Researchers seek the role of antioxidants in apoptosis. *J. Natl. Cancer Inst.* **89,** 413–414.

Hsu, S. Y., Kaipia, A., McGee, E., Lomeli, M., and Hsueh, A. J. (1997). Bok is a proapoptotic Bcl-2 protein with restricted expression in reproductive tissues and heterodimerizes with selective anti-apoptotic Bcl-2 family members. *Proc. Natl. Acad. Sci. U.S.A.* **94,** 12401–12406.

Inohara, N., Ding, L., Chen, S., and Nunez, G. (1997). Harakiri, a novel regulator of cell death, encodes a protein that activates apoptosis and interacts selectively with survival-promoting proteins Bcl-2 and Bcl-X(L). *EMBO J.* **16,** 1686–1694.

Inohara, N., Ekhterae, D., Garcia, I., Carrio, R., Merino, J., Merry, A., Chen, S., and Nunez, G. (1998). Mtd, a novel Bcl-2 family member activates apoptosis in the absence of heterodimerization with Bcl-2 and Bcl-XL. *J. Biol. Chem.* **273**, 8705–8710.

Inoue, M., Suzuki, R., Koide, T., Sakaguchi, N., Ogihara, Y., and Yabu, Y. (1994). Antioxidant, gallic acid, induces apoptosis in HL-60RG cells. *Biochem. Biophys. Res. Commun.* **204**, 898–904.

Jacobson, M. D. (1996). Reactive oxygen species and programmed cell death. *Trends Biochem. Sci.* **21**, 83–86.

Jacobson, M. D., and Raff, M. C. (1995). Programmed cell death and Bcl-2 protection in very low oxygen. *Nature (London)* **374**, 814–816.

Jee, S. H., Shen, S. C., Tseng, C. R., Chiu, H. C., and Kuo, M. L. (1998). Curcumin induces a p53-dependent apoptosis in human basal cell carcinoma cells. *J. Invest. Dermatol.* **111**, 656–661.

Johnson, T. M., Yu, Z. X., Ferrans, V. J., Lowenstein, R. A., and Finkel. T. (1996). Reactive oxygen species are downstream mediators of p53-dependent apoptosis. *Proc. Natl. Acad. Sci. U.S.A.* **93**, 11848–11852.

Kelekar, A., and Thompson, C. B. (1998). Bcl-2-family proteins: The role of the BH3 domain in apoptosis. *Trends Cell Biol.* **8**, 324–330.

King, K. L., and Cidlowski, J. A. (1998). Cell cycle regulation and apoptosis. *Annu. Rev. Physiol.* **60**, 601–617.

Kletsas, D., Barbieri, D., Stathakos, D., Botti, B., Bergamini, S., Tomasi, A., Monti, D., Malorni, W., and Franceschi, C. (1998). The highly reducing sugar 2-deoxy-D-ribose induces apoptosis in human fibroblasts by reduced glutathione depletion and cytoskeletal disruption. *Biochem. Biophys. Res. Commun.* **243**, 416–425.

Kubo, T., Enokido, Y., Yamada, M., Oka, T., Uchiyama, Y., and Hatanaka, H. (1996). Oxygen-induced apoptosis in PC12 cells with special reference to the role of Bcl-2. *Brain Res.* **733**, 175–183.

Kuo, M. L., Huang, T. S., and Lin, J. K. (1996). Curcumin, an antioxidant and anti-tumor promoter, induces apoptosis in human leukemia cells. *Biochim. Biophys. Acta.* **1317**, 95–100.

Lafon, C., Mathieu, C., Guerrin, M., Pierre, O., Vidal, S., and Valette, A. (1996). Transforming growth factor beta 1-induced apoptosis in human ovarian carcinoma cells: Protection by the antioxidant N-acetylcysteine and bcl- 2. *Cell Growth Differ.* **7**, 1095–1104.

Langer, C., Jurgensmeier, J. M., and Bauer, G. (1996). Reactive oxygen species act at both TGF-beta-dependent and -independent steps during induction of apoptosis of transformed cells by normal cells. *Exp. Cell Res.* **222**, 117–124.

Lee, Y., and Shacter, E. (1997). Bcl-2 does not protect Burkitt's lymphoma cells from oxidant-induced cell death. *Blood* **89**, 4480–4492.

Lin, E. Y., Orlofsky, A., Berger, M. S., and Prystowsky, M. B. (1993). Characterization of A1, a novel hemopoietic-specific early-response gene with sequence similarity to bcl-2. *J. Immunol.* **151**, 1979–1988.

Liu, M., Pelling, J. C., Ju, J., Chu, E., and Brash, D. E. (1998). Antioxidant action via p53-mediated apoptosis. *Cancer Res.* **58**, 1723–1729.

Liu, Y., Naumovski, L., and Hanawalt, P. (1997). Nucleotide excision repair capacity is attenuated in human promyelocytic HL60 cells that overexpress BCL2. *Cancer Res.* **57**, 1650–1653.

Lopes, S., Jurisicova, A., Sun, J. G., and Casper, R. F. (1998). Reactive oxygen species: Potential cause for DNA fragmentation in human spermatozoa. *Hum. Reprod.* **13**, 896–900.

Luo, Y., Umegaki, H., Wang, X., Abe, R., and Roth, G. S. (1998). Dopamine induces apoptosis through an oxidation-involved SAPK/JNK activation pathway. *J. Biol. Chem.* **273**, 3756–3764.

Manev, H., Cagnoli, C. M., Atabay, C., Kharlamov, E., Ikonomovic, M. D., and Grayson, D. R. (1995). Neuronal apoptosis in an in vitro model of photochemically induced oxidative stress. *Exp. Neurol.* **133,** 198–206.

Maroto, R., and Perez-Polo, J. R. (1997). BCL-2-related protein expression in apoptosis: Oxidative stress versus serum deprivation in PC12 cells. *J. Neurochem.* **69,** 514–523.

Masserano, J. M., Gong, L., Kulaga, H., Baker, I., and Wyatt, R. J. (1996). Dopamine induces apoptotic cell death of a catecholaminergic cell line derived from the central nervous system. *Mol. Pharmacol.* **50,** 1309–1315.

Maulik, N., Yoshida, T., and Das, D. K. (1998). Oxidative stress developed during the reperfusion of ischemic myocardium induces apoptosis. *Free Radical Biol. Med.* **24,** 869–875.

McLaughlin, K. A., Osborne, B. A., and Goldsby, R. A. (1996). The role of oxygen in thymocyte apoptosis. *Eur. J. Immunol.* **26,** 1170–1174.

Minn, A. J., Boise, L. H., and Thompson, C. B. (1996). Bcl-x(S) antagonizes the protective effects of Bcl-x(L). *J. Biol. Chem.* **271,** 6306–6312.

Minn, A. J., Swain, R. E., Ma, A., and Thompson, C. B. (1998). Recent progress on the regulation of apoptosis by Bcl-2 family members. *Adv. Immunol.* **70,** 245–279.

Mirkovic, N., Voehringer, D. W., Story, M. D., McConkey, D. J., McDonnell, T. J., and Meyn, R. E. (1997). Resistance to radiation-induced apoptosis in Bcl-2-expressing cells is reversed by depleting cellular thiols. *Oncogene* **15,** 1461–1470.

Newton, K., and Strasser, A. (1998). The Bcl-2 family and cell death regulation. *Curr. Opin. Genet. Dev.* **8,** 68–75.

Nguyen, M., Branton, P. E., Walton, P. A., Oltvai, Z. N., Korsmeyer, S. J., and Shore, G. C. (1994). Role of membrane anchor domain of Bcl-2 in suppression of apoptosis caused by E1B-defective adenovirus. *J. Biol. Chem.* **269,** 16521–16524.

O'Connor, L., Strasser, A., O'Reilly, L. A., Hausmann, G., Adams, J. M., Cory, S., and Huang, D. C. (1998). Bim: A novel member of the Bcl-2 family that promotes apoptosis. *EMBO J.* **17,** 384–395.

Offen, D., Ziv, I., Gorodin, S., Barzilai, A., Malik, Z., and Melamed, E. (1995). Dopamine-induced programmed cell death in mouse thymocytes. *Biochim. Biophys. Acta* **1268,** 171–177.

Offen, D., Ziv, I., Panet, H., Wasserman, L., Stein, R., Melamed, E., and Barzilai, A. (1997). Dopamine-induced apoptosis is inhibited in PC12 cells expressing Bcl-2. *Cell. Mol. Neurobiol.* **17,** 289–304.

Oltvai, Z. N., Milliman, C. L., and Korsmeyer, S. J. (1993). Bcl-2 heterodimerizes in vivo with a conserved homolog, Bax, that accelerates programmed cell death. *Cell (Cambridge, Mass.)* **74,** 609–619.

Oridate, N., Suzuki, S., Higuchi, M., Mitchell, M. F., Hong, W. K., and Lotan, R. (1997). Involvement of reactive oxygen species in N-(4-hydroxyphenyl)retinamide- induced apoptosis in cervical carcinoma cells. *J. Natl. Cancer Inst.* **89,** 1191–1198.

Packham, G., Ashmun, R. A., and Cleveland, J. L. (1996). Cytokines suppress apoptosis independent of increases in reactive oxygen levels. *J. Immunol.* **156,** 2792–2800.

Palozza, P., Maggiano, N., Calviello, G., Lanza, P., Piccioni, E., Ranelletti, F. O., and Bartoli, G. M. (1998). Canthaxanthin induces apoptosis in human cancer cell lines. *Carcinogenesis (London)* **19,** 373–376.

Pourzand, C., Rossier, G., Reelfs, O., Borner, C., and Tyrrell, R. M. (1997). Overxpression of Bcl-2 inhibits UVA-mediated immediate apoptosiinrat 6 fibroblasts: Evidence for the involvement of Bcl-2 as an antioxidant. *Cancer Res.* **57,** 1405–1411.

Salas-Vidal, E., Lomeli, H., Castro-Obregon, S., Cuervo, R., Escalante-Alcalde, D., and Covarrubias, L. (1998). Reactive oxygen species participate in the control of mouse embryonic cell death. *Exp. Cell Res.* **238,** 136–147.

Satoh, T., Sakai, N., Enokido, Y., Uchiyama, Y., and Hatanaka, H. (1996). Survival factor-insensitive generation of reactive oxygen species induced by serum deprivation in neuronal cells. *Brain Res.* **733,** 9–14.

Schulz, J. B., Bremen, D., Reed, J. C., Lommatzsch, J., Takayama, S., Wullner, U., Loschmann, P. A., Klockgether, T., and Weller, M. (1997). Cooperative interception of neuronal apoptosis by BCL-2 and BAG-1 expression: Prevention of caspase activation and reduced production of reactive oxygen species. *J. Neurochem.* **69,** 2075–2086.

Schulze-Osthoff, K., Los, M., and Baeuerle, P. A. (1995). Redox signalling by transcription factors NF-kappa B and AP-1 in lymphocytes. *Biochem. Pharmacol.* **50,** 735–741.

Schulze-Osthoff, K., Bauer, M. K., Vogt, M., and Wesselborg, S., (1997). Oxidative stress and signal transduction. *Int. J. Vitam. Nutr. Res.* **67,** 336–342.

Schulze-Osthoff, K., Ferrari, D., Los, M., Wesselborg, S., and Peter, M. E. (1998). Apoptosis signaling by death receptors. *Eur. J. Biochem.* **254,** 439–459.

Sen, C. K. (1998). Redox signaling and the emerging therapeutic potential of thiol antioxidants. *Biochem. Pharmacol.* **55,** 1747–1758.

Sen, C. K., and Packer, L. (1996). Antioxidant and redox regulation of gene transcription. *FASEB J.* **10,** 709–720.

Sen, C. K., Roy, S., and Packer, L. (1996). Involvement of intracellular Ca2+ in oxidant-induced NF-kappa B activation. *FEBS Lett.* **385,** 58–62.

Simizu, S., Imoto, M., Masuda, N., Takada, M., and Umezawa, K. (1996). Involvement of hydrogen peroxide production in erbstatin-induced apoptosis in human small cell lung carcinoma cells. *Cancer Res.* **56,** 4978–4982.

Singh, I., Pahan, K., Khan, M., and Singh, A. K. (1998). Cytokine-mediated induction of ceramide production is redox-sensitive. Implications to proinflammatory cytokine-mediated apoptosis in demyelinating diseases. *J. Biol. Chem.* **273,** 20354–20362.

Skulachev, V. P. (1998). Cytochrome c in the apoptotic and antioxidant cascades. *FEBS Lett.* **423,** 275–280.

Steinman, H. M. (1995). The Bcl-2 oncoprotein functions as a pro-oxidant. *J. Biol. Chem.* **270,** 3487–3490.

Stoian, I., Oros, A., and Moldoveanu, E. (1996). Apoptosis and free radicals. *Biochem. Mol. Med.* **59,** 93–97.

Straface, E., Santini, M. T., Donelli, G., Giacomoni, P. U., and Malorni, W. (1995). Vitamin E prevents UVB-induced cell blebbing and cell death in A431 epidermoid cells. *Int. J. Radiat. Biol.* **68,** 579–587.

Suzuki, K., Nakamura, M., Hatanaka, Y., Kayanoki, Y., Tatsumi, H., and Taniguchi, N. (1997). Induction of apoptotic cell death in human endothelial cells treated with snake venom: Implication of intracellular reactive oxygen species and protective effects of glutathione and superoxide dismutases. *J. Biochem. (Tokyo)* **122,** 1260–1264.

Tsai, J. C., Jain, M., Hsieh, C. M., Lee, W. S., Yoshizumi, M., Patterson, C., Perrella, M. A., Cooke, C., Wang, H., Haber, E., Schlegel, R., and Lee, M. E. (1996). Induction of apoptosis by pyrrolidinedithiocarbamate and N-acetylcysteine in vascular smooth muscle cells. *J. Biol. Chem.* **271,** 3667–3670.

Tyurina, Y. Y., Tyurin, V. A., Carta, G., Quinn, P. J., Schor, N. F., and Kagan, V. E. (1997). Direct evidence for antioxidant effect of Bcl-2 in PC12 rat pheochromocytoma cells. *Arch. Biochem. Biophys.* **344,** 413–423.

Veis, D. J., Sorenson, C. M., Shutter, J. R., and Korsmeyer, S. J. (1993). Bcl-2-deficient mice demonstrate fulminant lymphoid apoptosis, polycystic kidneys, and hypopigmented hair. *Cell (Cambridge, Mass.)* **75,** 229–240.

Vlachaki, M. T., and Meyn, R. E. (1998). ASTRO research fellowship: The role of BCL-2 and glutathione in an antioxidant pathway to prevent radiation-induced apoptosis. *Int. J. Radiat. Oncol. Biol. Phys.* **42,** 185–190.

Wang, H. G., and Reed, J. C. (1998). Mechanisms of Bcl-2 protein function. *Histol. Histopathol.* **13,** 521–530.

Wang, K., Yin, X. M., Chao, D. T., Milliman, C. L., and Korsmeyer, S. J. (1996). BID: A novel BH3 domain-only death agonist. *Genes Dev.* **10,** 2859–2869.

Watson, R. W., Rotstein, O. D., Nathens, A. B., Dackiw, A. P., and Marshall, J. C. (1996). Thiol-mediated redox regulation of neutrophil apoptosis. *Surgery* **120,** 150–157; discussion: pp. 157–158.

Yang, E., Zha, J., Jockel, J., Boise, L. H., Thompson, C. B., and Korsmeyer, S. J. (1995). Bad, a heterodimeric partner for Bcl-XL and Bcl-2, displaces Bax and promotes cell death. *Cell (Cambridge, Mass.)* **80,** 285–291.

Zamzami, N., Marchetti, P., Castedo, M., Decaudin, D., Macho, A., Hirsch, T., Susin, S. A., Petit, P. X., Mignotte, B., and Kroemer, G. (1995). Sequential reduction of mitochondrial transmembrane potential and generation of reactive oxygen species in early programmed cell death. *J. Exp. Med.* **182,** 367–377.

Zhang, P., Liu, B., Kang, S. W., Seo, M. S., Rhee, S. G., and Obeid, L. M. (1997). Thioredoxin peroxidase is a novel inhibitor of apoptosis with a mechanism distinct from that of Bcl-2. *J. Biol. Chem.* **272,** 30615–30618.

Zhou, P., Qian, L., Kozopas, K. M., and Craig, R. W. (1997). Mcl-1, a Bcl-2 family member, delays the death of hematopoietic cells under a variety of apoptosis-inducing conditions. *Blood* **89,** 630–643.

11 Role of Reactive Oxygen Species in Tumor Necrosis Factor Toxicity

Vera Goossens, Kurt De Vos, Dominique Vercammen, Margino Steemans, Katia Vancompernolle, Walter Fiers, Peter Vandenabeele, and Johan Grooten

Flanders Interuniversity Institute for Biotechnology and
University of Gent
B-9000 Gent, Belgium

I. INTRODUCTION

A. Cell Death by Tumor Necrosis Factor (TNF)

This chapter deals with the cytotoxic activity of the pleiotropic cytokine TNF on transformed cell lines. The initial emphasis on the direct cell-killing activity of TNF on tumor cells and its application in therapeutic strategies against cancer has evolved to the use of this cytokine as a model system for molecular mechanisms of cell death. Many tumor-derived established cell lines are susceptible to cell killing by TNF. The molecular mechanisms underlying TNF cytotoxicity vary between cell types, ranging from typical apoptosis characterized by membrane blebbing, chromatin condensation, and formation of apoptotic cell bodies to necrosis-like cell death (characterized mainly by swelling of the cell followed by lysis). Therefore, the elucidation of the signaling cascades activated by TNF in these different cell types will contribute to our knowledge of molecular execution pathways of apoptosis and necrosis, as well as of mechanisms affecting these cell death pathways. This chapter focuses on how oxidative stress participates in various stages of these cell death processes.

B. The TNF-R55 Death-Inducing Signaling Complex

Tumor necrosis factor exerts its function by binding specific membrane-anchored TNF receptors. There are two types of receptors: a 55-kDa receptor (TNF-R55) and a 75-kDa receptor (TNF-R75), which are found in nearly all cell types (Vandenabeele et al., 1995). Whereas TNF-R55-mediated signaling accounts for most TNF-induced effects, TNF-R75-mediated effects are largely restricted to T cells and lymphocyte-derived lines. However, at low ligand concentration, TNF-R75 can very efficiently assist TNF-R55-mediated effects by ligand passing (Tartaglia et al., 1993b). TNF receptors are members of the TNF receptor family, including, among others, Fas, lymphotoxin-β, and DR4/5; the corresponding ligands TNF, FasL, lymphotoxin-β, and TRAIL form the TNF family (Wallach et al., 1997). Clustering of several receptor molecules by the trimeric ligands or by cross-linking with antibodies results in activation of the receptors. Oligomerization of the intracellular C-terminal domain of TNF-R55 is sufficient for signaling to gene induction or cell killing, even when these domains are not anchored to the cell membrane (Vandevoorde et al., 1997). The intracellular region that is relevant for TNF-R55 signaling to cytotoxicity is the so-called death domain (DD) (Tartaglia et al., 1993a; Song et al., 1994; Boldin et al., 1995). Several other members of the TNF receptor family (Fas, DR4/5) and cytosolic proteins contain a DD. The latter exerts its effects via interactive properties; it was found to self-associate and to be capable of binding other (nonreceptor) DD-containing proteins (Varfolomeev et al., 1996). TNF-induced aggregation of the TNF-R55 DD is followed by recruitment of the DD adapter proteins TRADD (TNF-R55-associated DD; Hsu et al., 1995), FADD (Fas-associating protein with DD), and RIP (receptor-interacting protein; Boldin et al., 1996; Chinnaiyan et al., 1996). Next to its DD, which allows binding to TRADD, FADD contains a so-called death effector domain (DED). This DED also functions as a protein–protein interactive domain and is essential for the recruitment of caspase-8 to the receptor complex, thus forming the TNF-R55 death-inducing signaling complex (DISC). Caspase-8 is a member of the cysteine aspartic acid-specific protease family characterized by a conserved cysteine in the active site and specific substrate cleavage prior to an aspartate residue (Nicholson et al., 1995). Inactive procaspase-8 enters DISC by binding of its N-terminal DED to the DED of FADD, and is subsequently activated and released from DISC by autoproteolytic cleavage (Boldin et al., 1996; Medema et al., 1997; Muzio et al., 1998), thus initiating the death execution signal.

The receptor complex described links ligand binding to three main axes of signaling cascades: (i) to cell death initiated by caspase-8, (ii) to cell death initiated by caspase-independent signaling cascades, and (iii) to gene expression by activation of the nuclear factor NF-κB.

C. Caspase-Independent Signaling to Cell Death

Although caspase-8-initiated signaling to cell death has been confirmed in several cell types dying by apoptosis after triggering death receptors such as TNF-R55 or Fas, other TNF-induced death pathways occur independently of the activation of caspases (Jarvis *et al.*, 1994). Ceramide has been recognized as an important secondary messenger in membrane receptor signaling. The role, origin, and mode of action of this lipid messenger and its link to receptor-associated signal proteins are still unresolved. Distinct enzymes contribute to ceramide production after TNF treatment, ceramide synthase, and two forms of sphingomyelinases (Smases), a membrane-associated neutral (N-) Smase and an acidic (A-) Smase that resides in the endosomal–lysosomal compartment (Wiegmann *et al.*, 1994). Activation of N-Smase is an early signal transduction event initiated by recruitment of a WD-40 repeat containing protein FAN to the membrane-proximal N-Smase domain (NSD) of TNF-R55 (Adam *et al.*, 1996; Adam-Klages *et al.*, 1996). Signaling events downstream of N-Smase include the activation of ceramide-activated protein kinase and cytosolic phospholipase A_2 (PLA$_2$; Huwiler *et al.*, 1996). Activation of A-Smase requires the TNF-R55 DD-associating proteins TRADD and FADD and coincides with the occurrence of irreversible damage. Peptide inhibitors of caspase-8 only partially block A-Smase activation, indicating that it may involve a yet-to-define pathway distinct from caspase-8 and possibly independent of caspases. Events implicated in ceramide-mediated signaling to cell death may include the generation of reactive oxygen species (ROS), generated in the mitochondria at the ubiquinone site of the mitochondrial respiratory chain (García-Ruiz *et al.*, 1997; Quillet-Mary *et al.*, 1997).

D. Signal Transduction to Gene Expression by Activation of NF-κB

Tumor necrosis factor always induces activation of NF-κB. The latter was originally described as a B-cell-specific transcription factor essential for expression of the immunoglobulin light chain of the κ type. Currently, it is clear that NF-κB plays a central role in cellular gene expression induced by pathogens and inflammatory cytokines such as TNF and interleukin-1 (Baeuerle and Baltimore, 1996). TRAF-1 and TRAF-2 (TNF receptor-associated factor) were initially identified as proteins that are associated with the cytoplasmic domain of TNF-R75. TRAF-1 and TRAF-2 self-associate and also form heterodimers. They can bind the N-terminal part of the TRADD and RIP DD; TRAF-2 is recruited to the TNF-R55 complex by these interactions. TRAF-2 and RIP binding of TRADD are essential for optimal NF-κB activation (Hsu *et al.*, 1996a,b). Furthermore, TRAF-2 can associate with NIK (NF-κB-inducing kinase), a Ser/Thr kinase of the MAP-

KKK type (Malinin et al., 1997). NIK is the molecular link between TRAF-2 and the IκB kinases IKKα and IKKβ (Mercurio et al., 1997; Régnier et al., 1997; Zandi et al., 1997), which phosphorylate the inhibitory subunit IκB, leading to ubiquitinylation, degradation, and release of NF-κB, followed by translocation to the nucleus. Although the role of these proteins in NF-κB activation has been established in several models, other proteins and signal transduction pathways can be activated, resulting in NF-κB transactivation, including p38 MAP kinase (Beyaert et al., 1996).

Relevant to the scope of this chapter is the involvement of NF-κB-regulated genes in TNF-mediated cell death. Several genes that are under the control of NF-κB are involved in the protection of cells from cell death, such as mitochondrial manganese-dependent superoxide dismutase (MnSOD), an inhibitor of apoptosis proteins (IAPs) and A20. MnSOD is a general stress-induced protein that dismutates superoxide anion ($O_2^{\cdot-}$) formed by the leak of electrons of the respiratory chain carriers to hydrogen peroxide (H_2O_2). IAPs constitute an evolutionarily conserved family of homologous proteins that suppress apoptosis induced by multiple stimuli. Some IAPs directly bind and inhibit caspases, thereby preventing apoptosis (Liston et al., 1996). A20 is a zinc finger protein that has been identified as an antiapoptotic gene; it also strongly inhibits activation of NF-κB (Opipari et al., 1992; Jäättelä et al., 1996). A two-hybrid experiment showed that A20 can react with TRAF-1 and TRAF-2 (Song et al., 1996). However, its inhibitory mechanism remains unresolved.

II. REACTIVE OXYGEN SPECIES AS MEDIATORS OF CELL DEATH BY TNF

Reactive oxygen species, such as $O_2^{\cdot-}$, H_2O_2, organic peroxides, and hydroxyl radicals, are generated by all aerobic cells as by-products of metabolic reactions or in response to various stimuli. Electron transport chains in the plasma membrane, endoplasmatic reticulum, nuclear membranes, and especially in the inner mitochondrial membrane are the major sites of superoxide anion as a result of a single electron transfer to molecular oxygen at the level of electron carriers. Also, some fatty acid metabolites derived from arachidonic acid in the lipoxygenase pathway are free radicals.

When generated in a controlled manner, ROS play a role in various molecular processes. For example, a role for H_2O_2 as a secondary messenger in the activation mechanism of NF-κB has been demonstrated (Schreck et al., 1991; Schmidt et al., 1995). Although it was initially tempting to classify H_2O_2 as a general secondary messenger for the activation of this transcription factor, it subsequently became clear that this pathway may be limited to certain cell lines. In the murine fibrosarcoma cell line L929, activation of

NF-κB by TNF could not be inhibited by free radical scavengers or catalase; also, prooxidant molecules such as H_2O_2 could not induce activation of NF-κB (unpublished results).

Certain physiological stimuli result in an excessive production of ROS that play a role in cell killing and cell suicide. For example, factors released in an inflammatory environment (such as bacterial products, components of the complement, and cytokines) trigger an oxidative burst in leukocytes, which gives rise to the release of highly reactive radicals killing the pathogen, but also leading to cell suicide of the producing cell. Other physiological stimuli (such as TNF and FasL) often result in an increase in intracellular ROS levels that participate in the process of apoptosis or necrosis-like suicide. In these situations the fate of the cell depends on the balance between ROS levels and the level of endogenous intracellular antioxidant mechanisms. These mechanisms of defense against oxidative stress include enzymes dealing with $O_2^{\cdot-}$ (superoxide dismutases), H_2O_2 (catalase and glutathione peroxidase), and lipid hydroperoxides (phospholipid hydroperoxide glutathione peroxidase). Finally, unphysiological stimuli causing excessive oxidative stress result in primary necrosis as a consequence of lipid peroxidation and alterations of proteins and nucleic acids (Guénal et al., 1997).

The study of the contribution of ROS to TNF-induced cell suicide includes questions related to the nature and the source of ROS generated, the causal relationship between ROS and cell death, and the target(s) of ROS. As discussed later, not all questions have been answered yet. Furthermore, depending on the cell type, the answers vary considerably.

A. TNF-Induced Oxidative Burst

During inflammation, TNF released in the inflammatory environment will transiently activate neutrophils and macrophages, causing them to release $O_2^{\cdot-}$ as a consequence of activation of the plasma membrane NADPH oxidase (Shalaby et al., 1985; Klebanoff et al., 1986). This oxidative burst in leukocytes features a rapid but transient release of ROS and is part of the defense mechanism against microbial pathogens. Tumor cells may also be killed by a TNF-induced oxidative burst as the susceptibility of tumors to TNF-mediated regression is inversely correlated with their free radical buffering capacity (Zimmerman et al., 1989).

In primary human fibroblasts, TNF-mediated activation of a plasma membrane NADPH oxidase was demonstrated (Meier et al., 1989). In contrast to the rapid transient response of leukocytes, the oxidative burst caused by TNF in fibroblasts is characterized by a sustained release of relatively low amounts of reactive radicals. These ROS are not cytocidal to the producing cells; they respond to the increased oxidative stress by enhanced proliferation. In L929 cells, a TNF-induced oxidative burst has also been reported. Measurement of the release of superoxide anion from TNF-treated L929 cells by lucigenin revealed an immediate and sustained increase in ROS (Hennet et al., 1993). The membrane-impermeable nature of the lucigenin probe used (Mohazzab and Wolin, 1994) and of the superoxide anion generated makes it highly unlikely that the detected $O_2^{\cdot-}$ originated from the mitochondria. A more likely interpretation is that a fibroblast-type plasma membrane-associated NADPH oxidase is responsible for the observed ROS. Using the fluorogenic marker hydroethidin, a similar rapid and sustained ROS production in L929 cells after TNF treatment was observed; furthermore, overexpression of hsp27 in L929 cells protected the latter from TNF cytotoxicity by an hsp27-induced increase in GSH levels (Mehlen et al., 1996). In these transfectants, TNF still induced a rapid generation of ROS, but started from a lower basal level and apparently required more time to accumulate to a cytocidal threshold value. Thus an inverse correlation between free radical buffering capacity and susceptibility to TNF-mediated cell death was observed.

B. Intracellular ROS as Mediators of TNF-Mediated Cell Death

1. Intracellular ROS Are Involved in the Signaling Cascade to TNF Cytotoxicity

A first evidence for the involvement of ROS in the signaling cascade to TNF cytotoxicity was obtained in L929 cells, where culture under hypoxic conditions abrogated TNF cytotoxicity. It also became clear that certain free radical scavengers inhibited TNF cytotoxicity, whereas others had no effect (Matthews et al., 1987). Desferal and butylated hydroxytoluene (BHT) inhibited cell killing, whereas vitamin E or mercaptoethanol had no significant effect. Table I contains a list of antioxidant molecules that were tested on their effect on TNF cytotoxicity in L929 cells. The results demonstrate that iron chelators (desferral and desferrioxamine), hydroxyl radical scavengers (mannitol and DMSO), and lipophylic antioxidants [butylated hydroxyanisole (BHA) and BHT] inhibit cell killing by TNF (Higuchi et al., 1992; Schulze-Osthoff et al., 1992), which is not the case with the hydrophylic antioxidant enzymes catalase or superoxide dismutases (Goossens et al., 1999b). This indicates that free iron plays a major role, probably by promoting the formation of hydroxyl radicals that start a lipid peroxidation chain reaction. Contrary to L929 cells, overexpression of MnSOD was protective in ME180 cells (Wong et al., 1989). Apparently, in different cell types TNF induces ROS that exhibit differences in accessibility to radical scavengers. In L929 cells, TNF-induced free radicals are shielded from hydrophylic

Table I Effect of Low Molecular Weight Free Radical Scavengers, Iron Chelators, and Antioxidant Enzymes on TNF Cytotoxicity

	Antioxidant	Target molecule	Subcellular location	Effect on TNF cytotoxicity	Reference
Low molecular weight free radical scavengers	BHA	Lipid hydroperoxides	Membranes	Protection	Goossens et al. (1995)
	BHT	Lipid hydroperoxides	Membranes	Protection	Matthews et al. (1987)
					Goossens et al. (1995)
	Mannitol	·OH	Ubiquitous	Protection	Higuchi et al. (1992)
	Dimethyl sulfoxide	·OH	Ubiquitous	Protection	Unpublished results
	Vitamin C	ROS	Aqueous solution	No effect	Unpublished results
	Vitamin E	Lipid hydroperoxides	Membranes	No effect	Unpublished results
Iron chelators	Desferral	Free iron		Protection	Matthews et al. (1987)
	Desferrioxamine	Free iron		Protection	Schulze-Osthoff et al. (1992)
Antioxidant enzymes or enzyme mimics	MnSOD	$O_2^{·-}$	Mitochondria	No effect	Goossens et al. (1999b)
	Cu/ZnSOD	$O_2^{·-}$	Cytoplasm	No effect	Goossens et al. (1999b)
	CuDIPS	$O_2^{·-}$	Mitochondria + cytoplasm	No effect	Unpublished results
	Catalase	H_2O_2	Peroxisomes (cytoplasm)	No effect	Goossens et al. (1999b)
	Mit. catalase	H_2O_2	Mitochondria	No effect	Goossens et al. (1999b)

antioxidant enzymes (such as MnSOD and Mit. catalase in the mitochondrial matrix and Cu/ZnSOD and catalase in the cytoplasm), which points to their generation in a hydrophobic environment.

2. Cell Death-Inducing ROS of Mitochondrial Origin

During apoptosis, mitochondria often release apoptogenic factors from the mitochondrial intermembrane space, such as cytochrome c. In the presence of dATP, cytochrome c induces the Apaf-1-mediated processing of Apaf-3 or procaspase-9 to active caspase-9 (Zou *et al.*, 1997), resulting in activation of the executioner caspase-3, -6, and -7 (Li *et al.*, 1997). The release of cytochrome c is often (but not necessarily) accompanied by a disruption of the mitochondrial inner transmembrane potential ($\Delta\Psi_m$). This $\Delta\Psi_m$ collapse appears to be a general feature of early apoptosis in the sense that it is found in different cell types in response to different apoptosis stimuli, including TNF (Cossarizza *et al.*, 1995). The mechanism by which cells undergoing apoptosis lose their $\Delta\Psi_m$ is the so-called permeability transition (PT), caused by opening of the permeability transition pores or megachannels; this allows diffusion of solutes of <1500 kDa from the mitochondria and hence dissipation of $\Delta\Psi_m$. As far as we know, the PT pore-gating potential (i.e., the probability of PT pore opening at a given voltage) is regulated by cellular oxidative stress (Bernardi, 1992). This explains why ROS are among the most effective inducers of PT in isolated mitochondria. Furthermore, PT leads to generation of ROS as a result of uncoupling of the respiratory chain and disruption of thiol homeostasis. Thus ROS play a dual role in this apoptotic cascade; as an inducer of PT and/or as an obligate consequence of PT. Despite the evidence that exogenously added ROS induce PT, no detailed study showing a physiological death stimulus (TNF or Fas) inducing PT as a consequence of ROS generation has been reported yet. It has been demonstrated in lymphocytes undergoing apoptosis that $\Delta\Psi_m$ disruption is a cause rather than a consequence of ROS production (Castedo *et al.*, 1995; Zamzami *et al.*, 1995). So the exact role of ROS in cells dying due to a caspase-dependent apoptotic process has to be investigated in further detail. Especially in those cases in which ROS and PT participate in the cell death process, the causal relationship between ROS and PT should be verified.

L929 cells devoid of the oxidative burst-like fast ROS response exhibit a necrosis-like cell death phenotype after TNF treatment. In these cells, cell death is characterized by simultaneous membrane permeabilization and loss of clonogenic potential, as well as by the lack of DNA fragmentation (Grooten *et al.*, 1993). Using the fluorogenic ROS-specific marker dihydrorhodamine 123 (DHR123), an enhanced intracellular ROS generation was observed (Goossens *et al.*, 1995). ROS were detected late, between 2 hr to over 4 hr treatment of the cells with TNF, which was followed by cell death within approximately 1 hr after ROS production (Goossens et al., 1995). The causal relationship between these ROS and the cytotoxic mechanism of TNF was established on the basis of an immediate effect of lipophylic antioxidants (BHA and BHT), arresting the ongoing ROS increment and cell death, even when added 3 hr after TNF treatment. We have shown directly that mitochondria are at the basis of ROS production by comparing ROS-mediated DHR123 oxidation in untreated and TNF-treated cells permeabilized by digitonin. These permeabilized cells had lost their soluble cytoplasmic content as well as single membrane organelles, but had retained intact mitochondria. As shown in Fig. 1A, we observed a similar ROS-mediated oxidation of DHR123 in permeabilized TNF-treated cells and in intact counterparts; this confirms that mitochondria are the source of these ROS in TNF-treated cells (Goossens et al., 1999b). To estimate the contribution of $O_2^{\cdot-}$ and H_2O_2, the two major ROS generated in the mitochondria, to the cytotoxic phase of TNF signaling, we overexpressed MnSOD and catalase in the mitochondria (Goossens et al., 1999b). As already mentioned, both enzymes had no effect on TNF cytotoxicity, which correlated with a lack of effect on TNF-induced ROS levels. From this negative result we conclude that mitochondrial ROS are generated by a membrane-associated mechanism and exert their cytocidal activity on the site of production. Because inhibitors of the respiratory chain complex I and II block TNF-induced ROS generation, the electron transport chain in the inner mitochondrial membrane could be identified as the probable source of ROS.

3. Cell Death-Inducing ROS from the Lipoxygenase Pathway

In addition to the compelling evidence of involvement of mitochondrial ROS in TNF cytotoxicity in L929 cells, free radicals from the lipoxygenase pathway have also been shown to be mediators of cell death in the same cell line. Three modulators of this radical-generating signaling pathway, namely the lipoxygenase inhibitor nordihydroguaiaretic acid (NDGA), an arachidonate analog ETYA, and the lipoxygenase product 12(S)HETE, were significantly protective against TNF cytotoxicity (O'Donnell et al., 1995). Although these results are apparently contradictory to mitochondrial ROS as mediators of cell death, data suggest a mechanism that reconciles both lines of evidence. It was shown that damage to complex I, caused by hydroxyl radicals, and activation of PLA_2 are both involved in the cytolytic action of TNF in L.P3 cells (Higuchi et al., 1992). Hydroxyl radical formation was an early event and PLA_2 activation a late but necessary event in the cytotoxic process. Possibly the activated PLA_2 is the mitochondrial PLA_2 localized in the inner mitochondrial membrane and in the contact sites between inner and outer membrane, as demonstrated in WEHI-164 cells (Levrat and Louisot,

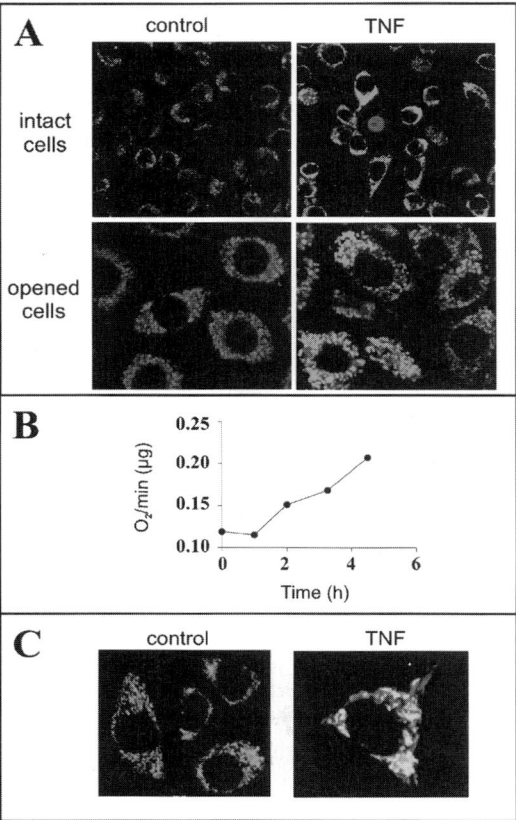

Figure 1 Analysis of TNF-induced mitochondrial changes. (A) TNF-induced ROS in intact and opened cells. ROS were detected by the fluorogenic marker DHR123. After oxidation by ROS, the uncharged membrane-permeable DHR123 is converted to the lipophylic cationic fluorescent marker R123, which accumulates in mitochondria on the basis of $\Delta\Psi_m$. DHR123-derived fluorescence is proportional to the amount of ROS in the cell. CLSM images of DHR123-derived fluorescence intensity (green) in untreated cells (control) and in TNF-treated intact or opened cells (3 hr) are shown. In intact cells, cell death was detected after uptake of the polar marker propidium iodide in the nucleus (red fluorescence). Opened cells are cells that were treated with digitonin, a mild detergent disrupting the plasma membrane, but leaving mitochondrial membranes and functionality intact. (B) TNF-induced increase in oxygen consumption. Oxygen consumption was measured with a Clark-type electrode. Viable cells were counted by trypan blue or propidium iodide exclusion; oxygen consumption determinations are expressed as $\mu g\ O_2/min/10^6$ viable cells. (C) TNF-induced increase in $\Delta\Psi_m$. TNF-induced changes in $\Delta\Psi_m$ were measured by $\Delta\Psi_m$-driven uptake of the charged, but still membrane-permeable JC-1. As a monomer, JC-1 emits green fluorescence, but at high concentrations aggregates are formed that display yellow-orange fluorescence emissions. An increase in $\Delta\Psi_m$ can be visualized by a change in the fluorescence wavelength. CLSM images of JC-1 fluorescence in untreated cells (control) and in TNF-treated cells in the lower right panels clearly show a TNF-induced increase in $\Delta\Psi_m$, visualized by J aggregates and yellow-orange fluorescence. (**See color reproduction in color plate section.**)

1996). Thus a link may exist between TNF-induced changes in the respiratory chain and the activation of mitochondrial PLA_2 and lipoxygenase.

C. Regulation of TNF-Induced Intracellular ROS Generation

1. Alteration of TNF-Induced Mitochondrial ROS Generation by Parameters Defining Oxidative Phosphorylation

We pinpointed the respiratory chain as the source for TNF-induced ROS generation in L929 cells. Therefore, we also wanted to analyze the impact of TNF on two other mitochondrial parameters, namely oxygen consumption and $\Delta\Psi_m$. Oxygen consumption in viable cells was measured with a Clark-type electrode. As shown in Fig. 1B, TNF treatment results in a time-dependent increase in oxygen consumption. $\Delta\Psi_m$ was estimated by a fluorescent marker JC-1 that accumulates in mitochondria on the basis of $\Delta\Psi_m$. As a monomer, JC-1 emits green fluorescence, but at high concentrations, J aggregates are formed that have yellow-orange fluorescence emission characteristics (Smiley et al., 1991). Differences in $\Delta\Psi_m$ can be visualized by changes in fluorescence wavelength. Representative images of JC-1 staining in TNF-treated (3 hr) and untreated cells are shown in Fig. 1C. The monomer form of JC-1 predominates in untreated cells (green fluorescence); J aggregates are formed in TNF-treated cells, which means that TNF treatment results in an increased $\Delta\Psi_m$. Comparison of the time kinetics of TNF-induced changes of mitochondrial parameters indicates that an increased ROS production is accompanied by an increased $\Delta\Psi_m$ and an enhanced oxygen consumption. These results support the notion that TNF-induced signal transduction pathways affect the overall rate of oxidative phosphorylation; this results in an increased electron flow sustaining a higher $\Delta\Psi_m$, which leads to increased oxygen consumption accompanied by higher ROS levels.

To further verify if TNF-induced mitochondrial changes are dependent on coupled mitochondria, we evaluated the effect of the uncoupling agent dinitrophenol (DNP) on TNF cytotoxicity. Based on the loss of accumulation of rhodamine 123, DNP treatment effectively uncoupled the mitochondria in L929 cells. However, this uncoupling did not change TNF-induced ROS formation or TNF-induced cell death (unpublished data). Thus the TNF-induced generation of ROS depends on an intact electron transport chain, but it is independent of its coupling to ATP synthesis.

Furthermore, we established that the bioenergetic status of L929 cells is an important modulator of the sensitivity of mitochondria to TNF signaling. Tumor cells, and also the L929 cell line, use the NADH-linked respiratory substrate glutamine instead of glucose. L929 cells deprived of glutamine (L929/Gln$^-$) are desensitized to TNF cytotoxicity (Goossens et al., 1996).

Desensitization could be mimicked in the presence of glutamine by inhibition of the mitochondrial enzymes glutaminase and transaminase, which are involved in glutamine oxidation, suggesting a direct link between glutamine oxidation and TNF-induced mitochondrial distress. Comparison between the induction of ROS by TNF in L929/Gln$^-$ and parental L929 cells suggested that the effect of glutamine involves a mechanism that renders the mitochondria more susceptible to TNF-induced mediators, resulting in accelerated ROS production and cytotoxicity (Goossens *et al.*, 1996). The oxygen consumption in both cell types was comparable, indicating a similar overall electron flow through the electron transport chain. We conclude that the difference in sensitivity was not due to the overall electron flow. Probably substrate-imposed differences in the electron flow through complex I vs complex II may affect the susceptibility to TNF-induced mitochondrial changes (Goossens *et al.*, 1999a).

2. Mitochondrial Protein Thiol Groups Regulate TNF-Induced Mitochondrial Dysfunction and Cytotoxicity

The main question remaining is how TNF-induced alterations in the respiratory activity of the mitochondria lead to the generation of these highly cytocidal ROS. The only defense mechanisms of mitochondria against ROS are MnSOD and the GSH redox cycle. The neutral effect of inhibitors of GSH synthesis and GSSG reduction on TNF-induced ROS formation, as opposed to the strongly enhancing effect of depletion of both cytosolic and mitochondrial GSH, suggested that GSH may be the major ROS scavenger (Goossens *et al.*, 1995). Thus perturbations of the mitochondrial thiol status as a consequence of GSH depletion may constitute the direct cause of cytocidal ROS activity.

To evaluate the role of mitochondrial GSH in TNF-mediated signaling to cell death, we directly quantified GSH levels in cellular and mitochondrial protein-free acid extracts of TNF-treated and untreated cells. We detected decreased GSH levels in mitochondria of TNF-treated cells, confirming the scavenging role for mitochondrial GSH (Goossens *et al.*, 1999b). However, it is unlikely that this relatively small decrease in GSH levels causes a shift in the thiol balance of sufficient amplitude to result in irreversible damage. Because the mitochondrial inner membrane contains several proteins with redox-regulated thiol groups, which are very susceptible targets of ROS modifications, we further analyzed if such proteins could be a target of TNF-induced oxidative stress. Using thiol-reactive reagents, we showed that formation of disulfides favors TNF-induced ROS generation and cell death, whereas impediment of disulfide bridge formation by treatment with monovalent sulfhydryl-blocking agents delays ROS generation and protects against cell death. Two possible mechanisms could explain these results.

First, disulfide formation in the critical protein(s) may be part of the ROS-producing system similarly activated by TNF or diamide. However, the absence of similar ROS production and cytotoxicity by diamide alone contradicts this possibility. Second, the protein inactivated by disulfide formation is a negative regulator of ROS formation and its inactivation no longer delays ROS generation. However, diamide enhances ROS production only in cooperation with TNF-imposed mitochondrial alterations. Probably, reduced NADH levels in TNF-treated cells determine this cooperative effect, as GSH and NADH levels are the major regulating systems of redox modification in protein thiol groups present in high concentrations in mitochondria. This is supported indirectly by the protective effect of inhibitors of the NADH-linked complex I and the attenuating effect of glucose oxidation vs glutamine oxidation because these treatments most likely result in changed available levels of NADH in mitochondria. Therefore, we propose that the critical protein is a negative regulator of ROS production that can be regulated by disulfide formation and by NADH levels. This mechanism would be similar to the regulation of the PT pore by oxidative agents through two regulatory sites (Chernyak and Bernardi, 1996). One site is linked to a critical dithiol and the other to pyridine nucleotides, and both sites independently modulate PT pore opening.

3. Involvement of Caspases in TNF-Induced Mitochondrial ROS Production

Two types of Fas-triggered caspase-dependent cell death cascades have been described (Scaffidi *et al.*, 1998). In type I cells, massive recruitment and activation of procaspase-8 at the DISC are closely followed by activation of the death executioner caspase-3. In these cells, mitochondrial PT and release of cytochrome c are observed; these are, however, not causally related to cell death, as blocking of mitochondria-related phenomena by Bcl-2 does not affect caspase-3 activation. In type II cells, neither caspase-8 nor caspase-3 is activated downstream of mitochondria; their activation upstream of mitochondria is blocked by Bcl-2 or Bcl-x_L. Although caspase-1, -2, -3, -4, and -6 are able to induce PT in intact mitochondria (Marzo *et al.*, 1998), the caspase activity remains undetectable in type II apoptosis until PT is induced. We have shown that all caspases at high concentrations are able to induce PT in isolated mitochondria (Steemans *et al.*, 1998). However, when tested in permeabilized cells in the presence of cytosolic factors, caspase-8 induces PT and cytochrome c release at very low concentrations. We have evidence that caspase-8 activates a cytosolic factor, called caspase-activated factor (CAF), that causes mitochondrial dysfunction. Preliminary purification demonstrated that CAF is a ~20-kDa protein that efficiently induces PT and cytochrome c release by a nonproteolytical mechanism

(Steemans et al., 1998). This factor might be partly involved in the regulation of Bcl-2 family members, moving the balance toward proapoptotic conditions. As mentioned in Section II,B,2, the involvement of ROS in this type of cell death has not been demonstrated directly. Release of cytochrome c, uncoupling of $\Delta\Psi_m$, and subsequent thiol imbalance will probably result in ROS that sustain the execution phase of cell death.

In some cell types, caspase-independent, but mitochondria-dependent pathways to cell death predominate (type III). As shown earlier, the generation of mitochondrial ROS in L929 cells determines TNF-induced cell death. We have reported the existence of a caspase-like activity in these cells that protects against rather than participates in TNF-induced cell death (Vercammen et al., 1998a). Overexpression of cytokine response modifier A (CrmA), a cowpox-encoded serpin-like caspase inhibitor, led to a 1000-fold enhancement of TNF-induced cytotoxicity. The same sensitizing effect was also observed when the cells were challenged with TNF in the presence of the broad-range caspase inhibitor zVAD-fmk. Further analysis revealed that this sensitization resulted directly from an accelerated ROS production after TNF treatment of the cells. Moreover, the nature of the ROS generated in the presence of caspase inhibitors was identical to that of the ROS produced after treatment with TNF alone; such ROS can indeed be scavenged by BHA, but cannot be neutralized by SOD or catalase, be it cytoplasmic or mitochondrial (unpublished results). Hence, it appears that caspases or related proteases in L929 cells are implicated in protection mechanisms against TNF-induced mitochondrial ROS production. TNF treatment of L929 cells does not result in any detectable caspase activity, suggesting that the caspase-mediated protection mechanism against TNF-induced ROS has a low, but constitutive expression profile. A surprising observation was that triggering of transfected Fas in L929 cells, a caspase-dependent apoptotic pathway of cell death, switched to necrosis-like cell death in the presence of caspase inhibitors (Vercammen et al., 1998b). In the presence of caspase inhibitors, this Fas-mediated cell death is also accompanied by elevated ROS levels and is inhibited by BHA. Because ROS scavenging had no protective effect on Fas-mediated apoptosis, we conclude that ROS formation does not play a causative role in Fas-mediated apoptosis. However, when the latter is blocked by caspase inhibitors, the ROS accumulated lead to necrosis-like cell death.

The exact nature of the protection mechanism of caspases or caspase-like proteases against ROS production has not been elucidated yet. Possibly, these proteases are involved in attenuating signals from the TNF receptor or Fas antigen to the mitochondria. Alternatively, these proteases are necessary to eliminate ROS-damaged mitochondria, which would accumulate in the presence of caspase inhibitors and may lead to an enhanced ROS production. Identification and further characterization of the caspase involved are needed to further analyze the exact molecular mechanism underlying this protection against elevated ROS.

4. TNF-Induced Signaling to Mitochondrial ROS Production

Our current knowledge of TNF-induced signaling to ROS generation and cytotoxicity includes membrane-proximal events, especially the role of receptor-associated proteins, in the initiation of cell death. Furthermore, at the other end of the pathway investigators are focusing on the mitochondrial events, either induction and regulation of cytochrome c release and PT or induction of cytocidal ROS. But how does the receptor-associated signal reach the mitochondria? As described earlier, activated caspases may form a molecular bridge to the mitochondria, either directly or by activating a cytosolic intermediate(s). A caspase-independent link from receptor to mitochondria may include accumulation of ceramide, a messenger molecule described to induce ROS generation in the mitochondria. However, another cell death-enhancing mitochondria-associated signaling pathway is activated in caspase-dependent and caspase-independent pathways, but is exclusive to TNF receptor triggering (De Vos et al., 1998). Whereas in untreated cells, mitochondria are typically distributed throughout the cytoplasm, they cluster asymmetrically in the vicinity of the nucleus in TNF-treated cells. A similar phenotype was obtained by inhibition of the molecular motor kinesin in the absence of TNF, suggesting a TNF-induced inhibition of kinesin as an underlying mechanism for this cluster formation. Such a response required the membrane-proximal region of TNF-R55 (i.e., the part of TNF-R55 between membrane and DD), which explains the absence of this response after Fas signaling. Interestingly, cell death induced by mutant TNF-R55 lacking the membrane-proximal region was delayed markedly when compared to full-length TNF-R55. Thus there seems to be a cell death-enhancing signaling pathway emanating from the membrane-proximal region of TNF-R55 that cooperates with the DD-dependent pathway. Because inhibition of kinesin by specific blocking of the tail domain of the molecule enhances cell death (besides causing clustering of the mitochondria), a functional behavior similar to that generated by the TNF-R55 membrane-proximal region was obtained. This similarity supports the notion that kinesin participates in TNF signal transduction and, due to its association with mitochondria, may in fact transduct the cytosolic signal to the organelles. The nature of the signaling pathway emanating from the membrane-proximal region of TNF-R55 remains to be analyzed.

III. CONCLUDING REMARKS

We conclude that ROS may act at different levels of the signaling and execution pathways to cell death. TNF induces the release of superoxide anion in the extracellular environment by the activation of membrane-bound NAD(P)H oxidases in leukocytes, primary fibroblasts and fibrosarcoma cells. Depending on the cellular surroundings, the oxidative stress created is used to kill pathogens or tumor cells and induces proliferation or suicide of the producing cell.

In transformed cell lines killed by TNF, this cytokine may induce the generation of intracellular ROS. Intracellular sources of TNF-induced ROS are mitochondria and the lipoxygenase pathway. How these ROS participate in cell death depends on the signaling and execution pathways activated. Caspase-dependent cell death is often accompanied by the release of cytochrome c from the mitochondria and/or by PT. These mitochondrial events can be induced by ROS; they do, however, also enforce the generation of ROS as a consequence of uncoupling of the respiratory chain and thiol imbalance. Although these ROS may not be necessary to cell death, they may accelerate the execution phase of cell death. The exact role of ROS in cells dying from caspase-dependent apoptosis has to be investigated in more detail. A direct causal relationship between intracellular ROS generation and TNF cytotoxicity has been established in L929 cells where a caspase-independent, but mitochondria-dependent pathway leads to cell death. The crucial role of mitochondria is the production of highly cytocidal ROS by the respiratory chain. Mitochondrial ROS are inaccessible to hydrophylic antioxidant enzymes, but are neutralized efficiently by iron chelators and lipophylic-free radical scavengers, indicating that they are not only produced in the mitochondrial membrane, but also exert cytocidal activity in a hydrophobic environment. The ROS-generating mechanism is regulated by a dithiol group(s) in a crucial protein(s), probably in cooperation with NADH. A caspase-like protection mechanism against mitochondrial ROS generation has now been defined, which needs to be characterized further.

ACKNOWLEDGMENTS

Research was supported by the Interuniversitaire Attractiepolen and the Vlaams Actiecomité voor Biotechnologie. K. De Vos and M. Steemans are fellows with the Vlaams Instituut voor de Bevordering van het Wetenschappelijk-technologisch Onderzoek in de Industrie. K. Vancompernolle and P. Vandenabeele are postdoctoral researchers with the Fonds voor Wetenschappelijk Onderzoek–Vlaanderen. Drs. D. Männel and P. Amstad are acknowledged for kindly donating the h-MnSOD gene as well as the h-Cu/ZnSOD and rat–human catalase genes, respectively. The authors are indebted to Dr. M. Davey for glutathione measurements.

REFERENCES

Adam, D., Wiegmann, K., Adam-Klages, S., Ruff, A., and Krönke, M. (1996). A novel cytoplasmic domain of the p55 TNF receptor initiates the neutral sphingomyelinase pathway. *J. Biol. Chem.* **271**, 14617–14622.

Adam-Klages, S., Adam, D., Wiegmann, K., Struve, S., Kolanus, W., Schneider-Mergener, J., and Krönke, M. (1996). FAN, a novel WD-repeat protein, couples the p55 TNF-receptor to neutral sphingomyelinase. *Cell (Cambridge, Mass.)* **86**, 937–947.

Baeuerle, P. A., and Baltimore, D. (1996). NF-κB: Ten years after. *Cell (Cambridge, Mass.)* **87**, 13–20.

Bernardi, P. (1992). Modulation of the mitochondrial cyclosporin A-sensitive permeability transition pore by the proton electrochemical gradient. Evidence that the pore can be opened by membrane depolarization. *J. Biol. Chem.* **267**, 8834–8839.

Beyaert, R., Cuenda, A., Vanden Berghe, W., Plaisance, S., Lee, J. C., Haegeman, G., Cohen, P., and Fiers, W. (1996). The p38/RK mitogen-activated protein kinase pathway regulates interleukin-6 synthesis in response to tumour necrosis factor. *EMBO J.* **15**, 1914–1923.

Boldin, M. P., Mett, I. L., Varfolomeev, E. E., Chumakov, I., Shemer-Avni, Y., Camonis, J. H., and Wallach, D. (1995). Self-association of the "death domains" of the p55 tumor necrosis factor (TNF) receptor and Fas/APO1 prompts signaling for TNF and Fas/APO1 effects. *J. Biol. Chem.* **270**, 387–391.

Boldin, M. P., Goncharov, T. M., Goltsev, Y. V., and Wallach, D. (1996). Involvement of MACH, a novel MORT1/FADD-interacting protease, in Fas/APO-1- and TNF receptor-induced cell death. *Cell (Cambridge, Mass.)* **85**, 803–815.

Castedo, M., Macho, A., Zamzami, N., Hirsch, T., Marchetti, P., Uriel, J., and Kroemer, G. (1995). Mitochondrial perturbations define lymphocytes undergoing apoptotic depletion in vivo. *Eur. J. Immunol.* **25**, 3277–3284.

Chernyak, B. V., and Bernardi, P. (1996). The mitochondrial permeability transition pore is modulated by oxidative agents through both pyridine nucleotides and glutathione at two separate sites. *Eur. J. Biochem.* **238**, 623–630.

Chinnaiyan, A. M., Tepper, C. G., Seldin, M. F., O'Rourke, K., Kischkel, F. C., Hellbardt, S., Krammer, P. H., Peter, M. E., and Dixit, V. M. (1996). FADD/MORT1 is a common mediator of CD95 (Fas/APO-1) and tumor necrosis factor receptor-induced apoptosis. *J. Biol. Chem.* **271**, 4961–4965.

Cossarizza, A., Franceschi, C., Monti, D., Salvioli, S., Bellesia, E., Rivabene, R., Biondo, L., Rainaldi, G., Tinari, A., and Malorni, W. (1995). Protective effect of N-acetylcysteine in tumor necrosis factor-α-induced apoptosis in U937 cells: The role of mitochondria. *Exp. Cell Res.* **220**, 232–240.

De Vos, K., Goossens, V., Boone, E., Vercammen, D., Vancompernolle, K., Vandenabeele, P., Haegeman, G., Fiers, W., and Grooten, J. (1998). The 55-kDa tumor necrosis factor receptor induces clustering of mitochondria through its membrane-proximal region. *J. Biol. Chem.* **273**, 9673–9680.

García-Ruiz, C., Colell, A., Marí, M., Morales, A., and Fernández-Checa, J. C. (1997). Direct effect of ceramide on the mitochondrial electron transport chain leads to generation of reactive oxygen species. Role of mitochondrial glutathione. *J. Biol. Chem.* **272**, 11369–11377.

Goossens, V., Grooten, J., De Vos, K., and Fiers, W. (1995). Direct evidence for tumor necrosis factor-induced mitochondrial reactive oxygen intermediates and their involvement in cytotoxicity. *Proc. Natl. Acad. Sci. U.S.A.* **92**, 8115–8119.

Goossens, V., Grooten, J., and Fiers, W. (1996). The oxidative metabolism of glutamine. A modulator of reactive oxygen intermediate-mediated cytotoxicity of tumor necrosis factor in L929 fibrosarcoma cells. *J. Biol. Chem.* **271**, 192–196.

Goossens, V., De Vos, K., Vercammen, D., Steemans, M., Vancompernolle, K., Fiers, W., Vandenabeele, P., and Grooten, J. (1999a). Redox regulation of TNF signaling. *BioFactors* **10**, in press.

Goossens, V., Stangé, G., Moens, K., Pipeleers, D., and Grooten, J. (1999b). Regulation of TNF-induced, mitochondria- and ROS-dependent cell death by the electron flux through the electron transport chain complex I. *Antioxidant Redox Signaling* **1**, in press.

Grooten, J., Goossens, V., Vanhaesebroeck, B., and Fiers, W. (1993). Cell membrane permeabilization and cellular collapse, followed by loss of dehydrogenase activity: Early events in tumour necrosis factor-induced cytotoxicity. *Cytokine* **5**, 546–555.

Guénal, I., Sidoti de Fraisse, C., Gaumer, S., and Mignotte, B. (1997). Bcl-2 and Hsp27 act at different levels to suppress programmed cell death. *Oncogene* **15**, 347–360.

Hennet, T., Richter, C., and Peterhans, E. (1993). Tumour necrosis factor-α induces superoxide anion generation in mitochondria of L929 cells. *Biochem. J.* **289**, 587–592.

Higuchi, M., Shirotani, K., Higashi, N., Toyoshima, S., and Osawa, T. (1992). Damage to mitochondrial respiration chain is related to phospholipase A_2 activation caused by tumor necrosis factor. *J. Immunother.* **12**, 41–49.

Hsu, H., Xiong, J., and Goeddel, D. V. (1995). The TNF receptor 1-associated protein TRADD signals cell death and NF-κB activation. *Cell (Cambridge, Mass.)* **81**, 495–504.

Hsu, H., Huang, J., Shu, H.-B., Baichwal, V., and Goeddel, D. V. (1996a). TNF-dependent recruitment of the protein kinase RIP to the TNF receptor-1 signaling complex. *Immunity* **4**, 387–396.

Hsu, H., Shu, H.-B., Pan, M.-G., and Goeddel, D. V. (1996b). TRADD-TRAF2 and TRADD-FADD interactions define two distinct TNF receptor 1 signal transduction pathways. *Cell (Cambridge, Mass.)* **84**, 299–308.

Huwiler, A., Brunner, J., Hummel, R., Vervoordeldonk, M., Stabel, S., van den Bosch, H., and Pfeilschifter, J. (1996). Ceramide-binding and activation defines protein kinase c-Raf as a ceramide-activated protein kinase. *Proc. Natl. Acad. Sci. U.S.A.* **93**, 6959–6963.

Jäättelä, M., Mouritzen, H., Elling, F., and Bastholm, L. (1996). A20 zinc finger protein inhibits TNF and IL-1 signaling. *J. Immunol.* **156**, 1166–1173.

Jarvis, W. D., Fornari, F. A., Jr., Browning, J. L., Gewirtz, D. A., Kolesnick, R. N., and Grant, S. (1994). Attenuation of ceramide-induced apoptosis by diglyceride in human myeloid leukemia cells. *J. Biol. Chem.* **269**, 31685–31692.

Klebanoff, S. J., Vadas, M. A., Harlan, J. M., Sparks, L. H., Gamble, J. R., Agosti, J. M., and Waltersdorph, A. M. (1986). Stimulation of neutrophils by tumor necrosis factor. *J. Immunol.* **136**, 4220–4225.

Levrat, C., and Louisot, P. (1996). Increase of mitochondrial PLA_2-released fatty acids is an early event in tumor necrosis factor α-treated WEHI-164 cells. *Biochem. Biophys. Res. Commun.* **221**, 531–538.

Li, P., Nijhawan, D., Budihardjo, I., Srinivasula, S. M., Ahmad, M., Alnemri, E. S., and Wang, X. (1997). Cytochrome c and dATP-dependent formation of Apaf-1/caspase-9 complex initiates an apoptotic protease cascade. *Cell (Cambridge, Mass.)* **91**, 479–489.

Liston, P., Roy, N., Tamai, K., Lefebvre, C., Baird, S., Cherton-Horvat, G., Farahani, R., McLean, M., Ikeda, J., MacKenzie, A., and Korneluk, R. G. (1996). Suppression of apoptosis in mammalian cells by NAIP and a related family of IAP genes. *Nature (London)* **379**, 349–353.

Malinin, N. L., Boldin, M. P., Kovalenko, A. V., and Wallach, D. (1997). MAP3K-related kinase involved in NF-κB induction by TNF, CD95 and IL-1. *Nature (London)* **385**, 540–544.

Marzo, I., Brenner, C., Zamzami, N., Susin, S. A., Beutner, G., Brdiczka, D., Rémy, R., Xie, Z.-H., Reed, J. C., and Kroemer, G. (1998). The permeability transition pore complex: A target for apoptosis regulation by caspases and Bcl-2-related proteins. *J. Exp. Med.* **187**, 1261–1271.

Matthews, N., Neale, M. L., Jackson, S. K., and Stark, J. M. (1987). Tumour cell killing by tumour necrosis factor: Inhibition by anaerobic conditions, free-radical scavengers and inhibitors of arachidonate metabolism. *Immunology* **62**, 153–155.

Medema, J. P., Scaffidi, C., Kischkel, F. C., Shevchenko, A., Mann, M., Krammer, P. H., and Peter, M. E. (1997). FLICE is activated by association with the CD95 death-inducing signaling complex (DISC). *EMBO J.* **16**, 2794–2804.

Mehlen, P., Kretz-Remy, C., Préville, X., and Arrigo, A.-P. (1996). Human hsp27, Drosophila hsp27 and human αB-crystallin expression-mediated increase in glutathione is essential for the protective activity of these proteins against TNFα-induced cell death. *EMBO J.* **15**, 2695–2706.

Meier, B., Radeke, H. H., Selle, S., Younes, M., Sies, H., Resch, K., and Habermehl, G. G. (1989). Human fibroblasts release reactive oxygen species in response to interleukin-1 or tumour necrosis factor-α. *Biochem. J.* **263**, 539–545.

Mercurio, F., Zhu, H., Murray, B. W., Shevchenko, A., Bennett, B. L., Li, J. W., Young, D. B., Barbosa, M., Mann, M., Manning, A., and Rao, A. (1997). IKK-1 and IKK-2: Cytokine-activated IκB kinases essential for NF-κB activation. *Science* **278**, 860–866.

Mohazzab, K. M., and Wolin, M. S. (1994). Sites of superoxide anion production detected by lucigenin in calf pulmonary artery smooth muscle. *Am. J. Physiol.* **267**, L815–L822.

Muzio, M., Stockwell, B. R., Stennicke, H. R., Salvesen, G. S., and Dixit, V. M. (1998). An induced proximity model for caspase-8 activation. *J. Biol. Chem.* **273**, 2926–2930.

Nicholson, D. W., Ali, A., Thornberry, N. A., Vaillancourt, J. P., Ding, C. K., Gallant, M., Gareau, Y., Griffin, P. R., Labelle, M., Lazebnik, Y. A., Munday, N. A., Raju, S. M., Smulson, M. E., Yamin, T.-T., Yu, V. L., and Miller, D. K. (1995). Identification and inhibition of the ICE/CED-3 protease necessary for mammalian apoptosis. *Nature (London)* **376**, 37–43.

O'Donnell, V. B., Spycher, S., and Azzi, A. (1995). Involvement of oxidants and oxidant-generating enzyme(s) in tumour-necrosis-factor-α-mediated apoptosis: Role for lipoxygenase pathway but not mitochondrial respiratory chain. *Biochem. J.* **310**, 133–141.

Opipari, A. W., Jr., Hu, H. M., Yabkowitz, R., and Dixit, V. M. (1992). The A20 zinc finger protein protects cells from tumor necrosis factor cytotoxicity. *J. Biol. Chem.* **267**, 12424–12427.

Quillet-Mary, A., Jaffrézou, J.-P., Mansat, V., Bordier, C., Naval, J., and Laurent, G. (1997). Implication of mitochondrial hydrogen peroxide generation in ceramide-induced apoptosis. *J. Biol. Chem.* **272**, 21388–21395.

Régnier, C. H., Song, H. Y., Gao, X., Goeddel, D. V., Cao, Z., and Rothe, M. (1997). Identification and characterization of an IκB kinase. *Cell (Cambridge, Mass.)* **90**, 373–383.

Scaffidi, C., Fulda, S., Srinivasan, A., Friesen, C., Li, F., Tomaselli, K. J., Debatin, K.-M., Krammer, P. H., and Peter, M. E. (1998). Two CD95 (APO-1/Fas) signaling pathways. *EMBO J.* **17**, 1675–1687.

Schmidt, K. N., Amstad, P., Cerutti, P., and Baeuerle, P. A. (1995). The roles of hydrogen peroxide and superoxide as messengers in the activation of transcription factor NF-κB. *Chem. Biol.* **2**, 13–22.

Schreck, R., Rieber, P., and Baeuerle, P. A. (1991). Reactive oxygen intermediates as apparently widely used messengers in the activation of the NF-κB transcription factor and HIV-1. *EMBO J.* **10**, 2247–2258.

Schulze-Osthoff, K., Bakker, A. C., Vanhaesebroeck, B., Beyaert, R., Jacob, W. A., and Fiers, W. (1992). Cytotoxic activity of tumor necrosis factor is mediated by early damage of mitochondrial functions. Evidence for the involvement of mitochondrial radical generation. *J. Biol. Chem.* **267**, 5317–5323.

Shalaby, M. R., Aggarwal, B. B., Rinderknecht, E., Svedersky, L. P., Finkle, B. S., and Palladino, M. A., Jr. (1985). Activation of human polymorphonuclear neutrophil functions by interferon-γ and tumor necrosis factor. *J. Immunol.* **135**, 2069–2073.

Smiley, S. T., Reers, M., Mottola-Hartshorn, C., Lin, M., Chen, A., Smith, T. W., Steele, G. D. Jr., and Chen, L. B. (1991). Intracellular heterogeneity in mitochondrial membrane potentials revealed by a J-aggregate-forming lipophilic cation JC-1. *Proc. Natl. Acad. Sci. USA* **88**, 3671–3675.

Song, H. Y., Dunbar, J. D., and Donner, D. B. (1994). Aggregation of the intracellular domain of the type 1 tumor necrosis factor receptor defined by the two-hybrid system. *J. Biol. Chem.* **269**, 22492–22495.

Song, H. Y., Rothe, M., and Goeddel, D. V. (1996). The tumor necrosis factor-inducible zinc finger protein A20 interacts with TRAF1/TRAF2 and inhibits NF-κB activation. *Proc. Natl. Acad. Sci. U.S.A.* **93**, 6721–6725.

Steemans, M., Goossens, V., Van de Craen, M., Van Herreweghe, F., Vancompernolle, K., De Vos, K., Vandenabeele, P., and Grooten, J. (1998). A caspase-activated factor (CAF) induces mitochondrial membrane depolarization and cytochrome c release by a nonproteolytic mechanism. *J. Exp. Med.* **188**, 2193–2198.

Tartaglia, L. A., Ayres, T. M., Wong, G. H. W., and Goeddel, D. V. (1993a). A novel domain within the 55 kd TNF receptor signals cell death. *Cell (Cambridge, Mass.)* **74**, 845–853.

Tartaglia, L. A., Pennica, D., and Goeddel, D. V. (1993b). Ligand passing: The 75-kDa tumor necrosis factor (TNF) receptor recruits TNF for signaling by the 55-kDa TNF receptor. *J. Biol. Chem.* **268**, 18542–18548.

Vandenabeele, P., Declercq, W., Beyaert, R., and Fiers, W. (1995). Two tumour necrosis factor receptors: Structure and function. *Trends Cell Biol.* **5**, 392–399.

Vandevoorde, V., Haegeman, G., and Fiers, W. (1997). Induced expression of trimerized intracellular domains of the human tumor necrosis factor (TNF) p55 receptor elicits TNF effects. *J. Cell Biol.* **137**, 1627–1638.

Varfolomeev, E. E., Boldin, M. P., Goncharov, T. M., and Wallach, D. (1996). A potential mechanism of "cross-talk" between the p55 tumor necrosis factor receptor and Fas/APO1: Proteins binding to the death domains of the two receptors also bind to each other. *J. Exp. Med.* **183**, 1271–1275.

Vercammen, D., Beyaert, R., Denecker, G., Goossens, V., Van Loo, G., Declercq, W., Grooten, J., Fiers, W., and Vandenabeele, P. (1998a). Inhibition of caspases increases the sensitivity of L929 cells to necrosis mediated by tumor necrosis factor. *J. Exp. Med.* **187**, 1477–1485.

Vercammen, D., Brouckaert, G., Denecker, G., Van de Craen, M., Declercq, W., Fiers, W., and Vandenabeele, P. (1998b). Dual signaling of the Fas receptor: Initiation of both apoptotic and necrotic cell death pathways. *J. Exp. Med.* **188**, 919–930.

Wallach, D., Boldin, M., Varfolomeev, E., Beyaert, R., Vandenabeele, P., and Fiers, W. (1997). Cell death induction by receptors of the TNF family: Towards a molecular understanding. *FEBS Lett.* **410**, 96–106.

Wiegmann, K., Schütze, S., Machleidt, T., Witte, D., and Krönke, M. (1994). Functional dichotomy of neutral and acidic sphingomyelinases in tumor necrosis factor signaling. *Cell (Cambridge, Mass.)* **78**, 1005–1015.

Wong, G. H. W., Elwell, J. H., Oberley, L. W., and Goeddel, D. V. (1989). Manganous superoxide dismutase is essential for cellular resistance to cytotoxicity of tumor necrosis factor. *Cell (Cambridge, Mass.)* **58**, 923–931.

Zamzami, N., Marchetti, P., Castedo, M., Zanin, C., Vayssière, J.-L., Petit, P. X., and Kroemer, G. (1995). Reduction in mitochondrial potential constitutes an early irreversible step of programmed lymphocyte death in vivo. *J. Exp. Med.* **181**, 1661–1672.

Zandi, E., Rothwarf, D. M., Delhase, M., Hayakawa, M., and Karin, M. (1997). The IκB kinase complex (IKK) contains two kinase subunits, IKKα and IKKβ, necessary for IκB phosphorylation and NF-κB activation. *Cell (Cambridge, Mass.)* **91**, 243–252.

Zimmerman, R. J., Chan, A., and Leadon, S. A. (1989). Oxidative damage in murine tumor cells treated in vitro by recombinant human tumor necrosis factor. *Cancer Res.* **49**, 1644–1648.

Zou, H., Henzel, W. J., Liu, X., Lutschg, A., and Wang, X. (1997). Apaf-1, a human protein homologous to C. elegans CED-4, participates in cytochrome c-dependent activation of caspase-3. *Cell (Cambridge, Mass.)* **90**, 405–413.

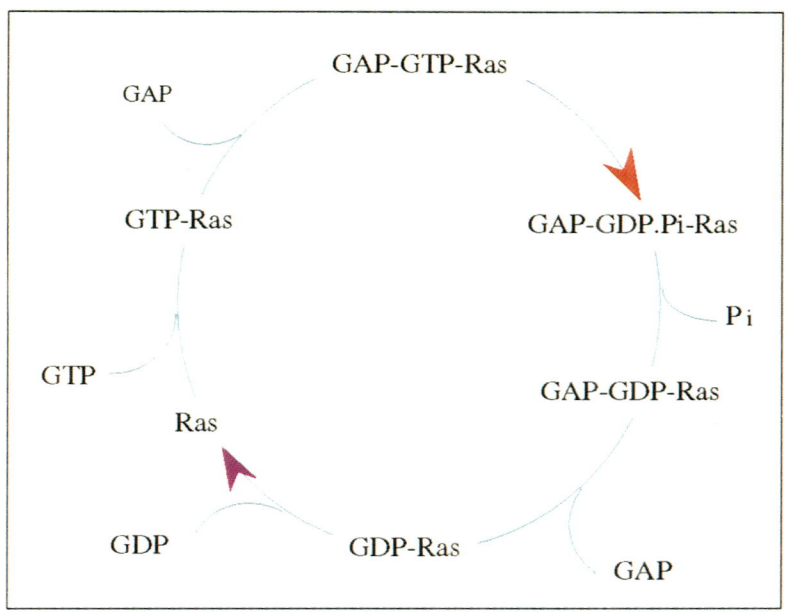

Figure 3-1

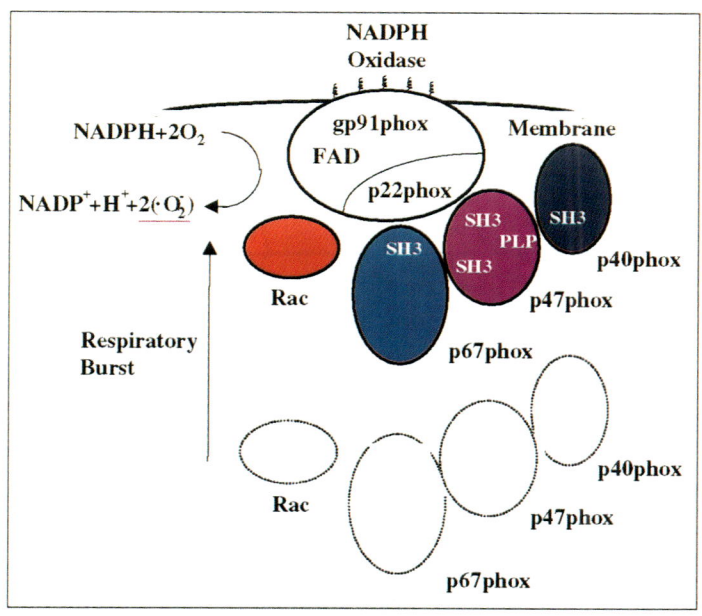

Figure 3-2

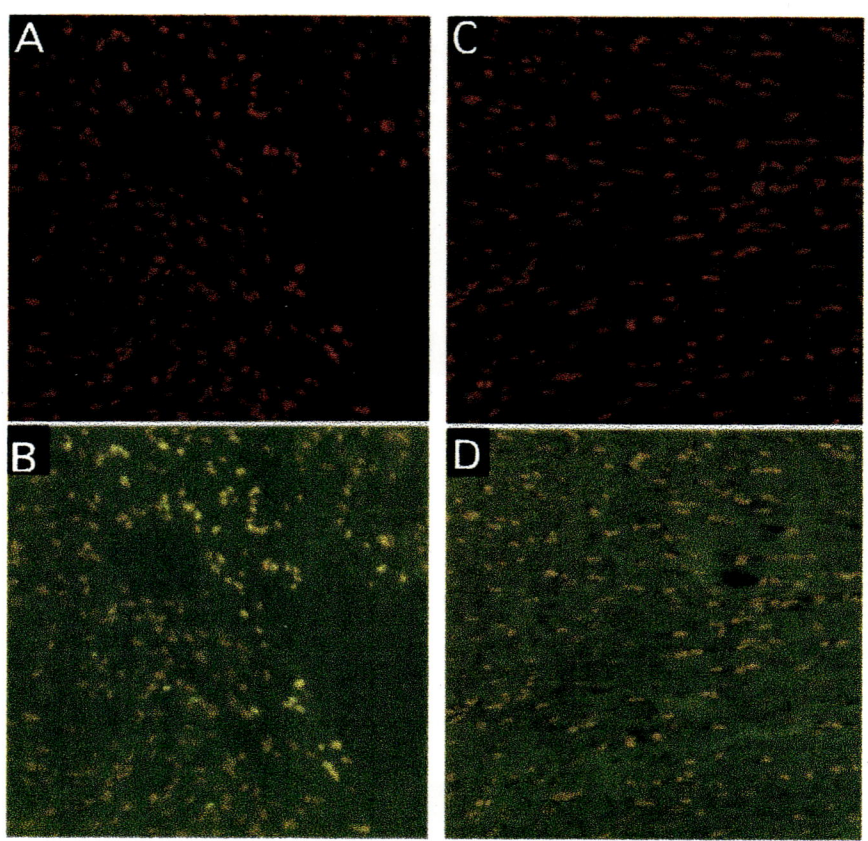

Figure 22-3

12 Redox Regulation of Cell Adhesion Processes

Sashwati Roy,* Chandan K. Sen,† Alexia Gozin,‡
Valèrie Andrieu,‡ and Catherine Pasquier‡

*Molecular and Cell Biology
University of California
Berkeley, California 94720
†Lawrence Berkeley National Laboratory
Berkeley, California 94720
‡INSERM U 479
CHU Xavier Bichat
75870 Paris, France

Cell adhesion processes play a major role in the regulation of immune functions as well as that of other vital biological processes, such as embryogenesis, cell growth, differentiation, and wound repair. Cell adhesion, a multistep process including rolling, firm attachment, and transmigration of leukocytes, is mediated by several classes of cell adhesion molecules. Cell adhesion molecule expression and adhesive properties of cells are modified greatly in several diseased conditions involving redox imbalances such as cancer, atherosclerosis, diabetes, chronic inflammation, and ischemia–reperfusion injury. Several stimuli such as cytokines, chemokines, and phorbol ester are known to activate the expression and function of cell adhesion molecules. The following two lines of evidence indicate that the overall cell adhesion process is redox regulated: (i) direct activation of cell adhesion processes by oxidants and (ii) inhibitory action of antioxidants on cell adhesion molecule expression and function. This chapter discusses the role of oxidants and antioxidants in the regulation of cell adhesion processes and the molecular mechanisms underlying such regulation.

I. INTRODUCTION

Adhesion of leukocytes to endothelial cells in postcapillary venules is one of the earliest and critical steps in acute and chronic inflammatory responses and is dependent on the expression of a variety of cell surface receptors also known as cell adhesion molecules (Albelda et al., 1994). Apart from immune functions, these cell adhesion molecules participate in orchestrating other vital biological processes, such as embryogenesis, cell growth, differentiation, and wound repair (Frenette and Wagner, 1996). The expression of cell adhesion molecules is known to be induced in response to several stimuli such as cytokines (tumor necrosis factor-α, TNF-α; interleukin-1α and β), phorbol 12-myristate 13-acetate (PMA), and lipopolysaccharide (LPS; Albelda et al., 1994). Oxidants and antioxidants have been shown to influence the expression of CAM and other cell–cell adhesion processes directly or indirectly (Aoki et al., 1996; Moynagh et al., 1994; Weber et al., 1994). This work mainly focuses on the role of oxidants and antioxidants in the regulation of cell adhesion molecule-related processes involved in inflammatory responses.

II. CELL ADHESION MOLECULES

Cell adhesion molecules are glycoproteins expressed on the cell surface. There are four main groups: the integrin family, the immunoglobulin superfamily, selectins, and cadherins.

A. Integrins

The integrins represent a family of heterodimeric, transmembrane glycoproteins that are formed by noncovalent association of α and β subunits. At present, 15α and 8β chains are known. The most widely studied subfamilies are $\beta 1$ [CD29, very late activation (VLA) members], $\beta 2$ (leukocyte integrins such as CD11a/CD18, CD11b/CD18, CD11c/CD18, and αd $\beta 2$), $\beta 3$ (CD61, cytoadhesions), and $\beta 7$ ($\alpha 4\beta 7$ and $\alpha E\beta 7$). As transmembrane proteins, integrins can interact with both extracellular molecules and intracellular proteins. On the cell exterior, domains of α and β subunits combine to form a ligand-binding site. Through this site, integrins can mediate cell binding to various substrates, including extracellular matrix molecules, other cell surface proteins (e.g., ICAM-1 and VCAM-1), and, in some cases, other integrins (Bevilacqua, 1993; Dejana et al., 1994; Gahmberg, 1997; Striz and Costabel, 1992).

B. Immunoglobulin Superfamily

The immunoglobulin (Ig) superfamily includes a large number of molecules with multiple Ig-like domains (Frenette and Wagner, 1996). Some of the major molecules of this group are intercellular adhesion molecules (ICAMs), vascular adhesion molecule-1 (VCAM-1), platelet-endothelial cell adhesion molecule-1 (PECAM-1), mucosal addressin cell adhesion molecule-1 (MAdCAM-1), and neuronal cell adhesion molecule (N-CAM).

Three forms of ICAMs have been identified: ICAM-1 (CD 54), ICAM-2 (CD 102), and ICAM-3 (CD 50). These ICAMs are not functionally redundant, their expression profiles are quite distinct, and they can bind to their counterreceptor, LFA1, differentially. ICAM-1 is basally expressed in low levels on leukocytes, fibroblasts, epithelial cells, and endothelial cells and consists of five Ig domains (Simmons, 1995). ICAM-1 is rapidly upregulated by inflammatory cytokines (Bevilacqua, 1993). ICAM-2 is a truncated form of ICAM-1, containing only two Ig-like extracellular domains. ICAM-3 is closely related to ICAM-1 and has five Ig domains. ICAM-3 is constitutively and abundantly expressed on "professional" antigen-presenting cells, such as Langerhans cells in the skin and dendritic cells in lympoid organs. A current working model for the interplay of the ICAMs is that ICAM-3 is the major ligand for LFA1 in the initiation of immune responses and that ICAM-1 takes over after these early stages. ICAMs not only function as an adhesion molecule, but also are emerging as a potent signaling molecules (Simmons, 1995).

VCAM-1 was originally identified as a cytokine-inducible adhesion molecule expressed on human endothelial cells (Osborn *et al.*, 1989). It is an important modulator of lymphocyte and monocyte trafficking. The major form of VCAM-1 in humans contains seven Ig-like domains. VCAM-1 binds the integrin VLA-4 through the first and fourth Ig-like domains. VCAM-1 is absent in nonstimulated endothelial cells; however, transcription-dependent upregulation can be elicited by cytokines and LPS (Collins *et al.*, 1995; Marui *et al.*, 1993; Weber *et al.*, 1995). Redox regulation of ICAM-1 and VCAM-1 expression has been discussed later in this chapter.

PECAM-1 (CD 31) is a 120 to 130-kDa integral membrane glycoprotein found on the surface of platelets, at endothelial intercellular junctions in culture, and on cells of myeloid lineage (Albelda *et al.*, 1991). PECAM-1 plays a major role in neutrophil recruitment at inflammatory sites and may be involved in the release of leukocytes from the bone marrow, in cardiovascular development, and in other interactive events taking place during thrombosis, wound healing, and angiogenesis (Albelda *et al.*, 1991). Various agents that cause oxidative stress have been shown to induce a multifold increase in the phosphorylation state of PECAM-1 (Kalra *et al.*, 1996; Rattan *et al.*,

1997; Shen et al., 1996; Sultana et al., 1996). An increase in the phosphorylation state of this protein on endothelial cells has been associated with enhanced transendothelial migration of leukocytes (Kalra et al., 1996; Rattan et al., 1997; Sultana et al., 1996).

The mucosal vascular addressin, MAdCAM-1, is an immunoglobulin superfamily adhesion molecule for lymphocytes that is expressed by mucosal venules and helps direct lymphocyte traffic into Peyer's patches and the intestinal lamina propria (Berlin et al., 1993). Neural-cell adhesion molecules (N-CAMs) are other major members of the immunoglobulin superfamily that exists in various isoforms generated by alternative splicing (Cremer et al., 1994). During embryonic development, N-CAMs are expressed in derivatives of all three germ layers, whereas in the adult animals they are predominantly present in neural and muscle tissues. Processes such as neurulation, axonal outgrowth, histogenesis of the retina, and development of the olfactory system are correlated with the regulated expression of N-CAMs (Cremer et al., 1994). The role of oxidants/antioxidants in the regulation of MAdCAM and N-CAM has not been well studied.

C. Selectins

The selectins are a family of three calcium-dependent lectins that mediate adhesive interactions between leukocytes and the endothelium during normal and abnormal inflammatory episodes (Erbe et al., 1993; Lasky, 1992). This family consists of three closely related glycoproteins: L-selectin (CD62L, LAM-1, MEL-14), E-selectin (CD62E, ELAM-1), and P-selectin (CD62P, PADGEM, GMP-140). Selectins belong to a single family because they share a common mosaic structure consisting of (i) a calcium-dependent lectin (sugar binding) domain at the NH_2 terminus, (ii) a single epidermal growth factor-like domain, (iii) two to nine short consensus repeat units homologous to domains found in complement-binding proteins, (iv) a transmembrane domain, and (v) a short C-terminal cytoplasmic domain (Tedder et al., 1995). The function of selectin is uniquely restricted to the vascular system. The expression of L-selectin is limited to hematopoietic cells. Lymphocytes and neutrophils exhibit a reversible loss of L-selectin expression after cellular activation. Concomitant with the decrease in L-selectin from the cell surface is an increase in expression of other adhesion receptors. P-selectin is constitutively found in Weibel-Palade bodies of endothelial cells and in α granules of platelets (Lasky, 1995). Reactive oxygen species have been known to stimulate the mobilization of P-selectins (Tedder et al., 1995). E-selectins are expressed within the vasculature at sites of inflammation. E-selectin mediates the adhesion of neutrophils to activated vascular endothelium (Bevilacqua et al., 1987) and may serve as a tissue-specific homing receptor for T-cell subsets (Shimizu et al., 1991). Regulation of expression and function of selectins by reactive oxygen species will be discussed in later sections.

D. Cadherins

The cadherins are a gene superfamily of calcium-dependent cell adhesion molecules and include epithelial (E), placental (P), and neural (N) subclasses (Takeichi, 1991; Tang and Honn, 1994). These are single-chain transmembrane proteins composed of a highly conserved cytoplasmic region and an extracellular domain containing Ca^{2+}-binding motifs (Bavisotto et al., 1990). Cadherins form cellular membrane contact also known as cell-to-cell adherens junctions. Cadherins are considered to be important regulators of morphogenesis. Decreased expression of this molecule has been associated with the invasiveness of tumor cells (Takeichi, 1991). Reactive oxygen species such as hydroperoxides have been shown to stimulate the internalization of cadherins in endothelial cells (Kevil et al., 1998).

E. Ligand–Receptor Complexes

Binding sites (ligands–receptors) are different for each cell adhesion molecule. Ligands for ICAM-1 and VCAM-1 on leukocyte are LFA-1 (CD11a/CD18) and VLA-4 ($\alpha_4\beta_1$), respectively. Interactions of PECAM-1 with its ligands are complex in that it is able to bind to itself (homophilic adhesion) as well as to non-PECAM-1 ligands (heterophilic adhesion) (Sun et al., 1996). MAdCAM-1 serves as a ligand for L-selectin and $\alpha_4\beta_7$-integrin (Panes and Granger, 1998). N-CAM mediates hetero- or homophilic adhesion and can associate laterally with the fibroblast growth factor receptor (FGFR). Following association with FGFR, N-CAM stimulates tyrosine kinase activity of the receptor to induce the outgrowth of neurites (Frenette and Wagner, 1996).

Selectins can bind to a variety of carbohydrate structures *in vitro*. The tetrasaccharide sialyl Lewis X (sLe^x, CD15s) and sialyl Lewis A (sLe^A) have been identified as a prototype ligand for both P- and E-selectin (Lasky, 1995). P-selectin also binds to P-selectin glycoprotein ligand-1 (PSGL-1) expressed by all blood neutrophils, monocytes, and lymphocytes (Tedder et al., 1995). L-selectin has been shown to bind with at least three different heavily glycosylated mucin-like proteins: GlyCam-1, CD34, and MAdCAM-1 (Tedder et al., 1995).

Cadherins are known to form complexes with catenins (α-, β-, and γ-catenin or plakoglobin). The assoaciation of cadherin with catenins is necessary for the adhesive functions of these molecules and for adherens junctions (Ranscht, 1994).

III. MULTISTEP MODEL OF CELL ADHESION

Lymphocyte recirculation and leukocyte extravasation involve a multistep process (Fig. 1). The initial adhesive interaction between neutrophils and the wall of postcapillary venules seems to involve the participation of selectins that mediate low-affinity binding that is manifested as leukocyte rolling (Lasky, 1995). These transiently bound leukocytes are then exposed to a low concentration of inflammatory mediators that result in integrin activation. The subsequent steps in the adhesion process involve firm adhesion and transendothelial migration.

A. Rolling

In the early phases of cell adhesion, leukocytes transiently adhere to the vessel wall in a process termed "rolling." Rolling of leukocytes is mediated by members of the selectin family, expressed both on the leukocyte and on the endothelial cell surface. The anatomic position of specific selectins may be crucial for their role in leukocyte migration. For example, L-selectin is concentrated on the tips of membranous projections present on the surface

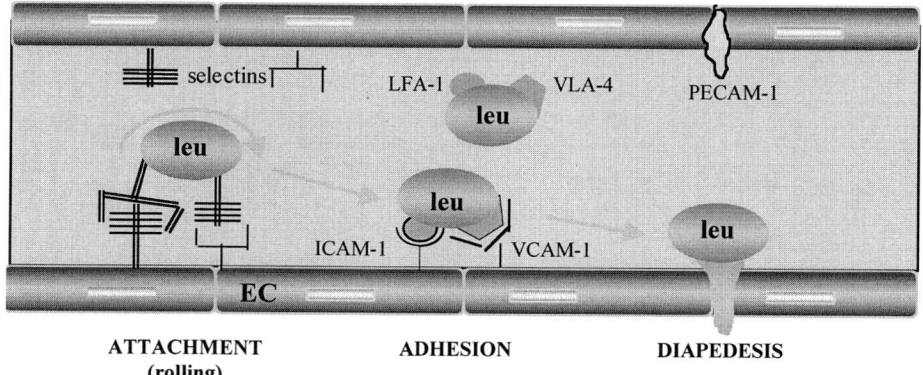

Figure 1 Schematic illustration of the multistep model of cell adhesion. In the initial phase, selectins mediate low-affinity binding that is manifested as leukocyte rolling. These transiently bound leukocytes then get attached firmly to endothelial cells, the process is mediated by adhesion molecules such as ICAM-I and VCAM-1 on endothelial cells and their ligands such as LFA-1 and VLA-4 on leukocytes. The transmigration of leukocytes through the endothelial cell lining is the final step mediated by adhesion molecules such as PECAM-1. EC, endothelial cells; Leu, leukocytes; LFA-1, lymphocyte function-associated antigen-1; ICAM-1, intercellular adhesion molecule-1; PECAM-1, platelet endothelial cell adhesion molecule-1; VCAM-1, vascular cell adhesion molecule-1; VLA-4, very late antigens-4.

of neutrophils and these projections have been considered to be the initial sites of contact between neutrophils and endothelial cells. In this position, L-selectin is able to bind favorably via its lectin domain to a newly expressed glycoprotein on the surface of stimulated endothelial cells. L-selectin on neutrophils has been suggested to be a predominant SLe^x bearing ligand (Picker et al., 1991).

B. Firm Adhesion

After "rolling", leukocytes firmly adhere to endothelial cells. This process is mediated by the binding of LFA-1 and VLA-4 expressed on leukocytes to ICAM-1 and VCAM-1 present on the endothelial cell surface, respectively. VLA-4 is expressed on all leukocytes except neutrophils. L-selectin binding to a newly expressed glycoconjugate on endothelial cells could lead to the activation of a $\beta 1$- or $\beta 3$-integrin. These integrins are responsible for the activation of intracellular signaling leading to structural modification of the membrane and particularly ICAM-1 expression.

C. Transmigration

The final step, transmigration into tissues, requires both a chemotactic stimulus and an engagement of PECAM-1. The surface distribution of ICAM-1 on endothelial cells (ICAM-1 is found on both apical and basal surfaces of endothelial cells and in intracellular junctions) and LFA-1/ICAM-1 interactions have been suggested to play an important role in the process of transendothelial migration of leukocytes (Vachula and Van Epps, 1992).

IV. DIRECT ACTIVATION OF CELL ADHESION PROCESSES BY REACTIVE OXYGEN SPECIES

Oxidants (e.g., H_2O_2, $O_2^{\cdot -}$ and $HO\cdot$) are known to induce the adherence of leukocytes to endothelial cells (Roy et al., 1999; Sellak et al., 1994). Treatment of human umbilical vein endothelial cells (HUVEC) with oxidant-generating systems such as hypoxanthine and xanthine oxidase ($O_2^{\cdot -}$ generating system) or oxidants such as H_2O_2 induced the adherence of neutrophils to HUVEC (Fig. 2) (Roy et al., 1999; Sellak et al., 1994). Catalase (200 IU ml^{-1}) completely abolished such an increase in adherence, whereas superoxide dismutase (100 IU ml^{-1}) had no effect, suggesting that H_2O_2, and not $O_2^{\cdot -}$, accounts for such effects of reactive oxygen species on cell adhesion (Sellak et al., 1994). Direct activation of HUVEC with H_2O_2 also increased adherence of leukocytes to these cells (Roy et al., 1999).

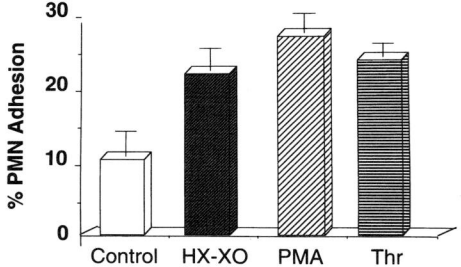

Figure 2 Adhesion of neutrophils to oxidant-stimulated endothelial cells. Confluent cultures of HUVEC were treated with (i) $O_2^{\cdot-}$ generating system (0.2 mM hypoxanthine + 4.5 mU/ml xanthine oxidase, HX-XO) for 15 min, (ii) PMA (100 ng/ml) for 10 min, or (iii) 2 U/ml thrombin (Thr) for 10 min. Following activation the HUVEC were cocultured with polymorphonuclear neutrophils (PMN) for 15 min. After coculture, nonadhered PMN were washed with phosphate-buffered saline, pH 7.4. The adherence of PMN to the HUVEC monolayer was evaluated by determining the myeloperoxidase activity of PMN associated with endothelial cells. The results are expressed as a percentage of the total PMN adhered to HUVEC. Values are the means ± SD of three experiments.

A. Effect of HO· Scavengers and Iron Chelators on Adherence of Leukocytes to Endothelial Cells

Pretreatment of HUVEC with hydroxyl radical scavengers such as dimethylthiourea or pentoxifylline or with iron chelators such as desferrioxamine or N,N'-di-(2-hydroxybenzyl)ethylenediamine-N,N'-diacetic acid dihydrochloride dihydrate inhibited the oxidant-induced adherence of neutrophils to HUVEC (Fig. 3) (DeForge *et al.*, 1992, 1993; Gutteridge, 1984; Pasquier *et al.*, 1991). These findings suggest that the oxidant-induced adherence could be attributed to scavenging of HO· generated from H_2O_2 inside the cells. These agents were not effective in downregulating PMA- or thrombin-induced adhesion of neutrophils to HUVEC.

B. Involvement of P-selectin, L-selectin, and ICAM-1

The relative contributions of P-selectin, L-selectin, and ICAM-1 in oxidant-induced cell adhesion processes has been studied by preincubating neutrophils with blocking antibodies against the respective adhesion molecules (Fig. 4). The P-selectin blocking antibody did not affect the adherence of neutrophils to HUVEC (data not shown), suggesting that P-selectin does not play any role in oxidant-induced cell adhesion. However, Patel *et al.* (1991) reported the involvement of P-selectin in the H_2O_2-induced adherence of neutrophils to endothelial cells. We observed that stimulation of HUVEC

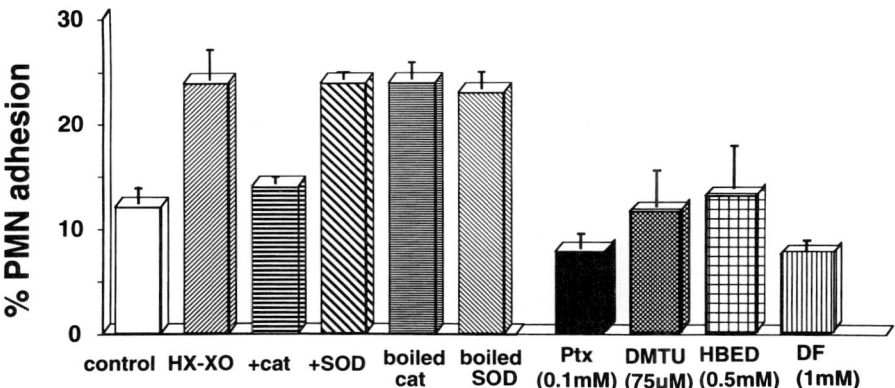

Figure 3 Effect of free radical scavengers and metal chelators on oxidant-induced adherence of neutrophils to HUVEC. Confluent cultures of HUVEC were pretreated with various free radical scavengers: 2000 U/ml catalase (cat); 750 U/ml superoxide dismutase (SOD); 75 μM dimethylthiourea (DMTU); 0.1 mM pentoxifylline (Ptx) or metal chelators; 1 mM desferrioxamine (DF); or 0.5 mM N,N'-di-(2-hydroxybenzyl)ethylenediamine-N,N'-diacetic acid dihydrochloride dihydrate (HBED) for 10 min. After pretreatment, HUVEC were activated by O_2^{-} generating system (0.2 mM hypoxanthine + 4.5 mU/ml xanthine oxidase, HX-XO) for 15 min. Following activation the HUVEC were cocultured with polymorphonuclear neutrophils (PMN) for 15 min (same as in Fig. 2). The results are expressed as a percentage of the total PMN adhered to HUVEC. Values are the means ± SD of three experiments.

with oxidants also had no influence on the surface expression of P-selectin. However, following oxidant treatment the total P-selectin levels, as detected in permeabilized HUVEC, were higher compared to that in nontreated cells. It appears that oxidant treatment releases P-selectin from intracellular stores such as Weibel-Palade bodies that are know to store P-selectin (Bonfanti et al., 1989). Adhesion of neutrophils to oxidant-treated endothelial cells involves ICAM-1, CD11b/CD18, and L-selectin (Sellak et al., 1994).

C. Protein Tyrosine Phosphorylation and Focal Adhesion Kinase

Phosphorylation of tyrosine residues on proteins plays an important role in the regulation of cell adhesion molecules expression and function (Bockholt and Burridge, 1993; Kornberg et al., 1992; Seufferlein and Rozengurt, 1994a). Figure 5 illustrates the time course of oxidant-induced protein tyrosine phosphorylation in HUVEC. All of these phosphorylated proteins were found to be translocated to the plasma membrane (Fig. 6). One of the proteins was identified as $pp125^{FAK}$ (FAK), a cytosolic tyrosine

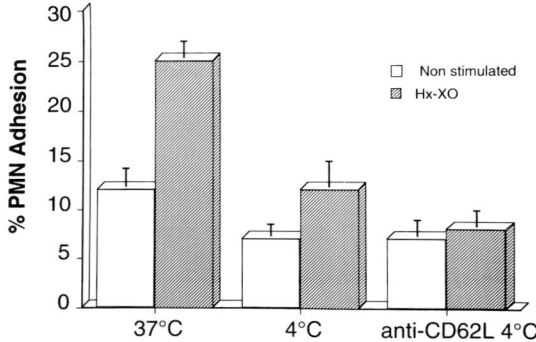

Figure 4 Role of L-selectin in adhesion of neutrophils to hypoxanthine–xanthine oxidase-activated HUVEC. Confluent cultures of HUVEC monolayers were activated by $O_2^{·-}$ generating system (0.2 mM hypoxanthine + 4.5 mU/ml xanthine oxidase, Hx-XO) for 15 min. Following activation, HUVEC were cocultured for 15 min with polymorphonuclear neutrophils (PMN) pretreated with or without anti-L selectin at 4°C. After coculture, nonadhered PMN were washed with phosphate-buffered saline, pH 7.4. The adherence of PMN to HUVEC monolayer was evaluated by determining the myeloperoxidase activity of PMN associated with endothelial cells. Results are expressed as a percentage of the total PMN adhered to HUVEC. Values are the means ± SD of three experiments.

kinase (Fig. 7A) (Schaller *et al.*, 1992). Tyrosine phosphorylation of the proteins of focal adhesions (focal adhesion kinase, FAK; paxillin; and p130cas) is an early event in mediating cell growth and differentiation

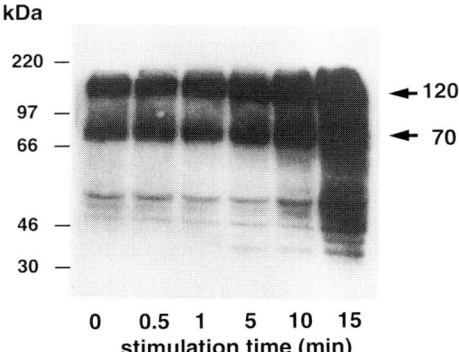

Figure 5 Time course of oxidant-induced tyrosine phosphorylation. HUVEC monolayers were activated by $O_2^{·-}$ generating system (0.2 mM hypoxanthine + 4.5 mU/ml xanthine oxidase) in a serum-free Hank's balanced salt solution (pH 7.4) for indicated time periods. The protein tyrosine phosphorylation of the cell lysate was evaluated by Western blot using the antiphosphotyrosine antibody.

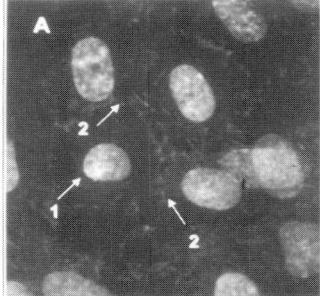 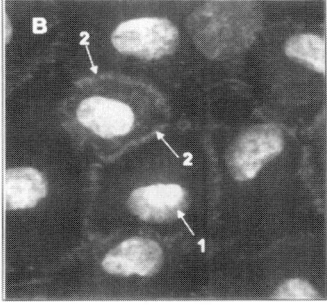

Figure 6 Oxidant-induced tyrosine phosphorylation. HUVEC cultured on glass coverslips were activated by $O_2^{\cdot-}$ generating system (0.2 mM hypoxanthine + 4.5 mU/ml xanthine oxidase, X-XO) in serum-free Hank's balanced salt solution (pH 7.4) for 5 min. The cells were fixed in 4% (w/v) paraformaldehyde in phosphate-buffered saline (pH 7.4, PBS) for 20 min and permeabilized with Triton X-100 [0.2% (v/v) in PBS] for 8 min at room temperature. Cells were then incubated with the antiphosphotyrosine antibody (marked as 1) for 1 hr followed by incubation in a secondary antibody conjugated to fluorescein isothiocyanate (FITC) for 1 hr. The nucleus was labeled with propidium iodide (marked as 2) after treatment with RNase for 30 min. Cells were visualized in a confocal imaging microscope using laser scanning. (A) Nonactivated cells. (B) X-XO-activated cells.

(Burridge *et al.*, 1992; Kornberg *et al.*, 1992; Lacerda *et al.*, 1996; Rankin and Rozengurt, 1994; Seckl *et al.*, 1995; Seufferlein and Rozengurt, 1994a,b, 1995; Sinnett-Smith *et al.*, 1993; Zachary *et al.*, 1992, 1993). FAK has also

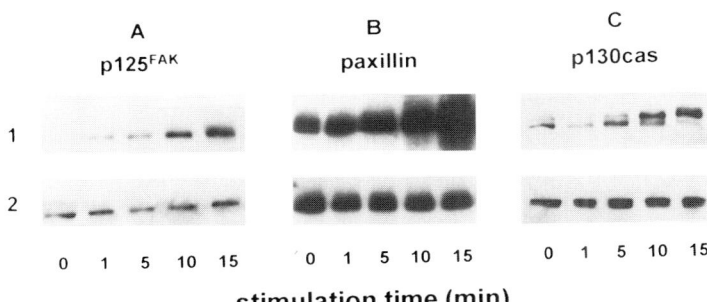

Figure 7 Effect of oxidant treatment on p125FAK, paxillin (PAX), and p130cas phosphorylation in HUVEC. HUVEC were activated by $O_2^{\cdot-}$ generating system (0.2 mM hypoxanthine + 4.5 mU/ml xanthine oxidase) and immunoprecipitated using (A) anti-p125FAK, (B) antipaxillin, and (C) anti-p130cas monoclonal antibodies. Tyrosine phosphorylation of the immunoprecipitated proteins was analyzed by Western blotting using the antiphosphotyrosine antibody (row 1). In order to normalize the amount of protein present in immunoprecipitates, the content of each protein was determined individually by Western blotting using the respective antibody, i.e., anti-p125FAK, antipaxillin, and anti-p130cas (row 2).

been observed to regulate the interaction of integrins with the cytoskeleton and/or with the extracellular matrix (Schaller et al., 1992). The other major proteins identified with increased phosphorylation following exposure to $O_2^{\cdot-}$ were paxillin (PAX) and p130cas (Figs. 7B and 7C). Both of these proteins are focal adhesion proteins and serve as substrates of FAK (Turner et al., 1990). An increase in the phosphorylation of FAK, paxillin, and p130cas was evident as early as 1 min following $O_2^{\cdot-}$ exposure. Phosphorylation levels continued to increase at least up to 15 min after exposure to $O_2^{\cdot-}$. The $O_2^{\cdot-}$ treatment of cells did not affect the total protein levels of FAK, paxillin, or p130cas (Figs. 7A–7C).

To evaluate the relative contributions of HO· and transition metals ions on the oxidant-mediated increase in the phosphorylation of focal adhesion proteins, HUVEC were treated with HO· radical scavengers (e.g., dimethylthiourea, pentoxifylline) or with metal chelators [e.g., desferrioxamine or N,N'-di-(2-hydroxybenzyl)ethylenediamine-N,N'-diacetic acid dihydrochloride dihydrate]. Western blots of HUVEC proteins immunoprecipitated with anti-FAK and antipaxillin antibodies after 1 min of stimulation with $O_2^{\cdot-}$ and probed with antiphosphotyrosine mAb demonstrate that the tyrosine phosphorylation of FAK and paxillin was inhibited by HO· scavengers (Fig. 8A) and chelators of metal ions (Fig. 8B). These data suggest that free metal ions and hydroxyl radicals play important roles in $O_2^{\cdot-}$-mediated cell adhesion as well as phosphorylation of proteins involved in focal adhesion. Direct evidence establishing a role of focal adhesion proteins in the oxidant-mediated adhesion of leukocytes to endothelial cells is still lacking.

V. IONIZING RADIATIONS AND ULTRAVIOLET RADIATION

Low dosages (5 Gy or higher) of ionizing radiation exposure are known to induce adhesion molecules expression (Hareyama et al., 1998; Heckmann et al., 1998). Reactive oxygen species have been suggested to be involved in upregulating the expression of adhesion molecules following ionizing radiation (Baeuml et al., 1997). Both in vivo and in vitro studies have shown that UV radiation is able to modulate cell adhesion molecule expression and function (Grether-Beck et al., 1996; Hallahan and Virudachalam, 1997; Krutmann and Grewe, 1995; Krutmann and Trefzer, 1992; Norris, 1993). A UVA radiation-mediated increase in ICAM-1 expression in keratinocytes has been observed to be inhibited by singlet oxygen quenchers (Grether-Beck et al., 1996). Furthermore, it has been shown that the effect of UVA on ICAM-1 expression could be mimicked in

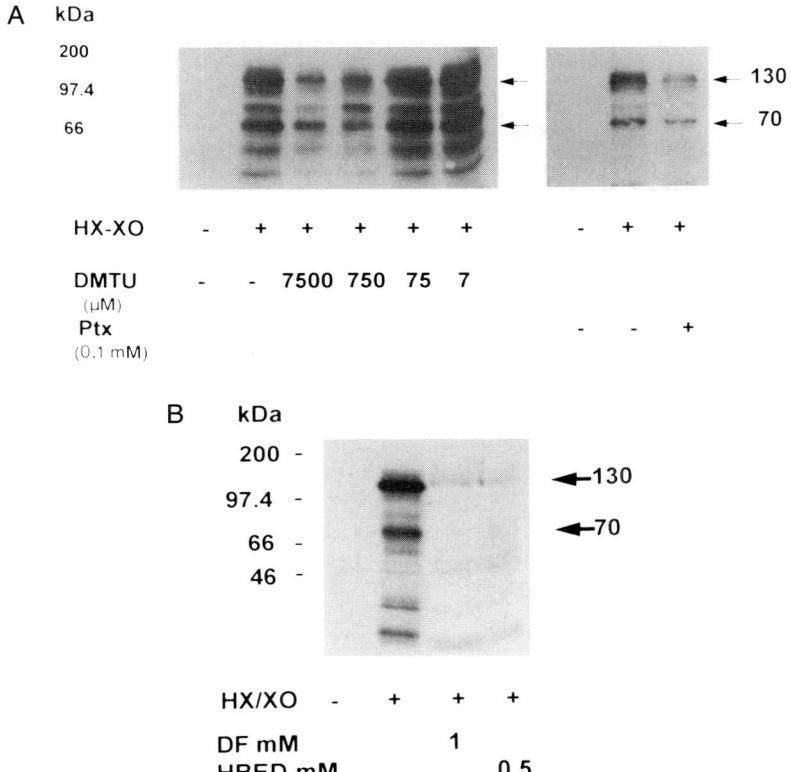

Figure 8 Effect of free radical scavengers and metal chelators on total protein tyrosine phosphorylation. Confluent cultures of HUVEC were pretreated with (A) hydroxyl radical scavengers: 7–7500 μM dimethylthiourea (DMTU) or 0.1 mM pentoxifylline (Ptx) or (B) metal chelators: 1 mM desferrioxamine (DF) or 0.5 mM N,N'-di-(2-hydroxybenzyl)ethylenediamine-N,N'-diacetic acid dihydrochloride dihydrate (HBED). Following pretreatment, HUVEC were activated by $O_2^{\cdot-}$ generating system (0.2 mM hypoxanthine + 4.5 mU/ml xanthine oxidase, HX-XO) for 15 min. Protein tyrosine phosphorylation was analyzed in the cell lysate by Western blot using the antiphosphotyrosine antibody.

unirradiated cells by a singlet oxygen-generating system, suggesting an important role of singlet oxygen in UV-induced cell adhesion processes (Grether-Beck et al., 1996). The AP-2 site has been identified as the UVA radiation- and singlet oxygen-responsive element of the human ICAM-1 gene (Grether-Beck et al., 1996).

VI. CIGARETTE SMOKE, HEAVY METALS, AND OTHER OXIDATIVE ENVIRONMENTAL POLLUTANTS

A common feature of cigarette smoke-associated diseases such as atherosclerosis and pulmonary emphysema is the activation, aggregation, and adhesion of leukocytes to micro- and macrovascular endothelium (Lehr *et al.*, 1994; Shen *et al.*, 1996). Cigarette smoke has been shown to induce the surface expression of ICAM-1, ELAM-1, and VCAM-1 in HUVEC (Shen *et al.*, 1996). Such increases were associated with an increase in the binding activity of NF-κB to the consensus motif common to cell adhesion molecule genes (Shen *et al.*, 1996). Furthermore, cigarette smoke condensate also increased the rate of transendothelial migration of vitamin D_3-differentiated monocyte and caused an approximately 10-fold increase in the phosphorylation of PECAM-1 (Shen *et al.*, 1996).

Heavy metal pollutants such as iron copper and zinc have been observed to activate cell adhesion molecule expression (Kennedy *et al.*, 1998; Martinotti *et al.*, 1995; Stal *et al.*, 1995). Among other oxidant-related environmental pollutants, ozone has been reported to increase P-selectin levels in bronchial vessels. Such a response may signify an early inflammatory response to ozone (Krishna *et al.*, 1997).

VII. NITRIC OXIDE

Decreased production of nitric oxide (NO), increased formation of reactive oxygen species, and increased endothelial expression of redox-sensitive adhesion molecules (e.g., ICAM-1 and VCAM-1) in the vessel wall are early and characteristic features of atherosclerosis, suggesting that these phenomenon are interrelated (Khan *et al.*, 1996). The NO· donor, diethylamine-NO, has been shown to downregulate VCAM-1 and ICAM-1 gene expression induced by TNF-α (Biffl *et al.*, 1996; Ikeda *et al.*, 1996; Khan *et al.*, 1996). E-selectin gene expression was not affected under such conditions (Khan *et al.*, 1996). Chronic blockade of NO production in rats by N(G)-nitro-L-arginine methyl ester administration has been shown to increase the expression of ICAM-1 and VCAM-1 (Luvara *et al.*, 1998).

VIII. ANTIOXIDANT REGULATION OF AGONIST-INDUCED CELL ADHESION PROCESS

Several classes of antioxidants (*e.g.*, thiol, phenolic, and flavonoid) have been reported to downregulate inducible cell adhesion molecule expression as well as cell–cell adhesion. The following section briefly describes some of the major categories of antioxidants and their effect on cell adhesion processes.

A. Thiols

Several studies have shown that thiol antioxidants downregulate the cytokine- or oxidant-induced expression of adhesion molecules (Aoki et al., 1996; Ikeda et al., 1994; Kawai et al., 1995; Marui et al., 1993; Moynagh et al., 1994; Offermann et al., 1996; Roy et al., 1998, 1999; Weber et al., 1994). At a concentration as high as 20 mM, N-acetylcysteine (NAC) has been shown to strongly inhibit ICAM-1 expression induced by H_2O_2 or cytokines in keratinocytes, whereas under the same conditions, pyrrolidine dithiocarbamate (PDTC) was less effective in preventing inducible ICAM-1 expression (Ikeda et al., 1994). However, both NAC and PDTC have been reported to modulate the induction of cell adhesion molecule expression by IL-1α in astrocytoma cells (Moynagh et al., 1994). In HUVEC, interleukin-1β-activated VCAM-1 gene expression has been observed to be repressed approximately 90% by the antioxidants PDTC (50 μM) and NAC (30 mM) (Marui et al., 1993). NAC (5–10 mM) and exogenous GSH (10–1000 μg/ml) have been shown to also decrease ICAM-1 expression in hyperoxia-exposed human pulmonary artery endothelial cells and HUVEC (Aoki et al., 1996). In contrast to the effect of NAC at high concentrations, lower concentrations of NAC (0.2 and 1 mM) have been observed to potentiate the IL-1α-induced expression of ICAM-1 and VCAM-1 (Moynagh et al., 1994).

Among thiol and related antioxidants, NAC and PDTC, which have been widely studied for their ability to regulate the expression of molecules and cell–cell adhesion, NAC is a clinically safe drug (Smilkstein et al., 1988). However, in all of the studies showing the efficacy of NAC to inhibit agonist-induced ICAM-1 or VCAM-1 expression, a high millimolar range (5–30 mM) of the drug was necessary. Pharmacokinetic studies of NAC in humans show that only up to 25 μM of NAC is available in human plasma following oral intake (Holdiness, 1991). PDTC, the other thiol antioxidant used widely in these studies, has never been tested for safety in humans.

We investigated the effects of α-lipoate on agonist (phorbol ester and TNF-α)-induced cell adhesion processes in human endothelial cells. α-Lipoate has been used safely for human therapy to treat complications associated with diabetes (Peter and Borbe, 1995; Ziegler et al., 1995). Pharmacokinetic studies of α-lipoate have shown that following a single orally administered dose (10 mg/kg body weight), the plasma concentration may reach up to 60–70 μM. Higher concentrations of α-lipoate can be achieved if the drug is administered intravenously (Peter and Borbe, 1995). At clinically relevant doses (50–100 μmM), lipoate downregulated the agonist-induced adhesion of T cells to endothelial cells, and the agonist induced ICAM-1 and VCAM-1 expression on endothelial cells (Fig. 9) (Roy et al., 1998, 1999).

The studies discussed in this section clearly indicate the presence of a thiol-sensitive mechanism for modulating ICAM-1 and VCAM-1 gene expression and suggest a potential novel therapeutic strategy for diseases characterized by cell adhesion-mediated tissue injury (Nathens et al., 1998).

B. Tocopherol

Low levels of α-tocopherol are related to a higher incidence of cardiovascular disease, and an increased intake of the vitamin appears to afford protection against cardiovascular disease (Devaraj et al., 1996). In addition to decreasing LDL oxidation, α-tocopherol has been suggested to exert intracellular effects on the adhesive properties of cells (e.g., monocytes and platelets) that are crucial in atherogenesis (Devaraj and Jialal, 1998; Devaraj et al., 1996). Pretreatment of HUVEC with α-tocopherol has been shown to downregulate agonist (interleukin-1, thrombin, or PMA)-induced monocytic adhesion to HUVEC (Faruqi et al., 1994). Such inhibition correlated with

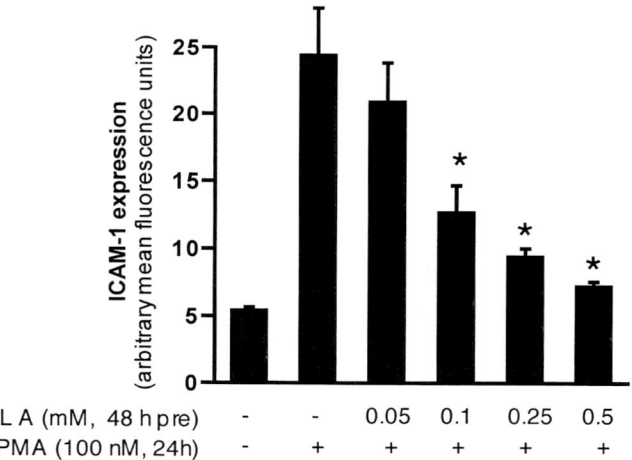

Figure 9 Dose relationship of α-lipoate pretreatment on PMA-induced ICAM-1 expression in human vascular endothelial (ECV) cells. ECV cells were pretreated with α-lipoate (0–500 μM, LA) for 48 hr and then activated with 100 nM PMA for 24 hr. Expression of ICAM-1 on ECV cells was determined using flow cytometry. Data are means ± SD of at least three experiments. *$p < 0.01$ when compared with PMA-induced ICAM-1 expression (Roy et al., 1998).

a decrease in steady-state levels of E-selectin mRNA and cell surface expression of E-selectin (Faruqi et al., 1994). Probucol also showed a similar effect as that of α-tocopherol on cell adhesion (Faruqi et al., 1994). Cotreatment of lipoate and α-tocopherol has been observed to be more effective in downregulating the agonist-induced adhesion of leukocytes to endothelial cells compared to the treatment of endothelial cells with either of the antioxidants alone (Roy et al., 1998). Cytokine-stimulated cell surface expression of VCAM-1 and E-selectin, but not of ICAM-1, has been reported to be inhibited by α-tocopherol succinate (Erl et al., 1997). However, such effects were not exerted either by α-tocopherol or its acetate derivative (Erl et al., 1997).

C. Ascorbate

Cigarette smoke-induced leukocyte adhesion to micro- and macrovascular endothelium and leukocyte-platelet aggregate formation is prevented by dietary or intravenous pretreatment with the water-soluble antioxidant ascorbate (Lehr et al., 1994). Consistently, restoration of reduced plasma vitamin C concentrations in smokers by oral supplementation has been shown to decrease cigarette smoke-induced monocyte adhesion (Weber et al., 1996).

D. Flavonoids

Treatment of human endothelial cells with certain hydroxyflavones and flavanols has been reported to inhibit cytokine-induced ICAM-1, VCAM-1, and E-selectin expression in human endothelial cells (Gerritsen et al., 1995). Apigenin is a potent flavone that has been reported to inhibit adhesion molecule expression in endothelial cells in a dose- and time-dependent manner. An effect of this flavone at the transcriptional level has been demonstrated (Gerritsen et al., 1995). Apigenin is also known to inhibit TNF-α-induced ICAM-1 expression *in vivo* (Panes et al., 1996). Cell adhesion regulatory effects of flavonoids have been consistently evident in other independent studies. The flavonoid delphinidin chloride (CAS 528-53-0, IdB 1056) inhibited acetylcholine- and sodium nitroprusside-induced adherence of leukocytes to the venular endothelium in diabetic hamsters (Bertuglia et al., 1995). 5-Methoxyflavanone, and more potently 5-methoxyflavone, downregulated indomethacin-induced leukocyte adherence to mesenteric venules (Blank et al., 1997). The flavonoid 2-(3-amino-phenyl)-8-methoxychromene-4-one (PD 098063) selectively blocks TNF-α-induced VCAM-1 expression in endothelial cells in a concentration-dependent manner but had no effect on ICAM-1 expression. This selective inhibition of agonist-induced VCAM-1 protein and gene expression by PD 098063 was through a NF-κB-independent mechanism(s) (Wolle et al., 1996). We have observed that quercetin, a potent antioxidant, suppresses agonist-induced ICAM-1 protein and gene expression in human endothelial cells (Kobuchi et al., 1998).

E. Molecular Mechanisms of Antioxidant Regulation

Because of the diverse chemical structure of the antioxidants, it is rather difficult to understand the exact mechanisms and sites of action of antioxidants downregulating agonist-induced cell adhesion (Fig. 10). Previous studies have suggested that activation of the agonist-induced transcription of cell adhesion molecules (e.g., E-selectin, ICAM-1, and VCAM-1) in cells are dependent, at least in part, on the activation of redox-sensitive transcription factors NF-κB and AP-1 (Marui et al., 1993; Voraberger et al., 1991). Among the thiol and related antioxidants studied, PDTC and high concentra-

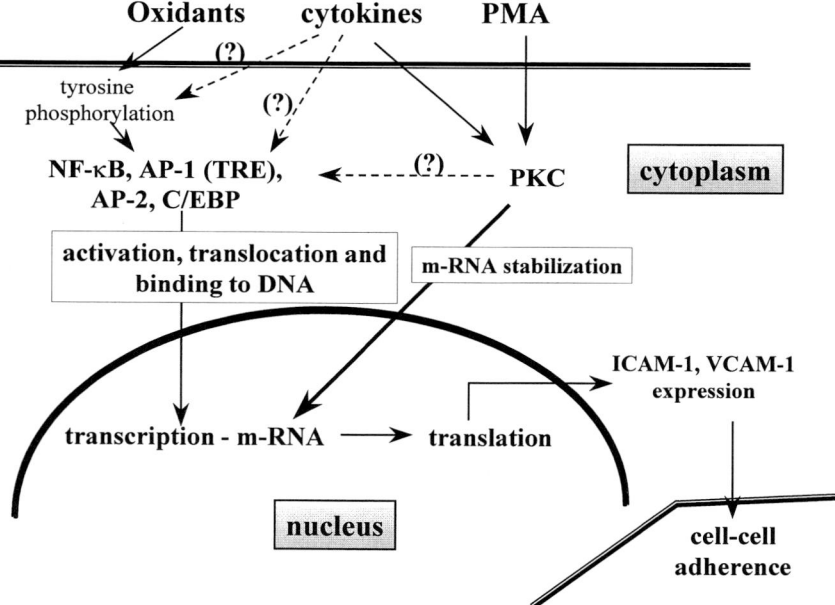

Figure 10 Signal transduction pathways of cell adhesion molecule expression. Several stimuli, such as cytokines, phorbol ester (PMA), and oxidant, induce the expression of cell adhesion molecules, such as ICAM-1 and VCAM-1. The expression of these adhesion molecules is regulated at transcriptional levels by several kinases (e.g., protein kinase C, PKC; and tyrosine kinases) and redox-sensitive transcription factors (e.g., NF-κB, AP-1, and AP-2). Posttranscriptional mechanisms such as stabilization of mRNA are also known to contribute to the induction of cell adhesion molecule expression.

tions (mM) of NAC suppress inducible adhesion molecule expression, at least in some cells, through NF-κB-dependent mechanisms (Marui et al., 1993; Sato et al., 1996; Weber et al., 1994). In astrocytoma cells, PDTC- and NAC-inhibited IL-1α-induced adhesion molecule expression, however, failed to block the IL-1α-mediated activation of NF-κB (Moynagh et al., 1994). The exact regulatory mechanisms of these thiol compounds on NF-κB activation are not yet known. Studies related to the understanding of mechanisms underlying the inhibitory effect of lipoate on inducible cell adhesion molecule expression clearly indicate that the mechanism of action is independent of NF-κB activity (Roy et al., 1998).

α-Tocopherol is known to inhibit protein kinase C (PKC) activity in some cell lines (Tasinato et al., 1995). PKC plays a major role in the expression of cell adhesion (Kvale and Holme, 1996). Faruqui et al. (1994) reported that inhibition of inducible cell adhesion by tocopherol was not dependent on the effects of tocopherol on PKC or NF-κB. In contrast, tocopherol-succinate has been observed to prevent monocytic cell adhesion to cytokine-stimulated endothelial cells by inhibiting the activation of NF-κB (Erl et al., 1997).

In most cases, the mechanisms that mediate the suppressive effect of flavonoids on cell adhesion have been reported not to be dependent on the activation of NF-κB (Gerritsen et al., 1995; Kobuchi et al., 1998). We have observed that the flavonoid quercetin significantly and dose dependently downregulates agonist-induced AP-1 activation in human endothelial cells. Such a decrease in inducible AP-1 activation was observed to be associated with the inhibitory effects of quercetin on the c-Jun NH$_2$-terminal kinase (JNK) pathway, but not on the extracellular signal-regulated protein kinase (ERK) pathway (Kobuchi et al., 1998).

IX. CELL ADHESION PROCESSES AND PATHOLOGIES WITH REDOX IMBALANCES

The expression of cell adhesion molecules and the adhesive properties of cells are modified greatly in several diseased conditions such as cancer, atherosclerosis, diabetes, chronic inflammation, and ischemia–reperfusion injury (Cutolo et al., 1993; Dosquet et al., 1992; Kukielka et al., 1994; Munro, 1993; Pantel et al., 1995; Thiery et al., 1996). Redox imbalances have been suggested to play a critical role in the etiology of such diseases. In the disorders mentioned, altered cell adhesion properties and redox imbalances occur simultaneously, suggesting that these two conditions may be interrelated. A brief review of the disease conditions where altered cell adhesion properties and redox imbalances have been observed simultaneously is presented in the following section.

A. Cancer and Metastasis

Most of the adhesion receptor families reported so far, including integrins, cadherins, selectins, immunoglobulins, and proteoglycans, have been implicated in various stages of tumor progression and metastasis. Integrins αIIb β3 and αL β2 and immunoglobulin family members PECAM-1, ICAM-1, and N-CAM in solid tumor cells are some examples of such adhesion molecules (Tang and Honn, 1994). Changes in the adhesive properties of solid tumour cells, such as downregulation of desmosomal proteins (e.g., plakoglobin) and neo-expression of ICAM-1 or MUC18, have also been suggested to be important determinants of the metastatic capability of individual malignant cells (Pantel *et al.*, 1995). The importance of adhesion molecules in tumor metastasis is also evidenced by their involvement in other important parameters of metastasis such as angiogenesis (Tang and Honn, 1994). An increase in oxidative stress markers such as oxidized DNA and compromised antioxidant have been reported in several types of cancer (Montagnier *et al.*, 1998). An elevated antioxidant status is also associated with decreased incidences of cancer (Montagnier *et al.*, 1998).

B. Ischemia–Reperfusion Injury

Injury associated with ischemia and reperfusion is a significant factor in a number of clinical diseases. Reactive oxygen species generated during the reperfusion of ischemic tissues is known to directly cause injury (Serrano *et al.*, 1996). Reperfusion of ischemic tissue is also associated with an acute inflammatory response that may exacerbate vascular and tissue damage further. Compelling evidence from a variety of animal models indicates that neutrophils are the principal effector cells of reperfusion injury and that the blockade of neutrophil adhesion to endothelium attenuates ischemia–reperfusion injury (Thiagarajan *et al.*, 1997). Monoclonal antibodies directed to CD18, P-selectin, and L-selectin have been shown to be effective in reducing ischemia–reperfusion injury to the rabbit ear and in reducing injury following hemorrhagic shock in both rabbits and nonhuman primates (Winn *et al.*, 1997, 1998). Antioxidants have been shown to suppress leukocyte adhesion and CD18 expression following ischemia–reperfusion, resulting in a marked suppression of leukocyte-mediated injury in the postischemic heart (Serrano *et al.*, 1996).

C. Diabetes

Abnormalities in the regulation of reactive oxygen species and transition metal metabolism are postulated to result in diabetes as well as its longer term complications (Baynes, 1991; Wolff, 1993). Diabetes is associated with an increased risk of premature vascular disease. Vascular research is currently focusing on a potential role of adhesion molecules in diabetes mellitus, as classical risk factors, including hyperlipidemia and hypertension, do not completely account for the increased incidence of atherosclerosis in diabetes. After the expression of adhesion molecules on the cell surface, they are shed into plasma (Roep *et al.*, 1994). Soluble adhesion molecules are detectable at low levels in healthy people but are increased in various disorders. Elevated concentrations of circulating ICAM-1 and L-selectin have been found in subjects at risk of developing insulin-dependent diabetes mellitus (IDDM) compared with recent-onset IDDM patients and healthy controls (Fasching *et al.*, 1996; Jilma *et al.*, 1996; Lampeter *et al.*, 1992; Roep *et al.*, 1994; Wagner and Jilma, 1997). Recombinant ICAM-1 has been shown to block T-cell proliferation in response to an islet cell autoantigen, suggesting that agents with inhibitory activity against inducible cell adhesion molecule expression may provide novel means to intervene in the pathogenesis of diabetes (Roep *et al.*, 1994).

D. Atherosclerosis

Invasion of the artery wall by leukocytes, particularly monocytes and T lymphocytes, is one of the earliest steps in the development of atherosclerotic lesion. Expression and activity of adhesion molecules are enhanced by inflammatory mediators released from leukocytes and endothelium; these mediators, in turn, are induced by high serum cholesterol levels or complement fragments (Campbell and Campbell, 1997; Yokota and Hansson, 1995). Fluid shear stress and circumferential stretch play important roles in maintaining the homeostasis of the blood vessel and can also contribute as pathophysiological factors in atherosclerosis. Shear stress has been reported to increase the production of reactive oxygen species, the activity of a number of kinases (e.g., focal adhesion kinases), and the expression of adhesion molecules such as ICAM-1 (Chien *et al.*, 1998).

E. Rheumatoid Arthritis

Rheumatoid arthritis is a systemic autoimmune disease and is characterized by chronic joint inflammation and infiltration of cells from the blood, especially activated T cells and macrophages, together with the formation of new blood vessels. Rheumatoid arthritis affects approximately 1% of the adult population in a female/male ratio ranging from 2:1 to 4:1 (Grossman and Brahn, 1997). Production of free radicals at inflammation sites has been suggested to play a significant role in the pathogenesis of rheumatoid arthritis (Araujo et al., 1998; Mapp et al., 1995; Maurice et al., 1998). Adhesion molecules facilitate the migration of cells to the joint as well as the attachment of synovium to bone and cartilage (Cunnane et al., 1998). Monocyte and lymphocyte traffic into the rheumatoid arthritis synovium is mediated by adhesion molecules such as ELAM-1, VCAM-1, ICAM-1, and ICAM-2, as well as by monocyte chemotactic protein 1 and $\beta 2$ integrins (CD11 a,b,c/CD18) (Cutolo et al., 1993; Veale and Maple, 1996). Anti-ICAM-1 therapy has shown beneficial effects in the treatment of rheumatoid arthritis (Rothlein and Jaeger, 1995).

F. Inflammatory Skin Disorders

During cutaneous inflammatory reactions, the recruitment of circulating leukocytes into the tissue critically depends on the regulated expression of endothelial cell adhesion molecules (Heckmann et al., 1998). Human epidermal keratinocytes have been observed to markedly express ICAM-1 during the course of inflammatory skin diseases (Ikeda et al., 1994). Manipulation of the intracellular redox state has been shown to modulate skin inflammation (Jones et al., 1997). There is accumulating evidence that the expression of certain adhesion molecules has important consequences for understanding patterns of cell movement in normal and pathologically altered skin (Walsh and Murphy, 1992). Agents that inhibit adhesion processes have been suggested to be useful in cutaneous inflammation and multistage carcinogenesis (Oberyszyn et al., 1998).

G. Infectious Diseases

The common response to infection is infiltration of the affected tissue by inflammatory cells. In terms of global health, two important infectious diseases where reactive oxygen species and cell adhesion molecules play important role are (i) protozoan parasites of the genus *Plasmodium,* particularly *Plasmodium falciparum* (tropical malaria), and (ii) human immunodeficiency virus (HIV).

Systemic endothelial activation, represented by the upregulation of inducible cell adhesion molecules on endothelium, and increased levels of soluble adhesion molecules have been observed in both severe and uncomplicated malaria when compared with uninfected controls (McCormick et al., 1997; Turner et al., 1998). A strong association between the expression of adhesion molecules (ICAM-1, LFA-1, and Mac-1) and injury in murine kidneys infected acutely with the fatal malaria parasite *Plasmodium berghei* ANKA has been observed (Rui-Mei et al., 1998). Acute malaria is associated with oxidative stress, mediated by redox-active iron released from the infected erythrocytes (Golenser et al., 1998).

The HIV-infected population is known to be oxidatively stressed and deficient in antioxidant micronutrients (Allard et al., 1998a,b; Montagnier et al., 1998). HIV-induced apoptosis of T cells has been shown to be preceded by an exponential increase in reactive oxygen species produced in mitochondria (Banki et al., 1998). Antioxidants have been suggested to play a role in the treatment of viral diseases (Peterhans, 1997). Monocytes are the major targets of HIV infection in AIDS patients. *In vitro* infection of monocytes with HIV has been shown to be associated with an increased expression of adhesion molecules, which increases both monocyte aggregation and monocyte/endothelial adhesion. The extracellular HIV-tat protein has been reported to activate monocytes, causing them to extravasate into underlying tissues and ultimately contribute to tissue damage as seen during the progression of AIDS (Lafrenie et al., 1996a,b).

REFERENCES

Albelda, S. M., Muller, W. A., Muller, C. A., Buck, C. A., and Newman, P. J. (1991). Molecular and cellular properties of PECAM-1 (endoCAM/ CD31): A novel vascular cell-cell adhesion molecule. *J. Biol. Chem.* **114**, 1059–1068.

Albelda, S. M., Smith, C. W., and Ward, P. A. (1994). Adhesion molecules and inflammatory injury. *FASEB J.* **8**, 504–512.

Allard, J. P., Aghdassi E., Chau J., Salit I., and Walmsley, S. (1998a). Oxidative stress and plasma antioxidant micronutrients in humans with HIV infection. *Am. J. Clin. Nutr.* **67**, 143–147.

Allard, J. P., Aghdassi E., Chau J., Tam, C., Kovacs, C. M., Salit, I. E., and Walmsley, S. L. (1998b). Effects of vitamin E and C supplementation on oxidative stress and viral load in HIV-infected subjects. *AIDS* **12**, 1653–1659.

Aoki, T., Suzuki, Y., Suzuki, K., Miyata, A., Oyamada, Y., Takasugi, T., Mori, M., Fujita, H., and Yamaguchi, K. (1996). Modulation of ICAM-1 expression by extracellular glutathione in hyperoxia-exposed human pulmonary artery endothelial cells. *Am. J. Respir. Cell. Mol. Biol.* **15**, 319–327.

Araujo, V., Arnal, C., Boronat, M., Ruiz, E., and Dominguez, C. (1998). Oxidant–antioxidant imbalance in blood of children with juvenile rheumatoid arthritis. *BioFactors* **8**, 155–159.

Baeuml, H., Behrends, U., Peter R. U., Mueller, S., Kammerbauer, C., Caughman, S. W., and Degitz, K. (1997). Ionizing radiation induces, via generation of reactive oxygen intermediates, intercellular adhesion molecule-1 (ICAM-1) gene transcription and NF kappa B-like binding activity in the ICAM-1 transcriptional regulatory region. *Free Radical Res.* **27**, 127–142.

Banki, K., Hutter, E., Gonchoroff N. J., and Perl, A. (1998). Molecular ordering in HIV-induced apoptosis. Oxidative stress, activation of caspases, and cell survival are regulated by transaldolase. *J. Biol. Chem.* **273,** 11944–11953.

Bavisotto, L. M., Schwartz, S. M., and Heimark, R. L. (1990). Modulation of Ca2(+)-dependent intercellular adhesion in bovine aortic and human umbilical vein endothelial cells by heparin-binding growth factors. *J. Cell. Physiol.* **143,** 39–51.

Baynes, J. W. (1991). Role of oxidative stress in development of complications in diabetes. *Diabetes* **40,** 405–412.

Berlin, C., Berg, E. L., Briskin, M. J., Andrew, D. P., Kilshaw, P. J., Holzmann, B., Weissman I. L., Hamann, A., and Butcher, E. C. (1993). Alpha 4 beta 7 integrin mediates lymphocyte binding to the mucosal vascular addressin MAdCAM-1. *Cell (Cambridge, Mass.)* **74,** 185–185.

Bertuglia, S., Malandrino, S., and Colantuoni A. (1995). Effects of the natural flavonoid delphinidin on diabetic microangiopathy. *Arzneimitt.-Forsch.* **45,** 481–485.

Bevilacqua, M. P. (1993). Endothelial-leukocyte adhesion molecules. *Annu. Rev. Immunol.* **11,** 767–804.

Bevilacqua, M. P., Pober J. S., Mendrick D. L., Cotran, R. S., and Gimbrone, M. A., Jr. (1987). Identification of an inducible endothelial-leukocyte adhesion molecule. *Proc. Natl. Acad. Sci. U.S.A.* **84,** 9238–9242.

Biffl, W. L., Moore, E. E., Moore, F. A., and Barnett, C. (1996). Nitric oxide reduces endothelial expression of intercellular adhesion molecule (ICAM)-1. *J. Surg. Res.* **63,** 328–332.

Blank, M. A., Ems, B. L., O'Brien, L. M., Weisshaar, P. S., Ares, J. J., Abel, P. W., McCafferty, D. M., and Wallace, J. L. (1997). Flavonoid-induced gastroprotection in rats: Role of blood flow and leukocyte adherence. *Digestion* **58,** 147–154.

Bockholt, S. M., and Burridge, K. (1993). Cell spreading on extracellular matrix proteins induces tyrosine phosphorylation of tensin. *J. Biol. Chem.* **268,** 14565–14567.

Bonfanti, R., Furie, B. C., Furie, B., and Wagner, D. D. (1989). PADGEM (GMP140) is a component of Weibel-Palade bodies of human endothelial cells. *Blood* **73,** 1109–1112.

Burridge, K., Turner, C. E., and Romer, L. H. (1992). Tyrosine phosphorylation of paxillin and pp125FAK accompanies cell adhesion to extracellular matrix: A role in cytoskeletal assembly. *J. Cell Biol.* **119,** 893–903.

Campbell, J. H., and Campbell, G. R. (1997). The cell biology of atherosclerosis—new developments. *Aust. N. Z. J. Med.* **27,** 497–500.

Chien, S., Li, S., Shyy, Y. J. (1998). Effects of mechanical forces on signal transduction and gene expression in endothelial cells. *Hypertension* **31,** 162–169.

Collins, T., Read, M. A., Neish, A. S., Whitley, M. Z., Thanos, D., and Maniatis, T. (1995). Transcriptional regulation of endothelial cell adhesion molecules: NF-kappa B and cytokine-inducible enhancers. *FASEB J.* **9,** 899–909.

Cremer, H., Lange, R., Christoph, A., Plomann, M., Vopper, G., Roes, J., Brown, R., Baldwin, S., Kraemer, P., Scheff, S., (1994). Inactivation of the N-CAM gene in mice results in size reduction of the olfactory bulb and deficits in spatial learning. *Nature (London)* **367,** 455–459.

Cunnane, G., Hummel, K. M., Muller-Ladner, U., Gay, R. E., and Gay, S. (1998). Mechanism of joint destruction in rheumatoid arthritis. *Arch Immunol. Ther. Exp.* **46,** 1–7.

Cutolo, M., Sulli, A., Barone, A., Seriolo, B., and Accardo, S. (1993). Macrophages, synovial tissue and rheumatoid arthritis. *Clin. Exp. Rheumatol.* **11,** 331–339.

DeForge, L. E., Fantone, J. C., Kenney, J. S., and Remick, D. G. (1992). Oxygen radical scavengers selectively inhibit interleukin 8 production in human whole blood. *J. Clin. Invest.* **90,** 2123–2129.

DeForge, L. E., Preston, A. M., Takeuchi, E., Kenney, J., Boxer, L. A., and Remick, D. G. (1993). Regulation of interleukin 8 gene expression by oxidant stress. *J. Biol. Chem.* **268,** 25568–25576.

Dejana, E., Breviario, F., and Caveda, L. (1994). Leukocyte-endothelial cell adhesive receptors. *Clin. Exp. Rheumatol.* **12** (Suppl. 10), S25–S28.

Devaraj, S., and Jialal, I. (1998). The effects of alpha-tocopherol on critical cells in atherogenesis. *Curr. Opin. Lipidol.* **9,** 11–15.

Devaraj, S., Li, D., and Jialal, I. (1996). The effects of alpha tocopherol supplementation on monocyte function. Decreased lipid oxidation, interleukin 1 beta secretion, and monocyte adhesion to endothelium. *J. Clin. Invest.* **98,** 756–763.

Dosquet, C., Weill, D., and Wautier, J. L. (1992). Molecular mechanism of blood monocyte adhesion to vascular endothelial cells. *Nouv. Rev. Fr. Hematol.* **34**(Suppl.), S55–S59.

Erbe, D. V., Watson, S. R., Presta, L. G., Wolitzky, B. A., Foxall, C., Brandley, B. K., and Lasky, L. A. (1993). P- and E-selectin use common sites for carbohydrate ligand recognition and cell adhesion. *J. Cell Biol.* **120,** 1227–1235.

Erl, W., Weber, C., Wardemann, C., and Weber, P. C. (1997). alpha-Tocopheryl succinate inhibits monocytic cell adhesion to endothelial cells by suppressing NF-kappa B mobilization. *Am. J. Physiol.* **273,** H634–H640.

Faruqi, R., de la Motte, C., and DiCorleto, P. E. (1994). Alpha-tocopherol inhibits agonist-induced monocytic cell adhesion to cultured human endothelial cells. *J. Clin. Invest.* **94,** 592–600.

Fasching, P., Waldhausl, W., and Wagner, O. F. (1996). Elevated circulating adhesion molecules in NIDDM—potential mediators in diabetic macroangiopathy. *Diabetologia* **39,** 1242–1244.

Frenette, P. S., and Wagner, D. D. (1996). Adhesion molecules. Part I. *N. Engl. J. Med.* **334,** 1526–1529.

Gahmberg, C. G. (1997). Leukocyte adhesion: CD11/CD18 integrins and intercellular adhesion molecules. *Curr. Opin. Cell Biol.* **9,** 643–650.

Gerritsen, M. E., Carley, W. W., Ranges, G. E., Shen, C. P., Phan, S. A., Ligon, G. F., and Perry, C. A. (1995). Flavonoids inhibit cytokine-induced endothelial cell adhesion protein gene expression. *Am. J. Pathol.* **147,** 278–292.

Golenser, J., Peled-Kamar, M., Schwartz, E., Friedman, I., Groner, Y., and Pollack, Y. (1998). Transgenic mice with elevated level of CuZnSOD are highly susceptible to malaria infection. *Free Radical Biol. Med.* **24,** 1504–1510.

Grether-Beck, S., Olaizola-Horn, S., Schmitt, H., Grewe, M., Jahnke, A., Johnson, J. P., Briviba, K., Sies, H., and Krutmann, J. (1996). Activation of transcription factor AP-2 mediates UVA radiation- and singlet oxygen-induced expression of the human intercellular adhesion molecule 1 gene. *Proc. Natl. Acad. Sci. U.S.A.* **93,** 14586–14591.

Grossman, J. M., and Brahn, E. (1997). Rheumatoid arthritis: Current clinical and research directions. *J. Women's Health* **6,** 627–638.

Gutteridge, J. M. (1984). Reactivity of hydroxyl and hydroxyl-like radicals discriminated by release of thiobarbituric acid-reactive material from deoxy sugars, nucleosides and benzoate. *Biochem. J.* **224,** 761–767.

Hallahan, D. E., and Virudachalam, S. (1997). Intercellular adhesion molecule 1 knockout abrogates radiation induced pulmonary inflammation. *Proc. Natl. Acad. Sci. U.S.A.* **94,** 6432–6437.

Hareyama, M., Imai, K., Oouchi, A., Takahashi, H., Hinoda, Y., Tsujisaki M., Adachi, M., Shonai, T., Sakata, K., and Morita, K. (1998). The effect of radiation on the expression of intercellular adhesion molecule-1 of human adenocarcinoma cells. *Int. J. Radiat. Oncol. Biol. Phys.* **40,** 691–696.

Heckmann, M., Douwes, K., Peter, R., and Degitz, K. (1998). Vascular activation of adhesion molecule mRNA and cell surface expression by ionizing radiation. *Exp. Cell Res.* **238**, 148–154.

Holdiness, M. R. (1991). Clinical pharmacokinetics of N-acetylcysteine. *Clin. Pharmacokinet* **20**, 123–134.

Ikeda, M., Schroeder, K. K., Mosher, L. B., Woods, C. W., and Akeson, A. L. (1994). Suppressive effect of antioxidants on intercellular adhesion molecule-1 (ICAM-1) expression in human epidermal keratinocytes. *J. Invest. Dermatol.* **103**, 791–796.

Ikeda, M., Ikeda, U., Takahashi, M., Shimada, K., Minota, S., and Kano, S. (1996). Nitric oxide inhibits intracellular adhesion molecule-1 expression in rat mesangial cells. *J. Am. Soc. Nephrol.* **7**, 2213–2218.

Jilma, B., Fasching, P., Ruthner, C., Rumplmayr, A., Ruzicka, S., Kapiotis, S., Wagner, O. F., and Eichler, H. G. (1996). Elevated circulating P-selectin in insulin dependent diabetes mellitus. *Thromb. Haemostasis* **76**, 328–332.

Jones, J. J., McGilvray, I. D., Nathens, A. B., Bitar, R., and Rotstein, O. D. (1997). Glutathione depletion prevents lipopolysaccharide-induced local skin inflammation. *Arch. Surg. (Chicago)* **132**, 1165–1169; discussion: p. 1170.

Kalra, V. K., Shen, Y., Sultana, C., and Rattan, V. (1996). Hypoxia induces PECAM-1 phosphorylation and transendothelial migration of monocytes. *Am. J. Physiol.* **271**, H2025–H2034.

Kawai, M., Nishikomori, R., Jung, E. Y., Tai, G., Yamanaka, C., Mayumi, M., and Heike, T. (1995). Pyrrolidine dithiocarbamate inhibits intercellular adhesion molecule-1 biosynthesis induced by cytokines in human fibroblasts. *J. Immunol.* **154**, 2333–2341.

Kennedy, T., Ghio, A. J., Reed, W., Samet, J., Zagorski, J., Quay J., Carter, J., Dailey, L., Hoidal, J. R., and Devlin, R. B. (1998). Copper-dependent inflammation and nuclear factor-kappaB activation by particulate air pollution. *Am. J. Respir. Cell Mol. Biol.* **19**, 366–378.

Kevil, C. G., Payne, D. K., Mire, E., and Alexander, J. S. (1998). Vascular permeability factor/vascular endothelial cell growth factor- mediated permeability occurs through disorganization of endothelial junctional proteins. *J. Biol. Chem.* **273**, 15099–15103.

Khan, B. V., Harrison, D. G., Olbrych, M. T., Alexander, R. W., and Medford, R. M. (1996). Nitric oxide regulates vascular cell adhesion molecule 1 gene expression and redox-sensitive transcriptional events in human vascular endothelial cells. *Proc. Natl. Acad. Sci. U.S.A.* **93**, 9114–9119.

Kobuchi, H., Roy, S., Sen, C. K., and Packer, L. (1999). Quercetin inhibits inducible ICAM-1 expression in human endothelial cells through the c-Jun NH2-terminal kinase pathway. *Am. J. Physiol.* **277** (*Cell Physiol.* **46**):C (in press).

Kornberg, L., Earp, H. S., Parsons, J. T., Schaller, M., and Juliano, R. L. (1992). Cell adhesion or integrin clustering increases phosphorylation of a focal adhesion-associated tyrosine kinase. *J. Biol. Chem.* **267**, 23439–23442.

Krishna, M. T., Blomberg, A., Biscione, G. L., Kelly, F., Sandstrom, T., Frew, A., and Holgate, S. (1997). Short-term ozone exposure upregulates P-selectin in normal human airways. *Am. J. Respir. Crit. Care Med.* **155**, 1798–1803.

Krutmann, J., and Grewe, M. (1995). Involvement of cytokines, DNA damage, and reactive oxygen intermediates in ultraviolet radiation-induced modulation of intercellular adhesion molecule-1 expression. *J. Invest. Dermatol.* **105**, 67S–70S.

Krutmann, J., and Trefzer, U. (1992). Modulation of the expression of intercellular adhesion molecule-1 (ICAM-1) in human keratinocytes by ultraviolet (UV) radiation. *Springer Semin. Immunopathol.* **13**, 333–344.

Kukielka, G. L., Youker, K. A., Hawkins, H. K., Perrard, J. L., Michael, L. H., Ballantyne, C. M., Smith, C. W., and Entman, M. L. (1994). Regulation of ICAM-1 and IL-6 in myocardial ischemia: Effect of reperfusion. *Ann. N.Y. Acad. Sci.* **723,** 258–270.

Kvale, D., and Holme, R. (1996). Intercellular adhesion molecule-1 (ICAM-1; CD54) expression in human hepatocytic cells depends on protein kinase C. *J. Hepatol.* **25,** 670–676.

Lacerda, H. M., Lax, A. J., and Rozengurt, E. (1996). Pasteurella multocida toxin, a potent intracellularly acting mitogen, induces p125FAK and paxillin tyrosine phosphorylation, actin stress fiber formation, and focal contact assembly in Swiss 3T3 cells. *J. Biol. Chem.* **271,** 439–445.

Lafrenie, R. M., Wahl, L. M., Epstein, J. S., Hewlett, I. K., Yamada, K. M., and Dhawan, S. (1996a). HIV-1-Tat modulates the function of monocytes and alters their interactions with microvessel endothelial cells. A mechanism of HIV pathogenesis. *J. Immunol.* **156,** 1638–1645.

Lafrenie, R. M., Wahl, L. M., Epstein, J. S., Hewlett, I. K., Yamada, K. M., and Dhawan, S. (1996b). HIV-1-Tat protein promotes chemotaxis and invasive behavior by monocytes. *J. Immunol.* **157,** 974–977.

Lampeter, E. R., Kishimoto, T. K., Rothlein, R., Mainolfi, E. A., Bertrams, J., Kolb, H., and Martin, S. (1992). Elevated levels of circulating adhesion molecules in IDDM patients and in subjects at risk for IDDM. *Diabetes* **41,** 1668–1671.

Lasky, L. A. (1992). Selectins: Interpreters of cell-specific carbohydrate information during inflammation. *Science* **258,** 964–969.

Lasky, L. A. (1995). Selectin-carbohydrate interactions and the initiation of the inflammatory response. *Annu. Rev. Biochem.* **64,** 113–139.

Lehr, H. A., Frei, B., and Arfors, K. E. (1994). Vitamin C prevents cigarette smoke-induced leukocyte aggregation and adhesion to endothelium in vivo. *Proc. Natl. Acad. Sci. U.S.A.* **91,** 7688–7692.

Luvara, G., Pueyo, M. E., Philippe, M., Mandet, C., Savoie, F., Henrion, D., and Michel, J. B. (1998). Chronic blockade of NO synthase activity induces a proinflammatory phenotype in the arterial wall: Prevention by angiotensin II antagonism. *Arterioscler. Thromb. Vasc. Biol.* **18,** 1408–1416.

Mapp, P. I., Grootveld, M. C., and Blake, D. R. (1995). Hypoxia, oxidative stress and rheumatoid arthritis. *Br. Med. Bull.* **51,** 419–436.

Martinotti, S., Toniato, E., Colagrande, A., Alesse, E., Alleva, C., Screpanti, I., Morrone, S., Scarpa, S., Frati, L., Hayday, A. C., Piovella, F., Gulino, A. (1995). Heavy-metal modulation of the human intercellular adhesion molecule (ICAM-1) gene expression. *Biochim. Biophys. Acta.* **1261,** 107–114.

Marui, N., Offermann, M. K., Swerlick, R., Kunsch, C., Rosen, C. A., Ahmad, M., Alexander, R. W., and Medford, R. M. (1993). Vascular cell adhesion molecule-1 (VCAM-1) gene transcription and expression are regulated through an antioxidant-sensitive mechanism in human vascular endothelial cells. *J. Clin. Invest.* **92,** 1866–1874.

Maurice, M. M., van der Voort, E., van Vilet, A. I., Tak, P. P., Breedveld, F. C., and Verweij, C. L. (1998). The rheumatoid joint: Redox-paradox? *In* "Oxidative Stress in Cancer, AIDS and Neurodegenerative Diseases" (L. Montagnier, R. Olivier and C. Pasquier, eds.), p. 517. Dekker, New York.

McCormick, C. J., Craig, A., Roberts, D., Newbold, C. I., and Berendt, A. R. (1997). Intercellular adhesion molecule-1 and CD36 synergize to mediate adherence of Plasmodium falciparum-infected erythrocytes to cultured human microvascular endothelial cells. *J. Clin. Invest.* **100,** 2521–2529.

Montagnier, L., Olivier, R., and Pasquier, C. (eds.) (1998). Oxidative stress in cancer, AIDS and neurodegenerative diseases. Marcel Dekker Inc., New York, NY.

Moynagh, P. N., Williams, D. C., and O'Neil, L. A. J. (1994). Activation of NF-kappa B and induction of vascular cell adhesion molecule-1 and intracellular adhesion molecule-1 expression in human glial cells by IL-1. Modulation by antioxidants. *J. Immunol.* **153**, 2681–2690.

Munro, J. M. (1993). Endothelial-leukocyte adhesive interactions in inflammatory diseases. *Eur. Heart J.* **14**(Suppl. K), 72–77.

Nathens, A. B., Bitar, R., Watson, R. W., Issekutz, T. B., Marshall, J. C., Dackiw, A. P., and Rotstein, O. D. (1998). Thiol-mediated regulation of ICAM-1 expression in endotoxin-induced acute lung injury. *J. Immunol.* **160**, 2959–2966.

Norris, D. A. (1993). Pathomechanisms of photosensitive lupus erythematosus. *J. Invest. Dermatol.* **100**, 58S–68S.

Oberyszyn, T. M., Conti, C. J., Ross, M. S., Oberyszyn, A. S., Tober, K. L., Rackoff, A. I., and Robertson, F. M. (1998). Beta2 integrin/ICAM-1 adhesion molecule interactions in cutaneous inflammation and tumor promotion. *Carcinogenesis (London)* **19**, 445–455.

Offermann, M. K., Lin, J. C., Mar, E. C., Shaw, R., Yang, J., and Medford, R. M. (1996). Antioxidant-sensitive regulation of inflammatory-response genes in Kaposi's sarcoma cells. *J. Acquired. Immune. Defic. Syndr. Hum. Retrovirol.* **13**, 1–11.

Osborn, L., Hession, C., Tizard, R., Vassallo, C., Luhowskyj, S., Chi-Rosso, G., and Lobb, R. (1989). Direct expression cloning of vascular cell adhesion molecule 1, a cytokine-induced endothelial protein that binds to lymphocytes. *Cell (Cambridge, Mass.)* **59**, 1203–1211.

Panes, J., and Granger, D. N. (1998). Leukocyte-endothelial cell interactions: Molecular mechanisms and implications in gastrointestinal disease. *Gastroenterology* **114**, 1066–1090.

Panes, J., Gerritsen, M. E., Anderson, D. C., Miyasaka, M., and Granger, D. N. (1996). Apigenin inhibits tumor necrosis factor-induced intercellular adhesion molecule-1 upregulation in vivo. *Microcirculation* **3**, 279–286.

Pantel, K., Schlimok, G., Angstwurm, M., Passlick, B., Izbicki, J. R., Johnson, J. P., and Riethmuller, G. (1995). Early metastasis of human solid tumours: Expression of cell adhesion molecules. *Ciba Found. Symp.* **189**, 157–170.

Pasquier, C., Franzini, E., Abedinzadeh, Z., Kaouadji, M. N., and Hakim, J. (1991). Gamma and pulse radiolysis study of pentoxifylline, a methylxanthine. *Int. J. Radiat. Biol.* **60**, 433–447.

Peter, G., and Borbe, H. O. (1995). Absorption of 7,8-^{14}C-rac-α-lipoic acid from *in situ* ligated segments of gastrointestinal tract of the rat. *Arzneim.-Forsch./Drug Res.* **45**, 293–329.

Peterhans, E. (1997). Oxidants and antioxidants in viral diseases: Disease mechanisms and metabolic regulation. *J. Nutr.* **127**, 962S–965S.

Picker, L. J., Warnock, R. A., Burns, A. R., Doerschuk, C. M., Berg, E. L., and Butcher, E. C. (1991). The neutrophil selectin LECAM-1 presents carbohydrate ligands to the vascular selectins ELAM-1 and GMP-140. *Cell (Cambridge, Mass.)* **66**, 921–933; erratum: **67** (6), 1267.

Rankin, S., and Rozengurt, E. (1994). Platelet-derived growth factor modulation of focal adhesion kinase (p125FAK) and paxillin tyrosine phosphorylation in Swiss 3T3 cells. Bell-shaped dose response and cross-talk with bombesin. *J. Biol. Chem.* **269**, 704–710.

Ranscht, B. (1994). Cadherins and catenins: Interactions and functions in embryonic development. *Curr. Opin. Cell Biol.* **6**, 740–746.

Rattan, V., Sultana, C., Shen, Y., and Kalra, V. K. (1997). Oxidant stress-induced transendothelial migration of monocytes is linked to phosphorylation of PECAM-1. *Am. J. Physiol.* **273**, E453–E461.

Roep, B. O., Heidenthal, E., de Vries, R. R., Kolb, H., and Martin, S. (1994). Soluble forms of intercellular adhesion molecule-1 in insulin-dependent diabetes mellitus. *Lancet* **343**, 1590–1593.

Rothlein, R., and Jaeger, J. R. (1995). Treatment of inflammatory diseases with a monoclonal antibody to intercellular adhesion molecule 1. *Ciba Found. Symp.* **189,** 200–208; discussion: pp. 208–211.

Roy, S., Sen, C. K., Kobuchi, H., and Packer, L. (1998). Antioxidant regulation of phorbol ester-induced adhesion of human jurkat T-cells to endothelial cells. *Free Radical Biol. Med.* **25,** 229–241.

Roy, S., Sen, C. K., and Packer, L. (1999). Determination of cell-cell adhesion in response to oxidants and antioxidants. *In* "Methods in Enzymology: Oxidants and Antioxidants" 300, pp. 395–401. Academic Press, San Diego, CA.

Rui-Mei, L., Kara, A. U., and Sinniah, R. (1998). In situ analysis of adhesion molecule expression in kidneys infected with murine malaria. *J. Pathol.* **185,** 219–225.

Sato, M., Miyazaki, T., Nagaya, T., Murata, Y., Ida, N., Maeda, K., and Seo, H. (1996). Antioxidants inhibit tumor necrosis factor-alpha mediated stimulation of interleukin-8, monocyte chemoattractant protein-1, and collagenase expression in cultured human synovial cells. *J. Rheumatol,* **23;** 432–438.

Schaller, M. D., Borgman, C. A., Cobb, B. S., Vines, R. R., Reynolds, A. B., and Parsons, J. T. (1992). pp125fak a structurally distinctive protein-tyrosine kinase associated with focal adhesions. *Proc. Natl. Acad. Sci. U.S.A.* **89,** 5192–5196.

Seckl, M. J., Morii, N., Narumiya, S., and Rozengurt, E. (1995). Guanosine 5'-3-O-(thio)triphosphate stimulates tyrosine phosphorylation of p125FAK and paxillin in permeabilized Swiss 3T3 cells. Role of p21rho. *J. Biol. Chem.* **270,** 6984–6990.

Sellak, H., Franzini, E., Hakim, J., and Pasquier, C. (1994). Reactive oxygen species rapidly increase endothelial ICAM-1 ability to bind neutrophils without detectable upregulation. *Blood* **83,** 2669–2677.

Serrano, C. V., Jr., Mikhail, E. A., Wang, P., Noble, B., Kuppusamy, P., and Zweier, J. L. (1996). Superoxide and hydrogen peroxide induce CD18-mediated adhesion in the post-ischemic heart. *Biochim. Biophys. Acta* **1316,** 191–202.

Seufferlein, T., and Rozengurt, E. (1994a). Lysophasphatidic acid stimulates tyrosine phosphorylation of focal adhesion kinase, paxillin, and p130. *J. Biol. Chem.* **269,** 9345–9351.

Seufferlein, T., and Rozengurt, E. (1994b). Sphingosine induces p125FAK and paxillin tyrosine phosphorylation, actin stress fiber formation, and focal contact assembly in Swiss 3T3 cells. *J. Biol. Chem.* **269,** 27610–27617.

Seufferlein, T., and Rozengurt, E. (1995). Sphingosylphosphorylcholine rapidly induces tyrosine phosphorylation of p125FAK and paxillin, rearrangement of the actin cytoskeleton and focal contact assembly. Requirement of p21rho in the signaling pathway. *J. Biol. Chem.* **270,** 24343–24351.

Shen, Y., Rattan, V., Sultana, C., and Kalra, V. K. (1996). Cigarette smoke condensate-induced adhesion molecule expression and transendothelial migration of monocytes. *Am. J. Physiol.* **270,** H1624–H1633.

Shimizu, Y., Newman, W., Gopal, T. V., Horgan, K. J., Graber, N., Beall, L. D., van Seventer, G. A., and Shaw, S. (1991). Four molecular pathways of T cell adhesion to endothelial cells: roles of LFA-1, VCAM-1, and ELAM-1 and changes in pathway hierarchy under different activation conditions. *J. Cell Biol.* **113,** 1203–1212.

Simmons, D. L. (1995). The role of ICAM expression in immunity and disease. *Cancer Surv.* **24,** 141–155.

Sinnett-Smith, J., Zachary, I., Valverde, A. M., and Rozengurt, E. (1993). Bombesin stimulation of p125 focal adhesion kinase tyrosine phosphorylation. Role of protein kinase C, Ca2+ mobilization, and the actin cytoskeleton. *J. Biol. Chem.* **268,** 14261–14268.

Smilkstein, M. J., Knapp, G. L., Kulig, K. W., and Rumack, B. H. (1988). Efficacy of oral N-acetylcysteine in the treatment of acetaminophen overdose. Analysis of the national multicenter study (1976 to 1985). *N. Engl. J. Med.* **319,** 1557–1562.

Stal, P., Broome, U., Scheynius, A., Befrits, R., and Hultcrantz, R. (1995). Kupffer cell iron overload induces intercellular adhesion molecule-1 expression on hepatocytes in genetic hemochromatosis. *Hepatology* **21**, 1308–1316.

Striz, I., and Costable, U. (1992). The role of integrins in the immune response. *Sarcoidosis* **9**, 88–94.

Sultana, C., Shen, Y., Rattan, V., and Kalra, V. K. (1996). Lipoxygenase metabolites induced expression of adhesion molecules and transendothelial migration of monocyte-like HL-60 cells is linked to protein kinase C activation. *J. Cell. Physiol.* **167**, 477–487.

Sun, J., Williams, J., Yan, H. C., Amin, K. M., Albelda, S. M., and DeLisser, H. M. (1996). Platelet endothelial cell adhesion molecule-1 (PECAM-1) homophilic adhesion is mediated by immunoglobulin-like domains 1 and 2 and depends on the cytoplasmic domain and the level of surface expression. *J. Biol. Chem.* **271**, 18561–18570.

Takeichi, M. (1991). Cadherin cell adhesion receptors as a morphogenetic regulator. *Science* **251**, 1451–1455.

Tang, D. G., and Honn, K. V. (1994). Adhesion molecules and tumor metastasis: An update. *Invasion Metastasis* **14**, 109–122.

Tasinato, A., Boscoboinik, D., Bartoli, G. M., Maroni, P., and Azzi, A. (1995). d-alpha-tocopherol inhibition of vascular smooth muscle cell proliferation occurs at physiological concentrations, correlates with protein kinase C inhibition, and is independent of its antioxidant properties. *Proc. Natl. Acad. Sci. U.S.A.* **92**, 12190–12194.

Tedder, T. F., Steeber, D. A., Chen, A., and Engel, P. (1995). The selectins: Vascular adhesion molecules. *FASEB J.* **9**, 866–873.

Thiagarajan, R. R., Winn, R. K., and Harlan, J. M. (1997). The role of leukocyte and endothelial adhesion molecules in ischemia-reperfusion injury. *Thromb Haemostasis* **78**, 310–314.

Thiery, J., Teupser, D., Walli, A. K., Ivandic, B., Nebendahl, K., Stein, O., Stein, Y., and Seidel, D. (1996). Study of causes underlying the low atherosclerotic response to dietary hypercholesterolemia in a selected strain of rabbits. *Atherosclerosis* **121**, 63–73.

Turner, C. E., Glenney, J. R., Jr., and Burridge, K. (1990). Paxillin: A new vinculin-binding protein present in focal adhesions. *J. Cell Biol.* **111**, 1059–1068.

Turner, G. D., Ly, V. C., Nguyen, T. H., Tran, T. H., Nguyen, H. P., Bethell, D., Wyllie, S., Louwrier, K., Fox, S. B., Gatter, K. C., Day, N. P., White, N. J., and Berendt, A. R. (1998). Systemic endothelial activation occurs in both mild and severe malaria. Correlating dermal microvascular endothelial cell phenotype and soluble cell adhesion molecules with disease severity. *Am. J. Pathol.* **152**, 1477–1487.

Vachula, M., and Van Epps, D. E. (1992). In vitro models of lymphocyte transendothelial migration. *Invasion Metastasis* **12**, 66–81.

Veale, D. J., and Maple, C. (1996). Cell adhesion molecules in rheumatoid arthritis. *Drugs Aging* **9**, 87–92.

Voraberger, G., Schafer, R., and Stratowa, C. (1991). Cloning of the human gene for intracellular adhesion molecule-1 and analysis of its 5'-regulatory region. *J. Immunol.* **147**, 2777–2786.

Wagner, O. F., and Jilma, B. (1997). Putative role of adhesion molecules in metabolic disorders. *Horm. Metab. Res.* **29**, 627–630.

Walsh, L. J., and Murphy, G. F. (1992). Role of adhesion molecules in cutaneous inflammation and neoplasia. *J. Cutaneous Pathol.* **19**, 161–171.

Weber, C., Erl, W., Pietsch, A., Strobel, M., Ziegler-Heitbrock, H. W., and Weber, P. C. (1994). Antioxidants inhibit monocyte adhesion by suppressing nuclear factor-kappa B mobilization and induction of vascular cell adhesion molecule-1 in endothelial cells stimulated to generate radicals. *Arterioscler. Thromb.* **14**, 1665–1673.

Weber, C., Negrescu, E., Erl, W., Pietsch, A., Frankenberger, M., Ziegler-Heitbrock, H. W., Siess, W., and Weber, P. C. (1995). Inhibitors of protein tyrosine kinase suppress TNF-stimulated induction of endothelial cell adhesion molecules. *J. Immunol.* **155,** 445–451.

Weber, C., Erl, W., Weber, K., and Weber, P. C. (1996). Increased adhesiveness of isolated monocytes to endothelium is prevented by vitamin C intake in smokers. *Circulation* **93,** 1488–1492.

Winn, R. K., Ramamoorthy, C., Vedder, N. B., Sharar, S. R., and Harlan, J. M. (1997). Leukocyte-endothelial cell interactions in ischemia-reperfusion injury. *Ann. N.Y. Acad. Sci.* **832,** 311–321.

Winn, R., Vedder, N., Ramamoorthy, C., Sharar, S., and Harlan, J. (1998). Endothelial and leukocyte adhesion molecules in inflammation and disease. *Blood Coagul. Fibrino.* **9** (Suppl. 2), S17–S23.

Wolff, S. P. (1993). Diabetes mellitus and free radicals. Free radicals, transition metals and oxidative stress in the aetiology of diabetes mellitus and complications. *Br. Med. Bull.* **49,** 642–652.

Wolle, J., Hill, R. R., Ferguson, E., Devall, L. J., Trivedi, B. K., Newton, R. S., and Saxena, U. (1996). Selective inhibition of tumor necrosis factor-induced vascular cell adhesion molecule-1 gene expression by a novel flavonoid. Lack of effect on transcription factor NF-kappa B. *Arterioscler. Thromb. Vasc. Biol.* **16,** 1501–1508.

Yokota, T., and Hansson, G. K. (1995). Immunological mechanisms in atherosclerosis. *J. Intern. Med.* **238,** 479–489.

Zachary, I., Sinnett-Smith, J., and Rozengurt, E. (1992). Bombesin, vasopressin, and endothelin stimulation of tyrosine phosphorylation in Swiss 3T3 cells. Identification of a novel tyrosine kinase as a major substrate. *J. Biol. Chem.* **267,** 19031–19034.

Zachary, I., Sinnett-Smith, J., Turner, C. E., and Rozengurt, E. (1993). Bombesin, vasopressin, and endothelin rapidly stimulate tyrosine phosphorylation of the focal adhesion-associated protein paxillin in Swiss 3T3 cells. *J. Biol. Chem.* **268,** 22060–22065.

Ziegler, D., Hanefeld, M., Ruhnau, K. J., Meissner, H. P., Lobisch, M., Schuete, K., and Gries, F. A. (1995). Treatment of symptomatic diabetic peripheral neuropathy with the antioxidant α-lipoic acid. *Diabetologia* **38,** 1425–1433.

13 Role of Thioredoxin and Redox Regulation in Oxidative Stress Response and Signaling

Hiroshi Masutani, Masaya Ueno, Shugo Ueda, and Junji Yodoi
Institute for Virus Research
Kyoto University
Sakyo, Kyoto 606-8397, Japan

Oxidative stress evokes various cellular events, including activation of transcription factors, apoptosis, and cell cycle arrest. Accumulating evidence shows that the reduction/oxidation regulating mechanism (redox regulation) plays an important role in the regulation of apoptosis/necrosis and cell cycle arrest elicited by oxidative stress. Thioredoxin (TRX) and related molecules maintain the cellular reducing environment, working in concert with the glutathione system. Several lines of evidence have indicated the active role of TRX on the regulation of apoptosis and cell cycle as an endogenous reducing molecule. TRX physiologically has cytoprotective effects against oxidative stress and is induced by various oxidative stresses through responsive elements in its promoter. In addition, TRX is translocated from the cytoplasm into the nucleus on oxidative stress, including ultraviolet light irradiation, and interacts physically with Ref-1 (redox factor 1)/APEX, an endoexonuclease located in the nucleus. Several reports showed that TRX and/or Ref-1 enhance the DNA-binding activity of AP-1, polyoma enhancer-binding protein-2 (PEBP2), and other transcription factors. Therefore, TRX nuclear translocation by oxidative stress may enhance the activity of nuclear transcription factors in cooperation with Ref-1. Based on these findings, in addition to a function in scavenging reactive oxygen species (ROS), we suggest that TRX is a regulator/modulator involved in various steps of cellular signaling against oxidative stress.

I. INTRODUCTION

Cells can be exposed to various stresses: oxidative stress, metals, chemicals, heat shock, osmotic stress, and hypoxia. Cells have multiple sensing and signaling mechanisms to each kind of stress to protect cells or individuals, although the detail of each mechanism remains unclear. Among them, the oxidative stress response has been studied intensively. Oxidative stress is largely mediated by ROS. Reactive Oxygen Species are generated by metal ion-catalyzed redox reactions, metabolism of chemicals, and normal physiological activities, including the inflammatory response to infection (Halliwell and Gutteridge, 1984). A series of studies have clarified the existence of a signaling pathway evoked by oxidative stress (Nakamura et al., 1997). Whereas ROS by itself may act as intracellular signaling mediators in physiological signal transduction (Schreck et al., 1991), excessive oxidation causes cellular death (Buttke and Sandstrom, 1994). Endogenous reducing molecules protect cells as antioxidants by scavenging excessive ROS. In addition to this function, evidence has accumulated suggesting that reducing molecules play important roles in cellular signaling through the reduction of cysteine residues of various important components of signal transduction pathways. This chapter discusses the role of reduction/oxidation regulation (redox regulation) in oxidative stress-evoked cellular signaling.

II. THIOREDOXIN AND RELATED MOLECULES

To protect cells from unwanted effects of oxygen, cells have developed defense mechanisms against oxidants. Thiol reduction is one of the mechanisms against oxidative stress. The TRX system, in concert with the glutathione and glutaredoxin system, constitutes a cellular reducing environment. TRX is a small protein with two redox-active cysteine residues in an active center (Cys-Gly-Pro-Cys-) and operates together with NADPH and TRX reductase as an efficient reducing system of exposed protein disulfides. TRX is present in many different prokaryotes and eukaryotes and appears to be present in all living cells (Holmgren, 1985; Holmgren and Björnstedt, 1995). We identified and cloned an adult T-cell leukemia (ATL)-derived factor (ADF) from the HTLV-I positive cell line ATL-2 that was found to be a human homolog of TRX (Tagaya et al., 1989; Wollman et al., 1988). Several cytokine-like factors, such as 3B6-IL-1, eosinophil cytotoxicity enhancing factor (ECEF), T-cell hybridoma MP6-derived B-cell growth factor, and early pregnancy factor, are identical or related to TRX, indicating that TRX plays a multifunctional role (Yodoi and Uchiyama, 1992). In the past years, important findings have been added to show that TRX participates in various cellular events. New members of TRX-related molecules in the mammalian system have also been newly identified. Thioredoxin reductase has a selenium-containing active center in the C terminus and several isoforms of thioredoxin reductase exist (Liu and Stadtman, 1997). Mammalian thioredoxin 2 (TRX2) has been cloned. TRX2 has high homology with TRX and has an active site C-G-P-C with thiol-reducing activity (Spyrou et al., 1997). It has a mitochondrial insertion signal and is specifically localized in mitochondria. TRX2 may be related to the protection of oxidative stress in mitochondria. Another family member of thioredoxin, namely nucleoredoxin, which is localized in the nucleus, has been identified. It also has thiol-reducing activity and may play a role in the regulation of transcription factors in the nucleus (Kurooka et al., 1997). Thioredoxin-dependent peroxidases (peroxiredoxin) are considered to be members of a family for intracellular hydrogen peroxide (Jin et al., 1997). Thus, the TRX system is composed of several related molecules forming a network of recognition and interaction with its active site cysteine residues.

III. THIOREDOXIN INDUCTION BY OXIDATIVE STRESS

Thioredoxin expression is induced by a variety of stresses, including virus infection, mitogens, phorbol myristate acetate (PMA), X-ray and ultraviolet irradiation, hydrogen peroxide, and ischemic reperfusion (Sachi et al., 1995). In the 5'-upstream sequence of the human TRX gene, there are putative binding sites for NF-κB, AP-1, CREB, and SP-1. We analyzed the 5'-upstream region of the TRX gene and newly identified an oxidative responsive element that is inducible by various oxidative agents. Using an electrophoretic mobility shift assay (EMSA), we identified in nuclear extracts from Jurkat cells DNA-binding activities specific to the sequences from -953 to -930. The binding was enhanced by the treatment of cells with hydrogen peroxide (Taniguchi et al., 1996). Isolation of the binding proteins to this sequence, as well as the further elucidation of the mechanism of the transduction of signals to this element, is underway. In the erythroleukemia cell line K562, TRX has been reported to be induced transcriptionally by hemin (Leppa et al., 1997). Although the responsive element for hemin has not yet been elucidated, it seems different from the oxidative responsive element described previously (our unpublished observation). In the TRX

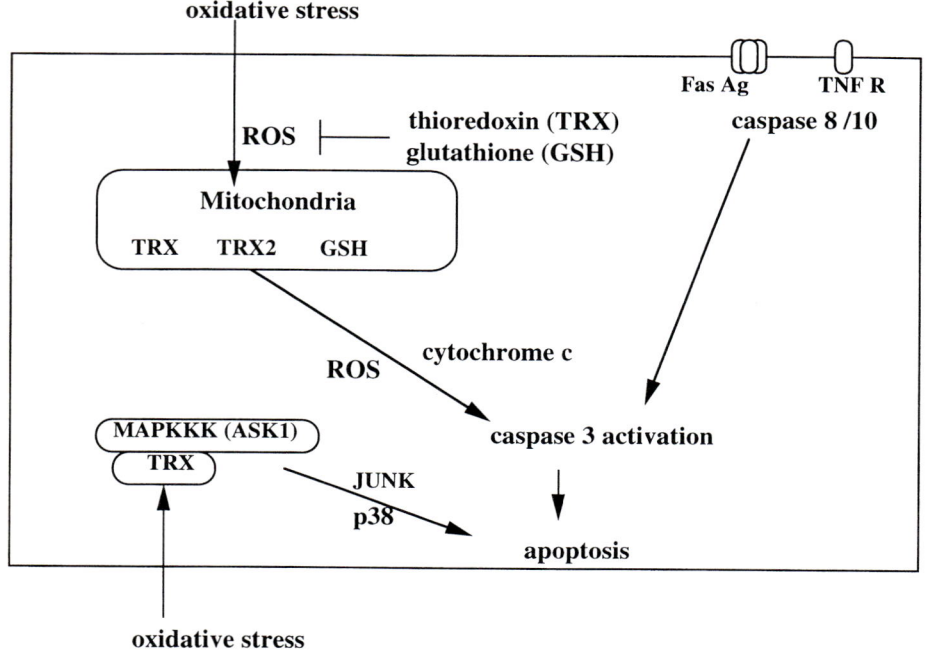

Figure 1 Redox regulation of apoptosis.

VI. REDOX REGULATION OF p53

Another important aspect of oxidative stress signaling is regulation via p53. p53 is induced rapidly by oxidative stress, including ultraviolet or ionic irradiation. p53 is thought to be a gatekeeper against DNA damage (Levine, 1997). One function of p53 is to activate the p21/WAF1 gene to arrest the cell cycle. p53 induces G1 arrest to give cells time to repair damaged DNA (Macleod et al., 1995). The function of p53 is dependent on its DNA-binding activity to its specific binding sequence in target genes. p53 has several functionally important cysteine residues (Cho et al., 1994). Mutagenesis of cysteine residues, especially around the metal center, abolished its sequence-specific DNA-binding activity (Rainwater et al., 1995). The DNA-binding activity of p53 is augmented by reducing agents and is diminished by oxidizing agents *in vitro* (Hainaut and Milner, 1993; Parks et al., 1997). Thus, the DNA-binding activity of p53 can be regulated by the redox state. Ref-1, which was reported to be a redox regulator of AP-1 (Demple et al., 1991; Seki et al., 1991; Xanthoudakis and Curran, 1992; Xanthoudakis et *al.*, 1992), augmented p53 function (Jayaraman *et al.*, 1997). We also showed that the p53 function was regulated by TRX and/or Ref-1 (Ueno *et al.*, 1999). In yeast, deletion of thioredoxin reductase gene inhibited p53-dependent reporter gene expression (Pearson and Merill, 1998). Therefore, a redox-regulating mechanism plays a role in the p53 transcriptional function and seems to be involved in the cell cycle regulation via p53 function (Fig. 2).

VII. NUCLEAR TRANSLOCATION OF THIOREDOXIN AND GENE REGULATION

Transcription factors are important sensing and signaling components of oxidative signaling. In bacteria, the OxyR transcription factor is activated through the formation of a disulfide bond and is deactivated by enzymatic reduction with glutaredoxin, the expression of which is regulated by OxyR (Storz *et al.*, 1990; Zheng *et al.*, 1998). Thus, a transcription factor bound stably to DNA serves as a redox sensor of oxidative stress. In higher organisms, oxidative stress signaling is more complex. In yeast, yAP-1, an AP-1 like transcription factor, and Skn7 cooperate on the yeast TRX2 promoter to induce transcription in response to oxidative stress (Kuge and Jones, 1994; Morgan *et al.*, 1997). yAP-1 is translocated to the nucleus on oxidative stress, and its cysteine residues are essential for translocation, indicating that the redox state of the cysteine residues is important in sensing the oxidative signal and yAP-1 activity (Kuge *et al.*, 1997).

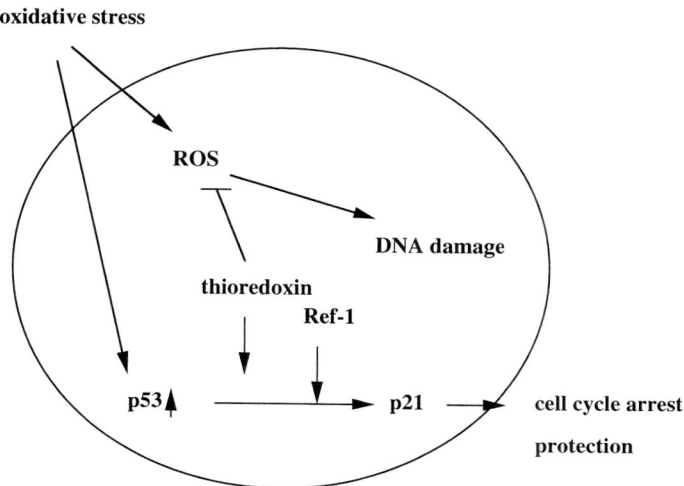

Figure 2   Cytoprotective action of TRX via p53 function.

Several transcription factors, including AP-1, NF-κB, c-Myb, Ets, p53, and zinc finger proteins, have been shown to be modulated by the cellular redox state (Abate *et al.*, 1990; Yodoi and Uchiyama, 1992; Nakamura *et al.*, 1997). There are other reports about the regulation of transcription factors by reducing agents. The interaction between dioxin receptor and xenobiotic responsive element (XRE) is modulated by TRX (Ireland *et al.*, 1995). Related bHLH transcription factors may be a target for redox regulation. Indeed, activation of the hypoxia-inducible transcription factor depends the redox-sensitive stabilization of its α subunit (Huang *et al.*, 1996). Nuclear receptors such as glucocorticoid receptors or estrogen receptors have also been shown to be activated by TRX (Makino *et al.*, 1996; Hayashi *et al.*, 1997). We showed previously that the *in vitro* DNA-binding activity of PEBP2 is enhanced by the addition of either TRX or Ref-1 (Akamatsu *et al.*, 1997). PEBP2/CBF belongs to a family (Runt family) of heterodimeric transcription factors. The Runt domain contains two conserved Cys residues. Site-directed mutagenesis of the Cys-115 and Cys-124 residues in the Runt domain from PEBP2aA revealed that the two Cys residues both contribute to the redox regulation of the Runt domain: Cys-115 is only moderately redox sensitive but is more critically involved in DNA binding than Cys-124, whereas the latter serves as a hypersensitive redox regulator that presumably acts through its conformational influence on the Runt domain. Data also show that the DNA-binding activity of p53 is regulated by TRX and/or Ref-1 *in vitro* and *in vivo* (Ueno *et al.*, 1999). Thus, the TRX–Ref-1 cascade may regulate at least some transcription factors.

In the nucleus, these transcription factors need to be reduced to be functional, whereas in the cytoplasm oxidative stress can activate NF-κB and AP-1. UV irradiation promptly induced the translocation of TRX into the nucleus. TRX was shown to be able to interact directly with Ref-1/APEX nuclease, a ubiquitous nuclear protein (Hirota *et al.*, 1997). Thus, the translocated TRX is beneficial to make a reduced environment in association with Ref-1 in the nucleus and may enhance the DNA binding of transcription factors. It is also possible that the translocation of TRX and the interaction between TRX and Ref-1 support the function of p53 against oxidative stress. TRX translocation may also be involved in the activation of the glucocorticoid receptor (Makino *et al.*, 1999). In the NF-κB system, direct binding between reduced TRX and NF-κB was demonstrated using nuclear magnetic resonance (Qin *et al.*, 1995). The association between NF-κB and TRX may be involved in the nuclear translocation of NF-κB (Hirota *et al.*, 1999). We have identified TRX-binding proteins that are localized in the nucleus by coimmunoprecipitation using an anti-TRX monoclonal antibody (Nishiyama *et al.*, 1998) or by a yeast two-hybrid assay (Nishiyama *et al.*, 1999), although the functions of the proteins are under investigation.

Nuclear localization of TRX is often observed in pathological tissue. In cervical tissue, TRX expression is observed in human papilloma virus-infected cells and TRX is localized in the nucleus (Fujii *et al.*, 1991). In renal proximal tubules, TRX is induced and translocated to nuclei by oxidative damage mediated by Fe-nitrilotriacetate (Tanaka *et al.*, 1997). Although the significance and mechanism of these observations are unclear, TRX translocation may be related to the cyotoprotection and pathogenesis of oxidative stress-related disorders.

VIII. CONCLUSION

The TRX system, composed of several related molecules, protects cells against oxidative stress. Redox regulation is deeply involved in the cell death mechanism as is p53 transcriptional function. Thus, in addition to a function of scavenging ROS, TRX appears to be a regulator/modulator involved in cellular signaling against oxidative stress. Enhancement of the thioredoxin system may be beneficial against oxidative stress-induced pathological conditions. Further elucidation of oxidative stress signaling and response may give a new therapeutic approach by the redox regulation of oxidative stress-mediating target molecules.

ACKNOWLEDGMENTS

We thank Dr. K. Hirota and other members of the laboratory for helpful discussion and Ms. Y. Kanekiyo for secretarial help. This work was supported by a grant-in-aid for Scientific Research and Special Project Research–Cancer Bioscience from the Ministry of Education, Science, and Culture of Japan and a grant-in-aid of Research for the Future from the Japan Society for the Promotion of Science, Japan.

REFERENCES

Abate, C., Patel, L., Rauscher, F. D., and Curran, T. (1990). Redox regulation of fos and jun DNA-binding activity in vitro. *Science* **249**, 1157–1161.

Akamatsu, Y., Ohno, T., Hirota, K., Kagoshima, H., Yodoi, J., and Shigesada, K. (1997). Redox regulation of the DNA binding activity in transcription factor PEBP2. The roles of two conserved cysteine residues. *J. Biol. Chem.* **272**, 14497–14500.

Buttke, T. M., and Sandstrom, P. A. (1994). Oxidative stress as a mediator of apoptosis. *Immunol. Today* **15**, 7–10.

Cho, Y., Gorina, S., Jeffrey, P. D., and Pavletich, N. P. (1994). Crystal structure of a p53 tumor suppressor–DNA complex: Understanding tumorigenic mutations. *Science* **265**, 346–355.

Clarke, F. M., Orozco, C., Perkins, A. V., Cock, I., Tonissen, K. F., Robins, A. J., and Wells, J. R. (1991). Identification of molecules involved in the 'early pregnancy factor' phenomenon. *J. Reprod. Fertil.* **93**, 525–539.

Demple, B., Herman, T., and Chen, D. S. (1991). Cloning and expression of APE, the cDNA encoding the major human apurinic endonuclease: Definition of a family of DNA repair enzymes. *Proc. Natl. Acad. Sci. U. S. A.* **88**, 11450–11454.

Fujii, S., Nanbu, Y., Nonogaki, H., Konishi, I., Mori, T., Masutani, H., and Yodoi, J. (1991). Coexpression of adult T-cell leukemia-derived factor, a human thioredoxin homologue, and human papillomavirus DNA in neoplastic cervical squamous epithelium. *Cancer (Philadelphia)* **68**, 1583–1591.

Gauntt, C. D., Ohira, A., Honda, O., Kigasawa, K., Fujimoto, T., Masutani, H., Yodoi, J., and Honda, Y. (1994). Mitochondrial induction of adult T cell leukemia derived factor (ADF/hTx) after oxidative stresses in retinal pigment epithelial cells. *Invest. Ophthalmol. Visual Sci.* **35**, 2916–2923.

Hainaut, P., and Milner, J. (1993). Redox modulation of p53 conformation and sequence-specific DNA binding in vitro. *Cancer Res.* **53**, 4469–4473.

Halliwell, B., and Gutteridge, J. M. C. (1984). Oxygen toxicity, oxygen radicals, transition metals and disease. *Biochem. J.* **219**, 1–14.

Hampton, M. B., and Orrenius, S. (1997). Dual regulation of caspase activity by hydrogen peroxide: Implications for apoptosis. *FEBS Lett.* **414**, 552–556.

Hayashi, S., Hajiro, N. K., Makino, Y., Eguchi, H., Yodoi, J., and Tanaka, H. (1997). Functional modulation of estrogen receptor by redox state with reference to thioredoxin as a mediator. *Nucleic Acids Res.* **25**, 4035–4040.

Heiden, M. G. V., Chandel, N. S., Williamson, E. K., Schumacker, P. T., and Thompson, C. B. (1997). Bcl-xL regulates the membrane potential and volume homeostasis of mitochondria. *Cell (Cambridge, Mass.)* **91**, 627–637.

Hirota, K., Matsui, M., Iwata, S., Nishiyama, A., Mori, K., and Yodoi, J. (1997). AP-1 transcriptional activity is regulated by a direct association between thioredoxin and Ref-1. *Proc. Natl. Acad. Sci. U. S. A.* **94**, 3633–3638.

Hirota, K. Murata, M., Sachi, Y., Nakamura, H., Takeuchi, J., Mori, K., and Yodoi, J. (1999). Distinct roles of thioredoxin in the cytoplasm and in the nucleus: A two step mechanism of redox regulation of transcription factor NF-κB. Submitted for publication.

Hirsh, T., Marchetti, P., Susin, S. S., Dallaporta, B., Zamzami, N., Marzo, I., Geuskins, M., and Kroemer, G. (1997). The apoptosis-necrosis paradox. Apoptogenic proteases activated after mitochondrial permeability transition determine the mode of cell death. *Oncogene* **15**, 1573–1581.

Holmgren, A. (1985). Thioredoxin. *Annu. Rev. Biochem.* **54**, 237–271.

Holmgren, A., and Björnstedt, M. (1995). Thioredoxin and thioredoxin reductase. *In* "Methods in Enzymology," (L. Packer, ed.), Vol. 252, pp. 199–208. Academic Press, San Diego, CA.

Hori, K., Katayama, M., Sato, N., Ishii, K., Waga, S., and Yodoi, J. (1994). Neuroprotection by glial cells through adult T cell leukemia-derived factor/human thioredoxin (ADF/TRX). *Brain Res.* **652**, 304–310.

Huang, L. E., Arany, Z., Livingston, D. M., and Bunn, H. F. (1996). Activation of hypoxia-inducible transcription factor depends primarily upon redox-sensitive stabilization of its alpha subunit. *J. Biol. Chem.* **271**, 32253–32259.

Ichijo, H., Nishida, E., Irie, K., ten Dijke, P., Saitoh, M., Moriguchi, T., Takagi, M., Matsumoto, K., Miyazono, K., and Gotoh, Y. (1997). Induction of apoptosis by ASK1, a mammalian MAPKKK that activates SAPK/JNK and p38 signaling pathways. *Science* **275**, 90–94.

Ireland, R. C., Li, S. Y. and Dougherty, J. J. (1995). The DNA binding of purified Ah receptor heterodimer is regulated by redox conditions. *Arch. Biochim. Biophys.* **319**, 470–480.

Iwata, S., Hori, T., Sato, N., Ueda, T. Y., Yamabe, T., Nakamura, H., Masutani, H., and Yodoi, J. (1994). Thiol-mediated redox regulation of lymphocyte proliferation. Possible involvement of adult T cell leukemia-derived factor and glutathione in transferrin receptor expression. *J. Immunol.* **152**, 5633–5642.

Jayaraman, L., Murthy, K. G., Zhu, C., Curran, T., Xanthoudakis, S., and Prives, C. (1997). Identification of redox/repair protein Ref-1 as a potent activator of p53. *Genes Dev.* **11**, 558–570.

Jin, D. Y., Chae, H. Z., Rhee, S. G., and Jeang, K. T. (1997). Regulatory role for a novel human thioredoxin peroxidase in NF-kappaB activation. *J. Biol. Chem.* **272**, 30952–30961.

Kawahara, N., Tanaka, T., Yokomizo, A., Nanri, H., Ono, M., Wada, M., Kohno, K., Takenaka, K., Sugimachi, K., and Kuwano, M. (1996). Enhanced coexpression of thioredoxin and high mobility group protein 1 genes in human hepatocellular carcinoma and the possible association with decreased sensitivity to cisplatin. *Cancer Res.* **56**, 5330–5333.

Kobayashi, F., Sagawa, N., Nanbu, Y., Kitaoka, Y., Mori, T., Fujii, S., Nakamura, H., Masutani, H., and Yodoi, J. (1995). Biochemical and topological analysis of adult T-cell leukaemia-derived factor, homologous to thioredoxin, in the pregnant human uterus. *Hum. Reprod.* **10**, 1603–1608.

Kroemer, G., Zamzami, N., and Susin, S. A. (1997). Mitochondrial control of apoptosis. *Immunol. Today* **18**, 44–51.

Kuge, S., and Jones, N. (1994). YAP1 dependent activation of TRX2 is essential for the response of Saccharomyces cerevisiae to oxidative stress by hydroperoxides. *EMBO J.* **13**, 655–664.

Kuge, S., Jones, N., and Nomoto, A. (1997). Regulation of yAP-1 nuclear localization in response to oxidative stress. *EMBO J.* **16**, 1710–1720.

Kurooka, H., Kato, K., Minoguchi, S., Takahashi, Y., Ikeda, J., Habu, S., Osawa, N., Buchberg, A. M., Moriwaki, K., Shisa, H., and Honjo, T. (1997). Cloning and characterization of the nucleoredoxin gene that encodes a novel nuclear protein related to thioredoxin. *Genomics* **39**, 331–339.

Leppa, S., Pirkkala, L., Chow, S. C., Eriksson, J. E., and Sistonen, L. (1997). Thioredoxin is transcriptionally induced upon activation of heat shock factor 2. *J. Biol. Chem.* **272**, 30400–30404.

Levine, A. J. (1997). p53, the cellular gatekeeper for growth and division. *Cell (Cambridge, Mass.)* **88**, 323–331.

Liu, S. Y., and Stadtman, T. C. (1997). Heparin-binding properties of selenium-containing thioredoxin reductase from HeLa cells and human lung adenocarcinoma cells. *Proc. Natl. Acad. Sci. U. S. A.* **94**, 6138–6141.

Macleod, K. F., Sherry, N., Hannon, G., Beach, D., Tokino, T., Kinzler, K., Vogelstein, B., and Jacks, T. (1995). p53-dependent and independent expression of p21 during cell growth, differentiation, and DNA damage. *Genes Dev.* **9**, 935–944.

Makino, Y., Okamoto, K., Yoshikawa, N., Aoshima, M., Hirota, K., Yodoi, J., Umesono, K., Makino, I., and Tanaka, H. (1996). Thioredoxin: A redox-regulating cellular cofactor for glucocorticoid hormone action. Cross talk between endocrine control of stress response and cellular antioxidant defense system. *J. Clin. Invest.* **98**, 2469–2477.

Makino, Y., Yoshikawa, N., Okamoto, K., Hirota, K., Yodoi, J., Makino, I., and Tanaka, H. (1999). Direct association with thioredoxin allows redox regulation of glucocorticoid receptor function. *J. Biol. Chem.* **274**, 3182–3188.

Marchetti, P., Decaudin, D., Macho, A., Zamzami, N., Hirsch, T., Susin, S. A., and Kroemer, G. (1997). Redox regulation of apoptosis: Impact of thiol oxidation status on mitochondrial function. *Eur. J. Immunol.* **27**, 289–296.

Matsuda, M., Masutani, H., Nakamura, H., Miyajima, S., Yamauchi, A., Yonehara, S., Uchida, A., Irimajiri, K., Horiuchi, A., and Yodoi, J. (1991). Protective activity of adult T cell leukemia-derived factor. (ADF) against tumor necrosis factor-dependent cytotoxicity on U937 cells. *J. Immunol.* **147**, 3837–3841.

Matsui, M., Oshima, M., Oshima, H., Takaku, K., Maruyama, T., Yodoi, J., and Taketo, M. M. (1996). Early embryonic lethality caused by targeted disruption of the mouse thioredoxin gene. *Dev. Biol.* **178**, 179–185.

Mitsui, A., Hirakawa, T. and Yodoi, J. (1992). Reactive oxygen-reducing and protein-refolding activities of adult T cell leukemia-derived factor/human thioredoxin. *Biochem. Biophys. Res. Commun.* **186**, 1220–1226.

Morgan, B. A., Banks, G. R., Toone, W. M., Raitt, D., Kuge, S., and Johnston, L. H. (1997). The Skn7 response regulator controls gene expression in the oxidative stress response of the budding yeast *Saccharomyces cerevisiae*. *EMBO J.* **16**, 1035–1044.

Nakamura, H., Matsuda, M., Furuke, K., Kitaoka, Y., Iwata, S., Toda, K., Inamoto, T., Yamaoka, Y., Ozawa, K., and Yodoi, J. (1994). Adult T cell leukemia-derived factor/human thioredoxin protects endothelial F-2 cell injury caused by activated neutrophils or hydrogen peroxide. *Immunol. Lett.* **42**, 75–80.

Nakamura, H., Nakamura, K., and Yodoi, J. (1997). Redox regulation of cellular activation. *Annu. Rev. Immunol.* **15**, 351–369.

Natsuyama, S., Noda, Y., Narimoto, K., Umaoka, Y., and Mori, T. (1992). Release of two-cell block by reduction of protein disulfide with thioredoxin from *Escherichia coli* in mice. *J. Reprod. Fertil.* **95**, 649–656.

Natsuyama, S., Noda, Y., Yamashita, M., Nagahama, Y., and Mori, T. (1993). Superoxide dismutase and thioredoxin restore defective p34cdc2 kinase activation in mouse two-cell block. *Biochim Biophys. Acta* **1176**, 90–94.

Nishiyama, A., Furuke, K., Hirota, K., Masutani, H., and Yodoi, J. (1998). Detection of a nuclear 60-kDa protein coimmunoprecipitated with human thioredoxin. In "Oxygen Homeostasis and its Dynamics" (Y. Ishimura. H., Shimada, and M. Suematsu, eds.), pp. 464–468. Springer-Verlag, Tokyo.

Nishiyama, A. Matsui, M., Iwata, S., Hirota, K., Masutani, H., Nakamura, H., Takagi, Y., Sono, H., Gon, Y., and Yodoi, J. (1999). Identification of thioredoxin binding protein-2/vitamin D_3 up-regulated protein 1 as a negative regulator of thioredoxin function and expression. *J. Biol. Chem.* in press.

Ohira, A., Honda, O., Gauntt, C. D., Yamamoto, M., Hori, K., Masutani, H., Yodoi, J., and Honda, Y. (1994). Oxidative stress induces adult T cell leukemia derived factor/thioredoxin in the rat retina. *Lab. Invest.* **70**, 279–285.

Parks, D., Bolinger, R., and Mann, K. (1997). Redox state regulates binding of p53 to sequence-specific DNA, but not to non-specific or mismatched DNA. *Nucleic Acids Res.* **25**, 1289–1295.

Pearson, G. D., Merill, G. F. (1998). Deletion of the *Saccharomyces cerevisiae* TRR1 gene encoding thioredoxin reductase inhibits p53-dependent reporter gene expression. *J. Biol. Chem.* **273**, 5431–5434.

Qin, J., Clore, G. M., Kennedy, W. M., Huth, J. R., and Gronenborn, A. M. (1995). Solution structure of human thioredoxin in a mixed disulfide intermediate complex with its target peptide from the transcription factor NF kappa B. *Structure* **3**, 289–297.

Rainwater, R., Parks, D., Anderson, M. E., Tegtmeyer, P., and Mann, K. (1995). Role of cysteine residues in regulation of p53 function. *Mol. Cell. Biol.* **15**, 3892–3903.

Reed, J. C. (1997). Cytochrome c: Can't live with it—can't live without it. *Cell (Cambridge, Mass.)* **91**, 559–562.

Sachi, Y., Hirota, K., Masutani, H., Toda, K., Okamoto, T., Takigawa, M., and Yodoi, J. (1995). Induction of ADF/TRX by oxidative stress in keratinocytes and lymphoid cells. *Immunol. Lett.* **44**, 189–193.
Saito, M., Nishitoh, H., Fujii, M., Takeda, K., Tobiume, K., Sawada, Y., Kawabata, M., Miyazono, K., and Ichijyo, H. (1998). Mammalian thioredoxin is a direct inhibitor of apoptosis signal-regulating kinase (ASK) 1. *EMBO J.* **17**, 2596–2606.
Sasada, T., Iwata, S., Sato, N., Kitaoka, Y., Hirota, K., Nakamura, K., Nishiyama, A., Taniguchi, Y., Takabayashi, A., and Yodoi, J. (1996). Redox control of resistance to cis-diamminedichloroplatinum (II) (CDDP): Protective effect of human thioredoxin against CDDP-induced cytotoxicity. *J. Clin. Invest.* **97**, 2268–2276.
Sato, N., Iwata, S., Nakamura, K., Hori, T., Mori, K., and Yodoi, J. (1995). Thiolmediated redox regulation of apoptosis. Possible roles of cellular thiols other than glutathione in T cell apoptosis. *J. Immunol.* **154**, 3194–3203.
Schreck, R., Rieber, P., and Baeuerle, P. A. (1991). Reactive oxygen intermediates as apparently widely used messengers in the activation of the NF-kappa B transcription factor and HIV-1. *EMBO J.* **10**, 2247–2258.
Seki, S., Akiyama, K., Watanabe, S., Hatsushika, M., Ikeda, S., and Tsutsui, K. (1991). cDNA and deduced amino acid sequence of a mouse DNA repair enzyme (APEX nuclease) with significant homology to *Escherichia coli* exonuclease III. *J. Biol. Chem.* **266**, 20797–20802.
Spyrou, G., Enmark, E., Miranda-Vizuete A., and Gustafsson, J. (1997). Cloning and expression of a novel mammalian thioredoxin. *J. Biol. Chem.* **272**, 2936–2941.
Storz, G., Tartaglia, L. A., and Ames, B. N. (1990). Transcriptional regulator of oxidative stress-inducible genes: Direct activation by oxidation. *Science* **248**, 189–194.
Tagaya, Y., Maeda, Y., Mitsui, A., Kondo, N., Matsui, H., Hamuro, J., Brown, N., Arai, K-I., Yokota, T., Wakasugi, H., and Yodoi, J. (1989). ATL-derived factor (ADF), an IL-2 receptor/Tac inducer homologous to thioredoxin; possible involvement of dithiol-reduction in the IL-2 receptor induction. *EMBO J.* **8**, 757–764.
Takagi, Y., Tokime, T., Nozaki, K., Gon, Y., Kikuchi, H., and Yodoi, J. (1998). Redox control of neuronal damage during brain ischemia after middle cerebral artery occlusion in the rat: Immunohistochemical and hybridization studies of thioredoxin. *J. Cereb. Blood Flow Metab.* **18**, 206–214.
Tanaka, T., Nishiyama, Y., Okada, K., Hirota, K., Matsui, M., Yodoi, J., Hiai, H., and Toyokuni, S. (1997). Induction and nuclear translocation of thioredoxin by oxidative damage in the mouse kidney: Independence of tubular necrosis and sulfhydryl depletion. *Lab. Invest.* **77**, 145–155.
Taniguchi, Y., Taniguchi, U. Y., Mori, K., and Yodoi, J. (1996). A novel promoter sequence is involved in the oxidative stress-induced expression of the adult T-cell leukemia-derived factor (ADF)/human thioredoxin (Trx) gene. *Nucleic Acids Res.* **24**, 2746–2752.
Tonissen, K. F., and Wells, J. R. E. (1991). Isolation and characterization of human thioredoxin-encoding genes. *Gene* **102**, 221–228.
Ueda, S., Nakamura, H., Masutani, H., Sasada, T., Yonehara, S., Takabayashi, A., Yamaoka, Y., and Yodoi, J. (1998). Redox regulation of caspase-3 (-like) protease activity: Regulatory roles of thioredoxin and cytochrome c. *J. Immunol.* **161**, 6689–6695.
Ueno, M., Masutani, H., Arai, R. J., Yamauchi, A., Hirota, K., Sakai, T., Inamoto, T., Yamaoka, Y., Yodoi, J., and Nikaidou, T. (1999). Redox regulation of P53 activity by thioredoxin. Submitted for publication.
Wada, H., Muro, K., Hirata, T., Yodoi, J., and Hitomi, S. (1995). Rejection and expression of thioredoxin in transplanted canine lung. *Chest* **108**, 810–814.

Wang, J., Kobayashi, M., Sakurada, K., Imamura, M., Moriuchi, T., and Hosokawa, M. (1997). Possible roles of an adult T-cell leukemia (ATL)-derived factor/thioredoxin in the drug resistance of ATL to adriamycin. *Blood* **89**, 2480–2487.

Wilson, K. P., Black, J. A., Thomson, J. A., Kim, E. E., Griffith, J. P., Navia, M. A., Murcko, S. P., Chambers, S. P., Aldape, R. A., Raybuck, S. A., and Livingston, D. J. (1994). Structure and mechanism of interleukin-1 beta converting enzyme. *Nature (London)* **370**, 270–275.

Wollman, E. E., d'Auriol, L., Rimsky, L., Shaw, A., Jacquot, J. P., *et al.* (1988). Cloning and expression of a cDNA for human thioredoxin. *J. Biol. Chem.* **263**, 15506–15512.

Xanthoudakis, S., and Curran, T. (1992). Identification and characterization of Ref-1, a nuclear protein that facilitates AP-1 DNA-binding activity. *EMBO J.* **11**, 653–665.

Xanthoudakis, S., Miao, G., Wang, F., Pan, Y. C., and Curran, T. (1992). Redox activation of Fos-Jun DNA binding activity is mediated by a DNA repair enzyme. *EMBO J.* **11**, 3323–3335.

Yamamoto, M., Ohira, A., Honda, O., Sato, N., Furuke, K., Yodoi, J., and Honda, Y. (1997). Analysis of localization of adult T-cell leukemia-derived factor in the transient ischemic rat retina after treatment with OP-1206 alpha-CD, a prostaglandin E1 analogue. *J. Histochem. Cytochem.* **45**, 63–70.

Yodoi, J., and Uchiyama, T. (1992). Diseases associated with HTLV-I: Virus, IL-2 receptor dysregulation and redox regulation. *Immunol. Today* **13**, 405–411.

Yokomizo, A., Ono, M., Nanri, H., Makino, Y., Ohga, T., Wada, M., Okamoto, T., Yodoi, J., Kuwano, M., and Kohno, K. (1995). Cellular levels of thioredoxin associated with drug sensitivity to cisplatin, mitomycin C, doxorubicin, and etoposide. *Cancer Res.* **55**, 4293–4296.

Zamzami, N., Marzo, I., Susin, S. S., Brenner, C., Larochette, N., Marchetti, P., Reed, J., Kofler, R., and Kroemer, G. (1998). The thiol crosslinking agent diamide overcomes the apoptosis-inhibitory effect of Bcl-2 by enforcing mitochondrial permeability transition. *Oncogene* **16**, 1055–1063.

Zhang, P., Liu, B., Kang, S. W., Seo, M. S., Rhee, S. G., and Obeid, L. M. (1997). Thioredoxin peroxidase is a novel inhibitor of apoptosis with a mechanism distinct from that of Bcl-2. *J. Biol. Chem.* **272**, 30615–30618.

Zheng, M., Aslund, F., and Storz, G. (1998). Activation of the OxyR transcription factor by reversible disulfide bond formation. *Science* **279**, 1718–1721.

14 Redox Regulation of p21, Role of Reactive Oxygen and Nitrogen Species in Cell Cycle Progression

Axel H. Schönthal,* Sebastian Mueller,†‡ and Enrique Cadenas‡

*Department of Molecular Microbiology and Immunology
School of Medicine
University of Southern California
Los Angeles, California 90089

†Department of Internal Medicine IV
University of Heidelberg
69115 Heidelberg, Germany

‡Department of Molecular Pharmacology and Toxicology
School of Pharmacy
University of Southern California
Los Angeles, California 90089

Damage to cell molecules by oxidants constitutes the basic tenet of the redox changes underlying signal transduction pathways and affecting cell growth control as well as several pathophysiological situations, such as aging, neurodegenerative diseases, and cancer. Depending on the interplay among intracellular signaling pathways, these processes may promote cell proliferation, induce apoptosis, or cause necrosis.

This chapter (a) provides a survey of some relevant aspects of cell cycle control and protein activity regulation by reactive species generated by different metabolic pathways, (b) emphasizes the role of oxidants in the expression of the cell cycle inhibitor p21 (WAF1, CIP1, or sdi1) by pathways involving either activation of the tumor suppressor p53 or signal transduction events, and (c) addresses the relative contribution of these genes in the sequence of events leading to the inhibition of cell proliferation. The emphasis on p21 is based on several features of this cell cycle inhibitor: (1) it is a focal point that integrates many types of signals that impact on cell division, (2) it is an attractive target for drug design because it can be induced by chemotherapeutic agents that elicit oxidative stress, and (3) its overexpression alone suppresses the growth of many different types of tumor cells in culture or *in vivo*. The oxidants discussed in the expression of p21 are those entailed in oxidative stress (metabolism of chemotherapeutic quinones, enzyme generation of hydrogen peroxide, hyperoxia, and related changes in the thiol/disulfide status of the cell), UV radiation, and systems generating nitric oxide.

I. INTRODUCTION

The formation of free radicals or oxidants is a well-established physiological event in aerobic cells, which convene enzymic and nonenzymic resources to remove these oxidizing species. An imbalance between oxidants and antioxidants, the two terms of the equation that define oxidative stress (Sies, 1986), and the consequent damage to cell molecules constitutes the basic tenet of the redox changes underlying signal transduction pathways and affecting cell growth control as well as several pathophysiological situations, such as aging, neurodegenerative diseases, and cancer.

The chemical stability of a variety of reactive species, whether of free radical character or not, varies substantially; regardless of their source, it could be stated that virtually all cell components—lipids, nucleic acids, proteins, and carbohydrates—in an appropriate setting are sensitive to damage by reactive species encompassing oxygen-, nitrogen-, carbon-, and sulfur-centered radicals and nonradical species, such as hydrogen peroxide, singlet oxygen, and peroxynitrite. That the cellular generation of oxidants overwhelms or bypasses the antioxidant defenses is attested by several lines of evidence: on the one hand, the participation of free radicals in several pathologies and the usefulness of certain antioxidant therapies indicate that these defenses are not entirely effective. On the other hand, the occurrence of fingerprints for oxidative damage has been found in DNA, proteins, carbohydrates, and lipids; this appears to stimulate repair enzymes, which are involved in the removal of nonfunctional cell components or in the actual repair of damaged biomolecules.

The effects of oxidative stress, however, are not all mediated through damage of cellular constituents, but may be the result of slight changes in the redox status of the cell. These changes may alter cell function by acting as, mimicking, or affecting second messengers in signal transduction pathways

(Forman and Cadenas, 1997). Oxidative stress is also involved in signaling the induction of antioxidant enzymes (as those controlled by *oxyR* and *soxR* loci) (Demple and Levin, 1991), activation protein phosphorylation (Schieven, 1997), and mimicking the effect of hormones, such as insulin (Richter and Laffranchi, 1997).

Thus, changes in the intracellular redox status elicited by the generation of oxidants can trigger selective gene expression, enzyme activation, DNA synthesis, and alter progression of the cell cycle. Depending on the interplay among intracellular signalling pathways, these processes may promote cell proliferation, induce apoptosis, or cause necrosis (Simon *et al.*, 1997a; Burdon, 1995). Hence, it may be surmised that the proliferative response to reactive oxygen species (or reactive species in general) is varied and dependent on the cell type and nature of the oxidant. It is of interest to understand how free radicals and the cellular redox status in general may alter or regulate these critical cell cycle regulators, such as cyclin-dependent kinases (Simon *et al.*, 1997a), by mechanisms involving signal transduction pathways or upstream genes induced by oxygen-free radicals.

II. p21 EXPRESSION AND CELL CYCLE CONTROL

The progression of eukaryotic cells through cell cycle transitions is regulated by cyclin-dependent protein kinases (CDKs); one of the roles of CDKs in the G_1 phase is to phosphorylate the retinoblastoma gene product (pRb) and to release the transactivator E_2F, which is essential for cell cycle progression ($G_1 \rightarrow S$) (Fig. 1). Both pRb and E_2F seem to play critical roles in the control of the cell cycle (Fig. 1).

The activity of CDKs is controlled positively by cyclins and negatively by CDK inhibitory proteins (CKIs), such as those in the p21, p27, and p57 family and the p16 (Ink4) family (Fig. 1). It may be surmised that impairment of the balance between these components may not just disrupt cell cycle progression, but also increase the susceptibility of cells to apoptosis.

Tumor suppressor p53 is activated in response to DNA damage (i.e., strand breaks) and can orchestrate a number of cellular responses; fundamental to understanding p53 function is the recognition that it regulates not only checkpoint control through the induction of other genes and resulting in cell cycle G_1 arrest (Fig. 2, pathway A), but also activates a pathway of apoptotic cell death that inhibits tumor development (Fig. 2, pathway B) (el-Deiry, 1998).

p53 exerts cell cycle control on induction of the cell cycle inhibitor p21, which plays a critical role consequential to its interaction with multiple cellular targets. p21 is a negative regulator of the cell cycle that not only

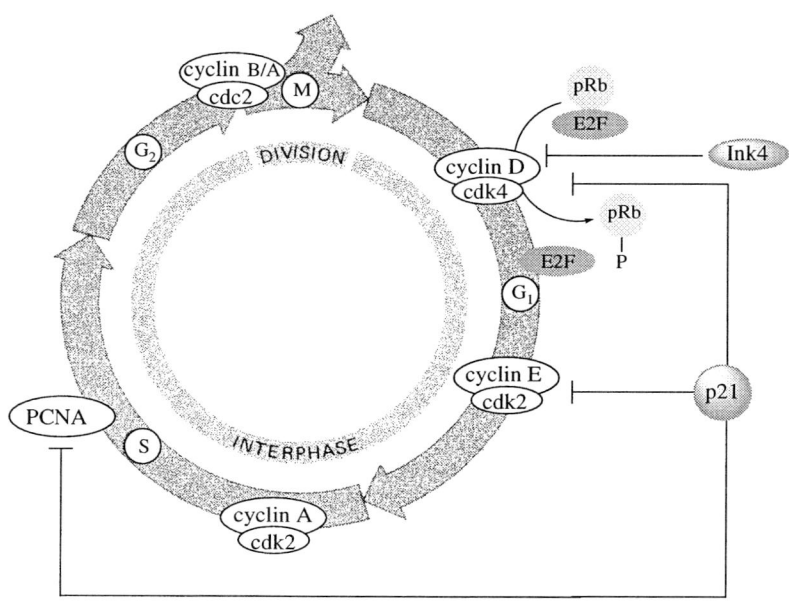

Figure 1 Schematic representation of the mammalian cell cycle. pRb, retinoblastoma gene product; E$_2$F, trans-activator; cdk, cyclin-dependent kinase; PCNA, proliferating cell nuclear antigen. Phosphorylation of pRb by the cyclin D–cdk4 complex releases E$_2$F, which activates genes involved in the S phase. Ink4 (p16) can inhibit the cyclin D–cdk4 complex. p21[WAF1/CIP1] can inhibit both cyclin D–cdk4 and cyclin E–cdk2 complexes, as well as form a complex with PCNA leading to the inhibition of DNA replication.

inhibits cyclin-dependent kinases by acting as a CKI, but also the proliferating cell nuclear antigen (PCNA), a subunit of DNA polymerase (Figs. 1 and 2). In the former instances, one potential downstream target of p21 inhibitory activity in G$_1$ is the cell cycle-dependent phosphorylation of Rb: p21 prevents the phosphorylation of Rb and, consequently, release of E$_2$F, which is required for the progression into the S phase (Fig. 2, pathway A). In the latter instances, p21 inhibits DNA replication (but not DNA repair synthesis) directly through binding to PCNA. Cells lacking p21 are defective in their ability to undergo G$_1$ arrest following DNA damage (Brugarolas *et al.*, 1995; Deng *et al.*, 1995) and are more sensitive to DNA damage (McDonald *et al.*, 1996).

p21 is, therefore, a focal point that integrates many types of signals that have an impact on cell division. Its ectopic expression suppresses the growth of different types of tumor cells in culture and *in vivo*, such as human brain, lung, prostate, osteosarcoma, and colon tumor cells. It may be surmised that upregulating p21 expression by anticancer agents could be effective in

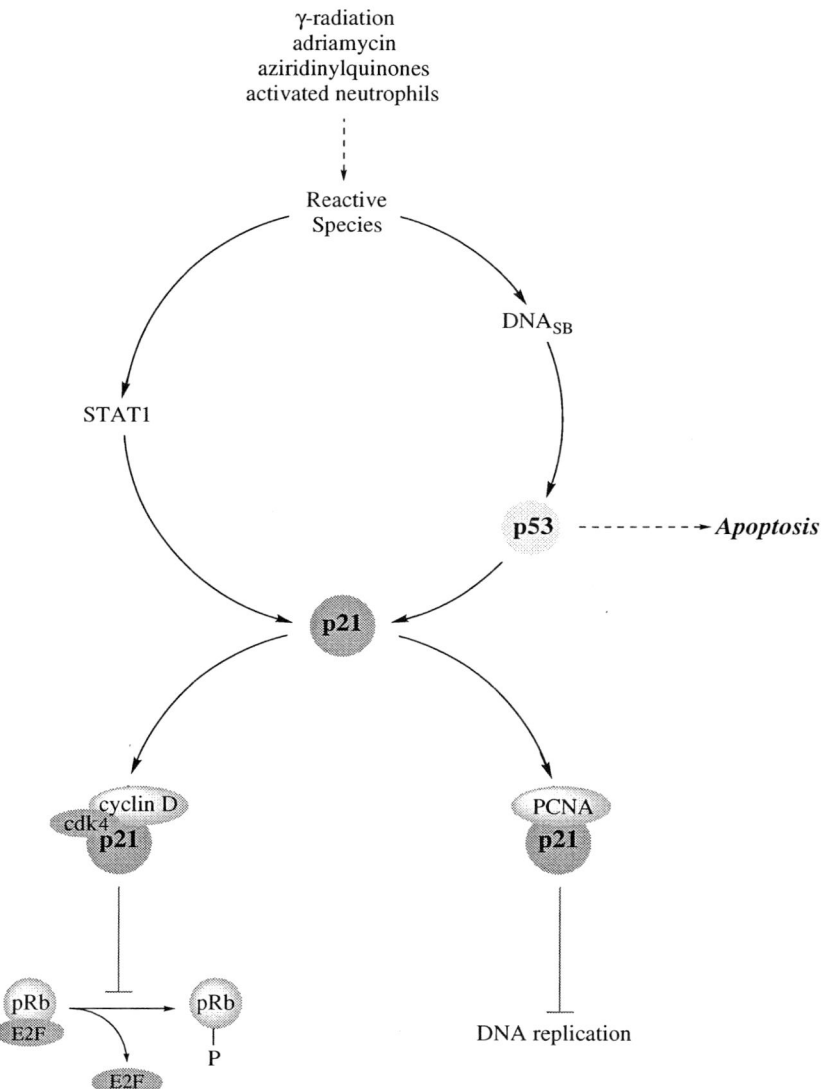

Figure 2 Induction of p21 by p53-dependent and -independent pathways. The scheme shows that reactive oxygen species (or reactive nitrogen species, see Fig. 7) are upstream of p53. These species, as generated during processes such as γ-radiation, metabolism of anticancer quinones, hyperoxia, or H_2O_2 metabolism, are involved in DNA oxidative damage (strand breaks) and subsequent expression of p53. The latter can exert cell cycle control on induction of p21 and activate a pathway of apoptotic cell death (Burdon, 1995). The p53-independent pathways for p21 induction also consider that reactive oxygen species are upstream of this gene: activation of the STAT system by oxidative stress (Sherr and Roberts, 1995) and induction of p21 by STAT proteins (el-deiry, et al., 1993) are key steps in this pathway.

arresting the cell cycle and be beneficial to cancer treatment. This view gains further significance when considering that the p53 tumor suppressor gene is mutated in more than 50% of human tumors and, accordingly, approaches to the molecular pharmacology of cancer are searching for anticancer drugs that are not dependent on intact p53 suppressor gene function for their activity.

III. INDUCTION OF p21: p53-DEPENDENT AND -INDEPENDENT PATHWAYS

The mechanism of p21 gene activation by exogenous agents remains unclear: p21 can be induced by a variety of external stimuli, such as growth factors, cytokines, and tumor promoters as well as by DNA-damaging agents and generators of reactive oxygen species, such as adriamycin, etoposide, and γ-radiation (Sheikh *et al.*, 1994; el-Deiry *et al.*, 1994).

The regulation of p21 is complex, involving transcriptional and post-transcriptional mechanisms. Dissection of the promoter region of the p21 gene (Fig. 3) revealed two binding sites for tumor suppressor p53, and it has been established that p53 indeed is a major regulator of p21 gene expression (el-Deiry *et al.*, 1993; 1995). In addition to binding sites for p53, there are binding sites for other transcription factors that may play a role in the regulation of p21 expression by p53-independent pathways (M.Liu *et al.*, 1996a, b; Chin *et al.*, 1996; Somasundaram *et al.*, 1997; Zeng *et al.*, 1997; Zeng and el-Deiry, 1996; Prowse *et al.*, 1997; Datto *et al.*, 1995). (Fig. 3). Hence, two separate pathways for the induction of p21 are currently recognized: a p53-dependent one activated by DNA damage (usually involving γ-irradiation models) (el-Deiry, 1998) and a p53-independent one activated by mitogens at the entry of the cell cycle (Michieli *et al.*, 1994) or during adriamycin metabolism (Russo *et al.*, 1995) or oxidative stress (Russo *et al.*, 1995) (Fig. 2).

With respect to p53-dependent pathways, it is generally assumed that the activation of p21 by p53 is a crucial event that secures the integrity of the genome in the defense against various adverse conditions. For instance, the p53 protein is activated in response to treatment of cells with different types of chemotherapeutic drugs or γ-radiation, assaults that are known to cause DNA damage (Harris, 1996; O'Connor *et al.*, 1997). The activation of p53 is followed by the increased expression of p21, a specific inhibitor of CDKs, and of PCNA (a subunit of DNA polymerase) (Hunter and Pines, 1994; Harper and Elledge, 1996; Sherr and Roberts, 1995), thus leading to cellular growth arrest. It is likely that the p53–p21-induced G_1 arrest involves yet another tumor suppressor, namely the product of the Rb tumor suppressor gene (Harrington *et al.*, 1998; Prives, 1998). Rb is active (cell cycle

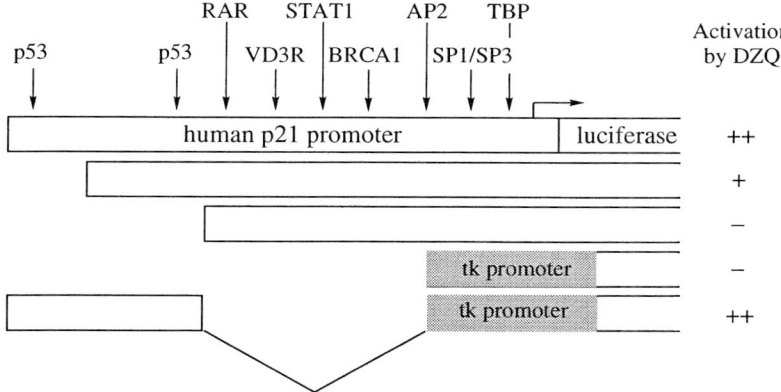

Figure 3 The p21 promoter, deletion mutants, and response to treatment with aziridinylbenzoquinone. (Left) The p21 promoter fused to the luciferase reporter gene. Various known transcription factor-binding sites are indicated. Different deletion mutants are represented below the full-length p21 promoter (2. 5 kb): the first mutant exhibits a 5' deletion, which removes the distal p53-binding site. In the second mutant, both p53-binding sites are removed. The bottom construct is a fusion between the p21 promoter region containing only the two p53-binding sites and a heterologous promoter, the thymidine kinase (tk) promoter. (Right) The response of each construct after transient transfection and treatment of cells with unsubstituted diaziridinylbenzoquinone (DZQ): ++, strong induction; +, intermediate induction; −, no induction. For experimental details, see Lane (1998). RAR, retinoic acid-binding site (el-Deiry *et al.*, 1994). VD3R, vitamin D_3-binding site (Sheikh *et al.*, 1994). STAT1, signal transducer and activator of transcription 1-binding site (el-Deiry *et al.*, 1993). BRCA1, breast cancer gene 1-binding site (el-Deiry *et al.*, 1995). AP2, activator protein 2-binding site (M. Liu *et al.*, 1996a) SP1/SP3, stimulator protein 1- and 3-binding site (Chin *et al.*, 1996; Somasundaram *et al.*, 1997).

inhibitory) when it is hypophosphorylated and becomes inactivated due to its hyperphosphorylation by CDKs (Hinds and Weinberg, 1994) (Fig. 1). The hypophosphorylated form of Rb sequesters the transcription factor E_2F, which is a positive regulator of several genes necessary for the execution of DNA replication (Fig. 2), such as thymidylate kinase, dihydrofolate reductase, thymidilate synthase, DNA polymerase α, and ribonucleotide reductase.

Two independent findings are of interest in the understanding of the p53-independent expression of p21: on the one hand, oxidative stress was reported to lead to the activation of signal transducer and activator of transcription (STAT) proteins (Simon *et al.*, 1997b), and on the other hand, STAT proteins have been shown to activate p21 expression in a p53-independent manner (Chin *et al.*, 1996) (Fig. 2). The response of the STAT family of transcription factors (particularly STAT1 and STAT3) to oxidative stress appears to be cell-type specific and occurs most likely via an oxidative stress-sensitive cellular tyrosine kinase (JAK or src kinases—both activated in response to H_2O_2—may be involved) (Schieven, 1997; Simon *et al.*, 1997b). Alternatively, it was suggested that the p53-independent expression of p21 in response to oxidative stress could be mediated by mechanisms involving activation of the mitogen-activated protein (MAP) kinase cascade (Russo *et al.*, 1995; Ihle, 1996), probably the oxidative stress-sensitive p38. Overall, maximal activation of transcription by STAT1 and STAT3 appears to require both tyrosine and serine phosphorylation (Wen *et al.*, 1995).

IV. ROLE OF FREE RADICALS IN THE REDOX REGULATION OF p21

Overall, the pause in proliferative activity allows the cells to repair their DNA damage and thus ensures that these defects are not replicated and transmitted to the daughter cells. In that sense, p53 and Rb were dubbed the *guardians of the genome,* and p21 is thought to be an essential executioner of these pathways (Lane, 1992, 1998). For this reason, several laboratories have investigated the underlying mechanisms by which free radicals impinge on the expression and activity of p21. The following is a brief survey of the potential role that oxygen- and nitrogen-centered radicals may play in the redox regulation of p21 and its implications for cell growth and apoptosis.

A. Oxidative Stress and p21 Induction

A role for oxygen-centered radicals in p21 expression is suggested by experiments involving bioactivation of aziridinylbenzoquinones in human cancer cells (Qiu *et al.*, 1996, 1998; Qiu and Cadenas, 1997; Wu *et al.*, 1998), oxidation of xanthine by xanthine oxidase in human fibroblasts, (F. Gansauge *et al.*, 1997), exposure of murine neonatal lung cells to hyperoxic conditions (McGrath, 1998), and changes in thiol/disulfide status of the cell (Russo *et al.*, 1995; Esposito *et al.*, 1997).

1. Bioactivation of Azyridinylbenzoquinones

Studies carried out in our laboratory were aimed at gathering information on the molecular mechanisms inherent in the activation of p21 during the metabolism of aziridinylbenzoquinones in different cancer cell lines. A comprehensive approach to elucidating this mechanism(s) involved the stepwise characterization of (a) the cellular activation pathways for the quinone, with emphasis on the overexpressed enzyme NAD(P)H:quinone oxidoreductase and the ensuing generation of reactive oxygen species; (b) the relationship between the latter and expression of p21; and (c) the effects of expression of p21—triggered by oxidative stress—on cell proliferation.

a. Cellular Activation of Aziridinylbenzoquinones and Oxidative Stress Most of the currently used antineoplastic agents elicit their effects by directly damaging cellular DNA or through mechanisms involving the inhibition of DNA synthesis, the mitotic apparatus, or topoisomerases (Hartwell and Kastan, 1994). The variability observed in the effectiveness of chemotherapeutic agents probably is an expression of the functional group chemistry of the compound, the mechanisms of bioactivation, and the cell-type-specific responses to DNA-damaging agents. Hence, biochemical distinctions between normal and cancer cells are of importance in the development of anticancer drugs. In this context, it is worth noting the potential role of NAD(PH) quinone oxidoreductase (NQOR: EC 1.6.99.2; also known as DT-diaphorase) in quinone bioactivation: (1) the enzyme is a unique flavoprotein that, at variance with other quinone reductases, catalyzes obligatory two-electron transfers (Cadenas, 1995; Tedeschi *et al.*, 1995; Li *et al.*, 1995), *i.e.*, it reduces quinones to hydroquinones (Fig. 4); and (2) this enzyme activity is highly expressed in most tumor cell lines and preneoplastic nodules, thereupon its significance in enzyme-directed bioreductive drug activation with emphasis on cancer chemotherapy (Riley and Workman, 1992). Chemically, this concept encompasses the NQOR-catalyzed formation of bioalkylating products that react effectively with DNA bases and, in some cases, leads to interstrand cross-links.

An alternative chemical view (Qiu *et al.*, 1996) of enzyme-directed bioreductive drug development purports that the enzyme-mediated activation of chemotherapeutic quinones with a particular functional group chemistry leads to the formation of redox labile hydroquinones, which decay with the formation of oxygen-free radicals (Cadenas, 1995). The mechanism underlying this effect probably involves a superoxide anion-dependent propagation of hydroquinone oxidation as illustrated in Fig. 4A (Ordoñez and Cadenas, 1992; Öllinger *et al.*, 1990) The oxidative and reductive decay pathways of the semiquinone form of aziridinylbenzoquinones—in connection with NQOR activity—may have profound implications in the subsequent steady-state concentration of free radicals, for the aziridinylbenzosemiquinones are key species in the propagation of hydroquinone autoxidation (Goin *et al.*, 1993) (Fig. 4B). Superoxide dismutase and glutathione, whose concentrations are similar in normal and cancer cells, facilitate the oxidative and reductive decay of the semiquinone, respectively. An increased cellular steady-state level of oxygen radicals, termed *oxidative stress* (by the mechanisms depicted in Fig. 4), elicits an array of effects on signal transduction processes, cell proliferation, and concomitant disruption of the cell cycle.

Aziridinylbenzoquinones are a group of antitumor agents that elicit cytotoxicity by generating either alkylating intermediates or reactive oxygen species (Ordoñez and Cadenas,1992; Moore, 1997; Lee *et al.*, 1992; Berar-

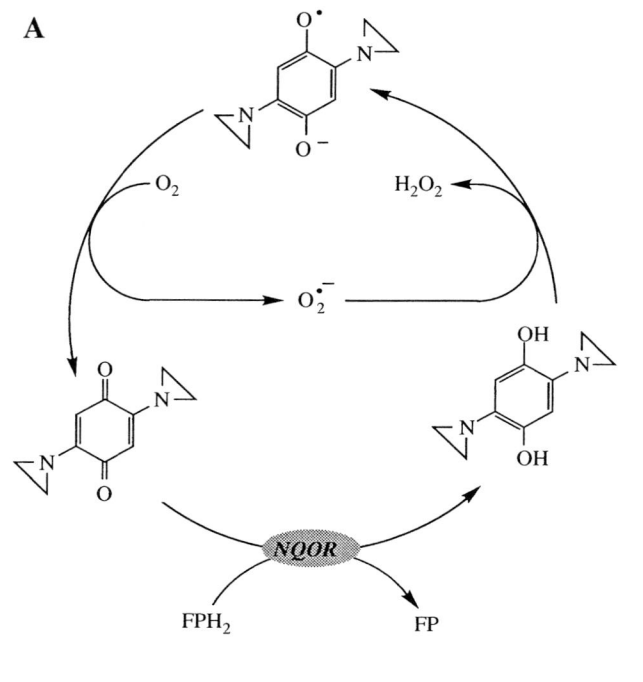

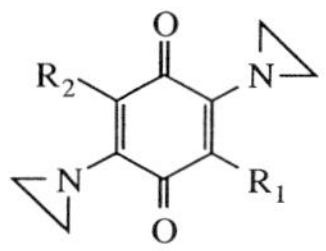

DZQ $R_3 = R_6 = H$
AZQ $R_3 = R_6 = NHCOOCH_2C_3$
F-DZQ $R_3 = R_6 = F, Cl, Br$

Figure 5 Structure of some aziridinylbenzoquinones. DZQ, 3,6-diaziridinyl-1,4-benzoquinone. AZQ, 2,5-bis(carboethoxyamino)-3,6-diaziridinyl-1,4-benzoquinone. F-DZQ, 2,5-bis-fluoro-3,6-diaziridinyl-1,4-benzoquinone. Cellular effects elicited by the aziridinylbenzoquinone are an expression of the functional group chemistry of these compounds and of the mode of bioactivation.

dini *et al.*, 1993; Gibson *et al.*, 1992; Fisher and Gutierrez, 1991a, b). The alkylating properties require protonation of the aziridinyl N followed by interaction with suitable neutrophiles, whereas the formation of reactive oxygen species is a consequence of redox-cycling properties of the quinone moiety (Fig. 5). The metabolism of aziridinylbenzoquinones with a different substitution pattern (Fig. 5) and its association with oxygen-free radical formation and how this impinged on p21 expression were examined in cells with different constitutive levels of NQOR: HT29 human colon carcinoma cells with an overexpressed NQOR activity (Qiu and Cadenas, 1997) HCT116 cells with intermediate activity of the enzyme, (Qiu *et al.*, 1996) and BE cells, which are devoid of NQOR activity (Qiu and Cadenas, 1997).

The metabolism of the unsubstituted aziridinylbenzoquinone DZQ (Fig. 6) in HT29 cells was accompanied by H_2O_2 formation, which was ~24 to 57-fold higher than that observed in the NQOR-lacking BE cells. Spin-trapping electron paramagnetic resonance (EPR) analysis of the metabolism of aziridinylbenzoquinones in cancer cells revealed the formation of spin adducts of $O_2^{\cdot-}$ and HO·; the contribution of these species to the overall EPR spectrum was dependent on the enzymic machinery with which the cell was endowed and the functional group chemistry of the aziridinylbenzoquinone. Thus, in BE cells—devoid of NQOR activity—the bioactivation

Figure 4 NAD(P)H:quinone oxidoreductase-catalyzed activation of aziridinylbenzoquinones and the oxidative and reductive decay of the semiquinone species. (A) Two-electron reduction of aziridinylbenzoquinones by NQOR and the redox transition of the hydroquinone product. $O_2^{\cdot-}$ acts as a propagating species, being generated during semiquinone autoxidation and consumed during hydroquinone oxidation (Cadenas, 1995). (B) Effect of superoxide dismutase and GSH on the oxidative and reductive decay of the semiquinone species of aziridinylbenzoquinones (Li *et al.*, 1995).

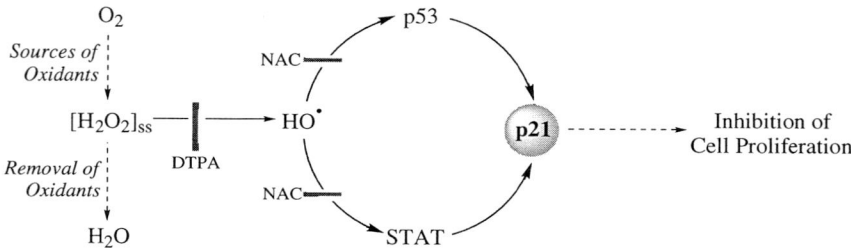

Figure 6 Cellular steady-state levels of H_2O_2 and p21 expression. The steady-state level of H_2O_2 in cells is determined by the (a) rate of formation and (b) removal of this species. It may be hypothesized that the HO·, generated by homolytic cleavage of the peroxide, is directly involved in the activation of p21 by p53-dependent and -independent pathways. NAC, N-acetylcysteine, scavenges HO·; DTPA, metal chelator, inhibits the transition metal-dependent cleavage of H_2O_2.

of AZQ and DZQ yielded spin adducts of $O_2^{·-}$ and HO·, with the former adduct prevailing on the latter, expectedly, these signals were insensitive to dicumarol. In HT29 cells, the metabolism of DZQ yielded mainly the spin adduct of HO· and a modest contribution of $O_2^{·-}$ adducts; dicumarol elicited a strong of HO· production associated with an increase in the accumulation of $O_2^{·-}$ adducts, consistent with a change in the bioactivation pattern of the quinone, i.e., from NQOR-catalyzed two-electron transfers to other flavoprotein-catalyzed one-electron transfers.

Experiments with several aziridinylbenzoquinones revealed that the unsubstituted quinone DZQ was a preferred substrate for the overexpressed NQOR activity in cancer cells and that the subsequent redox transitions of the hydroquinone product yielded mainly HO·. In agreement with this view, N-acetylcysteine, a powerful reductant used as an intracellular antioxidant in several experimental models, abolished the EPR signal observed during the metabolism of DZQ by HCT116 and HT29 cells, and a new signal was observed assigned to the N-acetylcysteinyl radical (Qiu et al., 1996; Qiu and Cadenas, 1997). This was consistent with the one-electron oxidation of the thiol by HO· and subsequent reaction of the thiyl radical with the spin trap.

b. Oxygen-Free Radicals and p21 Induction Treatment of the colon carcinoma cell lines (HT29, HCT116) with aziridinylbenzoquinones resulted in a concentration- and time-dependent increase of p21 mRNA levels. p21 induction was prevented partly by dicumarol, thereby suggesting the involvement of NAD(P)H:quinone oxidoreductase in quinone activation and, subsequently, p21 induction (Qiu et al., 1996; Qiu and Cadenas, 1997). A role for HO· in p21 induction is suggested by the effects of N-acetylcysteine and metal chelators (Fig. 6) (Qiu et al., 1996). The former, which displays a high reactivity toward HO·, inhibited p21 induction in HCT116 and HT29 cells, whereas the nonsulfur analog, N-acetylalanine, had no effect. In the later, the effect of metal chelators suggests the implication of redox-active metals as critical catalysts in the site-specific generation of HO·; DTPA, a metal chelator, inhibited p21 induction.

H_2O_2, a product of quinone redox cycling, and HO·, derived from its homolytic cleavage, appear central to understanding the role of oxidants in p21 expression. H_2O_2 elicited an increase of p21 mRNA levels in BE, HT29, HCT116, and K562 human chronic myelogenous leukemia cells; the latter lack p53, one of the activators of p21 transcription, thus suggesting that p21 expression can be accomplished in a p53-independent manner in these cells (Qiu et al., 1996; Qiu and Cadenas, 1997). Taken together, an enhancement of the cellular steady-state concentration of oxidants appears to be critical in the sequence depicted in Fig. 6. A high $[H_2O_2]_{ss}$ may be obtained despite a modest generation of this species, as, for example, during the metabolism of aziridinylbenzoquinones by BE cells, which have significant low levels of glutathione and glutathione-S-transferases and, consequently, a low rate of removal of H_2O_2. Although not conclusive, HO· may be suggested to be involved in the induction of p21 mRNA levels: this view is supported by the fact that p21 was highly induced under conditions encompassing the high production of H_2O_2 and HO· and the effects of DTPA and N-acetylcysteine on the induction of p21 mRNA levels (Fig. 6).

c. Oxidants, p21 Induction, and Cell Proliferation Under the experimental conditions described earlier and implying an increased cellular steady-state level of oxidants, the p21 protein level was elevated significantly, whereas the level of $p34^{cdc2}$ protein, one of the cyclin-dependent kinases (Fig. 1), remained unchanged following the treatment of HCT116 cells with the unsubstituted aziridinylbenzoquinone. The *cdc2*-associated H_1 kinase activity was strongly suppressed following the treatments described earlier; the histone H_1 kinase activities of cyclin-dependent kinases associated with cyclin A (immunoprecipitated by cyclin A antibody) were also inhibited substantially. These findings are in agreement with a role of p21 in inhibiting cyclin-dependent kinase activities (Qiu et al., 1996).

Cell proliferation was inhibited by aziridinylbenzoquinones at the same concentrations required to induce p21. The effect of these quinones on cell number appeared to be due to a decreased proliferation rather than outright toxicity. As mentioned earier p21 is also an upstream regulator of the retinoblastoma gene product (pRb) involved in G_1 cell cycle control. The cell cycle was arrested in the G_2/M phase following the supplementation with DZQ of human osteosarcoma Saos-2 cells (lacking both p53 and pRb) and HCT116 cells (Qiu *et al.*, 1998). This aziridinylbenzoquinone also induced p21 and apoptosis in Saos-2 cells. The transfection of the Rb gene into Saos-2 cells did not alter the level of p21 induction, but changed cell cycle arrest into the G_1 phase and prevented apoptosis (Qiu *et al.*, 1998). These findings suggest that quinones may lead to a p53-independent and pRb-preventable G_2/M arrest and apoptosis, which correlate with p21 induction (Qiu *et al.*, 1998).

i. H_2O_2 from the Xanthine/Xanthine Oxidase Reaction and Activated Neutrophils. Oxygen-centered radicals may be generated by sources other than the redox transitions of aziridinylbenzoquinones: oxidation of xanthine by xanthine oxidase was shown to lead to an enhanced expression of p21 in the human fibroblast cell line WI38, an effect suppressed by catalase. This p21 induction might be regulated by p53, whose protein was accumulated on exposure to xanthine/xanthine oxidase mixtures (F. Gansauge *et al.*, 1997; S. Gansauge *et al.*, 1997). p53 levels were increased in cultured human skin fibroblasts incubated with stimulated neutrophils by a mechanism that required the H_2O_2/myeloperoxidase-dependent generation of hypochlorous acid (Vile *et al.*, 1998). Although H_2O_2 alone elevated the levels of p53 and p21 in human fibroblasts, the fluxes required were 10-fold higher than those that were effective for hypochlorous acid; it may be surmised that the oxidant produced by activated neutrophils and responsible for the increase in p53 expression at the site of chronic inflammation is hypochlorous acid (Vile *et al.*, 1998).

ii. Hyperoxia. mRNA and protein p21 levels are induced in cells grown in culture and in murine neonatal and adult lung cells when exposed to hyperoxia (which may be considered an oxidative stress situation) (McGrath, 1998). Studies under *in vivo* conditions revealed that the localization of enhanced p21 was in peripheral airway cells. In these cells, hyperoxia-induced p21 expression appeared to be mediated by a p53 pathway, as suggested by using adult p53 mutant mice. The localization of p21 expression in the peripheral airway cells also suggests that p21 may act to inhibit growth of alveoli in neonatal lung and delay the repopulation of alveolar cells during hyperoxic conditions (McGrath, 1998).

iii. Thiol/Disulfide Status. Changes in thiols/disulfide status play a key role in redox signaling and the control of cell growth and death (Powis *et al.*, 1997). These changes may be associated with the generation of reactive oxygen species (or increase in the oxidation status of the cell), which, in turn, may alter the activity of transcription factors by direct interaction with these proteins. For example, the zinc-finger domain that is found in several transcription factors is especially sensitive to redox changes. The cysteine residues involved in finger formation interact with the Zn^{2+} only in their reduced form. Oxidation will induce the formation of disulfide bonds between the thiols invovled in finger stabilization, thereby disrupting the structure and preventing DNA binding. Examples of these types of transcription factors are SP-1, Egr-1, or members of the steroid-thyroid hormone receptor family (Kadonaga et al., 1987; Cao et al., 1990; Beato, 1989). Treatment of cells with oxidants, such as H_2O_2, N-ethylmaleimide, or diethylmaleate, results in reduced DNA-binding activity of these factors. These effects are reversible, as DNA-binding activity and transcriptional activation by these factors are restored on removal of the oxidant or through the addition of dithiothreitol and Zn^{2+} ions. Similarly, the pretreatment of cells with N-acetylcysteine significantly blunted the effects of subsequently added oxidants (Espositoa et al., 1994, 1995; Ammendola et al., 1994; Hutchison et al., 1991).

Although not a member of the group of Zn-finger transcription factors, the p53 protein also harbors reduced cysteines that are crucial for the formation of its DNA-binding domain. In this case, Zn^{2+} is associated with one histidine and three cysteine residues to contribute to the formation of the DNA-binding domain (Cho et al., 1994). Oxidizing and/or DNA-damaging agents such as N-ethylmaleimide (NEM), Methylmethane sulfonate (MMS), diamide, or various sources of radiation have been shown to greatly reduce the ability of p53 to bind to its DNA target site, and this effect is strictly correlated with the redox regulation of these three cysteine residues (Hainaut and Milner, 1993; Lu and Lane, 1993; Hupp et al., 1993; Zhan et al., 1993). The reduced DNA-binding activity of p53 correlates very closely with the impaired activation of its target genes, such as p21. However, under these conditions, p53-independent pathways for p21 induction are also possible: treatment of cultured human cells with diethylmaleimide has been found to induce the expression of p21, whereas under the same conditions the DNA-binding and transcriptional activity of p53 was inhibited (Russo et al., 1995). In fact, this induction was also observed in human tumor cells lacking the p53 gene (Russo et al., 1995). Because N-acetylcysteine was able to prevent p21 induction by diethylmaleate, this effect likely was a consequence of the ability of diethylmaleate to reduce intracellular GSH levels. Intriguingly, the increased expression of p21 in response to diethylmaleate treatment appeared to include regulation at the posttranscriptional level through the increased stability of its mRNA (Esposito et al., 1997), an effect apparently regulated via the activation of the MAP kinase pathway by reactive oxygen species (Esposito et al., 1997; Fialkow et al., 1994; Y. Liu et al., 1996).

B. Ultraviolet Radiation and p21 Induction

Ultraviolet light (UV) interacts with many types of cellular macromolecules and elicits a complex set of acute cellular responses, among them the activation of signal transduction pathways and changes in gene expression (Bender, *et al.*, 1997; Engelberg *et al.*, 1994; Griffiths *et al.*, 1998). Some of these processes are modulated by UV-induced damage to proteins and DNA. Misfolded proteins can be offset by molecular chaperones, whereas DNA photoproducts such as cyclobutane dimers and pyrimidine(6-4)pyrimidones can be corrected through nucleotide excision repair. In addition, UV is able to cause perturbations of the cellular redox equilibrium as a result of free radical release. The major reactive oxygen species generated during this process appears to be HO·.

A number of genes have been shown (Smith and Fornace, 1997; Maltzman and Czyzyk, 1984) to be activated in response to UV irradiation of cells, most notably for this review, p53 and p21. The induction of p53 by UV appears to be controlled by two events (Eller *et al.*, 1997; Renzing *et al.*, 1976; Martinez *et al.*, 1997) one is the process of nucleotide excision repair, which transiently generates free DNA ends (DNA breaks) and DNA-damage intermediates. The other is the production of oxygen-free radicals. The latter process is sensitive to N-acetylcysteine pretreatment, whereas the former is not (Renzing *et al.*, 1996). In fact, it has been demonstrated that N-acetylcysteine does not reduce the number of radiation-induced DNA strand breaks and therefore is not able to completely prevent the induction of p53 by short-wave UV (Renzing *et al.*, 1976; Ewing, 1991; Ghosh *et al.*, 1993). However, higher doses of UV, which induce more oxidative stress, result in an increase in the number of cells exhibiting highly increased levels of p53 protein (Renzing *et al.*, 1996).

It seems that different types of UV (A, B, C) may involve distinct pathways to generate cellular responses. For example, whereas DNA single strand breaks are involved predominantly in the UV-C-dependent increase in p53, the induction of p53 caused by solar UV (UV-A and UV-B, 290–380 nm) has been suggested to be regulated through the generation of reactive oxygen species (Vile, 1997). In the case of solar UV, the induction of p53 could be inhibited by scavengers of active oxygen species, such as N-acetylcysteine, ascorbate, and α-tocopherol (Vile, 1997). However, this issue has not been completely clarified, and many details remain to be established.

In general, the induction of p53 in response to UV leads to G_1 phase cell cycle arrest, which is mediated largely through the increased expression of the p21 gene (Smith and Fornace, 1997). It appears that under these conditions, cyclin–CDK complexes containing *cdk2* are the main targets of inhibition associated with p21 expression (Poon *et al.*, 1996). Other complexes containing *cdk4* are inhibited by cooperation of p21 with other CDK inhibitors, whereas *cdc2* is inactivated *via* phosphorylation of its threonine-14 and tyrosine-15 residues (Poon *et al.,* 1996). While these findings point to the involvement of several different mechanisms for UV-induced cell cycle arrest, it is evident that p21 plays a major role in these processes. It is less clear, however, whether the UV-induced stimulation of p21 expression is achieved through p53-dependent or p53-independent pathways, as examples for both are available.

For instance, the existence of p53-dependent activation of p21 in response to UV irradiation has been shown in cells expressing a mutant of p53 that lacks sequence-specific DNA-binding capacity (Bissonnette and Hunting, 1998). These cells were found to not undergo cell cycle arrest following exposure to UV because of their inability to induce p21. In similar studies by others (Liu and Pelling, 1995), p21 was induced in UV-A/B-irradiated keratinocytes containing wild-type p53, but not in cells deficient in p53. In contrast, p21 induction was found to be independent of p53 in several other studies using solar UV or UV-C at 254 nm. For example, UV-C was shown to induce p21 expression and concomitant cell cycle arrest in normal human skin fibroblasts as well as in cells derived from Li-Fraumeni syndrome patients, who are p53 deficient (Loignon *et al.,* 1997). Moreover, on exposure of volunteers to solar UV, induction of p53 as well as p21 could be detected in epidermal cells, whereas in mesenchymal cells of the upper dermis only p21, but not p53, could be detected (Pontin *et al.,* 1995).

The basis for these discrepancies is not clear, but in addition to cell type, genetic, or other biological differences, the experimental settings might contribute to the generation of different results. In this regard, the length of the experiment or the energy of UV used is likely to determine the respective results. For example, using UV-C-irradiated rat embryo fibroblasts, it was found that p21 was induced at 6 hr, but not at later times, after exposure to 10 J/m^2; however, irradiation of these cells with 50 J/m^2 did not stimulate p21 expression at any of the times examined (Reinke and Lozano, 1997).

Intriguingly, although p53 is frequently found mutated in a great variety of human tumors and tumor cell lines, inactivating mutations of p21 are found very rarely in tumor cells (Shiohara *et al.,* 1994; Kamb, 1998). In some tumor cell lines, however, a second form of p21 protein, with a faster mobility in denaturing gel electrophoresis, can be detected (Poon and Hunter, 1998). This same form, which appears to harbor a deletion of approximately 10 amino acids at its C terminus, which is expected to remove the PCNA-binding domain, a second cyclin binding domain, and the nuclear localization signal, was also found in cells treated with high doses of UV light (Poon and Hunter, 1998). Although the mechanism of this truncation has not been established yet, it is assumed that this shorter form of p21 could contribute to the deregulation of the cell cycle and could be involved in cellular transformation. In other experiments, where the cDNA of p21 was transfected into cells, it was established that UV irradiation resulted in point mutations of p21 that impaired its growth-arresting activity (Lu *et al.*, 1996). Although some of these mutations occurred repeatedly in independently isolated cellular clones, these types of mutations are not found in tumors, raising the intriguing possibility that such mutations do not confer selective advantage to cells. In support of such a hypothesis, we have observed that on homozygous deletion of the p21 gene in human colon carcinoma cells, these cells lose their tumorigenic phenotype (Mueller *et al.*, 1999). This latter finding was surprising and would argue for the requirement of p21 in the process of tumor development.

C. Nitric Oxide and p21 Induction

Nitric oxide (NO·; systematic name: nitrogen monoxide), a free radical formed in a variety of cell types by NO synthase and involved in an wide array of physiological and pathophysiological phenomena (Gross and Wolin, 1995) can induce apoptosis in different systems. In addition to the original two actions of nitric oxide on vascular smooth muscle and platelets, NO plays additional physiological roles as a modulator of cell proliferation and of transcriptional expression (Ignarro, 1996, 1997). The effects of nitric oxide (NO·) on p21 induction as it relates to inhibition of cell proliferation, decrease of neurite extension, and apoptosis were investigated in smooth muscle cells, pheochromocytoma PC12 cells, and human and rat cancer cells, respectively (Poluha *et al.*, 1997; Ishida *et al.*, 1997; Ho *et al.*, 1996). Experimental models leading to an increased cellular steady-state level of NO· consisted of activation of nitric oxide synthase, exposure to agents releasing NO·, or to an atmosphere of NO gas (Fig. 7).

NO·, generated on activation of nitric oxide synthase by nerve growth factor (NGF) in pheochromocytome PC12 cells acts as a second messenger activating the p21 promoter and inducing p21 expression. p21 expression appears to occur via p53-dependent and -independent pathways: inhibition of nitric oxide synthase, and thereby a decrease in the steady-state levels of NO·, resulted in a decreased accumulation of p53 and activation of the p21 promoter, effects that correlated with a decreased expression of neuronal markers and neurite extension. These effects reveal a sequence as that illustrated in Fig. 7; however, induction of p21 with isopropyl-1-thio-β-D-galactopyranoside, under conditions of inhibited nitric oxide synthase, led to a decreased neurite extension, thus indicating the occurrence of p53-independent pathways for p21 induction (Poluha *et al.*, 1997). Toxicity elicited

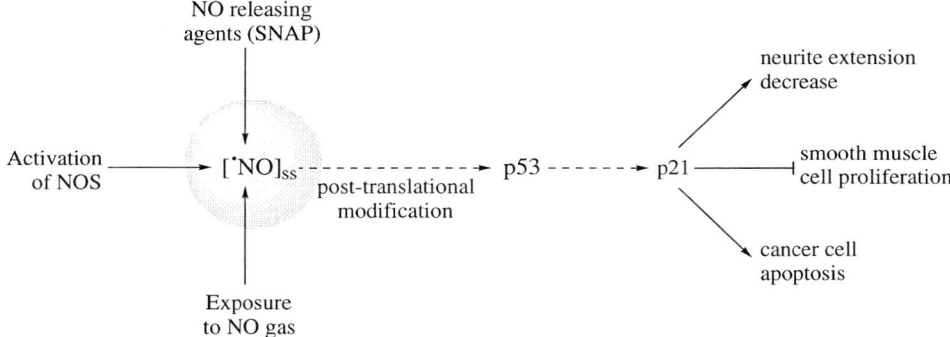

Figure 7 Nitric oxide and p21 induction. NO· generated either on activation of nitric oxide synthase by nerve growth factor (NGF) in pheochromocytome PC12 cells (Mueller *et al.*, 1999) or from the releasing agent S-nitroso-N-acetylpenicillamine in smooth muscle cells (Gross and Wolin, 1995) or on exposure of NO· gas in cancer cells (Ignarro, 1996) elicits p21 induction in a p53-dependent manner. See text for explanations.

by peroxynitrite (ONOO⁻), a major oxidative product resulting from the diffusion-controlled reaction of superoxide anion and NO·, is affected differently when cells are exposed to different growth factors: pretreatment of PC12 with NGF protected against ONOO− toxicity, whereas that with acidic and basic fibroblast growth factor (FGF-1 and FGF-2) exacerbated ONOO⁻ toxicity. This different action of growth factors was explained on the basis that NGF was protective in PC12 cells by activating a phosphoinositol-3-phosphate kinase pathway, whereas FGF increased toxicity by activating a p21 Ras-dependent pathway (Spear *et al.*, 1997).

NO·, generated by a releasing agent (S-nitroso-N-acetylpenicillamine), inhibited DNA synthesis (assessed by [^{3}H]thymidine incorporation and triggered by fetal bovine serum and basic fibroblast growth factor) in smooth muscle cells. NO·, also inhibited the activity of *cdk2* and Rb phosphorylation, whereas it had no effect on the activities of cyclin D-associated kinases (*cdk4* and *cdk6*). Under conditions involving an increased steady-state level of NO·, the mRNA and protein levels of p21 were high in the early G_1 phase as well as the levels of *cdk2*-associated p21. It may be concluded that NO· inhibits the G_1/S transition by inhibiting *cdk2*-mediated phosphorylation of Rb and that p21 induction is involved in the *cdk2* inhibition (Ishida *et al.*, 1997).

NO· induced nuclear accumulation of p53 in a dose- and time-dependent manner in human cancer cells, a process regulated by posttranslational modification. Furthermore, NO· induced apoptosis in COLO 205 and HepG2 cells containing wild-type p53; this was associated with an increased of p21 protein. Conversely, in cells lacking (Hep 3B) or with a mutated (HT29) p53, p21 levels were not increased on exposure to NO·. This suggests that the apoptosis resulting from the exposure of certain types of cancer cells to NO· requires a functional p53 leading to p21 expression (Ho *et al.*, 1996).

V. CONCLUDING REMARKS

Regulation of CDK activity by either downregulating cyclin expression or upregulating CKI expression could be effective in arresting cell cycle and could be beneficial to cancer treatment (Qiu *et al.*, 1996), especially when considered that cancer may be a consequence of either the overexpression of cyclins or the loss of CKIs. Hence, advances in our understanding of the mechanisms by which free radicals affect cell cycle progression and its checkpoint controls seem essential for identifying potential targets for improved cancer therapies.

An enhanced cellular steady-state concentration of oxygen radicals—as a consequence of either the enzymic activation of anticancer quinones (Qiu *et al.*, 1996; Qiu and Cadenas, 1997) or in hyperoxia or under certain conditions of UV irradiation (Renzing *et al.*, 1996; Vile, 1997)—appears to be of utmost significance for further signaling pathways controlling cell cycle progression via p53-dependent or -independent pathways. Likewise, DNA damage elicited by nitrogen-centered radicals, e.g., NO· or the products of its redox chemistry, may establish the basis for expression of the tumor suppressor gene p53 and further apoptosis (Messmer *et al.*, 1994).

Oxidative stress leads to the activation of signal transducer and activator of transcription (STAT) proteins, which have been shown to activate p21 expression in a p53-independent manner. Therefore, it is possible that oxyradicals may elicit p21 induction through STAT-dependent pathways. Concerning the p53-dependent pathways for p21 induction, free radicals were considered in this review as upstream mediators of p53 (via DNA oxidative damage) and apoptosis (Fig. 2). Conversely, it was proposed that reactive oxygen species were downstream mediators of p53-dependent apoptosis (Johnson *et al.*, 1996). This suggests that one function of p53 may be to regulate the intracellular redox state. Further elucidation of the specific mechanism by which p53 generates oxyradicals to induce apoptosis might offer new approaches for the development of therapeutic agents capable of selectively inducing apoptosis in normal or neoplastic cells (Johnson *et al.*, 1996).

The relative contribution of STAT- and p53-dependent pathways to p21 induction and of cell cycle arrest and/or apoptosis to growth inhibition depends largely on the cell genotype. Because p53 is mutated in 50% of the tumors, understanding of the p53-independent pathways for p21 activation is of utmost importance in cancer chemotherapy. p21 is an attractive target for drug design because it can be induced by chemotherapeutic agents that elicit oxidative stress, and its overexpression alone suppresses the growth of many different types of tumor cells in culture or *in vivo*.

REFERENCES

Ammendola, R., Mesuraca, M., Russo, T., and Cimino, F. (1994). The DNA-binding efficiency of Sp1 is affected by redox changes. *Eur. J. Biochem.* **225**, 483–489.

Beato, M. (1989). Gene regulation by steroid hormones. *Cell (Cambridge, Mass.)* **56**, 335–344.

Bender, K., Blattner, C., Knebel, A., Iordanov, M., Herrlich, P., and Rahmsdorf, H. J. (1997). UV-induced signal transduction. *J. Photochem. Photobiol. B* **37**, 1–17.

Berardini, M. D., Souhami, R. L., Lee, C. S., Gibson, N. W., Butler, J., and Hartley, J. A. (1993). Two structurally related diaziridinylbenzoquinones preferentially cross-link DNA at different sites upon reduction with DT-diaphorase. *Biochemistry* **32**, 3306–3312.

Bissonnette, N., and Hunting, D. J. (1998). p21-induced cycle arrest in G1 protects cells from apoptosis induced by UV-irradiation or RNA polymerase II blockage. *Oncogene* **16**, 3461–3469.

Brugarolas, J., Chandraskeran, C., Gordon, I., Beach, D., Jacks, T., and Hannon, G. (1995). Radiation-induced cell cycle arrest compromised by p21 deficiency. *Nature, (London)*, **377**, 552–556.

Burdon, R. H. (1995). Superoxide and hydrogen peroxide in relation to mammalian proliferation. *Free Radical Biol. Med.* **18**, 775–794.

Cadenas, E. (1995). Antioxidant and prooxidant functions of DT-diaphorase in quinone metabolism. *Biochem. Pharmacol.* **49**, 127–140.

Cao, X. M., Koski, R. A., Gashler, A., McKiernan, M., Morris, C. F., Gaffney, R., Hay, R. V., and Sukhatme, V. P. (1990). Identification and characterization of the Egr-1 gene product, a DNA-binding zinc finger protein induced by differentiation and growth signals. *Mol. Cell Biol.* **10**, 1931–1939.

Chin, Y. E., Kitagawa, M., Su, W.-C. S., You, Z.-H., Iwamoto, Y., and Fu, X.-Y. (1996). Cell growth arrest and induction of cyclin-dependent kinase inhibitor p21 WAF1/CIP1 mediated by STAT1. *Science* **272**, 719–722.

Cho, Y., Gorina, S., Jeffrey, P. D., and Pavletich, N. P. (1994). Crystal structure of a p53 tumor suppressor-DNA complex: Understanding tumorigenic mutations. *Science* **265**, 346–355.

Datto, M. B., Yu, Y., and Wang, X. F. (1995). Functional analysis of the transforming growth factor beta responsive elements in the WAF1/Cip1/p21 promoter. *J. Biol. Chem.* **270**, 28623–28628.

Demple, B., and Levin, J. D. (1991). Repair system for radical-damaged DNA. In "Oxidative Stress: Oxidants and Antioxidants" (H. Sies, ed.), pp. 119–154. Academic Press, London.

Deng, C., Zhang, P., Harper, J. W., Elledge, S. J., and Leder, P. (1995). Mice lacking p21$^{CIP1/WAF1}$ undergo normal development, but are defective in G_1 checkpoint control. *Cell (Cambridge, Mass)* **82**, 675–684.

el-Deiry, W. S. (1998). p21/p53, cellular growth control and genomic integrity. In "Current Topics in Microbiology and Immunology" (P. K. Vogt and S. I. Reed, eds.). pp. 121–137. Springer-Verlag, Berlin.

el-Deiry, W. S., Tokino, T., Velculescu, V. E., Levy, D. B., Parsons, R., Trent, J. M., Lin, D., Mercer, W. E., Kinzler, K. W., and Vogelstein, B. (1993). WAF1, a potential mediator of p53 tumor suppression. *Cell (Cambridge, Mass.)* **75,** 817–825.

el-Deiry, W. S. Harper, J. W., O'Connor, P. M., Velculescu, V. E., Canman, C. E., Jackman, J., Pietenpol, J. A., Burrell, M., Hill, D. E., Wang, Y., Wiman, K. G., Mercer, W. E., Kastan, M. B., Kohn, K. W., Elledge, S. J., Kinzler, K. W., and Vogelstein, B. (1994). WAF1/CIP1 is induced in p53-mediated G_1 arrest and apoptosis. *Cancer Res.* **54,** 1169–1174.

el-Deiry, W. S., Tokino, T., Waldman, T., Oliner, J. D., Velculescu, V. E., Burrell, M., Hill, D. E., Healy, E., Rees, J. L., Hamilton, S. R., Kinzler, K. W., and Vogelstein, B. (1995) Topological control of p21WAF1/CIP1 expression in normal and neoplastic tissues. *Cancer Res.* **55,** 2910–2919.

Eller, M. S., Maeda, T., Magnoni, C., Atwal, D., and Gilchrest, B. A. (1997). Enhancement of DNA repair in human skin cells by thymidine dinucleotides: Evidence for a p53-mediated mammalian SOS response. *Proc. Natl. Acad. Sci. U. S. A.* **94,** 12627–12632.

Engelberg, D., Klein, C., Martinetto, H., Struhl, K., and Karin, M. (1994). The UV response involving the Ras signaling pathway and AP-1 transcription factors is conserved between yeast and mammals. *Cell (Cambridge, Mass.)* **77,** 381–390.

Esposito, F., Agosti, V., Morrone, G., Morra, F., Cuomo, C., Russo, T., Venuta, S., and Cimino, F. (1994). Inhibition of the differentiation of human myeloid cell lines by redox changes induced through glutathione depletion. *Biochem. J.* **301,** 649–653.

Esposito, F., Cuccovillo, F., Morra, F., Russo, T., and Cimino, F. (1995). DNA binding activity of the glucocorticoid receptor is sensitive to redox changes in intact cells. *Biochim. Biophys. Acta* **1260,** 308–314.

Esposito, F., Cuccovillo, F., Vanoni, M., Cimino, F., Anderson, C. W., Appella, E., and Russo, T. (1997). Redox-mediated regulation of p21 (waf1/cip1) expression involves a posttranscriptional mechanism and activation of the mitogen-activated protein kinase pathway. *Eur. J. Biochem.* **245,** 730–737.

Ewing, D. (1991). Can ·OH scavengers protect against direct UV-C damage in vivo? *Int. J. Radiat. Biol.* **60,** 449–452.

Fialkow, L., Chan, C. K., Rotin, D., Grinstein, S., and Downey, G. P. (1994). Activation of the mitogen-activated protein kinase signaling pathway in neutrophils. Role of oxidants. *J. Biol. Chem.* **269,** 31234–31242.

Fisher, G. R., and Gutierrez, P. L. (1991a). The reductive metabolism of diaziquinone (AZQ) in the S9 fraction of MCF-7 cells: Free radical formation and NAD(P)H: quinone-acceptor oxidoreductase (DT-diaphorase) activity. *Free Radical Biol. Med.* **10,** 359–370.

Fisher, G. R., and Gutierrez, P. L. (1991b) Free radical formation and DNA strand breakage during metabolism of diaziquone by NAD(P)H quinone-acceptor oxidoreductase (DT-diaphorase) and NADPH cytochrome c reductase. *Free Radical Biol. Med.* **11,** 597–607.

Forman, H. J., and Cadenas, E. (1997). "Oxidative Stress and Signal Transduction." Chapman & Hall, New York.

Gansauge, F., Gansauge, S., Poch, B., Schoenberg, M. H., and Beger, H. G. (1997). Oxygen radical-induced apoptosis of proliferating human fibroblasts. *Langenbecks Arch. Chir., Suppl.* **114,** 359–362.

Gansauge, S., Gansauge, F., Gause, H., Poch, B., Schoenberg, M. H., and Beger, H. G. (1997). The induction of apoptosis in proliferating human fibroblasts by oxygen radicals is associated with a p53- and p21$^{\text{WAF1CIP1}}$ induction. *FEBS Lett.* **404,** 6–10.

Ghosh, R., Amstad, P., and Cerutti, P. (1993). UVB-induced DNA breaks interfere with transcriptional induction of c-fos. *Mol. Cell. Biol.* **13,** 6992–6999.

Gibson, N. W., Hartley, J. A., Butler, J., Siegel, D., and Ross, D. (1992). Relationship between DT-diaphorase-mediated metabolism of a series of aziridinylbenzoquinones and DNA damage and cytotoxicity. *Mol. Pharmacol.* **42,** 531–536.

Goin, J., Ordoñez, I., and Cadenas, E. (1993). Reductive and oxidative decay pathways of semiquinones: The effects of glutathione and superoxide dismutase. *In* "Active Oxygens, Lipid Peroxides, and Antioxidants" (K. Yagi, ed.), pp. 69–82. CRC Press, Boca Raton, FL.

Griffiths, H. R., Mistry, P., Herbert, K. E., and Lunec, J. (1998). Molecular and cellular effects of ultraviolet light-induced genotoxicity. *Crit. Rev. Clin. Lab. Sci.* **35**, 189–237.

Gross, S. S., and Wolin, M. S. (1995). Nitric oxide: Pathophysiological mechanisms. *Ann. Rev. Physiol.* **57**, 737–769.

Hainaut, P., and Milner, J. (1993). A structural role for metal ions in the "wild-type" conformation of the tumor suppressor protein p53. *Cancer Res.* **53** 1739–1742.

Harper, J. W., and Elledge, S. J. (1996). Cdk inhibitors in development and cancer. *Cur. Opin. Genet. Dev.* **6**, 56–64.

Harrington, E. A., Bruce, J. L., Harlow, E., and Dyson, N. (1998). pRB plays an essential role in cell cycle arrest induced by DNA damage. *Proc. Natl. Acad. Sci. U. S. A.* **95**, 11945–11950.

Harris, C. C. (1996). p53 Tumor suppressor gene: From the basic research laboratory to the clinic—an abridged historical perspective. *Carcinogenesis (London)*, **17**, 1187–1198.

Hartwell, L. H., and Kastan, M. B. (1994). Cell cycle control and cancer. *Science* **266**, 1821–1828.

Hinds, P. W., and Weinberg, R. A. (1994). Tumor suppressor genes. *Curr. Opin. Genet. Dev.* **4**, 135–141.

Ho, Y. S., Wang, Y. J., and Lin, J. K. (1996). Induction of p53 and p21/WAF1/CIP1 expression by nitric oxide and their association with apoptosis in human cancer cells. *Mol. Carcinog.* **16**, 20–31.

Hunter, T., and Pines, J. (1994). Cyclins and cancer II: Cyclin D and CDK inhibitors come of age. *Cell (Cambridge, Mass.)* **79**, 573–582.

Hupp, T. R., Meek, D. W., Midgley, C. A., and Lane, D. P. (1993). Activation of the cryptic DNA binding function of mutant forms of p53. *Nucleic Acids Res.* **21**, 3167–3174.

Hutchison, K. A., Matic, G., Meshinchi, S., Bresnick, E. H., and Pratt, W. B. (1991). Redox manipulation of DNA binding activity and BuGR epitope reactivity of the glucocorticoid receptor. *J. Biol. Chem.* **266**, 10505–10509.

Ignarro, L. J. (1996). Nitric oxide as a communication signal in vascular and neuronal cells. *In* "Nitric Oxide: Principles and Actions" (J. Lancaster, Jr. ed.), pp. 111–137. Academic Press, San Diego, CA.

Ignarro, L. J. (1997). Activation and regulation of the nitric oxide-cyclic GMP signal transduction pathway by oxidative stress. *In* "Oxidative Stress and Signal Transduction" (H. J. Forman, and E. Cadenas, eds.). pp. 3–31. Chapman & Hall, New York.

Ihle, J. N. (1996). STATs and MAPKs: Obligate or opportunistic partners in signaling. *BioEssays* **18**, 95–98.

Ishida, A., Sasaguri, T., Kosaka, C., Nojima, H., and Ogata, J. (1997). Induction of the cyclin-dependent kinase inhibitor p21(Sdi1/Cip1/Waf1) by nitric oxide-generating vasodilator in vascular smooth muscle cells. *J. Biol. Chem.* **272**, 10050–10057.

Johnson, T. M., Yu, Z. X., Ferrans, V. J., Lowenstein, R. A., and Finkel, T. (1996). Reactive oxygen species are downstream mediators of p53-dependent apoptosis. *Proc. Natl. Acad. Sci. U. S. A.* **93**, 11848–11852.

Kadonaga, J. T., Carner, K. R., Masiarz, F. R., and Tjian, R. (1987). Isolation of cDNA encoding transcription factor Sp1 and functional analysis of the DNA binding domain. *Cell (Cambridge, Mass.)* **51**, 1079–1090.

Kamb, A. (1998). Cyclin-dependent kinase inhibitors and human cancer. *Curr. Top. Microbiol. Immunol.* **227**, 139–148.

Lane, D. P. (1992). Cancer. p53, guardian of the genome. *Nature (London)* **358**, 15–16.

Lane, D. (1998). Awakening angels. *Nature (London)* **394**, 616–617.

Lee, C. S., Hartley, J. A., Berardini, M. D., Butler, J., Siegel, D., Ross, D., and Gibson, N. W. (1992). Alteration in DNA cross-linking and sequence selectivity of a series of aziridinyl-benzoquinones after enzymatic reduction by DT-diaphorase. *Biochemistry* **31**, 3019–3025.

Li, R., Bianchet, M. A., Talalay, P., and Amzel, L. M. (1995). The three-dimensional structure of NAD(P)H:quinone reductase, a flavoprotein involved in cancer chemoprotection and chemotherapy: mechanism of the two-electron reduction. *Proc. Natl. Acad. Sci. U. S. A.* **92**, 8846–8850.

Liu, M., and Pelling, J. C. (1995). UV-B/A irradiation of mouse keratinocytes results in p53-mediated WAF1/CIP1 expression. *Oncogene* **10**, 1955–1960.

Liu, M., Lee, M. H., Cohen, M., Bommakanti, M., and Freedman, L. P. (1996a). Transcriptional activation of the Cdk inhibitor p21 by vitamin D3 leads to the induced differentiation of the myelomonocytic cell line U937. *Genes Dev.* **10**, 142–153.

Liu, M., Iavarone, A., and Freedman, L. P. (1996b). Transcriptional activation of the human p21 (WAF1/CIP1) gene by retinoic acid receptor. Correlation with retinoid induction of U937 cell differentiation. *J. Biol. Chem.* **271**, 31723–31728.

Liu, Y., Martindale, J. L., Gorospe, M., and Holbrook, N. J. (1996). Regulation of p21$^{WAF1/CIP1}$ expression through mitogen-activated protein kinase signaling pathway. *Cancer Res.* **56**, 31–35.

Loignon, M., Fetni, R., Gordon, A. J., and Drobetsky, E. A. (1997). A p53-independent pathway for induction of p21$^{waf1/cip1}$ and concomitant G_1 arrest in UV-irradiated human skin fibroblasts. *Cancer Res.* **57**, 3390–3394.

Lu, X., and Lane, D. P. (1993). Differential induction of transcriptionally active p53 following UV or ionizing radiation: Defects in chromosome instability syndromes? *Cell (Cambridge, Mass.)* **75**, 765–778.

Lu, Y., Yamagishi, N., Miyakoshi, J., Noda, A., Yagi, T., and Takebe, H. (1996). Sites and types of UV-induced mutations leading to inactivation of the growth-arresting activity in p21 (sdi1/cip1/waf1) cDNA. *Carcinogenesis (London)* **17**, 2343–2345.

Maltzman, W., and Czyzyk, L. (1984). UV irradiation stimulates levels of p53 cellular tumor antigen in nontransformed mouse cells. *Mol. Cell. Biol.* **4**, 1689–1694.

Martinez, J. D., Pennington, M. E., Craven, M. T., Warters, R. L., and Cress, A. E. (1997). Free radicals generated by ionizing radiation signal nuclear translocation of p53. *Cell Growth Differ.* **8**, 941–949.

McDonald, R. E., Wu, G. S., Waldman, T., and el-Deiry, W. S. (1996). Repair defect in p21$^{WACF1/CIP1-/-}$ human cancer cells. *Cancer Res.* **56**, 2250–2255.

McGrath, S. A. (1998). Induction of p21$^{WAF/CIP1}$ during hyperoxia. *Am. J. Respir. Cell Mol. Biol.* **18**, 179–187.

Messmer, U. K., Ankarcrona, M., Nicotera, P., and Brune, B. (1994). p53 expression in nitric oxide-induced apoptosis. *FEBS Lett.* **355**, 23–26.

Michieli, P., Chedid, M., Lin, D., Pierce, J. H., Mercer, W. E., and Givol, D. (1994). Induction of WAF1/CIP1 by a p53-independent pathway. *Cancer Res.* **54**, 3391–3395.

Moore, H. W. (1977). Bioactivation as a model for drug design. Bioreductive alkylation. *Science* **197**, 527–532.

Mueller, S., Cadenas, E., and Schönthal, A. H. (1999). *Cancer Res.*, submitted.

O'Connor, P. M., Jackman, J., Bae, I., Myers, T. G., Fan, S., Mutoh, M., Scudiero, D. A., Monks, A., Sausville, E. A., Weinstein, J. N., Friend, S., Fornace, A. J., Jr., and Kohn, K. W. (1997). Characterization of the p53 tumor suppressor pathway in cell lines of the National Cancer Institute Anticancer Drug Screen and correlations with the growth-inhibitory potency of 123 anticancer drugs. *Cancer Res.* **57**, 4285–4300.

Öllinger, K., Buffinton, G. D., Ernster, L., and Cadenas, E. (1990). Effect of superoxide dismutase on the autoxidation of substituted hydro- and semi-naphthoquinones. *Chem. Biol. Interact.* **73,** 53–76.

Ordoñez, I. D., and Cadenas, E. (1992). Thiol oxidation coupled to DT-diaphorase-catalysed reduction of diaziquone. Reductive and oxidative pathways of diaziquone semiquinone modulated by glutathione and superoxide dismutase. *Biochem. J.* **286,** 481–490.

Poluha, W., Schonhoff, C. M., Harrington, K. S., Lachyankar, M. B., Crosbie, N. E., Bulseco, D. A., and Ross, A. H. (1997). A novel, nerve growth factor-activated pathway involving nitric oxide, p53, and p21^{WAF1} regulates neuronal differentiation of PC12 cells. *J. Biol. Chem.* **272,** 24002–24007.

Pontén, F., Berne, B., Ren, Z. P., Nister, M., and Pontén, J. (1995). Ultraviolet light induces expression of p53 and p21 in human skin: Effect of sunscreen and constitutive p21 expression in skin appendages. *J. Invest. Dermatol.* **105,** 402–406.

Poon, R. Y., and Hunter, T. (1998). Expression of a novel form of p21$^{Cip1/Waf1}$ in UV-irradiated and transformed cells. *Oncogene* **16,** 1333–1343.

Poon, R. Y. C., Jiang, W., Toyoshima, H., and Hunter, T. (1996). Cyclin-dependent kinases are inactivated by a combination of p21 and Thr-14/Tyr-15 phosphorylation after UV-induced DNA damage. *J. Biol. Chem.* **271,** 13283–13291.

Powis, G., Gasdaska, J. R., and Baker, A. (1997). Redox signaling and the control of cell growth and death. *In* "Antioxidants in Disease Mechanisms and Therapy" (H. Sies, ed.), pp. 329–359. Academic Press, San Diego, CA.

Prives, C. (1998). Signaling to p53: Breaking the MDM2-p53 circuit. *Cell (Cambridge, Mass.)* **95,** 5–8.

Prowse, D. M., Bolgan, L., Molnar, A., and Dotto, G. P. (1997). Involvement of the Sp3 transcription factor in induction of p21$^{CIP1/WAF1}$ in keratinocyte differentiation. *J. Biol. Chem.* **272,** 1308–1314.

Qiu, X., and Cadenas, E. (1997). The role of NAD(P)H:quinone oxidoreductase in quinone-mediated p21 induction in human colon carcinoma cells. *Arch. Biochem. Biophys.* **346,** 241–251.

Qiu, X., Schönthal, A. H., Forman, H. J., and Cadenas, E. (1996). Induction of p21 mediated by reactive oxygen species formed during the metabolism of aziridinylbenzoquinones by HCT116 cells. *J. Biol. Chem.* **271,** 31915–31922.

Qiu, X., Schönthal, A. H., and Cadenas, E. (1998). Anticancer quinones induce pRb-preventable G$_2$/M cell cycle arrest and apoptosis. *Free Radical Biol. Med.* **24,** 848–854.

Reinke, V., and Lozano, G. (1997). Differential activation of p53 targets in cells treated with ultraviolet radiation that undergo both apoptosis and growth arrest. *Radiat. Res.* **148,** 115–122.

Renzing, J., Hansen, S., and Lane, D. P. (1996). Oxidative stress is involved in the UV activation of p53. *J. Cell Sci.* **109,** 1105–1112.

Richter, C., and Laffranchi, R. (1997). Nitric oxide signaling through mitochondrial calcium release. *In* "Oxidative Stress and Signal Transduction" (H. J. Forman and E. Cadenas, eds.), pp. 52–76. Chapman & Hall, New York.

Riley, R. J., and Workman, P. (1992). DT-diaphorase and cancer chemotherapy. *Biochem. Pharmacol.* **43,** 1657–1669.

Russo, T., Zambrano, N., Esposito, F., Ammendola, R., Cimino, F., Fiscella, M., Jackman, J., O'Connor, P. M., Anderson, C. W., and Appella, E. (1995). A p53-independent pathway for activation of WAF1-CIP1 expression following oxidative stress. *J. Biol. Chem.* **270,** 29386–29391.

Schieven, G. L. (1997). Tyrosine phosphorylation in oxidative stress. *In* "Oxidative Stress and Signal Transduction" (H. J. Forman and E. Cadenas, eds.), pp. 200–238. Chapman & Hall, New York.

Sheikh, M. S., Li, X., Chen, J., Shao, Z., Ordoñez, J. V., and Foutana, J. A. (1994). Mechanisms of regulation of WAF1/CIP1 gene expression in human breast carcinoma: Role of p53-dependent and independent signal transduction pathways. *Oncogene* **9**, 3407–3415.

Sherr, C. J., and Roberts, J. M. (1995). Inhibitors of mammalian G_1 cyclin-dependent kinases. *Genes Dev.* **9**, 1149–1163.

Shiohara, M., el-Deiry, W. S., Wada, M., Nakamaki, T., Takeuchi, S., Yang, R., Chen, D. L., Vogelstein, B., and Koeffler, H. P. (1994). Absence of WAF1 mutations in a variety of human malignancies. *Blood* **84**, 3781–3784.

Sies, H. (1986). Biochemistry of oxidative stress. *Angew. Chem., Int. Ed. Engl.* **25**, 1058–1071.

Simon, A. R., Cochran, B. H., and Fanburg, B. L. (1997a) Oxidative stress and cell proliferation. *In* "Oxygen, Gene Expression, and Cellular Function" (D. Massaro and L. Clerch, eds.), pp. 123–138. Dekker, New York.

Simon, A. R., Fanburg, B. L., and Cochran, B. H. (1997b). STAT activation by oxidative stress. *In* "Oxidative Stress and Signal Transduction" (H. J. Forman and E. Cadenas, eds.), pp. 260–271. Chapman & Hall, New York.

Smith, M. L., and Fornace, A. J., Jr. (1997). p53-mediated protective responses to UV irradiation. *Proc. Natl. Acad. Sci. U. S. A.* **94**, 12255–12257.

Somasundaram, K., Zhang, H., Zeng, Y. X., Houvras, Y., Peng, Y., Wu, G. S., Licht, J. D., Weber, B. L., and el-Deiry, W. S. (1997). Arrest of the cell cycle by the tumour-suppressor BRCA1 requires the CDK-inhibitor p21WAF1/CiP1. *Nature (London)* **389**, 187–190.

Spear, N., Estévez, A. G., Radi, R., and Beckman, J. S. (1997). Peroxynitrite and cell signaling. *In* "Oxidative Stress and Cell Signaling" (H. J. Forman and E. Cadenas, eds.), pp. 32–51. Chapman & Hall, New York.

Tedeschi, G., Chen, S., and Massey, V. (1995). DT-Diaphorase: Redox potential, steady state, and rapid reaction studies. *J. Biol. Chem.* **270**, 1198–1204.

Vile, G. F. (1997). Active oxygen species mediate the solar ultraviolet radiation-dependent increase in the tumour suppressor protein p53 in human skin fibroblasts. *FEBS Lett.* **412**, 70–74.

Vile, G. F., Rothwell, L. A., and Kettle, A. J. (1998). Hypochlorous acid activates the tumor suppressor protein p53 in cultured human skin fibroblasts. *Arch. Biochem. Biophys.* **359**, 51–56.

Wen, Z., Zhong, Z., and Darnell, J. E., Jr. (1995). Maximal activation of transcription by Stat1 and Stat3 requires both tyrosine and serine phosphorylation. *Cell (Cambridge, Mass.)* **82**, 241–250.

Wu, R.-C., Hohenstein, A., Park, J. M., Qiu, X., Mueller, S., Cadenas, E., and Schönthal, A. H. (1998). Role of p53 in aziridinylbenzoquinone-induced $p21^{wafl}$ expression. *Oncogene* **17**, 357–365.

Zeng, Y.-X., and el-Deiry, W. S. (1996). Regulation of $p21^{WAF1/CIP1}$ expression by p53-independent pathways. *Oncogene* **12**, 1557–1564.

Zeng, Y.-X., Somasundaram, K., and el-Deiry, W. S. (1997). AP2 inhibits cancer cell growth and activates $p21^{WAF1/CIP1}$ expression. *Nat. Genet.* **15**, 78–82.

Zhan, Q., Carrier, F., and Fornace, A. J., Jr. (1993). Induction of cellular p53 activity by DNA-damaging agents and growth arrest. *Mol. Cell. Biol.* **13**, 4242–4250.

III Clinical Implications of Redox Signaling and Antioxidant Therapy

15 Effects of Lipoxygenases on Gene Expression in Mammalian Cells

Helena Viita and Seppo Ylä-Herttuala
A.I. Virtanen Institute
University of Kuopio
FIN-70211 Kuopio, Finland

Lipoxygenases are intracellular enzymes that oxidize either free or esterified polyunsaturated fatty acids. Their metabolites are lipid peroxides that can react with other biological compounds and alter the cellular redox state. Many of the physiological roles of lipoxygenases are still unknown, but 5-lipoxygenase is involved in leukotriene metabolism, 12-lipoxygenase in the angiotensin pathway, and 15-lipoxygenase in reticulocyte maturation. A common finding characterizing lipoxygenases is their involvement in the regulation of gene expression of transcription factors, cytokines, protooncogenes, and growth factors. Acting as secondary messengers in various signal transduction pathways, lipoxygenases have been linked to many pathological conditions, such as asthma, inflammation, immune response, and atherosclerosis. This chapter reviews literature about the effects of lipoxygenases on the regulation of gene expression in mammalian cells.

I. INTRODUCTION

A. Lipid Peroxidation by Lipoxygenases

Lipoxygenases (LO) are nonheme-containing dioxygenases that recognize the 1,4-pentadiene structure of polyunsaturated fatty acids and incorporate single molecules of oxygen at specific carbon atoms of their substrate fatty acids. Their primary reaction products are hydroperoxy fatty acids containing conjugated dienes (Fig. 1). *In vivo*, these fairly reactive compounds are usually reduced to their corresponding hydroxy fatty acids or are converted to other secondary products. The oxygenation site of the LO substrate differs from enzyme to enzyme. Each LO is referred to as arachidonate x-LO where x is the carbon predominantly oxygenated in arachidonic acid (AA). The numbering of the carbons starts from the carboxylic carbon (Yamamoto, 1992).

The substrate specificities of the different LOs vary remarkably. 5-LO reacts predominantly with C_{20} fatty acids (Yamamoto, 1992), and the mouse 8S-LO can use AA and linoleic acid (LA) as substrate (Jisaka *et al.*, 1997). There are two distinct types of 12-LOs that differ in their substrate specificity: the platelet-type 12-LO reacts mainly with AA, whereas the leukocyte-type 12-LO has a wide substrate specificity and can react with C_{18} and C_{22} fatty

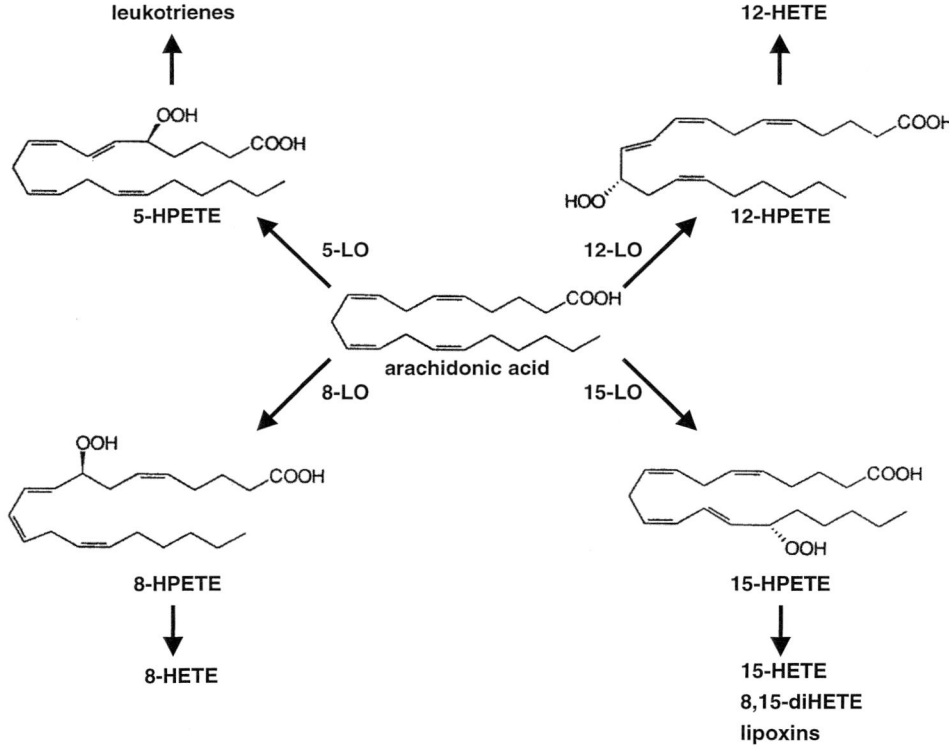

Figure 1 Arachidonic acid metabolism by mammalian lipoxygenases.

acids as fast as with AA. Reticulocyte 15-LO also shows wide substrate specificity in terms of carbon chain length: it can oxygenate unsaturated C_{18}, C_{20}, and C_{22} fatty acids in free or esterified form. The optimal substrate is LA, and reaction with AA produces mainly 15S-hydroperoxyeicosatetraenoic acid (15S-HPETE), but also 12S-HPETE as a side product (Yamamoto, 1992). On the contrary, the 15S-LO isolated from human hair roots produces only 12S-HPETE from AA, and LA is a very poor substrate (Brash et al., 1997).

B. Cloning of Lipoxygenase cDNAs and Genes

Mammalian LO cDNAs and genes have been cloned from various species and libraries (Table I) since 1987, when a partial cDNA clone of the rabbit reticulocyte 15-LO was isolated (Thiele et al., 1987). The LOs are clearly a family of enzymes that have a relatively high overall sequence similarity and some conserved regions responsible for iron binding and catalysis (Sigal, 1991).

C. Physiological Roles of Lipoxygenases

All LOs are intracellular enzymes involved in the regulated metabolism of AA (Fig. 1), which is a common constituent of cell membrane phospholipids. In response to various external stimuli, free AA is released from the membranes by the action of phospholipases. The released AA is consequently metabolized via the cyclooxygenase (CO) or the LO pathway (Sigal, 1991). The best known physiological role of 5-LO is the production of leukotrienes (LT) (A_4, B_4, C_4, D_4, and E_4), which have been indicated in immune reactions, hypersensitivity, inflammation, and asthma (Samuelsson et al., 1987). 5-LO activating protein (FLAP) is also essential for leukotriene synthesis (Mancini et al., 1993).

In contrast, the physiological roles of 8-, 12-, and 15-LO are still mostly unknown. Some roles for 12-LO metabolites have been reported in the angiotensin pathway (Natarajan et al., 1993, 1994), chemoattraction (Nakao et al., 1982), angiogenesis and mitogenesis (Setty et al., 1987), neurotransmission (Piomelli et al., 1987), regulation of adhesion molecules (Grossi et al., 1989), and cancer metastasis (Honn et al., 1994).

15-LO was first purified from rabbit reticulocytes and was shown to be involved in the degradation of reticulocyte mitochondria during red cell maturation (Rapoport et al., 1979). Since then, a growing body of evidence has implied the involvement of 15-LO in the oxidative modification of low-density lipoprotein (LDL) (Parthasarathy et al., 1989; McNally et al., 1990; Rankin et al., 1991) and in the inflammatory process of atherosclerosis

Table I Complementary and Genomic DNAs for Mammalian Lipoxygenases

	cDNA	Genomic DNA
5-LO	Human lung and placenta (Matsumoto et al., 1988) Human differentiated HL-60 cells (Dixon et al., 1988) Rat basophilic leukemia cells (Balcarek et al., 1988) Mouse macrophage (Chen et al., 1995) Syrian hamster embryo fibroblast (Kitzler and Eling, 1996)	Human genomic bacteriophage and cosmid libraries (Funk et al., 1989) Guinea pig genomic DNA library (Chopra et al., 1992)
8-LO	Mouse epidermis (Jisaka et al., 1997)	
L-12-LO[a]	Porcine leukocyte (Yoshimoto et al., 1990a) Bovine epithelium (De Marzo et al., 1992)[d] Rat brain/leukocyte/lung (Watanabe et al., 1993; Hada et al., 1994; Katoh et al., 1994)[d] Mouse leukocyte type (Chen et al., 1994a; Krieg et al., 1995) Mouse macrophage (Freire-Moar et al., 1995)[d]	Porcine EMBL3 genomic library (Arakawa et al., 1992) Mouse leukocyte type from 129 Sv genomic library (Chen et al., 1994a) Mouse macrophage from 129 Sv genomic library (Freire-Moar et al., 1995)[d]
P-12-LO[b]	Human platelet/erythroleukemia cells (Funk et al., 1990; Yoshimoto et al., 1990b) Human platelet (Izumi et al., 1990) Human epidermal cells (Hussain et al., 1994; Hagmann et al., 1996) Mouse platelet type (Chen et al., 1994a; Krieg et al., 1995) Mouse lung carcinoma (Hagmann et al., 1995)	Human genomic bacteriophage and cosmid (Funk et al., 1992) and λ libraries (Yoshimoto et al., 1992) Mouse platelet type from 129 Sv genomic library (Chen et al., 1994a)
E-12-LO[c]	Mouse epidermal (Aloxe) (van Dijk et al., 1995; Funk et al., 1996; Kinzig et al., 1997)	Mouse 12/15-LO (Aloxe) from 129 Sv genomic phage library (van Dijk et al., 1995)
15-LO	Rabbit reticulocyte (Thiele et al., 1987; Fleming et al., 1989) Human reticulocyte (Sigal et al., 1988) Human bronchus (Sigal et al., 1992)[d] Human breast carcinoma (Reddy et al., 1997)[d]	Rabbit genomic DNA library (O'Prey et al., 1989) Human monocyte (Kritzik et al., 1997)

(*continues*)

Table I (Continued)

cDNA	Genomic DNA
Human hair root (Brash et al., 1997) Human monocyte (Kritzik et al., 1997)[d]	

[a] Leukocyte-type 12-LO.
[b] Platelet-type 12-LO.
[c] Epidermal 12-LO.
[d] Homolog for human reticulocyte 15-LO.

development (Ylä-Herttuala et al., 1990, 1991, 1995; Kühn et al., 1994b, 1997; Hiltunen et al., 1995). 15-LO expression is regulated at the transcriptional level by a transcriptional silencer on the 5′ flanking DNA (O'Prey and Harrison, 1995) and at the translational level by a 3′-untranslated region-binding protein (Ostareck-Lederer et al., 1994).

There are two important differences between the reticulocyte 15-LO and other mammalian LOs in respect to atherogenesis. As mentioned previously (Section I,A), the optimal substrate for reticulocyte 15-LO is LA, which is rich in LDL. In addition to this, 15-LO (and the closely related porcine 12-LO) is the only known mammalian LO that can also oxidize complex lipids such as biomembranes and LDL (Kühn et al., 1994a).

D. Experimental Systems Used for Lipoxygenase Studies

Most of the *in vitro* LO studies have used various cell lines expressing LO activity and an array of different inhibitors. Some studies have taken advantage of transient transfection techniques (Wölle et al., 1996) or virus-mediated stable gene transfer (Benz et al., 1995; Ezaki et al., 1995; Viita et al., 1998) for generating cell lines expressing a specific LO. Virus-mediated gene transfer has been used for *in vivo* studies (Ylä-Herttuala et al., 1995). Also, transgenic LO mice and rabbits (Harats et al., 1995; Shen et al., 1995, 1996) and LO knockout mice (Chen et al., 1994b; Goulet et al., 1994; Sun and Funk, 1996) have been generated, but the mouse strains have not revealed any dramatically altered phenotypes.

II. EFFECTS OF LIPOXYGENASES ON GENE EXPRESSION

A new exciting area of LO research has emerged. Oxidized metabolites produced by different LOs can alter the redox balance within the cells and thus act as mediators in signal transduction pathways and in the regulation of gene expression (Fig. 2). Studies indicating the involvement of LOs in the regulation of gene expression are from several different fields of research. However, a common finding is LO involvement in the regulation of gene expression of transcription factors, cytokines, protooncogenes, and growth factors. In addition, LO reaction products have some direct physiological effects on cells and homeostasis. The effects on gene expression are summarized in the following sections according to the specific LO involved. Section II,D reviews studies in which LO involvement was shown, but not specified to a certain LO.

A. 5-Lipoxygenase

The primary product of 5-LO with AA is 5-HPETE which is metabolized further to LTs (Fig. 1). These have been implicated in immune reactions, hypersensitivity, inflammation, and asthma (Samuelsson *et al.*, 1987).

1. Regulation of Transcription Factor NF-κB

NF-κB is a well-defined transcription factor regulated by the intracellular redox state. NF-κB is involved in the induced expression of various genes that are sensitive to oxidative stress (Fig. 2) (Baeuerle and Henkel, 1994; Sen and Packer, 1996).

Tepoxalin, a dual inhibitor of CO and 5-LO, was shown to suppress NF-κB activation in various cell lines (Kazmi *et al.*, 1995). Also, NF-κB-

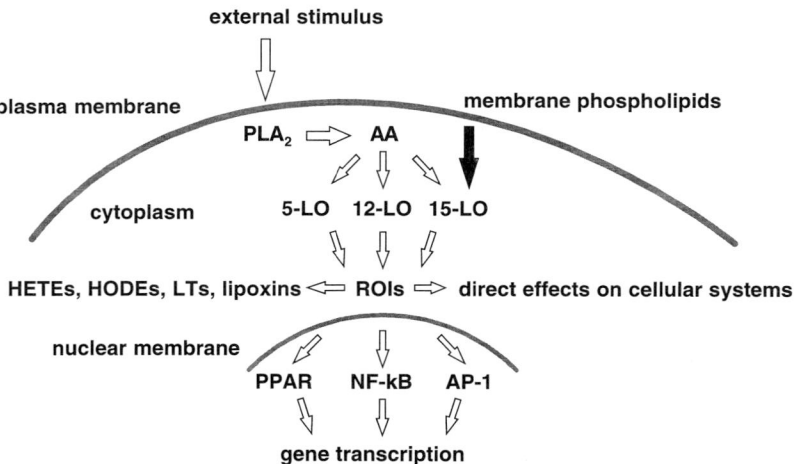

Figure 2   Possible pathways for effects of LOs on gene expression and cellular functions.

dependent gene expression of interleukin-6 (IL-6) and c-myc was suppressed. However, other CO and LO inhibitors did not inhibit NF-κB activity, so the inhibitory activity of tepoxalin was probably not related to the AA pathway. The actual mechanism of tepoxalin-induced inhibition of NF-κB remains unclear.

Los et al., (1995) have shown that ligation of the surface receptor CD28 in primary T lymphocytes leads to the rapid intracellular formation of reactive oxygen intermediates (ROIs), activation of NF-κB, and IL-2 gene expression. Their data show that 5-LO is involved in ROI formation and the following events.

No data of 5-LO involvement in the regulation of transcription factor activating protein 1 (AP-1) are yet available.

2. Regulation of Protein Kinase C

Protein kinase C (PKC) is a protein kinase family with multiple isoforms. It plays an integral part in signal transduction and can be activated by various lipid metabolites (Nishizuka, 1995). PKC, AA, and some of its LO products have been shown to be involved in gonadotropin-releasing hormone-induced gonadotropin secretion and synthesis. These relationships have been studied in more detail by Shraga-Levine et al. (1996) who show that the addition of AA or its 5-LO products 5-hydroxyeicosatetraenoic acid (5-HETE) or LTC$_4$ increase the mRNA expression of PKCβ in pituitary αT3-1 cells stimulated by the gonadotropin hormone analog. This could be abolished by the phospholipase A$_2$ (PLA$_2$) inhibitor 4-bromophenacyl bromide or the 5-LO inhibitor L-656,224.

3. Regulation of Cytokines

a. Tumor Necrosis Factor α Tumor necrosis factor (TNF)-α is a redox-sensitive cytokine with various biological effects that have been linked to the AA cascade. TNF-α can be activated transcriptionally by a wide variety of different stimuli, e.g., phorbol esters and lipopolysaccharide (LPS). Several studies have implicated the involvement of AA cascade, PLA$_2$, and 5-LO in the signaling events leading to the activation of TNF-α (Horiguchi et al., 1989; Mohri et al., 1990; Spriggs et al., 1990).

Horiguchi et al. (1989) have studied the phorbol ester-induced activation of TNF-α and AA release in human promyelocytic leukemia HL-60 cells. They showed that the induced mRNA expression of TNF-α was mediated by the activation of PKC and that PLA$_2$ inhibitors blocked both the induction of TNF-α expression and the increase in AA release. Furthermore, TNF-α induction could also be blocked by the 5-LO inhibitor ketoconazole and induced by the 5-LO metabolite LTB$_4$. They suggest that 12-O-tetradeca-noylphorbol-13-acetate (TPA)-induced TNF-α expression is mediated by

AA cascade and is regulated by metabolites of this pathway. The phorbol ester-mediated induction of TNF-α in HL-60 cells has also been shown to be blocked by ketoconazole (Meier et al., 1995).

The effects of LPS on PLA$_2$ activity and TNF-α expression in HL-60 cells were studied by Mohri et al. (1990). LPS stimulated PLA$_2$ activity and TNF-α expression in these cells, and the induction of TNF-α could be blocked by inhibitors of PLA$_2$ and 5-LO (ketoconazole). These results suggest that the LPS-mediated induction of TNF-α is regulated through the AA cascade, especially by PLA$_2$ and 5-LO.

5-LO has also been indicated in the regulation of both autoinduction and radiation induction of TNF-α gene expression (Spriggs et al., 1990; Hallahan et al., 1994).

b. Interleukins Interleukins belong to the hematopoietin family of cytokines, are produced by T lymphocytes in response to antigen-stimulated cellular activation, and function in the immune response by inducing biochemical signals (Paul and Seder, 1994).

5-LO has been implicated in the activation of NF-κB and IL-2 gene expression (Los et al., 1995). They propose that this signaling cascade involved protein tyrosine kinase (PTK) activity, which activated PLA$_2$ and 5-LO, leading to ROI formation via 5-LO metabolites. Increased oxidative stress then activated NF-κB and led to the induction of IL-2 gene expression. 5-LO inhibition has also been shown to inhibit the accumulation of IL-2 and IL-6 mRNA in human peripheral blood mononuclear cells (Atluru et al., 1993), and phorbol ester induced IL-1β mRNA expression in HL-60 cells (Meier et al., 1995). Also, a dual inhibitor of CO and 5-LO, tepoxalin, has been shown to suppress NF-κB regulated expression of IL-6 (Kazmi et al., 1995).

4. Protooncogene Regulation and Mitogenesis

Protooncogenes are cellular counterparts of the viral oncogenes and are involved in normal cell functions. Their association with tumor formation requires mutation or aberrant activation, e.g., constitutive activation of a regulated function, expression of the gene in an unusual cell type, or overexpression in the usual tissue (Lewin, 1994).

5-LO has been implicated in growth factor-induced mitogenesis. Inhibition of 5-LO suppressed platelet-derived growth factor (PDGF)-induced expression of the protooncogenes *fos* and *egr* and inhibited PDGF-induced mitogenesis (Beno et al., 1995).

5. Adhesion Molecule Expression

Inflammation is characterized by the recruitment of leukocytes from the circulation and subsequent migration to the subendothelial space. Endothelial cells expressing cell surface adhesion molecules play a major role in this event (Lee et al., 1997). Many of the adhesion molecules are under the regulation of redox-sensitive transcription factor NF-κB (Collins et al., 1995).

Lee et al. (1997) have studied the IL-1β-induced vascular cell adhesion molecule-1 (VCAM-1) and intercellular adhesion molecule-1 (ICAM-1) gene expression in human vascular endothelial cells (HUVEC). They have shown that IL-1β-induced VCAM-1 and ICAM-1 cell surface expression and VCAM-1 mRNA expression and promoter activity can be blocked by LO inhibitors [nonspecific nordihydroguairetic acid (NDGA) and specific 5-LO inhibitor AA861]. These inhibitors did not block the nuclear translocation of NF-κB nor the phosphorylation of p65. According to their studies, 5-LO inhibitors reduced the functional activity of NF-κB/Rel proteins in HUVECs and thus blocked the IL-1β-mediated induction of VCAM-1 gene expression. A possible role for the 5-LO inhibitor as an anti-inflammatory agent is also suggested.

6. Gonadotropin α-Subunit Expression

The gonadotropin-releasing hormone regulates the synthesis and release of gonadotropins that contain a common α subunit and a specific β subunit. The signaling events involved in the gonadotropin release include translocation and activation of PKC, release of AA, and formation of bioactive LO products. The regulation of α gene expression was studied in a transformed gonadotroph cell line that produces the common α subunit (Ben-Menahem et al., 1994). 5-LO metabolites 5-HETE and LTC$_4$ stimulated α-subunit mRNA levels, and gonadotropin-releasing hormone-induced α-subunit expression was inhibited by the specific 5-LO inhibitor L-656,224.

7. Type I Collagen Gene Expression

Chen et al. (1996) have studied the effects of 5-LO inhibitor ICI 230487 on the expression of type I collagen in rat hepatic stellate cells. The inhibitor resulted in a marked decrease in the type I collagen mRNA expression. Their studies further showed that the suppression of gene transcription was localized to a NF-1-binding domain and an AP-2-binding domain adjacent to it in the proximal promoter region. They concluded that the decreased gene expression due to the 5-LO inhibitor is partly a consequence of the AP-2 transmodulation of NF-1-dependent gene transcription.

This finding could lead to the development of therapeutic approaches for pathologic wound healing conditions where type I collagen is overexpressed, e.g., in hepatic fibrosis and cirrhosis.

In conclusion, several studies have implicated the involvement of 5-LO in the regulation of gene expression of transcription factors, cytokines, and various other genes. 5-LO has been implicated in pathologic situations, such as asthma, immune reactions, and inflammation. Specific 5-LO inhibitors could be potentially useful drugs in these pathologic conditions.

B. 12-Lipoxygenase

The physiological role of 12-LO and its reaction products is still mostly unknown. The primary product from AA is 12-HPETE (Fig. 1), which *in vivo* is reduced to 12-HETE. Some possible functions of 12-LO metabolites have been reported (see Section I,C).

1. Protooncogene Expression and DNA Synthesis

Neonatal rat lens epithelium has a high 12(S)-HETE synthetic capacity, which decreases with age as epithelial cell proliferation decreases. Incubation of rat lenses with epidermal growth factor and insulin stimulated 12-HETE production, induced c-*fos* and c-*myc* mRNAs, and DNA synthesis (Lysz *et al.*, 1994). The 12-LO inhibitor cinnamyl-3,4-dihydroxy-α-cyanocinnamate (CDC) inhibited endogenous 12-HETE production and growth factor-stimulated DNA synthesis. Also, the expression of the protooncogenes c-*fos* and c-*myc* was inhibited. These inhibitory effects were prevented by 12(S)-HETE, but not by 5(S)-HETE or 15(S)-HETE. The same effects were also observed in primary human lens epithelial cells (Arora *et al.*, 1996) where epidermal growth factor and insulin stimulated 12(S)-HETE release, DNA synthesis, and c-*fos* mRNA expression. Stimulation of DNA synthesis and c-*fos* mRNA expression were inhibited by CDC, and these inhibitory effects were completely reversed by exogenous 12(S)-HETE. No induction of DNA synthesis was observed in the absence of growth factors.

The authors suggest a role for 12(S)-HETE in the signal transduction events that couple receptor activation with the transcription of growth regulatory genes. They also suggest further studies for evaluating the potential use of LO inhibitors for limiting lens cell proliferation after cataract surgery.

2. Expression of Vascular Endothelial Growth Factor

Vascular endothelial growth factor (VEGF) can promote the growth of endothelial cells and increase vascular permeability and monocyte migration. It has been implicated in the pathogenic neovascularization associated with diabetes, atherosclerosis, and tumor angiogenesis (Connolly *et al.*, 1989; Keck *et al.*, 1989; Natarajan *et al.*, 1997).

Because 12-LO products of AA have angiogenic properties, Natarajan *et al.* (1997) studied the effect of 12-HETE on VEGF expression in human and porcine smooth muscle cells. 12-HETE significantly increased VEGF mRNA and protein expression in both these cell lines and in both normal or high glucose conditions. Other ROIs (superoxide and hydrogen peroxide) have previously been shown to increase VEGF gene expression (Kuroki et al., 1996). By increasing VEGF expression, 12-LO may have a role in diabetes and vascular diseases, neovascularization, ischemia, and tumor angiogenesis.

3. Integrin Induction

Vascular endothelial cells express several integrin receptors that play an essential role in the formation and maintenance of endothelial cell monolayers and in linking the endothelial cells to extracellular matrix. Tang et al. (1995) studied the expression of integrin $\alpha_v\beta_3$ in mouse endothelial cells from lung microvasculature and showed that 12(S)-HETE transcriptionally activated the expression of integrin αv in these cells in a PKC-dependent manner. They suggest a possible role for 12(S)-HETE in tumor cell adhesion, angiogenesis, hemostasis, and other vascular events.

Even though the physiological role of 12-LO and its products is still unknown, there are some studies showing 12-LO mediated regulation of gene expression as reviewed earlier. 12-LO may have a role in regulating protooncogene expression and cell proliferation in many vascular events and in tumor cell adhesion and tumor metastasis.

C. 15-Lipoxygenase

The main reaction products of reticulocyte 15-LO are 15-HPETE and 13-hydroperoxyoctadecadienoic acid (13-HPODE) from AA and LA, respectively. These are *in vivo* reduced to their hydroxy derivatives 15-HETE and 13-hydroxyoctadecadienoic acid (13-HODE). Some possible physiological roles of 15-LO are presented in Section II,C.

1. Adhesion Molecule Expression

15-Lipoxygenase has been indicated in the induction of VCAM-1 gene expression in a study where human 15-LO cDNA was transfected transiently to bovine aortic endothelial cells (Wölle et al., 1996). Transfection led to the expression of 15-LO protein and enzymatic activity, but no detectable VCAM-1 message was expressed. After stimulus with TNF-α, VCAM-1 expression was clearly induced in 15-LO cells compared to cells that had been transfected with a control plasmid.

We have studied the expression of adhesion molecules VCAM-1 and ICAM-1 in human ECV304 cells where human 15-LO cDNA was stably transduced by retrovirus-mediated gene transfer (Viita et al., 1999). Our results show that the gene transfer leads to 15-LO mRNA expression and enzymatic activity. Because 15-LO can affect the redox balance within the cells, we wanted to study the effect of 15-LO expression on the activation of NF-κB and on the expression of NF-κB-regulated adhesion molecules VCAM-1 and ICAM-1. The control cells in this study were ECV304 cells transduced with retrovirus carrying the *Escherichia coli lacZ* gene. Our results show that NF-κB activation is clearly potentiated in 15-LO cells compared to control cells in response to various stimuli. Also, VCAM-1 and ICAM-1 expression were induced in 15-LO cells compared to control cells, and the induction could be prevented by antioxidant α-lipoic acid in a dose-dependent manner. Both these studies indicate that 15-LO can be involved in the early phase of atherosclerosis by amplifying the effects of inflammatory stimuli on the activation of NF-κB and the induction of adhesion molecules.

2. Regulation of Macrophage Gene Expression through Peroxisome Proliferator-Activated Receptor γ

Peroxisome proliferator-activated receptors (PPARs) are a subfamily of the nuclear hormone receptor gene family. Activated PPARs heterodimerize with another nuclear receptor, retinoid X receptor, and after binding to specific peroxisome proliferator response elements, the complex alters the transcription of their target genes (Fig. 2). PPARs were known to be activated by various substances, but originally no direct binding of any of these compounds to the receptors could be demonstrated (Schoonjans *et al.*, 1997). Nagy *et al.* (1998) showed that oxidized LDL activates PPARγ-dependent transcription through a novel signaling pathway involving scavenger receptor-mediated cellular uptake. Two of the major oxidized lipid components of oxidized LDL, 9-HODE and 13-HODE, were endogenous activators and ligands of PPARγ. This finding and the demonstration of PPARγ expression at high levels in the foam cells of atherosclerotic lesions (Tontonoz *et al.*, 1998) have led to the conclusion that PPARγ may be a key regulator of gene expression of atherosclerotic foam cells.

As described earlier, very little is known so far about the possible role of 15-LO in regulating gene expression. Current data give new aspects about the role of 15-LO in atherosclerosis, giving evidence that in addition to its possible roles in the oxidation of LDL and macrophage differentiation, 15-LO may also have a major role in regulating gene expression in atherosclerotic lesions (Hiltunen *et al.*, 1995).

D. Effects Associated with Lipoxygenases without Demonstration of Positional Specificity

1. Protooncogene Expression, Mitogen-Activated Protein Kinase Activation, DNA Synthesis, and Growth Induction

Rao *et al.* (1995) studied the effects of LA and its LO products HPODEs on various parameters affecting mitogenesis in vascular smooth muscle cells

(VSMC) in order to determine their role in the modulation of cell growth in atherosclerosis. LA and 9- and 13-HPODE induced c-*fos*, c-*jun*, and c-*myc* protooncogene mRNA expression, mitogen-activated protein kinase (MAPK) activation, and DNA synthesis in VSMC. All these effects were blocked by the LO inhibitor NDGA, suggesting that LOs and their products were needed for these effects. In another study by Rao *et al.* (1996), hydrogen peroxide stimulated the production of 12- and 15-HPETE in VSMC, and both these HPETEs induced the expression of c-fos and c-jun protein. This induction was blocked by NDGA. Tebbey and Buttke (1993) have also shown the induction of c-*fos* gene transcription by AA and the suppression of this induction by the LO inhibitor NDGA. Thus, LO-mediated lipid peroxidation and altered redox status could play an important role in the modulation of VSMC growth in atherosclerosis.

2. Regulation of Transcription Factor-Activating Protein 1

AP-1 is a transcription factor regulated by the intracellular redox state. The major constituents of AP-1 are cFos and cJun proteins. DNA binding of Fos–Jun or Jun–Jun complexes to the TPA response element (TRE) of several genes is required for the transcriptional regulation induced by AP-1. Most of the induced AP-1 are Fos–Jun heterodimers (Sen and Packer, 1996). Since Rao *et al.* (1996) (Section II,D,1) showed that 12- and 15-HPETE induced the expression of the protooncogenes c-*fos* and c-*jun*, constituents of AP-1, they also studied the effects of these HPETEs on the DNA-binding activity of AP-1 and on AP-1-regulated transcription. Both these HPETEs, hydrogen peroxide, and AA stimulated AP-1 DNA-binding activity and also AP-1-dependent reporter gene transcription in VSMC. NDGA significantly inhibited AP-1 activity stimulated by both hydrogen peroxide and AA. They suggest that the LO metabolites 12- and 15-HPETE are involved in the activation of c-*fos* and c-*jun* protooncogenes, AP-1 activity, oxidative stress-induced AP-1-dependent gene transcription, and growth regulation in VSMC.

3. Stimulation of Nerve Growth Factor Synthesis

Cytokines are potent inducers of nerve growth factor (NGF) expression both in peripheral tissues and in the central nervous system. NGF synthesis is induced by inflammatory mediators IL-1, TNF-α, and transforming growth factor-β (TGF-β), and NGF and the NGF receptor system have been implicated in inflammatory and immune reactions. The simultaneous addition of IL-1β and TNF-α stimulated the synthesis of NGF in rat renal mesangial cells. Pretreatment of the cells with NDGA abolished this induction, whereas specific CO inhibitors had no effect. This suggested a regulatory role for a LO metabolite in mediating NGF gene expression in response to IL-1β and

TNF-α, and a possible role in the pathophysiology of inflammatory renal diseases (Steiner *et al.*, 1991).

III. CONCLUSIONS

Lipoxygenases and their metabolites have been indicated as possible mediators in signal transduction pathways and in the regulation of gene expression. In most of the experiments done so far, different LO inhibitors and LO metabolites have been used in studying the effects of LOs and their reaction products on gene expression. Some studies have already given evidence about the effects of LO activity on the cellular redox state and on the activation of redox-sensitive transcription factors regulating the expression of various genes.

In the future, studies with knockout animals and stable cell lines where specific LO cDNAs have been transfected should give more insights about the physiological roles of different LOs.

ACKNOWLEDGMENTS

This study was supported by grants from the Finnish Academy, Finnish Cultural Foundation, Finnish Foundation for Cardiovascular Research, Ida Montin Foundation, Sigrid Jusélius Foundation, and Aarne and Aili Turunen Foundation.

REFERENCES

Arakawa, T., Oshima, T., Kishimoto, K., Yoshimoto, T., and Yamamoto, S. (1992). Molecular structure and function of the porcine arachidonate 12-lipoxygenase gene. *J. Biol. Chem.* **267**, 12188–12191.

Arora, J. K., Lysz, T. W., and Zelenka, P. S. (1996). A role for 12(S)-HETE in the response of human lens epithelial cells to epidermal growth factor and insulin. *Invest. Ophthalmol. Visual Sci.* **37**, 1411–1418.

Atluru, D., Gudapaty, S., O'Donnell, M. P., and Woloschak, G. E. (1993). Inhibition of human mononuclear cell proliferation, interleukin synthesis, mRNA for IL-2, IL-6, and leukotriene B4 synthesis by a lipoxygenase inhibitor. *J. Leukocyte Biol.* **54**, 269–274.

Baeuerle, P. A., and Henkel, T. (1994). Function and activation of NF-κB in the immune system. *Annu. Rev. Immunol.* **12**, 141–179.

Balcarek, J. M., Theisen, T. W., Cook, M. N., Varrichio, A., Hwang, S.-M., Strohsacker, M. W., and Crooke, S. T. (1988). Isolation and characterization of a cDNA clone encoding rat 5-lipoxygenase. *J. Biol. Chem.* **263**, 13937–13941.

Ben-Menahem, D., Shraga-Levine, Z., Limor, R., and Naor, Z. (1994). Arachidonic acid and lipoxygenase products stimulate gonadotropin α-subunit mRNA levels in pituitary αT3-1 cell line: Role in gonadotropin releasing hormone action. *Biochemistry.* **33**, 12795–12799.

Beno, D. W. A., Mullen, J., and Davis, B. H. (1995). Lipoxygenase inhibitors block PDGF-induced mitogenesis: A MAPK-independent mechanism that blocks *fos* and *egr*. *Am. J. Physiol.* **268**, C604–C610.

Benz, D. J., Mol, M., Ezaski, M., Mori-Ito, N., Zelán, I., Miyanohara, A., Friedmann, T., Parthasarathy, S., Steinberg, D., and Witztum, J. L. (1995). Enhanced levels of lipoperoxides. In low density lipoprotein incubated with murine fibroblasts expressing high levels of human 15-lipoxygenase. *J. Biol. Chem.* **270**, 5191–5197.

Brash, A. R., Boeglin, W. E., and Chang, M. S. (1997). Discovery of a second 15S-lipoxygenase in humans. *Proc. Natl. Acad. Sci. U.S.A.* **94**, 6148–6152.

Chen, X. -S., Kurre, U., Jenkins, N. A., Copeland, N. G., and Funk, C. D. (1994a). cDNA cloning, expression, mutagenesis of C-terminal isoleucine, genomic structure, and chromosomal localizations of murine 12-lipoxygenases. *J. Biol. Chem.* **269**, 13979–13987.

Chen, X.-S., Sheller, J. R., Johnson, E. N., and Funk, C. D. (1994b). Role of leukotrienes revealed by targeted disruption of the 5-lipoxygenase gene. *Nature (London)* **372**, 179–182.

Chen, X.-S., Naumann, T. A., Kurre, U., Jenkins, N. A., Copeland, N. G., and Funk, C. D. (1995). cDNA cloning, expression, mutagenesis, intracellular localization, and gene chromosomal assignment of mouse 5-lipoxygenase *J. Biol. Chem.* **270**, 17993–17999.

Chen, A., Beno, D. W. A., and Davis, B. H. (1996). Suppression of stellate cell type I collagen gene expression involves AP-2 transmodulation of nuclear factor-1-dependent gene transcription. *J. Biol. Chem.* **271**, 25994–25998.

Chopra, A., Ferreira-Alves, D. L., Sirois, P., and Thirion, J. P. (1992). Cloning of the guinea pig 5-lipoxygenase gene and nucleotide sequence of its promoter. *Biochem. Biophys. Res. Commun.* **185**, 489–495.

Collins, T., Read, M. A., Neish, A. S., Whitley, M. Z., and Maniatis, T. (1995). Transcriptional regulation of endothelial cell adhesion molecules: NF-kappa B and cytokine-inducible enhancers. *FASEB J.* **9**, 899–909.

Connolly, D. T., Heuvelman, D. M., Nelson, R., Olander, J. V., Eppley, B. L., Delfino, J. J., Siegel, N. R., Leimgruber, R. M., and Feder, J. (1989). Tumor vascular permeability factor stimulates endothelial cell growth and angiogenesis. *J. Clin. Invest.* **84**, 1470–1478.

De Marzo, N., Sloane, D. L., Dicharry, S., Highland, E., and Sigal, E. (1992). Cloning and expression of an airway epithelial 12-lipoxygenase. *Am. J. Physiol.* **262**, L198–L207.

Dixon, R. A. F., Jones, R. E., Diehl, R. E., Bennett, C. D., Kargman, S., and Rouzer, C. A. (1988). Cloning of the cDNA for human 5-lipoxygenase. *Proc. Natl. Acad. Sci. U.S.A.* **85**, 416–420.

Ezaki, M., Witztum, J. L., and Steinberg, D. (1995). Lipoperoxides in LDL incubated with fibroblasts that overexpress 15-lipoxygenase. *J. Lipid Res.* **36**, 1996–2004.

Fleming, J., Thiele, B. J., Chester, J., O'Prey, J., Janetzki, S., Aitken, A., Anton, I. A., Rapoport, S. M., and Harrison, P. R. (1989). The complete sequence of the rabbit erythroid cell-specific 15-lipoxygenase mRNA: Comparison of the predicted amino acid sequence of the erythrocyte lipoxygenase with other lipoxygenases. *Gene* **79**, 181–188.

Freire-Moar, J., Alavi-Nassab, A., Ng, M., Mulkins, M., and Sigal, E. (1995). Cloning and characterization of a murine macrophage lipoxygenase. *Biochim. Biophys. Acta* **1254**, 112–116.

Funk, C. D., Hoshiko, S., Matsumoto, T., Rådmark, O., and Samuelsson, B. (1989). Characterization of the human 5-lipoxygenase gene. *Proc. Natl. Acad. Sci. U.S.A.* **86**, 2587–2591.

Funk, C. D., Furci, L., and FitzGerald, G. A. (1990). Molecular cloning, primary structure, and expression of the human platelet/erythroleukemia cell 12-lipoxygenase. *Proc. Natl. Acad. Sci. U.S.A.* **87**, 5638–5642.

Funk, C. D., Funk, L. B., FitzGerald, G. A., and Samuelsson, B. (1992). Characterization of human 12-lipoxygenase genes. *Proc. Natl. Acad. Sci. U.S.A.* **89,** 3962–3966.

Funk, C. D., Keeney, D. S., Oliw, E. H., Boeglin, W. E., and Brash, A. R. (1996). Functional expression and cellular localization of a mouse epidermal lipoxygenase. *J. Biol. Chem.* **271,** 23338–23344.

Goulet, J. L., Snouwaert, J. N., Latour, A. M., Coffman, T. M., And Koller, B. H. (1994). Altered inflammatory responses in leukotriene-deficient mice. *Proc. Natl. Acad. Sci. U.S.A.* **91,** 12852–12856.

Grossi, I. M., Fitzgerald, L. A., Umbarger, L. A., Nelson, K. K., Diglio, C. A., Taylor, J. D., and Honn, K. V. (1989). Bidirectional control of membrane expression and/or activation of the tumor cell IRGpIIb/IIIa receptor and tumor cell adhesion by lipoxygenase products of arachidonic acid and linoleic acid. *Cancer Res.* **49,** 1029–1037.

Hada, T., Hagiya, H., Suzuki, H., Arakawa, T., Nakamura, M., Matsuda, S., Yoshimoto, T., Yamamoto, S., Azekawa, T., Morita, Y., Ishimura, K., and Kim, H. -Y. (1994). Arachidonate 12-lipoxygenase of rat pineal glands: Catalytic properties and primary structure deduced from its cDNA. *Biochim. Biophys. Acta* **1211,** 221–228.

Hagmann, W., Gao, X., Zacharek, A., Wojciechowski, L. A., and Honn, K. V. (1995). 12-Lipoxygenase in Lewis lung carcinoma cells: Molecular identity, intracellular distribution of activity and protein, and Ca(2+)-dependent translocation from cytosol to membranes. *Prostaglandins* **49,** 49–62.

Hagmann, W., Gao, X., Timar, J., Chen, Y. Q., Strohmaier, A.-R., Fahrenkopf, C., Kagawa, D., Lee, M., Zacharek, A., and Honn, K. V. (1996). 12-Lipoxygenase in A431 cells: Genetic identity, modulation of expression, and intracellular localization. *Exp. Cell Res.* **228,** 197–205.

Hallahan, D. E., Virudachalam, S., Kufe, D. W., and Weichselbaum, R. R. (1994). Ketoconazole attenuates radiation-induction of tumor necrosis factor. *Int. J. Radiat. Oncol. Biol. Phys.* **29,** 777–780.

Harats, D., Kurihara, H., Belloni, P., Oakley, H., Ziober, A., Ackley, D., Cain, G., Kurihara, Y., Lawn, R., and Sigal, E. (1995). Targeting gene expression to the vascular wall in transgenic mice using the murine preproendothelin-1 promoter. *J. Clin. Invest.* **95,** 1335–1344.

Hiltunen, T., Luoma, J., Nikkari, T., and Ylä-Herttuala, S. (1995). Induction of 15-lipoxygenase mRNA and protein in early atherosclerotic lesions. *Circulation* **92,** 3297–3303.

Honn, K. V., Tang, D. G., Gao, X., Butovich, I. A., Liu, B., Timar, J., and Hagmann, W. (1994). 12-Lipoxygenases and 12(S)-HETE: Role in cancer metastasis. *Cancer Metastasis Rev.* **13,** 365–396.

Horiguchi, J., Spriggs, D., Imamura, K., Stone, R., Luebbers, R., and Kufe, D. (1989). Role of arachidonic acid metabolism in transcriptional induction of tumor necrosis factor gene expression by phorbol ester. *Mol. Cell. Biol.* **9,** 252–258.

Hussain, H., Shornick, L. P., Shannon, V. R., Wilson, J. D., Funk, C. D., Pentland, A. P., and Holtzman, M. J. (1994). Epidermis contains platelet-type 12-lipoxygenase that is overexpressed in germinal layer keratinocytes in psoriasis. *Am. J. Physiol.* **266,** C243–C253.

Izumi, T., Hoshiko, S., Rådmark, O., and Samuelsson, B. (1990). Cloning of the cDNA for human 12-lipoxygenase. *Proc. Natl. Acad. Sci. U.S.A.* **87,** 7477–7481.

Jisaka, M., Kim, R. B., Boeglin, W. E., Nanney, L. B., and Brash, A. R. (1997). Molecular cloning and functional expression of a phorbol ester-inducible 8S-lipoxygenase from mouse skin. *J. Biol. Chem.* **272,** 24410–24416.

Katoh, T., Lakkis, F. G., Makita, N., and Badr, K. F. (1994). Co-regulated expression of glomerular 12/15-lipoxygenase and interleukin-4 mRNAs in rat nephrotoxic nephritis. *Kidney Int.* **46,** 341–349.

Kazmi, S. M. I., Plante, R. K., Visconti, V., Taylor, G. R., Zhou, L., and Lau, C. Y. (1995). Suppression of NFκB activation and NFκB-dependent gene expression by tepoxalin, a dual inhibitor of cyclooxygenase and 5-lipoxygenase. *J. Cell. Biochem.* 57, 299–310.

Keck, P. J., Hauser, S. D., Krivi, G., Sanzo, K., Warren, T., Feder, J., and Connolly, D. T. (1989). Vascular permeability factor, an endothelial cell mitogen related to PDGF. *Science* 246, 1309–1312.

Kinzig, A., Fürstenberger, G., Bürger, F., Vogel, S., Müller-Decker, K., Mincheva, A., Lichter, P., Marks, F., and Krieg, P. (1997). Murine epidermal lipoxygenase (Aloxe) encodes a 12-lipoxygenase isoform. *FEBS Lett.* 402, 162–166.

Kitzler, J. W., and Eling, T. E. (1996). Cloning, sequencing and expression of a 5-lipoxygenase from Syrian hamster embryo fibroblasts. *Prostaglandins Leukotrines Essent. Fatty Acids* 55, 269–277.

Krieg, P., Kinzig, A., Ress-Loschke, M., Vogel, S., Vanlandingham, B., Stephan, M., Lehmann, W. D., Marks, F., and Fürstenberger, G. (1995). 12-Lipoxygenase isoenzymes in mouse skin tumor development. *Mol. Carcinog.* 14, 118–129.

Kritzik, M. R., Ziober, A. F., Dicharry, S., Conrad, D. J., and Sigal, E. (1997). Characterization and sequence of an additional 15-lipoxygenase transcript and of the human gene. *Biochim. Biophys. Acta* 1352, 267–281.

Kühn, H., Belkner, J., Suzuki, H., and Yamamoto, S. (1994a). Oxidative modification of human lipoproteins by lipoxygenases of different positional specificities, *J. Lipid Res.* 35, 1749–1759.

Kühn, H., Belkner, J., Zaiss, S., Fährenklemper, T., and Wohlfeil, S. (1994b). Involvement of 15-lipoxygenase in early stages of atherogenesis. *J. Exp. Med.* 179, 1903–1911.

Kühn, H., Heydeck, D., Hugou, I., and Gniwotta, C. (1997). In vivo action of 15-lipoxygenase in early stages of human atherogenesis. *J. Clin. Invest.* 99, 888–893.

Kuroki, M., Voest, E. E., Amano, S., Beerepoot, L. V., Takashima, S., Tolentino, M., Kim, R. Y., Rohan, R. M., Colby, K. A., Yeo, K.-T., and Adamis, A. P. (1996). Reactive oxygen intermediates increase vascular endothelial growth factor expression in vitro and in vivo. *J. Clin. Invest.* 98, 1667–1675.

Lee, S., Felts, K. A., Parry, G. C. N., Armacost, L. M., and Cobb, R. R. (1997). Inhibition of 5-lipoxygenase blocks IL-1β-induced vascular adhesion molecule-1 gene expression in human endothelial cells. *J. Immunol.* 158, 3401–3407.

Lewin, B. (1994). "Genes V." Oxford University Press, Oxford.

Los, M., Schenk, H., Hexel, K., Baeuerle, P. A., Dröge, W., and Schulze-Osthoff, K. (1995). IL-2 gene expression and NF-κB activation through CD28 requires reactive oxygen production by 5-lipoxygenase. *EMBO J.* 14, 3731–3740.

Lysz, T. W., Arora, J. K., Lin, C., and Zelenka, P. S. (1994). 12(S)-hydroxyeicosatetraenoic acid regulates DNA synthesis and protooncogene expression induced by epidermal growth factor and insulin in rat lens epithelium. *Cell Growth Differ.* 5, 1069–1076.

Mancini, J. A., Abramovitz, M., Cox, M. E., Wong, E., Charleson, S., Perrier, H., Wang, Z., Prasit, P., and Vickers, P. J. (1993). 5-Lipoxygenase-activating protein is an arachidonate binding protein. *FEBS Lett.* 318, 277–281.

Matsumoto, T., Funk, C. D., Rådmark, O., Höög, J.-O., Jörnvall, H., and Samuelsson, B. (1988). Molecular cloning and amino acid sequence of human 5-lipoxygenase. *Proc. Natl. Acad. Sci. U.S.A.* 85, 26–30.

McNally, A. K., Chisolm G. M., III, Morel, D. W., and Cathcart, M. K. (1990). Activated human monocytes oxidize low-density lipoprotein by a lipoxygenase-dependent pathway. *J. Immunol.* 145, 254–259.

Meier, R. W., Niklaus, G., Dewald, B., Fey, M. F., and Tobler, A. (1995). Inhibition of the arachidonic acid pathway prevents induction of IL-8 mRNA by phorbol ester and changes the release of IL-8 from HL 60 cells: Differential inhibition of induced expression of IL-8, TNF-α, IL1-α, and IL-1β. *J. Cell. Physiol.* **165**, 62–70.

Mohri, M., Spriggs, D. R., and Kufe, D. (1990). Effects of lipopolysaccharide on phospholipase A$_2$ activity and tumor necrosis factor expression in HL-60 cells. *J. Immunol.* **144**, 2678–2682.

Nagy, L., Tontonoz, P., Alvarez, J. G. A., Chen, H., and Evans, R. M. (1998). Oxidized LDL regulates macrophage gene expression through ligand activation of PPARγ. *Cell (Cambridge, Mass.)* **93**, 229–240.

Nakao, J., Ooyama, T., Ito, H., Chang, W.-C., and Murota, S. (1982). Comparative effect of lipoxygenase products of arachidonic acid on rat aortic smooth muscle cell migration. *Atherosclerosis* **44**, 339–342.

Natarajan, R., Gu, J. L., Rossi, J., Gonzales, N., Lanting, L., Xu, L., and Nadler, J. (1993). Elevated glucose and angiotensin II increase 12-lipoxygenase activity and expression in porcine aortic smooth muscle cells. *Proc. Natl. Acad. Sci. U.S.A.* **90**, 4947–4951.

Natarajan, R., Gonzales, N., Lanting, L., and Nadler, J. (1994). Role of the lipoxygenase pathway in angiotensin II-induced vascular smooth muscle cell hypertrophy. *Hypertension* **23**, 1142–1147.

Natarajan, R., Bai, W., Lanting, L., Gonzales, N., and Nadler, J. (1997). Effects of high glucose on vascular endothelial growth factor expression in vascular smooth muscle cells. *Am. J. Physiol.* **273**, H2224–H2231.

Nishizuka, Y. (1995). Protein kinase C and lipid signaling for sustained cellular responses. *FASEB J.* **9**, 484–496.

O'Prey, J., and Harrison, P. R. (1995). Tissue-specific regulation of the rabbit 15-lipoxygenase gene in erythroid cells by a transcriptional silencer. *Nucleic Acids Res.* **23**, 3664–3672.

O'Prey, J., Chester, J., Thiele, B. J., Janetzki, S., Prehn, S., Fleming, J., and Harrison, P. R. (1989). The promoter structure and complete sequence of the gene encoding the rabbit erythroid cell-specific 15-lipoxygenase. *Gene.* **84**, 493–499.

Ostareck-Lederer, A., Ostareck, D. H., Standart, N., and Thiele, B. J. (1994). Translation of 15-lipoxygenase mRNA is inhibited by a protein that binds to a repeated sequence in the 3' untranslated region. *EMBO J.* **13**, 1476–1481.

Parthasarathy, S., Wieland, E., and Steinberg, D. (1989). A role for endothelial cell lipoxygenase in the oxidative modification of low density lipoprotein *Proc. Natl. Acad. Sci. U.S.A.* **86**, 1046–1050.

Paul, W. E., and Seder, R. A. (1994). Lymphocyte responses and cytokines. *Cell (Cambridge, Mass.)* **76**, 241–251.

Piomelli, D., Volterra, A., Dale, N., Siegelbaum, S. A., Kandel, E. R., Schwartz, J. H., and Belardetti, F. (1987). Lipoxygenase metabolites of arachidonic acid as second messengers for presynaptic inhibition of *Aplysia* Sensory cells. *Nature (London)* **328**, 38–43.

Rankin, S. M., Parthasarathy, S., and Steinberg, D. (1991). Evidence for a dominant role of lipoxygenase(s) in the oxidation of LDL by mouse peritoneal macrophages. *J. Lipid Res.* **32**, 449–456.

Rao, G. N., Alexander, R. W., and Runge, M. S. (1995). Linoleic acid and its metabolites, hydroperoxyoctadecadienoic acids, stimulate *c-Fos*, *c-Jun*, and *c-Myc* mRNA expression, mitogen-activated protein kinase activation, and growth in rat aortic smooth muscle cells. *J. Clin. Invest.* **96**, 842–847.

Rao, G. N., Glasgow, W. C., Eling, T. E., and Runge, M. S. (1996). Role of hydroperoxyeicosatetraenoic acids in oxidative stress-induced activating protein 1 (AP-1) activity. *J. Biol. Chem.* **271**, 27760–27764.

Rapoport, S. M., Schewe, T., Wiesner, R., Halangk, W., Ludwig, P., Janicke-Höhne, M., Tannert, C., Hiebsch, C., and Klatt, D. (1979). The lipoxygenase of reticulocytes. Purification, characterization and biological dynamics of the lipoxygenase; its identity with the respiratory inhibitors of the reticulocytes. *Eur. J. Biochem.* **96**, 546–561.

Reddy, N., Everhart, A., Eling, T., and Glasgow, W. (1997). Characterization of a 15-lipoxygenase in human breast carcinoma BT-20 cells: Stimulation of 13-HODE formation by TGF$^\alpha$/EGF. *Biochem. Biophys. Res. Commun.* **231**, 111–116.

Samuelsson, B., Dahlén, S.-E., Lindgren, J. Å., Rouzer, C. A., and Serhan, C. N. (1987). Leukotrienes and lipoxins: Structures, biosynthesis, and biological effects. *Science* **237**, 1171–1176.

Schoonjans, K., Martin, G., Staels, B., and Auwerx, J. (1997). Peroxisome proliferator-activated receptors, orphans with ligands and functions. *Curr. Opin. Lipidol.* **8**, 159–166.

Sen, C. K., and Packer, L. (1996). Antioxidant and redox regulation of gene transcription. *FASEB J.* **10**, 709–720.

Setty, B. N. Y., Graeber, J. E., and Stuart, M. J. (1987). The mitogenic effect of 15- and 12-hydroxyeicosatetraenoic acid on endothelial cells may be mediated via diacylglycerol kinase. *J. Biol. Chem.* **262**, 17613–17622.

Shen, J., Kühn, H., Petho-Schramm, A., and Chan, L. (1995). Transgenic rabbits with the integrated human 15-lipoxygenase gene driven by a lysozyme promoter: Macrophage-specific expression and variable positional specificity of the transgenic enzyme. *FASEB J.* **9**, 1623–1631.

Shen, J., Herderick, E., Cornhill, J. F., Zsigmond, E., Kim, H.-S., Kühn, H., Guevara, N. V., and Chan, L. (1996). Macrophage-mediated 15-lipoxygenase expression protects against atherosclerosis development. *J. Clin. Invest.* **98**, 2201–2208.

Shraga-Levine, Z., Ben-Menahem, D., and Naor, Z. (1996). Arachidonic acid and lipoxygenase products stimulate protein kinase Cβ mRNA levels in pituitary αT3-1 cell line: role in gonadotropin-releasing hormone action. *Biochem. J.* **316**, 667–670.

Sigal, E. (1991). The molecular biology of mammalian arachidonic acid metabolism. *Am. J. Physiol.* **260**, L13–L28.

Sigal, E., Craik, C. S., Highland, E., Grunberger, D., Costello, L. L., Dixon, R. A. F., and Nadel, J. A. (1988). Molecular cloning and primary structure of human 15-lipoxygenase. *Biochem. Biophys. Res. Commun.* **157**, 457–464.

Sigal, E., Dicharry, S., Highland, E., and Finkbeiner, W. E. (1992). Cloning of human airway 15-lipoxygenase: Identity to the reticulocyte enzyme and expression in epithelium. *Am. J. Physiol.* **262**, L392–L398.

Spriggs, D. R., Sherman, M. L., Imamura, K., Mohri, M., Rodriguez, C., Robbins, G., and Kufe, D. W. (1990). Phospholipase A$_2$ activation and autoinduction of tumor necrosis factor gene expression by tumor necrosis factor. *Cancer Res.* **50**, 7101–7107.

Steiner, P., Pfeilschifter, J., Boeckh, C., Radeke, H., and Otten, U. (1991). Interleukin-1β and tumor necrosis factor-α synergistically stimulate nerve growth factor synthesis in rat mesangial cells. *Am. J. Physiol.* **261**, F792–F798.

Sun, D., and Funk, C. D. (1996). Disruption of 12/15-lipoxygenase expression in peritoneal macrophages. *J. Biol. Chem.* **271**, 24055–24062.

Tang, D. G., Diglio, C. A., Bazaz, R., and Honn, K. V. (1995). Transcriptional activation of endothelial cell integrin αv by protein kinase C activator 12(S)-HETE. *J. Cell Sci.* **108**, 2629–2644.

Tebbey, P. W., and Buttke, T. M. (1993). Independent arachidonic acid-mediated gene regulatory pathways in lymphocytes. *Biochem. Biophys. Res. Commun.* **194**, 862–868.

Thiele, B. J., Fleming, J., Kasturi, K., O'Prey, J., Black, E., Chester, J., Rapoport, S. M., and Harrison, P. R. (1987). Cloning of a rabbit erythroid-cell-specific lipoxygenase mRNA. *Gene* **57**, 111-119.

Tontonoz, P., Nagy, L., Alvarez, J. G. A., Thomazy, V. A., and Evans, R. M. (1998). PPARγ promotes monocyte/macrophage differentiation and uptake of oxidized LDL. *Cell (Cambridge, Mass.)* **93**, 241-252.
van Dijk, K. W., Steketee, K., Havekes, L., Frants, R., and Hofker, M. (1995). Genomic and cDNA cloning of a novel mouse lipoxygenase gene. *Biochim. Biophys. Acta* **1259**, 4–8.
Viita, H., Sen, C. K., Roy, S., Siljamäki, T., Nikkari, T., and Ylä-Herttuala, S. (1999). High expression of human 15-lipoxygenase induces nuclear factor kappa B-mediated expression of vascular cell adhesion molecule 1 and intercellular adhesion molecule 1 in and T-cell adhesion on human endothelial cells. *Antiox. Redox Signal.* **1**, 83–96.
Watanabe, T., Medina, J. F., Haeggström, J. Z., Rådmark, O., and Samuelsson, B. (1993). Molecular cloning of a 12-lipoxygenase cDNA from rat brain. *Eur. J. Biochem.* **212**, 605–612.
Wölle, J., Welch, K. A., Devall, L. J., Cornicelli, J. A., and Saxena, U. (1996). Transient overexpression of human 15-lipoxygenase in aortic endothelial cells enhances tumor necrosis factor-induced vascular cell adhesion molecule-1 gene expression. *Biochem. Biophys. Res. Commun.* **220**, 310–314.
Yamamoto, S. (1992). Mammalian lipoxygenases: Molecular structures and functions. *Biochim. Biophys. Acta* **1128**, 117–131.
Ylä-Herttuala, S., Rosenfeld, M. E., Parthasarathy, S., Glass, C. K., Sigal, E., Witztum, J. L., and Steinberg, D. (1990): Colocalization of 15-lipoxygenase mRNA and protein with epitopes of oxidized low density lipoprotein in macrophage-rich areas of atherosclerotic lesions. *Proc. Natl. Acad. Sci. U.S.A.* **87**, 6959–6963.
Ylä-Herttuala, S., Rosenfeld, M. E., Parthasarathy, S., Sigal, E., Särkioja, T., Witztum, J. L., and Steinberg, D. (1991). Gene expression in macrophage-rich human atherosclerotic lesions. 15-Lipoxygenase and acetyl low density lipoprotein receptor messenger RNA colocalize with oxidation specific lipid-protein adducts. *J. Clin. Invest.* **87**, 1146–1152.
Ylä-Herttuala, S., Luoma, J., Viita, H., Hiltunen, T., Sisto, T., and Nikkari, T. (1995). Transfer of 15-lipoxygenase gene into rabbit iliac arteries results in the appearance of oxidation-specific lipid-protein adducts characteristic of oxidized low density lipoprotein. *J. Clin. Invest.* **95**, 2692–2698.
Yoshimoto, T., Suzuki, H., Yamamoto, S., Takai, T., Yokoyama, C., and Tanabe, T. (1990a). Cloning and sequence analysis of the cDNA for arachidonate 12-lipoxygenase of porcine leukocytes. *Proc. Natl. Acad. Sci. U.S.A.* **87**, 2142–2146.
Yoshimoto, T., Yamamoto, Y., Arakawa, T., Suzuki, H., Yamamoto, S., Yokoyama, C., Tanabe, T., and Toh, H. (1990b). Molecular cloning and expression of human arachidonate 12-lipoxygenase. *Biochem. Biophys. Res. Commun.* **172**, 1230–1235.
Yoshimoto, T., Arakawa, T., Hada, T., Yamamoto, S., and Takahashi, E. (1992). Structure and chromosomal localization of human arachidonate 12-lipoxygenase gene. *J. Biol. Chem.* **267**, 24805–24809.

16 α-Tocopherol in the Pathogenesis of Atherosclerosis: One Molecule, Several Functions

Angelo Azzi

Institute of Biochemistry and Molecular Biology
University of Bern
3012 Bern, Switzerland

The annual incidence of coronary artery disease in the United States is estimated at 616,900 cases, with first-year costs of treatment totaling $5.54 billion. Five- and 10-year cumulative costs in 1995 dollars for patients who are initially free of coronary artery disease are estimated at $9.2 and $16.5 billion, respectively; for all patients with coronary artery disease, these costs are estimated to be $71.5 and $126.6 billion, respectively. The direct medical costs of coronary artery disease create a large economic burden for the United States health-care system (Russell *et al.*, 1998). A similar situation is present in other industrialized countries. Risk factors at the basis of the onset of coronary heart disease include smoking, high blood pressure, diabetes, and dietary factors (Jha *et al.*, 1995). High blood levels of cholesterol, especially bound with low-density lipoprotein, have been associated with an increased risk of atherosclerosis (Esterbauer *et al.*, 1991; Steinberg and Witztum, 1990). Dietary factors may also be important in addition to the control of high blood cholesterol levels in the prevention of atherosclerosis (Esterbauer *et al.*, 1991; Ulbricht and Southgate, 1991). Epidemiological data, as well as animal studies, indicate that a diminution of dietary antioxidants increases the risk of atherosclerosis and that an increased intake of antioxidants may prevent coronary heart disease (Gey *et al.*, 1991; Ulbricht and Southgate, 1991).

I. VITAMIN E AS AN ANTIOXIDANT

Vitamin E is an essential nutrient because the body cannot synthesize its own vitamin E, which must be provided by foods and supplements.

Vitamin E is a generic term that includes all entities that exhibit the biological activity of natural vitamin E, RRR-α-tocopherol. In nature, eight substances have been found to have vitamin E activity: RRR-α-, RRR-β-, RRR-γ-, and RRR-δ-tocopherol as well as RRR-α-, RRR-β- RRR-γ-, and RRR-δ-tocotrienol. The acetate and succinate derivatives of natural tocopherols have vitamin E activity, as do synthetic tocopherols and their acetate and succinate derivatives. Of these, RRR-α-tocopherol has the highest biopotency, and its activity is the standard against which all the others must be compared (Burton *et al.*, 1982, 1983, 1998).

The human liver has a protein, tocopherol transfer protein (Arita *et al.*, 1995; Traber *et al.*, 1993), capable of transferring specifically the dietary RRR-α-tocopherol to lipoproteins. The other natural vitamin E forms and the synthetic ones are transported much more poorly. Absence of a functional transport protein is associated with a neurological syndrome, ataxia with isolated vitamin E deficiency (AVED) (Gotoda *et al.*, 1995; Ouahchi *et al.*, 1995; Traber *et al.*, 1993).

Vitamin E is liposoluble and as such is mostly present in cell membranes and in low-density lipoproteins (LDL). Free radicals are created both in metabolic processes and as a result of environment pollution (e.g., superoxide, hydroxyl radicals, nitrogen dioxide, ozone, heavy metals, halogenated hydrocarbons, ionizing radiation, and cigarette smoke).

Oxygen-derived free radicals are particularly abundant, as the human organism produces them in a number of reactions (mitochondrial respiratory chain, drug hydroxylating reactions in the endoplasmatic reticulum, cytosolic enzymes, etc.). A conservative calculation may indicate the production of 100 kg of oxygen-free radicals in the life span of an individual. The highly reactive free radicals modify cell constituents, including DNA, proteins, and the polyunsaturated fatty acids of phospholipid. In the latter case, free radical production is amplified by chain reactions. Free radical damage of cell membranes, of nucleic acids, and of proteins occur. This event is followed by activation or inactivation of enzymes, cell deregulation, changes in cell–cell interactions, and in response to hormones and neurotransmitters. Important mutagenic events can also take place. Cell dysfunction, cell death, and carcinogenesis may ultimately take place.

Vitamin E is understood to be the major antioxidant in tissues and in blood as well as the primary defense against lipid peroxidation (Pryor et al., 1993). It neutralizes free radicals by terminating chain reactions, protecting against free radical/oxidative damage. The resulting vitamin E-oxidized products can be reduced back to native vitamin E by ascorbic acid (Wefers, 1988, Kagan et al., 1992; Packer et al., 1979; Packer, 1992) or lypoic acid (Scholich, 1989, Packer and Suzuki, 1993). It can be calculated that one molecule of vitamin E protects up to 1000 molecules of polyunsaturated fatty acids against oxidative damage.

Free radicals, lipid peroxidation, and oxidative modification of LDL play a role in the first steps of atherosclerosis (Esterbauer et al., 1992a,b; Sato et al., 1990). After penetration into the intima, LDL become oxidatively modified and originate foam cells, one of the earliest stages in the development of atherosclerosis (Morin and Peng, 1989). Macrophages do not internalize native LDL at any significant rate unless they are oxidatively modified; this process leads to accumulation inside the macrophage of cholesterol and cholesteryl esters (Mitchinson et al., 1990; Müller et al., 1996) and become foam cells, an early event in the development of atherosclerotic lesions.

Research results support the hypothesis that the oxidative modification of LDL results in the enhanced uptake by macrophages through a specific receptor, which may lead to the conversion of macrophages into lipid-laden foam cells. Oxidized LDL is highly toxic to cells and may be responsible for damage to the endothelial layer and destruction of smooth muscle cells (Hoff et al., 1989; Jurgens et al., 1987; Sambrano et al., 1994; Schwartz et al., 1991; Yla Herttuala et al., 1989). It is well accepted that fatty streaks, which are characterized by an accumulation of foam cells just beneath the arterial endothelium, are a precursor of later, clinically significant lesions. In experimental atherosclerosis, the natural progression of lesions is the development of fibrous plaques and more advanced lesions.

Frontiers in atherosclerosis research are thus moving from lipoprotein metabolism and control of hyperlipidemia to the cellular events in the artery wall. Emerging hypotheses, including the oxidative modification hypothesis of LDL, suggest a new approach, with antioxidants, that could complement and be additive to the control of hypercholesterolemia in the prevention of atherosclerosis (Steinberg and Witztum, 1990).

The following studies deal with the problem of LDL oxidation, accumulation in cells, and protection by antioxidants.

The effect of 6 months of α-tocopherol treatment on the susceptibility of low-density lipoproteins to oxidative modification and on established atherosclerotic lesions was studied in Watanabe heritable hyperlipidemic (WHHL) rabbits. At the end of the 6-month period, the mean±SD percentage area of aorta covered with plaques was 58.7 ± 10.1% in control animals,

62.7 ± 12.0% in probucol-treated animals, and 48.9 ± 13.8% in animals treated with vitamin E. This study demonstrates that at low dosage, vitamin E is a more effective antioxidant than probucol (Kleinveld et al., 1994).

The uptake of native and modified LDL in foam cells in atherosclerotic tissue was studied in an *in vitro* perfusion system for rabbit aorta. About 40 times more LDL per cell was accumulated in the foam cell fraction than in the smooth muscle cell fraction. The accumulation of radiolabeled LDL in plaques and in foam cells was reduced 30–55% by adding vitamin E (0.1 mg/ml) to the system. These studies show an uptake of LDL by foam cells in the atherosclerotic tissue. The observation that an antioxidant, vitamin E, may decrease this uptake suggests that the oxidative modification of LDL is of importance for this process (Wiklund et al., 1991).

The time course of the depletion of α-tocopherol in human low-density lipoprotein was measured during macrophage-mediated and cell-free oxidation. The formation of oxidatively modified, high uptake species of LDL in these systems was not detectable until all of the endogenous α-tocopherol had been consumed. Supplementation of the α-tocopherol content of LDL by 24 hr loading *in vivo* extended the duration of the lag period during which no detectable oxidative modification occurred (Jessup et al., 1990).

In another study in humans it has been shown that high-dose α-tocopherol supplementation decreased the susceptibility of LDL to oxidation. Hence, the aim of the present study was to ascertain the minimum dose of α-tocopherol that would decrease the susceptibility of LDL to oxidation. The effect of α-tocopherol in doses of 60, 200, 400, 800, and 1200 IU/day on copper-catalyzed LDL oxidation was tested in a randomized, placebo-controlled study over 8 weeks. There were eight subjects in each group. There was a dose-dependent increase in plasma and lipid-standardized α-tocopherol levels with increasing doses of α-tocopherol supplementation. LDL α-tocopherol appeared to follow a similar trend. Subjects supplemented with 400, 800, or 1200 IU vitamin E per day for 8 weeks showed a decreased susceptibility of LDL to oxidation. There was no significant effect of daily supplementation with 60 or 200 IU vitamin E for 8 weeks (Jialal et al., 1995).

Two groups of 12 male subjects were given either placebo or α-tocopherol (800 IU/day) for a period of 12 weeks. α-Tocopherol therapy did not result in any side effects or exert an adverse effect on the plasma lipid and lipoprotein profile. The supplemented group had 3.3 and 4.4-fold higher levels compared to placebo at 6 and 12 weeks, respectively. At both 6 and 12 weeks the mean rate of oxidation was lower in the α-tocopherol group (Jialal and Grundy, 1992).

The effect of α-tocopherol was tested both *in vitro* and *in vivo* on LDL oxidation and glycation. There was a progressive increase in the susceptibility of LDL to oxidation with increasing LDL glycation as evidenced by a reduced lag time of copper-catalyzed LDL oxidation. Data from the α-tocopherol supplementation study confirmed the *in vitro* findings that α-tocopherol significantly decreases the oxidative susceptibility of LDL (Li *et al.*, 1996).

In summary, it has been demonstrated clearly that vitamin E exerts antioxidant properties *in vitro*, in plasma, and at a cellular level in animals and humans. This property is of central importance in the onset of atherosclerosis. Part of the effects of vitamin E may, however, be referable to other properties of the substance.

II. VITAMIN E FUNCTION AT THE LEVEL OF PROTEIN KINASE C: SMOOTH MUSCLE CELL PROLIFERATION, MONOCYTE ADHESION, THROMBOCYTE AGGREGATION AND ADHESION, AND NEUTROPHIL ACTIVATION

Numerous biological activities of α-tocopherol have been described *in vivo*. However, its molecular mechanism of action and its direct cellular targets are still subject of current investigations.

Smooth muscle cell proliferation, monocyte adhesion, thrombocyte aggregation and adhesion, and neutrophil activation are important cellular events in the initiation and progression of atherosclerosis. At the basis of a number of cell, these cell effects appear to be the regulation of protein kinase C by α-tocopherol (Boscoboinik *et al.*, 1991).

Inhibition by α-tocopherol of protein kinase C, a key enzyme in the signaling pathways that regulates cellular events and gene expression, has been reviewed in Azzi *et al.*, (1992, 1995, 1996) and Newton (1995). Such an inhibition is not caused by a direct binding of α-tocopherol to the enzyme but by preventing its activation via phosphorylation (Tasinato *et al.*, 1995). Phosphorylation has emerged as an important mode of regulation of protein kinase C (Bornancin and Parker, 1996; Dutil *et al.*, 1994). Several reports have shown that phosphorylation at the activation loop activates the enzyme and that dephosphorylation by the catalytic subunit of protein phosphatase 2A leads to a loss of its activity.

By comparing α-tocopherol with analogous compounds exhibiting similar antioxidant properties such as β-tocopherol, it is concluded that α-tocopherol exerts its action independently of its free-radical scavenger capacity and most probably by interacting with a yet not characterized receptor molecule in smooth muscle cells (Boscoboinik *et al.*, 1991, 1994, 1995).

The effect of α-tocopherol on the different protein kinase C isoforms was also analyzed and it was found that α-tocopherol prevents uniquely protein kinase C-α phosphorylation and its functional activation. Because protein kinase C translocation from cytosol to membranes and the protein levels were not changed by α-tocopherol treatment during the transition G_0/G_1, it was investigated how α-tocopherol influenced the *de novo* synthesis of protein kinase C after its downregulation by phorbol esters. α-Tocopherol inhibits proliferation by inducing an arrest at the G_1/S boundary in several cell types (Azzi *et al.*, 1993, 1995), particularly in smooth muscle cells. Smooth muscle cell proliferation is an early event in the development of atherosclerosis (Raines and Ross, 1993; Ross, 1993).

Activation of monocytes by lipopolysaccharide results in the release of reactive oxygen species (superoxide anion, hydrogen peroxide), lipid oxidation, release of the potentially atherogenic cytokine, interleukin 1β, and monocyte–endothelial adhesion. α-Tocopherol inhibits these effects, crucial in atherogenesis, due to an inhibition of protein kinase C activity (Devaraj *et al.*, 1996)

The effect of vitamin E on platelet function has been reviewed by Steiner (1993). Although vitamin E inhibits platelet aggregation *in vitro*, it has no significant effect *in vivo* when administered in doses up to 1200 U/day. Platelet adhesion, however, is strongly inhibited by α-tocopherol. Doses of 400 IU/day provide greater than 75% inhibition of platelet adhesion to a variety of adhesive proteins when tested at a low shear rate in a laminar flow chamber. The antiadhesive effect of vitamin E appears to be related to a reduction in the number and size of pseudopodia on platelet activation. These effects may be related to the protein kinase C inhibition produced by α-tocopherol

Platelet adhesion was also measured in a randomized subgroup of both study populations using collagen III as the adhesive surface (Steiner *et al.*, 1995). One hundred patients with transient ischemic attacks, minor strokes, or residual ischemic neurologic deficits were enrolled in a double-blind, randomized study comparing the effects of aspirin plus vitamin E [0.4 g (400 IU)/day; $n = 52$] with aspirin alone (325 mg; $n = 48$). The patients received study medication for 2 years or until they reached a termination point. There was a highly significant reduction in platelet adhesiveness in patients who were taking vitamin E plus aspirin compared to those taking aspirin only. Measurement of α-tocopherol concentrations confirmed compliance of the patients with the medication schedule, showing a near doubling of serum concentrations of α-tocopherol. It is concluded that the combination of vitamin E and a platelet antiaggregating agent (e.g., aspirin) significantly enhances the efficacy of the preventive treatment regimen in patients with transient ischemic attacks and other ischemic cerebrovascular events. Preliminary results show a significant reduction in the incidence of ischemic events in patients in the vitamin E plus aspirin group compared to patients taking only aspirin. There was no significant difference in the incidence of hemorrhagic stroke, although both patients who developed it were taking vitamin E.

T-cell-mediated functions have also been studied (Meydani et al., 1997). Elderly subjects (at least 65 years of age) consuming 200 mg/day of vitamin E had a 65% increase in delayed-type hypersensitivity skin response and a 6-fold increase in antibody titer to hepatitis B compared to placebo (17% and 3-fold, respectively), 60-mg/day (41% and 3-fold, respectively), and 800-mg/day (49% and 2.5-fold, respectively) groups. The 200-mg/day group also had a significant increase in antibody titer to tetanus vaccine. Subjects in the upper tertile of serum α-tocopherol concentration [$>48.4\mu$mol/l (2.08 mg/dl)] after supplementation had a higher antibody response to hepatitis B and delayed-type hypersensitivity skin response. These results indicate that a level of vitamin E greater than currently recommended enhances certain clinically relevant *in vivo* indexes of T-cell-mediated function in healthy elderly persons. No adverse effects were observed with vitamin E supplementation.

Phorbol 12-myristate 13-acetate (PMA)-induced superoxide generation in neutrophils is α-tocopherol sensitive at a concentration much lower than that for the inhibition of PMA-activated phospholipid-dependent protein kinase (ID_{50} = 30 μM). The α-tocopherol-sensitive superoxide generation is also observed in neutrophils induced by dioctanoylglycerol and calcium ionophore A23187 but not by formylmethionylleucylphenylalanine, opsonized zymosan, and sodium dodecyl sulfate. The pattern of inhibition by α-tocopherol is quite similar to that of staurosporine, a specific inhibitor of protein kinase C. These results indicate that the neutrophil content of α-tocopherol is an important factor in superoxide generation (Kanno et al., 1996).

Ex vivo smooth muscle cells obtained from the aorta of cholesterol-fed rabbits exhibited a twofold increase of protein kinase C expression and activity. The increase of protein kinase C expression and activity was affected by vitamin E treatment (Özer et al., 1998; Sirikci et al., 1996).

In summary, the inhibition of protein kinase C elicited by vitamin E is a specific event at a cellular level, as well as at the level of *ex vivo* tissue.

III. CONSUMPTION OF VITAMIN E AND THE DEVELOPMENT OF ATHEROSCLEROTIC MODIFICATIONS OF THE ARTERIAL WALL

A. Protective Effects of Antioxidants on Development and Progression of Atherosclerosis in Animals

A number of animal studies have evaluated the protective effects of antioxidants, primarily vitamin E, on the development and progression of atherosclerosis.

Vitamin E supplementation in hens that develop hyperlipidemia and atherosclerotic lesions resulted in a reduction of plasma lipid peroxides. It is suggested that hyperlipidemia without lipid peroxidation does not promote the development of atherosclerosis or does so at a decreased rate (Smith and Kummerow, 1989).

The efficacy of vitamin E supplementation on atherosclerosis has been investigated in studies of rabbits fed a high fat diet. In nonsupplemented rabbits, atherosclerotic plaque coverage averaged 76%. In contrast, the area of intima covered by atherosclerotic plaques averaged only 28% in rabbits supplemented with selenium and vitamin E (Wojcicki et al., 1991).

Although the amount of plaque formation in the aorta of WHHL rabbits (undergoing spontaneous atherosclerosis) did not differ significantly in vitamin E-supplemented and nonsupplemented animals, the concentration of cholesterol in lesions of the aortic arch was approximately 25% lower in vitamin E-supplemented animals. Plasma levels of total cholesterol, LDL cholesterol, and triglycerides were 20–30% lower in the vitamin E-supplemented group compared to the control group. Blood and aortic tissue levels of lipid peroxides decreased in vitamin E-supplemented rabbits on high cholesterol diets or on regular rabbit chow. Atherosclerotic plaques were significantly smaller in cholesterol-fed rabbits on vitamin E supplementation than in nonsupplemented rabbits (Prasad and Kalra, 1993).

In another study with WHHL rabbits, lowering of blood cholesterol and triglyceride levels was observed with a 32% inhibition of early aortic lesion development in vitamin E-supplemented animals (Williams et al., 1992).

Vitamin E pretreatment (beginning 19 days before the procedure) in rabbits significantly inhibited restenosis after angioplasty (Lafont et al., 1995).

Prevention and regression of induced atherosclerosis by α-tocopherol were investigated in 24 male monkeys. One group received a basal diet, whereas three others consumed an atherogenic diet. Two of the latter groups also received tocopherol, one at the onset of the study (prevention) and the other after atherosclerosis was established by ultrasound evaluation (regression). Atherosclerosis was monitored over a 36-month period by duplex ultrasound imaging of the common carotid arteries. At termination, 24 arterial sites were examined for histopathology. In those animals receiving an atherogenic diet, mean percentage ultrasound stenosis at 36 months posttreatment was lower in tocopherol-supplemented groups (61 and 18%) than in the nonsupplemented group (87%). Plasma tocopherol concentration was negatively correlated with percentage ultrasound stenosis ($p < 0.002$). Percentage stenosis in the regression group decreased from 33 to 8% ($p < 0.05$) 8 months after tocopherol supplementation. These result indicate that α-tocopherol is prophylactically and therapeutically effective in atherosclerosis (Verlangieri and Bush, 1992).

B. Epidemiological Studies Showing Protective Effects of Antioxidants on Coronary Heart Disease in Humans

A number of studies (Table I) have investigated the relationship between plasma vitamin E concentrations or intakes and incidence of coronary heart disease.

In 12 European populations of middle-aged men (40–59 years of age) with "common" plasma cholesterol (5.7–6.2 mmol/liter) and blood pressure, both classical risk factors lacked significant correlation to ischemic heart disease mortality, whereas absolute levels of vitamin E showed a strong inverse correlation ($r^2 = 0.63$, $p = 0.002$). In all populations, cholesterol and diastolic blood pressure were moderately associated, but their correlation was inferior to that of vitamin E. Thus, in this study the cross-cultural differences of ischemic heart disease mortality are primarily attributable to the plasma status of vitamin E, which might have protective functions. In the study, vitamin C showed a moderate relation whereas vitamin A showed a weak relation to death from ischemic heart disease (Gey and Puska, 1989; Gey et al., 1991).

A study of middle-aged men in Switzerland (Basel Prospective Study) showed a significantly increased risk of ischemic heart disease among those in the lowest quartile of plasma levels of carotene and/or vitamin C (Gey et al., 1993).

Low plasma levels of vitamins A, C, E, and β-carotene were significantly related to risk of coronary artery disease in the urban population of India (Singh et al., 1995).

In two studies in Turkey, serum vitamin E and serum β-carotene levels were lower in patients with cardiovascular disease than in healthy controls (Torun et al., 1994, 1995).

Plasma vitamin E levels were significantly lower in both groups of angina patients investigated in Poland than in controls (Sklodowska et al., 1991).

The relation between risk of angina pectoris and plasma concentrations of vitamins A, C, and E and carotene was examined in a population case-control study of 110 cases of angina in men aged 35–54. Plasma concentrations of vitamins C and E and carotene were significantly inversely related to the risk of angina. The findings suggest that populations with a high incidence of coronary heart disease may benefit greatly from vitamin E intake (Riemersma et al., 1991).

A study in 19 western European countries and 5 non-European countries evaluated the association between dietary antioxidant intake and coronary

Table I Relationship between Vitamin E Concentrations and Incidence of Coronary Heart Disease

Study location	Sample/heart disease cases[a]	Correlation	Reference
Human epidemiological studies showing beneficial effects of dietary or high serum vitamin E and coronary heart disease			
Europe	12 populations	Inverse association between ischemic heart disease mortality and plasma vitamin E levels	Gey and Puska (1989)
Switzerland	2,974/132	Increased ischemic heart disease risk at lowest quartile of plasma carotene and/or vitamin C levels	Gey et al. (1993)
India	595/72	Low plasma levels of vitamins A, C, and E and β-carotene were associated with an increased risk of coronary artery disease	Singh et al. (1995)
Turkey	n.a./62	Serum vitamin E levels were lower in patients with cardiovascular disease	Torun et al. (1995)
Turkey	n.a./195	Serum β-carotene levels were lower in patients with cardiovascular disease	Torun et al. (1994)
Poland	n.a./53	Levels of vitamin E were lower in plasma in angina patients and in red blood cells in patients with unstable angina	Sklodowska et al. (1991)
United Kingdom	6,000/110	Inverse association between plasma vitamin C, vitamin E, and carotene levels and risk of angina	Riemersma et al. (1991)
19 western European and 5 non-European countries	n.a.	Inverse correlation between dietary intake of vitamin C, vitamin E, β-carotene, and phenolic antioxidant-rich foods and coronary heart disease mortality.	Bellizzi et al. (1994)
United States	25,802/123	Increasing risk for subsequent myocardial infarction with decreasing serum β-carotene levels and a suggestive trend with decreasing serum lutein levels	Street et al. (1994)

United States	1,899/282	Inverse association between serum carotenoid levels and risk of coronary heart disease	Morris et al. (1994)
Finland	5,133/244	Decreased risk of heart disease in men and women in the highest tertile of vitamin E intake and heart disease mortality in women with vitamin C, vitamin E, and carotenoid intakes in the highest tertile	Knekt et al. (1994)
United States	1,556/231	Lower risk of coronary heart disease mortality in the highest tertile of β-carotene and vitamin C intake	Pandey et al. (1995)
United States	34,486/242	Lower risk of death from coronary heart disease in the two highest quintiles of vitamin E intake	Kushi et al. (1996)
United States	11,178/1,101	Use of vitamin E supplements decreased the risk of coronary heart disease mortality	Losonczy et al. (1996)
United States	n.a./156	Coronary artery disease progression was lower in men who took vitamin E supplements	Hodis et al. (1995)
United States	39,910/667	Decreased risk of coronary heart disease in men who took vitamin E supplements	Rimm et al. (1993)
United States	87,245/552	Decreased risk of coronary heart disease in women who took vitamin E supplements	Stampfer et al. (1993)
	Human epidemiological studies showing no clear association between serum vitamin E and coronary heart disease		
Finland	12,000/92	No consistent association between serum selenium, vitamin A, or vitamin E concentrations and risk of death from coronary artery disease	Salonen et al. (1985)
Netherlands	10,532/106	No significant association between serum vitamin A, vitamin E, and selenium levels and risk of death from cardiovascular disease	Kok et al. (1987)

[a] n.a., not available.

heart disease in men under 65 years of age. There was a significant inverse correlation between α-tocopherol intake and coronary heart disease mortality (Bellizzi et al., 1994).

Serum antioxidant levels and the risk of myocardial infarction have been investigated in two studies in the United States. Their data demonstrate a significant increasing risk for subsequent myocardial infarction with decreasing serum carotenoid levels. A protective association with serum vitamin E levels was suggested only among individuals with high serum cholesterol concentrations (Morris et al., 1994; Street et al., 1994).

Numerous studies have also investigated the association between varying levels of antioxidant intake and risk or severity of coronary heart disease. The association between dietary intakes of vitamins C and E and carotenoids and the subsequent mortality from coronary heart disease was evaluated in 14 years of follow-up of Finnish men and women initially free from heart disease. The relative risk of heart disease was 32% lower for men and 65% lower for women in the highest tertile of vitamin E intake compared to the lowest tertile. The relative risk of coronary heart disease mortality was 84% lower in women with both dietary vitamin E and carotenoid intakes in the highest tertile and 83% lower in women with both vitamin C and vitamin E intakes in the highest tertile (Knekt et al., 1994). During 24 years of follow-up of middle-aged men in the United States, men in the highest tertile of combined β-carotene and vitamin C intake had a 21% lower risk of death from coronary heart disease than men with intakes in the lowest tertile (Pandey et al., 1995).

In 1986, 34,486 postmenopausal women in the United States with no cardiovascular disease completed a questionnaire that assessed, among other factors, their intake of vitamins A, E, and C from food sources and supplements. During approximately 7 years of follow-up (ending December 31, 1992), 242 of the women died of coronary heart disease. In analyses adjusted for age and dietary energy intake, vitamin E consumption appeared to be inversely associated with the risk of death from coronary heart disease. The relative risk of death from coronary heart disease was 58% lower in the highest quintiles of dietary vitamin E intake. These results suggest that the intake of vitamin E from food in postmenopausal women is inversely associated with the risk of death from coronary heart disease (Kushi et al., 1996).

Another study (Losonczy et al., 1996) examined vitamin E and vitamin C supplement use in relation to mortality risk and whether vitamin C enhanced the effects of vitamin E in 11,178 persons aged 67–105 years who participated in the Established Populations for Epidemiologic Studies of the Elderly in 1984–1993. Use of vitamin E at two points in time was also associated with a 47% reduced risk of coronary mortality. The simultaneous use of vitamins E and C was associated with a lower risk of coronary mortality (53%). These findings are consistent with those for younger persons and suggest protective effects of vitamin E supplements in the elderly.

Another study on the association between vitamin C and vitamin E intake and coronary artery disease progression was carried out in middle-aged men in the United States with previous coronary bypass graft surgery who were placed on a cholesterol-lowering diet. Overall, men who took at least 100 IU vitamin E per day from supplements showed significantly less coronary artery disease progression for all lesions than men who took less than 100 IU per day for mild to moderate lesions. However, in this study no benefit was found for supplementary vitamin C (Hodis et al., 1995).

Two large epidemiological studies in the United States examined the association between antioxidant intake and risk of coronary heart disease in men and women. In a group of 39,910 male health professionals, men who took vitamin E supplements in doses of at least 100 IU per day for at least 2 years had a 37% lower relative risk of coronary heart disease compared to men who did not take vitamin E supplements. A high intake of vitamin C was not associated with a lower risk of coronary heart disease (Hodis et al., 1995).

In a subsequent study of 87,245 female nurses, women who took vitamin E supplements for more than 2 years had a 41% lower relative risk of major coronary disease. Women who took vitamin E supplements for short periods had little apparent benefit, but those who took them for more than 2 years had a relative risk of major coronary disease of 0.59. High vitamin E intakes from dietary sources were not associated with a significant risk reduction, but even the highest dietary vitamin E intakes were far lower than intakes among supplement users (Stampfer et al., 1993). Researchers in both Harvard-based studies noted that these data and other evidence suggest that vitamin E supplements may decrease heart disease risk.

C. Human Intervention Trials on Effects of Antioxidant Supplements on Risk or Progression of Coronary Heart Disease

A number of studies show that vitamin E decreases the oxidative susceptibility of LDL. Animal investigations as well as human epidemiological data show that antioxidants decrease the risk or progression of coronary heart disease.

Intervention trials will be now discussed aiming to assess the effects of antioxidant supplements on risk or progression of heart disease in various groups.

A randomized nutrition intervention trial provided daily multiple vitamin–mineral supplementation in 29,584 adults from Linxian, China. The largest reductions were for cerebrovascular disease mortality (at doses of one to two times the U.S. recommended dietary allowances) mortality from cerebrovascular disease over a 5-year period. Mortality was 10% lower among subjects supplemented with β-carotene, vitamin E, and selenium (Blot et al., 1995).

A randomized, double-blind, placebo-controlled primary prevention trial (the ATBC study) was primarily aimed to determine whether daily supplementation with α-tocopherol, β-carotene, or both would reduce the incidence of lung cancer and other cancers. A total of 29,133 male smokers (21 cigarettes a day for 36 years) 50–69 years of age were given α-tocopherol (50 mg per day) and/or 20 mg synthetic β-carotene for 5–8 years. Thirty-five fewer deaths from ischemic heart disease and 11 fewer deaths from ischemic stroke, but 22 more deaths from hemorrhagic stroke among men who received vitamin E supplements were counted. Fewer cases of prostate cancer were diagnosed among those who received α-tocopherol than among those who did not. α-Tocopherol had no apparent effect on total mortality, although more deaths from hemorrhagic stroke were observed among men who received this supplement than among those who did not (Albanes et al., 1995).

A large double-blind trial (the CARET study) in the United States evaluated the effects of combined daily supplementation of 30 mg synthetic β-carotene and 25,000 IU vitamin A or placebo on the incidence of cancer and cardiovascular disease. Objects of the study were 18,314 smokers, former smokers, and workers exposed to asbestos over an average follow-up period of 4 years. In the supplemented group, the relative risk of death from cardiovascular disease was 26% higher than in the group on placebo. Researchers observed that favorable effects of antioxidants may be particularly difficult to achieve with continuing cancer-promoting and atherosclerosis-promoting assaults in smokers (Omenn, 1996).

In the Physicians' Health Study, 22,071 healthy male physicians 40–84 years of age in the United States received either 50 mg β-carotene or a placebo on alternate days for a period of 12 years. There were no significant differences in the overall incidence of deaths from cardiovascular disease or in the number of men who suffered a stroke or myocardial infarction in the β-carotene-supplemented group compared to the placebo group. In current and former smokers, those assigned to receive β-carotene supplements had no significant increase or decrease in the risk of cardiovascular disease (Hennekens et al., 1996). In a subgroup analysis of 333 physicians with a history of stable angina pectoris and/or coronary revascularization, the β-carotene-supplemented group had a significant 51% reduction in the risk of major coronary events. The beneficial effect of β-carotene first appeared during the second year of supplementation (Gaziano, 1994).

The last randomized trial studied the effect of vitamin E supplementation (400 or 800 IU per day) on the risk of myocardial infarction in 2002 patients with angiographic evidence of coronary atherosclerosis. Results of the Cambridge Heart Antioxidant Study (CHAOS) demonstrate that vitamin E supplementation significantly decreased the risk of cardiovascular disease and nonfatal myocardial infarction by 47%. After 200 days of treatment, vitamin E produced a 77% decrease in the risk of nonfatal myocardial infarction alone. Total mortality from cardiovascular disease was not significantly lower in the vitamin E-supplemented group, with most of the deaths occurring during the first 200 days of follow-up (Stephens et al., 1996).

D. Commentary to Intervention Studies

The intervention trial in Linxian, China, and the CHAOS study have shown protective effects of antioxidant supplements in the prevention of ischemic heart disease, whereas, the ATBC, CARET, and Physicians' Health Studies did not find a protective effect. However, both the ATBC and the CARET studied were conducted in high-risk groups, smokers and asbestos workers for a long period of time. The ATBC study included vitamin E doses of only 50 IU per day, although results of recent studies have clearly shown that doses of 100 to 400 IU per day are necessary to achieve protective effects against the development or the progression of coronary heart disease. Results of these intervention trials are also in apparent conflict with the large amount of evidence that has demonstrated an effect of vitamin E in protecting LDL from oxidation and diminishing the risk of ischemic heart disease in animals. An inverse correlation between the intake of antioxidants and of fruits and vegetables rich in antioxidants and the risk of coronary heart disease in human epidemiological studies is also very convincing.

The lack of a consistent association among serum selenium, vitamin A or vitamin E levels, and risk of death from coronary heart disease was reported in only epidemiological studies in Finland and the Netherlands (Kok et al., 1987; Salonen et al., 1985). However, it was also noted that these data must be considered with reservation due to several methodological problems, including lack of standardization for cholesterol and triglycerides (Gey et al., 1991).

In summary, both epidemiological and intervention studies, in animals and humans, have shown, in great majority, a strong protective influence of vitamin E in the development of atherosclerosis, especially measured in terms of ischemic art disease. Although a diet rich in fruit and vegetables may have positive effects, supplementation with relatively high doses of vitamin E has shown the most impressive results.

IV. OTHER NONANTIOXIDANT PROPERTIES OR EFFECTS OF VITAMIN E

Vitamin E, in the reduced form, shows very modest anticlotting activity. In contrast, vitamin E quinone (α-tocopheryl quinone) is an anticoagulant. This observation may have significance for field trials in which vitamin E is observed to exhibit beneficial effects on ischemic heart disease and stroke. Vitamin E quinone is an inhibitor of the vitamin K-dependent carboxylase that controls blood clotting (Dowd and Zheng, 1995). This result obtained *in vitro* may suggest caution in the usage of high doses of vitamin E in patients with blood coagulation anomalies.

To clarify whether vitamin E enhances the pharmacological effect of warfarin, a double-blind clinical trial was completed in which 21 subjects taking chronic warfarin therapy were randomized to receive either vitamin E or placebo. None of the subjects who received vitamin E had a significant change in the international normalized ratio, and thus it appears that vitamin E can safely be given to patients who require chronic warfarin therapy (Kim and White, 1996). This may be based on the notion that a conversion of α-tocopheryl quinone into α-tocopherol in humans has been demonstrated (Moore and Ingold, 1997).

In vitro results suggest that despite α-tocopherol action as an antioxidant, γ-tocopherol could also be required. Its function would be to effectively remove the peroxynitrite-derived nitrating species present in the atherosclerotic plaque that may participate in the development of the process (Christen *et al.*, 1997). Relevance of this report for cellular, animal, and human situations would be worth investigating, especially in the light of the report (Clément *et al.*, 1995) indicating a stimulation by γ-tocopherol of α-tocopherol uptake of in rat tissues.

In summary, the anticoagulant effect of the oxidized product of α-tocopherol, although demonstrated *in vitro*, does not appear to be important *in vivo*. This may be due to its rereduction or to prevention of α-tocopherol oxidation by other antioxidants, such as ascorbic acid.

REFERENCES

Albanes, D., Heinonen, O. P., Huttunen, J. K., Taylor, P. R., Virtamo, J., Edwards, B. K., Haapakoski, J., Rautalahti, M., Hartman, A. M., Palmgren, J., and Greenwald, P. (1995). Effects of α-tocopherol and β-carotene supplements on cancer incidence in the Alpha-Tocopherol Beta-Carotene Cancer Prevention Study. *Am. J. Clin. Nutr.* 62(Suppl.), 1427S–1430S.

Arita, M., Sato, Y., Miyata, A., Tanabe, T., Takahashi, E., Kayden, H. J., Arai, H., and Inoue, K. (1995). Human alpha-tocopherol transfer protein: cDNA cloning, expression and chromosomal localization. *Biochem. J.* 306, 437–443.

Azzi, A., Boscoboinik, D., and Hensey, C. (1992). The protein kinase C family. *Eur. J. Biochem.* **208**, 547–557.

Azzi, A., Boscoboinik, D., Chatelain, E., Özer, N. K., and Stäuble, B. (1993). d-alpha-tocopherol control of cell proliferation. *Mol. Aspects Med.* **14**, 265–271.

Azzi, A., Boscoboinik, D., Marilley D., Özer, N. K., Stäuble, B., and Tasinato, A. (1995). Vitamin E: A sensor and an information transducer of the cell oxidation state. *Am. J. Clin. Nutr.* **62**(Suppl.), 1337S–1346S.

Azzi, A., Cantoni, O., Özer, N. K., Boscoboinik, D., and Spycher, S. (1996). The role of hydrogen peroxide and RRR-α-tocopherol in smooth muscle cell proliferation. *Cell Death Differ.* **3**, 79–90.

Bellizzi, M. C., Franklin, M. F., Duthie, G. G., and James, W. P. (1994). Vitamin e and coronary heart disease: The European paradox. *Eur. J. Clin. Nutr.* **48**, 822–831; *erratum:* **49**(3), 230 (1995).

Blot, W. J., Li, J. Y., Taylor, P. R., Guo, W. D., Dawsey, S. M., and Li, B. (1995). The Linxian trials: Mortality rates by vitamin-mineral intervention group. *Am. J. Clin. Nutr.* **62**(Suppl.), 1424S–1426S.

Bornancin, F., and Parker, P. J. (1996). Phosphorylation of threonine 638 critically controls the dephosphorylation and inactivation of protein kinase Cα. *Curr. Biol.* **6**, 1114–1123.

Boscoboinik, D., Szewczyk, A., Hensey, C., and Azzi, A. (1991). Inhibition of cell proliferation by alpha-tocopherol. Role of protein kinase C. *J. Biol. Chem.* **266**, 6188–6194.

Boscoboinik, D. O., Chatelain, E., Bartoli, G. M., Stäuble, B., and Azzi, A. (1994). Inhibition of protein kinase C activity and vascular smooth muscle cell growth by d-alpha-tocopherol. *Biochim. Biophys. Acta.* **1224**, 418–426.

Boscoboinik, D., Özer, N. K., Moser, U., and Azzi, A. (1995). Tocopherols and 6-hydroxychroman-2-carbonitrile derivatives inhibit vascular smooth muscle cell proliferation by a nonantioxidant mechanism. *Arch. Biochem. Biophys.* **318**, 241–246.

Burton, G. W., Joyce, A., and Ingold, K. U. (1982). First proof that vitamin E is major lipid-soluble, chain- breaking antioxidant in human blood plasma (letter). *Lancet* **2**, 327.

Burton, G. W., Joyce, A., and Ingold, K. U. (1983). Is vitamin E the only lipid-soluble, chain-breaking antioxidant in human blood plasma and erythrocyte membranes? *Arch. Biochem. Biophys.* **221**, 281–290.

Burton, G. W., Traber, M. G., Acuff, R. V., Walters, D. N., Kayden, H., Hughes, L., and Ingold, K. U. (1998). Human plasma and tissue alpha-tocopherol concentrations in response to supplementation with deuterated natural and synthetic vitamin E. *Am. J. Clin. Nutr.* **67**, 669–684.

Christen, S., Woodall, A. A., Shigenaga, M. K., Southwell-Keely, P. T., Duncan, M. W., and Ames, B. N. (1997). gamma-Tocopherol traps mutagenic electrophiles such as NO_x and complements α-tocopherol: Physiological implications. *Proc. Natl. Acad. Sci. U.S.A.* **94**, 3217–3222.

Clément, M., Dinh, L., and Bourre, J.-M. (1995). Uptake of dietary RRR-α- and RRR-gamma-tocopherol by nervous tissues, liver and muscle in vitamin-E-deficient rats. *Biochim. Biophys. Acta Lipids Lipid Metab.* **1256**, 175–180.

Devaraj, S., Li, D., and Jialal, I. (1996). The effects of alpha tocopherol supplementation on monocyte function. Decreased lipid oxidation, interleukin 1 beta secretion, and monocyte adhesion to endothelium. *J. Clin. Invest.* **98**, 756–763.

Dowd, P., and Zheng, Z. B. (1995). On the mechanism of the anticlotting action of vitamin E quinone. *Proc. Natl. Acad. Sci. U.S.A.* **92**, 8171–8175.

Dutil, E. M., Keranen, L. M., DePaoli-Roach, A. A., and Newton, A. C. (1994). *In vivo* regulation of protein kinase C by trans-phosphorylation followed by autophosphorylation. *J. Biol. Chem.* **269**, 29359–29362.

Esterbauer, H., Dieber Rotheneder, M., Striegl, G., and Waeg, G. (1991). Role of vitamin E in preventing the oxidation of low-density lipoprotein. *Am. J. Clin. Nutr.* 53, 314S–321S.

Esterbauer, H., Puhl, H., Waeg, G., Krebs, A., Tatzber, F., and Rabl, H. (1992a). Vitamin E and atherosclerosis: An overview. *J. Nutr. Sci. Vitaminol., Spec No.*, pp. 177–182.

Esterbauer, H., Waeg, G., Puhl, H., Dieber Rotheneder, M., and Tatzber, F. (1992b). Inhibition of LDL oxidation by antioxidants. *EXS* 62, 145–157.

Gaziano, J. M. (1994). Antioxidant vitamins and coronary artery disease risk. *Am. J. Med.* 97, 18S–21S.

Gey, K. F., and Puska, P. (1989). Plasma vitamins E and A inversely correlated to mortality from ischemic heart disease in cross-cultural epidemiology. *Ann. N. Y. Acad. Sci.* 570, 268–282.

Gey, K. F., Puska, P., Jordan, P., and Moser, U. K. (1991). Inverse correlation between plasma vitamin E and mortality from ischemic heart disease in cross-cultural epidemiology. *Am. J. Clin. Nutr.* 53, 326S–334S.

Gey, K. F., Stahelin, H. B., and Eichholzer, M. (1993). Poor plasma status of carotene and vitamin C is associated with higher mortality from ischemic heart disease and stroke: Basel Prospective Study. *Clin. Invest.* 71, 3–6.

Gotoda, T., Arita, M., Arai, H., Inoue, K., Yokota, T., Fukuo, Y., Yazaki, Y., and Yamada, N. (1995). Adult-onset spinocerebellar dysfunction caused by a mutation in the gene for the alpha-tocopherol-transfer protein. *N. Engl. J. Med.* 333, 1313–1318.

Hennekens, C. H., Buring, J. E., Manson, J. E., Stampfer, M., Rosner, B., Cook, N. R., Bélanger, C., La Motte, F., Gaziano, J. M., Ridker, P. M., Willett, W., and Peto, R. (1996). Lack of effect of long-term supplementation with beta carotene on the incidence of malignant neoplasms and cardiovascular disease. *N. Engl. J. Med.* 334, 1145–1149.

Hodis, H. N., Mack, W. J., LaBree, L., Cashin-Hemphill, L., Sevanian, A., and Azen, S. P. (1995). Serial coronary angiographic evidence that antioxidant vitamin intake reduces progression of coronary artery atherosclerosis. *J. Am. Med. Assoc.* 273, 1849–1854.

Hoff, H. F., O'Neil, J., Chisolm, G. M., Cole, T. B., Quehenberger, O., Esterbauer, H., and Jurgens, G. (1989). Modification of low density lipoprotein with 4-hydroxynonenal induces uptake by macrophages. *Arteriosclerosis* 9, 538–549.

Jessup, W., Rankin, S. M., De Whalley, C. V., Hoult, U. R., Scott, J., and Leake, D. S. (1990). Alpha-tocopherol consumption during low-density-lipoprotein oxidation. *Biochem. J.* 265, 399–405.

Jha, P., Flather, M., Lonn, E., Farkouh, M., and Yusuf, S. (1995). Antioxidant vitamins and cardiovascular disease—A critical review of epidemiologic and clinical trial data. *Ann. Intern. Med.* 123, 860–872.

Jialal, I., and Grundy, S. M. (1992). Effect of dietary supplementation with alpha-tocopherol on the oxidative modification of low density lipoprotein. *J. Lipid Res.* 33, 899–906.

Jialal, I., Fuller, C. J., and Huet, B. A. (1995). The effect of alpha-tocopherol supplementation on ldl oxidation: A dose-response study. *Arterioscler. Thromb. Vasc. Biol.* 15, 190–198.

Jurgens, G., Hoff, H. F., Chisolm, G. M., and Esterbauer, H. (1987). Modification of human serum low density lipoprotein by oxidation—characterization and pathophysiological implications. *Chem. Phys. Lipids.* 45, 315–336.

Kagan, V. E., Serbinova, E. A., Forte, T., Scita, G., and Packer, L. (1992). Recycling of vitamin E in human low density lipoproteins. *J. Lipid Res.* 33, 385–397.

Kanno, T., Utsumi, T., Takehara, Y., Ide, A., Akiyama, J., Yoshioka, T., Horton, A. A., and Utsumi, K. (1996). Inhibition of neutrophil-superoxide generation by α-tocopherol and coenzyme Q. *Free Radical Res.* 24, 281–289.

Kim, J. M., and White, R. H. (1996). Effect of vitamin E on the anticoagulant response to warfarin. *Am. J. Cardiol.* 77, 545–546.

Kleinveld, H. A., Demacker, P. N. M., and Stalenhoef, A. F. H. (1994). Comparative study on the effect of low-dose vitamin E and probucol on the susceptibility of LDL to oxidation and the progression of atherosclerosis in Watanabe heritable hyperlipidemic rabbits. *Arterioscler. Thromb.* **14,** 1386–1391.

Knekt, P., Reunanen, A., Jarvinen, R., Seppanen, R., Heliovaara, M., and Aromaa, A. (1994). Antioxidant vitamin intake and coronary mortality in a longitudinal population study. *Am. J. Epidemiol.* **139,** 1180–1189.

Kok, F. J., de Bruijn, A. M., Vermeeren R., Hofman A., van Laar, A., de Bruin, M., Hermus, R. J., and Valkenburg, H. A. (1987). Serum selenium, vitamin antioxidants, and cardiovascular mortality: A 9-year follow-up study in the netherlands. *Am. J. Clin. Nutr.* **45,** 462–468.

Kushi, L. H., Folsom, A. R., Prineas, R. J., Mink, P. J., Wu, Y., and Bostick, R. M. (1996). Dietary antioxidant vitamins and death from coronary heart disease in postmenopausal women. *N. Engl. J. Med.* **334,** 1156–1162.

Lafont, A. M., Chai, Y. C., Cornhill, J. F., Whitlow, P. L., Howe, P. H., and Chisolm, G. M. (1995). Effect of alpha-tocopherol on restenosis after angioplasty in a model of experimental atherosclerosis. *J. Clin. Invest* **95,** 1018–1025.

Li, D., Devaraj, S., Fuller, C., Bucala, R., and Jialal, I. (1996). Effect of alpha-tocopherol on ldl oxidation and glycation: In vitro and in vivo studies. *J. Lipid Res.* **37,** 1978–1986.

Losonczy, K. G., Harris, T. B., and Havlik, R. J. (1996). Vitamin E and vitamin C supplement use and risk of all-cause and coronary heart disease mortality in older persons: The established populations for epidemiologic studies of the elderly. *Am. J. Clin. Nutr.* **64,** 190–196.

Meydani, S. N., Meydani, M., Blumberg, J. B., Leka, L. S., Siber, G., Loszewski, R., Thompson, C., Pedrosa, M. C., Diamond, R. D., and Stollar, B. D. (1997). Vitamin E supplementation and in vivo immune response in healthy elderly subjects—A randomized controlled trial. *JAMA, J. Am. Med. Assoc.* **277,** 1380–1386.

Mitchinson, M. J., Ball, R. Y., Carpenter, K. H., Enright, J. H., and Brabbs, C. E. (1990). Ceroid, macrophages and atherosclerosis. *Biochem. Soc. Trans.* **18,** 1066–1069.

Moore, A. N. J., and Ingold, K. U. (1997). α-Tocopheryl quinone is converted into vitamin E in man. *Free Radical Biol. Med.* **22,** 931–934.

Morin, R. J., and Peng, S. K. (1989). The role of cholesterol oxidation products in the pathogenesis of atherosclerosis. *Ann. Clin. Lab. Sci.* **19,** 225–237.

Morris, D. L., Kritchevsky, S. B., and Davis, C. E. (1994). Serum carotenoids and coronary heart disease: The Lipid Research Clinics Coronary Primary Prevention Trial and Follow-up Study. *JAMA. J. Am. Med. Assoc.* **272,** 1439–1441.

Müller, K., Hardwick, S. J., Marchant, C. E., Law, N. S., Waeg, G., Esterbauer, H., Carpenter, K. L. H., and Mitchinson, M. J. (1996). Cytotoxic and chemotactic potencies of several aldehydic components of oxidised low density lipoprotein for human monocyte-macrophages. *FEBS Lett.* **388,** 165–168.

Newton, A. C. (1995). Protein kinase C: Structure, function, and regulation. *J. Biol. Chem.* **270,** 28495–28498.

Omenn, G. S. (1996). Antioxidant vitamins, cancer, and cardiovascular disease. *N. Engl. J. Med.* **335,** 1067–1068.

Ouahchi, K., Arita, M., Kayden, H., Hentati, F., Ben Hamida, M., Sokol, R., Arai, H., Inoue, K., Mandel, J. L., and Koenig, M. (1995). Ataxia with isolated vitamin E deficiency is caused by mutations in the α-tocopherol transfer protein. *Nat. Genet.* **9,** 141–145.

Özer, N. K., Sirikci, O., Taha, S., San, T., Moser, U., and Azzi, A. (1998). Effect of vitamin E and probucol on dietary cholesterol-induced atherosclerosis in rabbits. *Free Radical Biol. Med.* **24,** 226–33.

Packer, J. E., Slater, T. F., and Willson, R. L. (1979). Direct observation of a free radical interaction between vitamin E and vitamin C. *Nature* (London) **278**, 737–738.

Packer, L. (1992). Interactions among antioxidants in health and disease: Vitamin E and its redox cycle. *Proc. Soc. Exp. Biol. Med.* **200**, 271–276.

Packer, L., and Suzuki, Y. J. (1993). Vitamin E and alpha-lipoate: Role in antioxidant recycling and activation of the NF-kappa B transcription factor. *Mol. Aspects. Med.* **14**, 229–239.

Pandey, D. K., Shekelle, R., Selwyn, B. J., Tangney, C., and Stamler, J. (1995). Dietary vitamin c and beta-carotene and risk of death in middle-aged men. The western electric study. *Am. J. Epidemiol.* **142**, 1269–1278.

Prasad, K., and Kalra, J. (1993). Oxygen free radicals and hypercholesterolemic atherosclerosis: Effect of vitamin E. *Am. Heart J.* **125**, 958–973.

Pryor, A. W., Cornicelli, J. A., Devall, L. J., Tait, B., Trivedi, B. K., Witiak, D. T., and Wu, M. (1993). A rapid screening test to determine the antioxidant potencies of natural and synthetic antioxidants. *J. Org. Chem.* **58**, 3521–3532.

Raines, E. W., and Ross, R. (1993). Smooth muscle cells and the pathogenesis of the lesions of atherosclerosis. *Br. Heart J.* **69**, S30–S37.

Riemersma, R. A., Wood, D. A., MacIntyre, C. C., Elton, R. A., Gey, K. F., and Oliver, M. (1991). Risk of angina pectoris and plasma concentrations of vitamins A, C, and E and carotene. *Lancet* **337**, 1–5.

Rimm, E. B., Stampfer, M. J., Ascherio, A., Giovannucci, E., Colditz, G. A., and Willett, W. C. (1993). Vitamin E consumption and the risk of coronary heart disease in men. *N. Engl. J. Med.* **328**, 1450–1456.

Ross, R. (1993). The pathogenesis of atherosclerosis: A perspective for the 1990s. *Nature* (London) **362**, 801–809.

Russell, M. W., Huse, D. M., Drowns, S., Hamel, E. C., and Hartz, S. C. (1998). Direct medical costs of coronary artery disease in the United States. *Am. J. Cardiol.* **81**, 1110–1115.

Salonen, J. T., Salonen, R., Penttila, I., Herranen, J., Jauhiainen, M., Kantola, M., Lappetelainen, R., Maenpaa, P. H., Alfthan, G., and Puska, P. (1985). Serum fatty acids, apolipoproteins, selenium and vitamin antioxidants and the risk of death from coronary artery disease. *Am. J. Cardiol.* **56**, 226–231.

Sambrano, G. R., Parthasarathy, S., and Steinberg, D. (1994). Recognition of oxidatively damaged erythrocytes by a macrophage receptor with specificity for oxidized low density lipoprotein. *Proc. Natl. Acad. Sci. U.S.A.* **91**, 3265–3269.

Sato, K., Niki, E., and Shimasaki, H. (1990). Free radical-mediated chain oxidation of low density lipoprotein and its synergistic inhibition by vitamin E and vitamin C. *Arch. Biochem. Biophys.* **279**, 402–405.

Scholich, H., Murphy, M. E., and Sies, H. (1989). Antioxidant activity of dihydrolipoate against microsomal lipid peroxidation and its dependence on alpha-tocopherol. *Biochim. Biophys. Acta* **1001**, 256–261.

Schwartz, C. J., Valente, A. J., Sprague, E. A., Kelley, J. L., and Nerem, R. M. (1991). The pathogenesis of atherosclerosis: An overview. *Clin. Cardiol.* **14**, 11–16.

Singh, R. B., Ghosh, S., Niaz, M. A., Singh, R., Beegum, R., Chibo, H., Shoumin, Z., and Postiglione, A., (1995). Dietary intake, plasma levels of antioxidant vitamins, and oxidative stress in relation to coronary artery disease in elderly subjects. *Am. J. Cardiol.* **76**, 1233–1238.

Sirikci, Ö., Özer, N. K., and Azzi, A. (1996). Dietary cholesterol-induced changes of protein kinase C and the effect of vitamin E in rabbit aortic smooth muscle cells. *Atherosclerosis* **126**, 253–263.

Sklodowska, M., Wasowicz, W., Gromadzinska, J., Miroslaw, W., Strzelczyk, M., Malczyk, J., and Goch, J. H. (1991). Selenium and vitamin E concentrations in plasma and erythrocytes of angina pectoris patients. *Trace Elem. Med.* **8**, 113–117.

Smith, T. L., and Kummerow, F. A. (1989). Effect of dietary vitamin E on plasma lipids and atherogenesis in restricted ovulator chickens. *Atherosclerosis* **75**, 105–109.
Stampfer, M. J., Hennekens, C. H., Manson, J. E., Colditz, G. A., Rosner, B., and Willett, W. C. (1993). Vitamin E consumption and the risk of coronary disease in women. *N. Engl. J. Med.* **328**, 1444–1449.
Steinberg, D., and Witztum, J. L. (1990). Lipoproteins and atherogenesis: Current concepts. *JAMA. J. Am. Med. Assoc.* **264**, 3047–3052.
Steiner, M. (1993). Vitamin E: More than an antioxidant. *Clin. Cardiol.* **16**, 116–118.
Steiner, M., Glantz, M., and Lekos, A. (1995). Vitamin E plus aspirin compared with aspirin alone in patients with transient ischemic attacks. *Am. J. Clin. Nutr.* **62**, 1381S–1384S.
Stephens, N. G., Parsons, A., Schofield, P. M., Kelly, F., Cheeseman, K., Mitchinson, M. J., and Brown, M. J. (1996). Randomised controlled trial of vitamin E in patients with coronary disease: Cambridge Heart Antioxidant Study (CHAOS). *Lancet* **347**, 781–786.
Street, D. A., Comstock, G. W., Salkeld, R. M., Schuep, W., and Klag, M. J. (1994). Serum antioxidants and myocardial infarction: Are low levels of carotenoids and alpha-tocopherol risk factors for myocardial infarction? *Circulation* **90**, 1154–1161.
Tasinato, A., Boscoboinik, D., Bartoli, G. M., Maroni, P., and Azzi, A. (1995). d-α-tocopherol inhibition of vascular smooth muscle cell proliferation occurs at physiological concentrations, correlates with protein kinase C inhibition, and is independent of its antioxidant properties. *Proc. Natl. Acad. Sci. U.S.A.* **92**, 12190–12194.
Torun, M., Yardim, S., Sargin, H., and Simsek, B. (1994). Evaluation of serum beta-carotene levels in patients with cardiovascular diseases. *J. Clin. Pharmacol. Ther.* **19**, 61–63.
Torun, M., Avci, N., and Yardim, S. (1995). Serum levels of vitamin E in relation to cardiovascular diseases. *J. Clin. Pharmacol. Ther.* **20**, 335–340.
Traber, M. G., Sokol, R. J., Kohlschutter, A., Yokota, T., Muller, D. P., Dufour, R., and Kayden, H. J. (1993). Impaired discrimination between stereoisomers of alpha-tocopherol in patients with familial isolated vitamin E deficiency. *J. Lipid Res.* **34**, 201–210.
Ulbricht, T. L., and Southgate, D. A. (1991). Coronary heart disease: Seven dietary factors. *Lancet* **338**, 985–992.
Verlangieri, A. J., and Bush, M. J. (1992). Effects of d-alpha-tocopherol supplementation on experimentally induced primate atherosclerosis. *J. Am. Coll. Nutr.* **11**, 131–138.
Wefers, H., and Sies, H. (1988). The protection by ascorbate and glutathione against microsomal lipid peroxidation is dependent on vitamin E. *Eur. J. Biochem.* **174**, 353–357.
Wiklund, O., Mattsson, L., Bjornheden, T., Camejo, G., and Bondjers, G. (1991). Uptake and degradation of low density lipoproteins in atherosclerotic rabbit aorta: Role of local LDL modification. *J. Lipid Res.* **32**, 55–62.
Williams, R. J., Motteram, J. M., Sharp, C. H., and Gallagher, P. J. (1992). Dietary vitamin E and the attenuation of early lesion development in modified watanabe rabbits. *Atherosclerosis* **94**, 153–159.
Wojcicki, J., Rozewicka, L., Barcew-Wiszniewska, B., Samochowiec, L., Juzwiak, S., Kadlubowska, D., Tustanowski, S., and Juzyszyn, Z. (1991). Effect of selenium and vitamin E on the development of experimental atherosclerosis in rabbits. *Atherosclerosis* **87**, 9–16.
Yla Herttuala, S., Palinski, W., Rosenfeld, M. E., Parthasarathy, S., Carew, T. E., Butler, S., Witztum, J. L., and Steinberg, D. (1989). Evidence for the presence of oxidatively modified low density lipoprotein in atherosclerotic lesions of rabbit and man. *J. Clin. Invest.* **84**, 1086–1095.

17 Peroxiredoxins in Cell Signaling and HIV Infection

Dong-Yan Jin and Kuan-Teh Jeang
Laboratory of Molecular Microbiology
National Institute of Allergy and Infectious Diseases
Bethesda, Maryland 20892

Living organisms constantly produce reactive oxygen species (ROS) such as H_2O_2 during normal cellular metabolism and in response to external stimuli such as UV radiation. ROS are evolutionarily ancient threats to biomolecules, including lipids, carbohydrates, nucleic acids, and proteins (Berlett and Stadtman, 1997; Henle and Linn, 1997). To combat this threat, cells have evolved antioxidant defenses (Sies, 1993; Scandalios, 1997). Antioxidants such as superoxide dismutase, catalase, glutathione, glutathione peroxidase, glutaredoxin, and thioredoxin are ubiquitously present in organisms ranging from bacteria to human. A delicate balance of these defense functions is important for survival.

In addition to well-documented antioxidant enzymes, a novel family of peroxidases, designated peroxiredoxins (Chae *et al.*, 1994c), has been characterized (Chae *et al.*, 1994a). Peroxiredoxins are found in all organisms and their primary sequences are highly conserved (Chae and Rhee, 1994; Rhee and Chae, 1994). This ubiquity and high degree of conservation suggest a biological importance of this family of enzymes. Several lines of evidence support that peroxiredoxins are critically involved in cellular defense, receptor signaling, protein phosphorylation, transcriptional regulation, and apoptosis (Ichimiya *et al.*, 1997; Jin *et al.*, 1997; Zhang *et al.*, 1997; Wen and Van Etten, 1997; Chen *et al.*, 1998; Kang *et al.*, 1998a). In addition, peroxiredoxins have crucial roles in cellular immunity and infection (Sauri *et al.*, 1996; Sherman *et al.*, 1996; Haridas *et al.*, 1998; McGonigle *et al.*,

1998). Thus, peroxiredoxins can suppress HIV infection through inhibition of NF-κB signaling (Jin et al., 1997).

This chapter summarizes the existing evidence and provides perspectives on the structure of peroxiredoxins and their functions in signal transduction and in HIV-1 infection. Readers should also refer to two papers by Kang et al., (1998a) and McGonigle et al., (1998), which provide comprehensive descriptions of individual peroxiredoxins.

I. PEROXIREDOXINS: A LARGE FAMILY OF ANTIOXIDANT ENZYMES

The first identified peroxiredoxin was a 25-kDa thiol-specific antioxidant (TSA) from budding yeast, which reduces peroxides using thioredoxin as the immediate electron donor (K. Kim et al., 1988; I.-H. Kim et al., 1989; Chae et al., 1993, 1994a,c; Netto et al., 1996). This "protector" activity, which specifically prevents enzymes from damage by thiol-dependent metal-catalyzed oxidation, was initially described by Kim and co-workers (1988). One commonly used biological assay for TSA and many other peroxiredoxins is the protection of glutamine synthetase from oxidation by dithiothreitol (DTT)/Fe^{3+}/O_2. It is now well understood that these antioxidant properties of TSA and peroxiredoxins are ascribed, at least in part, to their ability to scavenge hydrogen peroxide (Netto et al., 1996). The peroxidatic center of peroxiredoxins contains one highly conserved cysteine residue (Chae et al., 1994a,c). The thiol specificity of their antioxidant activities, as described in earlier studies, is explained by the preservation of the active sulfhydryl form of the enzymes by thiols such as dithiothreitol (DTT). Peroxiredoxins are thought to be active peroxidases, but some of them are supported by electron donors other than thioredoxin. However, an antioxidant function of peroxiredoxins, which directly protects bacterial and human cells from reactive nitrogen species (RNS), has also been suggested in a more recent study (Chen et al., 1998). This function is apparently independent of the peroxidase activity. It remains to be seen whether peroxiredoxins can generally scavenge RNS in addition to reactive oxygen species.

The TSA sequence was first reported in 1994 (Chae et al., 1993, 1994a). Currently there are several hundred members in the peroxiredoxin family. While the rapid accumulation of novel cDNAs makes it difficult to compile an exhaustive alignment of all peroxiredoxins, sequence information does provide a path for tracing the evolution of peroxiredoxins and has paved the way toward a better understanding of the structure–function relationship of various peroxiredoxins. Major subfamilies of peroxiredoxins are summarized in Table I. Most peroxiredoxins currently deposited in data bases are

Table I Major Subfamilies of Peroxiredoxins

Acronym	Name	First identified source	Described function	Reference
p20	Thiol peroxidase p20 Scavengase p20	*E. coli* (periplasmic space)	Thioredoxin peroxidase	Cha *et al.* (1995) Cha *et al.* (1996) Wan *et al.* (1997) Zhou *et al.* (1997)
BCP	Bacterioferritin comigratory protein	*E. coli*	Unknown	Chae *et al.* (1994a)
TSA	Thiol-specific antioxidant	Yeast	Thioredoxin peroxidase	Chae *et al.* (1993, 1994a,c)
AhpC	Alkyl hydroperoxide reductase C	*S. typhimurium*	Flavin-dependent peroxidase	Chae *et al.* (1994a) Poole (1996) Ellis and Poole (1997a)
ICPrx	One cysteine peroxiredoxin	Bovine Human	Glutathione peroxidase Peroxidase linked to unidentified source of electron	Shichi and Demar (1990) Munz *et al.* (1997) Frank *et al.* (1997) Kang *et al.* (1998b)

well represented by these major subfamilies. Two phylogenetic trees based on parsimony (Fig. 1) or distance matrix (Fig. 2) are also shown for a larger subset of prototypic peroxiredoxins discussed here.

A. Microbial Peroxiredoxins

The complete genome sequences of a dozen bacteria and one yeast strain have been determined. This information has permitted a thorough comparison of microbial peroxiredoxins, yielding novel insights into the molecular evolution and the biology of these antioxidant enzymes.

There are three peroxiredoxins in *Escherichia coli*. The alkyl hydroperoxide reductase colorless subunit (AhpC) is a prototype peroxiredoxin that reduces hydroperoxide using electrons transferred from AhpF, a thioredoxin reductase-like flavoenzyme (Poole, 1996; Ellis and Poole, 1997a,b). Scavengase p20 (thiol peroxidase) is a thioredoxin-dependent peroxidase found in the periplasmic space (Cha *et al.*, 1995, 1996; Wan *et al.*, 1997; Zhou *et al.*, 1997). The third *E. coli* enzyme, misleadingly named bacterioferritin comigratory protein (BCP), is poorly understood.

AhpC, p20, and BCP from *E. coli* conserve a peroxiredoxin motif surrounding a cysteine residue, which is the putative redox center (Zhou *et al.*, 1997). However, the three peroxiredoxins are so diverged that *E. coli* AhpC is more closely related to human AOE372 (30% identity and 41% similarity)

Figure 1 Consensus parsimony tree of peroxiredoxins. Phylogenies were inferred from protein sequences using parsimony. Parsimony tree was constructed by the PROTPARS program in the PHYLIP package (version 3.572c; Felsenstein, 1996). SEQBOOT and CONSENSUS programs in the same package were used to perform bootstrap replication and to produce the majority rule consensus tree from 100 replicates. GenBank (GB/ or gi/) or SwissProt (SP/) accession/identification numbers of the sequences are TSA-yeast, SP/P34760, gi/575691; TSA2-yeast, SP/Q04120, gi/927720; BAS1-thale cress, GB/X94218, gi/1498198; AOE372-mouse, GB/U96746, gi/3024715; AOE372-human, SP/Q13162, gi/3024727; MER5-mouse, GB/M28723; MER5-human, GB/D49396; TPx-mouse, GB/U20611; TPx-human, GB/Z22588; PAG-mouse, GB/D16142; PAG-human, GB/X67951; TSA-amoeba, GB/X70996; TXNPx-C. fasciculata, GB/AF020947; YkuU-B. subtilis, GB/AJ222587, gi/2632242; AhpC1-Aquifex, GB/AE000692, gi/2983132; AhpC-B. subtilis, GB/D78193, SP/P80239; AhpC-E. coli, GB/U82598, SP/P26427; 1CPrx-mouse, GB/Y12883; 1CPrx-human, GB/D14662; 1CPrx-cyanobacteria, GB/D90905, gi/1652457; 1CPrx-yeast, GB/Z23261, SP/P34227; 1CPrx-archaea, GB/AE001087, gi/2650373; AhpC2-Aquifex, GB/AE000711, gi/2983402; Per1-thale cress, GB/Y12089, gi/1926269; TSA-cyanobacteria, GB/D64000, gi/1001510; BCP-Aquifex, GB/AE000691, gi/2983147; BCP-yeast, SP/P40553, gi/558392; BCP2-cyanobacteria, GB/D90900, gi/1651777; BCP-E.coli, GB/M63654,SP/P23480; BCP-cyanobacteria, GB/D90900, gi/1651799; YgaF-B. subtilis, GB/Z82044, gi/1673395; p20-E. coli, GB/U93212, SP/P37901; p20-Aquifex, GB/AE000692, gi/2983133; YtgI-B. subtilis, GB/AF008220, gi/2293238.

17. Peroxiredoxins

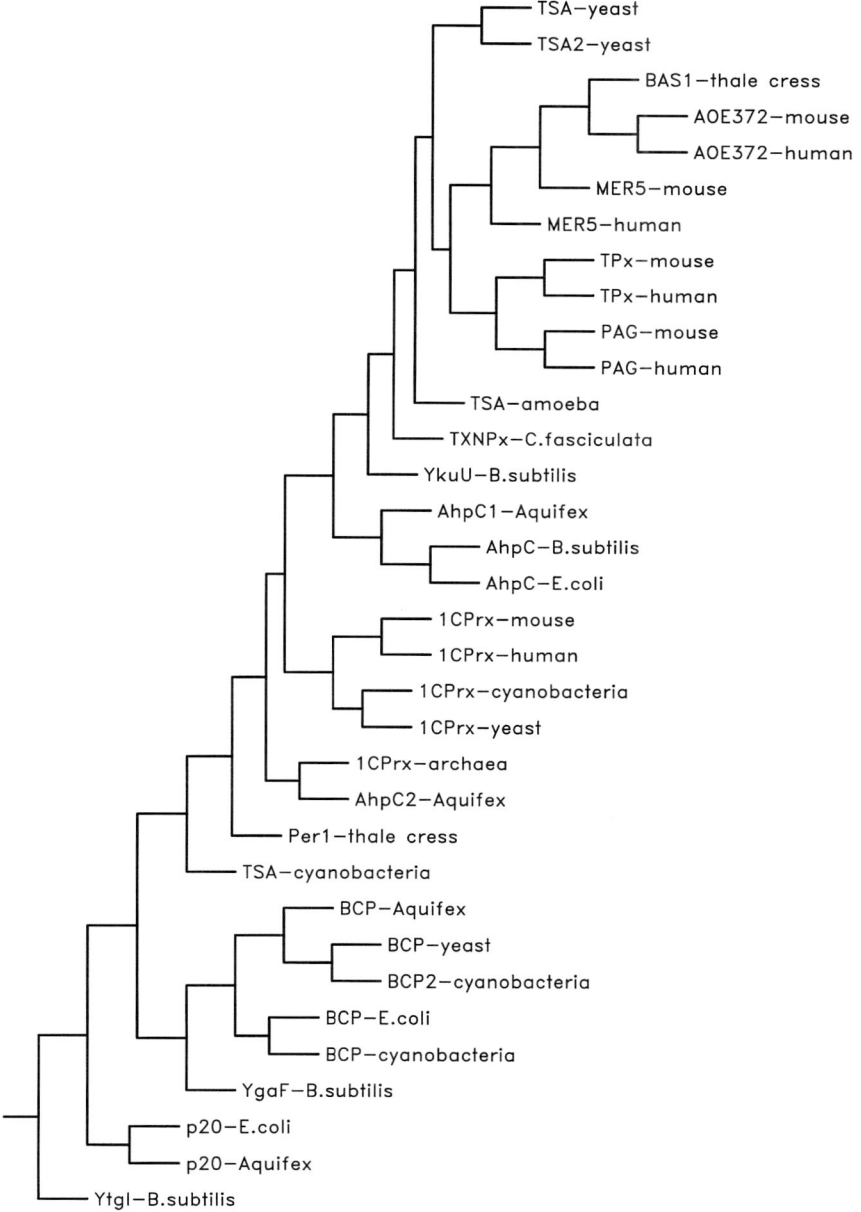

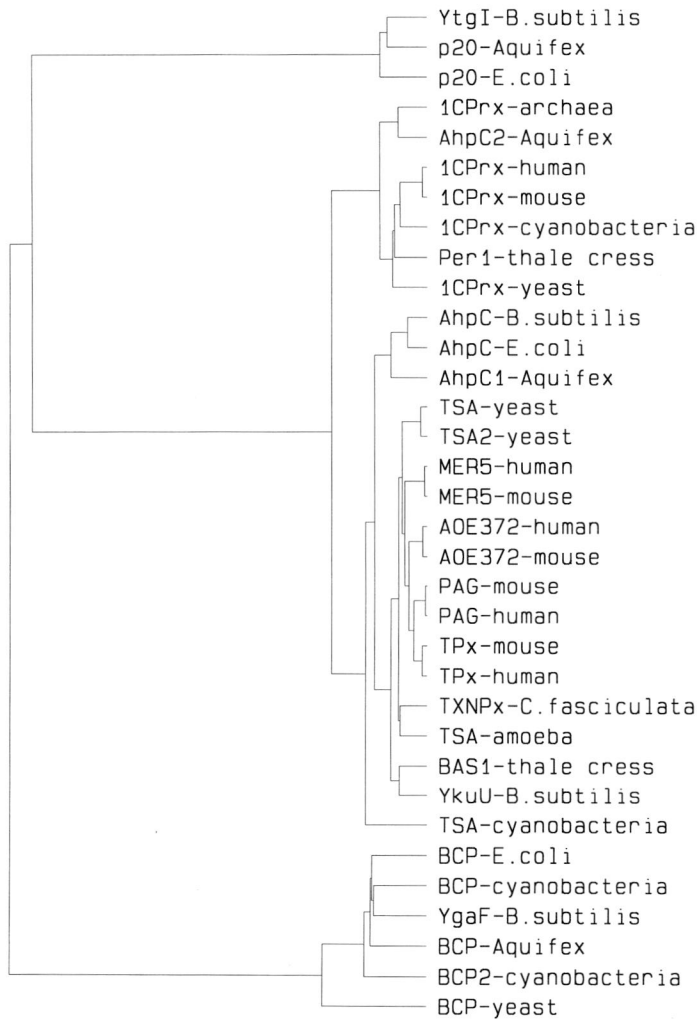

Figure 2 Distance matrix tree of peroxiredoxins. Phylogenies were inferred from aligned protein sequences (generated by the PILEUP program in the GCG package) using a distance matrix calculated by the DISTANCES program. Tree reconstruction was performed by the GROWTREE program using an UPGMA algorithm.

than to *E. coli* p20 (17% identity and 31% similarity) or *E. coli* BCP (21% identity and 34% similarity). Plausibly, they represent three distinct subfamilies of peroxiredoxins (Table I).

Proteins strikingly homologous (30–70% identity) to *E. coli* AhpC, p20, or BCP are commonly found in other bacteria (not discussed here) such as *Helicobacter pylori* and *S. typhimurium*. Bacterial BCP and p20 contain only one cysteine. By this criterion, they are more similar to the monocysteine peroxiredoxins (1CPrx) found in higher organisms (Kang *et al.*, 1998b). In contrast, AhpC proteins have two conserved cysteines and are more closely related to the TSA subfamily. Notably, not all bacteria have the three types of peroxiredoxins; several have only one or two. For example, only one peroxiredoxin (1CPrx) is found in the three archaea (*Methanococcus jannaschii*, *Methanobacterium thermoautotrophicum*, and *Archaeoglobus fulgidus*) with complete genomes sequence. In contrast, *Bacillus subtilis*, the cyanobacterium *Synechocystis* sp., and the hyperthermophilic bacterium *Aquifex aeolicus* encode four peroxiredoxins in their genome (Figs. 1 and 2). Conceivably, the presence of multiple peeroxiredoxins in a genome reflects the selection pressure from oxidative stress. Consistent with this, both cyanobacteria and thermophilic bacteria have greater needs for detoxification of hydrogen peroxide.

The budding yeast genome encodes four peroxiredoxins. TSA and TSA2 are closely related members in the TSA subfamily of eukaryotic enzymes. 1CPrx belongs to another subfamily of enzymes found in cyanobacteria, archaea, and eukaryotes. However, yeast BCP is currently the only eukaryotic member in the BCP subfamily (Figs. 1 and 2).

A closer examination of the evolutionary pathways of peroxiredoxins reveals several unexpected connections. *Aquifex* AhpC1, p20, and BCP are closely linked to their bacterial counterparts. *Aquifex* AhpC2 is diverged from cyanobacterial 1CPrx, but clusters with archaeal 1CPrx in phylogenetic trees (Figs. 1 and 2). Surprisingly, a BAS1- or TSA-like YkuU protein was found in *B. subtilis*. Although the two trees place YkuU in different arms, it is clear that YkuU has no counterpart in other bacteria but is related to plant BAS-1 (Fig. 2) and to members in the TSA subfamily. These similarities between peroxiredoxins from disparate sources thought to be unrelated suggest that gene swapping between ancient species might have occurred more frequently than expected.

Another interesting finding from an analysis of the genomic organization of microbial peroxiredoxin genes is the frequent clustering of redox-related genes. In *H. pylori*, p20 and the superoxide dismutase genes share an overlapping promoter region, but are transcribed in opposite directions (Wan *et al.*, 1997). In *A. aeolicus*, the BCP and sulfur oxygenase reductase (SOR) genes are similarly oriented. In addition, a head-to-tail gene cluster of p20 and AhpC1 was also observed in *A. aeolicus* (Deckert *et al.*, 1998). The close proximity of these genes suggests that they might be coregulated by the same set of transcription factors. In addition, the orientation of the overlapping genes allows them to be regulated at the RNA level by antisense inhibition or by related mechanisms.

Expression of *E. coli* AhpC is governed by redox-regulated transcription factor OxyR (Storz *et al.*, 1989; Tartaglia *et al.*, 1989). OxyR responds to exposure of H_2O_2 and is activated through the reversible formation of disulfide bonds. The reactive intermediate is a cysteine sulfenic acid. Activation of OxyR is balanced by reduction through the thiol donors glutaredoxin and GSH (Zheng *et al.*, 1998). The expression of bacterial AhpC is also induced by oxidants (peroxides and superoxide anions), thiols, Cd^{2+}, osmotic shock, and iron depletion (Armstrong-Buisseret *et al.*, 1995; Tai and Zhu, 1995; Mongkolsuk *et al.*, 1997). The promoter region upstream of the *E. coli* p20 gene is also responsive to oxidative stress (Kim *et al.*, 1996). The expression of yeast TSA is induced by O_2, Fe^{3+}, and 2-mercaptoethanol (Kim *et al.*, 1989).

Mutations in the mycobacterial *ahpC* promoter, which result in increased expression of AhpC, have been isolated. Interestingly, these mutations contribute to the emergence of isoniazid-resistant mycobacteria (Dhandayuthapani *et al.*, 1996; Sherman *et al.*, 1996; Wilson and Collins, 1996). Isoniazid, a front-line antituberculosis agent, is a prodrug that requires activation by catalase-peroxidase KatG, a component of the mycobacterial antioxidant defense. Inactivation of KatG prevents activation of isoniazid and is a major mechanism of drug resistance. To survive the loss of KatG, mutant bacteria generate a second compensatory mutation that activates AhpC expression. These studies on drug resistance provide novel and important insights into the cooperation of peroxiredoxins with other peroxidases in the bacterial defense system against oxidative stress.

One way to explore the functions of peroxiredoxins is through the phenotype of null mutants in bacteria and yeasts. An *E. coli* strain carrying a deletion of the entire *ahp* locus was first described by Storz *et al.*, (1989). This strain exhibited hypersensitivity to cumene hydroperoxide under aerobic conditions. This sensitivity can be complemented with *E. coli aphC*, as well as a mouse MER5 plasmid (Tsuji *et al.*, 1995). Interestingly, *S. typhimurium* disrupted in *ahpC* became hypersusceptible not only to ROS, but also to RNS (Chen *et al.*, 1998). This defect can be complemented with *M. tuberculosis* or *S. typhimurium ahpC*. These observations suggest that AhpC is the enzyme that confers resistance to RNS. OxyR, which controls AhpC expression in many bacteria, is activated by both oxidative and nitrosative stress (Hausladen *et al.*, 1996). Hence, the OxyR–AhpC axis likely represents a bacterial defense system against both oxidation and nitrosation. Point mutants of bacterial AhpC have also been isolated or constructed (Ferrante *et al.*, 1995; Ellis and Poole, 1997a,b). The phenotype of some mutants is instructive. For example, a substitution of valine for glycine at position 142 confers a three-fold increase in the activity to detoxify tetralin hydroperoxide, an organic solvent (Ferrante et al., 1995). A p20 null mutant (*tpx*) strain of *E. coli* has also been constructed. This strain was viable in aerobic culture but grew slowly and with an increased sensitivity to peroxides and paraquat. A six-fold greater induction of Mn-superoxide dismutase was also documented in the *tpx* mutant treated with paraquat (Cha et al., 1996). Additionally, both C94S and C61S mutants of *E. coli* p20 lost peroxidase activity (Cha et al., 1996; Wan et al., 1997). A haploid *tsa* mutant of budding yeast was indistinguishable from wild type under anaerobic conditions (Chae et al., 1993), but aerobically the mutant grew significantly more slowly. This mutant also exhibited a higher level of DNA damage (Lee and Park, 1998) and greater sensitivity to peroxides, menadione, and paraquat (Chae et al., 1993; Lee and Park, 1998).

As revealed by an examination of their complete genomic sequences, *E. coli* has three different peroxiredoxins whereas budding yeast has four. Hence, it would be of great interest to construct and test multiple simultaneous mutations in these genes. True peroxiredoxinnull strains would be useful for the elucidation of biological functions of peroxiredoxins.

B. Mammalian Peroxiredoxins

Peroxiredoxins are found ubiquitously in organisms ranging from plants, bacteria, invertebrates, to mammals. There are two major groups of peroxiredoxins in higher plants, 1CPrx and BAS1. Plant 1CPrx, originally called rehydrin and specifically expressed in the aleurone and the embryos, is a dormancy-related protein associated with rehydration events (Stacy et al., 1996; Haslekas et al., 1998). Plant BAS1, a two cysteine peroxiredoxin, is a nuclear-encoded chloroplast protein (Baier and Dietz, 1997). Peroxiredoxins are also found in helminth and protozoan parasites. Some of the parasite peroxiredoxins are secreted or expressed on the cell surface (McGonigle et al., 1998). One peroxiredoxin from pathogenic *Entamoeba histolytica*, TSA-ameba, has been studied by several groups (Bruchhaus et al., 1997; Poole et al., 1997).

Mammalian peroxiredoxins are classified into six distinct groups, Prx I–Prx VI, based on amino acid sequences and immunological reactivities (Jin et al., 1997; Kang et al., 1998a,b). Prx I and Prx II, represented by thioredoxin peroxidase (TPx) and proliferation-associated gene product (PAG), respectively, are primarily cytosolic proteins (Jin et al., 1997; Kang et al., 1998a), although a nuclear localization of PAG was reported by one group of investigators (Wen and Van Etten, 1997). Prx III, represented by MER5, contains an N-terminal signal sequence and is targeted to mitochondria (Watabe et al., 1994; Kang et al., 1998a). Prx IV, represented by

AOE372, also has a signal peptide and is found abundantly in the cytosol (Jin *et al.*, 1997). One report describes AOE372 as a secreted cytokine (Haridas *et al.*, 1998). However, because AOE372 is expressed abundantly inside cells, it is difficult to rule out that the AOE372 protein recovered from culture supernatants is from lysed or apoptotic cells. Prx V represents an unpublished group of protein (S.G. Rhee, personal communication). Prx VI, represented by 1CPrx (or hORF6), is exclusively cytoplasmic (Frank *et al.*, 1997; Kang *et al.*, 1998b).

All mammalian peroxiredoxins conserve a motif surrounding a cysteine residue that corresponds to Cys-47 in TPx. In addition to this, Prx I, Prx II, Prx III, and Prx IV have a second peroxiredoxin motif surrounding the second conserved cysteine that corresponds to Cys-180 at the C terminus of TPx. This C-terminal cysteine is absent in Prx VI. Mammalian peroxiredoxins in the same group generally share more than 90% homology in their amino acid sequences. For example, 91% or residues in human and mouse AOE372, and 95% in human and mouse PAG are identical. Peroxiredoxins in Prx I, Prx II, Prx III, and Prx IV groups belong to the same TSA family. The cross-group sequence homology is in the range of 60–80%. Thus, human AOE372 and human PAG share 65% identity and 83% similarity. Prx VI is more distantly related to other groups of mammalian peroxiredoxins. Human 1CPrx shares only 26% identity and 43% similarity with human AOE372, but 89% identity and 93% similarity with mouse 1CPrx.

Mammalian peroxiredoxins are highly abundant in various tissues and cells (Jin *et al.*, 1997; Kang *et al.*, 1998a). Prx I, Prx II, and Prx III together account for 0.2–0.8% of total soluble protein in cultured cells (Kang *et al.*, 1998a). However, different peroxiredoxins do show distinct tissue distribution profiles (Ishii *et al.*, 1993; Prosperi *et al.*, 1993; Iwahara *et al.*, 1995; Watabe *et al.*, 1994; Ichimiya *et al.*, 1997; Jin *et al.*, 1997; Kim *et al.*, 1997; Munz *et al.*, 1997). Thus, it is interpreted that peroxiredoxins serve specific functions in a subcellular and tissue-restricted manner.

Inducible expression has been demonstrated for several mammalian peroxiredoxins. Inducers include hydrogen peroxide (for induction of human TSA; Kim *et al.*, 1997), heme/heavy metal(Cd^{2+}, Co^{2+}, or ArO_2^-)/metalloporphyrins/protop IX/diethyl maleate/glucose/glucose oxidase (induction of mouse, rat, and human PAG; Ishii *et al.*, 1993; Immenschuh *et al.*, 1995, 1997; Prosperi *et al.*, 1998), and keratinocyte growth factor (induction of 1CPrx; Munz *et al.*, 1997). The increased expression of individual peroxiredoxins has also been correlated with murine erythroleukemia (MEL) cells in a chemically differentiated state (TPx; Ichimiya *et al.*, 1997) or prior to the accumulation of globin (MER5; Yamamoto *et al.*, 1989) or hemoglobin (TPx; Rabilloud *et al.*, 1995), areas of the brain susceptible to ischemic injury (TPx; Ichimiya *et al.*, 1997), keratinocytes of the hyperproliferative epithilium at the wound edge (1CPrx; Munz *et al.,* 1997), psoriatic skin (1CPrx; Frank *et al.,* 1997), transformed cells (PAG; Prosperi *et al.,* 1993), and cells entering S phase (PAG; Prosperi *et al.,* 1998). Taken together, peroxiredoxins are suggested to be involved in cell differentiation, proliferation, and defense against oxidative stress.

The chromosomal localization of several mammalian peroxiredoxin genes has been determined. The human TPx gene maps to chromosome 13q12, a region containing BRCA2 and a gene associated with a form of muscular dystrophy (Pahl *et al.,* 1995; Jacob *et al.,* 1996). Human PAG gene maps to 1p34, and PAG pseudogene maps to chromosome 9q22 (Prosperi *et al.,* 1994). Mouse 1CPrx localizes to chromosome 1 (Iakoubova *et al.,* 1997). The map position of mouse 1CPrx corresponds to *Atherosclerosis 1 (Ath1),* a locus that is associated with an increased susceptibility to atherosclerotic plaque formation. Because the link between antioxidants and atherosclerotic heart disease is well established (Diaz *et al.,* 1997), 1CPrx should be considered as a candidate atherosclerosis gene. Human and mouse AOE372 gene maps to chromosome X (Xp21-22.1 in human, Haridas *et al.,* 1998; GenBank U96746).

II. PEROXIREDOXINS: STRUCTURE AND FUNCTION

All peroxiredoxins conserve at least one cysteine motif at the N terminus. In addition to this N-terminal cysteine motif, many peroxiredoxins possess a second similar motif at the C terminus. These peroxiredoxin motifs have been implicated as being important for the catalysis of peroxide reduction. This peroxidase activity, which accounts for the antioxidant properties of peroxiredoxins, is modulated by the microenvironment affecting the redox center residues. The redox status of the reactive center cysteine dictates the specificity and potency of peroxiredoxin activity. The determination of a crystal structure of human 1CPrx C91S at 2 Å resolution (Choi *et al.,* 1998) provides a framework for understanding the mechanism and regulation of peroxiredoxin functions.

A. Mechanism of Peroxide Reduction

Peroxiredoxins have neither tightly bound metal ions nor prosthetic groups such as heme or flavin (Kim *et al.,* 1988). The ability of peroxiredoxins to reduce peroxides is attributed to their cysteine residues. However, one peroxiredoxin, PAG, is a heme-binding protein (Immenschuh *et al.,* 1995). On treatment of cells with a low concentration of hemin, the antioxidant activity of PAG is inhibited (Ishii *et al.,* 1993). These observations suggest that heme binding is physiologically relevant to PAG activities. Further investigations are required to clarify the interaction of heme with peroxiredoxins.

The mechanisms by which peroxiredoxins reduce peroxides have only been studied using a limited number of models. These studies include bacterial AphC (Poole, 1996; Ellis and Poole, 1997a,b), yeast and rat TPx (Chae et al., 1994c; Netto et al., 1996), and human 1CPrx (Netto et al., 1996; Kang et al., 1998b). In two-cysteine peroxiredoxins such as AhpC and TPx, only one cysteine (Cys-46 in AhpC and Cys-47 in TPx) is absolutely required for peroxidase activity (Netto et al., 1996; Ellis and Poole, 1997a,b). 1CPrx, as well as AhpC C165S and TPx C170S mutants, is fully active in comparison with wild-type AhpC and TPx. The peroxidatic center cysteine in these peroxiredoxins is sufficient to support peroxide reduction. When a peroxiredoxin is reduced, a single cysteine thiolate (Cys-S-) attacks the $-O-O-$ bond, releasing ROH and forming a cysteine sulfenic acid (Cys-SOH) within the enzyme. This sulfenic acid intermediate can be trapped by the electrophilic reagent 7-chloro-4-nitrobenz-2-oxa-1,3-diazole (Ellis and Poole, 1997a). The formation of cysteine sulfenic acid has also been demonstrated conclusively in the X-ray crystal structure of 1CPrx (Choi et al., 1998). This catalytic mechanism has several precedents in redox-active proteins. For example, a cysteine sulfenic acid is the intermediate in the activation of OxyR (Zheng et al., 1998) and NADH peroxidase.

The physiological electron donors in the peroxidatic reaction have only been documented for a few peroxiredoxins. These peroxiredoxin partners usually associate with peroxiredoxins through protein–protein interaction. In many bacteria, the electron donor for AphC is AhpF, a flavoenzyme closely related to thioredoxin reductase. AhpF transfers electrons from NADH to AphC via the flavin and the direct dithiol–disulfide interchange, followed by the transfer of electrons from the dithiol center of AhpC to the hydroperoxide substrates (Poole, 1996). Interestingly, AhpF is not required for the protection conferred by AhpC against RNS (Chen et al., 1998). Thus, this protection is likely independent of the peroxidase activity of AhpC.

The physiological hydrogen donor for TPx (Prx I) has been identified as thioredoxin (Chae et al., 1994b). Thioredoxin and thioredoxin reductase mediate the flow of electrons from NADPH to oxidized TPx. The thioredoxin system can support the peroxidase activity of bacterial p20 (Cha et al., 1995; Wan et al., 1997) and several other mammalian peroxiredoxins, including PAG (Prx II; Kang et al., 1998a), MER5 (Prx III; Kang et al., 1998a), AOE372 (Prx IV; Jin et al., 1997), and Prx V. In the reduction of hydrogen peroxide by Prx I, Prx II, and Prx III, the K_m value for thioredoxin is approximately 3–6 μM (Kang et al., 1998a). Thus, these peroxiredoxins bind thioredoxin with high affinity.

One report has suggested that glutathione supports the reduction of hydrogen peroxide and organic hydroperoxides by bovine 1CPrx (Shichi and Demar, 1990). However, this has not been substantiated (Frank et al., 1997). A recent report demonstrates the peroxidase activity of human 1CPrx in the presence of a nonphysiological electron donor, DTT; however, neither thioredoxin nor glutathione was able to reduce oxidized 1CPrx (Kang et al., 1998b).

TXNPx, a novel peroxiredoxin from the trypanosomatid *Crithidia fasiculata*, has been identified (Montemartini et al., 1998). The unique feature of TXNPx is its dependence on tryparedoxin, a thioredoxin-related protein found only in trypanosomatids (Gommel et al., 1997). Tryparedoxin supports the TXNPx activity at the expense of the unique trypanothione, N^1, N^8-bis(glutathionyl)spermidine (Nogoceke et al., 1997).

The microenvironment of the peroxidatic center cysteine residue in peroxiredoxins and the interaction of peroxiredoxins with their partners are important in the regulation of the potency and specificity of peroxidase activity. In two-cysteine peroxiredoxins, the cysteine sulfenic acid intermediate formed in the reaction of peroxide reduction is unstable due to the presence of a nearby cysteine thiol. This C-terminal cysteine (Cys-165 in AhpC and Cys-170 in TPx) reacts rapidly with the nascent cysteine sulfenic acid to generate a disulfide bridge, releasing a molecular of water. In the oxidized enzyme, interchain disulfide bonds are usually formed between two different cysteines (Cys-46 and Cys-165 in AhpC, or Cys-47 and Cys-170 in TPx). The formation of intrachain disulfide bridges has also been suggested for human Prx V. Thus, the C-terminal cysteine (Cys-165 in AhpC or Cys-170 in TPx), albeit not involved directly in peroxide reduction, has an important role in the prevention of the further oxidation of redox center cysteines (Cys-46 in AhpC or Cys-47 in TPx) by substrate and the preservation of peroxiredoxin activity.

Homodimerization of two-cysteine peroxiredoxins AhpC, TPx, PAG, and AOE372 has been documented (Chae et al., 1994b; Jin et al., 1997; Schroder and Ponting, 1998). Additionally, heterodimerization between PAG and AOE372 has also been reported (Jin et al., 1997). Seemingly, intracellular antioxidant activities of peroxiredoxins can be regulated through homo- and heterodimer formation. These subunit assortments between different forms of peroxiredoxins influence potency and perhaps specificity of action.

Reduction of hydrogen peroxide by 1CPrx is through an as yet unidentified physiological electron donor (Kang et al., 1998b). The reaction is rapid at pH 7. It has been postulated that the pK_a value of Cys-47-SH would be low as a result of the electrostatic interaction between cysteine thiolate (Cys-S$^+$) and the nearby basic residues. The low pK_a microenvironment would contribute to the stabilization of the oxidized enzyme in the absence of interchain disulfide bonds. 1CPrx exists as a dimer (Choi et al., 1998). However, existence of the 1CPrx monomer under certain conditions has not been excluded (Kang et al., 1998b; Choi et al., 1998).

B. Structural Basis of Peroxiredoxin Activity

A high-resolution X-ray crystal structure of a human 1CPrx C91S has been published (Choi et al., 1998). In this structure of dimeric 1CPrx, the Val-179 residue, which corresponds to Cys-170 in TPx, is very close (8.33 Å) to Cys-47 of the second subunit. An interchain disulfide bridge could be formed easily. Thus, this 2.0-Å structure is illustrative of many, if not all, peroxiredoxins and provides useful information on the structural basis of peroxiredoxin activity.

1CPrx C91S is a tightly bound dimer in the crystal. This dimer is stabilized by a network of hydrogen bonds and hydrophobic interactions. Each monomer contains two discrete domains. The larger N-terminal domain has a thioredoxin fold, which is also found in other redox proteins such as thioredoxin, glutaredoxin, glutathione peroxidase, glutathione S-transferase, and DsbA. The smaller C-terminal domain is used for dimerization.

The redox center Cys-47 is located at the bottom of a narrow pocket with a diameter of approximately 4 Å and a depth of approximately 7 Å. The two Cys-47 residues in the dimer are separated and likely function independently. Interestingly, both Cys-47 residues exist as cysteine sulfenic acid in the crystal. Several positively charged residues, including His-39, Arg-132, and Arg-155, are in close proximity to Cys-47. The interaction of His-39/Arg-132 with Cys-47 likely contributes to the decrease of the pK_a of the cysteine thiol. The increased nucleophilicity due to pK_a shift serves as a structural basis for the reactivity of Cys-47.

III. PEROXIREDOXINS IN CELL SIGNALING

Intracellular signaling pathways are complex networks that modulate nuclear gene expression. Intermediating proteins utilized in signal transduction are unmerous and incompletely understood. Kinases and phosphatases modify function by phosphorylation and dephosphorylation. Acetylases and deacetylases regulate activity by acetylation and deacetylation. Oxidants and antioxidants represent a different set of signal transducers that act through redox. Similar to other signaling events, redox can serve as the critical switch in many signaling processes. Physiologically relevant ROS and RNS that serve as secondary messengers have been documented for bacteria, plants, and mammals. Balancing these ROS are antioxidants such as peroxiredoxins. This delicate balance inside cells between oxidants and antioxidants ultimately determines the activity of many transcription factors.

In bacteria, stress conditions, such as exposure to oxygen, initiate genetic programs that coordinate the activation of transcription factors controlling the expression of defense-related regulons. For example, *E. coli ahpC*, together with other antioxidant genes *katG* (catalase-hydroperoxidase), *ahpF*, and *gorA* (glutathione reductase), is under the control of OxyR, the prototypic redox-regulated transcription factor. OxyR is regulated through the reversible formation of disulfide bonds using cysteine sulfenic acid as an intermediate (Zheng *et al.*, 1998). In plants, various pathogens stimulate an oxidative burst generating O_2^- and H_2O_2. These ROS are used as secondary messengers to establish a systematic signal network necessary for plant immunity (Alvarez *et al.*, 1998). In mammals, the involvement of oxidants and antioxidants in receptor signaling, protein phosphorylation, gene transcription, and apoptosis is under intense investigation (Nakamura *et al.*, 1997; Finkel, 1998). The section briefly reviews some of the recent developments as related to peroxiredoxins.

A. Receptor Signaling and Protein Phosphorylation

Stimulation of various types of mammalian cells with cytokines, growth factors, or phorbol esters results in an increase of ROS (reviewed in Finkel, 1998). This increase in intracellular ROS can be measured using peroxide-sensitive fluorophores such as 2′,7′-dichlorofluorescin (DCF). The production of ROS and the contribution of ROS to receptor signaling have been studied in two models. In one system, vascular smooth muscle cells were stimulated with platelet-derived growth factor (PDGF; Sundaresan *et al.*, 1995). In the second, A431 epidermoid carcinoma cells were induced with epidermal growth factor (EGF; Bae *et al.*, 1997). In both studies, a rapid increase in ROS peaked within 5 min of stimulation and returned to baseline after 20–30 min. This time course correlated with the kinetics of the ligand-stimulated tyrosine phosphorylation. In addition, treatments that altered the intracellular ROS concentration also correlated with the changes in tyrosine phosphorylation. Interestingly, the transient expression of human peroxiredoxins TPx or PAG in NIH 3T3 and A431 cells abolished the PDGF- and EGF-dependent increase of intracellular ROS (Kang *et al.*, 1998a). Plausibly, the PDGF- and EGF-induced tyrosine phosphorylation is inhibited. These observations support that the intracellular level of functional peroxiredoxins is important for the balance between oxidants and antioxidants and the redox regulation of ligand-stimulated receptor signaling.

A report by Wen and Van Etten (1997), however, suggests a more direct regulatory function of human peroxiredoxin PAG in tyrosine phosphorylation. In that study, PAG was demonstrated to be a binding partner and a SH3-specific physiological inhibitor of nonreceptor tyrosine kinase c-Ab1. Surprisingly, an N-terminal truncated mutant of PAG without the conserved Cys-52 residue was still functional as a c-Ab1 inhibitor. Hence, the peroxidase activity of PAG appears to be dispensable for c-Ab1 inhibition. Alternatively, because truncated PAG can interact with full-length PAG, AOE372, or other peroxiredoxins, the described observation may be explained by the recruitment of partners or interference with partner formation. c-Ab1 is a nuclear protein; however, the subcellular localization of PAG is controversial. Two independent groups demonstrated PAG as a cytoplasmic protein (Jin *et al.*, 1997; Kang *et al.*, 1998a), but a nuclear localization of PAG has also been suggested (Wen and Van Etten, 1997). The interaction between PAG and c-Ab1 was identified in yeast two-hybrid screening and was verified by coimmunoprecipitation and functional assays. However, the molecular mechanisms underlying this interaction remain to be seen.

More recently, another binding partner for mammalian peroxiredoxin(s) has been identified. The initial identification was through yeast two-hybrid screening, and the interaction was verified by *in vitro* protein affinity chromatography. The bait protein was human T-cell cyclophilin 18 and the partner peroxiredoxin was human MER-5 (Jaschke *et al.*, 1998). Cyclophilin 18 is the primary receptor for cyclosporin A in T cells. A complex of cyclosporin A and cyclophilin 18 inhibits the protein phosphatase calcineurin, a key regulator in T-cell activation. The physical interaction between MER5 and cyclophilin 18 resulted in the stimulation of MER5 antioxidant activity, as measured by the glutamine synthetase protection assay. Because the antioxidant properties of peroxiredoxins are due to their peroxidase activity (Netto *et al.*, 1996), the MER5 peroxidase activity should also be stimulated. It was proposed that cyclophilin 18, a peptidylprolyl *cis-trans* isomerase, can assist MER5 folding. However, both peroxiredoxins and cyclophilins are multifunctional proteins. It remains open whether MER5 or other peroxiredoxins have roles either in the regulation of calcineurin-dependent protein dephosphorylation or in the modulation of other cyclophilin-dependent functions. We note that MER5 is primarily a mitochondrial protein (Watabe *et al.*, 1994; Kang *et al.*, 1998a). If so, cytoplasmic cyclophilin 18 might not be its primary physiological interactive partner. A possibility is that MER5 binds to mitochondrial cyclophilin and that an alternative interaction is that between cyclophilin 18 and another cytoplasmic peroxiredoxin. The next section discusses a model for the regulation of apoptosis, which is based on the interaction of peroxiredoxin and cyclophilin.

B. Redox Regulation of NF-κB Activation

One redox-regulated transcription factor in higher eukaryotes is NF-κB. NF-κB is a member of the Rel family of proteins that exists ambiently in the cytosol in association with the inhibitor protein IκB (Baeuerle and Baltimore, 1988, 1996; Sen and Packer, 1996). A wide variety of inducers, including phorbol esters, tumor necrosis factor (TNF)-α, and viral infection, can activate NF-κB. Studies have implicated ROS as a common signal transducer for these diverse inducers (Anderson et al., 1994).

One pathway of NF-κB activation involves the site-specific phosphorylation of IκB-α on Ser-32 and Ser-36 by IκB-α kinases (reviewed in Stancovski and Baltimore, 1997; Verma and Stevenson, 1997). It has been suggested that serine phosphorylation targets IκB for ubiquitination and degradation. IκB inactivation, without ubiquitination and proteolysis, has also been reported to occur as a consequence of phosphorylation on Tyr-42 (Imbert et al., 1996). In both instances, phosphorylation results in the nuclear translocation of NF-κB. For stimuli such as oxidants, which potently and rapidly modulate nuclear NF-κB activity, IκB kinases and IκB-α may represent critical activation targets.

Human peroxiredoxins AOE372 and PAG have been documented to regulate NF-κB activation (Jin et al., 1997; Kang et al., 1998a). The transient expression of AOE372 (and/or PAG) plasmid in HeLa cells potently suppressed NF-κB activity induced by various stimuli, including TNF-α, phorbol ester, and HIV-1 Tat. This suppression correlates with the hypophosphorylation of cytoplasmic IκB-α, as well as lower level and lower DNA-binding activity of nuclear NF-κB. These results suggest that AOE372 modulates IκB-α phosphorylation in the cytosol and thus affects a peroxiredoxin-dependent redox step. It is noteworthy that this activity of the thioredoxin peroxidases AOE372 and PAG mirrors that described previously for thioredoxin (Schulze-Osthoff et al., 1995). These effects of peroxiredoxins are consistent with reports ascribing an activity to glutathione peroxidase on IκB-α (Kretz-Remy et al., 1996). Considered with other studies on thioredoxin (Schulze-Osthoff et al., 1995; Nakamura et al., 1997), we believe that thioredoxin and thioredoxin peroxidase are potent antioxidants for NF-κB activation.

C. Apoptosis

Apoptosis, or programmed cell death, is a genetically controlled process in which cells commit suicide in response to various stimuli. It has been known that apoptosis can be activated by at least two cross-communicating pathways: a mitochondrial pathway and a death receptor pathway (Cryns and Yuan, 1998). At the core of the first route is the release of cytochrome c from mitochondria to the cytoplasm. This event is inhibited by Bcl-2 and Bcl-2-related proteins. In contrast, the second route is triggered by the direct recruitment of caspase proenzymes to the ligand-stimulated death receptor through the homophilic interaction with adaptor proteins, such as FADD. There is ample evidence that both pathways are regulated by redox (Hockenbery et al., 1993; Polyak et al., 1997; Bauer et al., 1998; Saitoh et al., 1998). The regulatory roles for peroxiredoxins in cellular apoptosis have begun to be characterized.

Murine TPx (or Prx I) is expressed more abundantly in areas of the brain most sensitive to hypoxic and ischemic injury. In the model of neuronal survival, an antiapoptotic effect of mouse TPx has been documented (Ichimiya et al., 1997). Thus in PC12 pheochromocytoma cells, TPx overexpression promotes cell survival on withdrawal of nerve growth factor and serum. The antiapoptotic mechanisms of peroxiredoxins and Bcl-2 were compared in another study by Zhang and co-workers (1997). These authors demonstrated a protection from apoptosis induced by serum starvation, ceramide, or ectoposide in MOLT-4 leukemia cells stably expressing human PAG (Prx II). They also found that PAG functions upstream of Bcl-2 by preventing the accumulation of hydrogen peroxide and inhibiting the release of mitochondrial cytochrome c into the cytosol.

Two studies that describe the prevention of mitochondrial permeability transition (MPT) by yeast TSA (Kowaltowski et al., 1998) and the physical interaction of human mitochondrial peroxiredoxin MER5 with cyclophilin (Jaschke et al., 1998) may provide novel and important insight into the physiological functions of peroxiredoxins in the mitochondrial pathway of apoptosis. MPT is thought to be one of the initial events in apoptosis. The MPT pore, a dynamic multiprotein channel formed between the inner and the outer mitochondrial membranes, is an important target of apoptosis regulation by caspase, Bcl-2 and related proteins, ROS, and calcium flux (Marzo et al., 1998). The opening of the MPT pore, induced by calcium and inorganic phosphate and evidenced by mitochondrial swelling and membrane protein thiol oxidation, can be inhibited by the addition of recombinant peroxiredoxin. Although experimental data were presented with yeast TSA, similar results were also obtained using mammalian MER5 (Kowaltowski et al., 1998), a physiologically relevant peroxiredoxin that resides in mitochondria and is degraded by ATP-dependent proteases (Yamamoto et al., 1989; Watabe et al., 1994; Kang et al., 1998a). The regulation of MPT by peroxiredoxins was thought to be mediated by redox. Thus, the inhibition of MPT correlates with the scavenge of hydrogen peroxide by TSA, which is fully reduced by mitochondrial components, probably the mitochondrial thioredoxin system. The TSA mutant C47S, which is defective of the peroxidase activity, lost the ability to inhibit MPT. However, it remains open as to whether peroxiredoxin can modulate MPT through mechanisms other than redox. Notably, MPT is also inhibited by nanomolar concentrations of cyclosporin A. This inhibition is mediated, at least in part, through the interaction of mitochondrial cyclophilin with the inner membrane of the organelle (Nicolli *et al.*, 1996). Mutational analyses of a series of cyclosporin A derivatives have suggested that immunosuppression through cyclophilin and calcineurin is separable from MPT inhibition. Taken together with the documentation of an interaction of human mitochondrial peroxiredoxin MER5 with cyclophilin 18 (Jaschke *et al.*, 1998), and presumably with mitochondrial cyclosporin as well, these studies led us to a model in which peroxiredoxins modulate MPT and apoptosis via protein–protein interaction with cyclophilin. In this regard, it would be of interest to see whether this regulation by peroxiredoxins is independent of the peptidylprolyl *cis-trans* isomerase activity of cyclophilin.

Another model of apoptosis regulated by peroxiredoxins has been suggested in a study by Chen and co-workers (1998). They stably expressed mycobacterial AhpC in human cells and checked for protection against RNS. AhpC not only increased the resistance to exogenously added nitric oxide donor S-nitrosoglutathione (GSNO), but also prevented cytotoxicity and apoptosis caused by the overexpression of inducible nitric oxide synthetase. Thus, peroxiredoxins can modulate apoptotic pathways by countering the effects of RNS.

IV. PEROXIREDOXINS IN HIV INFECTION

Several lines of evidence suggest that oxidants and antioxidants affect the progression of HIV infection and the pathogenesis of AIDS. First, oxidants can activate HIV through NF-κB (Schreck *et al.*, 1991). Thus HIV-1 LTR DNA stably transformed into human lymphoid cells lines can be activated on low-dose treatment with H_2O_2 (Kurata, 1996; Okamoto *et al.*, 1992). Second, HIV infection is associated with defects in antioxidants such as thioredoxin, glutathione, superoxide dismutase, peroxidases, and catalase (Piedimonte *et al.*, 1997). Glutathione deficiency was found in HIV-infected individuals and was shown to be associated with impaired survival in AIDS (Herzenberg *et al.*, 1997). Abnormally low glutathione peroxidase activity in both symptomatic AIDS patients and chronically infected cell lines has also been documented (Sandstrom *et al.*, 1994). Third, antioxidants such as N-acetyl-L-cysteine (NAC) suppress HIV expression and replication in infected cells (Kalebic *et al.*, 1991), and administration of NAC to HIV-infected people may be associated with the prolongation of survival (Staal et al., 1993; Herzenberg et al., 1997). Finally, AIDS is manifested by the depletion of CD4$^+$ T lymphocytes as a result of apoptosis and ROS are critically involved in this dysregulation.

It is noteworthy that redox is a double-edged sword. While in certain circumstances, antioxidant inhibits HIV transcription (Staal et al., 1993; Sappey et al., 1994), it has also been reported that the constitutively elevated expression of antioxidant such as glutathione peroxidase accelerated HIV replication and the associated cytopathic effects (Sandstrom et al., 1998). Interestingly, glutathione and glutathione peroxidase exhibited opposite effects in this system. These studies revealed the complexity with which different antioxidants, or even different isoforms of the same antioxidant, subserve different functions at different stages of infection.

In principle, peroxiredoxins can be involved in the redox regulation of HIV infection. Analysis of the steady-state amount of AOE372 protein in HIV-infected cells compared to uninfected cells demonstrated a marked decrease (Jin et al., 1997). This decrease suggests that the aberrant expression of AOE372 is associated with HIV infection. However, because there are several subfamilies of human peroxiredoxins, it would be of interest to document the expression patterns of other peroxiredoxins and to shed light on how they cooperate to dysregulate intracellular redox status.

AOE372 and PAG have been shown to inhibit IκB-α phosphorylation and NF-κB activation (Jin et al., 1997; Kang et al., 1998a). A biological important corollary of the regulatory activity of peroxiredoxins on NF-κB is an effect on HIV infection. Indeed, the forced overexpression of AOE372 in cells transfected with an infectious molecular clone of HIV-1 reduced the production of viral proteins as assayed by either p24 or viral reverse transcriptase (Jin et al., 1997). These findings are compatible with a model in which peroxiredoxins exert an effect on HIV infection through NF-κB. Thus, peroxiredoxins are implicated as an additional antioxidant enzyme involved in HIV infection and AIDS pathogenesis. Similar to glutathione peroxidase, these enzymes could serve as new targets for the therapeutic regulation of intracellular redox levels in HIV-infected individuals.

V. FUTURE DIRECTIONS

Research into peroxiredoxins has advanced rapidly. Nevertheless, many questions remain unanswered. In the large family of peroxiredoxins, only a limited number of enzymes have been characterized for function. Coordinated efforts using different biological systems remain necessary to address the basic questions as to how peroxiredoxins scavenge RNS and ROS. For peroxiredoxins in higher eukaryotes, it would be of importance to elucidate whether some peroxiredoxins are secreted by cells and whether peroxiredoxins interact with cyclophilin(s) and/or c-Ab1. Given that defects in antioxidant enzymes are frequent causes of inherited diseases (e.g., one defect in superoxide dismutase is linked to familial amyotrophic lateral sclerosis), the genomic definition of mammalian peroxiredoxins, including genomic mapping and disease association, merits further investigation. With the high-resolution crystal structure of human 1CPrx, it remains of interest to obtain crystals for other peroxiredoxins more distantly related to 1CPrx, as well as heterodimeric peroxiredoxins. Concerning the regulation of cell signaling by peroxiredoxins, more detailed molecular mechanisms, including the definition of main intracellular targets of peroxiredoxins, the modulation of cellular targets by peroxiredoxins, and the specificity and regulation of peroxiredoxin activity by homo- and heterodimerization, should be characterized. We believe that a better understanding of the biological functions of peroxiredoxins will reveal novel opportunities for therapeutic interventions in human diseases.

REFERENCES

Alvarez, M. E., Pennell, R. I., Meijer, P. J., Ishikawa, A., Dixon, R. A., and Lamb, C. (1998). Reactive oxygen intermediates mediate a systemic signal network in the establishment of plant immunity. *Cell (Cambridge, Mass.)* **92**, 773–784.

Anderson, M. T., Staal, F. J. T., Gilter, C., Herzenberg, L. A., and Herzenberg, L.A. (1994). Separation of oxidant-initiated and redox-regulated steps in the NF-κB signal transduction pathway. *Proc. Natl. Acad. Sci. U.S.A.* **91**, 11527–11531.

Armstrong-Buisseret, L., Cole, M. B., and Stewart, G. S. (1995). A homologue to the *Escherichia coli* alkyl hydroperoxide reductase AhpC is induced by osmotic upshock in *Staphylococcus aureus*. *Microbiology* **141**, 1655–1661.

Bae, Y. S., Kang, S. W., Seo, M. S., Baines, I. C., Tekle, E., Chock, P. B., and Rhee, S. G. (1997). Epidermal growth factor (EGF)-induced generation of hydrogen peroxide: Role in EGF receptor-mediated tyrosine phosphorylation. *J. Biol. Chem.* **272**, 217–221.

Baeuerle, P. A., and Baltimore, D. (1988). IκB: A specific inhibitor of the NF-κB transcription factor. *Science* **242**, 540–546.

Baeuerle, P. A., and Baltimore, D. (1996). NF-κB: Ten years after. *Cell (Cambridge, Mass.)* **87**, 13–20.

Baier, M., and Dietz, K. J. (1997). The plant 2-Cys peroxiredoxin BAS1 is a nuclear-encoded chloroplast protein: Its expressional regulation, phylogenetic origin, and implications for its specific physiological function in plants. *Plant J.* **12**, 179–190.

Bauer, M. K. A., Vogt, M., Los, M., Siegel, J., Wesselborg, S., and Schulze-Osthoff, K. (1998). Role of reactive oxygen intermediates in activation-induced CD95 (APO-1/Fas) ligand expression. *J. Biol. Chem.* **273**, 8048–8055.

Berlett, B. S., and Stadtman, E. R. (1997). Protein oxidation in aging, disease, and oxidative stress. *J. Biol. Chem.* **272**, 20313–20316.

Bruchhaus, I., Richter, S., and Tannich, E. (1997). Removal of hydrogen peroxide by the 29 kDa protein of *Entamoeba histolytica*. *Biochem. J.* **326**, 785–789.

Cha, M.-K., Kim, H.-K., and Kim, I.-H. (1995). Thioredoxin-linked "thiol peroxidase" from periplasmic space of *Escherichia coli*. *J. Biol. Chem.* **270,** 28635–28641.

Cha, M.-K., Kim, H.-K., and Kim, I.-H. (1996). Mutation and mutagenesis of thiol peroxidase of *Escherichia coli* and a new type of thiol peroxidase family. *J. Bacteriol.* **178,** 5610–5614.

Chae, H. Z., and Rhee, S. G. (1994). A thiol-specific antioxidant and sequence homology to various proteins of unknown function. *BioFactors* **4,** 177–180.

Chae, H. Z., Kim, I.-H., Kim, K., and Rhee, S. G. (1993). Cloning, sequencing, and mutation of thiol-specific antioxidant gene of *Saccharomyces cerevisiae*. *J. Biol. Chem.* **268,** 16815–16821.

Chae, H. Z., Robison, K., Poole, L. B., Church, G., Storz, G., and Rhee, S. G. (1994a). Cloning and sequencing of thiol-specific antioxidant from mammalian brain: Alkyl hydroperoxide reductase and thiol-specific antioxidant define a large family of antioxidant enzymes. *Proc. Natl. Acad. Sci. U.S.A.* **91,** 7017–7021.

Chae, H. Z., Uhm, T. B., and Rhee, S. G. (1994b). Dimerization of thiol-specific antioxidant and the essential role of cysteine 47. *Proc. Natl. Acad. Sci. U.S.A.* **91,** 7022–7026.

Chae, H. Z., Chung, S. J., and Rhee, S. G. (1994c). Thioredoxin-dependent peroxide reductase from yeast. *J. Biol. Chem.* **269,** 27670–27678.

Chen, L., Xie, Q.-W., and Nathan, C. (1998). Alkyl hydroperoxide reductase subunit C (AhpC) protects bacterial and human cells against reactive nitrogen intermediates. *Mol. Cell* **1,** 795–805.

Choi, H.-J., Kang, S. W., Yang, C.-H., Rhee, S. G., and Ryu, S.-E. (1998). Crystal structure of a novel human peroxidase enzyme at 2.0 angstrom resolution. *Nat. Struct. Biol.* **5,** 400–406.

Cryns, V., and Yuan, J. (1998). Proteases to die for. *Genes Dev.* **12,** 1551–1570.

Deckert, G., Warren, P. V., Gaasterland, T., Young, W. G., Lenox, A. L., Graham, D. E., Overbeek, R., Snead, M. A., Keller, M., Aujay, M., Huber, R., Feldman, R. A., Short, J. M., Olsen, G. J., and Swanson, R. V. (1998). The complete genome of the hyperthermophilic bacterium *Aquifex aeolicus*. *Nature (London)* **392,** 353–358.

Dhandayuthapani, S., Zhang, Y., Mudd, M. H., and Deretic, V. (1996). Oxidative stress response and its role in sensitivity to isoniazid in mycobacteria: Characterization and inducibility of *ahpC* by peroxides in *Mycobacterium smegmatis* and lack of expression in *M. aurum* and *M. tuberculosis*. *J. Bacteriol.* **178,** 3641–3649.

Diaz, M. N., Frei, B., Vita, J. A., and Keaney, J. F., Jr. (1997). Antioxidants and atherosclerotic heart disease. *N. Engl. J. Med.* **337,** 408–416.

Ellis, H. R., and Poole, L. B. (1997a). Roles for the two cysteine residues of AhpC in catalysis of peroxide reduction by alkyl hydroperoxide reductase from *Salmonella typhimurium*. *Biochemistry* **36,** 13349–13356.

Ellis, H. R., and Poole, L. B. (1997b). Novel application of 7-chloro-4-nitrobenzo-2-oxa-1,3-diazole to identify cysteine sulfenic acid in the AhpC component of alkyl hydroperoxide reductase. *Biochemistry* **36,** 15013–15018.

Felsenstein, J. (1996). Inferring phylogenies from protein sequences by parsimony, distance and likelihood methods. *In* "Methods in Enzymology" (R. E. Doolittle, ed.), Vol. 266, pp. 418–427. Academic Press, San Diego, CA.

Ferrante, A. A., Augliera, J., Lewis, K., and Klibanov, A. M. (1995). Cloning of an organic solvent-resistance gene in *Escherichia coli*: The unexpected role of alkylhydroperoxide reductase. *Proc. Natl. Acad. Sci. U.S.A.* **92,** 7617–7621.

Finkel, T. (1998). Oxygen radicals and signaling. *Curr. Opin. Cell Biol.* **10,** 248–253.

Frank, S., Munz, B., and Werner, S. (1997). The human homologue of a bovine non-selenium glutathione peroxidase is a novel keratinocyte growth factor-regulated gene. *Oncogene* **14,** 915–921.

Gommel, D. U., Nogoceke, E., Morr, M., Kiess, M., Kalisz, H. M., and Flohe, L. (1997). Catalytic characteristics of tryparedoxin. *Eur. J. Biochem.* **248**, 913–918.
Haridas, V., Ni, J., Meager, A., Su, J., Yu, G.-L., Zhai, Y., Kyaw, H., Akama, K. T., Hu, J., Van Eldik, L. J., and Aggarwal, B. B. (1998). TRANK, a novel cytokine that activates NF-κB and c-Jun N-terminal kinase. *J. Immunol.* **161**, 1–6.
Haslekas, C., Stacy, R. A., Nygaard, V., Culianez-Macia, F. A., and Aalen, R. B. (1998). The expression of a peroxiredoxin antioxidant gene, AtPer1, in *Arabidopsis thaliana* is seed-specific and related to dormancy. *Plant Mol. Biol.* **36**, 833–845.
Hausladen, A., Privalle, C. T., Keng, T., DeAngelo, J., and Stamler, J. S. (1996). Nitrosative stress: Activation of the transcription factor OxyR. *Cell (Cambridge, Mass.)* **86**, 719–729.
Henle, E. S., and Linn, S. (1997). Formation, prevention, and repair of DNA damage by iron/hydrogen peroxide. *J. Biol. Chem.* **272**, 19095–19098.
Herzenberg, L. A., De Rosa, S. C., Dubs, J. G., Roederer, M., Anderson, M. T., Ela, S. W., Deresinski, S. C., and Herzenberg, L. A. (1997). Glutathione deficiency is associated with impaired survival in HIV disease. *Proc. Natl. Acad. Sci. U.S.A.* **94**, 1967–1972.
Hockenbery, D. M., Oltvai, Z. N., Yin, X.-M., Milliman, C. L., and Korsmeyer, S. J. (1993). Bcl-2 functions in an antioxidant pathway to prevent apoptosis. *Cell (Cambridge, Mass.)* **75**, 241–251.
Iakoubova, O. A., Pacella, L. A., Her, H., and Beier, D. R. (1997). LTW4 protein on mouse chromosome 1 is a member of a family of antioxidant proteins. *Genomics* **42**, 474–478.
Ichimiya, S., Davis, J. G., O'Rourke, D. M., Katsumata, M., and Greene, M. I. (1997). Murine thioredoxin peroxidase delays neuronal apoptosis and is expressed in areas of the brain most susceptible to hypoxic and ischemic injury. *DNA Cell Biol.* **16**, 311–321.
Imbert, V., Rupec, R. A., Livolsi, A., Pahl, H. L., Traenckner, E. B.-M., Mueller-Dieckmann, C., Farahifar, D., Rossi, B., Auberger, P., Baeuerle, P. A., and Peyron, J.-F. (1996). Tyrosine phosphorylation of IκB-α activates NF-κB without proteolytic degradation of IκB-α. *Cell (Cambridge, Mass.)* **86**, 787–798.
Immenschuh, S., Iwahara, S., Satoh, H., Nell, C., Katz, N., and Müller-Eberhard, U. (1995). Expression of the mRNA of heme-binding protein 23 is coordinated with that of heme oxygenase-1 by heme and heavy metals in primary rat hepatocytes and hepatoma cells. *Biochemistry* **34**, 13407–13411.
Immenschuh, S., Nell, C., Iwahara, S., Katz, N., and Müller-Eberhard, U. (1997). Gene regulation of HBP 23 by metalloporphyrins and protoporphyrin IX in liver and hepatocyte cultures. *Biochem. Biophys. Res. Commun.* **231**, 667–670.
Ishii, T., Yamada, M., Sato, H., Matsue, M., Taketani, S., Nakayama, K., Sugita, Y., and Bannai, S. (1993). Cloning and characterization of a 23-kDa stress-induced mouse peritoneal macrophage protein. *J. Biol. Chem.* **268**, 18633–18636.
Iwahara, S., Satoh, H., Song, D.-X., Webb, J., Burlingame, A. L., Nagae, Y., and Muller-Eberhard, U. (1995). Purification, characterization, and cloning of a heme-binding protein (23 kDa) in rat liver cytosol. *Biochemistry* **34**, 13398–13406.
Jacob, A. N., Kandpal, G., and Kandpal, R. P. (1996). Isolation of expressed sequences that include a gene for familial breast cancer (BRCA2) and other novel transcripts from a five megabase region on chromosome 13q12. *Oncogene* **13**, 213–221.
Jaschke, A., Mi, H., and Tropschug, M. (1998). Human T cell cyclophilin18 binds to thiol-specific antioxidant protein aop1 and stimulates its activity. *J. Mol. Biol.* **277**, 763–769.
Jin, D.-Y., Chae, H. Z., Rhee, S. G., and Jeang, K.-T. (1997). Regulatory role for a novel human thioredoxin peroxidase in NF-κB activation. *J. Biol. Chem.* **272**, 30952–30961.
Kalebic, T., Kinter, A., Poli, G., Anderson, M. E., Meister, A., and Fauci, AS. (1991). Suppression of human immunodeficiency virus expression in chronically infected monocytic cells by

glutathione, glutathione ester and N-acetylcysteine. *Proc. Natl. Acad. Sci. U.S.A.* **88**, 986–990.

Kang, S. W., Chae, H. Z., Seo, M. S., Kim, K., Baines, I. C., and Rhee, S. G. (1998a). Mammalian peroxiredoxin isoforms can reduce hydrogen peroxide generated in response to growth factors and tumor necrosis factor-α. *J. Biol. Chem.* **273**, 6297–6302.

Kang, S. W., Baines, I. C., and Rhee, S. G. (1998b). Characterization of a mammalian peroxiredoxin that contains one conserved cysteine. *J. Biol. Chem.* **273**, 6303–6311.

Kim, A. T., Sarafian, T. A., and Shau, H. (1997). Characterization of antioxidant properties of natural killer-enhancing factor-B and induction of its expression by hydrogen peroxide. *Toxicol. Appl. Pharmacol.* **147**, 135–142.

Kim, H.-K., Kim, S.-J., Lee, J.-W., Lee, J.-W., Cha, M.-K., and Kim, I.-H. (1996). Identification of promoter in the 5'-flanking region of the *E. coli* thioredoxin-linked thiol peroxidase gene: Evidence for the existence of oxygen-related transcriptional regulatory protein. *Biochem. Biophys. Res. Commun.* **221**, 641–646.

Kim, I.-H., Kim, K., and Rhee, S. G. (1989). Induction of an antioxidant protein of *Saccharomyces cerevisiae* by O_2, Fe^{3+}, or 2-mercaptoethanol. *Proc. Natl. Acad. Sci. U.S.A.* **86**, 6018–6022.

Kim, K., Kim, I. H., Lee, K.-Y., Rhee, S. G., and Stadtman, E. R. (1988). The isolation and purification of a specific "protector" protein which inhibits enzyme inactivation by a thiol/Fe(III)/O_2 mixed-function oxidation system. *J. Biol. Chem.* **263**, 4704–4711.

Kowaltowski, A. J., Netto, L. E., and Vercesi, A. E. (1998). The thiol-specific antioxidant enzyme prevents mitochondrial permeability transition: Evidence for the participation of reactive oxygen species in this mechanism. *J. Biol. Chem.* **273**, 12766–12769.

Kretz-Remy, C., Mehlen, P., Mirault, M.-E., and Arrigo, A.-P. (1996). Inhibition of IκB-α phosphorylation and degradation and subsequent NF-κB activation by glutathione peroxidase overexpression. *J. Cell Biol.* **133**, 1083–1093.

Kurata, S.-I. (1996). Sensitization of the HIV-1-LTR upon long term low dose oxidative stress. *J. Biol. Chem.* **271**, 21798–21802.

Lee, S. M., and Park, J. W. (1993). A yeast mutant lacking thiol-dependent protector protein is hypersensitive to menadione. *Biochim. Biophys. Acta* **1382**, 167–175.

Marzo, I., Brenner, C., Zamzam., N., Susin, S. A., Beutner, G., Brdiczka, D., Remy, R., Xie, Z. H., Reed, J. C., and Kroemer, G. (1998). The permeability transition pore complex: A target for apoptosis regulation by caspases and bcl-2-related proteins. *J. Exp. Med.* **187**, 1261–1271.

McGonigle, S., Dalton, J. P., and James, E. R. (1998). Peroxidoxins: A new antioxidant family. *Parasitol. Today* **14**, 139–145.

Mongkolsuk, S., Loprasert, S., Whangsuk, W., Fuangthong, M., and Atichartpongkun, S. (1997). Characterization of transcription organization and analysis of unique expression patterns of an alkyl hydroperoxide reductase C gene (*ahpC*) and the peroxide regulator operon *ahpF-oxyR-orfX* from *Xanthomonas compestris* pv. phaseoli. *J. Bacteriol.* **179**, 3950–3955.

Montemartini, M., Nogoceke, E., Singh, M., Steinert, P., Flohe, L., and Kalisz, H. M. (1998). Sequence analysis of the tryparedoxin peroxidase gene from *Crithidia fasciculata* and its functional expression in *Escherichia coli*. *J. Biol. Chem.* **273**, 4864–4871.

Munz, B., Frank, S., Hubner, G., Olsen, E., and Werner, S. (1997). A novel type of glutathione peroxidase: Expression and regulation during wound repair. *Biochem. J.* **326**, 579–585.

Nakamura, H., Nakamura, K., and Yodoi, J. (1997). Redox regulation of cellular activation. *Annu. Rev. Immunol.* **15**, 351–369.

Netto, L. E. S., Chae, H. Z., Kang, S.-W., Rhee, S. G., and Stadtman, E. R. (1996). Removal of hydrogen peroxide by thiol-specific antioxidant enzyme (TSA) is involved with its

antioxidant properties: TSA possesses thiol peroxidase activity. *J. Biol. Chem.* **271**, 15315–15321.
Nicolli, A., Basso, E., Petronilli, V., Wenger, R. M., and Bernardi, P. (1996). Interactions of cyclophilin with the mitochondrial inner membrane and regulation of the permeability transition pore, and cyclosporin A-sensitive channel. *J. Biol. Chem.* **271**, 2185–2192.
Nogoceke, E., Gommel, D. U., Kiess, M., Kalisz, H. M., and Flohe, L. (1997). A unique cascade of oxidoreductases catalyses trypanothione-mediated peroxide metabolism in *Crithidia fasciculata*. *Biol. Chem. Hoppe-Seyler* **378**, 827–836.
Okamoto, T., Ogiwara, H., Hayashi, T., Mitsui, A., Kawabe, T., and Yodoi, J. (1992). Human thioredoxin/adult T cell leukemia-derived factor activates the enhancer binding protein of human immunodeficiency virus type 1 by thiol redox control mechanism. *Int. Immunol.* **4**, 811–819.
Pahl, P., Berger, R., Hart, I., Chae, H. Z., Rhee, S. G., and Patterson, D. (1995). Localization of TDPX1, a human homologue of the yeast thioredoxin-dependent peroxide reductase gene (TPX), to chromosome 13q12. *Genomics* **26**, 602–606.
Piedimonte, G., Guetard, D., Magnani, M., Corsi, D., Picerno, I., Spataro, P., Kramer, L., Montroni, M., Silvestri, G., Torres, R. J. F., and Montagnier, L. (1997). Oxidative protein damage and degradation in lymphocytes from patients infected with human immunodeficiency virus. *J. Infect. Dis.* **176**, 655–664.
Polyak, K., Xia, Y., Zweier, J. L., Kinzler, K. W., and Vogelstein, B. (1997). A model for p53-induced apoptosis. *Nature (London)* **389**, 300–305.
Poole, L. B. (1996). Flavin-dependent alkyl hydroperoxide reductase from *Salmonella typhimurium*. 2. Cystine disulfides involved in catalysis of peroxide reduction. *Biochemistry* **35**, 65–75.
Poole, L. B., Chae, H. Z., Flores, B. M., Reed, S. L., Rhee, S. G., and Torian, B. E. (1997). Peroxidase activity of a TSA-like antioxidant protein from a pathogenic amoeba. *Free Radical Biol. Med.* **23**, 955–959.
Prosperi, M.-T., Ferbus, D., Karczinski, I., and Goubin, G. (1993). A human cDNA corresponding to a gene overexpressed during cell proliferation encodes a product sharing homology with amoebic and bacterial proteins. *J. Biol. Chem.* **268**, 11050–11056.
Prosperi, M.-T., Apiou, F., Dutrillaux, B., and Goubin, G. (1994). Organization and chromosomal assignment of two human PAG gene loci: PAGA encoding a functional gene and PAGB a processed pseudogene. *Genomics* **19**, 236–241.
Prosperi, M.-T., Ferbus, D., Rouillard, D., and Goubin, G. (1998). The pag gene product, a physiological inhibitor of c-Ab1 tyrosine kinase, is overexpressed in cells entering S phase and by contact with agents inducing oxidative stress. *FEBS Lett.* **423**, 39–44.
Rabilloud, T., Berthier, R., Vincon, M., Ferbus, D., Goubin, G., and Lawrence, J. J. (1995). Early events in erythroid differentiation: Accumulation of the acidic peroxidoxin (PRP/TSA/NKEF-B). *Biochem. J.* **312**, 699–705.
Rhee, S. G., and Chae, H. Z. (1994). Thioredoxin peroxidase and peroxiredoxin family. *Mol. Cells* **4**, 137–142.
Saitoh, M., Nishitoh, H., Fujii, M., Takeda, K., Tobiume, K., Sawada, Y., Kawabata, M., Miyazono, K., and Ichijo, H. (1998). Mammalian thioredoxin is a direct inhibitor of apoptosis signal-regulating kinase (ASK) 1.*EMBO J.* **17**, 2596–2606.
Sandstrom, P. A., Tebbey, P. W., Van Cleave, S., and Buttke, T. M. (1994). Lipid hydroperoxides induce apoptosis in T cells displaying a HIV-associated glutathione peroxidase deficiency. *J. Biol. Chem.* **269**, 798–801.
Sandstrom, P. A., Murray, J., Folks, T. M., and Diamond, A. M. (1998). Antioxidant defenses influence HIV-1 replication and associated cytopathic effects. *Free Radical Biol. Med.* **24**, 1485–1491.

Sappey, C., Legrand-Poels, S., Best-Belpomme, M., Favier, A., Rentier, B., and Piette, J. (1994). Stimulation of glutathione peroxidase activity decreases HIV type 1 activation after oxidative stress. *AIDS Res. Hum. Retroviruses* **10,** 1451–1461.

Sauri, H., Ashjian, P. H., Kim, A. T., and Shau, H. (1996). Recombinant natural killer enhancing factor augments natural killer cytotoxicity. *J. Leukocyte Biol.* **59,** 925–931.

Scandalios, J. G. (1997). "Oxidative Stress and the Molecular Biology of Antioxidant Defenses." Cold Spring Harbor Lab. Press, Cold Spring Harbor, NY.

Schreck, R., Rieber, P., and Baeuerle, P. A. (1991). Reactive oxygen intermediates as apparently widely used messengers in the activation of the NF-κB transcription factor and HIV-1. *EMBO J.* **10,** 2247–2258.

Schroder E., and Ponting, C. P. (1998). Evidence that peroxiredoxins are novel members of the thioredoxin fold superfamily. *Protein Sci.* **7,** 2465–2468.

Schulze-Osthoff, K., Schenk, H., and Droge, W. (1995). Effects of thioredoxin on activation of transcription factor NF-κB. *In* "Methods in Enzymology" (L. Packer, ed.), Vol. 252, pp. 253–264. Academic Press, San Diego, CA.

Sen, C. K., and Packer, L. (1996). Antioxidant and redox regulation of gene transcription. *FASEB J.* **10,** 709–720.

Sherman, D. R., Mdluli, K., Hickey, M. J., Arain, T. M., Morris, S. L., Barry, C. E., 3rd, and Stover, C. K. (1996). Compensatory *ahpC* gene expression in isoniazid-resistant *Mycobacterium tuberculosis Science* **272,** 1641–1643.

Shichi, H., and Demar, J. C. (1990). Non-selenium glutathione peroxidase without glutathione S-transferase activity from bovine ciliary body. *Exp. Eye Res.* **50,** 513–520.

Sies, H. (1993). Strategy of antioxidant defense. *Eur. J. Biochem.* **215,** 213–219.

Staal, F. J. T., Roederer, M., Raju, P. A., Anderson, M. T., Ela, S. W., Herzenberg, L. A., and Herzenberg, L. A. (1993). Antioxidants inhibit stimulation of HIV transcription. *AIDS Res. Hum. Retroviruses* **9,** 299–306.

Stacy, R. A., Munthe, E., Steinum, T., Sharma, B., and Aalen, R. B. (1996). A peroxiredoxin antioxidant is encoded by a dormancy-related gene, Perl, expressed during late development in the aleurone and embryo of barley grains. *Plant Mol. Biol.* **31,** 1205–1216.

Stancovski, I., and Baltimore, D. (1997). NF-κB activation: The IκB kinase revealed? *Cell (Cambridge, Mass.)* **91,** 299–302.

Storz, G., Jacobson, F. S., Tartaglia, L. A., Morgan, R. W., Silveira, L. A., and Ames, B. N., (1989). An alkyl hydroperoxide reductase induced by oxidative stress in *Salmonella typhimurium* and *Escherichia coli*: Genetic characterization and cloning of *ahp*. *J. Bacteriol.* **171,** 2049–2055.

Sundaresan, M., Yu, Z.-X., Ferrans, V. J., Irani, K., and Finkel, T. (1995). Requirement for generation of H_2O_2 for platelet-derived growth factor signal transduction. *Science* **270,** 296–299.

Tai, S. S., and Zhu, Y. Y. (1995). Cloning of a *Corynebacterium diphtheriae* ironrepressible gene that shares sequence homology with the AhpC subunit of alkyl hydroperoxide reductase of *Salmonella typhimurium*. *J. Bacteriol.* **177,** 3512–3517.

Tartaglia, L. A., Storz, G., and Ames, B. N. (1989). Identification and molecular analysis of *oxyR*-regulated promoters important for the bacterial adaptation to oxidative stress. *J. Mol. Biol.* **210,** 709–719.

Tsuji, K., Copeland, N. G., Jenkins, N. A., and Obinata, M. (1995). Mammalian antioxidant protein complements alkylhydroperoxide reductase (ahpC) mutation in *Escherichia coli*. *Biochem. J.* **307,** 377–381.

Verma, I. M., and Stevenson, J. (1997). IκB kinase: Beginning, not the end. *Proc. Natl. Acad. Sci. U.S.A.* **94,** 11758–11760.

Wan, X.-Y., Zhou, Y., Yan, Z.-Y., Wang, H.-L., Hou, Y.-D., and Jin, D.-Y. (1997). Scavengase p20: A novel family of bacterial antioxidant enzymes. *FEBS Lett.* **407,** 32–36.

Watabe, S., Kohno, H., Kouyama, H., Hiroi, T., Yago, N., and Nakazawa, T. (1994). Purification and characterization of a substrate protein for mitochondrial ATP-dependent protease in bovine adrenal cortex. *J. Biochem. (Tokyo)* **115,** 648–654.

Wen, S.-T., and Van Etten, R. A. (1997). The PAG gene product, a stress-induced protein with antioxidant properties, is an Ab1 SH3-binding protein and a physiological inhibitor of c-Ab1 tyrosine kinase activity. *Genes Dev.* **11,** 2456–2467.

Wilson, T. M., and Collins, D. M. (1996). *ahpC,* a gene involved in isoniazid resistance of the *Mycobacterium tuberculosis* complex. *Mol. Microbiol.* **19,** 1025–1034.

Yamamoto, T., Matsui, Y., Natori, S., and Obinata, M. (1989). Cloning of a housekeeping-type gene (MER5) preferentially expressed in murine erythroid cells. *Gene* **80,** 337–343.

Zhang, P., Liu, B., Kang, S. W., Seo, M. S., Rhee, S. G., and Obeid, L. M. (1997). Thioredoxin peroxidase is a novel inhibitor of apoptosis with a mechanism distinct from that of Bcl-2. *J. Biol. Chem.* **272,** 30615–30618.

Zheng, M., Aslund, F., and Storz, G. (1998). Activation of the OxyR transcription factor by reversible disulfide bond formation. *Science* **279,** 1718–1721.

Zhou, Y., Wan, X.-Y., Wang, H.-L., Yan, Z.-Y., Hou, Y.-D., and Jin, D.-Y. (1997). Bacterial scavengase p20 is structurally and functionally related to peroxiredoxins. *Biochem. Biophys. Res. Commun.* **233,** 848–852.

18 Enhanced Activity of an Oxidation Product of Lycopene Found in Tomato Products and Human Serum Relevant to Cancer Prevention

*John S. Bertram, Timothy King, Laurie Fukishima, and Frederick Khachik**

Cancer Research Center of Hawaii
University of Hawaii
Honolulu, Hawaii 96813
*Department of Chemistry and Biochemistry
Joint Institute for Food Safety and Applied Nutrition
University of Maryland
College Park, Maryland 20742

There is abundant epidemiological evidence that the consumption of dietary carotenoids reduces the risk of cancer; however, supplementation with β-carotene has been shown not to be effective. It is unclear whether this failure is due to the choice of carotenoid or the dose employed, whether supplementation with mixtures of carotenoids as found in the diet is required for effectiveness, or whether the epidemiological associations are in fact incorrect. It has been shown previously that three major dietary carotenoids—β-carotene, lutein, and lycopene—can increase connexin 43 gene expression in 10T1/2 cells and in human keratinocytes in organotypic culture. Connexin 43 codes for a major gap junctional protein that has been implicated in growth control. We now demonstrate that an oxidation product of lycopene, 2,6-cyclolycopene-1,5-diol, is found in tomato products and increases in human serum after ingestion of these products. This compound has been shown to possess greater activity than the parent molecule in its ability to increase connexin 43 levels in mouse and human cells. This suggests that lycopene may function both as an antioxidant, protecting cells from oxidative damage, and as the source of a molecule with biological actions that may additionally protect cells from neoplasia.

I. INTRODUCTION

With our increasing understanding of the causes of cancer at the environmental and genetic level there has grown an increasing realization that many cancers may be preventable. Whereas primary prevention, i.e., the avoidance or elimination of carcinogenic agents such as tobacco, is clearly desirable, epidemiological research has indicated that the consumption of many foods, principally green and yellow vegetables and fruits, results in a lower risk of cancer at many sites. Among the constituents of such foods, their content of carotenoids correlates most strongly with a decreased risk (Bertram et al., 1987). Because of the known importance of the pro-vitamin A carotenoids, principally β-carotene, in human nutrition and the known role of the retinoids as cancer preventive agents (Moon et al., 1994), initial studies focused almost exclusively on this carotenoid. In response to the highly persuasive epidemiological evidence, a number of clinical intervention trials were initiated utilizing purified, synthetic β-carotene administered at doses approximately 10-fold higher than that found in a normal Western diet. The results of three of the largest trials are now available (Heinonen and Albanes, 1994; Omenn et al., 1994; Hennekens et al., 1996). In all three trials, β-carotene failed to decrease cancer rates and, more disturbingly, in two of these trials, both conducted in current or former smokers and/or asbestos-exposed workers, there was a suggestion that lung cancer rates were actually increased among smokers. During the conduct of these trials, additional quantitative information became available regarding the presence of other carotenoids in foods associated with a lower cancer risk (Micozzi et al., 1990). High-performance liquid chromatography (HPLC) analysis has revealed a total of about 34 carotenoids in human serum (Khachik et al., 1997a). These include seven metabolites of lutein and two of lycopene, a carotenoid associated with a decreased risk of prostate cancer (Gann et al., 1999). These new studies presented the opportunity to determine if these oxidation products have biological activity, which may explain some of the actions attributed to carotenoids. In previous research we had shown that carotenoids were capable of inhibiting neoplastic transformation in a model cell system regardless of their pro-vitamin A activity (Bertram *et al.*, 1991). Moreover, we discovered that activity strongly correlated with their ability to increase expression of a gene, connexin 43, coding for a gap junctional protein (Zhang *et al.*, 1991). Because gap junctional communication (GJC) has, in many studies, been linked to increased growth control (Neveu and Bertram, 1999) and because the chemopreventive retinoids also increase expression of this gene (Merriman and Bertram, 1979), these studies suggested a mechanism for the effects of carotenoids. This chapter utilizes the expression of this gene as an intermediate marker of response in immortalized mouse fibroblasts and human keratinocytes.

II. MATERIALS AND METHODS

A. Clinical Studies

1. Bioavailability of Lycopene

Fifteen, nonsmoking, healthy Caucasian volunteers consisting of seven females and eight males, ages 35–62, participated in this study. The subjects were divided into four groups who received 70–75 mg/day of lycopene or placebo for 4 weeks from four different treatments. These were (1) tomato juice (475 g/day), providing 75 mg/day of lycopene; (2) lycopene soft gels, providing 75 mg/day of lycopene; (3) lycopene Beadlets (Lycobeads), providing 70 mg/day of lycopene; and (4) placebo consisting of beadlets with no added lycopene. Lycopene soft gels and beadlets are two different formulations of lycopene that are prepared from a concentrated tomato extract known as tomato oleoresin. Whereas lycopene soft gels are prepared directly from tomato oleoresin, lycopene beadlets are formulated from a mixture of tomato oleoresin, collagen, disaccharides, and food starch according to a proprietary technology developed by H. Reisman Corporation (Orange, New Jersey). The beadlet formulation of carotenoids, in general, is known to greatly enhance the absorption and bioavailability of these compounds in humans. The control group in the study received beadlets with no added lycopene. All subjects were on a self-selected diet throughout the study but avoided foods that were rich sources of lycopene (tomatoes and tomato products, pink grapefruits, apricots). The study was a randomized crossover design in which each subject received all of the four treatments and served as their own control. The subjects entered the study 1 week before receiving the 4-week treatment and during this time restricted the consumption of lycopene-containing foods. At the end of each 4-week treatment, each group underwent 6 weeks of "wash out" to allow for the blood lycopene concentration of the subjects to decline and return to baseline. At various intervals throughout the study, the serum concentration of lycopene and 33 other carotenoids was determined for each subject by HPLC (Paetau *et al.*, 1998).

B. Chemicals

Lycopene and 2,6-cyclolycopene 1,5-diol (referred to in the text as cyclolycopene) were prepared and separated by HPLC as described (Khachik *et al.*, 1998b). Tetrahydofuran (THF, 99.9% + 0.025% BHT) was purchased from Aldrich Chemicals (Milwaukee, WI). Structures of the tested carotenoids are shown in Fig. 2.

C. Cell Culture Conditions

Two immortalized cell lines were used in this study: the mouse embryonal fibroblast cell line C3H/10T1/2 (10T1/2) (Reznikoff *et al.*, 1973) and the human keratinocyte cell line HaCaT (Fitzgerald *et al.*, 1994; Ryle *et al.*, 1989). All cultures were incubated at 37°C, 5% carbon dioxide, and 95% humidity. 10T1/2 cells were cultured as monolayers as described previously (Bertram *et al.*, 1991). 10T1/2 cells were treated when confluent for the indicated time periods. HaCaT cells were cultured in low-calcium (0.1 mmol/liter) serum-free keratinocyte medium (GIBCO-BRL, Grand Island, NY) supplemented with epidermal growth factor (5 μg/ml), bovine pituitary extract (35 μg/ml), and insulin (5 μg/ml) in monolayer culture until subconfluent. They were then harvested and placed on Millicell-CM collagen-coated culture plate inserts (Millipore, Bedford, MA) for approximately 7 days when high-calcium medium (1.15 mmol/liter), supplemented as described earlier, was placed below the filters while the surface of the filters was exposed to the atmosphere. During this time a multilayered differentiating organotypic culture was produced (Asselineau and Darmon, 1995). Cells were treated with carotenoids or retinoic acid dissolved in THF or acetone, respectively, using the precautions and procedures discussed previously (Cooney *et al.*, 1993) and added to culture medium concurrent with exposure to the air interface. Cells were refed and retreated every 2 days.

D. Protein Electrophoresis and Western Blotting

HaCaT cells and 10T1/2 cells were harvested and Cx43 solubilized as described previously (Rogers *et al.*, 1990). Proteins were electrophoresed on 10% sodium dodecyl sulfate (SDS)–polyacrylamide gels and were subsequently incubated with a rabbit polyclonal antibody raised against a synthetic peptide corresponding to the 15 residues of the C-terminal domain of Cx43 (Rogers *et al.*, 1990). This sequence is 100% homologous in the mouse, rat, and human. Bound antibody was visualized using a chemiluminescent detection system (Tropix, Bedford, MA) and recorded on X-ray film.

III. RESULTS

A. Evidence for Preferential Absorption of Cyclolycopene over Dietary Lycopene Obtained from Tomato Products

The bioavailability of lycopene in humans from consumption of lycopene-rich foods has been reported previously (Gärtner et al., 1997; Stahl and Sies, 1992). However, in a study involving 15 healthy human subjects, Paetau et al. (1998) examined the bioavailability of lycopene from purified supplements as well as foods. In this study the serum concentration of lycopene in all groups who received the three lycopene treatments increased significantly in comparison to the controls who were on placebo (beadlets with no added lycopene). The mean serum lycopene concentration of all groups supplemented with Lycobeads increased by 1.5-fold in comparison to the mean of baseline values (0.601 μmol/liter $\pm$ 0.055) after 1 week and remained at a steady concentration during the last 3 weeks of treatment. Treatment with lycopene soft gels also resulted in a 1.5-fold increase in mean lycopene serum concentration in the subjects after 2 weeks of supplementation in comparison to mean baseline values (0.579 μmol/liter $\pm$ 0.052) and remained unchanged during the last 2 weeks of supplementation. However, the mean serum concentration of lycopene in all groups who were on tomato juice treatment increased very slowly with time and after 4 weeks of ingestion it was only elevated by 1.3-fold in comparison to mean baseline values (0.581 μmol/liter $\pm$ 0.057) (Fig. 1, left) (Paetau et al., 1998). Although lycopene appeared to be bioavailable from all the lycopene sources, the poor bioavailability of lycopene from tomato juice was particularly noticeable, which confirms previous reports (Gartner et al., 1997).

Concurrently with the increase in serum lycopene, there was an increase in a compound subsequently identified as cyclolycopene; this increase was most evident in the case of tomato juice supplementation (Fig. 1, right). Analysis of all lycopene preparations fed to human subjects revealed the presence of this compound at a level of approximately 0.5% of the lycopene content. However, analysis of plasma after supplementation with lycopene revealed that cyclolycopene was present at the level of approximately 50% of that of lycopene (for tomato juice, 0.41 and 0.21 μmol/liter, respectively) (Paetau et al., 1998). The structure of 2,6-cyclolycopene 1,5-diol has been established by partial synthesis (Khachik et al., 1998b), and several mechanisms that account for the presence of this metabolite in tomato-based foods

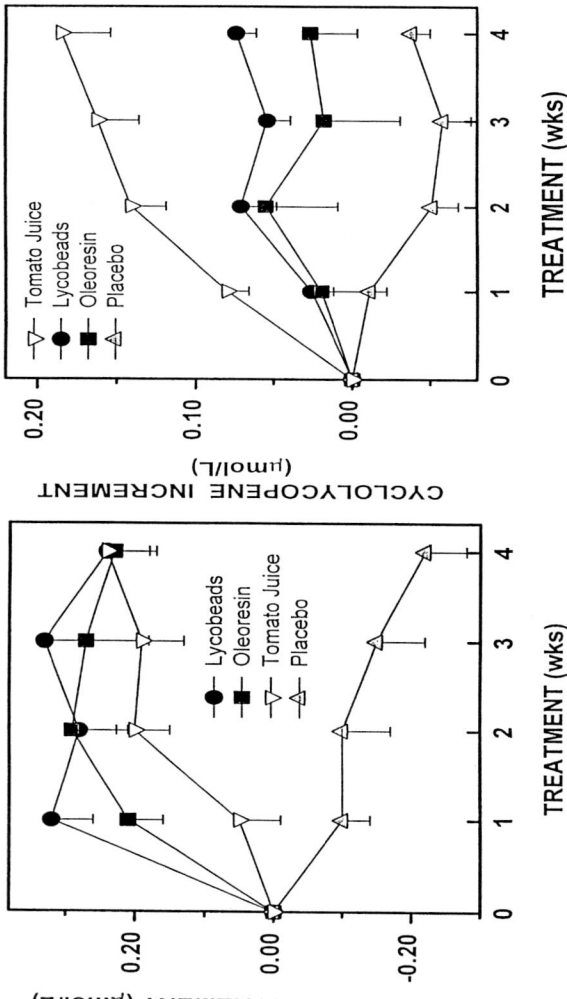

Figure 1 Ingestion of lycopene by human volunteers leads to increases in plasma lycopene (left) and its oxidized derivative cyclolycopene (right). Fifteen, nonsmoking, healthy Caucasian volunteers consisting of seven females and eight males, ages 35–62, participated in this study. The subjects were divided into four groups who received 70–75 mg/day of lycopene or placebo for 4 weeks from four different treatments. These were (1) tomato juice (475 g/day), providing 75 mg/day of lycopene; (2) lycopene soft gels, providing 75 mg/day of lycopene; (3) lycopene Beadlets, providing 70 mg/day of lycopene; and (4) placebo consisting of beadlets with no added lycopene. From Paetau et al., (1998), with permission.

and in human serum have been proposed (Khachik *et al.*, 1998a). It has been suggested that although this oxidative metabolite of lycopene is present in foods at very low concentrations, the preferential uptake of this compound from foods could account for its disproportionately large presence in serum. Alternatively the authors proposed that the *in vivo* oxidation of lycopene humans may also lead to this metabolite. Without further studies, e.g., utilizing stable isotopes, we cannot discriminate between these two alternatives. The structures of these compounds are shown in Fig. 2.

As shown previously, lycopene possesses biological activities in the 10T1/2 mouse transformation system: these activities include the ability to inhibit chemically induced neoplastic transformation (Bertram *et al.*, 1991) and to regulate the expression of connexin 43, a gene coding for gap junctions (Zhang *et al.*, 1991). Because we did not expect biological activity to be associated with a straight-chain hydrocarbon and because it seemed possible that this oxidized derivative of lycopene could be formed under conditions of cell culture, we decided to investigate whether this metabolite possessed biological activity. We thus compared the activity of the two compounds

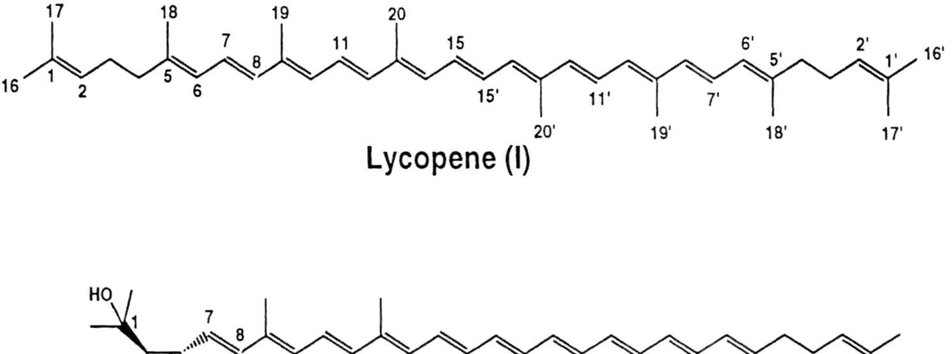

Lycopene (I)

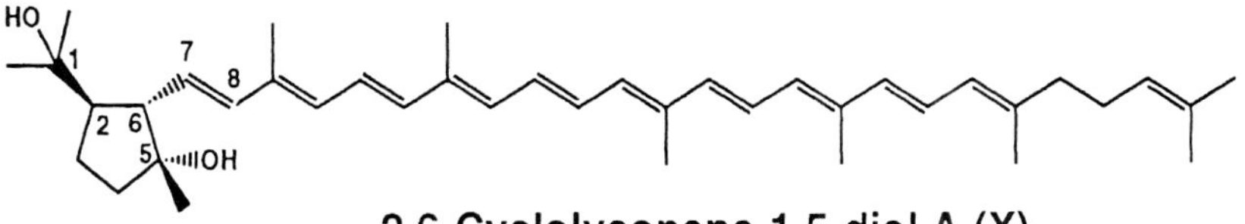

2,6-Cyclolycopene-1,5-diol A (X)

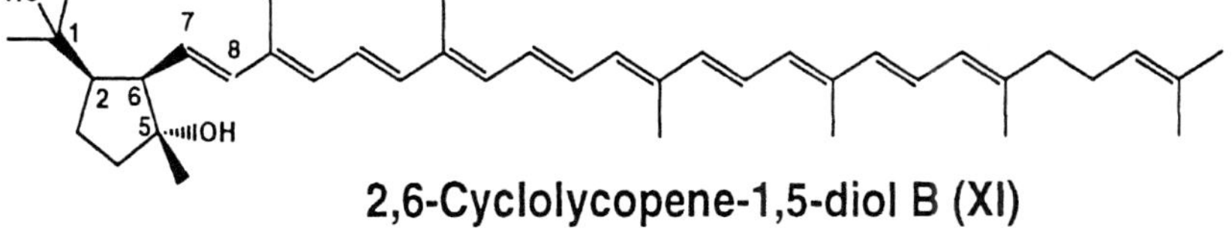

2,6-Cyclolycopene-1,5-diol B (XI)

Figure 2 Structures of lycopene and cyclolycopenes.

B. Effects of Lycopene and Cyclolycopene on 10T1/2 Murine Embryo Cells: Induction of Connexin 43

To investigate if lycopene and its oxidized metabolite cyclolycopene modulate Cx43 expression in confluent 10T1/2 cells, cultures were treated for 7 days, harvested, and analyzed by Western blotting. Lycopene caused small increases in Cx43 at 10^{-6} and $10^{-5} M$ (Fig. 3A, lanes 4 and 5), whereas the lycopene oxidation product cyclolycopene exhibited weak activity at $10^{-7} M$ (lane 6) and exhibited greater activity than lycopene itself at a concentration of $10^{-5} M$ (lane 8). Figure 3B shows a Coomassie-stained membrane of the same region to demonstrate equal protein loading and transfer. Although these data strongly suggest that cyclolycopene is a more potent molecule than lycopene in its ability to upregulate connexin 43, we must add the caveat that the stability of this and the other metabolites in cells and in culture medium has yet to be investigated. When added under the conditions used in the molecular studies, the stability of lycopene is depicted in Fig. 4. It will be seen that lycopene is surprisingly stable in serum-containing medium, with this concentration decreasing little over the 7-day assay period. After rising during the first 24 hr of incubation, however,

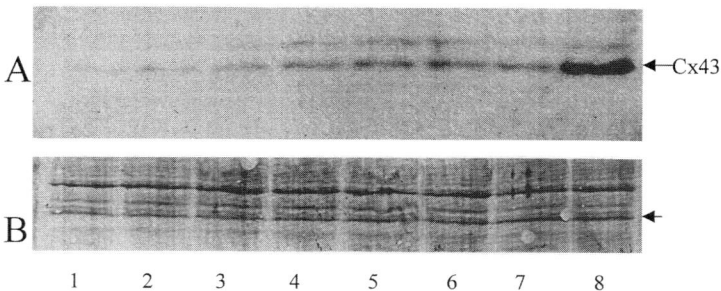

Figure 3 Induction of connexin 43 by lycopene and its metabolite in 10T/1/2 cells. Confluent 10T1/2 cells were treated with carotenoids for 7 days. Cells were then harvested, lysed, and the extracts solubilized. Gel electrophoresis, Western transfer, and detection of Cx43 were performed as described previously (Rogers et al., 1990). Equal amounts of protein (25 μg) were loaded in each lane. (A) Immunoblot probed with a rabbit polyclonal antibody to connexin 43. (B) The same region of the blot stained with Coomassie blue to indicate equal protein loading. Lanes: 1, no treatment; 2, THF solvent control; 3–5, lycopene 10^{-7}, 10^{-6}, and $10^{-5}M$; and 6–8, cyclolycopene, 10^{-8}, 10^{-6}, and $10^{-5}M$. The arrow in B represents the approximate position of connexin 43.

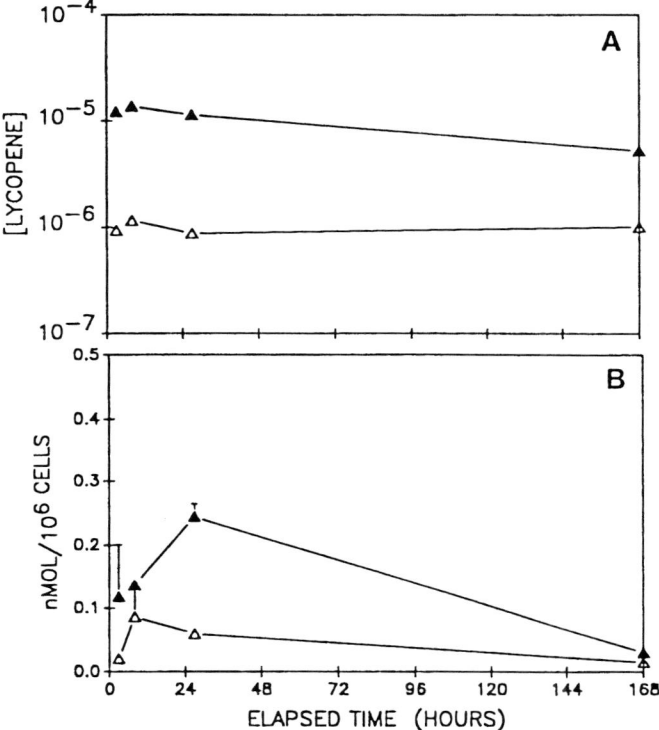

Figure 4 Chemical stability of lycopene under the conditions utilized for molecular studies. Lycopene, at a concentration of $10^{-5}M$ (▲) or $10^{-6}M$ (△), was added to confluent cultures of 10T/1/2 cells as a THF solution. At increasing times after addition, media and cells were harvested, and lycopene contents were analyzed by HPLC as described previously. From Bertram et al. (1991), with permission.

cellular levels, as shown in Fig. 4 B, decreased markedly over the remaining time period.

As reported previously, lycopene possesses the ability, along with many other carotenoids, to suppress carcinogen-induced neoplastic transformation in these 10T/1/2 cells (Bertram et al., 1991). For retinoids and carotenoids, the potency of this suppression was correlated with the ability to induce connexin 43 (Zhang et al., 1991). Because of a limited supply of cyclolycopene, we have so far been unable to test its ability to suppress transformation, thus its chemopreventive properties are at present unknown. However, if the correlation between the induction of Cx43 and the ability to inhibit transformation is maintained with this molecule, we predict strong activity.

C. Effects of Lycopene and Cyclolycopene in HaCaT Cells

Connexin 43 is expressed in differentiating human keratinocytes, and it has been demonstrated that the treatment of intact human skin with *all-trans* retinoic acid causes the induction of connexin 43 in suprabasal cells of the epidermis (Guo *et al.*, 1992). Because of the requirement for differentiation for connexin inducibility (Bertram and Bortkiewicz, 1995; Zhang *et al.*, 1995), keratinocytes were cultured as an organotypic culture in order to allow differentiation to proceed. For reasons of reproducibility, availability, and ease of use, these new studies did not utilize primary keratinocytes as before (Bertram and Bortkiewicz, 1995), instead we used the immortalized human keratinocyte cell line HaCaT. Studies by others had shown this line to closely resemble normal human keratinocytes in its differentiation pattern in organotypic culture (Ryle *et al.*, 1989).

As shown in Fig. 5, treatment of HaCaT cells with lycopene and cyclolycopene increased the expression of connexin 43. As with 10T/1/2 cells, cyclolycopene (Fig. 5, lanes 6–8) was more active than lycopene (Fig. 5, lanes 3–5) at concentrations of 10^{-7}–10^{-5} M, respectively. As can be seen, both compounds increased the level of connexin 43 in a dose-dependent manner. We had reported previously that retinoic acid will also upregulate connexin 43 under these conditions. Comparison of these data with previous results showed that retinoic acid was, as expected, a far more potent inducer than carotenoids (Bertram and Bortkiewicz, 1995). Nevertheless, the ability of both molecules to upregulate this gene further underlines the similarity between the actions of carotenoids and retinoids in cultured keratinocytes.

We have not determined what proportion, if any, of added lycopene becomes converted to cyclolycopene under conditions of cell culture. If substantial conversion occurs, the apparent activity of lycopene may be a consequence of this conversion.

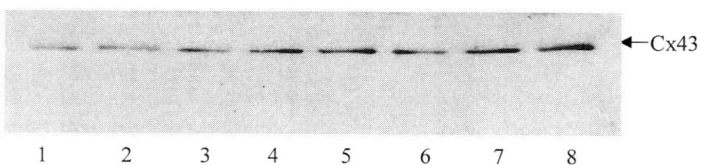

Figure 5 Induction of connexin 43 (Cx43) by lycopene and its metabolite in organotypic HaCaT cell cultures. HaCaT cells were grown in organotypic culture as described previously (Bertram and Bortkiewicz, 1995). Cells were treated with retinoic acid or carotenoids for 7 days. Gel electrophoresis, Western blotting, and detection of Cx43 were performed as described previously (Rogers *et al.*, 1990). Lane 1, control; lane 2, THF 0.5% solvent control; lanes 3–5, lycopene 10^{-7}, 10^{-6}, and 10^{-5} M, respectively; and lanes 6–8, cyclolycopene 10^{-7}, 10^{-6}, and 10^{-5} M, respectively.

IV. DISCUSSION

We have shown that certain retinoids (Merriman and Bertram, 1979) and carotenoids, including β-carotene (Bertram et al., 1991), can inhibit the carcinogen-induced neoplastic transformation of 10T1/2 cells and that this inhibition correlates with increased gap junctional intercellular communication (GJIC) (Hossain et al., 1989; Zhang et al., 1991). In both cases the observed increase in GJIC is mediated through an increase in Cx43 at the mRNA and protein level (Zhang et al., 1992; Rogers et al., 1990). Because β-carotene is converted readily in mammals to retinoids, many of its effects have been considered to be mediated through breakdown, either spontaneously or enzymatically, to retinoids. However, our demonstration that non-pro-vitamin A carotenoids, such as lycopene, also mediate responses similar to retinoic acid (Zhang et al., 1992) has led to a reevaluation of this concept, particularly the question of whether the conversion of classical non-pro-vitamin A carotenoids to retinoid-like molecules can occur. Most of the actions of retinoic acid appear to be mediated by the nuclear retinoic acid receptors RAR-α, -β, and -γ and RXR-α, -β, and -γ through interactions with their cognate responsive elements in retinoid responsive genes (Pfahl, 1993). One such gene, RAR-β, has been shown to be induced in 10T1/2 cells at the transcriptional level by the synthetic retinoid TTNPB (Zhang et al., 1992). However, canthaxanthin (CTX), a non-pro-vitamin A carotenoid, which had been shown to induce connexin 43 mRNA, did not induce RAR-β mRNA (Zhang et al., 1992), indicating that an independent mechanism for mRNA induction by carotenoids may exist. Studies utilizing liarozole, an inhibitor of cytochrome P450-mediated catabolism of retinoic acid, also failed to indicate the conversion of CTX to the expected retinoid, 4-oxoretinoic acid (Acevedo and Bertram, 1995). Others, however, have demonstrated in different experimental systems, by chemical (Hanusch et al., 1995) and molecular techniques (Nikawa et al., 1995), the conversion of CTX to 4-oxoretinoic acid. The reason for this discrepancy is not known. Similar studies with lycopene or its derivatives have not been performed.

While conversion of many dietary carotenoids to compounds shown to have retinoid-like properties is certainly feasible on a chemical basis, it is more difficult to imagine such conversion in the case of the straight-chain carotenoid lycopene, which would require cyclization, oxidation, and cleavage. It, however, exhibits chemopreventive activity (Bertram et al., 1991) and, as demonstrated here, induces connexin 43 in mouse 10T1/2 cells (Fig. 2) and human keratinocytes (Fig. 4). The results presented here suggest a possible explanation for this surprising activity of lycopene: its conversion to the cyclic compound 2,6-cyclolycopene-1,5-diol as a consequence of oxidation and subsequent rearrangement. Because this conversion may occur under conditions of cell culture and because small amounts are probably always present in lycopene when exposed to oxygen, as indicated by the presence of this molecule in tomato products, the activity of lycopene in our systems may be entirely due to the activity of this oxidation product. We have not yet had the opportunity to determine if this cyclic compound possesses retinoid-like activity. The potential of lycopene as a biologically important carotenoid in the human diet has been the subject of several reviews (Stahl and Sies, 1996; Gerster, 1997; Clinton, 1998).

Retinoids are known to be potent modifiers of differentiation in human keratinocytes in culture and in human skin (Fisher and Voorhees, 1996). This is the basis of their extensive and successful use in dermatology. This effect can be observed as a decrease in expression of mature keratins such as K1 (Kopan et al., 1987) and in markers of terminal differentiation such as loricrin (Brown et al., 1994). This regulation is known to occur at the transcriptional level and to be mediated through retinoic acid receptors (RARs) and retinoid X receptors (RXRs) (Tomic et al., 1990). In the case of loricrin, functional interactions between these nuclear receptors and the promoter region appear to involve AP-1 sites (DiSepio et al., 1995). We have shown previously that certain carotenoids also alter differentiation in organotypic culture (Bertram and Bortkiewicz, 1995). HaCaT cells and early passage human keratinocytes undergo pseudo-normal differentiation in organotypic culture and express connexin 43 in suprabasal cells in a manner similar to that observed *in vivo* (Guo et al., 1992; Salomon et al., 1994). Under these conditions, expression of this gene is increased when exposed to retinoic acid and to dietary carotenoids found in human serum (Fig. 4). This response to retinoic acid was predicted by previous clinical studies in which it was demonstrated that topical retinoic acid would induce connexin 43 in intact human skin (Guo et al., 1992). The present results with carotenoids predict that these molecules should produce dermatological effects in sufficient concentration; whether these effects would mirror those seen after retinoic acid treatment is uncertain given the known disparities between responses of intact human skin and keratinocytes in organotypic culture (Fisher and Voorhees, 1996). However, the ability of retinoic acid to elevate levels of connexin 43 under both situations (i.e., *in vivo* and in culture) suggests that carotenoids may also have this ability *in vivo*. If so, we would suggest that this action may be significant in terms of cancer chemoprevention and in the proposed role of carotenoids as micronutrients that reduce cancer risk (Mayne, 1996). However, we are unaware of reports that dietary carotenoids or β-carotene supplementation have had retinoid-like effects on skin or other organs. On the contrary, β-carotene has not been reported to exert any clinical toxicity, even after supplementation at levels up to 180 mg/day, concentrations which cause appreciable yellowing of skin (Heywood et al., 1985). In vivo, a low concentration of retinoic acid causes increased thickening of the epidermis, de novo synthesis of collagen in the dermis, and inhibition of UV-induced proteases: effects beneficial to photo-aged skin (Fisher and Voorhees, 1996). Moreover, the concentrations of carotenoids used in these studies were in the range of $10^{-6}M$; concentrations that can be reached by dietary supplementation with lycopene-rich foods as shown here. At present it is unclear why carotenoids have not been reported to produce changes in skin, one of the most sensitive organs to retinoids.

The research presented here has two implications: (1) that a product of the oxidation of lycopene found in tomato-based foods increases the expression of connexin 43 to higher levels than that obtained after treatment with the parent carotenoid and (2) that both lycopene and this oxidation product possess activity in cultured human keratinocytes, suggesting that they will possess similar activity in vivo. We have demonstrated that increased GJC is associated with increased growth control (Hossain et al., 1989; Mehta et al., 1986, 1989; Hossain and Bertram, 1994), and others have shown that the transfection of connexin genes into human tumor cells reduces their neoplastic potential (Mehta et al., 1991). Thus, if the in vitro models are predictive of in vivo behavior, an increased connexin expression in epithelium as a consequence of ingestion of, or supplementation with dietary carotenoids, or therapeutic treatment of retinoids could decrease the proliferation of carcinogen-initiated cells, thereby inhibiting their neoplastic conversion. Support for this hypothesis of an inverse relationship between connexin 43 expression and proliferation comes from data conducted in head and neck cancer patients in which we demonstrated an association between loss of expression of connexin 43 in oral mucosa and abnormal proliferation in dysplastic and malignant tissue (Sakr et al., 1996).

ACKNOWLEDGMENTS

This work was supported by Grant 95-34135-1775 from the USDA to JSB. FK thanks H. Reisman Corp. (Orange, New Jersey) and LycoRed Natural Products Industries (Beer Sheva, Israel) for partial support of the lycopene supplementation studies involving humans.

REFERENCES

Acevedo, P., and Bertram, J. S. (1995). Liarozole potentiates the cancer chemopreventive activity of and the up-regulation of gap junctional communication and connexin43 expression by retinoic acid and -carotene in 10T1/2 cells. Carcinogenesis (London) **16**, 2215–2222.

Asselineau, D., and Darmon, M. (1995). Retinoic acid provokes metaplasia of epithelium formed in vitro by adult human epidermal keratinocytes. *Differentiation (Berlin)* **58**, 297–306.

Bertram, J. S., and Bortkiewicz, H. (1995). Dietary carotenoids inhibit neoplastic transformation and modulate gene expression in mouse and human cells. *Am. J. Clin. Nutr.* **62**, 1327S–1336S.

Bertram, J. S., Kolonel, L. N., and Meyskens, F. L. (1987). Rationale and strategies for chemoprevention of cancer in humans. *Cancer Res.* **47**, 3012–3031.

Bertram, J. S., Pung, A., Churley, M., Kappock, T. J. I., Wilkins, L. R., and Cooney, R. V. (1991). Diverse carotenoids protect against chemically induced neoplastic transformation. *Carcinogenesis (London)* **12**, 671–678.

Brown, L. J., Geesin, J. C., Rothnagel, J. A., Roop, D. R., and Gordon, J. S. (1994). Retinoic acid suppression of loricrin expression in reconstituted human skin cultured at the liquid-air interface. *J. Invest. Dermatol.* **102**, 886–890.

Cooney, R. V., Kappock, T. J., Pung, A., and Bertram, J. S. (1993). Solubilization, cellular uptake, and activity of β-carotene and other carotenoids as inhibitors of neoplastic transformation in cultured cells. In "Methods in Enzymology" (L. Packer, ed.), Vol. 214, pp. 55–68. Academic Press, San Diego, CA.

DiSepio, D., Jones, A., Longley, M. A., Bundman, D., Rothnagel, J. A., and Roop, D. R. (1995). The proximal promoter of the mouse loricrin gene contains a functional AP-1 element and directs keratinocyte-specific but not differentiation-specific expression. *J. Biol. Chem.* **18**, 10792–10799.

Fisher, G. J., and Voorhees, J. J. (1996). Molecular mechanisms of retinoid actions in skin. *FASEB J.* **10**, 1002–1013.

Fitzgerald, D. J., Fusenig, N. E., Boukamp, P., Piccoli, C., Mesnil, M., and Yamasaki, H. (1994). Expression and function of connexin in normal and transformed human keratinocytes in culture. *Carcinogenesis (London)* **15**, 1859–1866.

Gann, P. H., Ma, J., Giovannucci, E., Willett, W., Sacks, F. M., Hennekens, C. H., and Stampfer, M. J. (1999). Lower prostate cancer risk in men with elevated plasma lycopene levels: results of a prospective analysis. *Cancer Research*, **59**, 1225–1230.

Gärtner, C., Stahl, W., and Sies, H. (1997). Lycopene is more bioavailable from tomato paste than from fresh tomatoes. *Am. J. Clin. Nutr.* **66**(1), 116–122.

Gerster, H. (1997). The potential role of lycopene for human health. *J. Am. Coll. Nutr.* **16**, 109–126.

Guo, H., Acevedo, P., Parsa, D. F., and Bertram, J. S. (1992). The gap-junctional protein connexin 43 is expressed in dermis and epidermis of human skin: differential modulation by retinoids. *J. Invest. Dermatol.* **99**, 460–467.

Hanusch, M., Stahl, W., Schulz, W. A., and Sies, H. (1995). Inductin of gap junctional communication by 4-oxoretinoic acid generated from its precursor canthaxanthin. *Arch. Biochem. Biophys.* **317**, 423–428.

Heinonen, O. P., and Albanes, D. (1994). The effect of vitamin E and beta carotene on the incidence of lung cancer and other cancers in male smokers. *N. Engl. J. Med.* **330**, 1029–1035.

Hennekens, C. H., Buring, J. E., Manson, J. E., Stampfer, M., Rosner, B., Cook, N. R., Bélanger, C., LaMotte, F., Gaziano, J. M., Ridker, P. M., Willett, W., and Peto, R. (1996). Lack of effect of long-term supplementation with beta carotene on the incidence of malignant neoplasms and cardiovascular disease. *N. Engl. J. Med.* **334**, 1145–1149.

Heywood, R., Palmer, A. K., Gregson, R. L., and Hummler, H. (1985). The toxicity of beta-carotene. *Toxicology* **36**, 91–100.

Hossain, M. Z., and Bertram, J. S. (1994). Retinoids suppress proliferation, induce cell spreading, and up-regulate connexin43 expression only in postconfluent 10T1/2 cells: Implications for the role of gap junctional communication. *Cell Growth Differ.* **5,** 1253–1261.

Hossain, M. Z., Wilkens, L. R., Mehta, P. P., Loewenstein, W. R., and Bertram, J. S. (1989). Enhancement of gap junctional communication by retinoids correlates with their ability to inhibit neoplastic transformation. *Carcinogenesis (London)* **10,** 1743–1748.

Khachik, F., Pfander, H., Traber, B. (1998a). Proposed mechanisms for the formation of the synthetic and naturally occurring metabolites of lycopene in tomato products and human serum *J. Agric. Food Chem.,* **46,** 4885–4890.

Khachik, F., Spangler, C. J., Smith, J. C., Jr., Canfield, L. M., Pfander, H., and Steck, A. (1997a). Identification, quantification, and relative concentrations of carotenoids, their metabolites in human milk and serum. *Anal. Chem.* **69,** 1873–1881.

Khachik, F., Steck, A., and Pfander, H., (1997b). Bioavailability, metabolism, and possible mechanism of chemoprevention by lutein and lycopene in humans. *In* "Food Factors for Cancer Prevention" (H. Ohigashi, T. Osawa, J. Tereo, S. Watanabe, and T. Yoshikawa, eds.), pp. 542–547. Springer-Verlag, Tokyo.

Khachik, F., Steck, A., Niggli, U. A., Pfander, H. (1998b). Partial synthesis and structural elucidation of the oxidative metabolites of lycopene identified in tomato paste, tomato juice and human serum *J. Agric. Food Chem.,* **46,** 4874–4884.

Kopan, R., Traska, G., and Fuchs, E. (1987). Retinoids as important regulators of terminal differentiation: Examining keratin expression in individual epidermal cells at various stages of keratinization. *J. Cell Biol.* **105,** 427–440.

Mayne, S. T. (1996). Beta-carotene, carotenoids, and disease prevention in humans. *FASEB J.* **10,** 690–701.

Mehta, P. P., Bertram, J. S., and Loewenstein, W. R. (1986). Growth inhibition of transformed cells correlates with their junctional communication with normal cells. *Cell (Cambridge, Mass.)* **44,** 187–196.

Mehta, P. P., Bertram, J. S., and Loewenstein, W. R. (1989). The actions of retinoids on cellular growth correlate with their actions on gap junctional communication. *J. Cell Biol.* **108,** 1053–1065.

Mehta, P. P., Hotz Wagenblatt, A., Rose, B., Shalloway, D., and Loewenstein, W. R. (1991). Incorporation of the gene for a cell-cell channel protein into transformed cells leads to normalization of growth. *J. Membr. Biol.* **124,** 207–225.

Merriman, R., and Bertram, J. S. (1979). Reversible inhibition by retinoids of 3-methylcholanthrene-induced neoplastic transformation in C3H10T1/2 cells. *Cancer Res.* **39,** 1661–1666.

Micozzi, M. S., Beecher, G. R., Taylor, P. R., and Khachik, F. (1990). Carotenoid analyses of selected raw and cooked foods associated with a lower risk for cancer. *J. Natl. Cancer Inst.* **82,** 282–285.

Moon, R. C., Mehta, R. G., and Rao, K. V. N. (1994). Retinoids and cancer in experimental animals. *In* The Retinoids (M. B. Sporn, A. B. Roberts, and D. S. Goodman, eds.), pp. 573–596. Raven Press, New York.

Neveu, M., and Bertram, J. S. (1999). Gap junctions and neoplasia. *In* "Gap Junctions: Advances in Cellular and Molecular Biology" (E. L. Hertzberg, ed.). JAI Press, Greenwich, CT (in press).

Nikawa, T., Schulz, W. A., Van den Brink, C. E., Hanusch, M., Van der Saag, P., Stahl, W., and Sies, H. (1995). Efficacy of all-*trans*-β-carotene, canthaxanthin, and all-*trans*-, 9-*cis*-, and 4-oxoretinoic acids in inducing differentiation of an F9 embryonal carcinoma RAR-β-*lacZ* reporter cell line. *Arch. Biochem. Biophys.* **316,** 665–672.

Omenn, G. S., Goodman, G., Thornquist, M., Grizzle, J., Rosenstock, L., Barnhart, S., Balmes, J., Cherniack, M. G., Cullen, M. R., Glass, A., Keogh, J., Meyskens, F., Jr., Valanis, B., and Williams, J., Jr. (1994). The β-Carotene and Retinol Efficacy Trial (CARET) for chemoprevention of lung cancer in high risk populations: Smokers and asbestos-exposed workers. *Cancer Res.* **54**(Suppl.), 2038s–2043s.

Paetau, I., Khachik, F., Brown, E. D., Beecher, G. R., Kramer, T. R., Chittams, J., and Clevidence, B. A. (1998). Chronic ingestion of lycopene-rich tomato juice or lycopene supplements significantly increases plasma concentrations of lycopene and related tomato carotenoids in humans. *Am. J. Clin. Nutr.* **68**, 1187–1195.

Pfahl, M. (1993). Signal transduction by retinoid receptors. *Skin Pharmacol.* **6** (Suppl. 1), 8–16.

Reznikoff, C. A., Bertram, J. S., Brankow, D. W., and Heidelberger, C. (1973). Quantitative and qualitative studies of chemical transformation of cloned C3H mouse embryo cells sensitive to postconfluence inhibition of cell division. *Cancer Res.* **33**, 2339–2349.

Rogers, M., Berestecky, J. M., Hossain, M. Z., Guo, H. M., Kadle, R., Nicholson, B. J., and Bertram, J. S. (1990). Retinoid-enhanced gap junctional communication is achieved by increased levels of connexin 43 mRNA and protein. *Mol. Carcinog.* **3**, 335–343.

Ryle, C. M., Breitkreutz, D., Stark, H. J., Leigh, I. M., Steinert, P. M., Roop, D., and Fusenig, N. E. (1989). Density-dependent modulation of synthesis of keratins 1 and 10 in the human keratinocyte line HACAT and in ras-transfected tumorigenic clones. *Differentiation (Berlin)* **40**, 42–54.

Sakr, W., Tabaska, P., Kucuk, O., and Bertram, J. S. (1996). Differential expression of connexin 43 in normal, preneoplastic and neoplastic squamous epithelium of the upper aerodigestive tract. *Proc. AACR* **37**, 269(abstr.).

Salomon, D., Masgrau, E., Vischer, S., Ullrich, S., Dupont, E., Sappino, P., Saurat, J. -H., and Meda, P. (1994). Topography of mammalian connexins in human skin. *J. Invest. Dermatol.* **103**, 240–247.

Stahl, W., and Sies, H. (1992). Uptake of lycopene and its geometrical isomers is greater from heat-processed than from unprocessed tomato juice in humans. *J. Nutr.* **122**, 2161–2166.

Stahl, W., and Sies, H. (1996). Lycopene: A biologically important carotenoid for humans? *Arch. Biochem. Biophys.* **336**, 1–9.

Tomic, M., Jiang, C. K., Epstein, H. S., Freedberg, I. M., Samuels, H. H., and Blumenberg, M. (1990). Nuclear receptors for retinoic acid and thyroid hormone regulate transcription of keratin genes. *Cell Regul.* **1**, 965–973.

Zhang, L. -X., Cooney, R. V., and Bertram, J. S. (1991). Carotenoids enhance gap junctional communication and inhibit lipid peroxidation in C3H/10T1/2 cells: Relationship to their cancer chemopreventive action. *Carcinogenesis (London)* **12**, 2109–2114.

Zhang, L., Cooney, R. V., and Bertram, J. S. (1992). Carotenoids up-regulate connexin43 gene expression independent of their pro-vitamin A or antioxidant properties. *Cancer Res.* **52**, 5707–5712.

Zhang, L. -X., Acevedo, P., Guo, H., and Bertram, J. S. (1995). Upregulation of gap junctional communication and connexin43 gene expression by carotenoids in human dermal fibroblasts but not in human keratinocytes. *Mol. Carcinog.* **12**, 50–58.

19 Antioxidant Genes and Reactive Oxygen Species in Down's Syndrome

Cécile Bladier, Judy B. de Haan, and Ismail Kola
Centre for Functional Genomics and Human Disease
Institute of Reproduction and Development
Monash Medical Centre
Monash University
VIC 3168, Australia

Down's syndrome (DS) results from the overexpression of genes found on chromosome 21 or part thereof. Research has focused on understanding how additional copies of these genes contribute to the phenotype. In particular, much attention has been given to the Cu/Zn superoxide dismutase (*SOD1*) gene since this gene encodes an antioxidant enzyme that plays a key role in reactive oxygen species (ROS) homeostasis. This chapter presents the perturbations in the ROS metabolism, resulting from elevated levels of SOD1, and some of the disorders and pathologies of DS thought to be associated with these perturbations. It focuses on the premature aging process observed in DS and the associated Alzheimer's type dementia as there is strong evidence for an involvement of altered ROS in these diseases. It also proposes a possible role for SOD1 in the various neuropathological aspects of DS, in some of the disorders of the immune system, and in the stimulation of inflammatory processes. Finally, other genes on chromosome 21 that may be regulated or have a redox function in the cell are investigated in the light of their possible involvement in some of the pathologies associated with DS.

I. INTRODUCTION

Down's syndrome is a genetic disorder resulting from either total or partial trisomy of chromosome 21 (Lejeune *et al.*, 1959). It is well recognized in the community due to its prevalence, occurring at a frequency of 1 in 700–1000 live births, and due to the characteristic morphological features such as shortened stature, flattened nose, and epicanthic eye folds (Patterson, 1987; Kola, 1997). Patients with DS suffer from a wide range of clinical pathologies, including mental retardation, microcephaly, bone and skeletal abnormalities, congenital malformations of most organs (especially the heart), high susceptibility to infections, cataracts, and increased incidence of leukemia and diabetes (Epstein, 1986; Pueschel, 1990). Individuals with DS also display features of premature aging and those who survive past their third decade usually develop Alzheimer's disease (AD) (Kesslak et al., 1994; Korenberg, 1995).

The main challenge for researchers working on DS is to clarify the mechanism(s) by which the additional copy of chromosome 21, or part thereof, contributes to the different pathologies (Kola, 1997). Several advances have been achieved in the past years, which has led to the current view that DS is a gene dosage disorder. It is believed that the expression of proteins that are encoded by the genes of the extra copy of chromosome 21 result in an imbalance in various biochemical pathways, which in turn leads to the development of the pathologies seen in DS (Anneren and Edman, 1993; Groner, 1995; Sumarsono et al., 1996). The number of functional genes that map to the entire chromosome 21 has been estimated to range between 250 and 1000. The majority of these genes are concentrated on the distal part of the long arm of chromosome 21 (region 21q22.2–21q22.3). Currently, approximately 70 known genes have been mapped to this region. The elucidation of the biological function of these genes is under investigation, and various studies have already given important insight into their contribution to the Down's syndrome phenotype (Kola and Hertzog, 1997).

The *SOD1* gene product is thought to play a prominent role in several of the clinical features of DS. *SOD1* is localized at 21q22.1 (Sinet et al., 1976) and codes for the antioxidant enzyme Cu/Zn-superoxide dismutase-1 (SOD1), which plays a key role in the metabolism of ROS (Fridovich, 1986). Reactive oxygen species are molecules formed spontaneously during the course of normal cellular processes involving oxygen. These species have the potential to cause, at high concentrations, oxidative damage to molecules and organelles, either directly or by serving as a substrate for the formation of other ROS (Kehrer, 1993). However, they are also, when present at relatively low concentrations, important signaling molecules and regulators of the expression of various genes (Khan and Wilson, 1995; Winyard and Blake, 1997). In order to protect against ROS and to maintain their cellular concentration at a steady-state level, aerobic organisms have developed enzymatic and nonenzymatic antioxidant systems. Superoxide dismutases are part of the enzymatic antioxidant defense: they catalyze the dismutation of superoxide anions to hydrogen peroxide. Hydrogen peroxide is then neutralized to water and oxygen by either glutathione peroxidases (GPXs) and/or catalase (Yu, 1994). This concerted action of SODs and GPXs and/or catalase prevents the accumulation of hydrogen peroxide and the subsequent formation of hydroxyl radicals (·OH) via the Fenton's reaction (Baker and Gebicki, 1984) and via H_2O_2 inactivation of SOD1 (Yim *et al.*, 1990). Hydroxyl radicals are the most reactive and noxious radical species. They react with practically any macromolecule and initiate chains of peroxidation, especially lipid peroxidation, resulting in damage to membranes and organelles (Kappus, 1985).

In Down's syndrome, the balance between the first and the second step of the antioxidant enzymatic pathway is perturbed: SOD1 is elevated in a variety of cell types and organs as a result of increased gene dosage (Feaster *et al.*, 1977; Brooksbank and Balazs, 1984; Anneren and Epstein, 1987; de Haan *et al.*, 1995, 1997). Consequently, it has been hypothesized that an increased generation of H_2O_2 occurs that causes a disequilibrium in the steady-state level of ROS. In addition, side reactions catalyzed by SOD1 also increase, e.g., the formation of free radicals in the presence of H_2O_2 and anionic scavengers (Yim *et al.*, 1993) and the nitration of protein tyrosine residues by peroxynitrite (Beckman *et al.*, 1993). These in turn could lead to damage to macromolecules and organelles, inappropriate signaling and gene expression, and therefore metabolic impairment and loss of cellular function (Fig. 1).

This hypothesis has been investigated extensively and has led to some important advances in the understanding of several of the perturbations and clinical features of DS. The aim of this chapter is to present some of the disorders and pathologies of DS in which the overexpression of *SOD1* and the subsequent perturbation of ROS metabolism are thought to play a significant role. In addition, this chapter addresses the possible involvement of other genes on chromosome 21 that may function in an antioxidant capacity or may be regulated by an altered redox state in the phenotype of the syndrome.

II. PREMATURE AGING

The phenomenon of premature aging observed in DS is probably the pathophysiology where a role for *SOD1* has been the most extensively investigated. The normal aging process is associated with an increased generation of ROS (Harman, 1994), an altered antioxidant defense, and increased cellular oxidative stress and damage (Matsuo, 1993). Therefore, it has been proposed that the overexpression of *SOD1*, resulting in a perturbation in

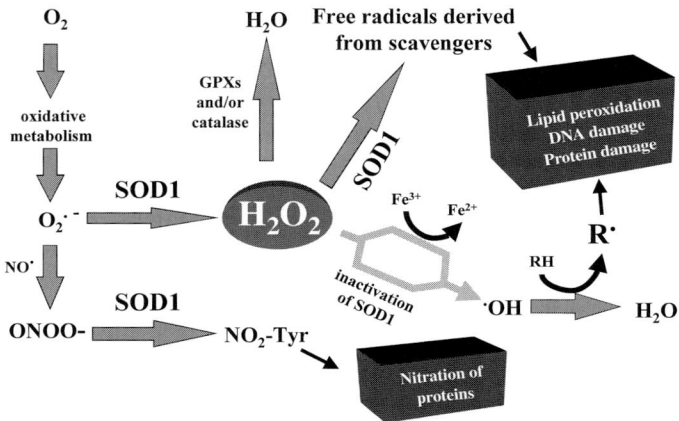

Figure 1 Schematic representation of the antioxidant pathway and some consequences of an overexpression of SOD1. The antioxidant pathway is a two-step process beginning with dismutation of $O_2^{·-}$ via SOD1 to H_2O_2 and the subsequent conversion of H_2O_2 by either GPXs or catalase into H_2O. An imbalance in the SOD-to-GPXs/catalase ratio results in increased H_2O_2 production, which in turn leads to increased ·OH formation either via inactivation of SOD1 itself or through interaction with transition metals in the Fenton reaction. ·OH radicals rapidly interact and damage DNA, lipids, and/or proteins. Elevated H_2O_2, together with anionic scavengers and SOD1, results in more radicals and subsequent damage to molecules. Elevated SOD1 also leads to the increased nitration of proteins though interactions with $OONO^{·-}$.

ROS metabolism, could contribute to the premature aging process observed in patients with DS.

Various approaches and strategies have been undertaken to test this hypothesis. First, studies investigating the levels of antioxidant enzyme activity in DS cells and organs found that SOD1 activity was elevated by 1.5-fold in erythrocytes and platelets (Sinet *et al.*, 1975), fibroblasts (Anneren and Epstein, 1987), lymphocytes (Feaster *et al.*, 1977), and fetal brain (Brooksbank and Balazs, 1984) in agreement with a gene dosage effect. Catalase activity, however, was found to be normal in cells of patients with DS as compared to control cells, whereas GPX activity was either slightly increased or unchanged. Importantly, in organs and cells where GPX activity was unchanged (fetal brain) or slightly increased (fibroblasts), an enhancement in lipid peroxidation was observed (Brooksbank and Balazs, 1984; Anneren and Epstein, 1987). These results indicate that an alteration in the ratio of SOD1 to GPX and catalase activity exists in some DS cells and organs and that this is correlated with oxidative cellular damage. This notion of oxidative damage, caused by an altered SOD1 to GPX and catalase ratio, was investigated further in normal diploid tissues as a function of age by de Haan et al. (1992) and Cristiano et al. (1995) on the basis that if this altered ratio was etiological (or contributory) to premature aging, then it has to fulfil the important postulate of being involved in the "normal" aging process per se. These authors showed that SOD1 activity increases with age in all organs investigated, with a concomitant increase in catalase and/or GPX activities in most organs. Thus the SOD1/(GPX and catalase) ratio is unchanged. In the brain, however, such an adaptive rise of GPX and catalase activities does not occur, resulting in an altered SOD1 to GPX and catalase ratio; interestingly, the brain is the only organ that shows an increase in lipid peroxidation with age. These data therefore strongly suggest a correlation between an altered SOD1 to GPX and catalase ratio and oxidative damage, which in turn may contribute to aging changes.

A further line of evidence that an altered antioxidant balance may contribute to oxidative damage and aging changes comes from data on cultured cell lines, where an elevation in SOD1 activity was achieved through transfection with the *SOD1* gene. Stably transfected cells overexpressing *SOD1* were shown to exhibit increased lipid peroxidation as compared with controls cells (Elroy-Stein et al., 1986), as well as increased sensitivity to DNA damage following H_2O_2 exposure (Amstad et al., 1991; Teixeira and Meneghini, 1996). In addition, cells that had an altered ratio of SOD1 to GPX and catalase activity (as a consequence of *SOD1* transfection) were shown to have higher intracellular levels of H_2O_2 and to display a senescent phenotype, as manifested by a slower growth rate, an altered morphology, and an increased expression of the senescence marker Cip1 (Fig. 2) (de Haan et al., 1996). This senescent phenotype was also observed in cells derived from children with DS, as well as in cells exposed directly to H_2O_2 (Fig. 3) (de Haan et al., 1996; Bladier et al., 1997). However, cells that had "adapted" to elevated SOD1 levels by upregulating GPX (such that the SOD1/GPX ratios were unchanged) were biochemically, morphologically, and genetically indistinguishable from the parental controls (Fig. 2). These findings provide strong evidence that the overexpression of *SOD1*, resulting in an altered antioxidant balance, can cause at least *in vitro*, oxidative stress injury leading to earlier senescence.

A third line of evidence that an altered antioxidant balance may contribute to oxidative damage and aging comes from the establishment of *SOD1* transgenic mice. Avraham et al. (1988, 1991) and Yarom et al. (1988) reported that *SOD1* transgenic mice undergo premature aging with respect to neuromuscular junction morphology (NMJ). The NMJ of the tongue and leg muscles from these mice exhibits pathological changes (e.g., atrophy, degeneration, and withdrawal of the terminal axons of the tongue muscle; mitochondrial accumulation and membranous degenerated structures of the end plates of the leg muscle) that are similar to those observed in muscles

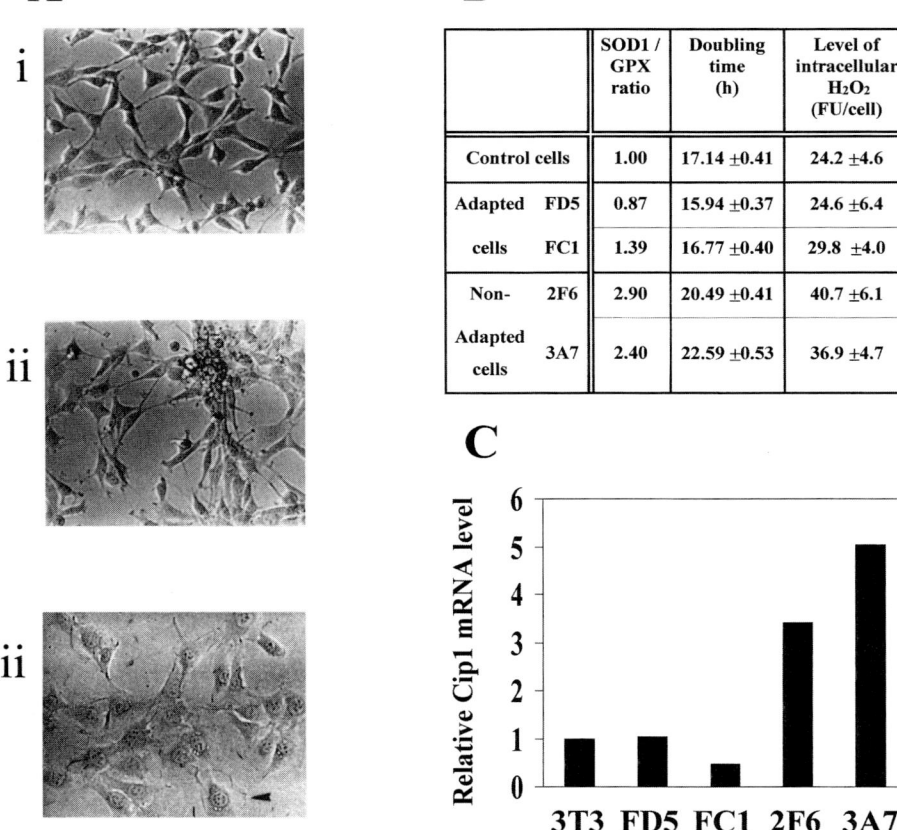

Figure 2 An altered SOD1-to-GPX (and catalase) ratio leads to aging changes in cultured cells. Two cell populations were obtained after transfection with human *SOD1* cDNA in NIH 3T3 cells. (A) Cellular morphology of (i) control cells, (ii) "adapted cells," and (iii) "nonadapted" cells. Adapted cells (cells that upregulate GPX in response to elevated SOD1 levels) are indistinguishable from control cells, whereas "nonadapted cells" (cells that did not upregulate GPX in response to elevated SOD1 levels) show features of senescence in culture, e. g., increased adherent surface area, and larger cell and nucleus volume (magnification ×250). (B) The SOD1/GPX ratio, doubling times, and levels of intracellular H_2O_2 in control and *SOD1*-overexpressing cells. A greatly elevated SOD1/GPX ratio in nonadapted cells was accompanied by increased doubling times, i.e., slower cell growth and elevated levels of intracellular H_2O_2 in two "nonadapted" cell clones, 2F6 and 3A7. Conversely, no change in the SOD1/GPX ratio (for cell clone FD5) and an only slightly elevated ratio (for cell clone FC1) resulted in no significant difference in doubling time or levels of H_2O_2 for the two "adapted" cell clones. (C) A diagrammatic representation of *Cip1* (a known marker of senscence) mRNA levels in controls, "nonadapted," and "adapted" cells. Nonadapted cells (2F6 and 3A7) show higher constitutive levels of *Cip1* mRNA than adapted or control cells. h, hour; FU/cell, fluorescent units per cell.

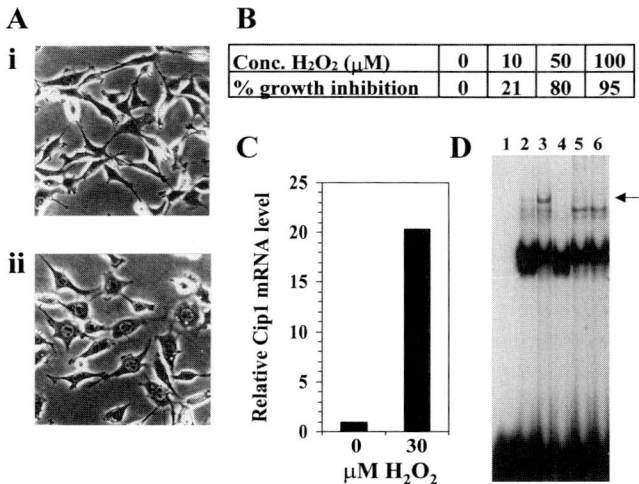

Figure 3 Effect of H_2O_2 treatment on cultured cells. (A) Cellular morphology of (i) NIH 3T3 cells and (ii) NIH 3T3 cells after exposure to 25 μM H_2O_2. H_2O_2 treatment induces features of senescence in culture, e.g., increased adherent surface area, and larger cell and nucleus volume (magnification ×250). (B) The percentage growth inhibition of NIH 3T3 cells after exposure to increasing concentrations of H_2O_2. (C) A diagrammatic representation of *Cip1* mRNA levels of NIH 3T3 cells and NIH 3T3 cells after a 24-hr exposure to 30 μM H_2O_2. (D) A electrophoretic mobility shift assay of lysate from NIH 3T3 cells (with and without exposure to 150 μM H_2O_2) incubated with a ^{32}P-radiolabeled NF-κB consensus oligonucleotide. Lane 1, migration of unbound ^{32}P-radiolabeled NF-κB oligonucleotide; lane 2, migration of nuclear proteins from untreated NIH 3T3 cells together with ^{32}P-radiolabeled NF-κB oligonucleotide; and lane 3, migration of nuclear proteins from NIH 3T3 cells after exposure to 150 μM H_2O_2 for 5 hr, together with ^{32}P-radiolabeled NF-κB oligonucleotide. Note that H_2O_2 treatment of these cells enhances the binding and/or availability of a complex to the NF-κB oligonucleotide (arrow). Lane 4, migration of nuclear proteins from NIH 3T3 cells after exposure to 150 μM H_2O_2 for 5 hr, together with ^{32}P-radiolabeled NF-κB oligonucleotide in the presence of excess unlabeled NF-κB oligonucleotide; lane 5, migration of nuclear proteins from NIH 3T3 cells after exposure to 150 μM H_2O_2 for 5 hr, together with ^{32}P-radiolabeled NF-κB oligonucleotide and p50 antibody; lane 6, migration of nuclear proteins from NIH 3T3 cells after exposure to 150 μM H_2O_2 for 5 hr, together with ^{32}P-radiolabeled NF-κB oligonucleotide and p65 antibody. Note that the complex induced by H_2O_2 is competed off by the unlabeled NF-κB oligonucleotide (lane 4), p50 antibody (lane 5), and p65 antibody (lane 6), indicating that this complex is NF-κB.

of aging rats and mice (Fahim and Robbins, 1982; Cardasis, 1983), as well as in tongue muscles of individuals with DS (Yarom *et al.*, 1986, 1987). In addition, Barkats *et al.* (1993) have reported neuromorphological abnormalities usually associated with premature aging (e.g., the diminution of the Mossy fiber innervation in the hippocampus) in the brain of *SOD1* transgenic mice. Furthermore, elevated levels of thiobarbituric acid (TBA)-reactive material were observed in transgenic brains as compared with controls, strongly suggesting increased levels of peroxidation and oxidative damage *in vivo* (Ceballos-Picot *et al.*, 1992). Finally, Nabarra *et al.* (1996) demonstrated that *SOD1* transgenic mice display a premature involution of the thymus, an organ that has been described as a clock in the process of aging due to the close relationship with its involution, the dysfunction of the immune system, and the aging process (Kay and Makinodan, 1981).

Taken together, the observations that an altered SOD1/(GPX + catalase) ratio (i) exists in DS cells/organs as well as in the aging brain and correlates with increased oxidative damage; (ii) leads to a senescence phenotype *in vitro*; and (iii) causes signs of premature aging *in vivo* strongly argue in favor of a role for an altered antioxidant balance due to the overexpression of *SOD1* in the acceleration of the aging process in DS.

III. ALZHEIMER'S DISEASE

Alzheimer's disease is a neurodegenerative disorder characterized by a progressive loss of memory, intellectual function, and cognitive abilities (Wisniewski *et al.*, 1994). In the normal population, AD usually arises during the fifth or later decades of life and affects approximately 10% of individuals aged 65 and above (Keefover, 1996). In DS, however, almost 100% of patients develop Alzheimer's-type neuronal pathology by the third to fourth decade of life (Dalton and Wisniewski, 1990). There is now considerable evidence that both amyloid β-peptide (Aβ) and oxidative stress are implicated in the pathogenesis of AD (Friedlich and Butcher, 1994; Mattson, 1997a). It has been shown that Aβ, which is a normal product of cell metabolism derived from the β-amyloid precursor protein (APP), is overproduced in the brain of individuals with AD and forms insoluble fibrillar aggregates, called "senile" or "neuritic" plaques, that promote neuronal degeneration (Mattson, 1997b). This neurotoxic effect of Aβ deposits may involve the generation of ROS, which disrupts cellular calcium homeostasis and renders neurons more susceptible to excitotoxicity and apoptosis (Behl *et al.*, 1994, Behl, 1997). Three genes have been identified that, when mutated, lead to aberrant APP processing and increased levels of Aβ: *APP*, presenilin 1 (*PS1*), and presenilin 2 (*PS2*) (Lendon *et al.*, 1997). Mutations in *APP* cause a shift in APP processing, leading to a decreased production of the secreted neuroprotective form of APP and an increased liberation of Aβ (Mattson, 1997b). Mutations in presenilin genes may lead to increased generation of Aβ by disrupting the calcium homeostasis of the endoplasmic reticulum, which results in mitochondrial impairment, increased oxidative stress, and consequently aberrant APP processing (Guo et al., 1997). In addition, the ε4 allele of the apolipoprotein E gene has been shown to increase the risk of AD, and some data suggest that this could occur via a perturbation in the level of oxidative stress in the brain parenchyma and vasculature and a stimulation of Aβ fibril formation (Mattson, 1997a).

Individuals with DS have three copies of both *APP* and *SOD1* genes (the gene coding for APP is also localized on chromosome 21). Therefore, individuals with DS might be exposed, from the beginning of life, to a concomitant excess of free radicals and APP levels. It is believed that the prooxidant conditions, together with enhanced APP formation, may be responsible for the early and invariable β-amyloid accumulation in these patients. This hypothesis is supported by several lines of evidence. Ceballos *et al.* (1991) have demonstrated that *SOD1* is expressed predominantly in the pyramidal neurons of the brain, which are the neurons that degeneate in AD. The elevated levels of SOD1 in DS might therefore perturb the ROS balance in these neurons, leading to oxidative damage and aberrant APP processing. In addition, Busciglio and Yankner (1995) reported clear evidence that oxidative stress and neuronal damage begin accumulating *in utero* in the DS brain: cortical neurons from aborted DS conceptuses show increased lipid peroxidation and apoptosis in culture as compared with normal cortical neurons. This effect seems to be mediated via excess levels of hydrogen peroxide as it can be prevented with compounds such as N-acetylcysteine and catalase (but not by SOD1). These data indicate that a defect in the metabolism of ROS exists in neurons of DS fetuses that places the developing brain in an environment of increased oxidative stress. This increased oxidative stress might cause aberrant APP processing, leading to early cerebral Aβ deposits. Furthermore, Bar-Peled *et al.* (1996) demonstrated that *SOD1* overexpression in *SOD1* transgenic mice leads to a chronic prooxidant state in the brain, as manifest by increased levels of oxidized glutathione and altered calcium homeostasis. This chronic oxidant state renders neurons more susceptible to apoptotic death when subjected to kainic acid and may also render neurons more susceptible to β-amyloid toxicity. Moreover, we also demonstrated, using *GPX1* knockout mice, that an imbalance in the SOD1 to GPX1 ratio in the brain leads to increased susceptibility of neurons to H_2O_2-mediated toxicity (de Haan/Bladier *et al.*, 1998). Interestingly, it has been shown that the promoter of the *APP* gene contains a heat shock element (HSE) (Salbaum *et al.*, 1988; Dewjii and Do, 1996) and that oxidative stress can activate gene transcription via the HSE (Liu and Thiele, 1996). This would suggest the following scenario: oxidative stress, due to the elevated SOD1/GPX ratio, leads to higher levels of APP via the transcriptional induction of the *APP* gene. This is turn causes an increased formation of Aβ (due to aberrant APP processing) and fibrillar amyloid deposits that result in further oxidative stress, disruption of calcium homeostasis, mitochondrial dysfunction, and consequently increased apoptosis and neurodegeneration (Fig. 4).

Finally, further evidence that oxidative stress, together with excess APP, may be involved in the early onset of AD in patients with DS comes from the fact that elevated levels of APP are insufficient to produce amyloid deposition by the age at which it occurs in DS. Stably transfected cells overexpressing *APP* exhibit increased levels of Aβ but do not produce any

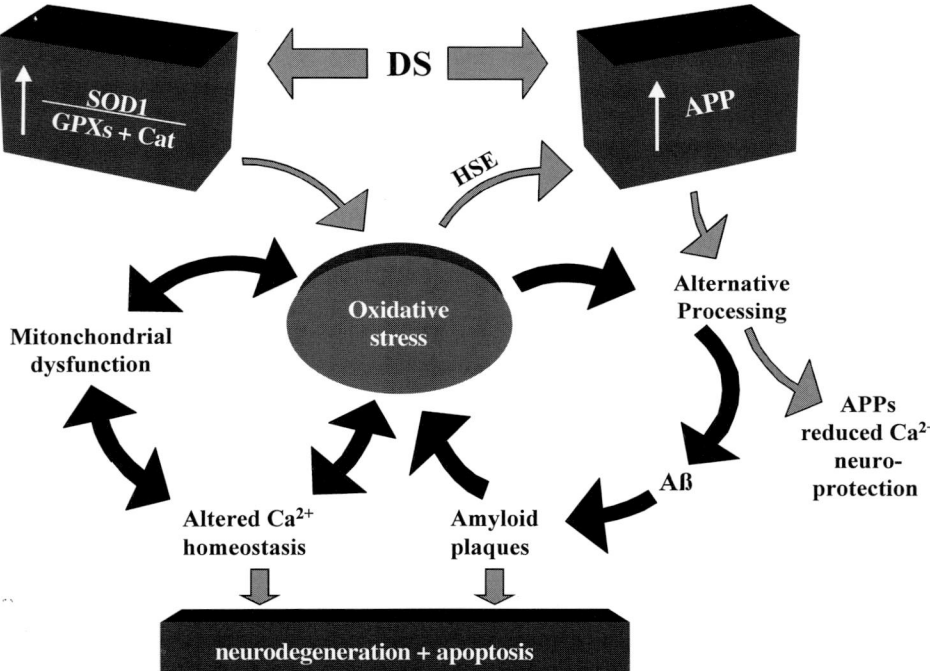

Figure 4 A possible role for SOD1 and APP in the early onset of Alzheimer's disease in patients with DS. Elevated levels of SOD1/(GPXs+catalase) ratio in the brain of patients with DS causes oxidative stress, which, in conjunction with elevated APP levels, leads to an alternative APP processing; this results in decreased production of the neuroprotective form of APP and increased liberation of toxic Aβ. In addition, oxidative stress further increases APP levels through the induction of *APP* transcription via a heat shock element (HSE) in the promoter of the *APP* gene. This leads to the elevated formation of Aβ and amyloid plaques, which, in turn, cause further oxidative stress, mitochondrial dysfunction, disruption of calcium homeostasis, and increased apoptosis and neurodegeneration. Increased amyloid deposition, together with apoptosis and neurodegeneration in the brain, may contribute to the early onset of AD-type pathology in DS.

extracellular Aβ deposits (Maruyama et al., 1990), and mice overexpressing *APP* do not completely develop the features of AD as they do not display extracellular deposits of fibrillar Aβ and/or neuronal degeneration (Mattson, 1997b). Nevertheless, data support the notion that the trisomy of *APP* may be necessary for the development of AD dementia in patients with DS. Indeed a case has been reported of a 78-year-old woman with DS but no Alzheimer's disease—trisomy 21 was partial and the gene for APP was present in only two copies (Prasher et al., 1998). Therefore these data indicate that other factors, such as oxidative stress (in conjunction with high APP levels), are needed to exacerbate β-amyloidosis and lead to the early onset of AD seen in DS. Further experiments using transgenic mouse models are needed to confirm these hypotheses. The generation of mice overexpressing both *APP* and *SOD1* genes, for example, should provide a good model system in which to test the notion that a synergistic interaction of *APP* and *SOD1* predispose patients with DS to the pathogenesis of AD.

IV. OTHER NEUROPATHOLOGIES

In addition to AD, other neuropathologies and neurological disorders of DS are also believed to result partly from ROS-induced cellular damage and enhanced apoptosis caused by elevated levels of SOD1 and ROS. Elroy-Stein and Groner (1988) observed that rat PC12 cells overexpressing the human *SOD1* gene have altered neurotransmitter (serotonin and catecholamine) uptake resulting from an impaired chromaffin granule transport mechanism. This impairment appears to be caused by lipid peroxidation of the membrane due to oxidative stress. These data are similar to some of the neurophysiological abnormalities seen in patients with DS, e.g., the reduction in neurotransmitter content in the nerve terminals of DS brains (Yates et al., 1983) and the defective membrane K^+ permeability of cultured neurons from individuals with DS (Scott et al., 1983). In addition, Schickler et al. (1989) found that serotonin levels are reduced in blood platelets of *SOD1* transgenic mice, which is similar to the reduction observed in patients with DS (Coyle et al., 1986). This effect is due to an alteration of the platelet uptake process in the dense granules. Because neurotransmitter uptake is an important determinant of various processes of the nervous system, these authors suggest a possible association of their data with the mental retardation and hypotonia observed in individuals with DS. This hypothesis is supported by the observation that an elevation of blood serotonin, following the administration of 5-hydroxytryptophan (the precursor of serotonin), improves muscular tone, motor activity, and sleep disorders in children with DS (Bazelon et al., 1968). Further support for the involvement of elevated

SOD1 activity in some of the neurological abnormalities of patients with DS comes from data of Gahtan et al. (1998) who demonstrate that *SOD1* transgenic mice have an increase in the tetanic stimulation-evoked formation of H_2O_2, resulting in cognitive deficits and impaired hippocampal long-term potentiation. However, it remains unclear which target of the brain is specifically altered by the excess of H_2O_2. Interestingly, these mice also show impairment of muscle function, as manifest by altered electromyography and reduced ability in the rope-grip test (Peled-Kamar et al., 1997), thus suggesting that overexpression of *SOD1* might contribute to the characteristic hypotonia of patients with DS and may also predispose individuals with DS to familial amyotrophic lateral sclerosis (FALS). However, FALS-like anomalies have not been reported in patients with DS as yet.

In addition to *SOD1*, other genes on chromosome 21 have been shown to contribute to learning defects. The recently cloned human homolog of *Drosophila minibrain* (*MNB*) leads to learning and memory impairment when overexpressed in mice (Smith et al., 1997). The newly developed partial trisomy 16 mouse TS108 Cje (trisomic for the region between *SOD1* and *MX2* genes; *SOD1* is inactivated and is present in only two copies) also displays learning and behavioral defects (Sago et al., 1998). Therefore, in addition to *SOD1*, other genes on chromosome 21 may contribute to learning impairment. The challenge now is to understand the specific defect(s) caused by these genes when present in three copies and to elucidate how elevated levels give rise to the mental retardation seen in DS.

V. IMMUNOLOGICAL DEFICIENCY

Individuals with DS present with various abnormalities of their immune system. For example, they display an early thymic involution and various alterations in the distribution and function of T and B lymphocytes and of thymocyte subpopulations, as well as bone marrow abnormalities (Burgio et al., 1983; Philip et al., 1986; Larocca et al., 1988). These defects are thought to be responsible, partly at least, for the high susceptibility of patients with DS to bacterial and viral infections (Epstein, 1986, 1989; Levin, 1987). Interestingly, a possible involvement of increased SOD1 activity in some of these defects has been proposed. Peled-Kamar et al. (1995) reported that thymuses of *SOD1* transgenic mice display histological disorganization (e.g., diminution in the thickness of the cortex, enlargement of the medulla, an increase in Hassal's corpuscles and amount of keratin) that are reminiscent of those seen in patients with DS and in trisomy 16 mice. Nabarra et al. (1996) also described structural modifications of the thymus in *SOD1* transgenic mice: premature involution characterized by a retracted cellular network leaving large spaces filled with lymphocytes and cellular fragments. These features strongly resemble those seen in thymuses of patients with DS (Fonseca et al., 1989). In addition, Peled-Kamar et al. (1995) showed that elevated levels of SOD1 in the thymuses of transgenic mice lead to the increased susceptibility of thymocytes to the bacterial endotoxin lipopolysaccharide (LPS), as manifested by premature and enhanced apoptosis. Importantly, the susceptibility to LPS was associated with elevated levels of H_2O_2 and lipid peroxidation, suggesting that LPS-induced apoptosis might result from oxidative damage. Taken together, these data support the notion that elevated levels of SOD1, which result in an increase in ROS, oxidative damage, and apoptosis, may be involved in the immunological disorders of individuals with DS and consequently in their increased susceptibility to infections.

Furthermore, data of Peled-Kamar et al. (1995) also indicate that the additional copy of the *SOD1* gene might play a role in the hematopoietic disorders seen in patients with DS. Indeed, cultured bone marrow myeloid progenitor cells of *SOD1* transgenic mice produce less granulocyte and macrophage colonies than those of control mice at a suboptimal concentration of interleukin 3, and it has been shown that individuals with DS display a decreased number of hematopoietic colony-forming cells in their peripheral blood. Moreover, this diminished ability to form colonies is further reduced when *SOD1* transgenic bone marrow cells are exposed to tumour necrosis factor-α (TNF-α). Because TNF-α is a cytokine known to generate ROS, this suggests that oxidative stress might be responsible for the increased sensitivity of *SOD1* transgenic hematopoietic progenitors to TNF-α.

VI. INFLAMMATION

There is now considerable evidence that the activation of inflammatory mechanisms in the central nervous system contributes to the pathogenesis of Alzheimer's disease (for a review, see Aisen and Davis, 1994; McGeer and McGeer, 1995). Indeed, numerous markers of inflammation have been reported in the brains of AD patients: acute-phase proteins such as α_1-antichymotrypsin and C-reactive protein are elevated in the cerebrospinal fluid of AD patients, and α_1-antichymotrypsin is a component of amyloid plaques; cytokines [particularly interleukin-1 (IL-1), interleukin-6 (IL-6), and TNF-α], which are the primary mediators of the acute-phase response, are elevated; activated microglia that generate ROS and that produce cytokines are present around AD lesions; and components of the complement pathway are found in the area of dystrophic neurites and neurofibrillar tangles. In addition, clinical trials indicate that anti-inflammatory drugs reduce the risk and slow the rate of progression of AD. Rogers *et al.* (1993), for example, found that patients with AD who received indomethacin over a 6-month period had significant cognitive improvement as compared to placebo-treated patients; and in a 15-year study, Stewart *et al.* (1997) demonstrated that ibuprofen administration for as little as 2 years lowered the risk of developing AD.

Reactive oxygen species can cause inflammation (for a review, see Winyard and Blake, 1997). They act either directly via the oxidative modification of molecules such as α_1-antichymotrypsin, collagen, and low-density proteins or indirectly via the activation of transcription factors such as NF-κB or AP-1, which regulate the transcription of numerous genes involved in cellular inflammatory responses. NF-κB, for example, has been shown to modulate the expression of genes coding for TNFα, IL-1, IL-6, MHC class I antigens, and serum amyloid A precursor, and the AP-1 DNA-binding site has been found in a wide range of genes such as transforming growth factors α and β, interleukin-2, and human collagenase.

In light of these data, it is tempting to propose that the altered ROS metabolism in DS, due to the elevated SOD1/(GPXs+catalase) ratio, may exacerbate the inflammatory processes in the brain of patients with DS, thus contributing to the early onset of AD. Indeed, oxidative stress leads, as mentioned earlier to the increased liberation of amyloid β-peptide, and Aβ has been shown to promote inflammation. For example, Aβ activates microglia directly (Meda *et al.*, 1995), which make a variety of inflammatory cytokines, including IL-1α, IL-1β, IL-6, and TNFα. Moreover, NF-κB may be activated by the prooxidant environment in the brain, leading to the induction of genes involved in inflammation; it may also increase APP levels, and consequently Aβ liberation, as the APP promoter contains NF-κB-binding sequences (Grilli *et al.*, 1995). Furthermore, oxidative stress is involved in the disruption of calcium homeostasis, and elevated levels of calcium might promote the activation of phospholipase A$_2$ (PLA$_2$). PLA$_2$ might then lead to the release of arachidonic acid and to the subsequent formation of leukotrienes and prostaglandins, which are known to play an important role in the inflammation process (Clark *et al.*, 1995; Stephenson *et al.*, 1996) (Fig. 5). Interestingly, data of Kalman *et al.* (1997) show signs of inflammation in the serum of patients with DS, as manifest by elevated levels of IL-6, which correlates with the severity of the dementia in these patients.

The exacerbation of inflammatory processes by the prooxidant state in DS can also be proposed to contribute to the development of diabetes and the edema of the neck, which are prevalent in patients with DS and have been shown to have an inflammatory component. Indeed, as mentioned previously, NF-κB, which is induced by ROS, upregulates IL-1 and TNFα

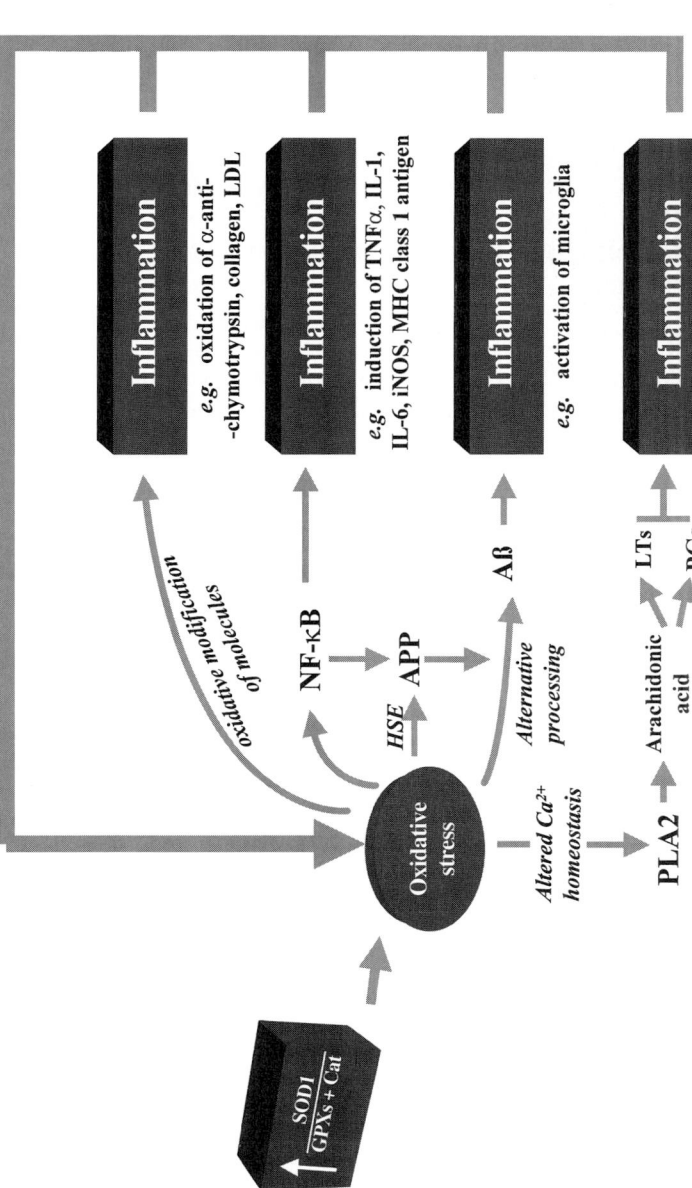

Figure 5 Exacerbation of the inflammatory process by the prooxidant state in the brain of patients with DS. The elevated SOD1/(GPXs+catalase) ratio in the brain of patients with DS causes oxidative stress, which stimulates inflammatory processes via the oxidative modification of molecules or via the activation of NF-κB, which induces genes involved in inflammation. In addition, oxidative stress causes an increased liberation of Aβ, which promotes inflammation; NF-κB also stimulates Aβ formation through the induction of the APP transcription. Furthermore, oxidative stress disrupts calcium homeostasis, which leads to elevated levels of calcium and the induction of phospholipase A_2 (PLA2); this in turn activates the arachidonic acid cascade and leads to the formation of leukotrienes (LTS) and prostaglandins (PGS), which also stimulate the inflammatory processes.

and, interestingly, these cytokines have been found to contribute to the destruction of pancreatic β-cells in insulin-dependent diabetes mellitus (Yamada et al., 1996; Iwahashi et al., 1996). In addition, diabetic complications lead to chronic inflammatory states such as nephropathy, due to the generation of ROS, which may be exacerbated in DS due to the already elevated levels of oxidative stress.

VII. OTHER ANTIOXIDANT GENES OR GENES REGULATED BY AN ALTERED REDOX STATE ON CHROMOSOME 21

A number of other genes localized on chromosome 21 may have an antioxidant function and/or be regulated by the redox state of the cell. The *CBS* gene, for example, which is localized at 21q22.3 and codes for the enzyme cystathionine-β-synthase, may play a role in the antioxidant protection of the vascular endothelial cell. Indeed, cystathionine-β-synthase, together with serine, catalyzes the conversion of homocysteine (Hcy) to cystathionine, and elevated levels of Hcy have been proposed as an important risk factor for coronary artery disease (Clarke et al., 1991), possibly via the reduction of GPX1 activity, the increased generation of H_2O_2, and the subsequent decrease in NO· bioavaibility and increased LDL oxidation (Upchurch et al., 1997; Nishio and Watanabe, 1997). Interestingly, patients with DS are protected against the development of atherosclerosis (Yla-Herttuala et al., 1989) and have lower levels of plasma homocyst(e)ine than control individuals with various types of mental retardation (Chadefaux et al., 1988). It is tempting to speculate whether the elevated levels of *CBS*, due to increased gene dosage, lead to decreased homocysteine levels and consequently protection against atherosclerosis.

The *CBR* gene, coding for the enzyme carbonyl reductase, is another gene on chromosome 21 (at 21q22.12) that may have an antioxidant function. Carbonyl reductase is a NADPH-dependent oxidoreductase with a broad specificity for carbonyl compounds (Wermuth, 1982). It may therefore play a role in the detoxification of oxidized compounds. Indeed, cells overexpressing carbonyl reductase display an increased resistance to paraquat, a potent herbicide whose toxicity is thought to result from increased lipid peroxidation and toxic carbonyl products (Kelner et al., 1997). However, the significance of the overexpression of this antioxidant enzyme in DS still remains to be elucidated.

A member of the heat shock protein family, *STCH*, has been localized at 21q11.1 (Brodsky et al., 1995). Its expression may be activated by the prooxidant state in DS, as it is well known that heat shock proteins (HSPs) are induced in response to numerous types of injury, including oxidative stress. The newly described gene, *DSCR1*, which maps to 21q22.2–21q22.3, may also be regulated by the altered cellular redox state in DS. Indeed, the transcription of its hamster homolog, *adapt78*, has been shown to be induced by hydrogen peroxide (Crawford *et al.*, 1997). Both *STCH* and *DSCR1* may be involved in the protection of cells against oxidative stress. Indeed, HSPs have been reported to enhance the survival of cells exposed to oxidative stress (Arrigo, 1998), and it is reasonable to speculate that DSCR1 may also participate in the protection against ROS, as induction of *adapt78* mRNA following a pretreatment exposure to H_2O_2 correlates with an adaptive response (Crawford *et al.*, 1997). However, whether the overexpression of both genes plays a protective role in DS or, on the contrary, gives rise to perturbations and contributes to some of the pathophysiologies of the syndrome still remains to be answered. It has been shown that HSPs protect cells against ROS-triggered necrosis or apoptosis via a decrease of the intracellular level of ROS in a glutathione-dependent way. Furthermore, HSPs also protect against apoptosis induced independently of ROS and play an important role during early cell differentiation by modulating the apoptotic process (for a review, see Arrigo, 1998). It is therefore possible that elevated levels of STCH in DS lead to perturbations in the intracellular redox state and in the apoptotic process, thereby causing various metabolic impairments.

Finally, the gene *GABPα*, which maps to 21q21–21q22.1 and codes for the α subunit of the GA-binding protein transcription complex (GABP), is also redox regulated. GABP is a tetramer composed of two subunits; GABPα and GABPβ. The α subunit of GABP is a member of the ETS family of transcription factors and mediates specific DNA binding. Chinenov *et al.* (1998) have demonstrated that GABP is subject to redox regulation through the oxidation/reduction of COOH-terminal cysteine residues in the α subunit. In particular, the DNA binding of GABPα is inhibited by prooxidant conditions. In addition, GABP has been shown to regulate the transcription of nuclear-encoded mitochondrial proteins such as the cytochrome c oxidase subunits IV and Vb (Virbasius *et al.*, 1993; Carter and Avadhani, 1994), the mitochondrial transcription factor 1 (Virbasius and Scarpulla, 1994), and the ATP synthase β subunit (Villena *et al.*, 1994). Importantly, mitochondrial enzyme deficiencies have been reported in patients with DS (Prince *et al.*, 1994). It is therefore tempting to hypothesize that the prooxidant conditions in DS may result in a reduction of GABP DNA-binding activity and consequently may lead to the altered transcription of various mitochondrial proteins.

VIII. CONCLUSION

Extensive work has been done in the past years to understand the consequences of the overexpression of the antioxidant gene *SOD1* in DS.

This has revealed the prominent role of ROS and oxidative stress in various pathologies of the syndrome. In particular, it has shown that the disruption of the ROS homeostasis in DS, as a consequence of elevated levels of the SOD1/(GPX+catalase) ratio, results in damage to molecules, inappropriate gene expression, and increased sensitivity of cells to excitotoxicity and apoptosis. This in turn leads to cellular dysfunctions and diseases.

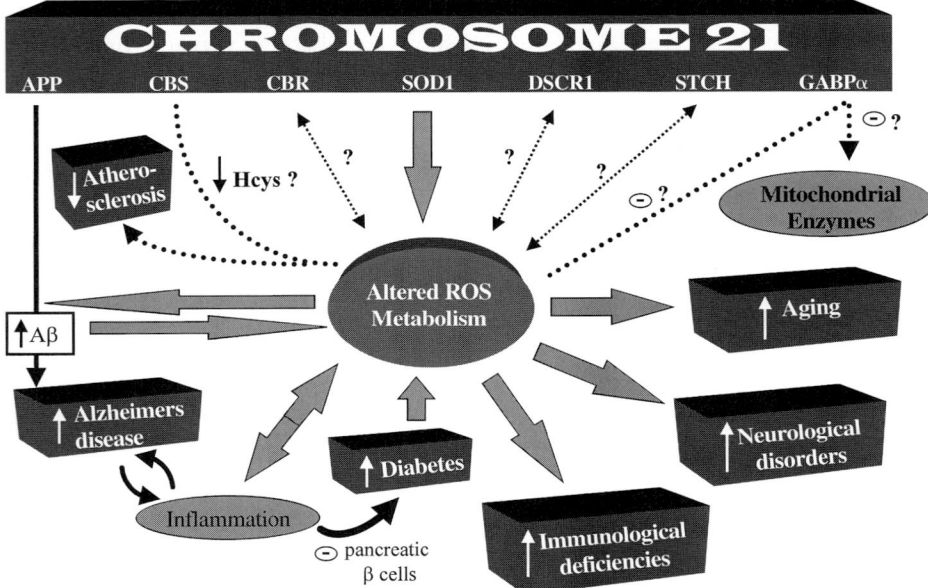

Figure 6 Genes on chromosome 21 that alter/might alter the ROS metabolism and whose expression is/may be perturbed by an altered ROS balance. Overexpression of *SOD1* in DS leads to an elevated SOD1/(GPXs+catalase) ratio that results in the increased generation of H_2O_2 and altered ROS metabolism. This in turn contributes to accelerated aging, increased neurological disorders, and immunological deficiencies in patients with DS. In addition, oxidative stress, in conjunction with elevated APP levels, leads to alternative APP processing, increased formation of Aβ, and early onset of Alzheimer's disease in patients with DS. Increased formation of Aβ also leads to the increased generation of ROS and further oxidative stress. Furthermore, oxidative stress exacerbates the inflammatory processes that contribute to the development of AD and diabetes. Overexpression of *CBS* in DS may lead to decreased level of plasma homocyst(e)ine, which in turn might lower the level of ROS in vascular endothelial cells, thus resulting in the decreased incidence of atherosclerosis. Elevated levels of CBR may alter ROS metabolism through the reduction of carbonyl compounds. The prooxidant state in DS might lead to the elevated expression of *STCH* and *DSCR1*, which in turn may affect the redox state of the cells. Finally, the prooxidant state in DS may also downregulate the binding activity of the transcription factor GABPα, thus resulting in the altered transcription of various mitochondrial proteins.

Further work will probably reveal that other genes on chromosome 21, such as *CBS, CBR, STCH, DSCR1,* and *GABPα,* also contribute to perturbations in the ROS metabolism and/or have their expression affected by the altered ROS balance in DS (Fig. 6). The generation of knockout and transgenic mice for each of these genes will be useful in elucidating the biological function of these genes and the specific defect(s) caused by their overexpression. In addition, the generation of transgenic mice with various combinations of these genes should also be undertaken in order to understand the consequences of the interactions of these genes in DS.

It is hoped that this research will provide new avenues for the prevention and treatment of several of the disorders occurring in patients with DS. A clinical trial by Sano *et al.* (1997) has demonstrated that the antioxidants vitamin E and selegiline significantly delay the time of death, institutionalization, and severe dementia of patients with AD. Because the prooxidant state in DS may contribute to the early onset of AD in DS, it is probable that these antioxidants will also help delay the development of this disease in patients with DS. In addition, this may also improve other disorders of the syndrome where oxidative stress plays a central role.

ACKNOWLEDGMENTS

The authors thank Peter Griffiths and Juliet Taylor for technical assistance and Sika Ristevski for critical reading of the manuscript. This work is funded by grants from the National Health and Medical Research Council of Australia.

REFERENCES

Aisen, P. S., and Davis, K. L. (1994). Inflammatory mechanisms in Alzheimer's disease: Implication for therapy. *Am. J. Psychiatry* **151,** 1105–1113.

Amstad, P., Peskin, A., Shah, G., Mirault, M. E., Moret, R., Zibinden, I., and Cerutti, P. (1991). The balance between CuZn-superoxide dismutase and catalase affects the sensitivity of mouse epidermal cells to oxidative stress. *Biochemistry,* **30,** 9305–9313.

Anneren, G., and Edman, B. (1993). Down's syndrome—a gene dosage disease caused by trisomy of genes within a small segment of the long arm of chromosome 21, exemplified by the study of effects from the superoxide-dismutase type 1 *(SOD-1)* gene. *APMIS Suppl.* **40,** 71–79.

Anneren, G., and Epstein, C. J. (1987). Lipid peroxidation and superoxide dismutase-I and glutathione peroxidase activities in trisomy 16 fetal mice and human trisomy 21 fibroblasts. *Pediatr. Res.* **21,** 88–92.

Arrigo, A. P. (1998). Small stress proteins: Chaperones that acts as regulators of intracellular redox state and programmed cell death. *Biol. Chem. Hoppe-Seyler* **379,** 19–26.

Avraham, K., Schickler, M., Sapoznikow, R., Yarom, R., and Groner, Y. (1988). Down's syndrome: Abnormal neuromuscular junction in tongue of transgenic mice with levels of human CuZn superoxide dismutase. *Cell. (Cambridge, Mass.)* **54**, 823–829.

Avraham, K., Sugaman, H., Rotshenker, S., and Groner, Y. (1991). Down's syndrome: Morphological remodelling and increased complexity in the neuromuscular junction of transgenic CuZn-superoxide dismutase mice. *J. Neurocytol.* **20**, 208–215.

Baker, M. S., and Gebicki, J. M. (1984). The effect of pH on the conversion of superoxide to hydroxyl free radicals. *Arch. Biochem. Biophys.* **234**, 258–264.

Barkats, M., Bertholet, J. Y., Venault, P., Ceballot-Picot, I., Nicole, A., Phillips, J., Moutier, R., Roubertoux, P., Sinet, P. M., and Cohen-Salmon, C. (1993). Hippocampal mossy fiber changes in mice transgenic for the human copper-zinc superoxide dismutase gene. *Neurosci. Lett.* **160**, 24–28.

Bar-Peled, O., Korkotian, E., Segal, M., and Groner, Y. (1996). Constitutive overexpression of Cu/Zn superoxide dismutase exacerbates kainic acid-induced apoptosis of transgenic-Cu/Zn superoxide dismutase neurons. *Proc. Natl. Acad. Sci. U.S.A.* **93**, 8530–8535.

Bazelon, M., Barnet, A., Lodge, A., and Shelbourne, S. A. (1968). The effect of high doses of 5-hydroxytryptophan on a patient with trisomy 21. Clinical, chemical and EEG correlations. *Brain Res.* **11**, 397–411.

Beckman, J. S., Carson, M., Smith, C. D., and Koppenol, W. H. (1993). ALS, Sod and peroxinitrite. *Nature (London)* **364**, 584–587.

Behl, C (1997). Amyloid beta-protein toxicity and oxidative stress in Alzheimer's disease. *Cell Tissue Res.* **290**, 471–480.

Behl, C., Davis, J. B., Lesley, R., and Schubert, D. (1994). Hydrogen peroxide mediates amyloid beta protein toxicity. *Cell (Cambridge, Mass.)* **77**, 817–827.

Bladier, C., Wolvetang, E. J., Hutchinson, P., de Haan, J. B., and Kola, I. (1997). Response of a primary human fibroblast cell line to H_2O_2: Senescence-like growth arrest or apoptosis? *Cell Growth Differ.* **8**, 589–598.

Brodsky, G., Otterson, G. A., Parry, B. B., Hart, I., Patterson, D., and Kaye, F. J. (1995). Localization of *STCH* to human chromosome 21q11.1. *Genomics* **30**, 627–628.

Brooksbank, B. W. L., and Balazs, R. (1984). Superoxide dismutase, glutathione peroxidase and lipid peroxidation in Down's syndrome fetal brain. *Dev. Brain Res.* **16**, 37–44.

Burgio, G. R., Ugazio, A., Nespoli, L., and Maccario, R. (1983). Down's syndrome: A model of immunodeficiency. *Birth Defects, Orig. Artic. Ser.* **19**, 325–327.

Busciglio, J., and Yankner, B. A. (1995). Apoptosis and increased generation of reactive oxygen species in Down's syndrome neurons *in vitro*. *Nature (London)* **378**, 777–779.

Cardasis, C. A. (1983). Ultrastructural evidence of continued reorganization at the aging (11-26 months) rat soleus neuromuscular junction. *Anat. Rec.* **207**, 399–415.

Carter, R. S., and Avadhani, N. G. (1994). Cooperative binding of GA-binding protein transcription factors to duplicated transcription initiation region repeats of the cytochrome c oxidase IV gene. *J. Biol. Chem.* **269**, 4381–4387.

Ceballos, I., Javoy-Agid, F., Delacourte, A., Defossez, A., Lafon, M., Hirsch, E., Nicole, A., Sinet, P. M., and Agid, Y. (1991). Neuronal localization of copper-zinc superoxide dismutase protein and mRNA within the human hippocampus from control and Alzheimer's disease brains. *Free Radical Res. Commun.* **12–13**, 571–580.

Ceballos-Picot, I., Nicole, A., Clement, M., Bourre, J. M., and Sinet, P. M. (1992). Age-related changes in antioxidant enzymes and lipid peroxidation in brains of control and transgenic mice overexpressing copper-zinc superoxide dismutase. *Mutat. Res.* **275**, 281–293.

Chadefaux, B., Ceballos, I., Hamet, M., Coude, M., Poissonnier, M., Kamoun, P., and Allard, D. (1988). Is absence of atheroma in Down's syndrome due to decreased homocysteine levels? *Lancet* **2**, 741.

Chinenov, Y., Schmidt, T., Yang, X. Y., and Martin, M. E. (1998). Identification of redox-sensitive cysteines in GA-binding protein-alpha that regulate DNA binding and heterodimerization. *J. Biol. Chem.* 273, 6203–6209.

Clark, J. D., Schievella, A. R., Nalefski, E. A., and Lin, L. L. (1995). Cytosolic phospholipase A2. *J. Lipid Mediators Cell Signall.* 12, 83–117.

Clarke, R., Daly, L., Robinson, K., Naughter, E., Cahalane, S., Fowler, B., and Graham, I. (1991). Hyperhomocysteinemia: An independent risk factor for vascular disease. *N. Engl. J. Med.* 324, 1149–1155.

Coyle, J. T., Oster-Granite, M. L., and Gearhart, J. D. (1986). The neurobiologic consequences of Down's syndrome. *Brain Res. Bull.* 16, 773–787.

Crawford, D. R., Leahy, K. P., Abramova, N., Lan, L., Wang, Y., and Davies, K. J. A. (1997). Hamster *adapt78* mRNA is a Down's syndrome critical region homologue that is inducible by oxidative stress. *Arch. Biochem. Biophys.* 342, 6–12.

Cristiano, F., de Haan, J. B., Iannello, R., and Kola, I. (1995). Changes in the levels of enzymes which modulate the antioxidant balance occur during aging and correlate with cellular damage. *Mech. Ageing Dev.* 80, 93–105.

Dalton, A., and Wisniewski, H. (1990). Down's syndrome and dementia of Alzheimer's disease. *Int. Rev. Psychiatry* 2, 41–50.

de Haan, J. B., Newman, J. D., and Kola, I. (1992). Cu/Zn superoxide dismutase mRNA and enzyme activity, and susceptibility to lipid peroxidation, increases with aging in murine brains. *Mol. Brain Res.* 13, 179–186.

de Haan, J. B., Cristiano, F., Iannello, R. C., and Kola, I. (1995). Cu/Zn-superoxide dismutase and glutathione peroxidase during aging. *Biochem. Mol. Biol. Int.* 35, 1281–1297.

de Haan, J. B., Cristiano, F., Iannello, R. C., Bladier, C., Kelner, M. J., and Kola, I. (1996). Elevation in the ratio of Cu/Zn-superoxide dismutase to glutathione peroxidase activity induces features of cellular senescence and this effect is mediated by hydrogen peroxide. *Hum. Mol. Genet.* 5, 283–292.

de Haan, J. B., Wolvetang, E. J., Cristiano, F., Iannello, R. C., Bladier, C., Kelner, M. J., and Kola, I. (1997). Reactive oxygen species and their contribution to pathology in Down's syndrome. *Adv. Pharmacol.* 38, 379–402.

de Haan,* J. B., Bladier,* C., Griffiths, P., Kelner, M., O'Shea, R. D., Cheung, N. S., Bronson, R. T., Silvestro, M. J., Wild, S., Zheng, S. S., Beart, P. M., Hertzog, P. J., and Kola, I. (1998). Mice with a homozygous null mutation for the most abundant glutathione peroxidase, Gpx1, show increased susceptibility to the oxidative stress-inducing agents, paraquat and hydrogen peroxide. *J. Biol. Chem.* 273, 22528–22536. *both authors contributed equally to this work.

Dewji, N. N., and Do, C. (1996). Heat shock factor-1 mediates the transcriptional activation of Alzheimer's beta-amyloid precursor protein gene in response to stress. *Brain Res. Mol. Brain Res.* 35, 325–328.

Elroy-Stein, O., and Groner, Y. (1988). Impaired neurotransmitter uptake in PC12 cells overexpressing human Cu/Zn-superoxide dismutase—Implication for gene dosage effects in Down's syndrome. *Cell (Cambridge, Mass.)* 52, 259–267.

Elroy-Stein, O., Bernstein, Y., and Groner, Y. (1986). Overproduction of human Cu/Zn-superoxide dismutase in transfected cells: Extenuation of paraquat-mediated cytotoxicity and enhancement of lipid peroxidation. *EMBO J.* 5, 615–622.

Epstein, C. J. (1986). "The Consequences of Chromosome Imbalance: Principles, Mechanisms and Models." Cambridge University Press, New York.

Epstein C. J. (1989). *In* "The Metabolic Basis of Inherited Disease" (C. R. Scriver, A. L. Beaudet, W. S. Sly, and D. Valle, eds.), pp 291–326. McGraw-Hill, New York.

Fahim, M. A., and Robbins, N. (1982). Ultrastructural studies of young and old mouse neuromuscular junctions. *J. Neurocytol.* **11,** 641–656.

Feaster, W. W., Kwok, L. W., and Epstein, C. J. (1977). Dosage effects for superoxide dismutase-1 on nucleated cells aneuploid for chromosome 21. *Am. J. Hum. Genet.* **29,** 563–570.

Fonseca, E., Lannes-Viera, J., Villa-Verde, D., and Savino, W. (1989). Thymic extracellular matrix in DS. *Braz. J. Med. Biol. Res.,* **22,** 971–974.

Fridovich, I. (1986). Superoxide dismutases. *Adv. Enzymol. Relat. Areas Mol. Biol.* **58,** 61–97.

Friedlich, A. L., and Butcher, L. L. (1994). Involvement of free oxygen radicals in β-amyloidosis: An hypothesis. *Neurobiol. Aging* **15,** 443–455.

Gahtan, E., Auerbach, J. M., Groner, Y., and Segal, M. (1998). Reversible impairment of long-term potentiation in transgenic Cu/Zn-SOD mice. *Eur. J. Neurosci.* **10,** 538–544.

Grilli, M., Ribola, M., Alberici, A., Valerio, A., Memo, M., and Spano, P. (1995). Identification and characterization of a kappa B/Rel binding site in the regulatory region of amyloid precursor protein gene. *J. Biol. Chem.* **270,** 26774–26777.

Groner, Y. (1995). Transgenic models for chromosome 21 gene dosage effects. *Prog. Clin. Biol. Res.* **393,** 193–212.

Guo, Q., Sopher, B. L., Furukawa, K., Pham, D. G., Robinson, N., Martin, G. M., and Mattson, M. P. (1997). Alzheimer's presenilin mutation sensitizes neural cells to apoptosis induced by trophic factor withdrawal and amyloid beta-peptide: Involvement of calcium and oxyradicals. *J. Neurosci.* **17,** 4212–4222.

Harman, D. (1994). Free-radical theory of aging: Increasing the functional life span. *Ann. N. Y. Acad. Sci.* **717,** 1–15.

Iwahashi, H., Hanafusa, T., Eguchi, Y., Nakajima, H., Miyagawa, J., Itoh, N., Tomita, K., Namba, M., Kuwajima, M., Noguchi, T., Tsujimoto, Y., and Matsuzawa, Y. (1996). Cytokine-induced apoptotic cell death in a mouse pancreatic beta-cell line: Inhibition by Bcl-2. *Diabetologia* **39,** 530–536.

Kalman, J. Juhasz, A. Laird, G., Dickens, P., Jardanhazy, T., Rimanoczy, A., Boncz, I., Parry-Jones, W. L., and Janka, Z. (1997). Serum interleukin-6 levels correlates with the severity of dementia in Down's syndrome and in Alzheimer's disease. *Acta Neurol. Scand.* **96,** 236–240.

Kappus, H. (1985). Lipid peroxidation: Mechanism, analysis, enzymology and biological relevance. In "Oxidative Stress" (H. Sies, ed.), pp. 273–311, Academic Press, London.

Kay, M. M., and Makinodan, T. (1981). Relationship between aging and the immune system. *Prog. Allergy* **29,** 134–181.

Keefover, R. W. (1996). The clinical epidemiology of Alzheimer's disease. *Neuroepidemiology* **14,** 337–351.

Kehrer, J. P. (1993). Free radicals, mediators of tissue injury and disease. *Crit. Rev. Toxicol.* **23,** 21–48.

Kelner, M. J., Estes, L., Rutherford, M., Uglik, S. F., and Peitzke, J. A. (1997). Heterologous expression of carbonyl reductase: Demonstration of protasglandin 9-ketoreductase activity and paraquat resistance. *Life Sci.* **61,** 2317–2322.

Kesslak, J. P., Nagata, S. F., Lott, I., and Nalcioglu, O. (1994). Magnetic resonnance imaging analysis of age related changes in the brains of individuals with Down's syndrome. *Neurology* **44,** 1039–1045.

Khan, A. U., and Wilson, T. (1995). Reactive oxygen species as cellular messengers. *Chem. Biol.* **2,** 437–445.

Kola, I. (1997). Simple minded mice from "*in vivo*" libraries. *Nat. Genet.* **16,** 8–9.

Kola, I., and Hertzog, P. J. (1997). Animal models in the study of the biological function of genes on human chromosome 21 and their role in the pathophysiology of Down's syndrome. *Hum. Mol. Genet.* **6,** 1713–1727.

Korenberg, J. R. (1995). Mental modelling. *Nat. Genet.* **11**, 109–111.
Larocca, L. M., Piatelli, M., Valitutti, S., Castellino, F., Maggiano, N., and Musiani, P. (1988). Alterations in thymocyte subpopulations in Down's syndrome (trisomy 21). *Clin. Immunol. Immunopathol.* **49**, 175–186.
Lejeune, J., Gauthier, M., and Turpin, R. (1959). Etude des chromosomes somatiques de neufs enfants mongoliens. *C. R. Hebd. Seances Acad. Sci.* **248**, 1721–1722.
Lendon, C. L., Ashall, F., and Goate, A. M. (1997). Exploring the etiology of Alzheimer's disease using molecular genetics. *JAMA, J. Am. Med. Assoc.* **277**, 825–883.
Levin, S. (1987). The immune system and susceptibility to infections in Down's syndrome. *Prog. Clin. Biol. Res.* **246**, 143–162.
Liu, X. D., and Thiele, D. J. (1996). Oxidative stress induced heat shock factor phosphorylation and HSF-dependent activation of yeast metallothionein gene transcription. *Genes Dev.* **10**, 592–603.
Maruyama, K., Terakado, K., Usami, M., and Yoshikawa, K. (1990). Formation of amyloid-like fibrils in COS cells overexpressing part of the Alzheimer's amyloid protein precursor. *Nature (London)* **347**, 566–569.
Mattson, M. P. (1997a). Advances fuel Alzheimer's conundrum. *Nat. Genet.* **17**, 254–256.
Mattson, M. P. (1997b). Cellular actions of β-amyloid precursor protein and its soluble and fibrillogenic derivatives. *Physiol. Rev.* **77**, 1081–1132.
Matsuo, M. (1993). Age-related alterations in antioxidative defense. In "Free Radicals in Aging" (B. P. Yu, ed.), pp. 143–181. CRC Press, Boca Raton, FL.
McGeer, P. L., and McGeer, E. G. (1995). Chronic inflammatory mechanisms of the brain: Implications for the treatment of Alzheimer's disease. *Dementia* **9**, 111–123.
Meda, L., Cassatella, M., Szendrei, G., Otvos, L., Baron, P., Villalba, M., Ferrari, D., and Rossi, F. (1995). Activation of microglial cells by β-amyloid protein and interferon-gamma. *Nature (London)*, **374**, 647–650.
Nabarra, B., Casanova, M., Paris, D., Nicole, A., Toyama, K., Sinet, P. M., Ceballos, I., and London, J. (1996). Transgenic mice overexpressing the human Cu/Zn-SOD gene: Ultrastructural studies of a premature thymic involution model of Down's syndrome (Trisomy 21). *Lab. Invest.* **74**, 617–626.
Nishio, E., and Watanabe, Y. (1997). Homocysteine as a modulator of platelet derived growth factor action in vascular smooth muscle cells: A possible role for hydrogen peroxide. *Br. J. Pharmacol.* **122**, 269–274.
Patterson, D. H. (1987). The cause of Down's syndrome. *Sci. Am.* **257**, 42–49.
Peled-Kamar, M., Lotem, J., Okon, E., Sachs, L., and Groner, Y. (1995). Thymic abnormalities and enhanced apoptosis of thymocytes and bone marrow cells in transgenic mice overexpressing Cu/Zn-superoxide dismutase: Implications for Down's syndrome. *EMBO J.* **14**, 4985–4993.
Peled-Kamar, M., Lotem, J., Wirguin, I., Weiner, L., Hermalin, A., and Groner, Y. (1997). Oxidative stress mediates impairment of muscle function in transgenic mice with elevated level of wild-type Cu/Zn superoxide dismutase. *Proc. Natl. Acad. Sci. U.S.A.* **94**, 3883–3887.
Philip, R., Berger, A., McManus, N., Warner, N., Peacock, M., and Epstein, L. B. (1986). Abnormalities of the *in vitro* cellular and humoral responses to tetanus and influenza antigens with concomitant numerical alterations in lymphocyte subsets in Down's syndrome. *J. Immunol.* **136**, 1661–1667.
Prasher, V. P., Farrer, M. J., Kessling, A. M., Fisher, E. M. C., West, R. J., Barber, P. C., and Butler, A. C. (1998). Molecular mapping of Alzheimer-type dementia in Down's syndrome. *Ann. Neurol.* **43**, 380–383.

Prince, J., Jia, S., Bave, U., Anneren, G., and Oreland, L. (1994). Mitochondrial enzyme deficiencies in Down's syndrome. *J. Neural. Transm. Park. Dis. Dement. Sect.* **8**, 171–181.

Pueschel, S. (1990). Clinical aspects of Down's syndrome from infancy to adulthood. *Am. J. Med. Genet.* **7**, 52–56.

Rogers, J., Kirby, L. C., Hempelman, S. R., Berry, D. L., McGeer, P. L., Kaszniak, A. W., Zalinski, J., Cofield, M., Mansukhani, L., and Wilson, P. (1993) Clinical trial of indomethacin in Alzheimer's disease. *Neurology,* **43**, 1609–1611.

Sago, H., Carlson, E. J., Smith, D. J., Kilbridge, J., Rubin, E. M., Mobley, W. C., Epstein, C. J., and Huang, T. T. (1998), *Ts 108 Cje,* a new partial trisomy 16 mouse model for Down's syndrome, exhibits learning and behavioural abnormalities. *Proc. Natl. Acad. Sci. U.S.A.* **95**, 6256–6261.

Salbaum, J. M., Weidemann, A., Lemaire, H., Masters, C. L., and Beyreuther, K. (1988). The promoter of Alzheimer's disease amyloid A4 precursor gene. *EMBO J.* **7**, 2807–2813.

Sano, M., Ernesto, C., Thomas, R. G., Klauber, M. R., Schafer, K., Grundman, M., Woodbury, P., Growdon, J., Cotman, C. W., Pfeiffer, E., Schneider, L. S., and Thal, L. J. (1997). A controlled trial of selegiline, alpha-tocopherol, or both as treatment for Alzheimer's disease. *N. Engl. J. Med.* **336**, 1216–1222.

Schickler, M., Knobler, H., Avraham, K. B., Elroy-Stein, O., and Groner, Y. (1989). Diminished serotonin uptake in platelets of transgenic mice with increased Cu/Zn-superoxide dismutase activity. *EMBO J.* **8**, 1385–1392.

Scott, B. S., Becker, L. E., and Petit, J. L. (1983). Neurobiology of Down's syndrome. *Prog. Neurobiol.* **21**, 199–237.

Sinet, P. M., Michelson, A. M., Bazin, A., Lejeune, J., and Jerome, H. (1975). Increase in glutathione peroxidase activity in erythrocytes from trisomy 21 subjects. *Biochem. Biophys. Res. Commun.* **67**, 910–915.

Sinet, P. M., Coutrier, J., and Dutillaux, B. (1976). Trisomie 21 et superoxyde dismutase-1 (IPO-A): Tentative de localisation sur la sous-bande 21q22. 1. *Exp. Cell Res.* **97**, 47–55.

Smith, D. J., Steven, M. E., Sudanagunta, S. B., Bronson, R. T., Makhinson, M., Watabe, A. M., O'Dell, T. J., Fung, J., and Weier, H. U. (1997). Functional screening of 2Mb of human chromosome 21q22. 2 in transgenic mice implicates *minibrain* in learning defects associated with Down's syndrome. *Nat. Genet.* **16**, 28–36.

Stephenson, D. T., Lemere, C. A., Selkoe, D. J., and Clemens, J. E. (1996). Cytosolic phospholipase A(2) (CPLA(2)) immunoreactivity is in Alzheimers disease brain. *Neurobiol. Dis.* **3**, 51–63.

Stewart, W. F., Kawas, C., Corrada, M., and Metter, E. J. (1997). Risk of Alzheimer's disease and duration of NSAID use. *Neurology* **48**, 626–632.

Sumarsono, S. H., Wilson, T. J., Tymms, M. J., Venter, D. J., Corrick, C. M., Kola, R., Lahoud, M. H., Papas, T. S., Seth, A., and Kola, I. (1996). Down's syndrome-like skeletal abnormalities in Ets2 transgenic mice. *Nature (London)* **379**, 534–537.

Teixeria, H. D., and Meneghini, R. (1996). Chinese hamster fibroblasts overexpressing CuZn-superoxide dismutase undergo a global reduction in antioxidants and an increasing sensitivity of DNA to oxidative damage. *Biochem. J.* **315**, 821–825.

Upchurch, G. R., Welch, G. N., Fabrain, A. J., Freedman, J. E., Johnsen, J. L., Keaney, J. F., and Loscalzo, J. (1997). Homocyst(e)ine decreases bioavailable nitric oxide by a mechanism involving gluthathione peroxidase. *J. Biol. Chem.* **272**, 17012–17017.

Villena, J. A., Martin, I., Vinas, O., Cormand, B., Iglesias, R., Mampel, T., Giralt, M., and Villarroya, F. (1994). ETS transcription factors regulate the expression of the gene for the human mitochondrial ATP sythase beta-subunit. *J. Biol. Chem.* **269**, 32649–32654.

Virbasius, J. V., and Scarpulla, R. C. (1994). Activation of the human mitochondrial transcription factor A gene by nuclear respiratory factors: A potential regulatory link between

nuclear and mitochondrial gene expression in organelle biogenesis. *Proc. Natl. Acad. Sci. U.S.A.* **91,** 1309–1313.
Virbasius, J. V., Virbasius, C. A., and Scarpulla, R. C. (1993). Identity of GABP with NRF-2, a multisubunit activator of cytochrome oxidase expression, reveals a cellular role for an ETS domain activator of viral promoters. *Genes Dev.* **7,** 380–382.
Wermuth, B. (1982). "Enzymology of Carbonyl Metabolism," pp. 261–274. Liss, New York.
Winyard, P. G., and Blake, D. R. (1997). Antioxidants, redox-regulated transcription factors, and inflammation. *Adv. Pharmacol.* **38,** 403–421.
Wisniewski, H. M., Silverman, W., and Wegiel, J. (1994). Aging, Alzheimer's disease and mental retardation. *J. Intell. Disabil. Res.* **38,** 233–239.
Yamada, K., Takane-Gyotoku, N., Yuan, X., Ichikawa, F., Inada, C., and Nonaka, K. (1996). Mouse islet cell lysis mediated by interleukin-1-induced Fas. *Diabetologia* **39,** 1306–1312.
Yarom, R., Sagher, U., Havivi, Y., Peled, I. J., and Wexler, M. R. (1986). Myofibers in tongues of Down's syndrome. *J. Neurol. Sci.* **73,** 279–287.
Yarom, R., Sherman, Y., Sagher, U., Peled, I. J., and Wexler, M. R. (1987). Elevated concentrations of elements and abnormalities of neuromuscular junctions in tongues muscles of Down's syndrome. *J. Neurol. Sci.* **79,** 315–326.
Yarom, R., Sapoznikov, D., Havivi, Y., Avraham, K. B., Schickler, M., and Groner, Y. (1988). Premature aging changes in neuromuscular junctions of transgenic mice with an extra human CuZnSOD gene: A model for tongue pathology in Down's syndrome. *J. Neurol. Sci.* **88,** 41–53.
Yates, C. M., Simpson, J., Gordon, A., Maloney, A. F., Allison, Y., Ritchiem, I. M., and Urguhart A. (1983). Catecholamine and cholinergic enzymes in presenile and senile Alzheimer-type dementia and Down's syndrome. *Brain Res.* **280,** 119–126.
Yim, M. B., Chock, P. B., and Stadtman, E. R. (1990). Copper, zinc superoxide dismutase catalyses hydroxyl radical production from hydrogen peroxide. *Proc. Natl. Acad. Sci. U.S.A.* **87,** 5006–5010.
Yim, M. B. Chock, P. B., and Stadtman, E. R. (1993). Enzyme function of copper, zinc superoxide as a free radical generator. *J. Biol. Chem.* **268,** 4099–4105.
Yla-Herttuala, S., Luoma, J., Nikkari, T., and Kivimaki, T. (1989). Down's syndrome and atherosclerosis. *Atherosclerosis* **76,** 269–172.
Yu, B. P. (1994). Cellular defenses against damage from reactive oxygen species. *Physiol. Rev.* **74,** 139–162.

20 Oxidant-Mediated Repression of Mitochondrial DNA Transcription

Bruce S. Kristal* and Byung P. Yu[†]

*Dementia Research Service
Burke Medical Research Institute
White Plains, New York 10605
and Department of Biochemistry
Cornell University Medical College

[†]Department of Physiology
University of Texas Health Science Center
San Antonio, Texas 78284

Proper mitochondrial function is essential for cellular homeostasis and is dependent on the continued, properly related expression of the mitochondrial genome, mtDNA, which encodes subunits of four of the five enzymatic complexes that comprise the respiratory chain. *In vitro* studies, however, have demonstrated that mtDNA transcription is extremely sensitive to inhibition by specific oxidants, notably peroxyl radicals. Abnormalities in mtDNA expression have been observed in aging and in pathological conditions, including ischemia–reperfusion and Alzheimer's disease. These observations, coupled with evidence of decreases in mtDNA expression in diabetes that precede the overt loss in respiratory capacity that accompanies this disease, suggest a role for this type of inhibition *in vivo* and enabled us to propose a model for progressive mitochondrial dysfunction. Data that the endogenous defense systems that protect the mitochondrial gene expression system appear regulated and are amenable to manipulation suggest potential targets for intervention, but the nature of these defenses is unknown.

I. INTRODUCTION

Mitochondrial energy production is essential for cellular and systemic homeostasis. Oxidative phosphorylation, which takes place primarily in and at the inner mitochondrial membrane, is the primary source of energy for most mammalian tissues and organs (Linnane et al., 1989; Wallace, 1992). Indeed, the coupling of the consumption of oxygen by the mitochondrial electron transport chain to the phosphorylation of ADP provides 85–90% of the ATP used in higher eukaryotes and represents one of the most fundamental metabolic properties of higher organisms. Repression of oxidative phosphorylation with respiratory inhibitors (e.g., cyanide, sodium azide) and/or uncouplers (e.g., dinitrophenol) can lead to numerous pathologies (e.g., blindness, deafness, and ataxia) (Wallace, 1992). Abnormalities in oxidative phosphorylation function are now known to be associated with numerous human pathologies, including mitochondrial DNA (mtDNA)-mediated disorders [e.g., mitochondrial myopathies (Shoffner et al., 1990; Wallace, 1992; Chomyn et al., 1992)], acute pathological processes [e.g., ischemia–reperfusion injury (Allen and Stone, 1994)], and chronic disorders ranging from diabetes (Hall et al., 1960; Mokhtar et al., 1993) to neurodegenerative disease [e.g., Parkinson's disease (Mizuno et al., 1995)]. Clearly, the ability to maintain mitochondrial respiratory function is essential for the maintenance of cellular homeostasis, function, and survival.

Oxidative phosphorylation, in turn, requires several components, including an intact mitochondrial inner membrane, the tricarboxylic acid cycle, a series of transporters, and the five enzymatic complexes involved in converting the energy stored in NADH and $FADH_2$ to energy stored in the high energy phosphate bonds of ATP. Four of these five multisubunit enzyme complexes require components encoded by mtDNA, and mitochondrial function is therefore dependent on proper mtDNA expression.

Following a brief description of the nature of the mitochondrial genome and the mechanisms underlying its expression, this chapter summarizes evidence from *in vitro* studies that suggest that mtDNA transcription is sensitive to specific forms of oxidative stress. Evidence from studies of liver mitochondria isolated from diabetic rats will then be presented showing that dysfunctional mtDNA transcription may contribute to diabetes-associated mitochondrial abnormalities. This evidence is consistent with the hypothesis that the oxidant-mediated inhibition of transcription may be relevant for disease processes *in vivo*. Building on data from these and other studies, we will discuss what is known about the nature of the defenses that protect mtDNA transcription from oxidants and the ability to manipulate these defenses. The chapter then concludes with a discussion of other systems in which oxidant-mediated inhibition of mtDNA expression might be involved.

II. MITOCHONDRIAL GENOME

Mammalian mitochondrial genomes are compact, double-stranded, circular DNAs that encode polypeptides that are constituents of the respiratory chain (see Fig. 1). The copy number appears tissue specific and highly variable, with estimates for cells containing mitochondria ranging from as few as four (platelets, Shuster et al., 1988) to as many as ~7900 (HeLa cells; King and Attardi, 1989). The mitochondrial genetic code is slightly different than that used by the nuclear genome. The human mitochondrial genome has 16,569 bp and, like at least those of mice, rats, and cows, has no introns and few base pairs not involved directly in coding or regulation. These genomes encode 13 known polypeptide chains (see Fig. 1), including 7 subunits of complex I (NADH dehydrogenase), 1 of complex III (cytochrome bc_1 complex), 3 of complex IV (cytochrome oxidase), and 2 of complex V (ATP synthase). mtDNA also encodes the rRNAs and tRNAs required for translation of these 13 polypeptides (see Fig. 1). mtDNA function, including transcription, replication, and their regulation, is dependent on proteins initially encoded in the nucleus and synthesized on cytoplasmic ribosomes.

Replication requires mtDNA transcription, which generates the primer required to initiate DNA synthesis. Replication may be linked to the cell cycle in some dividing tissues/cultured cells, but is clearly independent of such regulation in nondividing cells (i.e., neurons). Replication involves two origins, the origin of heavy strand replication, which is located within the D-loop region, and the origin of light strand replication, which is located approximately one-third of the genome away from the D-loop region (see Fig. 1). Replication is carried out by polymerase γ, which has been shown to be homologous to phage polymerases. The importance of mtDNA integrity and proper mtDNA expression is readily seen by considering the number of described human diseases whose origin lies in mtDNA abnormalities, including both depletion and various mutations, including deletions. These diseases, and the abnormalities that underlie them, have been well described elsewhere (see reviews such as Wallace, 1992; DiMauro and Wallace, 1993). Although this chapter focuses on oxidant-mediated loss of transcription, it is worth noting that the obligate requirement for transcription to generate the primer needed for replication suggests that the oxidant-mediated inhibition of mtDNA transcription may well contribute to mitochondrially based diseases by preventing adequate replication as well as by preventing adequate gene expression.

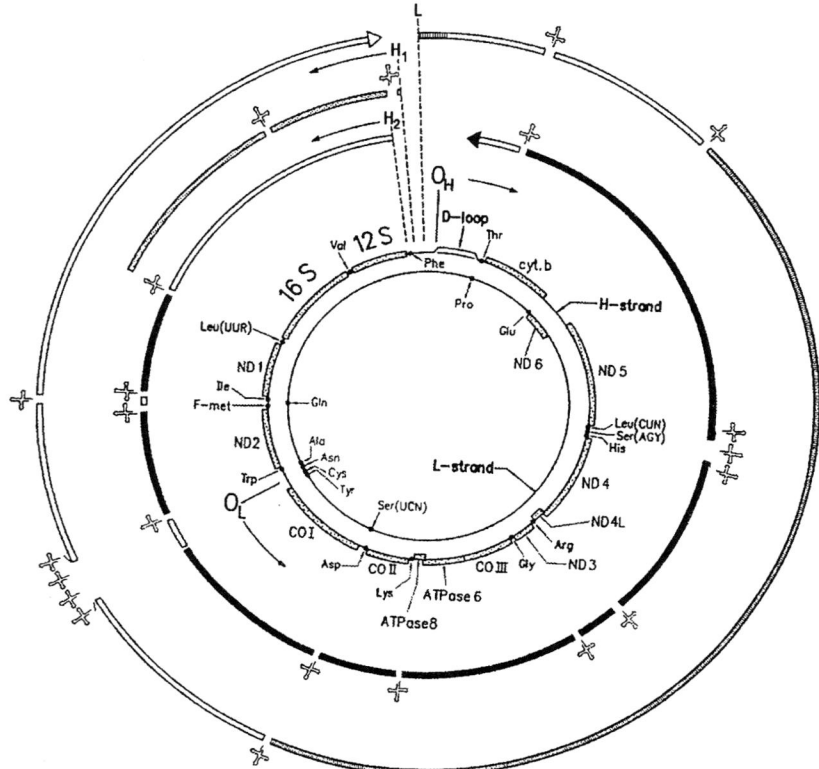

Figure 1 The human mitochondrial genome. The two inner circles show the positions of the two rRNA genes and the reading frames (stippled bars) and tRNA genes (filled circles). In the outer portion of the diagram, the identified functional RNA species other than tRNAs are represented by stippled bars (rRNA species), filled bars (mRNAs transcribed from the H-strand) or a hatched bar (ND6 mRNA, transcribed from the L-strand). tRNAs are represented by cloverleaf structures. The open bars represent unstable, presumably nonfunctional by-products. H_1 and H_2, initiation sites of the rDNA and, respectively, whole H-strand transcription units; L, initiation sites of the L-strand transcription unit. CoI, CoII, and COIII, subunits I, II, and III of cytochrome c oxidase; CYT b, apocytochrome b; ATPase 6 and ATPase 8, subunits 6 and 8 of H^+-ATPase; ND1, ND2, ND3, ND4, ND4L, ND5, and ND6, subunits of NADH dehydrogenase; O_H, O_L, origin of H-strand and, respectively, L-strand synthesis. From Attardi et al. (1990), modified from Attardi (1986). Reproduced with permission from Attardi et al. (1990).

III. mtDNA EXPRESSION AND REPLICATION

A. Molecular Biology of mtDNA Transcription

mtDNA is transcribed bidirectionally from the heavy and light strand promoters (see Fig. 1 and its legend). Cleavage of the polycistronic RNAs yields individual mRNA transcripts. The light strand RNA transcript includes the ND6 structural gene and eight tRNAs (see Fig. 1). All other genes are encoded on one of the two heavy strand transcripts (termed H_1 and H_2). An excellent overview of this process may be found in Attardi et al. (1990).

Regulation of the mitochondrial promoters is mediated by the interaction of nuclear-encoded *trans*-acting factors with the *cis*-regulatory elements for both heavy and light strand promoters located within the D-loop region (Chang and Clayton, 1986a,b; Topper and Clayton, 1989). Interactions of these *cis* sites with *trans*-acting protein factors have been studied with a combination of *in vivo, in situ*, and *in vitro* techniques, including *in organello* footprinting and reconstitution approaches (Walberg and Clayton, 1983; Gaines and Attardi, 1984; Fisher and Clayton, 1985; Fisher et al., 1989, 1992; Clayton, 1991; Shadel and Clayton, 1993; Ghivizzani et al., 1993, 1994; Cantatore et al., 1995; Dairaghi et al., 1995; Enríquez et al., 1996). The proteins involved in transcription include the core polymerase, mitochondrial transcription factor A (mtTFA also called mtTF1), an accessory protein that recognizes sites just upstream of the initiation site, and a species-specific protein involved in promoter recognition (e.g., Shadel and Clayton, 1993). The core polymerase has amino acid homology to T3 and T7 phage polymerases, but not *Escherichia coli* RNA polymerase (Masters et al., 1987; Schinkel and Tabak, 1989). The sites recognized by mtTFA appear heavily divergent among species (e.g., mouse/human), but binding requirements are apparently highly flexible, as human and mouse proteins will recognize each other's promoter (Fisher et al., 1989). mtTFA, which has homology to the high mobility group proteins, can wrap and bend DNA, requires correct positioning to function (Fisher et al., 1992; Ghivizzani et al., 1994; Dairaghi et al., 1995), and has been proposed as a limiting factor for mtDNA transcription (Garstka et al., 1994). Termination, both at the end of mature transcripts as well as to make replication primers, also appears to involve *trans*-acting factors (Kruse et al., 1989; Cantatore et al., 1995). Posttranscriptional processing (e.g., cleavage) and stability also play roles in expression. As one example, processing and stability of rRNAs are inhibited by the loss of interaction with nucleus/cytosol, whereas transcription of these messages, as well as synthesis, processing, and turnover of mt-mRNA, is unaffected (Enríquez et al., 1991, 1996). An example of translational control comes from the work of Polosa and Attardi (1991), who demonstrated that the developmental regulation of the ND5 gene product is not correlated with RNA levels in brain synaptic mitochondria.

B. Regulation of mtDNA Expression

mtDNA transcription seems capable of responding to both local and global (system level) signals. A rationale for the latter is readily appreciated when one considers that proper mitochondrial function is dependent on the successful coordination of signals received by nuclear and mitochondrial genomes. Responsiveness can be tissue specific, as response to vitamin D though its metabolite 1α, 25-dihydroxyvitamin D3 [1α,25-(OH)2D3] differs in the kidney (mtDNA transcription decreases) and intestine (mtDNA transcription increases; Chou et al., 1995). Regulation involves at least three levels (examples follow): (1) mtDNA copy number, (2) mtTFA levels, and (3) posttranscriptional regulation (e.g., stability, translational control).

1. mtDNA Copy Number

The oxidative capacity of rabbit muscle cells is correlated with oxidative capacity of the muscle, which in turn is apparently mediated by an effect on mtDNA copy number (gene dosage, Williams, 1986). Changes in copy number are insufficient, however, to account for changes of expression in all tissues [e.g., Bogert et al., 1993 (human tissues)] or explain the response to environmental stimuli (e.g., cold exposure in rats, Martin et al., 1993) or transformation (Glaichenhaus et al., 1986, rat fibroblasts).

2. mtTFA Levels

A major aspect of the coordinate regulation of nuclear and mitochondrial genomes is that mtTFA expression is modulated substantially by nuclear respiratory factors 1 and 2 (NRF-1 and 2, respectively), *trans*-acting factors previously linked to the expression of nuclear genes involved in regulation (Virbasisu and Scarpulla, 1994). Evidence from studies of the ability of the thyroid hormone to stimulate mitochondrial function (and mtDNA expression) implicates mtTFA in the integration of extracellular signals to mtDNA transcription (Garstka et al., 1994) and function. It is worth noting, however, that the thyroid hormone stimulation of mitochondrial function, which has been studied primarily in isolated cells and tissues from rats, may be complex (Mutvei et al., 1989; Van Itallie, 1990; Garstka et al., 1994; Katyare et al., 1994) and may involve mitochondrial dehydrogenases as well as electron transport (Katyare et al., 1994).

3. Posttranscriptional Regulation

Alteration of mtDNA expression by corticotrophin in bovine adrenocortical cells shows apparent modulation of both the ratio of expression from the H_1 and H_2 sites as well as mRNA transcript-specific effects on stability (Raikhinstein and Hanukoglu, 1993). Stability may also be affected by specific antisense transcripts (Tullo et al., 1994). Finally, it is worth noting that the mechanisms underlying the transcriptional increases following the exposure to estrogen [GH_4C_1 rat pituitary tumor cells (Van Itallie and Dannies, 1988)] and the overall and message-specific increases in rat liver mtDNA expression following chronic alcohol consumption (Enríquez et al., 1992) currently remain unknown.

Thus, as shown in the previous three paragraphs, investigations from multiple laboratories have demonstrated a series of modulators that can increase or modulate mtDNA expression at both transcriptional and post-transcriptional levels. These studies have greatly aided our understanding of the regulation of mtDNA expression and replication under normal, physiological, and some pathophysiological conditions, such as alcoholism and transformation. What is less well understood, however, is what happens under other conditions in which cells are stressed. In particular, the well-accepted interaction between oxidative stress and mitochondrial dysfunction, and the implied threat to mtDNA, led us to begin to test whether mtDNA transcription was sensitive to oxidative stress.

IV. MITOCHONDRIA–OXIDANT INTERACTIONS

Mitochondria are both important targets, and important sources, of reactive species. Because of the potential for the univalent transfer of electrons from the electron transport chain to oxygen (electron leak), mitochondria are likely to be the major source of reactive species in eukaryotes (Loschen et al., 1974; Nohl and Hegner, 1978; Nohl et al., 1981, 1993; Turrens et al., 1985; Bandy and Davison, 1990; Dykens, 1994; Barja et al., 1994; Taylor et al., 1995; Packer et al., 1996). Generation of reactive oxygen species (ROS) appears to be increased in damaged mitochondria and in cells with compromised mitochondrial function. Conversely, mitochondrial structure and function are extremely sensitive to oxidants. For example, an acute exposure to relatively high levels of oxidants, especially in the presence of calcium, can induce the mitochondrial permeability transition, uncouple oxidative phosphorylation, and contribute to cytotoxicity via necrosis and/or apoptosis (through release of cytochrome c). The chemical composition of the mitochondria and its constituents contributes to its oxidant sensitivity. Mitochondrial membranes are highly polyunsaturated, making them excellent targets for peroxidation. Iron–sulfur proteins (e.g., aconitase, rhodanese) are abundant, essential, and highly susceptible to oxidant-mediated damage. The potential combination of free iron or copper and hydrogen peroxide formed from the dismutation of $O_2^{\cdot-}$ increases the odds of production of $\cdot OH$ through Fenton chemistry. Potential by-products of $\cdot OH$-mediated damage are indeed observed. These damage by-products include the aromatic hydroxylation products o- and m-tyrosine, which are seen in some but not all mitochondrial preparations (Kristal et al., 1998), and the DNA damage product 8-hydroxydeoxyguanosine, which may reach concentrations ~15 times that found in nuclear DNA (Richter et al., 1988; Chung et al., 1992).

Chemical considerations are not the only factors that contribute to the oxidant sensitivity of mitochondria: spatial considerations compound the problem. The major spatial consideration is, of course, the proximity of the electron transport chain to the proteins, lipids, and nucleic acids that comprise the mitochondria. A second consideration is that mtDNA is bound to the inner mitochondrial membrane, thus making it readily accessible to lipid peroxyl radicals and lipid peroxidation by-products (Hillar et al., 1979). The extent to which mtDNA is susceptible to oxidant-mediated damage is highlighted by the studies of Hruszkewycz (1988) and Hruszkewycz and Bergtold (1990). They demonstrated that the peroxidation of membrane lipids leads rapidly to cross-linking of mitochondrial DNA and increased generation of 8-hydroxydeoxyguanosine. Believing that loss of mtDNA function would precede the severe cross-linking damage observed by Hruszkewycz (1988) and Hruszkewycz and Bergtold (1990), we tested the hypothesis that peroxyl radicals and/or the by-products of lipid peroxidation reactions would inhibit mtDNA transcription.

V. OXIDANT-MEDIATED REPRESSION OF mtDNA TRANSCRIPTION

A. Oxidant-Species Specificity *in Vitro*

The ability of peroxyl radicals and/or the by-products of lipid peroxidation reactions to inhibit mtDNA transcription was addressed using the *in situ* system developed by Gaines and Attardi (1984). In this technique, isolated mitochondria are incubated (30 min, 37°C) in the presence of ^{32}P-UTP (Gaines and Attardi, 1984; Kristal *et al.*, 1994a, 1997a). Mitochondrial RNA is then isolated and analyzed by polyacrylamide gel electrophoresis. The system allows elongation and *de novo* initiation and can transcribe at about 10–15% of the *in vivo* rate, but only for relatively limited periods of time (maximum yield at 30 min; Gaines and Attardi, 1984). This system enables transcription of mitochondria isolated from the heart, liver, and kidney as well as HeLa cells (Gaines and Attardi, 1984; Kristal *et al.*, 1994a,c). Transcription appears independent of internal mitochondrial energetics, which has two notable advantages and one notable disadvantage. One major advantage for us, which may be less critical for other groups, is that the yield of freshly transcribed RNA is essentially independent of whether the mitochondrial sample had been previously frozen. Thus, this system enabled the use of frozen samples, a necessity given the requirement for the simultaneous analysis of multiple samples from different animals that needed to be collected at different times. Another advantage is simplicity; use of a respiration-independent system enabled reduction in the number of variables in our study of potential mechanisms. This simplicity does, however, come at the disadvantage that it precluded meaningful study of the involvement of energetics in our system. Such an involvement is supported by the observations that moderate levels of ATP seem to stimulate transcription, while higher levels repress it (Kristal et al., 1994a; Enríquez et al., 1996). This is also consistent with data of Greco et al., (1989), who observed an increase in transcription in isolated mitochondria following exposure to a helium–neon laser, a treatment that also appears to increase mitochondrial metabolism. Further studies are needed, however, to confirm these observations and to address mechanisms.

Initial studies revealed that mtDNA transcription is extremely sensitive to inhibition by peroxyl radicals (Kristal et al., 1994a, see Fig. 2). The oxidant sensitivity of mtDNA transcription was evaluated by challenging mitochondria with increasing doses of the azo compounds AAPH [2,2′-azobis-(2-amidinopropane) hydrochloride] and AMVN [2,2′-azobis-(2,4-dimethyl valeronitrile)] (Sato et al., 1990; Niki et al., 1991; Takenaka et al., 1991) prior to allowing in situ transcription using the method developed by Gaines and Attardi (1984). Thermal decomposition of these compounds

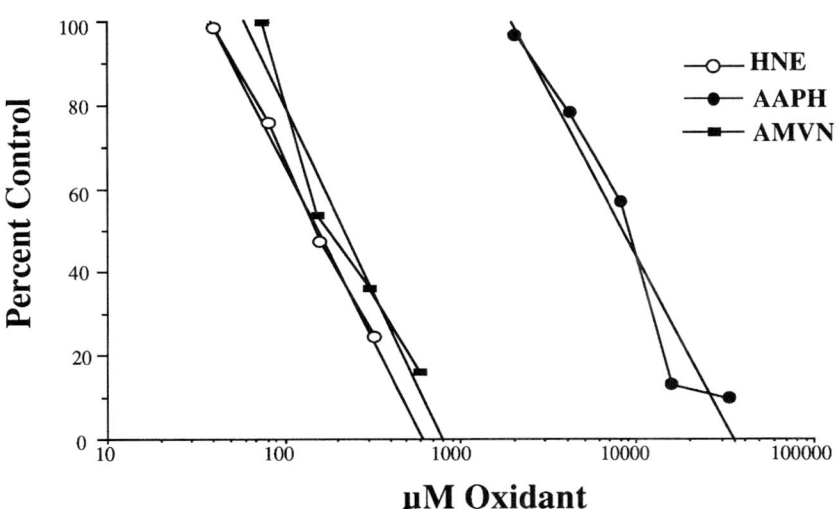

Figure 2 AAPH, AMVN, and 4-hydroxynonenal inhibit liver mtDNA transcription in a dose-dependent manner. Control levels of transcription were defined as those that occurred in a mitochondrial sample that was not exposed to oxidant. Data adapted from Kristal et al. (1994a). Copyright 1994, with permission from Elsevier Science.

initially yields carbon-centered free radicals, which then rapidly undergo further reactions with molecular oxygen, leading to the production of peroxyl radicals. The hydrophilic (AAPH) and hydrophobic (AMVN) nature of these chemicals suggests that they would reach different sites in the mitochondria and thus generate free radical challenges that differ in their location as well as, potentially, in the secondary radicals that they generate. Pretreatment of isolated mitochondria with peroxyl radicals generated by AAPH or AMVN impairs transcription at lower concentrations than those levels needed to induce detectable levels of lipid peroxidation. Thus, both hydrophilic (AAPH) and hydrophobic (AMVN) peroxyl radicals, or the secondary radicals they generate, appear capable of repressing transcription (Kristal *et al.*, 1994a). The sensitivity of the transcription system is seen most clearly in evidence that levels of AAPH or AMVN that decreased mtDNA transcription by 50–100% failed to induce detectable evidence of lipid peroxidation (by examination of malondialdehyde, 4-hydroxynonenal, and diene conjugation, Kristal *et al.*, 1994a), which had previously been considered the most sensitive marker.

Further studies revealed that the sensitivity of mtDNA transcription is limited to specific species of oxidants. In contrast to the ability of both hydrophilic (AAPH-generated) and hydrophobic (AMVN-generated) peroxyl radicals to impair transcription, preincubation with ADP/Fe/NADPH, which is known to generate substantial lipid peroxides, does not impair transcription (Kristal *et al.*, 1994a). Furthermore, concentrations of malondialdehyde as high as 10 mM had no effect on transcription, but exposure to 4-hydroxynonenal (4-HNE), a reactive aldehydic by-product of lipid peroxidation, did inhibit transcription (Kristal *et al.*, 1994a). [Note, however, that the dose required, 180 μM, is high compared to concentrations of this compound known to affect other systems (Esterbauer *et al.*, 1991; Kristal *et al.*, 1996), suggesting that, unless driven by spatial considerations, mtDNA transcription is unlikely to be the primary intracellular target of the toxicity of hydroxynonenal.] Thus, it appears that mtDNA transcription is sensitive only to specific species of oxidants (i.e., peroxyl radicals but not to those generated by ADP/Fe/NADPH). Additional support for the argument of the sensitivity of mtDNA transcription to specific species of oxidants comes from further analysis of the AAPH and AMVN data described earlier. Calculations show that about 1.2 nmol peroxyl radicals (per mg protein) are generated in reactions that prevent transcription, whereas the respiratory activities of this quantity of mitochondria present in these reactions would be expected to generate between 13 and 50 nmol superoxide radicals during the same period of time. This suggests that the sensitivity of mtDNA transcription to superoxide is low. As a further comparison, it may be considered that experiments using submitochondrial particles suggest that the sensitivity of the respiratory proteins to superoxide radicals is about 500-fold lower (ID_{50}s ~600 nM) than the sensitivity of transcription to peroxyl radicals.

Even without considering the possibility that the components of the transcription system itself are subject to cumulative damage, the sensitivity of mtDNA transcription to inhibition suggests that chronic, low-level oxidative stress would lead to a chronic state of partial (or complete) repression and thus lead to pathology. This problem would be exacerbated if the defense systems that protect the transcription system gradually weaken over time. Transcription of the mitochondrial genome could thus be impaired, perhaps chronically, by progressively lower levels of oxidative stress. Frequent inhibition may lead to impaired respiration due to an inability to replace respiratory complexes. Although not yet directly confirmed experimentally, this hypothesis is consistent with our data from streptozotocin-induced diabetic rats (see later). Interestingly, subtle impairments in energy metabolism can lead to excitotoxic cell death in the nervous system (Albin and Greenamyre, 1992; Beal et al., 1993). This suggests that if transcriptional inhibition leads to even slight respiratory problems, other biochemical pathways may amplify the pathogenic effect.

B. A Role for Decreased mtDNA Transcription in Diabetes-Associated Mitochondrial Dysfunction?

Studies of liver mitochondrial function in diabetic animals provide evidence for chronic, progressive impairment of the mitochondrial gene expression system *in vivo*. It has been known since the 1950s that uncontrolled type I diabetes can lead to frank defects in respiration rates and respiratory control ratios (Hall et al., 1960; Mackerer et al., 1971; Harano et al., 1972; Rinehart et al., 1982; Churchill et al., 1983; Pierce and Dhalla, 1985; Rogers et al., 1986; Bedetti et al., 1987; Grinblat et al., 1988; Brignone et al., 1991; Tanaka et al., 1992; Mokhtar et al., 1993; Memon et al., 1995). A role for the loss of transcription in overt mitochondrial respiratory failure is supported by observations of decreased protein synthesis in diabetic mitochondria (Rinehart et al., 1982; Memon et al., 1995). The diabetic model was therefore used to test whether mtDNA transcription decreases at or before the stage of disease in which overt respiratory dysfunction begins.

Studies indicate that the transcription of rat liver mtDNA is impaired greatly by chronic (3 month) streptozotocin-induced diabetes in rats (see Fig. 3; Kristal et al., 1997a). Interestingly, levels of the mitochondrially encoded RNAs increase whereas mtDNA levels decrease in skeletal muscle samples taken from human type I and II diabetics or from the NOD diabetic mouse and in liver samples taken from short-term (1 week) streptozotocin-

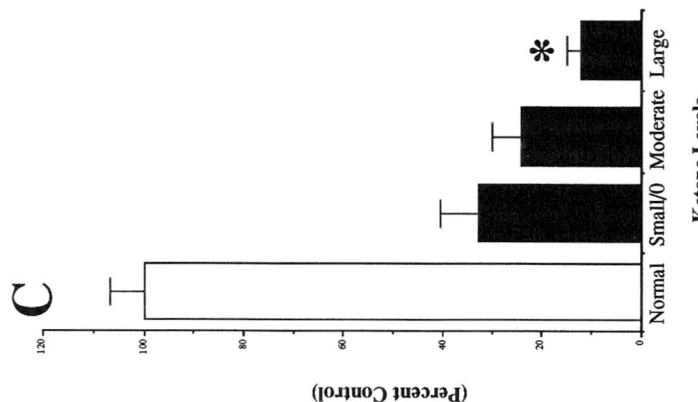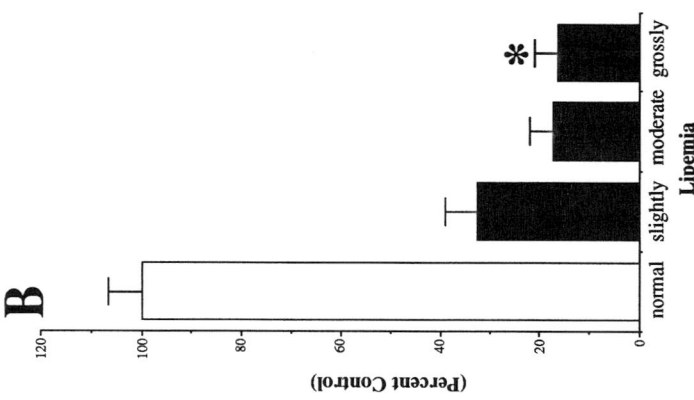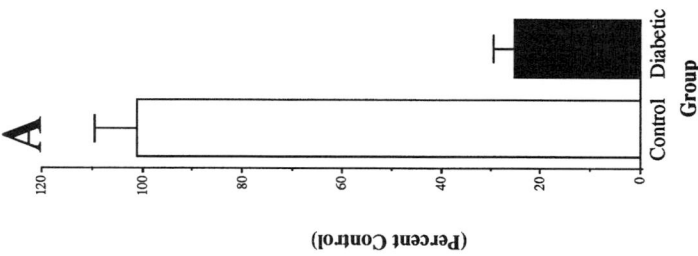

induced diabetic rats (Antonetti *et al.*, 1995). The loss in transcriptional capacity observed in liver mitochondria isolated from chronically diabetic animals is progressive throughout the course of the disease (as assessed by the rise in serum ketones) and is associated with an alteration in several parameters of mitochondrial oxidant resistance, most notably with resistance to the oxidant-mediated inhibition of transcription (Fig. 4; Kristal *et al.*, 1997b). mtDNA transcription eventually falls to <5% of control levels in the most severely affected animals (mean: 85% loss of function in animals with clinically defined "large" serum ketone levels, see Fig. 3). This observation is notable because an 85% loss of mitochondrial DNA has been estimated to represent the threshold for many mitochondrial myopathies.

The oxidant sensitivity of mtDNA transcription was also affected by diabetes. Resistance to AAPH remains nearly constant during the stage in disease processes in which as much as 75% of transcriptional capacity is lost ("early stage," see Fig. 4). Later in disease processes, mitochondrial transcription drops an additional 10–20% ("late stage," 5–15% of control levels remaining) and resistance to AAPH increases approximately twofold. We believe that this late-stage induction of mitochondrial resistance represents a secondary genetic response to life-threatening decreases in mitochondrial function. In contrast to resistance to AAPH, resistance to AMVN increases rapidly following disease induction and progresses throughout disease processes, suggesting that an increase in the defense system that protects against AMVN is a primary response to an oxidative stress generated in diabetic mitochondria.

A causative link between increased oxidative stress and decreased liver mtDNA transcription in diabetic rats is suggested by, and consistent with, the aforementioned *in vitro* studies in which mtDNA transcription was

---

Figure 3 Transcriptional capacity in diabetic animals. Total mtDNA transcription was measured in mitochondria isolated from diabetic or control (nondiabetic) animals. Data are expressed as the percentage of transcription in a group of samples (e.g., mitochondria from diabetics) vs that in the group of samples from control (i.e., nondiabetic) rats. (A) Total transcription from liver mitochondria in control and diabetic animals. Means ± SEM plotted, $N = 10$ (controls), $N = 17$ (diabetic), $p < 0.0001$ for all diabetic animals versus controls. (B) Total transcriptional capacity in diabetic animals as classified by serum lipemia levels. Lipemia levels follow clinical definition. Means ± SEM plotted, $N = 10$ (controls), $N = 6$ (slightly lipemic), $N = 3$ (moderate lipemia), $N = 7$ (grossly lipemic), $p < 0.0001$ for all diabetic animals versus controls, *p ~0.06 versus slightly lipemic. (C) Total transcriptional capacity in diabetic animals as classified by serum ketone levels. Ketone levels follow clinical definition. Means ± SEM plotted, $N = 10$ (controls), $N = 4$ (small/0), $N = 7$ (moderate), $N = 5$ (large), $p < 0.0001$ for all diabetic animals versus controls, *$p < 0.05$ versus small/0 ketone levels. Figure and portions of legend from Kristal *et al.* (1997a). Copyright 1997, with permission from Elsevier Science.

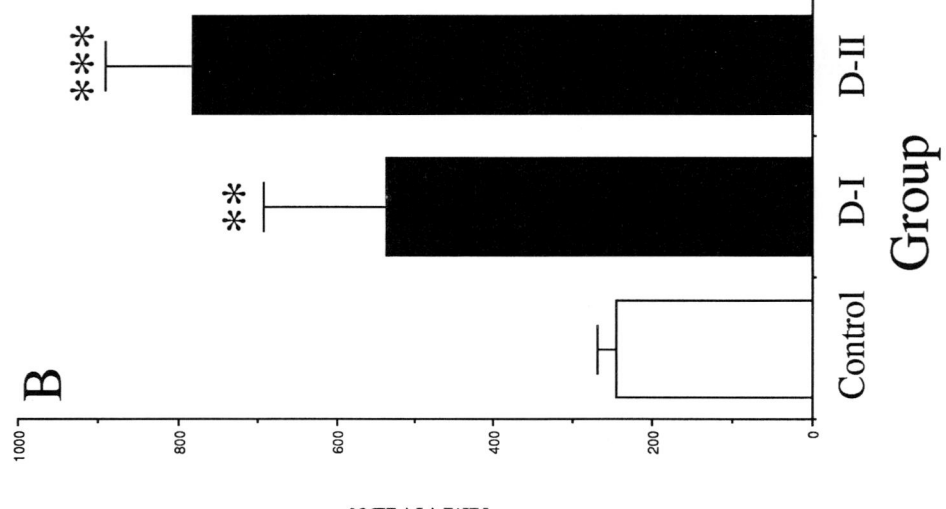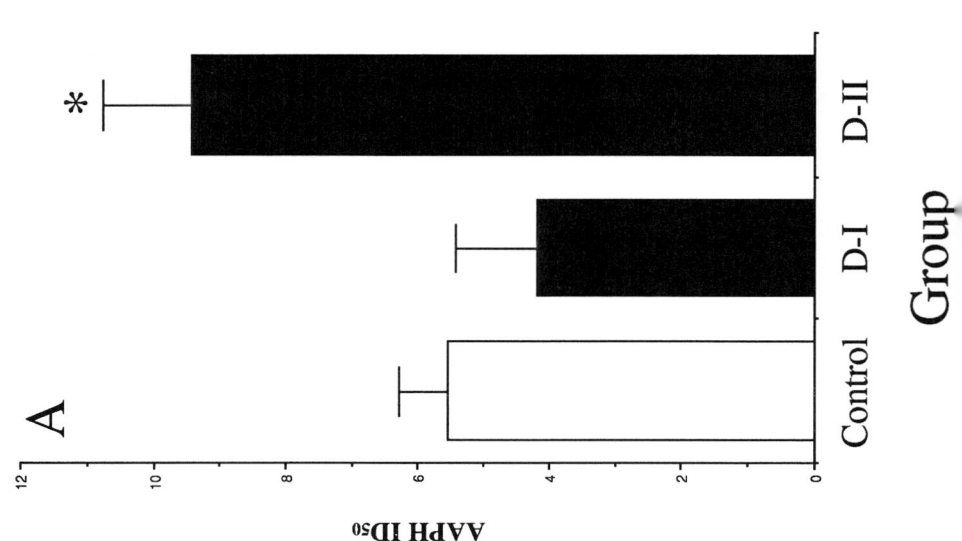

found to be sensitive to oxidants. The potential linkage is strengthened by the observed changes in the defense systems that protect mtDNA transcription from repression by oxidants (see Fig. 4) and by the concomitant appearance of changes in several other parameters that can be associated with oxidant stress, such as changes in the susceptibility of the mitochondrial lipids to peroxidation (Kristal et al., 1997a). To explore this possible linkage, we began to trace the source of this oxidative challenge.

A detailed analysis of mitochondrial respiration was carried out because dysfunctional mitochondria can, as noted earlier, generate ROS through an electron leak within the respiratory chain. To test the hypothesis that the loss of transcriptional competence is a relatively early effect in diabetes, studies concentrated on animals whose ketones levels suggested less severe disease. Mitochondria isolated from these animals had normal respiratory control ratios (RCRs) and had no significant changes in respiratory function in complexes I, II, III, or IV whether assayed in the presence (state 3) or absence (state 4) of ADP (Kristal et al., 1997b). Thus, mitochondria used in this study had no overt respiratory deficits.

Mitochondria isolated from animals with early stage diabetes did, however, already have one substantial defect, a decrease in the ADP/O ratio. This ratio is an index of the ability to couple oxygen utilization to ATP production. As described in detail in the original paper (Kristal et al., 1997b), analysis of these data suggests that approximately 3 to 10 times more free radicals are generated in mitochondria (at center P of the electron transport chain) from diabetic animals than those generated from mitochondria from control animals. Alternative rationales for the decrease in ADP/O ratio (e.g., proton leak mechanisms, altered activity of the Ca^{2+}/H^+ exchanger) are unlikely, as they would not display the specific increase in antimycin A-insensitive oxygen consumption that was observed. Data cited are consistent with a role for this increased ROS generation in the disease-associated loss of mitochondrial transcription and the eventual collapse of mitochondrial function observed in late-stage diabetes (Hall et al., 1960; Harano et al.,

---

Figure 4 Resistance to oxidant-induced repression of liver mtDNA transcription. Levels of AAPH (in mM, Fig. 2A) or AMVN (in μM, Fig. 2B) required to inhibit 50% of transcription expressed relative to the amount of transcriptional capacity remaining in the same sample. Control, control animals; group D-I, diabetic animals having 25–45% of control levels of transcription remaining; and group D-II, diabetic animals having 5–15% of control levels of transcription remaining. Means ± SEM plotted, $N = 10$ (control), $N = 6$ (group D-I), $N = 9$ (group D-II). $*p < 0.05$ versus control, $p < 0.01$ versus group D-I, $**p < 0.05$ versus control, $p \sim 0.10$ versus group D-II, $***p < 0.001$ versus control, $p \sim 0.10$ versus group D-I. Figure and legend from Kristal, et al. (1997a). Copyright 1997, with permission from Elsevier Science.

1972; Rinehart *et al.*, 1982; Churchill *et al.*, 1983; Rogers *et al.*, 1986; Bedetti *et al.*, 1987; Grinblat *et al.*, 1988; Brignone *et al.*, 1991; Tanaka *et al.*, 1992; Mokhtar *et al.*, 1993). This interpretation is supported by evidence that both the defect at center P and the loss of as much as 75% of mtDNA transcription appear to precede the loss of respiratory capacity (based on comparison of data in Kristal, 1997a,b). Together, these data suggest that the well-established diabetes-associated loss of mitochondrial respiratory function is predominantly due to chronic exposure of the mitochondrial transcription system to levels of oxidants that even the compensatory increase in defense systems cannot detoxify. This analysis is consistent with our working hypothesis that either decreases in mtDNA transcription or increases in oxidative stress can create a positive feedback cycle, leading to progressively greater levels of oxidative stress and lowered mtDNA transcription (Fig. 5). The model implies that entry to the pathway may occur at any site, suggesting the possibility that any systemic abnormality that leads to either a loss of mtDNA transcription or to increased oxidative stress has the potential to mediate dysfunction through this pathway.

These studies also highlight important distinctions between mitochondrial deletions and inhibition of mitochondrial transcription. Mitochondrial DNA deletions, once they occur, represent permanent, nonrepairable damage. However, the low levels of these deletions occurring in people without overt myopathy may indicate that they represent relatively rare events. In contrast, the oxidant-mediated inhibition of mitochondrial transcription appears to occur at very low levels of specific forms of oxidative stress. Although this suggests that mitochondrial gene expression might be suppressed in acute situations of high oxidative stress, it also implies that the oxidant-induced lowering of transcription might be most likely to cause problems when mitochondria are subjected to long-term, low-level oxidative stress, such as might occur during aging or, as shown earlier, during diseases such as diabetes. This view suggests the importance of protecting mtDNA expression.

C. Protecting Transcription: Tissue Specificity and Antioxidant Protection

Mitochondria rely on a complex network of defenses to prevent both short- and long-term consequences of exposure to oxidants. Mitochondrial antioxidant defenses are complex, including both hydrophilic (e.g., glutathione) and hydrophobic low molecular weight antioxidants (e.g., vitamin E, coenzyme Q), enzymatic free radical scavengers (e.g., Mn-superoxide dismutase and glutathione peroxidase), and organelle-specific defenses such as the proton gradient. We have taken several approaches in attempting to

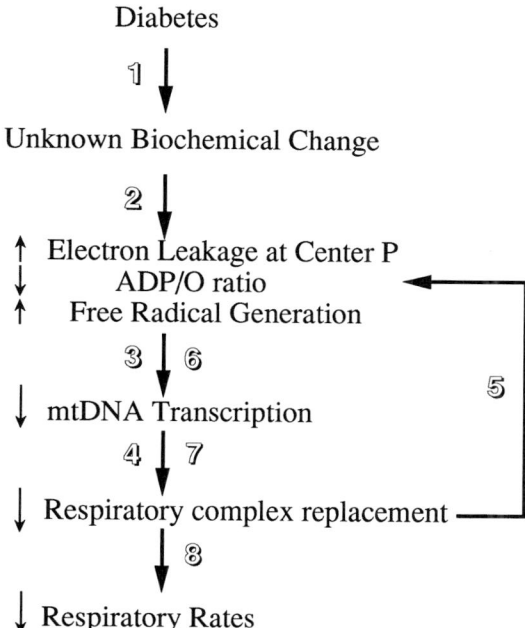

Figure 5 Model for progression of diabetes-induced mitochondrial dysfunction. (Steps 1 and 2) The onset of diabetes leads to some as yet uncharacterized biochemical event that uncouples a fraction of electron transport at center P. This would be predicted to decrease the ADP/O ratio and increase free radical generation. (Step 3) Increased free radicals cause the observed decrease in mitochondrial transcription. (Step 4) Transcription decreases to the point where it is insufficient to maintain respiratory protein complex levels. (Step 5) Loss of respiratory complex components, particularly cytochrome b, worsens the leak at center P by stabilizing $Q_p^{\cdot-}$ (the P-side ubisemiquinone). (Steps 6 and 7) Progressive loss of transcription leads to an equally progressive loss of functional respiratory complexes. (Step 8) Loss of respiratory complexes proceeds to a level where they are insufficient to retain respiration.

determine the nature of the defenses that protect mtDNA transcription from oxidative stress.

Data obtained in our experiments on diabetic animals suggest that different defenses, or sets of defenses, protect against AAPH and AMVN (Kristal *et al.*, 1997a; see Fig. 4). As noted earlier, resistance to AMVN-mediated inhibition increases shortly after an animal becomes diabetic (Kristal *et al.*, 1997a, and unpublished data), whereas resistance to AAPH-mediated inhibition increases only at relatively late stages of disease. This conclusion is supported by more recently completed studies on the effects of aging and dietary manipulation on these defenses (Kristal and Yu, 1998).

Data from diabetes experiments also appear to rule out the enzymatic free radical scavengers as the limiting factors in resistance to inhibition by AAPH and AMVN (Kristal et al., 1997a). This interpretation follows from observations that MnSOD does not change and that glutathione peroxidase and (peroxisomal) catalase decrease at stages of disease in which resistance to AMVN and AAPH is increasing. Likewise, Sukalski et al. (1993) have shown previously that glutathione reductase activity decreases in liver mitochondria from an essentially identical diabetic rat model, suggesting that this enzyme is also not limiting factors in the resistance of mtDNA transcription to inhibition by AAPH or AMVN.

Attempts to block AAPH- and AMVN-mediated inhibition with α-tocopherol, ascorbate, or glutathione suggest that these low molecular weight antioxidants are also not the limiting factors (Kristal et al., 1994b). Although each can attenuate inhibition by AAPH and AMVN, none substantially prevents inhibition. Although not conclusive, these data suggest that the endogenous factors that protect against oxidant-mediated inhibition of transcription are either stronger antioxidants and/or that there are spatial considerations not yet identified.

Examination of the tissue specificity revealed that resistance to the oxidant-mediated inhibition of mtDNA transcription is greater in both the brain and the heart than in the liver (Table I; Kristal et al., 1994c). The observations that brain defenses are high and that mtDNA transcription drops in the aging brain (Gadaleta et al., 1990; Fernandez-Silva et al., 1991) may reflect the necessity of protecting mtDNA transcription in the face of the high levels of oxidative phosphorylation conducted in the brain and the inability to continue this protection adequately during the aging process. Defenses in different brain regions have yet to be studied, but may be of interest because, while mtDNA transcription is qualitatively equivalent across different brain regions, evidence from *in situ* studies shows a more than fivefold variation in transcription rates (Enríquez et al., 1993).

Table I Resistance to Oxidant-Mediated Inhibition Is Tissue Specific[a]

Tissue	ID_{50} AAPH[b]	ID_{50} AMVN[b]
Liver	7.8 mM	220 μM
Brain	45 mM	770 μM
Heart	Not determined	460 μM

[a] From Kristal et al. (1994c).
[b] Dose of AAPH or AMVN that inhibited 50% of mtDNA transcription

Thus, the biochemical identities of the major defense systems that prevent the oxidant-mediated inhibition of transcription remain unknown, although some insight into these defenses has been obtained. Data from both diabetes and tissue specificity experiments suggest that at least two defense mechanisms exist and that these defenses are, at least to some extent, subject to modification and/or physiological regulation. Data from diabetes experiments and antioxidant supplementation experiments are most consistent with a model in which the defenses are nonenzymatic, potent, low molecular weight antioxidants. To address the nature of physiological defense against the oxidant-mediated inhibition of mtDNA transcription further, we chose to examine mitochondria isolated from aged and dietary restricted rats, complementary models in which free radical defenses and damage show divergent relationships.

D. Manipulating Resistance to Oxidants: Dietary Restriction

Dietary restriction regimens display a well-documented ability to increase longevity and decrease morbidity in rodents (Kristal and Yu, 1994; Shimokawa and Higami, 1994). One of the most striking effects of dietary restriction (DR) is its ability to decrease oxidative damage to critical targets, including mitochondria (Weindruch *et al.*, 1980; Laganiere and Yu, 1989; Yu *et al.*, 1992; Sohal *et al.*, 1994; Yu, 1996). Mitochondria from dietary restricted animals have lower basal levels of lipid peroxidation, generate less lipid peroxidation on challenge, and detoxify harmful lipid peroxidation by-products at greater rates than those from *ad libitum*-fed rats (Laganiere and Yu, 1987; Chen and Yu, 1996). Mitochondria from dietary restricted animals also maintain membrane structure and function (e.g., fluidity and lipid composition) later in life than those from *ad libitum*-fed rats (Yu *et al.*, 1992). Further work shows that steady-state levels of 8-hydroxydeoxyguanosine (Chung *et al.*, 1992) and mitochondrial DNA deletions (Yu *et al.*, 1996) are lower in the livers of dietary restricted rats, indicating that restriction protects the mitochondrial genome as well as the lipid bilayer. Part of the underlying mechanism may be the reduced production of free radicals from these mitochondria as compared to mitochondria isolated from *ad libitum*-fed controls (Sohal *et al.*, 1994). Together, these and other studies indicate that dietary restriction can protect mitochondrial lipids, proteins, and nucleic acids from free radical attack. This raises the questions as to whether mtDNA transcription is also well protected and as to whether the dietary-restricted model could be used to manipulate these defenses (Kristal and Yu, 1998).

Resistance to AAPH and AMVN was evaluated in liver mitochondria from five to seven Fischer 344 rats of each age (6, 12, 18, and 24 months) maintained on *ad libitum* or dietary-restricted regimens (Kristal and Yu, 1998). Resistance to AAPH-mediated inhibition was not affected significantly by either age or diet, although the high variability in the resistance to this agent limited the conclusions that we could draw. In contrast, dietary-restricted rats maintained a greater resistance to AMVN-mediated inhibition than their *ad libitum*-fed counterparts in three of the four ages tested (Fig. 6). This augmented resistance was U-shaped, with dietary-restricted samples at 6 and 24 months of age having resistances four- and seven-fold higher than *ad libitum*-fed rats of the same ages. These results show that dietary restriction augments the defense systems that protect one of the mitochondria's most vulnerable systems. Furthermore, the differential modulation of resistance to AAPH and AMVN with age and diet highlights distinctions between these agents and the ability of an organism to independently modulate the resistance to their effects.

E. Oxidant-Mediated Inhibition of mtDNA Transcription in Other Systems?

To our knowledge, there is no experimental evidence for the *direct* inhibition of mtDNA expression by oxidants in other systems. Nonetheless,

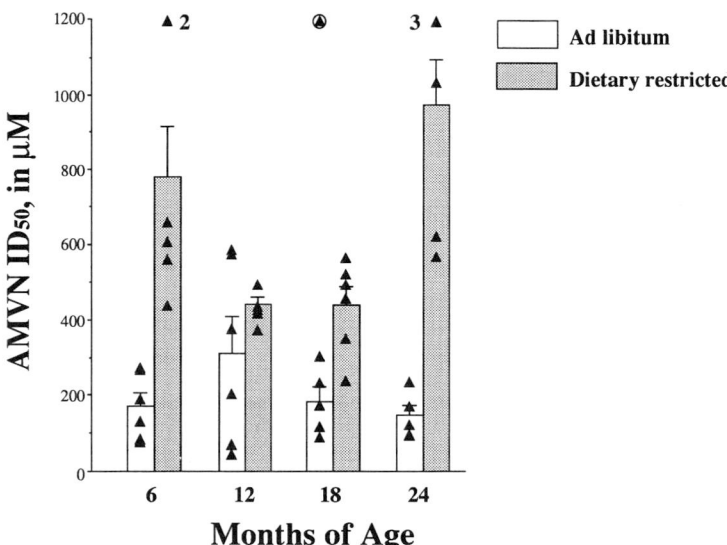

Figure 6 Resistance to AMVN-mediated inhibition of liver mtDNA transcription. ID_{50} values for AMVN were determined by dose–response analysis and are plotted by age and dietary group. Bars represent mean ± SEM. Numbers indicate multiple overlapping points. Figure and legend from Kristal and Yu (1998), with permission. See reference for details of statistical analysis.

data do exist suggesting the possibility that oxidant-mediated inhibition occurs under several situations. For example, mtDNA expression is reduced following partial obstruction of the rabbit bladder (Zhao et al., 1994) and in animal models of Duchenne muscular dystrophy and mitochondrial myopathy (Gannoun-Zaki et al., 1995; Vijayasarathy et al., 1994). Expression is also reduced in renal ischemia–reperfusion in the rat (Van Itallie et al., 1993) and after transient forebrain ischemia in the gerbil, where it has been hypothesized to contribute to the selective vulnerability of Ca1 pyramidal neurons in the hippocampus (Abe et al., 1993, 1995). Furthermore, mtDNA expression is reduced in cultured cells exposed to oxidants as well as in the aging and Alzheimer's brain.

Two reports reveal inhibition of mtDNA transcription in cultured cells exposed to oxidants. In the first, Vincent et al. (1994) showed that chondrocytes exposed to oxidant stress generated by hypoxanthine and xanthine oxidase decreased mtDNA transcripts by as much as 60%. In the second, Crawford et al. (1997) showed that the exposure of cultured HA-1 (hamster fibroblasts) to hydrogen peroxide led to a specific decrease in intact mtDNA-encoded transcripts, apparently due to increased degradation. The authors suggest that this is a protective mechanism. This possibility, along with the possibility that it serves as a general regulatory mechanism (Allen et al., 1995) remains a promising line of study that has yet to be addressed in detail. In addition, the treatment of cultured human mammary carcinoma cells with the antitumor drug mofarotene (Ro 40-8757) reduces the level of one mtDNA transcript through an apparently indirect mechanism (Uchida et al., 1994).

Age-associated decreases of mtDNA transcription have been noted in Drosophilia (Calleja et al., 1993) and in the brain and heart, but not in the liver, of aging rodents (Fernandez-Silva et al., 1991; Gadaleta et al., 1990). The defect in brain transcription can be reversed by treatment with L-carnitine (Gadaleta et al., 1990) and may explain the noted defects in mitochondrial protein synthesis and respiratory function noted in the aging mouse brain (Takai et al., 1995). Interestingly, Chandrasekaran et al. (1994) have determined that mtDNA transcripts are reduced in Alzheimer's disease, a result supported and expanded by further work from the same group (Rapoport et al., 1996; Chandrasekaran et al., 1997; Hatanpää et al., 1996). The current hypothesis suggested by these researchers is that this represents a regulated event.

VI. SUMMARY

In vitro studies have demonstrated that mtDNA transcription is extremely sensitive to specific oxidants. The defenses that protect the mitochondrial gene expression system appear to be nonenzymatic, potent, low molecular weight antioxidants, and it appears that these defenses are tissue specific and are subject to modulation by both endogenous (e.g., disease) and exogenous (e.g., diet) factors. Evidence of decreases in mtDNA expression in diabetes has enabled us to suggest a role for this type of inhibition *in vivo* and to build a model for progressive mitochondrial dysfunction. Growing evidence supports altered mtDNA transcription in other systems, but the exact mechanisms by which oxidants play roles in these effects remain to be defined.

REFERENCES

Abe, K., Kawagoe, J., Aoki, M., and Kogure, K. (1993). Changes of mitochondrial DNA and heat shock protein gene expressions in gerbil hippocampus after transient forebrain ischemia. *J. Cereb. Blood Flow Metab.* **13**, 773–780.
Abe, K., Aoki, M., Kawagoe, J., Yoshida, T., Hattori, A., Kogure, K., and Itoyama, Y. (1995). Ischemic delayed neuronal death, a mitochondrial hypothesis. *Stroke* **26**, 1478–1489.
Albin, R. L., and Greenamyre, J. T. (1992). Alternative excitotoxic hypotheses. *Neurology* **42**, 733–738.
Allen, C. A., Hakansson, G., and Allen, J. F. (1995). Redox conditions specify the proteins synthesized by isolated choloroplasts and mitochondria. *Redox Rep.* **1**, 119–123.
Allen, S. P., and Stone, D. (1994). Mitochondrial function in ischaemia/reperfusion in the heart. *In* "Mitochondria, Proteins and Disease" (V. Darley-Usmar and A. H. V. Shapira, eds.), pp. 143–156. Portland Press, London.
Antonetti, D. A., Reynet, C., and Kahn, C. R. (1995). Increased expression of mitochondrial-encoded genes in skeletal muscle of humans with diabetes mellitus. *J. Clin. Invest.* **95**, 1383–1388.
Attardi, G. (1986). The elucidation of the human mitochondrial genome: A historical perspective. *BioEssays* **5**, 34–39.
Attardi, G., Chomyn, A. King, M. P., Kruse, B., Polosa, P. L., and Murdter, N. N. (1990). Biogenesis and assembly of the mitochondrial respiratory chain: Structural, genetic, and pathological aspects. *Biochem. Soc. Trans.* **18**, 509–513.
Bandy, B., and Davison, A. J. (1990). Mitochondrial mutations may increase oxidative stress: Implications for carcinogenesis and aging? *Free Radical Biol. Med* **8**, 523–539.
Barja, G., Cadenas, S., Rojas, C., Perez-Campo, and R. Lopez-Torres, M. (1994). Low mitochondrial free radical production per unit O_2 consumption can explain the simultaneous presence of high longevity and high aerobic metabolic rate in birds. *Free Radical Res.* **21**, 317–328.
Beal, M. F., Hyman, B. T., and Koroshetz, W. J. (1993). Do defects in mitochondrial energy metabolism underlie the pathology of neurodegenerative diseases? *Trends Neurosci.* **16**, 125–131.
Bedetti, C. D., Montero, G. A., and Stoppani, A. O. M. (1987). Effect of diabetes, adrenalectomy, and hypophysectomy on D-3-hydroxybutyrate dehydrogenase activity and substrate oxidation by rat mitochondria. *Biochem. Int.* **14**, 45–54.
Bogert, C. V., DeVries, H., Holtrop, M., Muus, P., Dekker, H. L., Van Galen, M. J. M., Bolhius, P. A., and Taanman, J.-W. (1993). Regulation of the expression of mitochondrial

proteins: Relationship between mtDNA copy number and cytochrome c oxidase activity in human cells and tissues. *Biochim. Biophys. Acta* **1144,** 177–183.

Brignone, J. A., de Brignone, C. M. C., de Mignone, I. R., Ricci, C.R., Susemihl, M. C., and Rodriguez, R. R. (1991). Improving effects obtained by the ovariectomy or treatment with tamoxifen of female diabetic rats over the function and enzyme activities of liver mitochondria. *Horm. Metab. Res.* **23,** 51–61.

Calleja, M., Pena, P., Ugalde, C. Ferreiro, C., Marco, R., and Garesse, R. (1993). Mitochondrial DNA remains intact during Drosophilia aging, but the levels of mitochondrial transcripts are significantly reduced. *J. Biol. Chem.* **268,** 18891–18897.

Cantatore, P., Daddabbo, L., Fracasso, F., and Gadaleta, M. N. (1995). Identification by *in organello* footprinting of protein contact sites and of single-stranded DNA sequences in the regulatory region of rat mitochondrial DNA. *J. Biol. Chem.* **270,** 25020–25027.

Chandrasekaran, K., Giordano, T., Brady, D. R., Stoll, J., Martin, L. J., and Rapoport, S. I. (1994). Impairment in mitochondrial cytochrome oxidase gene expression in Alzheimer disease. *Mol. Brain Res.* **24,** 336–340.

Chandrasekaran, K., Hatanpää, K., Rapoport, S. I., and Brady, D. R. (1997). Decreased expression of nuclear and mitochondrial DNA-encoded genes of oxidative phosphorylation in association neocortex in Alzheimer disease. *Brain Res. Mol. Brain Res.* **44,** 99–104.

Chang, D. D., and Clayton, D. A. (1986a). Precise assignment of the light-strand promoter of mouse mitochondrial DNA: A functional promoter consists of multiple upstream domains. *Mol. Cell. Biol.* **6,** 3253–3261.

Chang, D. D., and Clayton, D. A. (1986b). Precise assignment of the heavy strand promoter of Mouse mitochondrial DNA: Cognate start sites are not required for transcriptional initiation. *Mol. Cell. Biol.* **6,** 3262–3267.

Chen, J. J., and Yu, B. P. (1996). Detoxification of reactive aldehydes in mitochondria: Effect of age and dietary restriction. *Aging, Clin. Exp. Res.* **8,** 334–340.

Chomyn, A., Martinuzzi, A., Yoneda, M., Hurko, O., Johns, D., Lai, S. T., Nonaka, I., Angelini, C., Attaroi, G. (1992). MELAS mutation in mtDNA binding site for transcriptional termination factor causes defects in protein synthesis and in respiration but no changes in mature upstream and downstream transcripts. *Proc. Natl. Acad. Sci. U.S.A.* **89,** 4221–4225.

Chou, S. Y., Hannah, S. S., Lowe, K. E., Norman, A. W., and Henry, H. L. (1995). Tissue-specific regulation by vitamin D status of nuclear and mitochondrial gene expression in kidney and intestine. *Endocrinology (Baltimore)* **136,** 5520–5526.

Chung, M. H., Kasai, H., Nishimura, S., and Yu, B. P. (1992). Protection of DNA damage by dietary restriction. *Free Radical Biol. Med.* **12,** 523–525.

Churchill, P., McIntyre, J. O., Vidal, J. C., and Fleischer, S. (1983). Basis for decreased D-β-hydroxybutyrate dehydrogenase activity in liver mitochondria from diabetic rats. *Arch. Biochem. Biophys.* **224,** 659–670.

Clayton, D. A. (1991). Nuclear gadgets in mitochondrial DNA replication and transcription. *Trends Biochem. Sci.* **16,** 107–111.

Crawford, D. R., Wang, Y. H., Schools, G. P., Kochheiser, J., and Davies, K.J.A. (1997). Down-regulation of mammalian mitochondrial RNAs during oxidative stress. *Free Radical Biol. Med.* **22,** 551–559.

Dairaghi, D. J., Shadel, G. S., and Clayton, D. A. (1995). Human mitochondrial transcription factor A and promotor spacing integrity are required for transcription initiation. *Biochim Biophys. Acta* **1271,** 127–134.

DiMauro, S., and Wallace, D. C., eds. (1993). "Mitochondrial DNA in Human Pathology." Raven Press, New York.

Dykens, J. A. (1994). Isolated cerebral and cerebellar mitochondria produce free radicals when exposed to elevated Ca^{2+} and Na^+: Implications for neurodegeneration. *J. Neurochem.* **63**, 584–591.

Enríquez, J. A., López-Pérez, M. J., and Montoya, J. (1991). Saturation of the processing of newly synthesized rRNA in isolated brain mitochondria. *FEBS Lett.* **280**, 32–36.

Enríquez, J. A., Pérez-Martos, A., Fernández-Silva, P., López-Pérez, M. J., and Montoya, J. (1992). Specific increase of a mitochondrial RNA transcript in chronic ethanol-fed rats. *FEBS Lett.* **304**, 285–288.

Enríquez, J. A., Pérez-Martos, A., Fernández-Silva, P., López-Pérez, M. J., and Montoya, J. (1993). RNA synthesis in isolated mitochondria from brain cortex, cerebellum and stem: Evidence of different transcription rates. *Eur. J. Biochem.* **25**, 1951–1956.

Enríquez, J. A., Fernández-Silva, P., Pérez-Martos, A., López-Pérez, M. J., and Montoya, J. (1996). The synthesis of mRNA in isolated mitochondria can be maintained for several hours and is inhibited by high levels of ATP. *Eur. J. Biochem.* **237**, 601–610.

Esterbauer, H., Schaur, R. J., and Zollner, H. (1991). Chemistry and biochemistry of 4-hydroxynonenal, malondialdehyde, and related aldehydes. *Free Radical Biol. Med.* **11**, 81–128.

Fernandez-Silva, P., Petruzzella, V., Fracasso, F., Gadaleta, M. N., and Cantatore P. (1991). Reduced synthesis of mtRNA in isolated mitochondria or senescent rat brain. *Biochem. Biophys. Res. Commun.* **176**, 645–653.

Fisher, R. P., and Clayton, D. A. (1985). A transcription factor required for promotor recognition by human mitochondrial RNA polymerase. *J. Biol. Chem.* **260**, 11330–11338.

Fisher, R. P., Parisi, M. A., and Clayton, D. A. (1989). Flexible recognition of rapidly evolving promotor sequences by mitochondrial transcription factor 1. *Genes Dev.* **3**, 2202–2217.

Fisher, R. P., Lisowsky, T., Parisi, M. A., and Clayton, D. A. (1992). DNA wrapping and bending by a mitochondrial high mobility group-like transcriptional activator protein. *J. Biol. Chem.* **267**, 3358–3367.

Gadaleta, M. N., Petruzzella, V., Renis, M., Fracasso, F., and Cantatore, P. (1990). Reduced transcription of mitochondrial DNA in the senescent rat. Tissue dependence and effect of L-carnitine. *Eur. J. Biochem.* **187**, 501–506.

Gaines, G., and Attardi, G. (1984). Highly efficient RNA-synthesizing system that uses isolated human mitochondria: New initiation events and *in vivo*-like processing patterns. *Mol. Cell. Biol.* **4**, 1605–1617.

Gannoun-Zaki, L., Fournier-Bidoz, S., LeCam, G., Chambon, C., Millasseau, P.. Léger, J. J., and Dechesne, C. A. (1995). Down-regulation of mitochondrial mRNAs in the *mdx* mouse model for Duchenne muscular dystrophy. *FEBS Lett.* **375**, 268–272.

Garstka, H. L., Facke, M., Escribano, J. R., and Wiesner, R. J. (1994). Stoichiometry of mitochondrial transcripts and regulation of gene expression by mitochondrial transcription factor A. *Biochem. Biophys. Res. Commun.* **200**, 619–626.

Ghivizzani, S. C., Madsen, C. S., and Hauswirth, W. W. (1993). *In organello* footprinting. *J. Biol. Chem.* **268**, 8675–8682.

Ghivizzani, S. C., Madsen, C. S., Nelen, M. R., Ammini, C. V., and Hauswirth, W. W. (1994). *In Organello* footprint analysis of human mitochondrial DNA: Human mitochondrial transcription factor A interactions at the origin of replication. *Mol. Cell. Biol.* **14**, 7717–7730.

Glaichenhaus, N., Léopold, P., and Cuzin, F. (1986). Increased levels of mitochondrial gene expression in rat fibroblast cell immortalized or transformed by viral and cellular oncogenes. *EMBO J.* **5**, 1261–1265.

Greco, M., Guida, G., Perlino, E., Marra, E., and Quagliariello, E. (1989). Increase in RNA and protein synthesis by mitochondria irradiated with helium-neon laser. *Biochem. Biophys. Res. Commun.* **163**, 1428–1434.

Grinblat, L., Bedetti, C. D., and Stoppani, A. O. M. (1988). Calcium transport and energy coupling in diabetic rat liver mitochondria. *Biochem. Int.* **17,** 329–335.
Hall, J. C., Sordahl, L. A., and Stefko, P. L. (1960). The effect of insulin on oxidative phosphorylation in normal and diabetic mitochondria. *J. Biol. Chem.* **235,** 1536–1539.
Harano, Y., DePalma, R. G., Lavine, L., and Miller, M. (1972). Fatty acid oxidation, oxidative phosphorylation and ultrastructure of mitochondria in the diabetic liver. *Diabetes* **21,** 257–270.
Hatanpää, K., Brady, D. R., Stoll, J., Rapoport, S. I., and Chandrasekaran, K. (1996). Neuronal activity and early neurofibrillary tangles in Alzheimer's disease. *Ann. Neurol.* **40,** 411–420.
Hillar, M., Rangayya, V., Jafar, B. B., Chambers, D., Vitzu, M., and Wyborny, L. E. (1979). Membrane bound mitochondrial DNA: Isolation, transcription and protein composition, *Arch. Int. Physiol. Biochim.* **87,** 29–49.
Hruszkewycz, A. M. (1988). Evidence for mitochondrial DNA damage by lipid peroxidation, *Biochem. Biophys. Res. Commun.* **153,** 191–197.
Hruszkewycz, A. M., and Bergtold, D. S. (1990). The 8-hydroxyguanine content of isolated mitochondria increases with lipid peroxidation. *Mutat. Res.* **244,** 123–128.
Katyare, S. S., Bangur, C. S., and Howland, J. L. (1994). Is respiratory activity in the brain mitochondria responsive to thyroid hormone action?: A critical re-evaluation. *Biochem. J.* **302,** 857–860.
King, M., and Attardi, G. (1989). Human cells lacking mtDNA: Repopulation with exogenous mitochondria by complementation. *Science* **246,** 500–503.
Kristal, B. S., and Yu, B. P. (1994). Aging and its modulation by caloric restriction. *In* "Modulation of Aging Processes by Dietary Restriction" (B. P. Yu, ed.), pp. 1–35. CRC Press, Boca Raton, FL.
Kristal, B. S., and Yu, B. P. (1998). Dietary restriction enhances resistance to oxidant-mediated inhibition of mitochondrial transcription. *Age* **21,** 1–6.
Kristal, B. S., Chen, J. J., and Yu, B. P. (1994a). Sensitivity of mitochondrial transcription to different free radical species. *Free Radical Biol. Med.* **16,** 323–329.
Kristal, B. S., Park, B.-J., and Yu, B. P. (1994b). Anti-oxidants attenuate peroxyl-radical mediated inhibition of mitochondrial transcription. *Free Radical Biol. Med.* **16,** 653–660.
Kristal, B. S., Kim, J. D., and Yu, B. P. (1994c). Tissue specific peroxyl radical mediated inhibition of mitochondrial transcription. *Redox Rep.* **1,** 51–55.
Kristal, B. S., Park, B. K., and Yu, B. P. (1996). 4-hydroxyhexenal is a potent inducer of the mitochondrial permeability transition. *J. Biol. Chem.* **271,** 6033–6038.
Kristal, B. S., Koopmans, S. J., Jackson, C. T., Ikeno, Y., Park, B.-J., and Yu, B. P. (1997a). Oxidant-mediated repression of mitochondrial transcription in diabetic rats. *Free Radical Biol. Med.* **22,** 813–822.
Kristal, B. S., Jackson, C. T., Chung, H.-Y., Matsuda, M., Nguyen, H. C., and Yu, B. P. (1997b). Defects at center P in the electron transport chain underlie diabetes-associated mitochondrial dysfunction. *Free Radical Biol. Med.* **22,** 823–834.
Kristal, B. S., Vigneau-Callahan, K. E., and Matson, W. R. (1998). Simultaneous analysis of the majority of low-molecular weight, redox-active compounds from mitochondria. *Anal. Biochem.* **263,** 18–25.
Kruse, B., Narasimhan, N., and Attardi, G. (1989). Termination of transcription in human mitochondria: Identification and purification of a DNA binding protein factor that promotes termination. *Cell (Cambridge, Mass.)* **58,** 391–397.
Laganiere, S., and Yu, B. P. (1987). Anti-lipoperoxidation action of food restriction. *Biochem. Biophys. Res. Commun.* **145,** 1185–1191.
Laganiere, S., and Yu, B. P. (1989). Effect of chronic food restriction in aging rats. I. liver subcellular membranes. *Mech. Ageing Dev.* **48,** 207–219.

Linnane, A. W., Marzuki, S., Ozawa, T., and Tanaka, M. (1989). Mitochondrial DNA mutations as an important contributor to aging and degenerative diseases. *Lancet* **1**, 642–645.

Loschen, G., Azzi, A., Richter, C., and Flohé, L. (1974). Superoxide radicals as precursors of mitochondrial hydrogen peroxide. *FEBS Lett.* **42**, 68–72.

Mackerer, C. R., Paquet, R. J., Mehlmann, M. A., and Tobin, R. B. (1971). Oxidation and phosphorylation in liver mitochondria from alloxan and streptozotocin diabetic rats. *Proc. Soc. Exp. Biol. Med.* **137**, 992–995.

Martin, I., Vinas, O., Mampel, T, Iglesias, R., and Villarroya, F. (1993). Effects of cold environment on mitochondrial genome expression in the rat: Evidence for a tissue-specific increase in the liver, independent of changes in mitochondrial gene abundance. *Biochem. J.* **296**, 231–234.

Masters, B. S., Stohl, L. L., and Clayton, D. A. (1987). Yeast mitochondrial RNA polymerase is homologous to those encoded by bacteriophages T3 and T7. *Cell (Cambridge, Mass.)* **51**, 89–99.

Memon, R. A., Mohan, C., and Bessman, S. P. (1995). Impaired mitochondrial protein synthesis in streptozotocin diabetic rat hepatocytes. *Biochem. Mol. Biol. Int.* **37**, 627.

Mizuno, Y., Ikebe, S.-I., Hattori, N., Nakagawa-Hattori, Y., Mochizuki, H., Tanaka, M., and Ozawa, T. (1995). Role of mitochondria in the etiology and pathogenesis of Parkinson's disease. *Biochim. Biophys. Acta* **1271**, 265–274.

Mokhtar, N., Lavoie, J. P., Rousseau-Migneron, S., and Nadeau, A. (1993). Physical training reverses defect in mitochondrial energy production in heart of chronically diabetic rats. *Diabetes* **42**, 682–687.

Mutvei, A., Kuzela, S., and Nelson, B. D. (1989). Control of mitochondrial transcription by thyroid hormone. *Eur. J. Biochem.* **180**, 235–240.

Niki, E., Yamamoto, Y., Komuro, E., and Sato, K. (1991). Membrane damage due to lipid oxidation. *Am. J. Clin. Nutr.* **53**, 201S–205S.

Nohl, H., and Hegner, D. (1978). Do mitochondria produce oxygen radicals in vivo? *Eur. J. Biochem.* **82**, 563–567.

Nohl, H., Jordan, W., and Hegner, D. (1981). Identification of free hydroxyl radicals in respiring rat heart mitochondria by spin trapping with the nitrone DMPO. *FEBS Lett.* **123**, 241–244.

Nohl, H., Koltover, V., and Stolze, K. (1993). Ischemia/reperfusion impairs mitochondrial energy conservation and triggers $O_2^{\cdot-}$ release as a byproduct of respiration. *Free Radical Res. Commun.* **18**, 127–137.

Packer, M. A., Porteous, C. M., and Murphy, M. P. (1996). Superoxide production by mitochondria in the presence of nitric oxide forms peroxynitrite. *Biochem. Mol. Biol. Int.* **40**, 527–534.

Pierce, G. N., and Dhalla, N. S. (1985). Heart mitochondrial function in chronic experimental diabetes in rats. *Can. J. Cardiol.* **1**, 48–54.

Polosa, P. L., and Attardi, G. (1991). Distinctive pattern and translational control of mitochondrial protein synthesis in rat brain synaptic endings. *J. Biol. Chem.* **266**, 10011–10017.

Raikhinstein, M., and Hanukoglu, I. (1993). Mitochondrial-genome-encoded RNAs: Differential regulation by corticotropin in bovine adrenocortical cell. *Proc. Natl. Acad. Sci. U.S.A.* **90**, 10509–10513.

Rapoport, S. I., Hatanpää, K., Brady, D. R., and Chandrasekaran, K. (1996). Brain energy metabolism, cognitive function, and down-regulated oxidative phosphorylation in Alzheimer disease. *Neurodegeneration* **5**, 473–476.

Richter, C., Park., J.-W., and Ames, B. N. (1988). Normal oxidative damage to mitochondrial and nuclear DNA is extensive. *Proc. Natl. Acad. Sci. U.S.A.* **85**, 6465–6467.

Rinehart, R. W., Roberson, J., and Beattie, D. S. (1982). The effect of diabetes on protein synthesis and the respiratory chain of rat skeletal muscle and kidney mitochondria. *Arch. Biochem. Biophys.* **213**, 341–352.

Rogers, K. S., Friend W. H., and Higgins, E. S. (1986). Metabolic and mitochondrial disturbances in streptozotocin-treated Sprague-Dawley and Sherman rats. *Proc. Soc. Exp. Biol. Med.* **182**, 167–175.

Sato, K., Niki, E., and Shimasaki, H. (1990). Free radical-mediated chain oxidation of low density lipoprotein and its synergistic inhibition by vitamin E and C. *Arch Biochem. Biophys.* **279**, 403–405.

Schinkel, A. H., and Tabak, H. F. (1989). Mitochondrial RNA polymerase: Dual role in transcription and replication. *Trends Genet.* **5**, 149–154.

Shadel, G. S., and Clayton, D. A. (1993). Mitochondrial transcription initiation. *J. Biol. Chem.* **268**, 16083–16086.

Shimokawa, I., and Higami, Y. (1994). Effect of dietary restriction on pathological processes. In "Modulation of Aging Processes by Dietary Restriction" (B. P., Yu, ed.), pp. 247–266. CRC Press, Boca Raton, FL.

Shoffner, J. M., Lott, M. T., Lezza, A. M., Seibel, P., Ballinger, S. W., Wallace, D. C. (1990). Myoclonic epilepsy and ragged-red fiber disease (MERRF) is associated with a mitochondrial DNA tRNA(Lys) mutation. *Cell (Cambridge, Mass.)* **61**, 931–937.

Shuster, R. C., Rubenstein, A. J., and Wallace, D. C. (1988). Mitochondrial DNA in anucleate human blood cells. *Biochem. Biophys. Res. Commun.* **155**, 1360–1365.

Sohal, R. S., Ku, H. H., Agarwal, S., Forster, M. J., and Lal, H. (1994). Oxidative damage, mitochondrial oxidant generation and antioxidant defense during aging and in response to food restriction in the mouse. *Mech Ageing Dev.* **74**, 121–133.

Sukalski, K. A., Pinto, K. A., and Berntson, J. L. (1993). Decreased susceptibility of liver mitochondria from diabetic rats to oxidative damage and associated increase and α-tocopherol. *Free Radical Biol. Med.* **14**, 57–65.

Takai, D., Inoue, K., Shisa, H., Kagawa, Y., and Hayashi, J.-I. (1995). Age-associated changes of mitochondrial translation and respiratory function in mouse brain. *Biochem. Biophys. Res. Commun.* **217**, 668–674.

Takenaka, Y., Miki, M., and Yasuda, H. (1991). The effect of α-tocopherol as an antioxidant on the oxidation of membrane protein thiols induced by free radicals generated at different sites. *Arch. Biochem. Biophys.* **285**, 344–350.

Tanaka, Y., Konno, N., and Kako, K. J. (1992). Mitochondrial dysfunction observed in situ in cardiomyocytes of rats in experimental diabetes. *Cardiovasc. Res.* **26**, 409–414.

Taylor, D. E., Ghio, A. J., and Piantadosi, C. A. (1995). Reactive oxygen species produced by liver mitochondria of rats in sepsis. *Arch. Biochem. Biophys.* **316**, 70–76.

Topper, J. N., and Clayton, D. A. (1989). Identification of transcriptional regulatory elements in human mitochondrial DNA by linker substitution analysis. *Mol. Cell. Biol.* **9**, 1200–1211.

Tullo, A., Tanzariello, F., D'Erchia, A. M., Nardelli, M., Papeo, P. A., Sbisa, E., and Saccone, C. (1994). Transcription of rat mitochondrial NADH-dehydrogenase subunits: Presence of antisense and precursor RNA species. *FEBS Lett.* **354**, 30–36.

Turrens, J. F., Alexandre, A., and Lehninger, A. L. (1985). Ubisemiquinone is the electron donor for superoxide formation by complex III of heart mitochondria. *Arch. Biochem. Biophys.* **237**, 408–414.

Uchida, T., Inagaki, N., Furuichi, Y., and Eliason, J. F. (1994). Down-regulation of mitochondrial gene expression by the anti-tumor arotinoid Mofarotene (Ro 40-8757). *Int. J. Cancer* **58**, 891–897.

Van Itallie, C. M. (1990). Thyroid hormone and dexamethasone increase the levels of a messenger ribonucleic acid for a mitochondrially encoded subunit but not for a nuclear-encoded subunit of cytochrome c oxidase. *Endocrinology (Baltimore)* **127**, 55–62.

Van Itallie, C. M., and Dannies, P. S. (1988). Estrogen induces accumulation of the mitochondrial ribonucleic acid for subunit II of cytochrome oxidase in pituitary tumor cells. *Mol. Endocrinol.* **2**, 332–337.

Van Itallie, C. M., Van Why, S., Thulin, G., Kashgarian, M., and Siegel, N. J. (1993). Alterations in mitochondrial RNA expression after renal ischemia. *Am. J. Physiol.* **265**, C712–C719.

Vijayasarathy, C., Giger, U., Prociuk, U., Patterson, D. F., Breitschwerdt, E. B., and Avadhani, N. G. (1994). Canine mitochondrial myopathy associated with reduced mitochondrial mRNA and altered cytochrome c oxidase activities in fibroblasts and skeletal muscle. *Comp. Biochem. Physiol. A* **109A**, 887–894.

Vincent, F., Corral-Debrinski, M., and Adolphe, M. (1994). Transient mitochondrial transcript level decay in oxidative stressed chondrocytes. *J. Cell. Physiol.* **158**, 128–132.

Virbasius, J. V., and Scarpulla, R. C. (1994). Activation of the human mitochondrial transcription factor A gene by nuclear respiratory factors: A potential regulatory link between nuclear and mitochondrial gene expression in organelle biosynthesis. *Proc. Natl. Acad. Sci. U.S.A.* **91**, 1309–1313.

Walberg, M. W., and Clayton, D. A. (1983). *In vitro* transcription of human mitochondrial DNA. *J. Biol. Chem.* **258**, 1268–1275.

Wallace, D. C. (1992). Mitochondrial genetics: A paradigm for aging and degenerative diseases. *Science* **256**, 628–632.

Weindruch, R. H., Cheung, M. K. Verity, M. A., and Walford R. L. (1980). Modification of mitochondrial respiration by aging and dietary restriction. *Mech. Ageing. Dev.* **12**, 375–395.

Williams, R. S. (1986). Mitochondrial gene expression in mmmalian striated muscle. *J. Biol. Chem.* **261**, 12390–12394.

Yu, B. P. (1996). Aging and oxidative stress: Modulation by dietary restriction. *Free Radical Biol. Med.* **21**, 651–658.

Yu, B. P., Suescun, E. A., and Yang, S. Y. (1992). Effect of age-related lipid peroxidation on membrane fluidity and phospholipase A2: Modulation by dietary restriction. *Mech. Ageing Dev.* **65**, 17–33.

Yu, B. P., Chen, J. J., Kang, C. M., Choe, M., Maeng, Y. S., and Kristal, B. S. (1996). Mitochondrial aging and lipoperoxidative products. *Ann. N. Y. Acad. Sci.* **786**, 44–56.

Zhao, Y., Levin, R.M., Levin, S. S., Nevel, C.A., Haugaard, N., Hsu, T. H.-S., and Hudson, A. P. (1994). Partial outlet obstruction of the rabbit bladder results in changes in the mitochondrial genetic system. *Mol. Cell. Biochem.* **141**, 47–55.

21 Suppression of Insulin Gene Promoter Activity by Oxidative Stress

Yoshitaka Kajimoto, Yoshimitsu Yamasaki, Taka-aki Matsuoka, Hideaki Kaneto, and Masatsugu Hori

Department of Internal Medicine and Therapeutics (A8)
Osaka University Graduate School of Medicine
Suita 565-0871, Japan

We have found that the insulin gene promoter is rather sensitive to oxidative stress caused by the induction of glycation. Results of reporter gene analyses using β-cell-derived HIT-T15 cells revealed that approximately 50 and 80% of the insulin gene promoter activity was lost when the cells were kept for 3 days in the presence of 40 and 60 mM D-ribose, respectively. The addition of 1 mM aminoguanidine or 10 mM N-acetylcysteine prevented this phenomenon. In agreement with the suppression of promoter activity, a decrease in the insulin mRNA and insulin content was observed in glycation-induced cells. Such a decrease in promoter activity was not observed at all with the control β-actin gene. Because protein glycation occurs in pancreatic β cells when kept under chronic hyperglycemia, the oxidative stress-mediated suppression of the insulin gene transcription may cause a decline in insulin secretion from pancreatic β cells accompanied by a decrease in the insulin content of the cells and thus explain part of the β-cell glucose toxicity.

I. INTRODUCTION

Chronic hyperglycemia is not only a marker of poor glycemic control in diabetes but is itself a cause of impairment of both insulin secretion and biosynthesis: prolonged exposure of pancreatic β-cells to high glucose levels is known to cause β-cell dysfunction, despite the fact that glucose is the major physiologic stimulator of insulin secretion and biosynthesis *in vivo* (Unger and Grundy, 1985; Robertson, 1989) (Fig. 1). Histologically, such a damaged β cell often reveals extensive degranulation and is clinically associated with the development of diabetes in some model animals for noninsulin-dependent diabetes mellitus (NIDDM) (Eizirik *et al.*, 1992; Tokuyama *et al.*, 1995).

In order to enable effective prevention and treatment of glucose toxicity to β cells, it would be essential to understand the biochemical aspects of the phenomenon. Whereas technical difficulties have prevented extensive *in*

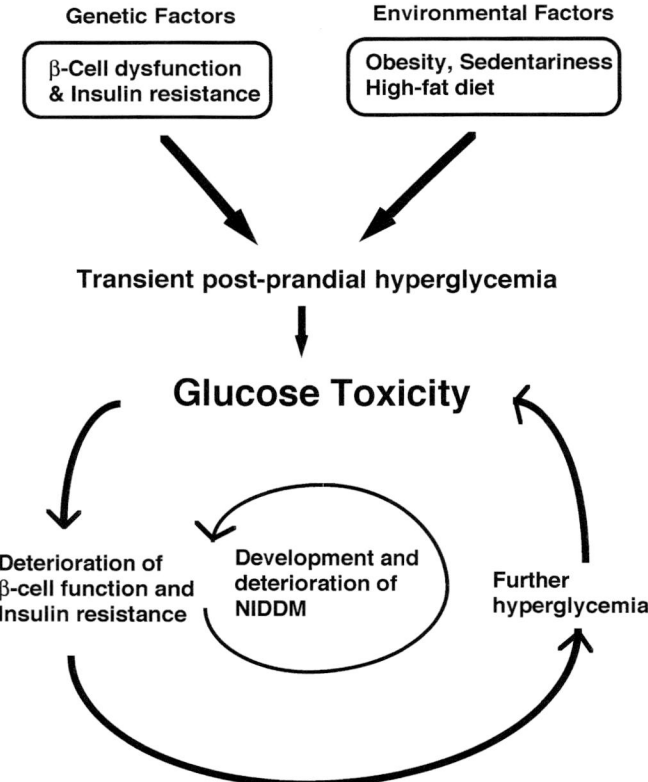

Figure 1 Development of NIDDM and glucose toxicity.

vivo research, a line of *in vitro* studies utilizing β-cell-derived HIT-T15 cells, which were kept under a high glucose condition for a very long period (≥ 25 passages), has provided useful information as to the possible mechanisms underlying the decrease in insulin biosynthesis (Robertson *et al.*, 1994; Olson *et al.*, 1993, 1995; Sharma *et al.*, 1995). According to the results of those studies, long-term exposure to a high glucose concentration caused the HIT-T15 cells to suppress insulin gene promoter activity and led to decreases in insulin mRNA amount, insulin content, and capacity for insulin secretion. These observations support the idea that reduction of the promoter activity may play a primary role in the impairment of the insulin biosynthesis and suggest the potential usefulness of such *in vitro* systems for the study of the molecular aspects of β-cell glucose toxicity.

As a step toward understanding the molecular background for β-cell glucose toxicity, we established an *in vitro* model system with HIT-T15 cells that allows us to evaluate the potential effects of glycation and the consequent increase of reactive oxygen species (ROS) on the β-cell function (Matsuoka *et al.*, 1997). The use of D-ribose could induce glycation in HIT-T15 cells during a relatively short period (<72 hr) between the transfection of the reporter gene plasmids and harvest of the cellular extracts. This made it possible to show that glycation, at least when induced in HIT-T15 cells suppresses the insulin gene transcription (Table I). Data summarized in this chapter thus support the idea that glycation and ROS-dependent suppression of the insulin gene promoter caused impairment of insulin synthesis in HIT-T15 cells.

Table I Effects of D-Ribose on Insulin, β-Actin, and RSV-LTR Promoters

Medium condition	Relative luciferase activity (%)		
	Insulin	β-Actin	RSV-LTR
0 mM D-ribose	100	100	100
40 mM D-ribose	48 ± 19	112 ± 12	252 ± 60
60 mM D-ribose	18 ± 6	87 ± 14	258 ± 95
40 mM D-ribose + 1 mM aminoguanidine	108 ± 10	98 ± 16	113 ± 20
60 mM D-ribose + 1 mM aminoguanidine	67 ± 18	88 ± 9	139 ± 52
40 mM D-ribose + 10 mM NAC	104 ± 10	97 ± 9	100 ± 19
60 mM D-ribose + 10 mM NAC	63 ± 23	86 ± 10	101 ± 20
1 mM aminoguanidine	100 ± 15	110 ± 11	96 ± 16
10 mM NAC	101 ± 18	103 ± 19	96 ± 18

II. CHRONIC HYPERGLYCEMIA INDUCES GLYCATION REACTION IN PANCREATIC β CELLS

Several possible mechanisms have been proposed to date as underlying the glucose-induced cell damage. They include increased osmolarity, an activated sorbitol pathway, and an activated glycation reaction (Clements, 1986; Eble et al., 1983). Among them, glycation seems to have broad pathological significance in diabetic complications. Under hyperglycemia, the production of various reducing sugars such as glucose, glucose 6-phosphate, and fructose increases through glycolysis and the polyol pathway. All of these reducing sugars are known to promote glycation reactions of various proteins (Franklin and Higgins, 1981; Kashimura et al., 1979). In diabetic animals, glycation is observed extensively in various organs and tissues, including kidney, liver, brain, and lung (Brownlee et al., 1984; Myint et al., 1995). Various glycated proteins, such as glycosylated hemoglobin, albumin, and lens crystalline, are produced nonenzymatically through the Maillard reaction (Brownlee et al., 1984; Njoroge and Monnier, 1989; Duhaiman et al., 1990). During the Maillard reaction, which in turn produces Schiff base, Amadori product, and advanced glycosylation end products (AGE), ROS are also produced and this is suggested to be involved in tissue damage (Sakurai and Tsuchiya, 1988; Baynes, 1991).

Pancreatic β cells express the high K_m glucose transporter GLUT2 abundantly and thereby display highly efficient glucose uptake when exposed to a high glucose concentration. As support for the idea that the glycation reaction may underlie the glucose toxicity to β cells, β cells kept under high glucose concentration were shown to contain AGE (Tajiri et al., 1997). Also, it has been shown that pancreatic β cells express relatively low levels of antioxidant enzymes (Lenzen et al., 1996). Therefore, once glycation-mediated oxidative stress is charged on β cells under diabetic conditions, it is likely to cause profound cell damage due to low levels of antioxidant enzyme expression in pancreatic β cells.

III. CIS- AND TRANS-ACTING FACTORS REGULATING INSULIN GENE TRANSCRIPTION

The 5'-flanking region of the insulin gene and the transcription factor binding to this region determine the cell-specific expression of the insulin gene in pancreatic β cells (Fig. 2; reviewed in Sander and German, 1997). Approximately 350-bp-long sequences located upstream of the transcription start sites are well conserved among mammalian insulin genes (Steiner et al., 1985), and this region of the rat insulin I gene was shown to be sufficient

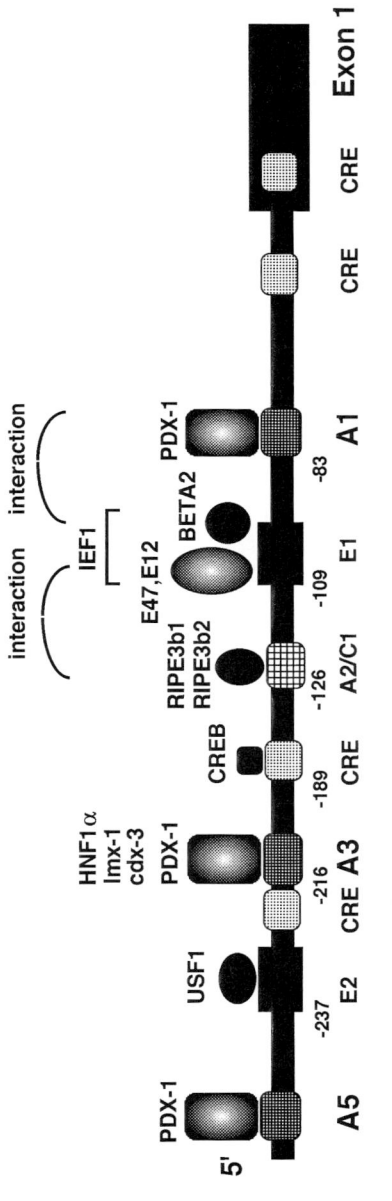

Figure 2 *cis*-acting elements and *trans*-acting factors for the human insulin gene.

for β-cell-specific expression in transgenic mice (Alpert *et al.*, 1988). As shown in Fig. 2, the 350-bp sequence of the insulin gene promoter contains multiple *cis*-acting motifs that act as recognition sites for DNA-binding transcription factors. Some of those transcription factors bind as a monomer and others as a heterodimer or homodimer. Expression of some of the factors is restricted to β cells or to a small subset of cell types and it is likely that the β-cell-specific expression of the insulin gene is achieved by the cooperative combination of those transcription factors (Sander and German, 1997).

Among the *cis*-acting motifs of the insulin gene, two *cis*-acting motifs, E and A elements, and their binding transcription factors play a primary role in generating the strong promoter activity of the insulin genes in β cells (Sander and German, 1997; Ohlsson and Edlund, 1986; Crowe and Tsai, 1989). The E elements, which can be classified, in a broad sense, as an E box (Ephrussi *et al.*, 1985), bind IEF1, a transcription factor complex composed of two different helix–loop–helix proteins, in β cells. IEF1 cannot be identified in most of the non β-cell tissues or cells but is present in α-cell-derived αTC1 clone 6 (αTC1.6) cells and thus is unlikely to be involved in the differentiation between β and α cells.

The A elements, however, are known as binding sites for a group of proteins that belong to the homeodomain family of transcription factors (German *et al.*, 1992; Ohlsson *et al.*, 1993). Among the A element-binding proteins, the one with the evident physiological significance is the homeodomain-containing transcription factor PDX-1 in mouse originally isolated as IPF1 (Ohlsson *et al.*, 1993) and in rat as STF-1/IDX-1 (Leonard *et al.*, 1993; Miller *et al.*, 1994). PDX-1 appears before insulin during the ontogeny of the mouse pancreas and its expression eventually becomes restricted to β cells in the adult (Guz *et al.*, 1995) and is essential for pancreas development (Jonsson *et al.*, 1984). We have shown that PDX-1 functions as a common transcription factor for the expression of at least three β-cell-specific genes: islet amyloid polypeptide (IAPP), insulin, and glucokinase (Watada *et al.*, 1996a,b).

IV. SENSITIVITY OF INSULIN GENE PROMOTER TO GLYCATION AND OXIDATIVE STRESS

Among reducing sugars that potentially induce glycation, D-ribose is outstanding for its very potent activity and thus is often used in *in vitro* studies as an inducer of glycation (Franklin and Higgins, 1981; Hasegawa *et al.*, 1995). We therefore used D-ribose to evaluate the effects of glycation on insulin gene transcription (Matsuoka *et al.*, 1997).

According to the results of reporter gene analyses, the insulin gene promoter but not the β-actin gene promoter or RSV-LTR promoter was sensitive to the induction of glycation. When HIT-T15 cells were kept for 72 hr with D-ribose, the activity of the insulin gene promoter was suppressed down to 48–18% of the control (Table I; Matsuoka et al., 1997). No significant changes were observed with the β-actin gene promoter, and an increase was observed with the RSV-LTR promoter in the presence of 40 and 60 mM D-ribose (Table I). Interestingly, the promoter-suppressing effects were neutralized, at least in part, with the addition of aminoguanidine (Table I). This suggested that the suppressive effect of D-ribose on the insulin gene promoter is mediated by the induction of glycation.

As described earlier, glycation often causes an increase of ROS, which are suggested to play a primary role in glycation-induced tissue damage (Baynes, 1991; Hunt et al., 1988). We therefore inhibited H_2O_2 production using 10 mM NAC. Then, similar to the results obtained with aminoguanidine, the insulin gene promoter activity was recovered (Table I). This indicated that ROS induction mediates the glycation-dependent suppression of the insulin gene promoter as well as the activation of the RSV-LTR promoter. We would like to note that the RSV-LTR promoter, which was activated by D-ribose in a glycation-dependent manner (Table I), is known to be a target of ROS-induced activation through the cis-acting motif CArG box (Trouche et al., 1993; Nose et al., 1991).

V. GLYCATION REDUCES INSULIN mRNA AMOUNT AND INSULIN CONTENT

The glycation-dependent suppression of the insulin gene promoter described earlier also caused the reduction of its transcripts. When insulin mRNA levels were measured by Northern blot analyses, the amounts of insulin mRNA decreased in HIT-T15 cells that had been kept for 5 days with 40 or 60 mM D-ribose (Fig. 3; Matsuoka et al., 1997). In agreement with the observations on promoter activities, 1 mM aminoguanidine and 10 mM NAC neutralized the suppressive effects of D-ribose.

A similar phenomenon was also observed with the insulin content. In agreement with the decrease in insulin mRNA, the insulin content of cells cultured in the presence of 40 mM D-ribose for 5 days was 41 ± 6 pmol/mg protein (average $\pm$ SD; $n = 5$), which was only 18% of that of the control cells kept for the same period without D-ribose (229 ± 52 pmol/mg protein). Thus the glycation-dependent suppression of the insulin gene promoter seemed to have reduced the amount of its transcript and protein in the cells.

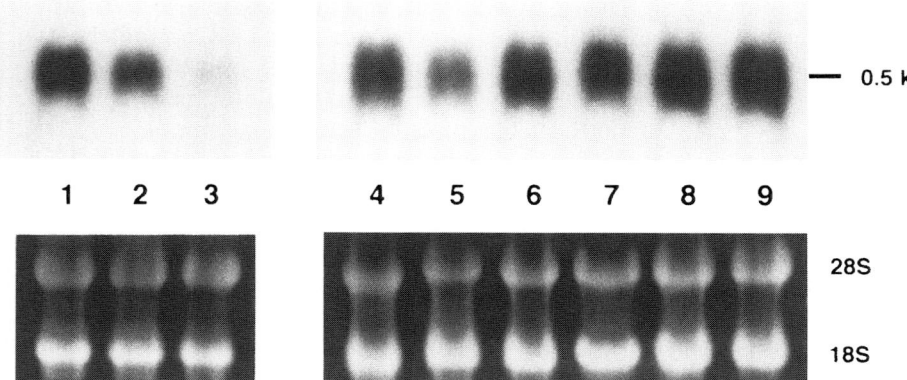

Figure 3 Oxidative stress-mediated reduction in insulin mRNA in HIT-T15 cells. HIT-T15 cells grown for 5 days under various conditions. (Left) Lane 1, control (0 mM D-ribose); lane 2, 40 mM D-ribose; and lane 3, 60 mM D-ribose. (Right) Lane 4, control (0 mM D-ribose); lane 5, 40 mM D-ribose; lane 6, 40 mM D-ribose plus 10 mM NAC; lane 7, 40 mM D-ribose plus 1 mM aminoguanidine; lane 8, 10 mM NAC (0 mM D-ribose); and lane 9, 1 mM aminoguanidine (0 mM D-ribose).

In terms of reversibility of the impairment of insulin gene transcription, the changes that occurred in HIT-T15 cells due to D-ribose-induced glycation and ROS are almost, if not totally, irreversible: when the cells were kept under 40 mM D-ribose for 5 days and then moved into media containing no D-ribose, there was only slight recovery from the decrease in the insulin mRNA amounts caused by 40 mM D-ribose (Matsuoka et al., 1997).

VI. GLYCATION-DEPENDENT REDUCTION OF DNA-BINDING ACTIVITY OF PDX-1

As a possible cause of the reduction in the insulin gene promoter activity, we found that the DNA-binding activity of PDX-1 is rather sensitive to glycation and the resulting oxidative stress. When HIT-T15 cells were kept for 5 days with 40 or 60 mM D-ribose, a marked reduction of PDX-1 binding to the human A3 probe was observed (Matsuoka et al., 1997). Coexistence of 1 mM aminoguanidine or 10 mM NAC in the media prevented this decrease. These results indicated that D-ribose suppressed the DNA-binding activity of PDX-1 in a manner dependent on glycation and ROS.

Is this decrease in the PDX-1-binding activity the cause of the suppression of insulin gene transcription or just a coincidence? The answer for this is not yet known. However, we have obtained some data that may support the latter possibility (Kajimoto *et al.*, 1997). We suppressed PDX-1 expression in β-cell-derived MIN6 cells using an antisense oligodeoxynucleotide (ODN) and searched for possible changes in β-cell-specific gene expression. Although the PDX-1 expression was suppressed to 14 ± 4% of the control and there was also a decrease in its DNA binding to the insulin gene A element, Northern blot analyses revealed no decrease in the amount of insulin mRNA in MIN6 cells. Similarly, no changes were detected in the transcription of the glucokinase or islet amyloid polypeptide gene, for which PDX-1 was shown to function as a transcription factor (Kajimoto *et al.*, 1997). Therefore, PDX-1 may not function as a rate determinant of insulin gene transcription. Although posttranslational suppression of the PDX-1 activity rather than the suppression of PDX-1 protein expression may possibly be important for the oxidative stress-mediated suppression of the insulin gene expression, we assume that other components constituting the transcription-controlling machinery also need to be examined to understand the molecular basis of impaired insulin biosynthesis (Fig. 4).

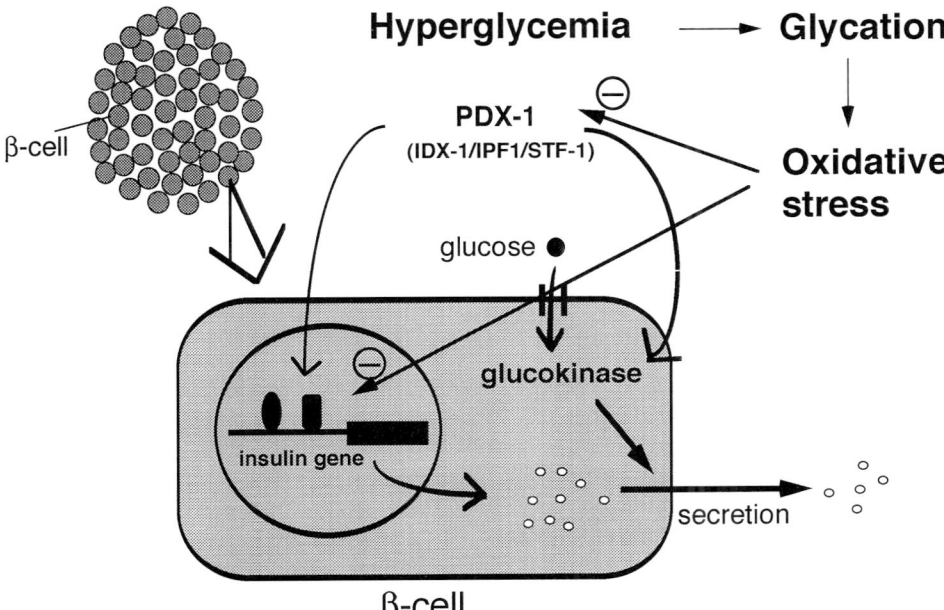

Figure 4 Glycation-dependent, oxidative stress-mediated suppression of insulin biosynthesis: a proposed model.

VII. CONCLUSIONS

Using an *in vitro* model cell system, we evaluated the potential effects of glycation and the consequent increase of ROS on β-cell function (Matsuoka *et al.*, 1997). Results indicated that glycation and the consequent increase of ROS are potential inhibitors of insulin gene transcription. The clear difference in the responses to D-ribose between the insulin gene and β-actin gene promoters suggested that the suppression of the promoter activity can be attributed to the impairment of specific transcription factors. Meanwhile, we found that the DNA-binding activity of PDX-1 is decreased in accordance with the decrease in insulin gene transcription; however, the pathophysiological significance of this phenomenon remained unknown. Several other factors, including RIPE3b1 whose activity also seemed to be inhibited after chronic exposure of HIT cells to high glucose concentrations (Sharma *et al.*, 1995), are also involved in regulation of the insulin gene and therefore their physiological roles in determining the transcription rate of the insulin gene need to be characterized. Because of the irreversibility of the phenomena, it is likely that the suppressive effects of glycation or ROS on insulin gene transcription, if they keep occurring *in vivo*, would accumulate in cells and inhibit insulin synthesis. Thus, because increased glycation in various tissues was identified in diabetes in model animals and in humans (Brownlee *et al.*, 1984; Myint *et al.*, 1995; Njoroge and Monnier, 1989; Sell and Monnier, 1990; Monnier *et al.*, 1984), a similar phenomenon is likely to occur *in vivo* in islets.

REFERENCES

Alpert, S., Hanahan, D., and Teitelman, G. (1988). Hybrid insulin genes reveal a developmental lineage for pancreatic endocrine cells and imply a relationship with neurons. *Cell (Cambridge, Mass.)* **53**, 295–308.

Baynes, J. W. (1991). Role of oxidative stress in development of complications in diabetes. *Diabetes* **40**, 405–412.

Brownlee, M., Vlassara, H., and Cerami, A. (1984). Nonenzymatic glycosylation and the pathogenesis of diabetic complications. *Ann. Intern. Med.* **101**, 527–537.

Bunn, H. F., and Higgins, P. J. (1981). Reaction of monosaccharides with proteins: possible evolutionary significance. *Science* **213**, 222–224.

Clements, R. S., Jr. (1986). The polyol pathway. A historical review. *Drugs* **32**(Suppl 2), 3–5.

Crowe, D. T., and Tsai, M. J. (1989). Mutagenesis of the rat insulin II 5′s-flanking region defines sequences important for expression in HIT cells. *Mol. Cell. Biol.* **9**, 1784–1789.

Duhaiman, A. S., Rabbani, N., and Cotlier, E. (1990). Camel lens crystallins glycosylation and high molecular weight aggregate formation in the presence of ferrous ions and glucose. *Biochem. Biophys. Res. Commun.* **173**, 823–832.

Eble, A. S., Thorpe, S. R., and Baynes, J. W. (1983). Nonenzymatic glucosylation and glucose-dependent cross-linking of protein. *J. Biol. Chem.* **258**, 9406–9412.

Eizirik, D. L., Korbutt, G. S., and Hellerström, C. (1992). Prolonged exposure of human pancreatic islets to high glucose concentration in vitro impairs the β-cell function. *J. Clin. Invest.* **90**, 1263-1268.

Ephrussi, A., Church, G. M., Tonegawa, S., and Gilbert, W. (1985). B lineage-specific interactions of an immunoglobulin enhancer with cellular factors in vivo. *Science* **227**, 134-140.

German, M. S., Wang, J., Chadwick, R. B., and Rutter, W. J. (1992). Synergistic activation of the insulin gene by a LIM-homeo domain protein and a basic helix-loop-helix protein: Building a functional insulin minienhancer complex. *Genes Dev.* **6**, 2165-2176.

Guz, Y., Montminy, M. R., Stein, R., Leonard, J., Gamer, L. W., Wright, C. V., and Teitelman, G. (1995). Expression of murine STF-1, a putative insulin gene transcription factor, in β cells of pancreas, duodenal epithelium and pancreatic exocrine and endocrine progenitors during ontogeny. *Development (Cambridge, UK)*, **121**, 11-18.

Hasegawa, G., Hunter, A. J., and Charonis, A. S. (1995). Matrix nonenzymatic glycosylation leads to altered cellular phenotype and intracellular tyrosine phosphorylation. *J. Biol. Chem.* **270**, 3278-3283.

Hunt, J. V., Dean, R. T., and Wolff, S. P. (1988). Hydroxyl radical production and autoxidative glycosylation. *Biochem. J.* **256**, 205-212.

Jonsson, J., Carlsson, L., Edlund, T., and Edlund, H. (1984). Insulin-promoter factor 1 is required for pancreas development in mice. *Nature (London)* **371**, 606-609.

Kajimoto, Y., Watada, H., Matsuoka, T., Kaneto, H., Fujitani, Y., Miyazaki, J., and Yamasaki, Y. (1997). Suppression of transcription factor PDX-1/IPF1/STF-1/IDX-1 causes no decrease in insulin mRNA in MIN6 cells. *J. Clin. Invest.* **100**, 1840-1846.

Kashimura, N., Morita, J., and Komano, T. (1979). Autoxidation and phagocidal action of some reducing sugar phosphate. *Carbohydr. Res.* **70**, C3-C7.

Lenzen, S., Drinkgern, J., and Tiedge, M. (1996). Low antioxidant enzyme gene expression in pancreatic islets compared with various other mouse tissues. *Free Radical Biol. Med.* **20**, 463-470.

Leonard, M., Peers, B., Johnson, T., Ferreri, K., Lee, S., Montminy, M. R. (1993). Characterization of somatostatin transactivator-1, a novel homeobox factor that stimulates somatostatin expression in pancreatic islet cells. *Mol. Endocrinol.* **7**, 1275-1283.

Matsuoka, T., Kajimoto, Y., Watada, H., Kaneto, H., Kishimoto, M., Umayahara, Y., Fujitani, Y., Kamada, T., Kawamori, R., and Yamasaki, Y. (1997). Glycation-dependent, reactive oxygen species-mediated suppression of the insulin gene promoter activity in HIT cells. *J. Clin. Invest.* **99**, 144-150.

Miller, C. P., McGehee, R. E., and Habener, J. F. (1994). IDX-1: A new homeodomain transcription factor expressed in rat pancreatic islets and duodenum that transactivates the somatostatin gene. *EMBO J.* **13**, 1145-1156.

Monnier, V. M., Kohn, R. R., and Cerami, A. (1984). Accelerated age-related browning of human collagen in diabetes mellitus. *Proc. Natl. Acad. Sci. U.S.A.* **81**, 583-587.

Myint, T., Hoshi, S., Ookawara, T., Miyazawa, N., Suzuki, K., and Taniguchi, N. (1995). Immunological detection of glycated proteins in normal and streptozotocin-induced diabetic rats using anti hexitol-lysine IgG. *Biochim. Biophys. Acta.* **1272**, 73-79.

Njoroge, F. J., and Monnier, V. M. (1989). The chemistry of the Maillard reaction under physiological conditions: A review. *Prog. Clin. Biol. Res.* **304**, 85-107.

Nose, K., Shibanuma, M., Kikuchi, K., Kageyama, H., Sakiyama, S., and Kuroki, T. (1991). Transcriptional activation of early-response genes by hydrogen peroxide in a mouse osteoblastic cell line. *Eur. J. Biochem.* **201**, 99-106.

Ohlsson, H., and Edlund, T. (1986). Sequence-specific interactions of nuclear factors with the insulin gene enhancer. *Cell (Cambridge, Mass.)* **45**, 35-44.

Ohlsson, H., Karlsson, K., and Edlund, T. (1993). IPF1, a homeodomain-containing transactivator of the insulin gene. *EMBO J.* **12**, 4251–4259.

Olson, L. K., Redmon, J. B., Towle, H. C., and Robertson, P. (1993). Chronic exposure of HIT cells to high glucose concentrations paradoxically decreases insulin gene transcription and alters binding of insulin gene regulatory protein. *J. Clin. Invest.* **92**, 514–519.

Olson, L. K., Sharma, A., Peshavaria, M., Wright, C. V. E., Towle, H. C., Robertson, R. P., and Stein, R. (1995). Reduction of insulin gene transcription in HIT-T15 β-cells chronically exposed to a supraphysiologic glucose concentration is associated with loss of STF-1 transcription factor expression. *Proc. Natl. Acad. Sci. U.S.A.* **92**, 9127–9131.

Robertson, R. P. (1989). Type II diabetes, glucose "non-sense," and islet desensitization. *Diabetes* **38**, 1501–1505.

Robertson, R. P., Olson, L. K., and Zhang, H. J. (1994). Differentiating glucose toxicity from glucose desensitization: A new message from the insulin gene. *Diabetes* **43**, 1085–1089.

Sakurai, T., and Tsuchiya, S. (1988). Superoxide production from nonenzymatically glycation protein. *FEBS Lett.* **236**, 406–410.

Sander, M., and German, M. S. (1997). The β cell transcription factors and development of the pancreas. *J. Mol. Med.* **75**, 327–340.

Sell, D. R., and Monnier, V. M. (1990). End-stage renal disease and diabetes catalyze the formation of a pentose-derived crosslink from aging human collagen. *J. Clin. Invest.* **85**, 380–384.

Sharma, A., Olson, L. K., Robertson, R. P., and Stein, R. (1995). The reduction of insulin gene transcription in HIT-T15 β cells chronically exposed to high glucose concentration is associated with loss of RIPE3b1 and STF-1 transcription factor expression. *Mol. Endocrinol.* **9**, 1127–1134.

Steiner, D. F., Chan, S. J., Welsh, J. M., and Kwok, S. C. M. (1985). Structure and evolution of the insulin gene. *Annu. Rev. Genet.* **19**, 463–484.

Tajiri, Y., Möller, C., and Grill, V. (1997). Long-term effects of aminoguanidine on insulin release and biosynthesis: Evidence that the formation of advanced glycosylation end products inhibits β cell function. *Endocrinology (Baltimore)* **138**, 273–280.

Tokuyama, Y., Sturis, J., DePaoli, A. M., Takeda, J., Stroffel, M., Tang, J., Sun, X., Polonsky, K. S., and Bell, G. I. (1995). Evolution of β-cell dysfunction in the male Zucker diabetic fatty rat. *Diabetes* **44**, 1447–1457.

Trouche, D., Grigoriev, M., Robin, P., and Bellan, A. H. (1993). The serum unresponsive Rous sarcoma virus promoter sustains a high serum response factor-dependent transcription in vitro. *Biochem. Biophys. Res. Commun.* **196**, 611–618.

Unger, R. H., and Grundy, S. (1985). Hyperglycemia as an inducer as well as a consequence of impaired islets cell function and insulin resistance: Implications for the management of diabetes. *Diabetologia* **28**, 119–121.

Watada, H., Kajimoto, Y., Umayahara, Y., Matsuoka, T., Kaneto, H., Fujitani, Y., Kamada, T., Kawamori, R., and Yamasaki, Y. (1996a). The human glucokinase gene beta-cell-type promoter: An essential role of insulin promoter factor 1 (IPF1)/PDX-1 in its activation in HIT-T15 cells. *Diabetes* **45**, 1478–1488.

Watada, H., Kajimoto, Y., Miyagawa, J., Hanafusa, T., Hamaguchi, K., Matsuoka, T., Yamamoto, K., Matsuzawa, Y., Kawamori, R., and Yamasaki, Y. (1996b). PDX-1 induces insulin and glucokinase gene expressions in αTC1 clone 6 cells in the presence of betacellulin. *Diabetes* **45**, 1826–1831.

22 Redox Regulation of Ischemic Adaptation

Nilanjana Maulik and Dipak K. Das
University of Connecticut School of Medicine
Farmington, Connecticut 06030

Ischemia and reperfusion cause injury to the heart, which is manifested by myocardial infarction, postischemic ventricular functional dysfunctions, arrhythmias, and cardiomyocyte apoptosis. Hearts can be adapted to ischemic–reperfusion injury by subjecting them to nonlethal cyclic episodes of short-term ischemia and reperfusion. The adapted myocardium becomes resistant to subsequent lethal ischemic injury. Reactive oxygen species (ROS) and oxidative stress play a crucial role in the pathophysiology of ischemic–reperfusion injury. When subjected to subsequent ischemia and reperfusion, adapted hearts generate a reduced amount of oxygen-free radicals compared to nonadapted hearts. The number of cardiomyocytes undergoing apoptotic cell death is reduced in adapted hearts subjected to ischemia and reperfusion. In concert, the adapted myocardium is associated with the increased antioxidant gene, Bcl-2, increased induction of the expression of the nuclear transcription factor NF-κB, and downregulation of AP-1 compared to nonadapted hearts. On the contrary, when nonadapted hearts are subjected to ischemia and reperfusion, Bcl-2 is downregulated whereas both NF-κB and AP-1 are upregulated. Ischemic adaptation triggers a tyrosine kinase-regulated signal transduction pathway involving p38 MAP kinase and MAPKAP kinase 2 leading to the phosphorylation and activation of heat shock protein (HSP) 27. The cardioprotective effects of adaptation can be blocked by pretreating the hearts with a hydroxyl radical scavenger, dimethyl thiourea, which also blocks the induction of transcription factors and HSP 27 gene expression. Taken together, it appears that myocardial adaptation to ischemia is precisely controlled by a redox switch and that oxygen-free radicals appear to function as a signaling molecule in this process.

I. INTRODUCTION

It is becoming increasingly apparent that redox signaling plays an essential role in myocardial ischemic reperfusion injury. The generation of oxygen-derived free radicals and the development of oxidative stress have been found to be important causative factors in the cellular injury associated with ischemia and reperfusion (Otani et al., 1986a,b). The presence of reactive oxygen species in the reperfused myocardium has been detected directly using electron spin-resonance (ESR) and high-performance liquid chromatography (HPLC) techniques (Tosaki et al., 1993a) and indirectly by estimating malonaldehyde formation and DNA damage (Cordis et al., 1995, 1998). Our laboratory has demonstrated that the reperfusion of ischemic myocardium undergoes apoptosis cell death, an oxidative stress-regulated process, further supporting the role of redox regulation in ischemic–reperfusion injury (Maulik et al., 1998a,b).

We and others have demonstrated a role of nitric oxide in the pathophysiology of ischemic–reperfusion injury. Numerous evidence exists in the literature indicating that exogenous nitric oxide can provide myocardial preservation in the face of ischemia (Maulik et al., 1995a, 1996; Engelman et al., 1996). Ischemic preconditioning and myocardial adaptation to ischemia have been found to be associated with the induction of the expression of inducible nitric oxide synthase (iNOS) mRNA (Tosaki et al., 1998).

A redox mechanism has been directly implicated in the activation of MAP kinases or JNK/SAPKs by H_2O_2, UV radiation, ischemia–reperfusion, and cytokines (Rouse et al., 1994). This pathway is controlled by Ras, Raf, MEK, and ERK kinases, and a small guanine nucleotide-binding protein links receptor tyrosine kinase activation to a cytosolic protein kinase cascade. This Ras/Raf/MEK/ERK pathway is generally known as the MAP kinase pathway for the stress signal. The signal is transmitted sequentially through three kinases: serine/threonine protein kinase (MAPKKK), which phosphorylates and activates MAPKK, which in turn phosphorylates and activates another serine/threonine protein kinase (MAPK) (Das, 1998). In this pathway, Raf corresponds to MAPKKK, MEK corresponds to MAPKK, and ERK corresponds to MAPK. Current evidence indicates that stress responses in mammalian cells are mediated by this signaling pathway (Rouse et al., 1994).

Evidence indicates that reduction/oxidation and oxidative stress/free radicals lead to the activation of NF-κB, which in turn induces the expression of genes (Duh et al., 1989; Toledano and Leonard, 1991; Meyer et al.,

1993). Interestingly, H_2O_2 was found to activate DNA binding of NF-κB *in vivo*, but not *in vitro* (Schreck *et al.*, 1991), suggesting that a by-product of H_2O_2 and not H_2O_2 itself may be responsible for the activation of NF-κB. Another related study using transient catalase overexpression in cos-1 cells showed that H_2O_2 may not serve as a messenger for tissue necrosis factor-α (TNF-α) or phorbol ester-induced NF-κB activation (Suzuki *et al.*, 1995). It is possible that the ·OH radical formed by the transient metal-catalyzed Fenton reaction during the reperfusion of ischemic heart (Bagchi *et al.*, 1990) can induce NF-κB activation. Inhibition of NF-κB induction by antioxidants further supports a role of free radicals in NF-κB activation (Ghosh *et al.*, 1990).

II. REDOX REGULATION OF ISCHEMIA–REPERFUSION

The cellular reduction/oxidation status (redox switch) regulates various aspects of cellular function. It also involves changes in the expression of particular genes including immediate early stress genes, induction of the genes coding for superoxide dismutase (SOD), heme oxygenase, thioredoxin, and heat shock proteins as a crucial step in the cellular response to oxidative stress (Jacquier-Sarlin and Polla, 1996). Thus, the "redox switch" may be viewed as the regulation of the intracellular reduction/oxidation state by an *invisible switch* that is involved in the pathogenesis of several disorders, including viral infections, cardiovascular diseases, immunodeficiency, malignant transformation, atherosclerosis, and diabetic abnormalities.

A. Generation of Reactive Oxygen Species

Myocardial cellular injury associated with the reperfusion of ischemic myocardium has been attributed to many interrelated factors, including intracellular Ca^{2+} overloading, loss of sarcolemmal phospholipids, and oxygen-free radical generation (Liu *et al.*, 1992). The participation of free radicals has been demonstrated from the beneficial effects of antioxidants and antioxidant enzymes (Tosaki *et al.*, 1993b; Das and Maulik, 1994) and free radical scavengers (Otani *et al.*, 1986b).

The presence of reactive oxygen species has been detected directly by HPLC using the electrochemical detection technique (Das *et al.*, 1989) and by ESR spectroscopy (Tosaki *et al.*, 1993b) as well as indirectly by the formation of malonaldehyde and 8-hydroxydeoxyguanosine (Cordis *et al.*, 1995, 1998). Tosaki *et al.* (1993b) compared the presence of ·OH using two different methods. Isolated buffer-perfused rat hearts were subjected to 30 min of normothermic global ischemia followed by 30 min of reperfusion.

5,5-Dimethyl-pyrroline-N-oxide (DMPO) was used as a spin-trap agent to detect ·OH radicals. In additional HPLC studies, salicylic acid was infused into the heart for the detection of ·OH radicals. In all studies, the effects of SOD and catalase on ·OH formation were examined. Irrespective of the methods used, the ·OH concentration was found to be increased drammatically between 60 and 90 sec of reperfusion, peaked between 180 and 210 sec, and then progressively decreased. In all cases, SOD and catalase were able to reduce the formation of ·OH radicals. The results further demonstrated that ·OH was produced only in fibrillating hearts, but not in nonfibrillating hearts.

B. Lipid Peroxidation

Reactive oxygen species generated in ischemic-reperfused myocardium attack most of the biomolecules. For example, polyunsaturated fatty acids in sarcolemmal phospholipids, proteins, and DNA molecules are potential targets of free radicals. After interaction with the lipids, free radicals cause lipid peroxidation, which can be detected as malonaldehyde formation. Numerous studies have demonstrated the increased formation of malonaldehyde in the ischemic-reperfused myocardium (Cordis *et al.*, 1995; Meerson *et al.*, 1982). A HPLC method has been described to monitor the malonaldehyde in heart that utilized the 2,4-dinitrophenylhydrazine derivatization of lipid metabolites (Cordis *et al.*, 1995). Results of the study demonstrated the progressive increase of malonaldehyde and other lipid metabolites, supporting previous reports concerning the increased formation of lipid peroxidation products during reperfusion.

C. DNA Damage

Although DNA is a well-known target for free radical attack, little attention had been paid to the injury of DNA molecules associated with ischemia and reperfusion. Damage of DNA by oxygen-free radicals results in the production of a large number of lesions that can be grouped into strand breaks and base modification products. At least three modified bases, 8-hydroxyguanine, 5-hydroxymethyluracil, and thymine glycol, are formed when ·OH attacks a DNA molecule (Cordis *et al.*, 1998). Following an oxidative insult on DNA, guanine bases rapidly undergo 8-hydroxylation, followed by the repairment of the damaged DNA by exonucleases. The formation of 8-hydroxydeoxyguanosine (8-OHDG), a product of hydroxyl radical (·OH)-DNA interaction, has been monitored in the postischemic myocardium. A simple HPLC with UV detection detected picomole levels of 8-OHDG in the preischemic heart that increased steadily and progressively as a function of reperfusion time. A similar rise in 8-OHDG was noticed when isolated hearts were perfused with a ·OH-generating system. Corroborating with the increased 8-OHDG formation, an increased amount of creatine kinase was released from the coronary effluent, indicating increased tissue injury. The formation of 8-OHDG was blocked completely when hearts were preperfused with the oxygen-free radical scavenger 1,3-dimethyl-2-thiourea (DMTU), which also reduced the appearance of CK significantly in the coronary effluent, suggesting that oxidative DNA damage plays a role in the pathophysiology of ischemic–reperfusion injury.

D. Thioredoxin and Glutaredoxin

Thioredoxin and glutaredoxin have been recognized as key players for redox regulation. Thioredoxin is a heat-stable protein (12 kDa) containing a redox-active disulfide bridge. It is present ubiquitously in the cells. The oxidized form of thioredoxin is reduced by NADPH and reductase. The two Cys residues can be reversibly oxidized to participate in redox reactions (Wieles *et al.*, 1995). Thus, thioredoxin is an important cellular factor possessing thiol-mediated redox activity and plays a crucial role in the regulation of cellular functions, programed cell death, and gene expression.

A study demonstrated the presence of glutaredoxin and thioredoxin in human plasma from patients undergoing open heart surgery. Plasma glutaredoxin levels did not change significantly during cardiopulmonary bypass, but thioredoxin levels in arterial plasma were elevated during reperfusion of the postcardioplegic heart (Nakamura *et al.*, 1998). These results indicate that oxidized thioredoxin is released into plasma during cardiopulmonary bypass, suggesting that these redox-active proteins may be important components in the plasma defense against oxidative stress.

E. Myocardial Adaptation to Ischemia

Myocardial adaptation to ischemic stress, which is the manifestation of the earlier stress response that occurs during repeated episodes of brief ischemia and reperfusion, can render the myocardium more tolerant to a subsequent potential lethal ischemic injury (Flack *et al.*, 1991). This transient adaptive response has been demonstrated to be associated with decreased reperfusion-induced arrhythmias (Tosaki *et al.*, 1994), increased the recovery of postischemic contractile functions (Kimura *et al.*, 1992; Cave *et al.*, 1993; Asimakis *et al.*, 1992), and reduction of the infarct size (Cave *et al.*, 1993; Schott *et al.*, 1990)). The adaptive protection has been found to be mediated by gene expression and their transcriptional regulation (Das *et al.*, 1993, 1995).

Ischemic preconditioning provides a powerful anti-ischemic protection of the heart. For example, ischemic preconditioning induced by repeated short-term ischemia and reperfusion has been found to reduce postischemic left ventricular functional abnormalities, ventricular arrhythmias, infarct size, and cell damage (Tosaki et al., 1994, 1998) and increased the recovery of postischemic contractile functions (Kimura et al., 1992; Cave et al., 1993). Although preconditioning is a powerful tool in protecting the myocardium from ischemic–reperfusion injury, considerable debate remains regarding its mechanism of action. Most of the existing studies have focused on the role of adenosine A1 receptors in preconditioning (Murry et al., 1990). In addition, α1-adrenergic receptors (Tosaki et al., 1995), muscarinic receptors (Kida et al., 1991), ATP-dependent potassium channels (Gross and Auchampach, 1992), multiple receptors, including bradykinin and angiotensin II receptors (Goto et al., 1995), and G-protein (Lawson et al., 1993) have been implicated in ischemic preconditioning. Irrespective of the signaling pathways involved, it is more or less accepted that the intracellular signaling leads to the translocation and activation of protein kinase C (PKC) (Tosaki et al., 1994). Involvement of PKC has been demonstrated directly by biochemical and immunological studies and indirectly by experiments demonstrating that PKC inhibitors block the beneficial effects of preconditioning (Tosaki et al., 1996; Bugge and Ytrehus, 1995).

Studies have demonstrated a tyrosine kinase-regulated MAP kinase signaling leading to the activation of MAPKAP kinase 2 and increased phosphorylation followed by the activation of HSP 27 (Maulik et al., 1996c). It is believed that reactive that reactive oxygen species play a crucial role in the signal transduction process for HSP 27 activation (Zu et al., 1997).

III. REDOX REGULATION OF ISCHEMIC ADAPTATION

As mentioned earlier, mammalian heart can be adapted for a longer period of ischemia by repeatedly subjecting the myocardium to short-term reversible ischemia followed by another brief period of reperfusion. Such an intermittent short duration of ischemia imposes redox/oxidative stress on cardiac tissue. Several lines of evidence show that oxidative stress is one of the most important mediators of molecular switches modulating candidate gene expression for adaptation. Prolong ischemia causes necrosis/infarction and ultimately heart failure. Strong similarities exist among the patterns of acute stress response induced by hypoxia/reoxygenation, ischemic/reperfusion, and reactive oxygen species (Herrlich et al., 1992; Schreck et al., 1992). Reactive oxygen species play an important role in the pathogenesis of a variety of human diseases. Thus, the cellular reduction/oxidation status (redox switch) regulates various aspects of cellular function. It also involves changes in the expression of particular genes, including immediate early stress genes. Induction of the genes coding for superoxide dismutase, heme oxygenase, thioredoxin, and heat shock proteins is a crucial step in the cellular response to oxidative stress (Matsui *et al.*, 1995).

A. Role of Free Radicals/Oxidative Stress

The development of oxidative stress plays a major role in the pathogenesis of myocardial ischemic–reperfusion injury. Ischemic adaptation, which involves repeated episodes [generally four times (4×PC) of ischemia and reperfusion initially, i.e., after the first episode of ischemia–reperfusion (1×PC)], produces oxygen-free radicals and develops oxidative stress, as expected. However, the amount of free radical generation does not increase progressively with each consecutive episode of ischemia–reperfusion. At the end of the fourth episode of ischemia–reperfusion (4×PC), the amount of free radicals actually begins to decline as compared to that produced after 1×PC. After acute ischemia and prolonged reperfusion following ischemic adaptation, the amount of free radicals/oxidative stress in the heart becomes significantly less compared to those in ischemic-reperfused hearts (without adaptation) (Cordis *et al.*, 1998). In concert, a number of oxidative stress-inducible genes as well as heat shock protein genes are induced in the adapted myocardium, which may comprise the cellular defense system (Das *et al.*, 1993). This suggests that oxidative stress developed in the ischemic-reperfused myocardium is translated into the defense system after subsequent ischemia–reperfusion.

The production of malonaldehyde (MDA) is an indicator for the development of oxidative stress. We have estimated the level of MDA formation to monitor the extent of lipid peroxidation (Cordis *et al.*, 1995). After 30 min of ischemia followed by 120 min of reperfusion, the MDA content was increased significantly compared to baseline values. At the end of a 2-hr reperfusion, the MDA content in adapted hearts was significantly less compared to that found in nonadapted hearts, demonstrating that adaptation lowers the oxidative stress in the heart. It appears that ROS generated during the early ischemia–reperfusion function as signaling molecules leading to the upregulation of antioxidants/antioxidant enzymes as well as other protective proteins such as heat shock and antioxidant proteins.

B. Signal Transduction

A study from our laboratory demonstrated that myocardial adaptation to ischemia could potentiate a signal transduction pathway leading to the activation of phospholipase D, MAP kinases, and MAPKAP kinase 2 activities that were inhibited by genistein, suggesting for the first time tyrosine kinase PLD as a potential signaling pathway for ischemic adaptation (Maulik et al., 1996c; Cohen et al., 1996). Many oxidants, such as H_2O_2, can increase protein tyrosine phosphorylation in conjunction with PLD activation (Ito et al., 1997). In neutrophils and HL-60 cells, oxygen-free radicals generated by f-Met-Leu-Phe resulted in increased protein tyrosine phosphorylation and PLD activation (Watson and Edwards, 1998). Genistein not only blocked the oxidant-mediated protein tyrosine phosphorylation in endothelial cells, but a correlation between tyrosine kinase inhibition and PLD activation was also observed with genistein. Although evidence supports a role of tyrosine kinases in PLD activation, the mechanisms of activation remain unknown. It is possible that PLD activation is the direct manifestation of protein tyrosine kinase phosphorylation. It is also possible that tyrosine kinase generates other intermediate proteins that are instrumental in PLD activation. The role of tyrosine kinases was further substantiated from the observations that the phosphorylation of protein tyrosine kinases was enhanced after ischemic preconditioning.

Many studies indicate that mitogen-activated protein (MAP) kinases, a novel serine/threonine protein kinase family, play an essential role in mediating intracellular signal transduction events (Seger and Krebs, 1995). In response to extracellular stimulation, MAP kinases are activated rapidly and, in turn, regulate cellular functions by inducing the phosphorylation of proteins, such as oncogene product c-*jun*, S6 ribosomal protein kinase, and MAP kinase-activated protein kinase 2 (Anderson et al., 1990). MAPKAP kinase 2 has been implicated in a novel mammalian stress-activated signal transduction pathway initiated by a variety of mitogens, proinflammatory cytokines, or environmental stresses, where it regulates its substrate molecules by serine/threonine phosphorylation. Stimulation of cultured cardiomyocytes with A_1 selective adrenergic analogs, endothelin 1, fibroblast growth factors, and mechanical stress activates the MAP kinase signaling cascade (Yamazaki et al., 1996). This kinase cascade includes MAP kinases, MAP kinase kinases, and MAP kinase kinase kinases. Three distinct MAP kinases have been identified: the ERK group, the JNK/SAPK group, and the p38 MAP kinase isoform of which the last one can activate MAPKAP kinase 2 by phosphorylation.

In case of rat heart, a mitogen-activated protein kinase cascade has already been identified (Lazou et al., 1994). These authors have demonstrated that MAPK isoforms p42MAPK and P44 MAPK and two peaks of MEK were activated by more than 10-fold in perfused hearts or ventricular myocytes exposed to PMA for 5 min. In our own study, we identified the participation of MAP kinase cascades in the ischemic preconditioning of rat hearts. The results of our study demonstrated that a kinase cascade involving tyrosine kinase–phospholipase D–MAP kinases–MAPKAP kinase 2 is triggered after ischemic stress.

MAP kinase activation has been found to be essential during the bombesin-induced, PKC-mediated sustained contraction in smooth muscle cells, and the redistribution of MAP kinases was colocalized with the redistribution of HSP 27 in smooth muscle cells (Yamada et al., 1995). In view of the evidence that the HSP 27 gene is induced after ischemic preconditioning (Das and Maulik, 1995), it seems likely that MAP kinases are involved in signal transduction leading to this gene expression. Indeed, in another related study, the activation of cardiac gene expression during phenylephrine-induced hypertrophy seemed to require ERK activation (Thorburn et al., 1995).

A new member of the MAP kinase family, p38 MAP kinase, has been identified. This MAP kinase seems to possess a dual phosphorylation motif (Thr-Gly-Tyr) in place of the Thr-Pro-Tyr motif present in Jnk and the Thr-Glu-Tyr motif present in Erk. Our laboratory has demonstrated that the P38 MAP kinase is translocated and activated after ischemic preconditioning. Additionally, an inhibitor of p38 MAP kinase blocked the effects of ischemic preconditioning.

The MAP kinase signal transduction pathway is likely to involve the activation of Ras or Raf-1, which in turn induces mitogen-activated protein kinase kinase (MKK) and MAP kinases. It is also known that the Raf-1 kinase possesses MAPKKK activity and lies upstream from MAPKK and MAP kinases in various cell types (Force et al., 1994). Seko et al., (1996) have demonstrated that hypoxia and hypoxia/reoxygenation activated Raf-1, and MKK as well as MAP kinases in cultured rat cardiomyocytes. Raf-1 operates downstream from cell surface-associated tyrosine kinases and upstream from MAP kinases. Raf is not strictly a member of the MEKK family, but is functionally analogous (Macdonald et al., 1993). Ras is part of the signal transduction chain extending from extracellular signals to transcriptional regulation in the nucleus (Morrison, 1995; Cleghon and Morrison, 1994).

MAPKAP kinase 2 is a downstream protein kinase in the stress-activated signal transduction pathway. Northern blot analysis indicated that this enzyme is highly expressed in heart tissues, suggesting that MAPKAP kinase 2 may function in myocardium to stress or mitogenic stimulation. In addition, the MAPKAP kinase 2 activity of cardiomyocytes is stimulated by a myocardial hypertrophic factor and oxidative stress as well as heat shock. Recombinant MAPKAP kinase 2 phosphorylates HSP 27 of cardiomyocytes in an in vitro phosphorylation assay. The known association between MAPKAP kinase 2 and heat shock proteins gives rise to the possibility that the kinase may be involved in a stress-related and redox-regulated mechanism of myocardium that leads to cardioprotection. Our previous studies indicate that environmental stresses, including heat shock and oxidative stress, and phorbol ester treatment of cultured cardiac myoblast cells resulted in a rapid increase in cellular MAPKAP kinase 2 activity (Zu et al., 1997).

C. Role of Oxygen Radicals as Second Messenger

Oxygen-derived free radicals have been implicated in the transmembrane signaling process (Schieven et al., 1993). In this study, the authors provided evidence that a tyrosine kinase inhibitor, herbimycin A, and a free radical scavenger, N-acetylcysteine, inhibit free radical-induced activation of NF-κB, indicating that activation triggered by ROS is dependent on tyrosine kinase activity. A large number of studies from different laboratories, including our own, showed the induction of the expression of antioxidant genes during preconditioning (Das et al., 1994). We have demonstrated nuclear translocation and activation of NF-κB in response to preconditioning (Maulik et al., 1998c). An increased binding of NF-κB was found to be dependent on both tyrosine kinase and p38 MAP kinase.

The nuclear transcription factor, NF-κB, has been found to play a role in the signaling process. A study from our laboratory demonstrated inhibition of the enhanced tyrosine kinase phosphorylation during ischemic adaptation by DMTU (Das et al., 1999). DMTU also inhibited the preconditioning-mediated increased phosphorylation of p38 MAP kinase and MAPKAP kinase 2 activity. However, DMTU had no effect on the translocation and activation of PKC resulting from preconditioning. Preconditioning reduced myocardial infarct size as expected. This cardioprotective effect of preconditioning was abolished by both DMTU and SN 50. Preconditioning resulted in the nuclear translocation and activation of NF-κB. Increased NF-κB binding was blocked by both DMTU and SN 50. The results of this study demonstrate that ROS play a crucial role in signal transduction mediated by preconditioning. This signaling process appears to be potentiated by tyrosine kinase phosphorylation, resulting in the activation of p38 MAP kinase and MAPKAP kinase 2 leading to the activation of NF-κB, suggesting a role of oxygen-free radicals as second messengers. Free radical signaling seems to be independent of PKC, although PKC is activated during the preconditioning process, suggesting a role of two separate signaling pathways in ischemic preconditioning.

D. Oxidative Stress-Inducible Genes

Studies from our laboratory documented the induction of BCL-2 in the ischemically adapted rat myocardium (Maulik et al., 1997). The same gene is being downregulated by the oxidative stress developed during the reperfusion of ischemic myocardium (Maulik et al., 1998d). Ebselen, a glutathione peroxidase mimic, can preserve Bcl-2 by eliminating oxidative stress in the postischemic-reperfused myocardium. Our laboratory has also documented that a sequential upregulation of energy metabolism genes is induced by ischemia (Moraru et al., 1994). The central role of mitochondria is oxidative ATP synthesis, which leads to the deprivation of oxygen, thus triggering the stimulus to synthesize some of the mitochondrial genes (ATPase 6, ATPase 8, ATPase $F_{1\alpha}$) that are associated with preconditioning, ultimately leading to myocardial adaptation. Both $^1O_2$ and ·OH have been implicated in the induction of the expression of heme oxygenase. The fact that the expression of heme oxygenase can be induced directly by oxygen-derived free radicals (Maines and Kappas, 1997) or by the depletion of glutathione (Ewing and Maines, 1993) raises the interesting possibility that oxidative stress developed during reperfusion potentiates the induction of heme oxygenase in heart. The inhibition of HO-1 induction by SOD and catalase demonstrates conclusively the involvement of oxygen-free radicals. This is supported further by the induction of heme oxygenase mRNA in heart by ·OH radicals (Maulik et al., 1996d). The role of heme oxygenase in cellular protection remains controversial.

E. Genes for Heat Shock Proteins and Cross-Talk with Oxidative Stress-Inducible Proteins

Both ischemia–reperfusion and ischemic adaptation result in the induction of the expression of several HSPs, including HSP 27, HSP 32, HSP 70, and HSP 89. After a heat shock, HSP is phosphorylated rapidly and then its synthesis is enhanced. Such enhancement not only confers the cells resistant to subsequent heat stress, but also makes them more resistant to oxidative stress. $TGF\beta_1$ was found to inhibit DNA synthesis in mouse osteoblastic cells in parallel with the phosphorylation of HSP 27 (Shibanuma et al., 1992) and the effects of both events were abolished by catalase (Nose et al., 1994). It has been demonstrated that a constitutively high expression of HSP 27 in the immortalized human fibroblast cell line KMST-6 made them susceptible to oxidative stress, resulting in growth arrest, and further suggested that this mechanism could involve the phosphorylation of HSP 27 (Arata et al., 1995). HSP 27 gene expression was found when isolated rat hearts were subjected to oxidative stress induced by interleukin-1α (IL-1α) (Maulik et al., 1993). In a biological system including the heart, cytokines such as IL-1α and TNF potentiate the generation of oxygen-free radicals. In this study, IL-1α induced the expression of HSP 27 mRNA within 2 hr and HSP protein after 48 hr. Hearts pretreated with IL-1α for 48 hr and then subjected to 30 min of ischemia followed by 30 min of reperfusion showed increased tolerance to ischemia–reperfusion injury. In another related study, heat shock was induced by injecting the pigs with an amphetamine and, after 40 hr, the isolated *in situ* pig hearts were subjected to 1 hr of LAD occlusion followed by 1 hr of global hypothermic cardioplegic arrest and 1 hr of normothermic reperfusion (Maulik et al., 1994). Such heat shock was associated with the enhanced induction of the HSP 27 mRNA in the hearts in concert with the reduction of postischemic myocardial injury (Maulik et al., 1995b). These hearts demonstrated a reduction in oxidative stress in conjunction with an increased tolerance to ischemia–reperfusion injury.

Oxidative stress was found to serve as redox signals for the induction of the heat shock response. For example, when HepG2 and V79 cells were exposed to a redox-cycling agent, menadione, the oxidative stress resulted in the oxidation of protein thiols in a dose-dependent manner (McDuffce et al., 1997). Cells treated with menadione exhibited activation of heat shock factor (HSF-1) in concert with the induction of the expression of HSP 70 mRNA.

F. Immediate Early Genes

Alteration in the expression patterns of many "early response" genes has been reported to be expressed in mammalian systems in response to oxidative stress. These include c-*fos*, *egr-1*, c-*jun*, and c-*myc*. Because these immediate early genes encode transcription factors, they have the power to further modulate gene expression (Das and Maulik, 1997). Numerous evidence exists to indicate that most protooncogenes are involved in the transcription regulation of a variety of genes, including stress-inducible genes. The protooncogene c-*fos* is known to be expressed under any kind of stress that leads to the alteration of the redox state within the system (Gu and Matlashewski, 1995). We have already established the expression of two protooncogenes, c-*fos* and c-*myc*, in the isolated, perfused, ischemic-preconditioned rat heart (Das and Maulik, 1997). In this experiment, isolated buffer-perfused rat hearts were made ischemic either 5 min followed by 10 min of reperfusion (1×PC) or for four 5-min episodes of ischemia each separated by 10 min of reperfusion (4×PC). The Northern hybridization technique detected c-*fos* and c-*myc* in 4×PC hearts after 60 min of reperfusion, and the signal for c-*fos* was much stronger than c-*myc*. 1×PC hearts showed some expression of c-*fos* mRNA, but only after 60 min of reperfusion. Studies have suggested that mild redox alterations can alter cell function via several mechanisms, one of which is definitely by the alteration in both early and late gene expression. Protein kinase C is instrumental for the activation of variety of genes, including protooncogenes.

G. Antioxidants, Oxidative Stress-Inducible Genes, and Intracellular Defense

There is general agreement that the amount of several antioxidants and antioxidant enzymes are reduced significantly after ischemia and reperfusion. For example, reduced amounts of SOD, catalase, and glutathione peroxidase enzymes, as well as α-tocopherol and ascorbic acid, have been found in the ischemic-reperfused myocardium. The loss of key antioxidant enzymes and antioxidants thus reduces the overall antioxidant reserve of the heart and makes the heart susceptible to ischemia–reperfusion injury. One of the major functions of antioxidants is to block free radical formation. Thus, the reduced antioxidant reserve in conjunction with increased free radical generation leads to the cellular injury associated with ischemia–reperfusion.

It is believed that antioxidants such as SOD, catalase, and glutathione peroxidase enzymes, as well as α-tocopherol, comprise the first line of defense against many degenerative diseases, including myocardial ischemic reperfusion injury (Das and Maulik, 1995). Additionally, induction of the expression of the oxidative stress-inducible genes can be viewed as a stress survival signal for the heart and may be viewed as another line of defense for the ischemic myocardium (Das *et al.*, 1995).

IV. ACTIVATION OF MULTIPLE TRANSCRIPTION FACTORS

It has been documented that ischemic adaptation translocated and increased the binding of nuclear transcription factor NF-κB in the heart (Maulik *et al.*, 1998c). NF-κB is a member of the Rel transcription factor family, which is involved in the regulation of stress defense mechanisms. Because ischemic adaptation was also found to reduce apoptosis, we speculated a direct role of NF-κB in apoptosis. AP-1 is another redox-sensitive signaling molecule that also plays an important regulatory role in cellular responses to stress induced by external factors, including UV radiation, phorbol esters, and TNF-α (McMahan and Monroe, 1992). The binding site of AP-1 is recognized by Jun family member homodimers and Jun/Fos family member heterodimers. The balance between Jun and Fos is very critical for gene expression. The induction of apoptosis by elevated levels of c-Jun is a crucial event in growth factor-deprived nerve cells. Stress induced by ischemia–reperfusion was shown previously to induce the activation of c-Jun (Diamond *et al.*, 1990; Ham *et al.*, 1995; Artuc *et al.*, 1997; Ishikawa *et al.*, 1997).

A. NF-κB

NF-κB is a critical regulator for gene expression induced by diverse stress signals, including mutagenic, oxidative, and hypoxic stresses. Activation of NF-κB is likely to be involved in the induction of gene expression associated with ischemic adaptation. Studies from our laboratory demonstrated ischemic adaptation translocated and increased the binding activity of NF-κB in heart. NF-κB-binding activity was found to be very low in nonischemic control hearts (Maulik et al., 1998c) (Fig. 1). Reperfusion of ischemic myocardium significantly increased the translocation of NF-κB from cytosol to nucleus. Perfusion of the heart with DMTU inhibited NF-κB translocation from cytosol to nucleus. The binding increased slightly for 1xPC hearts and appreciably for 4xPC hearts. Supershift assays were performed with polyclonal antibodies recognizing NF-κB, P50, and P65 subunit proteins. Both antibodies partially shifted the major ischemia/reperfusion and preconditioned mediated induction of NF-κB oligonucleotide complex.

B. AP-1

As mentioned earlier, AP-1 is a redox-sensitive signaling molecule that plays an important regulatory role in cellular responses to stress induced by external factors. The AP-1 transcription factor complex is composed of

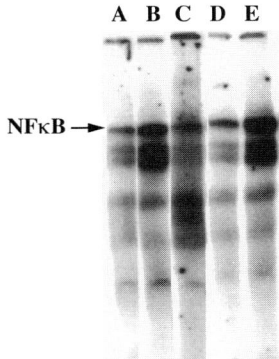

Figure 1 NF-κB-binding activity in rat heart. Nuclear translocation of NF-κB by ischemia–reperfusion and PC. Nuclear extracts were isolated from control and experimental hearts. These extracts were then used for electrophoretic moblity shift assay. Lane A, control; lane B, ischemia–reperfusion; lane C, DMTU + ischemia–reperfusion; lane D, 1×PC followed by ischemia–reperfusion; and lane E, 4×PC followed by ischemia–reperfusion. Results are representative of six animals per group. Each assay was run in duplicate. Data reproduced with permission from Maulik et al. (1998c).

a group of proteins encoded by the Jun and Fos families that bind to AP-1 consensus sequences. The regulation of AP-1 in response to external stimuli is mediated by members of the MAPK family (Macho et al., 1998). AP-1 regulates the activation of transcription of a variety of genes.

Electrophoretic mobility shift assays indicated increased AP-1-binding activity in the ischemic–reperfused rat heart compared to the control perfused group (Fig. 2). DMTU inhibited this binding activity significantly. Similar effects were also observed for ischemically adapted groups (both 1×PC and 4×PC). Myocardial adaptation induced by cyclic episodes of short-term ischemia and reperfusion results in the activation of NF-κB but not AP-1.

V. PROGRAMED CELL DEATH IN ISCHEMIA–REPERFUSION

Apoptosis or programmed cell death is a genetically controlled response for cells to commit suicide and is associated with DNA fragmentation or laddering. Common inducers of apoptosis include oxygen-free radicals/oxidative stress and Ca^{2+}, which are also implicated in the pathogenesis of myocardial ischemic reperfusion injury. Although cardiomyocyte death and infarction associated with ischemia–reperfusion injury are traditionally be-

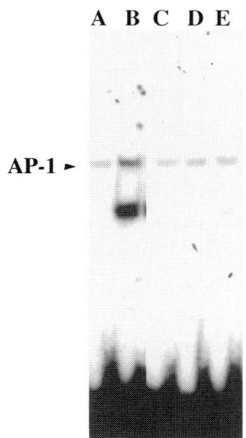

Figure 2 AP-1-binding activity in rat heart. Nuclear translocation of AP-1 by ischemia–reperfusion and PC. Nuclear extracts were isolated from control and experimental hearts. These extracts were then used for electrophoretic moblity shift assay. Lane A, control; lane B, ischemia–reperfusion; lane C, DMTU + ischemia–reperfusion; lane D, 1×PC followed by ischemia–reperfusion; and lane E, 4×PC followed by ischemia–reperfusion. Results are representative of six animals per group. Each assay was run in duplicate.

lieved to be induced via necrosis, which is a clear-cut mechanism of accidental cell death, its precise mechanism remains unclear. Studies have implicated apoptotic cell death in ischemic brain (Tominaga *et al.*, 1993) and ischemic liver (Fukuda *et al.*, 1993). Reperfusion of ischemic renal tissues was found to be associated with apoptopic cell death (Schumer *et al.*, 1992). Cardiomyocytes exposed to hypoxia revealed apoptotic cell death as evidenced by DNA fragmentation in conjunction with the expression of Fas mRNA (Tanaka *et al.*, 1994). More recently, evidence has been furnished in support of reperfusion injury-mediated apoptosis in cardiomyocytes. Based on pathologic evaluation, these investigators concluded that apoptosis might be a specific feature of reperfusion injury in cardiac myocytes. In another study, apoptotic and necrotic myocyte cell deaths associated with ischemia–reperfusion were shown to be independent contributing variables of infarct size in rats (Kajstura *et al.*, 1996). Apoptosis is a feature of human vascular pathology, including restenotic lesions, and, to a lesser extent, atherosclerotic lesions, suggesting that apoptosis may modulate the cellularity of lesions that produce human vascular obstruction (Kajstura *et al.*, 1996).

It has been demonstrated that apoptotic cell death is a function of the duration of reperfusion and that up to 2 hr of ischemia does not induce apoptosis (Isner *et al.*, 1995; Maulik *et al.*, 1998e). In this study, isolated perfused rat hearts were subjected to 15, 30, or 60 min of ischemia as well as to 15 min of ischemia followed by 30, 60, or 120 min of reperfusion. At the end of each experiment, hearts were processed for the evaluation of apoptosis and DNA laddering. Apoptosis was studied by visualizing the apoptotic cardiomyocytes by direct fluorescence detection of digoxigenin-labeled genomic DNA using APOPTAG in an *in situ* apoptosis detection kit. DNA laddering was evaluated by subjecting DNA obtained from the hearts to 1.8% agarose gel electrophoresis and photographing under UV illumination. Results of this study revealed apoptotic cells in only 90- and 120-min reperfused hearts as demonstrated by the intense fluorescence of the immunostained digoxigenin-labeled genomic DNA when observed under fluorescence microscopy. None of the ischemic hearts showed any evidence of apoptosis. These results were corroborated with the findings of DNA fragmentation, which showed increased ladders of DNA bands in the same reperfused hearts representing integer multiples of the internucleosomal DNA length (about 180 bp). The presence of apoptotic cells and DNA fragmentation in the myocardium were completely abolished by pretreating the myocardium with ebselen, a glutathione peroxidase mimic that also reduces the ischemic reperfusion injury as evidenced by the better recovery of left ventricular performance in the preconditioned myocardium (Maulik *et al.*, 1998d). In concert, ebselen reduced oxidative stress developed in the ischemic-reperfused myocardium. In another related study, SOD and catalase were found to ameliorate apoptotoic cell death (Maulik et al., 1999). Results of these studies indicate that the reperfusion of ischemic heart, but not ischemia, induces apoptotic cell death and DNA fragmentation, which can be inhibited by a glutathione peroxidase mimic such as ebselen or free radical scavengers SOD plus catalase.

VI. ISCHEMIC ADAPTATION AND APOPTOSIS

Our laboratory demonstrated a reduction of apoptotic cell death by myocardial adaptation to ischemia (Maulik et al., 1998d). The number of apoptotic cells was significantly higher (23%) in the ischemic–reperfused myocardium compared to control hearts. DMTU significantly reduced the number of apoptotic cells compared to ischemic-reperfused hearts. Myocardial adaptation to ischemia reduced the number of apoptotic cells (Fig. 3).

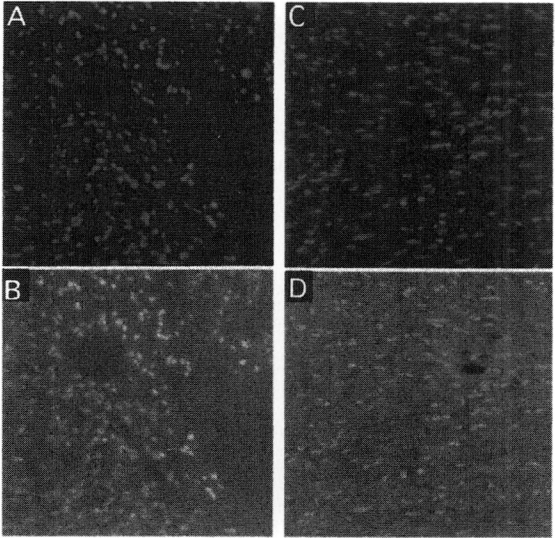

Figure 3 Effects of preconditioning on apoptotic cell death. Results for propidium iodide (PI) staining are shown in red (A and C). PI and fluorescein channels are superimposed to show apoptotic cells as glowing yellow-green cells under the microscope (B and D) on a green background. All normal cells are stained in orange when superimposed. A and B: 15 min ischemia followed by 120 min of reperfusion. C and D: PC followed by 15 min of ischemia and 120 min of reperfusion. Data reproduced with permission from Maulik et al. (1998b). (See color reproduction in color plate section.)

For 1×PC hearts, these number were reduced by 18%, whereas almost no apoptotic cells were noticed in 4×PC hearts.

Corroborating with these results, DNA fragmentation was clearly visualized in hearts subjected to 30 min of ischemia followed by 2 hr of reperfusion (Fig. 4). DNA ladders were not apparent in control hearts. DMTU significantly reduced DNA fragmentations when applied to the heart prior to ischemic insult. Ischemic adaptation was also associated with a decrease in DNA fragmentation (Fig. 5). The extent of inhibition was more in 4×PC hearts than in 1×PC hearts.

VII. SUMMARY AND CONCLUSION

The heart possesses the remarkable ability to adapt itself against any stressful situation by increasing resistance to the adverse consequences. Creating a stress reaction by repeated ischemia and reperfusion or subjecting the hearts to heat or oxidative stress enables the heart to meet the future stress challenge by upregulating its cellular defense through the direct accumulation of intracellular mediators that constitute the material basis of increased adaptation to stress. Thus, the powerful cardioprotective effect of adaptation is likely to originate at cellular and molecular levels (Das, 1996).

Ischemic preconditioning is the manifestation of the earlier stress response that occurs during repeated episodes of brief ischemia and reperfusion and can render the myocardium more tolerant to a subsequent potential lethal ischemic injury. This transient adaptive response has been demon-

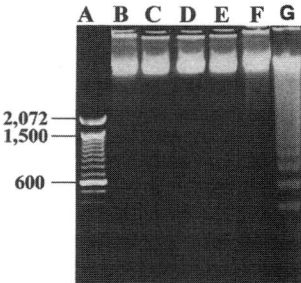

Figure 4 Effects of ischemia and reperfusion on DNA fragmentation. DNA was isolated from hearts and subjected to 1.8% agarose gel electrophoresis. A 100-bp DNA ladder was used as a molecular weight marker (lane A). Lane B, baseline; C, 15 min ischemia; D, 15 min ischemia followed by 30 min reperfusion; E, 15 min ischemia followed by 60 min reperfusion; F, 15 min ischemia followed by 90 min reperfusion; and G, 15 min ischemia followed by 120 min reperfusion. Data reproduced with permission from Maulik et al. (1998e).

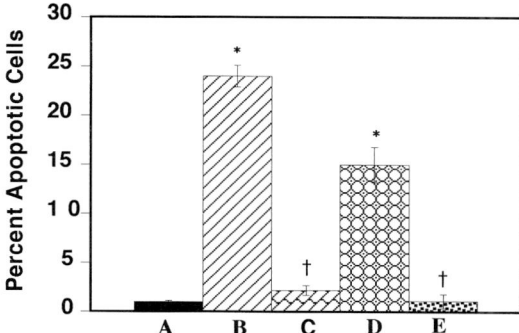

Figure 5 Estimation of the number of apoptotic cells (%) using double antibody staining. Evaluation of apoptosis revealed increased number of apoptotic cells in the ischemic–reperfused myocardium. Sections of control and experimental heart tissues were analyzed, and the number of apoptotic cells was counted under the microscope. The percentage of apoptotic cells is shown by a bar graph. (A) Control, (B) ischemia–reperfusion, (C) DMTU + ischemia–reperfusion, (D) 1 × PC followed by ischemia–reperfusion, and (E) 4×PC followed by ischemia–reperfusion. Results are shown as means ± SEM of six animals per group. Each assay was run in duplicate.

strated to be associated with decreased reperfusion-induced arrhythmias, increased recovery of postischemic contractile functions, and reduction of infarct size. The adaptive protection is believed to be mediated by gene expressions and their transcriptional regulation. Findings indicate that multiple kinases, including MAP kinases and MAPKAP kinase 2, are likely to be involved in the adaptive signaling process. The acutely developed adaptive effect is short-lived, lasting for only up to 2–3 hr. Hearts can subsequently undergo a secondary and delayed adaptation to stress presumably through the induction of the expression of new genes and their subsequent translation into proteins. A number of genes and proteins have been indentified as being possibly involved in the development of delayed preconditioning, including heat shock proteins, superoxide dismutase, catalase, and nitric oxide synthase as well as ATPase 6 and cytochrome b subunits. Such an adaptive response becomes evident only after approximately 24 hr of stress treatment and may include stress induced by heat shock, oxidant, or other stress-inducible agents. MAPKAP kinase 2 appears to link the early preconditioning effect to the delayed adaptative response (Fig. 6).

Evidence is rapidly accumulating to support the notion that oxygen-derived free radicals are involved in the transmembrane signaling process. A large number of studies from different laboratories, including our own, showed the induction of the expression of antioxidant genes during preconditioning. Our study on the role of NF-κB in tyrosine kinase signaling of

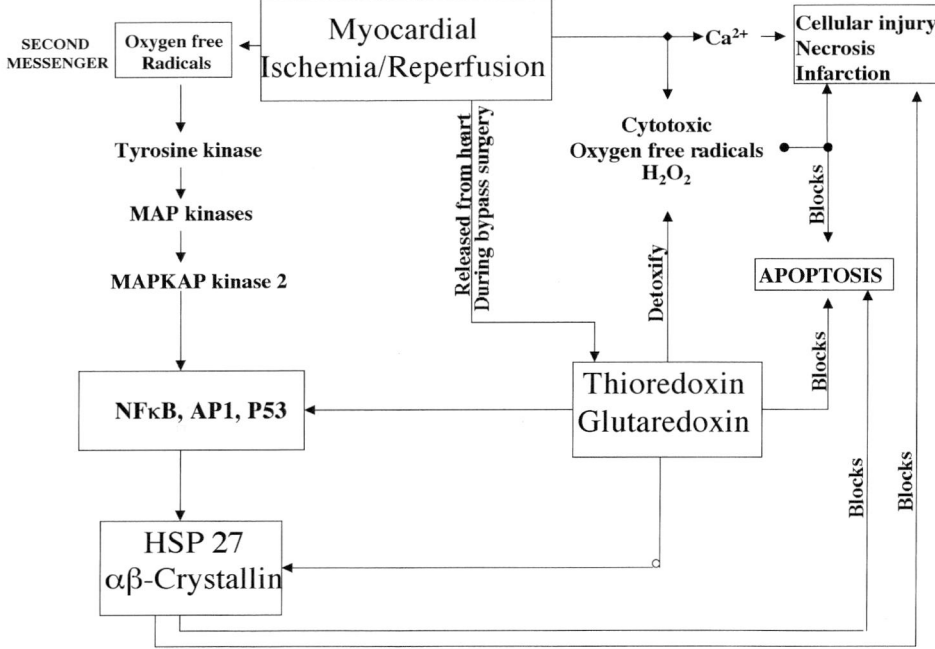

Figure 6 Proposed model for redox regulation of ischemic adaptation of heart.

ischemic adaptation and MAPKAP kinase 2 phosphorylation supports these earlier findings. Existing evidence that oxidative stress/free radicals lead to the activation of NF-κB, thereby inducing the expression of genes, also supports the role of ROS as messenger molecules.

The results of our study demonstrated that both AP-1 and NF-κB are activated after 30 min of ischemia followed by 2 hr of reperfusion. On the contrary, myocardial adaptation induced by cyclic episodes of short-term ischemia and reperfusion results in the activation of NF-κB, but not AP-1. Interestingly, the Bcl-2 gene is downregulated after 30 min of ischemia and 2 hr of reperfusion, but is induced significantly in the adapted heart. MDA formation, a presumptive marker of oxidative stress, is increased in ischemic–reperfused hearts, but is decreased in the adapted myocardium. In concert, cardiomyocyte apoptosis and DNA fragmentation increase in the ischemic–reperfused myocardium, but attenuate in adapted hearts. Taken together, these results suggest that while activation of NF-κB is necessary for myocardial ischemic adaptation, it does not control apoptosis. However, the expression of AP-1 is directly proportional to cardiomyocyte apoptosis, further indicating a redox regulation of myocardial ischemic adaptation.

REFERENCES

Anderson, N. G., Maller, J. L., Tonks, N. K., and Sturgill, T. W. (1990). Requirement for integration of signals from two distinct phosphorylation pathways for activation of Mapkinase. *Nature (London)* **343**, 651–653.

Arata, S., Hamaguchi, S., and Nose, K. (1995). Effects of the overexpression of the small heat shock protein, HSP 27, on the sensitivity of human fibroblast cells exposed to oxidative stress. *J. Cell. Physiol.* **163**, 458–465.

Artuc, M., Karman, D., Jurgovsky, K., and Schadendorf, D. (1997). Human melonoma cell lines selected in vitro displaying various levels of drug resistance against cisplatin, fotemustine, vindestine or etoposide: modulation of protooncogene expression. *Anticancer* **17**, 4359–4370.

Asimakis, G. K., Inners-McBride, K., Medellin, G., and Conti, V. R. (1992). Ischemic preconditioning attenuates acidosis and postischemic dysfunction in isolated rat heart. *Am. J. Physiol.* **263**, H887–H894.

Bagchi, D., Das, D. K., Engelman, R. M., Prasad, M. R., and Subramanian, R. (1990). Polymorphonuclear leucocytes an potential source of free radicals in the ischemic-reperfused myocardium. *Eur. Heart J.* **11**, 800–813.

Brand, T., Sharma, H. S., Fleischmann, K. E., Duncker, D. J., McFalls, E. O., Verdouw, P. D., and Schaper, W. (1992). Proto-oncogene expression in porcine myocardium subjected to ischemia/reperfusion. *Circulation* **71**, 1351–1360.

Bugge, E., and Ytrehus, K. (1995). Ischemic preconditioning is protein kinase c dependent but not through stimulation of a adrenergic or adenosine receptors in the isolated rat heart. *Cardiovasc. Res.* **29**, 401–406.

Cave, A. C., Collis, C. S., Downey, J. M., and Hearse, D. J. (1993). Improved functional recovery by ischemic preconditioning is not mediated by adenosine in the globally ischemic isolated rat heart. *Cardiovasc. Res.* **27**, 663–668.

Cerutti, P., Shah, G., Peskin, A., and Amstad, P. (1992). Oxidan carcinogenesis and antioxidant defense. *Ann. N. Y. Acad. Sci.* **663**, 158–166.

Cleghon, V., and Morrison, D. K. (1994). Raf-1 interacts with Fyn and Src in a non-phosphotyrosine-dependent manner. *J. Biol. Chem.* **269**, 17749–17755.

Cohen, M. V., Liu, G. S., Wang, P., Cordis, G. A., Das, D. K., and Downey, J. M. (1996). Phospholipase D plays a major role in ischemic preconditioning in rabbit heart. *Circulation* **94**, 1713–1718.

Cordis, G. A., Maulik, N., and Das, D. K. (1995). Detection of oxidative stress in heart by estimating the dinitrophenyl-hydrazine derivative of malonaldehyde. *J. Mol. Cell. Cardiol.* **27**, 1645–1653.

Cordis, G. A., Maulik, G., Bagchi, D., Riedel, W., and Das, D. K. (1998). Detection of oxidative DNA damage to ischemic reperfused rat hearts by 8-hydroxydeoxyguanosine formation. *J. Mol. Cell. Cardiol.* **30**, 1939–1944.

Das, D. K. (1996). Preconditioning potentiates molecular signaling for myocardial adaptation to ischemia. *Ann. N.Y. Acad Sci.* **793**, 191–209.

Das, D. K. (1998). Ischemic preconditioning: Role of multiple kinases in signal amplification and modulation. *Adv. Org. Biol.* **6**, 97–120.

Das, D. K., and Maulik, N. (1994). Evaluation of antioxidant effectiveness in ischemia reperfusion tissue injury methods. *In* "Methods in Enzymology" (L. Packer, ed.), Vol. 233, pp. 601–610. Academic Press, San Diego, CA.

Das, D. K., and Maulik, N. (1995). Cross talk between heat shock and oxidative stress inducible genes during myocardial adaptation to ischemia. *In* "Cell Biology of Trauma" (J. J. Lemasters and C. Oliver, eds.), pp. 193–211. CRC Press, Boca Raton, FL.

Das, D. K., and Maulik, N. (1997). Reprogramming of gene expression during myocardial adaptation to stress. In "Adaptation Biology and Medicine" (B. K. Sharma, N. Takeda, P. K. Ganguly, and P. K. Singal, eds.). Narosa Publishing House, pp. 82–101.

Das, D. K., George, A., Liu, X., and Rao, P. S. (1989). Detection of hydroxyl radicals in the mitochondria of ischemic-reperfused myocardium by trapping with salicylate. *Biochem. Biophys. Res. Commun.* **165**, 1004–1009.

Das, D. K., Engelman, R. M., and Kimura, Y. (1993). Molecular adaptation of cellular defenses following preconditioning of the heart by repeated ischemia. *Cardiovasc. Res.* **27**, 578–584.

Das, D. K., Moraru, I. I., Maulik, N., and Engelman, R. M. (1994). Gene expression during myocardial adaptation to ischemia and reperfusion. *Ann. N.Y. Acad. Sci.* **723**, 292–307.

Das, D. K., Maulik, N., and Moraru, I. I. (1995). Gene expression in acute myocardial stress. *J. Mol. Cell. Cardiol.* **27**, 181–193.

Das, D. K., Maulik, N., Sato, M., and Ray, P. (1999). Reactive oxygen species function as second messenger during ischemic preconditioning of heart. *Mol. Cell. Biochem* **196**, 59–67.

Diamond, M. I., Miner, J. N., Yoshinaga, S. K., and Yamamoto, K. R. (1990). Transcription factor interactions: selectors of positive or negative regulation from a single DNA element. *Science* **249**, 1266–1272.

Duh, E. J., Maury, W. J., Folks, T. M., Fauci, A. S., and Rabson, A. B. (1989). Tumor necrosis factor alpha activates human immunodeficiency virus type 1 through induction of nuclear factor binding to the NF-Kappa B sites in the long terminal repeat. *Proc. Natl. Acad. Sci. U.S.A.* **86**, 5974–5978.

Engelman, D. T., Watanabe, M., Maulik, N., Engelman, R. M., Rousou, J. A., Flack, J. E., Deaton, D. W., and Das, D. K. (1996). Critical timing of nitric oxide supplementation in cardioplegic arrest and reperfusion. *Circulation* **94**(II), 407–411.

Ewing, J. F., and Maines, M. D. (1993). Glutathione depletion induces heme-oxygenase 1 (HSP 32) mRNA and protein in rat brain. *J. Neurochem.* **60**, 1512–1519.

Flack, J. E., Kimura, Y., Engelman, R. M., Rousou, J. A., Iyengar, J., Jones, R., and Das, D. K. (1991). Preconditioning the heart by repeated stunning improves myocardial salvage. *Circulation* **84**, III369–III374.

Force, T., Bonventrem J. V., Heidecker, G., Rapp, U., Avruch, J., and Kyriakis, L. M. (1994). Enzymatic characteristics of the Raf-1 protein kinase. *Proc. Natl. Acad. Sci. U.S.A.* **91**, 1270–1274.

Fukuda, K., Kojiro, M., and Chiu, J. F. (1993). Demonstration of extensive chromatin cleavage in transplanted Morris Hepatoma 7777 tissue: Apoptosis or necrosis. *Am. J. Pathol.* **142**, 935–946.

Ghosh, S., Gifford, A. M., Riviere, L. R., Tempst, P., Nolan, G. P., and Baltimore, D. (1990). Cloning of the p50 DNA binding subunit of NF-kappa B: Homology to rel and dorsal. *Cell (Cambridge, Mass.)* **62**, 1019–1029.

Goto, M., Liu, Y., Yang, X.-M., Ardell, J. L., Cohen, M. V., and Downey, J. M. (1995). Role of bradykinin in protection of ischemic preconditioning in rabbit hearts. *Circ. Res.* **77**, 611–621.

Gross, G. J., and Auchampach, J. A. (1992). Blockade of ATP-sensitive potassium channels prevents myocardial preconditioning in dogs. *Circ. Res.* **73**, 656–670.

Gu, Z., and Matlashewski, G. (1995). Effect of human papillomavirus type 16 oncogenes on MAP kinase activity. *J. Virol.* **69**, 8051–8056.

Ham, J., Babij, C., Whitfield, J., Pfarr, C. M., Lallemand, D., Yaniv, M., and Rubin, L. L. (1995). A c-Jun dominant negative mutant protects sympathetic neurons against programmed cell death. *Neuron* **14**, 927–939.

Herrlich, P., Ponta, H., and Rahmsdorff, H. J. (1992). DNA damage-induced gene expression; signal transduction and relation to growth factor signaling. *Rev. Physiol. Biochem. Pharmacol.* **119**, 187–223.

Ishikawa, Y., Yokoo, T., and Kitamura, M. (1997). C-Jun/AP-1, but not NF-κB, is a mediator for oxidant-initiated apoptosis in glomerular mesangial cells. *Biochem. Biophys. Res. Commun.* **240**, 496–501.

Isner, J. M., Kearney, M., Bortman, S., and Passeri, J. (1995). Apoptosis in human atherosclerosis and restenosis. *Circulation* **91**, 2703–2711.

Ito, Y., Nakashima, S., and Nozawa, Y. (1997). Hydrogen peroxide-induced phospholipase D activation in rat pheochromocytoma PC12 cells: Possible involvement of Ca2+ dependent protein tyrosine kinase. *J. Neurochem.* **69**, 729–736.

Jacquier-Sarlin, M., and Polla, B. S. (1996). Dual regulation of heat-shock transcription factor (HSF) activation and DNA-binding activity by H2O2: Role of thioredoxin. *Biochem. J.* **318**, 187–193.

Kajstura, J., Cheng, W., Reiss, K., Clark, W. A., Sonnenblick, E. H., Krajewski, S., Reed, J. C., Olivetti, G., and Anversa, P. (1996). Apoptotic and necrotic myocyte cell deaths are independent contributing variables of infarct size in rats. *Lab. Invest.* **74**, 86–107.

Kida, M., Fujiwara, H., Ishida, M., Kawai, C., Ohura, M., Miura, I., and Yabuuchi, Y. (1991). Ischemic preconditioning preserved creatine phosphate and intracellular pH. *Circulation* **84**, 2495–2503.

Kimura, Y., Iyengar, J., Subramanian, R., Cordis, G. A., and Das, D. K. (1992). Myocaradial adaptation by repeated short term ischemia reduces post-ischemic dysfunction. *Basic Res. Cardiol.* **87**, 128–138.

Lawson, C. S., Coltart, D. J., and Hearse, D. J. (1993). The antiarrhythmic action of ischemic preconditioning in rat hearts does not involve functional Gi proteins. *Cardiovasc. Res.* **27**, 681–687.

Lazou, A., Bogoyevitch, M. A., Clerk, A., Fuller, S. J., Marshall, C. J., and Sugden, P. H. (1994). Regulation of mitogen-activated protein kinase cascade in adult rat heart preparations in vitro. *Circ. Res.* **75**, 932–941.

Liu, X., Engelman, R. M., Rousou, J. A., and Das, D. K. (1992). Reperfusion injury during cardiac surgery of adults. *In* "Pathophysiology of Reperfusion Injury" (D. K. Das, ed.), pp. 263–293. CRC Press, Boca Raton, FL.

Macdonald, S. G., Crews, C. M., Wu, L., Driller, J., Clark, R., Erikson, R. L., and McCormick, F. (1993). Reconstruction of the Raf-1-MEK-ERK signal transduction pathway *in vitro*. *Mol. Cell. Biol.* **13**, 6615–6620.

Macho, A., Blazquez, M. V., Navas, P., and Munoz, E. (1998). Induction of apoptosis by vanilloid compounds does not require de novo gene transcription and activator protein 1 activity. *Cell Growth Differ.* **9**, 277–286.

Maines, M. D., and Kappas, A. (1997). Metals as regulators of heme oxygenase. *Science* **198**, 1215–1221.

Matsui, M., Taniguchi Y., Hirota K., Taketo, M., and Yodoi, J. (1995). Structure of the mouse thioredoxin-encoding gene and its processed pseudogene. **152**, 165–171.

Maulik, N. (1999). Apoptotic cell death during ischemia/reperfusion and its attenuation by antioxidant therapy. *Am. J. Pathophysiol.* (in press).

Maulik, N., Engelman, R. M., Wei, Z., Lu, D., Rousou, J. A., and Das, D. K. (1993). Interleukin-1α preconditioning reduces myocardial ischemia reperfusion injury. *Circulation* **88**(Pt. II), 387–394.

Maulik, N., Wei, Z., Liu, X., Engelman, R. M., Rousou, J. A., Flack, J. E., Deaton, D. W., and Das, D. (1994). Improved postischemic ventricular functional recovery by amphetamine is linked with its ability to induce heat shock. *Mol. Cell. Biochem.* **137**, 17–24.

Maulik, N., Engelman, R. M., Wei, Z., Liu, X., Rousou, J. A., Flack, J. E., Deaton, D. W., and Das, D. K. (1995a). Drug-induced heat-shock preconditioning improves post-ischemic ventricular recovery after cardiopulmonary bypass. *Circulation* 92(Suppl. II), 381–388.

Maulik, N., Engelman, D. T., Watanabe, M., Engelman, R. M., and Das, D. K. (1995b). Nitric oxide signaling in ischemic heart. *Cardiovasc. Res.* 30, 593–601.

Maulik, N., Engelman, D., Watanabe, M., Engelman, R. M., Rousou, J. A., Flack, J., Deaton, D., and Das, D. K. (1996a). Nitric oxide/carbon monoxide: A free radical-dependent molecular switch for myocardial preservation during ischemic arrest. *Circulation* 94(II), 398–406.

Maulik, N., Engelman, R. M., Deaton, D. Flack, J. E., Rousou, J. A., and Das, D. K. (1996b). Reperfusion of ischemic myocardium induces apoptosis and DNA laddering with enhanced expression of protooncogene c-myc mRNA. *Circulation* 94(Suppl.) I415.

Maulik, N., Watanabe, M., Zu, Y. L., Huang, C. K., Cordis, G. A., Schley, J. A., and Das, D. K. (1996c). Ischemic preconditioning triggers the activation of MAP kinases and MAPKAP kinase 2 in rat hearts. *FEBS Lett.* 396, 233–237.

Maulik, N., Sharma, H. S., and Das, D. K. (1996d). Induction of the haem oxygenase gene expression during the reperfusion of ischemic rat myocardium. *J. Mol. Cell. Cardiol.* 28, 1261–1270.

Maulik, N., Yoshida, T., Engelman, R. M., Rousou, J. A., Flack, J. E., Deaton, D., and Das, D. K. (1997). Oxidative stress developed during reperfusion of ischemic myocardium downregulates the bcl-2 gene and induces apoptosis and DNA laddering. *Surg. Forum* 48, 245–248.

Maulik, N., Engelman, R. M., Rousou, J. A., Flack, J. E., Deaton, D. W., and Das, D. K. (1998a). Ischemic preconditioning suppresses apoptosis by upregulating the antideath gene Bcl-2. *Surg. Forum* 49, 209–211.

Maulik, N., Yoshida, T., Engelman, R. M., Deaton, D., Flack, J. E., Rousou, J. A., and Das, D. K. (1998b). Ischemic preconditioning attenuates apoptotic cell death associated with ischemia/reperfusion. *Mol. Cell. Biochem.* 186, 139–145.

Maulik, N., Sato, M., Price, B. D., and Das, D. K. (1998c). An essential role of NFκB in tyrosine kinase signaling of p38 MAP kinase regulation of myocardial adaptation to ischemia. *FEBS Lett.* 429, 365–369.

Maulik, N., Yoshida, T., and Das, D. K. (1998d). Oxidative stress developed during the reperfusion of ischemic myocardium induces apoptosis. *Free Radical Biol. Med.* 24, 869–875.

Maulik, N., Kagan, V.E., Vladimirm, A. T., and Das, D. K. (1998e). Redistribution of phosphatidylserine and phosphatidyl-ethanolamine precedes reperfusion-induced apoptosis. *Am. J. Physiol.* 274, H242–H248.

Maulik, N., Yoshida, T., and Das, D. K. (1999). Regulation of cardiomyocyte apoptosis in ischemic reperfused mouse heart by glutathione peroxidase. *Mol. Cell. Biochem.* 196, 13–21.

McDuffce, A. T., Senisterra, G., Huntley, S., Lepock, J. R., Sekhar, K. R., Meredith, M. J., Borrelli, M. J., Morrow, J. D., and Freeman, M. L. (1997). Proteins containing nonnative disulfide bonds generated by oxidative stress can act as signals for the induction of the heat shock response. *J. Cell. Physiol.* 171, 143–151.

McMahan, S. B., and Monroe, J. G. (1992). Role of primary response genes in generating cellular responses to growth factors. *FASEB J.* 62, 2707–2715.

Meerson, F. Z., Kagan, V. E., Kozlov, P., Beekina, L. M., and Khipenko, Y. V. (1982). The role of lipid peroxidation in pathogenesis of ischemic damage and the antioxidant protection of the heart. *Basic Res. Cardiol.* 77, 465–472.

Meyer, M., Schreck, R., and Baeuerle, P. A. (1993). H2O2 and antioxidants have opposite effects on activation of NF-kappa B and AP-1 in intact cells: AP-1 as secondary antioxidant-responsive factor. *EMBO J.* **12**, 2005–2015.

Moraru, I. I., Engelman, R. M., Rousou, J. A., Flack, J. E., Deaton, D. W., and Das, D. K. (1994). Myocardial ischemia triggers rapid expression of mitochondrial genes. *Surg. Forum* **45**, 315–317.

Morrison, D. K. (1995). Mechanisms regulating Raf-1 activity in signal transduction pathways. *Mol. Reprod. Dev.* **42**, 507–514.

Murry, C. E., Richard, V. J., Reimer, K. A., and Jennings, R. B. (1990). Ischemic preconditioning slows energy metabolism and delays ultrastructural damage during a sustained ischemic episode. *Circ. Res.* **66**, 913–931.

Nakamura, H., Vaage, J., Valen, G., Padilla, C. A., Bjornstedt, M., and Holmgren, A. (1998). Measurements of plasma glutaredoxin and thioredoxin in healthy volunteers and during open-heart surgery. *Free Radical Biol. Med.* **24**, 1176–1186.

Nose, K., Ohba, M., Shibanuma, M., and Kuroki, T. (1994). Involvement of hydrogen peroxide in the actions of TGFB1. *In* "Oxidative Stress, Cell Activation and Viral Infection" (C. Pasquier, R. Y. Oliver, C. Auclair, and L. Packer, eds.), pp. 21–34. Birkhaeuser, Basel.

Otani, H., Engelman, R. M., Rousou, J. A., Breyer, R. H., and Das, D. K. (1986a). Enhanced prostaglandin synthesis due to phospholipase breakdown in ischemic-reperfused myocardium. Control of its production by a phospholipase inhibitor or free radical scavengers. *J. Mol. Cell Cardiol.* **18**, 953–961.

Otani, H., Engelman, R. M., Rousou, J. A., Breyer, R. H., Lemeshow, S., and Das, D. K. (1986b). Cardiac performance during reperfusion improved by pretreatment with oxygen-free radical scavengers. *J. Thorac. Cardiovasc. Surg.* **91**, 290–295.

Rouse, J., Cohen, P., Trigon, S., Morange, M., Alonso-Llamazares, A., Zamanillo, D., Hunt, T., and Nebreda, A. R. (1994). A novel kinase cascade triggered by stress and heat shock that stimulates MAPKAP kinase 2 and phosphorylation of the small heat shock proteins. *Cell (Cambridge, Mass.)* **78**, 1027–1037.

Schieven, G. L., Kirihara, J. M., Myers, D. E., Ledbetter, J. A., and Uckun, F. M. (1993). Reactive oxygen intermediates activate NF-Kappa B in a tyrosine kinase-dependent mechanism and in combination with vanadate activate the p56 lck and p59 fyn tyrosine kinases in human lymphocytes. *Blood* **82**, 1212–1220.

Schott, R. J., Rohmann, S., Braun, E. R., and Schaper, W. (1990). Ischemic preconditioning reduces infarct size in swine myocardium. *Circ. Res.* **66**, 1133–1142.

Schreck, R., Rieber, P., and Baeuerle, P. A. (1991). Reactive oxygen intermediates as apparently widely used messengers in the activation of the NF-kappa B transcription factor and HIV-1. *EMBO J.* **10**, 2247–2258.

Schreck, R., Meier, B., Mannel, D., Droge, W., and Baeuerle, P. A. (1992). Dithiocarbamates as potent inhibitors of nuclear factor kappa B activation in intact cells. *J. Exp. Med.* **175**, 1181–1194.

Schumer, M., Colombel, M. C., Sawczuk, I. S., Gobe, G., Connor, J., O'Toole, K. M., Wise, G. J., and Buttyan, R. (1992). Morphologic, biochemical and molecular evidence of apoptosis during the reperfusion phase after brief periods of renal ischemia. *Am. J. Pathol.* **140**, 831–838.

Seger, R., and Krebs, E. G. (1995). The MAPK signaling cascade. *FASEB J.* **9**, 726–735.

Seko, Y., Tobe, K., Ueki, K., Kadowaki, T., and Yazaki, Y. (1996). Hypoxia and hypoxia/reoxygenation activate raf-1, mitogen-activated protein kinase kinase, mitogen-activated protein kinases, and S6 kinase in cultured rat cardiac myocytes. *Circ. Res.* **78**, 82–90.

Shibanuma, M., Kuroki, T., and Nose, K. (1992). Cell cycle dependent phosphorylation of HSP 28 by TGFβ_1 and H$_2$O$_2$ in normal mouse osteoblastic cells (MC$_3$T$_3$-E$_1$), but not in their ras-transformants. *Biochem. Biophys. Res. Commun.* **187**, 1418–1425.

Suzuki, Y. J., Mizuno, M., and Packer, L. (1995). Transient overexpression of catalase does not inhibit TNF- or PMA induced NF-KB activation. *Biochem. Biophys. Res. Commun.* **210**, 537–541.

Tanaka, M., Ito, H., Adachi, S., Akimoto, H., Nishikawa, T., Kasajima, T., Marumo, F., and Hiroe, M. (1994). Hypoxia induces apoptosis with enhanced expression of Fas antigen messenger RNA in cultured neonatal rat cardiomyocytes. *Circ. Res.* **75**, 426–433.

Thorburn, J., Carlson, M., Mansour, S. J., Chien, K. R., Ahn, N. G., and Thorburn, A. (1995). Inhibition of a signaling pathway in cardiac muscle cells by active mitogen-activated protein kinase kinase. *Mol. Biol. Cell* **6**, 1479–1490.

Toledano, M. B., and Leonard, W. J. (1991). Modulation of transcription factor NF-kappa B binding activity by oxidation-reduction in vitro. *Proc. Natl. Acad. Sci. U.S.A.* **88**, 4328–4332.

Tominaga, T., Kure, S., Narisawa, K., and Yoshimoto, T. (1993). Endonuclease activation following focal ischemic injury in the rat brain. *Brain Res.* **608**, 21–26.

Tosaki, A., Bagchi, D., Hellegouarch, A., Pali, T., Cordis, G. A., and Das, D. K. (1993a). Comparisons of ESR and HPLC methods for the detection of hydroxyl radicals in ischemic/reperfused hearts. A relationship between the genesis of oxygen-free radicals and reperfusion-induced arrhythmias. *Biochem. Pharmacol.* **45**, 961–969.

Tosaki, A., Droy-Lefaix, M. T., Pali, T., and Das, D. K. (1993b). Effects of SOD, catalase and a novel antiarrhythmic drug, EGB 671, on reperfusion-induced arrhythmias in isolated rat hearts. *Free Radical Biol. Med.* **14**, 361–370.

Tosaki, A., Cordis, G. A., Szerdahelyi, P., Engelman, R. M., and Das, D. K. (1994). Effects of preconditioning on reperfusion arrhythmias, myocardial functions, formation of free radicals, and ion shifts in isolated ischemic/reperfused rat hearts. *J. Cardiovasc. Pharmacol.* **23**, 365–373.

Tosaki, A., Engelman, D. T., Engelman, R. M., and Das, D. K. (1995). α-1 adrenergic receptor agonist-induced preconditioning in isolated working rat hearts. *J. Pharmacol. Exp. Ther.* **273**, 689–694.

Tosaki, A., Cordis, G. A., Szerdahelyi, P., Engelman, R. M., and Das, D. K. (1996). Role of protein kinase C in ischemic preconditioning of rat hearts. *J. Cardiovasc. Pharmacol.* **28**, 723–731.

Tosaki, A., Maulik, N., Elliott, G. T., Blasig, I. E., Engelman, R. M., and Das, D. K. (1998). Preconditioning of rat heart with monophosphoryl lipid A: A role for nitric oxide. *J. Pharmacol. Exp. Ther.* **285**, 1274–1279.

Watson, F., and Edwards, S. W. (1998). Stimulation of primed neutrophils by soluble immune complexes: Priming leads to enhanced intracellular $Ca2+$ elevations, activation of PLD and activation of the NADPH oxidase. *Biochem. Biophys. Res. Commun.* **247**, 819–826.

Wieles, B., Van Soolingen, D., Holmgren, A., Offringa, R., Ottenhoff, T., and Thole, J. (1995). Unique gene organization of thioredoxin and thioredoxin reductase in *Mycobacterium leprae*. *Mol. Microbiol.* **16**(5), 921–929.

Yamada, H., Strahler, J., Welsh, M. J., and Bitar, K. N. (1995). Activation of MAP kinase and translocation with HSP 27 in bombesin-induced contraction of rectosigmoid smooth muscle. *Am. J. Physiol.* **269**, G683–G691.

Yamazaki, T., Komuro, I., Kudoh, S., Zou, Y., Shiojima, I., Hiroi, Y., Mizuno, T., Maemura, K., Kurihara, H., Aikawa, R., Takano, H., and Yazaki, Y. (1996). Endothelin-1 is involved in mechanical stress-induced cardiomyocyte hypertrophy. *J. Biol. Chem.* **271**, 3221–3228.

Zu, Y. L., Ai, Y., Gilchrist, A., Maulik, N., Watras, J., Sha'afi, R. I., Das, D. K., and Huang, C. K. (1997). High expression and activation of MAP kinase-activated protein kinase 2 in myocardium. *J. Mol. Cell. Cardiol.* **29**, 2150–2168.

23 Oxidative Stress as a Governing Factor in Physiological Aging

William C. Orr and Rajindar S. Sohal
Department of Biological Sciences
Southern Methodist University
Dallas, Texas 75275

I. INTRODUCTION

The origin of the idea that the rate of energy utilization, and by implication, the rate of oxygen consumption, is a determinant of longevity can be traced to the studies of Rubner (1908) and later of Pearl (1928), who proposed the "rate of living" theory, which essentially predicts that longevity of an organism is inversely related to its rate of energy utilization. However, the nature of the biochemical link between the rate of metabolism or oxygen consumption and longevity remained largely an enigma until the discovery of the enzyme superoxide dismutase (SOD) and its substrate, superoxide anion radical, by McCord and Fridovich (1969), albeit the actual hypothesis, proposing the causal involvement of free radicals in aging, was advanced earlier by Harman (1956). Because of the tremendous expansion of knowledge in free radical biochemistry in the past three decades, a more contemporary concept, often referred to as the oxidative stress hypothesis of aging, has emerged.

The term oxidative stress, originally coined by Sies (1986), connotes the redox state in the cell where prooxidants outstrip the ability of the antioxidants to eliminate them, thereby resulting in the presence of steady-state amounts of products of the attacks of reactive oxygen species (ROS) on macromolecules. Because the concentration of oxidatively modified macromolecules is detectable under normal physiological conditions, even in young and healthy animals, it has been surmised that a certain level of oxidative stress prevails in normal cells. Because the concentration of oxidatively damaged molecules increases with age in tissues of various species, often in an exponential fashion, it can be inferred that the overall level of oxidative stress also increases with age. Oxidatively modified macromolecules almost invariably lose structural and functional integrity. The correlation between the age-associated increase in oxidative molecular damage and the concomitant losses in the efficiency of homeostatic mechanisms has indeed been used as a rationale for the oxidative stress hypothesis of aging. An excellent comprehensive review of studies bearing on the relationship between oxidative stress and aging has been authored by Beckman and Ames (1998). The present authors have also previously provided a detailed discussion of the work related to the role of oxidative stress in aging conducted in their laboratory (Sohal and Orr, 1995; Sohal and Weindruch, 1996). It is therefore deemed more profitable to focus here on more current issues relevant to the validation of the hypothesis.

II. REFORMULATION OF THE OXIDATIVE STRESS HYPOTHESIS OF AGING

Oxidative damage to macromolecules is widely believed to occur randomly, where any particular molecule in close proximity to ROS can be attacked stochastically. This view seems to be based on the notion that ROS, particularly the hydroxyl radical (HO·), believed to be the main agent of damage, are in general so highly reactive that their interactions with macromolecules cannot occur in a specific manner, akin to the interplay between an enzyme and its substrate. In addition, an age-related increase in oxidative damage has often been documented in tissue homogenates and not in individual molecular species, e.g., a specific protein or a specific gene. Significant age-associated increases in oxidatively damaged macromolecules, detected in tissue homogenates, have been considered as an indication of ubiquitous rather than specific damage. Hence, the oxidative stress hypothesis of aging is widely interpreted to mean that the physiological deterioration occurring during senescence is due to the progressive accumulation of randomly inflicted oxidative damage. The random damage model would predict that almost all the physiological and biochemical functions should be affected deleteriously, at least to some degree, a prediction that does not accord with the empirical observations; for instance, activities of most of the enzymes reportedly do not decline with age. In fact, loss of the functional capacity of cells during aging cannot be explained on the basis of a generalized decrease in enzyme activities.

Studies in this laboratory have indicated that oxidative damage to proteins in aged organisms is not random but highly specific. For example, in flight muscle mitochondria in the housefly, only adenine nucleotide translocase and aconitase were found to exhibit *ex vivo* an age-related increase in the accumulation of oxidative damage (Yan *et al.*, 1997; Yan and Sohal, 1998). Thus, we propose that the oxidative stress hypothesis should be modified to accommodate the specificity of damage during aging. The amended hypothesis can be stated as follows: "There is an intrinsic imbalance between prooxidants and antioxidants in cells resulting in a certain level of oxidative stress and macromolecular oxidative damage. Such damage is specific and not random and increases with age, causing senescence-associated losses in function."

If oxidative stress is a causal factor in aging, it would follow that the aging process may be governed by genes involved in (1) the regulation of the redox state, i.e., antioxidative defenses and ROS generation, (2) the repair or elimination of oxidized macromolecules, and (3) the specific targets of oxidative damage. Furthermore, the hypothesis would predict that attenuation of oxidative stress should result in the retardation of the rate of aging and prolongation of the life span.

III. GENETIC MANIPULATIONS TO ATTENUATE OXIDATIVE DAMAGE

Ample evidence exists that oxidative stress increases with age. Studies in our laboratory and others consistently support this to be true for aerobic organisms ranging from flies to mice to humans. Parameters used to measure oxidative stress include measures of protein damage (loss of sulfhydryls, protein carbonyl accumulation), DNA damage (8-hydroxydeoxyguanosine), membrane damage, the generation of ROS such as superoxide radicals, changes in oxidation state as measured by redox couples (GSH/GSSG, NADH/NAD, NADPH/NADP), and other parameters such as the activities of several enzymes that are particularly susceptible to oxidative damage (Sohal and Weindruch, 1996).

The oxidative stress hypothesis would predict that slowing down the overall accumulation of oxidative stress/damage would provide beneficial effects to the organism. Some efforts have been made in this direction by means of transgenic studies, primarily carried out in *Drosophila*. In these studies (Orr and Sohal, 1994; Sohal *et al.*, 1995), two stalwarts of the antioxidative defense system (Cu,Zn superoxide dismutase and catalase) were introduced into flies by microinjection and multiple lines were established. These overexpressor lines exhibited a reduction in the accumulation rate of oxidative damage that corresponded to an increase in life span that was both chronological and physiological. In other words, there was a "true" extension of life span as measured by metabolic potential (in this case the overall lifetime consumption of O_2 of the experimental groups relative to the controls) as well as an up to 34% increase in mean and maximum life span. Surprisingly, it appears that overexpression of Cu,Zn SOD alone in motor neurons has a similar effect, increasing life span by up to 40% with a comparable effect on the metabolic potential (Parkes et al., 1998). This very intriguing result suggests that there are certain tissues that are "limiting" with respect to life span and that increasing antioxidative defenses in these targets that are particularly labile to oxidative stress may have dramatic effects. The other side of this coin is that altering the redox state by overexpressing antioxidative genes may interfere with some tissue-specific function or some specific developmental event that requires a more prooxidizing state; the result might then be some functional deficit that could actually have a negative impact on survivorship. We and others (Reveillaud et al., 1991) have noted, for example, that overexpression of Cu,ZnSOD by more than 50% interferes with development and has an overall negative effect on viability.

In the studies just described, the increase in antioxidative defenses in transgenic flies resulted in an increase in both chronological and physiological life spans. Because simple environmental manipulations such as dropping the temperature from 25°C to 18°C can increase the chronological life span by up to twofold in cold-blooded animals without having any impact on total lifetime oxygen consumption (Miquel et al., 1979), this distinction is not a minor one. Such considerations are also relevant for studies involving genetic manipulations. For instance, a mutation resulting in the generation of a hypometabolic phenotype may mistakenly be represented as a longevity-promoting mutation when, in fact, it may well represent a simple reduction in the rate of oxygen consumption and thus no real increase in the physiological life span. These concerns may well apply to the long-lived *Caenorhabditis elegans* mutants that have been identified and characterized by several groups over the years (reviewed by Martin et al., 1996). W. Van Voorhies and S. Ward (personal communication) carefully determined that the metabolic rates of all the long-lived mutants that they examined (*age-1, clk-1, clk-2*, and *daf-2*) were lower than wild type. To corroborate this study further, they showed that in the *daf-2;daf-16* double mutant, which has a normal lifespan, the normal metabolic rate was restored. Thus any claims that these genes are specifically involved in aging are at best premature, as life span extension occurs under unphysiological conditions of semihibernation.

However, it is also possible that reliance solely on life span measures could erroneously discount the possible significance of certain genes in aging. We noted, for example, that in one study, flies containing no detectable catalase activity may live almost as long as controls possessing normal levels of catalase (Orr et al., 1992). However, these flies exhibit minimal physical activity and reduced oxygen consumption and, as a consequence, have a significantly reduced metabolic potential. In a similar vein, transgenics overexpressing candidate longevity genes may exhibit life spans similar to those of controls, but may in fact have a greater metabolic potential evinced in a higher level of activity. Reliance solely on survivorship data would then result in premature dismissal of a bona fide longevity-promoting gene.

As a general experimental strategy, bolstering selected antioxidative defenses has permitted some inroads to be made in reducing the rate of accumulation of oxidative damage with age, and extending life span, at least in *Drosophila*. Such studies provide compelling evidence supporting oxidative stress as a causal factor in the aging process, although similar transgenic experiments have not yet been duplicated in mammalian models. Efforts in this area have certainly been undertaken and there are now multiple mouse transgenics bearing the various antioxidative genes, including Cu,ZnSOD (Epstein et al., 1987), catalase (Li et al., 1997; Van Remmen et al., 1998), glutathione peroxidase (Mirochnitchenko et al., 1995), and MnSOD (Wispé et al., 1992). Although some interesting phenotypes have been noted for these transgenics (e.g., the reduction in tissue damage in response to acute stress conditions noted for Cu,ZnSOD transgenics; Kinouchi et al., 1991) overall, there is no obvious impact on life span. In fact, the only reports presented to date that do not include metabolic potential data indicate that the life span is reduced. This result has been attributed, at least in part, to the possible imbalance caused by the relatively high levels of overexpression obtained in these transgenics (>2-fold) or alternatively inappropriate temporal or spatial expression driven by heterologous promoters. The fact that some Cu,ZnSOD transgenics exhibit clear defects (Avraham et al., 1988) supports this hypothesis and suggests that optimal expression levels for mouse transgenics have not yet been obtained. In planning future mammalian transgenic studies, it might be advisable to take into account some of the lessons learned using *Drosophila*. In the case of *Drosophila* transgenics involving the integration of genomic fragments with "native" cis-regulatory controls, single antioxidative gene transgenics containing either Cu,ZnSOD or catalase alone were observed to exhibit little or no effect (Orr and Sohal, 1992, 1993), whereas transgenics carrying both together showed a significant amelioration (Orr and Sohal, 1994). Thus the more effective strategy might be to overexpress multiple antioxidative genes together to maintain a balanced antioxidative defense network and also consider focusing on transgenics where expression patterns and levels do not depart so dramatically from the normal situation. Alternatively, following the lead of the Parkes et al. (1998) study, it might be useful to focus antioxidant gene overexpression in tissues that are particularly susceptible to the deleterious effects of oxidative damage.

Further insights into the significance of specific antioxidative genes in controlling oxidative stress and in aging may be obtained by the study of antioxidative gene mutants. *Drosophila* mutants that are null or hypomorphic for Cu,ZnSOD (Phillips *et al.*, 1989) or catalase (Mackay and Bewley, 1989; Orr *et al.*, 1992) exhibit a marked reduction in survivorship and/ or metabolic potential and are hypersensitive to acute oxidative stress. In mouse knockouts for Cu,ZnSOD (Réaume *et al.*, 1996) and MnSOD (Li *et al.*, 1995; Melov *et al.*, 1998), hypersensitivity to acute stress is also observed; however, there are dramatic differences in survivorship. Surprisingly, Cu,ZnSOD knockout mice exhibit a relatively normal life span, although without data on metabolic potential it is difficult to assess these results as yet. In contrast, MnSOD knockout mice have a severely reduced mean life span (3–15 days, depending on the genetic background), suggesting a critical role for MnSOD in the antioxidative defense network; the fact that MnSOD localizes to mitochondria points further to the significance of this organelle as a primary source of oxidative stress in the cell. The severity of the knockout phenotype also suggests that the development of the knockout heterozygote or other mouse models with reduced MnSOD expression, as suggested by Przedborski and Schon (1998), may be of benefit in exploring further the various roles of MnSOD in modulating oxidative stress and aging.

If oxidative stress is in fact a universal factor promoting or controlling the rate of aging in aerobic organisms, then it is predictable that various interventions that have an impact on aging may do so by modulating the rate of ROS generation and/or the levels of antioxidative defenses, which in turn should be reflected in the rate of accrual of oxidative damage. One regimen that dependably extends the life span of multiple animal species, including mammals, is that of dietary or caloric restriction. It was first noted in the mid-1930s that a reduction in caloric intake in laboratory animals, where adequate nutrition is maintained, results in a consistent increase in maximum as well as mean life span (McCay *et al.*, 1935). More recent studies have shown that intermediate dietary regimens between full restriction and *ad libitum* have predictable effects on life span, i.e., an inverse relationship exists between caloric intake and life span (Weindruch *et al.*, 1986). Such studies have rendered the paradigm more robust, countering the criticism that the beneficial effects are simply due to removal of the animals from an unnatural *ad libitum* diet that induces obesity. In addition, it appears that applying caloric restriction in adults can still have a life span-extending effect (Weindruch and Walford, 1982), suggesting that the cause of this effect is not simply a reduced developmental rate in young animals.

What then mediates this life span-extending effect of caloric restriction? One clue may lie in the reduction in body temperature observed in mice and rats that are under caloric restriction. This is particularly clear in the case of mice who, over a 24-hr period, can show a drop of up to 13°C (Koizumi et al., 1992) and would suggest that over at least part of the calorically restricted animal's day respiration is reduced. If so, caloric restriction may well act by inducing a hypometabolic state and this apparent reduction in oxygen consumption could result in reduced levels of oxidative stress and damage. Evidence to date is consistent with this scenario. There is indeed a significant decrease in both mitochondrial O_2^- and H_2O_2 generation in tissues isolated from animals that are calorically restricted relative to those that are fed *ad libitum* at the same chronological age (Sohal et al., 1994a). Moreover, there is a clear reduction in relative levels of oxidative damage in calorically restricted animals, as measured by both protein carbonyls and 8-OHdG (Sohal et al., 1994a,b). This reduction in oxidative stress/damage may be explained solely by a reduction in oxyradical generation that might be obtained in a hypometabolic state. However, it is also plausible that caloric restriction, via an as yet undetermined mechanism, bolsters antioxidative gene expression and thus reduces levels of oxidative stress. It is, however, not yet possible to identify a consistent pattern of specific antioxidant gene upregulation in these animals.

IV. FACTORS AFFECTING THE GENERATION OF REACTIVE OXYGEN SPECIES

Cellular components that are involved directly in the generation of ROS can be reasonably included among possible candidates for governing the aging process. Although there are several intracellular sites of ROS generation, the mitochondrial electron transport chain is widely recognized to be the main source. Studies in this (Kwong and Sohal, 1998) and in other laboratories support the view that ubiquinone or coenzyme Q is the main site of mitochondrial O_2^-/H_2O_2 (Boveris and Chance, 1973; Turrens and McCord, 1990). This view is based on the following evidence: (1) rates of O_2^-/H_2O_2 by submitochondrial particles (smps) are highest in the presence of rotenone and antimycin A with succinate as the substrate (Turrens *et al.*, 1985; Kwong and Sohal, 1998); (2) bovine heart smps, depleted of endogenous CoQ and reconstituted with variable amounts of exogenous CoQ, exhibit a linear relationship with the amount of quinone added (Boveris and Chance, 1973; Boveris *et al.*, 1976); (3) rates of O_2^- generation by isolated NADH-ubiquinone reductase particles, supplemented with different CoQ homologs, are linearly dependent on the amount of exogenous CoQ (Cadenas *et al.*, 1977); and (4) in reconstituted smps, activities of succinate dehydrogenase and succinate-cytochrome c reductase reach a plateau at relatively low concentrations of reducible CoQ, whereas the rate of H_2O_2 generation is linearly related to a higher range of CoQ concentration (Boveris et al., 1976).

Previous studies in this laboratory have indicated that the rate of O_2^-/H_2O_2 generation varies greatly, even in the same type of tissue among different mammalian species, and is inversely related to the maximum life span of the species (Sohal et al., 1989, 1990; Ku et al., 1994). More recent studies have indicated that the nature of the CoQ homolog may play a role in determining the rates of mitochondrial O_2^-/H_2O_2 generation (Lass et al., 1997). For example, in a sample of nine different mammalian species, namely mouse, rat, guinea pig, rabbit, pig, goat, sheep, cow, and horse, which vary from 3.5 to 46 years in their maximum longevity, the rate of O_2^- generation in cardiac smps was directly related to the relative amount of CoQ9 and inversely related to the amount of CoQ10 extracted from mitochondria (Lass et al., 1997). The relationship between CoQ homologs and rate of O_2^- generation was tested experimentally in rat heart smps, naturally containing mainly CoQ9, and ox heart smps, with high natural CoQ10 content. Repeated extractions of rat heart smps with pentane exponentially depleted both CoQ homologs whereas the corresponding rates of O_2^- generation and oxygen consumption were lowered linearly. Reconstitution of both rat and ox heart smps with different amounts of CoQ9 or CoQ10 resulted in a sharp initial rise in O_2^- generation, which was followed by a plateau at high concentrations. At comparable concentrations, smps reconstituted with CoQ9 exhibited higher rates of O_2^- generation than those obtained with CoQ10. Because these differences were observed at CoQ concentrations that are supraphysiological and because the smps were exposed to harsh depletion/reconstitution procedures, unambiguous conclusions about the *in vivo* relevance of such studies cannot be drawn. Nevertheless, the experimental findings are consistent with those based on *ex vivo* studies on different species. It therefore seems that manipulations of CoQ homologs could provide an important tool for experimentally varying the rates of O_2^- generation. Future transgenic studies might explore this possibility.

V. WHAT ARE THE TARGETS OF OXIDATIVE STRESS?

Although the oxidative stress hypothesis has garnered considerable supporting evidence, it would be premature and unreasonable to claim that its validity has been fully established or that it can satisfactorily explain all of the various facets of the aging process. What is particularly lacking at present is a rational mechanistic explanation of how oxidative molecular damage causes specific attenuations in cellular functions during aging. Although it has been demonstrated convincingly that oxidative damage to various macromolecules increases during aging, such damage has nevertheless been mainly detected in tissue homogenates, which is an overall rather than specific measure of the damage to a particular type of molecular species. Such findings have led to the prevailing notion that oxidative damage occurs randomly and that age-associated losses in physiological functions are a collective reflection of such ubiquitous and random damage. This widely held view based on biochemical measurements of oxidative damage in tissue homogenates is, however, at fundamental odds with the actual cell physiological alterations observed during aging. Studies on the phenomenology of age-related changes in various cellular systems have demonstrated unambiguously that the magnitude of the age-related functional losses ranges from insignificant to a marked decline. For instance, specific activities of most of the enzymes do not decline during aging (Kanungo, 1980; Rothstein, 1982; Gafni, 1990), which is contradictory to the predictions of the random damage model. To be explicit, the point here is that the current view of random molecular oxidative damage is incompatible with the known facts of physiological aging of the cell.

A key issue to be addressed then is how molecular oxidative damage may be linked to specific losses in cellular function. On the basis of studies in our laboratory, the idea has emerged that the age-associated increase in protein oxidative damage is selective, involving only a small percentage of proteins, implying that oxidative molecular damage to proteins is a targeted and not a random phenomenon. Furthermore, it is plausible that such targeted protein damage may be responsible for the age-related attenuations in specific cellular functions.

Although oxidative modifications clearly affect nuclear and mitochondrial DNA and membrane lipids, they are also observed readily in proteins (Stadtman, 1992; Stadtman and Berlett, 1997). Oxidatively modified proteins are generally dysfunctional, losing catalytic or structural integrity. Proteins undergo a variety of oxidative modifications, including loss of sulfhydryl groups and conversion of tyrosine to dityrosine and histidine to oxohistidine, among others; however, the most frequently encountered modification is the addition of carbonyl groups (Stadtman, 1993; Dean *et al.*, 1997). Indeed, the protein carbonyl content has proven to be a robust general indicator of oxidative damage, and substantial increases have been reported during aging in the homogenates of a variety of tissues and species (summarized in Stadtman and Berlett, 1997). Such findings have contributed to the prevailing view that random modifications of protein molecules cause a generalized decrease in metabolic efficiency occurring during senescence.

Using insect flight muscle mitochondria, we tested the hypothesis that the oxidative modification of a particular protein target or targets occurs during aging and that the loss of activity of this key target initiates an amplification of oxidative damage, which becomes particularly evident during the later part of the life span of an organism. The reason for the selection of insect flight muscle mitochondria was that this tissue has an extremely high rate of O_2^-/H_2O_2 production (Sohal and Sohal, 1991; Sohal and Dubey, 1994) and they also exhibit an age-associated increase in DNA (Agarwal and Sohal, 1994) and protein oxidative damage (Sohal et al., 1993) and a decline in oxidative phosphorylation (Tribe, 1967). The flight ability of the insect also declines during aging (Sohal and Buchan, 1981). It was thus reasoned that if specific mitochondrial protein oxidative damage indeed plays a functional role in senescence, such a protein(s) should be detectable in these mitochondria.

Our approach entailed the isolation of mitochondria from the housefly flight muscle of both young and old flies, treatment of mitochondrial protein species with DNPH, a carbonyl-specific reagent, and immunochemical detection using polyclonal anti-DNP antibodies. This analysis revealed two protein species that differentially accumulated oxidative damage. To identify these species, the oxidized proteins were purified by isoelectric focusing, concentrated by sodium dodecyl sulfate–polyacrylamide gel electrophoresis, and blotted on Immobilon-P membranes for microsequencing. This allowed the identification of mitochondrial aconitase (Yan et al., 1997) and adenine nucleotide translocase (ADP/ATP exchange transporter), (Yan and Sohal, 1998) as being particularly susceptible molecular targets of oxidative stress, a result that was substantiated further by the relative loss of enzyme activities with age. Is there any evidence that aconitase and adenine nucleotide translocase could play a causal role in age-associated muscle degeneration? Although there is a need to test this hypothesis in the context of transgenics, suggestive evidence for this possibility exists. Not only do these two protein species exhibit age-associated accumulation of oxidative damage, they both exhibit a higher degree of oxidation in flies that have lost the ability to fly (crawlers) compared to the so-called fliers; in other words, their state of oxidation parallels their physiological age closely. Also, when fluoroacetate, an inhibitor of aconitase, was administered to houseflies at multiple concentrations, the life span was reduced significantly; moreover, this life span reduction was dose dependent. Finally, it is instructive to consider the known functions of these enzymes. Aconitase is thought to function principally in the tricarboxylic acid (TCA) cycle, catalyzing the conversion of citrate to isocitrate. A loss of aconitase activity would likely lead to a reduction in the efficiency of the TCA cycle, which would lead to a decreased efficiency of oxidative phosphorylation and a consequent energy deficit. A deficit in membrane-associated adenine nucleotide translocase activity may also have global deleterious functional consequences in flight muscle tissue, as an efficient supply of ADP is necessary to maintain an optimal rate of mitochondrial state 3 respiration as well as the adequate release of ATP to the cytosol.

VI. CONCLUDING REMARKS

The oxidative stress hypothesis continues to serve as a useful paradigm in understanding how the progressive functional deterioration that accompanies the aging process might occur. If the assumption that oxidative stress/damage accumulation is a fundamental causal factor in the aging of aerobic organisms is valid, it would follow that factors controlling the level of oxidative stress, as well as the specific molecular targets that are differentially susceptible to oxidative damage, would provide the framework for a fundamental understanding of physiological aging. Experimental manipulation of the antioxidative defenses and ROS generation, which together determine the level of oxidative stress, may eventually provide a mechanistic understanding of factors governing the age-associated increase in oxidative stress. Furthermore, it is our contention that an examination of those factors that are differentially susceptible to oxidative damage should provide an invaluable approach for making the connection between oxidative stress and the specific functional deterioration associated with aging. In this light, it is possible that strategies to render these targets less susceptible to oxidative damage may provide an important tool for intervention in the aging process.

REFERENCES

Agarwal, S., and Sohal, R. S. (1994). DNA oxidative damage and life expectancy in houseflies. *Proc. Natl. Acad. Sci. U.S.A.* **91**, 12332–12335.
Avraham, K. B., Schickler, M., Sapoznikov, D., Yarom, R., and Groner, Y. (1988). Down's syndrome: Abnormal neuromuscular junction in tongue of transgenic mice with elevated levels of human Cu, Zn-superoxide dismutase. *Cell (Cambridge, Mass.)* **54**, 823–829.
Beckman, K. B., and Ames, B. N. (1998). The free radical theory of aging matures. *Physiol. Rev.* **78**(2), 547–581.
Boveris, A., and Chance, B. (1973). The mitochondrial generation of hydrogen peroxide. *Biochem. J.* **134**, 707–716.
Boveris, A., Cadenas, E., and Stoppani, A. O. M. (1976). Role of ubiquinone in the mitochondrial generation of hydrogen peroxide. *Biochem. J.* **156**, 435–444.
Cadenas, E., Boveris, A., Ragan, C. I., and Stoppani, A. O. M. (1977). Production of superoxide radicals and hydrogen peroxide by NADH-ubiquinone reductase and ubiquinol cytochrome c reductase from beef heart mitochondria. *Arch. Biochem. Biophys.* **180**, 248–257.
Dean, R. T., Fu, S., Stocker, R., and Davies, M. J. (1997). Biochemistry and pathology of radical-mediated protein oxidation. *Biochem. J.* **324**, 1–18.

Epstein, C. J., Avraham, K. B., Lovett, M., Smith, S., Elroy-Stein, O., Rotman, G., Bry, C., and Groner, Y. (1987). Transgenic mice with increased Cu/Zn-superoxide dismutase activity: Animal model of dosage effects in Down syndrome. *Proc. Natl. Acad. Sci. U.S.A.* **84,** 8044–8048.

Gafni, A. (1990). Age-related effects in enzyme metabolism and catalysis. *Rev. Biol. Res. Aging* **4,** 315–336.

Harman, D. (1956). Aging: A theory based on free radical and radiation chemistry. *J. Gerontol.* **11,** 298–300.

Kanungo, M. S. (1980). "Biochemistry of Ageing." Academic Press, London.

Kinouchi, H., Epstein, C. J., Mizui, T., Carlson, E., Chen, S. F., and Chan, P. H. (1991). Attenuation of focal cerebral ischemic injury in transgenic mice overexpressing CuZn superoxide dismutase. *Proc. Natl. Acad. Sci. U.S.A.* **88,** 11158–11162.

Koizumi, A., Tsukada, M., Wada, Y., Masuda, H. and Weindruch, R. (1992). Mitotic activity in mice is suppressed by energy restriction-induced torpor. *J. Nutr.* **122,** 1446–1453.

Ku, H.-H., Brunk, U. T., and Sohal, R. S. (1994). Relationship between mitochondrial superoxide and hydrogen peroxide production and longevity of mammalian species. *Free Radical Biol. Med.* **15,** 621–627.

Kwong, L. K., and Sohal, R. S. (1998). Substrate and site specificity of hydrogen peroxide generation in mouse mitochondria. *Arch. Biochem. Biophys.* **350,** 118–126.

Lass, A., Agarwal, S., and Sohal, R. S. (1997). Mitochondrial ubiquinone homologues, superoxide radical generation and longevity in different mammalian species. *J. Biol. Chem.* **272,** 19199–19204.

Li, G., Chen, Y., Saari, J. T., and Kang, Y. J. (1997). Catalase-overexpressing transgenic mouse heart is resistant to ischemia-reperfusion injury. *Am. J. Physiol.* **273,** H1090–H1095.

Li, Y., Huang, T.-T., Carlson, E. J., Melov, S., Ursell, P. C., Olson, J. L., Noble, L. J., Yoshimura, M. P., Berger, C., Chan, P. H., Wallace, D. C., and Epstein, C. J. (1995). Dilated cardiomyopathy and neonatal lethality in mutant mice lacking manganese superoxide dismutase. *Nat. Genet.* **11,** 376–381.

Mackay, W. J., and Bewley, G. C. (1989). The genetics of catalase in *Drosophila melanogaster*: Isolation and characterization of acatalasemic mutants. *Genetics* **122,** 643–652.

Martin, G. M., Austad, S. N., and Johnson, T. E. (1996). Genetic analysis of ageing: Role of oxidative damage and environmental stresses. *Nat. Genet.* **13,** 25–34.

McCay, C. M., Crowell, M. F., and Maynard, L. A. (1935). The effect of retarded growth upon the length of lifespan and upon ultimate body size. *J. Nutr.* **10,** 63–79.

McCord, J. M., and Fridovich, I. (1969). Superoxide dismutase: An enzymic function for erythrocuprein (hemocuprein). *J. Biol. Chem.* **244,** 6049–6055.

Melov, S., Schneider, J. A., Day, B. J., Hinerfield, D., Coskun, P., Mirra, S. S., Crapo, J. D., and Wallace, D. C. (1998). A novel neurological phenotype in mice lacking mitochondrial manganese superoxide dismutase. *Nat. Genet.* **18,** 159–163.

Miquel, J., Lundgren, P. R., Bensch, K. J., and Attan, H. (1979). Effect of temperature on the lifespan, vitality and fine structure of *Drosophila melanogaster*. *Mech. Ageing Dev.* **10,** 93–99.

Mirochnitchenko, O., Palnitkar, U., Philbert, M., and Inouye, M. (1995). Thermosensitive phenotype of transgenic mice overproducing human glutathione peroxidases. *Proc. Natl. Acad. Sci. U.S.A.* **92,** 8120–8124.

Orr, W. C., and Sohal, R. S. (1992). The effects of catalase gene overexpression on life span and resistance to oxidative stress in transgenic *Drosophila melanogaster*. *Arch. Biochem. Biophys.* **297,** 35–41.

Orr, W. C., and Sohal, R. S. (1993). Effects of overexpression of Cu, Zn-superoxide dismutase on life span and response to oxidative stress in *Drosophila melanogaster*. *Arch. Biochem. Biophys.* **301,** 34–40.

Orr, W. C., and Sohal, R. S. (1994). Extension of life-span by overexpression of superoxide dismutase and catalase in *Drosophila melanogaster*. *Science* **263**, 1128–1130.

Orr, W. C., Arnold, L. A., and Sohal, R. S. (1992). Relationship between catalase activity, life span and some parameters associated with antioxidant defenses in *Drosophila melanogaster*. *Mech. Ageing Dev.* **63**, 287–296.

Parkes, T. L., Elia, A. J., Dickinson, D., Hilliker, A. J., Phillips, J. P., and Boulianne, G. L. (1998). Extension of Drosophila lifespan by overexpression of human SOD1 in motorneurons. *Nat. Genet.* **19**, 171–174.

Pearl, R. (1928). "The Rate of Living." Alfred A. Knopf, New York.

Phillips, J. P., Campbell, S. D., Michaud, D., Charbonneau, M., and Hilliker, A. (1989). Null mutation of copper/zinc superoxide dismutase in Drosophila confers hypersensitivity to paraquat and reduced longevity. *Proc. Natl. Acad. Sci. U.S.A.* **86**, 2761–2765.

Przedborski, S., and Schon, E. A. (1998). Loss of ROS—a radical response. *Nat. Genet.* **18**, 99–100.

Réaume, A. G., Elliott, J. L., Hoffman, E. K., Kowall, N. W., Ferrante, R. J., Siwek, D. F., Wilcox, H. M., Flood, D. G., Beal, M. F., Brown, R. H., Jr., Scott, R. W., and Snider, W. D. (1996). Motorneurons in Cu/Zn superoxide dismutase-deficient mice develop normally but exhibit enhanced cell death after axonal injury. *Nat. Genet.* **13**, 43–47.

Reveillaud, I., Niedzwiecki, A., Bensch, K. G., and Fleming, J. E. (1991). Expression of bovine superoxide dismutase in *Drosophila melanogaster* augments resistance to oxidative stress. *Mol. Cell. Biol.* **11**, 632–640.

Rothstein, M. (1982). "Biochemical Approaches to Aging." Academic Press, New York.

Rubner, M. (1908). "Das Problem der Lebensdauer und seine Beziehunger zum Wachstum und Ernaibrung." Oldenburg, Munchen.

Sies, H. (1986). Biochemistry of oxidative stress. *Angew. Chem., Int. Ed. Engl.* **25**, 1058–1071.

Sohal, R. S., and Buchan, P. B. (1981). Relationship between physical activity and life span in the adult housefly, *Musca domestica*. *Exp. Gerontol.* **16**, 157–162.

Sohal, R. S., and Dubey, A. (1994). Mitochondrial oxidative damage, hydrogen peroxide release and aging. *Free Radical Biol. Med.* **16**, 621–626.

Sohal, R. S., and Orr, W. C. (1995). Is oxidative stress a causal factor in aging. *In* "Molecular Aspects of Aging" (K. Esser and G. M. Martin, eds.), pp. 109–127. Wiley, New York.

Sohal, R. S., and Sohal, B. H. (1991). Hydrogen peroxide release by mitochondria increases during aging. *Mech. Ageing Dev.* **57**(2), 187–202.

Sohal, R. S., and Weindruch, R. (1996). Oxidative stress, caloric restriction, and aging. *Science* **273**, 59–63.

Sohal, R. S., Svensson, I., Sohal, B. H., and Brunk, U. T. (1989). Superoxide anion radical production in different animal species. *Mech. Ageing Dev.* **49**, 129–135.

Sohal, R. S., Svensson, I., and Brunk, U. T. (1990). Hydrogen peroxide production by liver mitochondria in different species. *Mech. Ageing Dev.* **53**, 209–215.

Sohal, R. S., Agarwal, S., Dubey, A., and Orr, W. C. (1993). Protein oxidative damage is associated with life expectancy. *Proc. Natl. Acad. Sci. U.S.A.* **90**, 7255–7259.

Sohal, R. S., Ku, H.-H., Agarwal, S., Forster, M. J., and Lal, H. (1994a). Oxidative damage, mitochondrial oxidant generation and antioxidant defenses during aging and in response to food restriction in the mouse. *Mech. Ageing Dev.* **74**, 121–133.

Sohal, R. S., Agarwal, S., Candas, M., Forster, M., and Lal, H. (1994b). Effect of age and caloric restriction on DNA oxidative damage in different tissues of C57BL/6 mice. *Mech. Ageing Dev.* **76**, 215–224.

Sohal, R. S., Agarwal, A., Agarwal, S., and Orr, W. C. (1995). Simultaneous overexpression of Cu, Zn-superoxide dismutase and catalase retards age-related oxidative damage and increases metabolic potential in *Drosophila melanogaster. J. Biol. Chem.* **270,** 15671–15674.

Stadtman, E. R. (1992). Protein oxidation and aging. *Science* **257,** 1220–1224.

Stadtman, E. R. (1993). Oxidation of free amino acids and amino acid residues in proteins by radiolysis and by metal-catalyzed reactions. *Annu. Rev. Biochem.* **62,** 797–821.

Stadtman, E. R., and Berlett, B. S. (1997). Reactive oxygen-mediated protein oxidation in aging and disease. *Chem. Res. Toxicol.* **10**(5), 485–494.

Tribe, M. A. (1967). Changes taking place in the respiratory efficiency of isolated flight muscle sarcosomes, associated with the age of the blowfly, *Calliphora erythrocephala. Comp. Biochem. Physiol.* **23,** 607–620.

Turrens, J. F., and McCord, J. M., (1990). Mitochondrial generation of reactive oxygen species. *In* "Free Radicals, Lipoproteins, and Membrane Lipids" (A. C. Paulet, L. Douste-Blazy, and R. Paoletti, eds.), pp. 203–212. Plenum, New York.

Turrens, J. F., Alexandre, A., and Lehninger, A. L. (1985). Ubisemiquinone is the electron donor for superoxide formation by complex III of heart mitochondria. *Arch. Biochem. Biophys.* **237,** 408–414.

Van Remmen, H., Williams, M. D., Yang, H., Walter, C. A., and Richardson, A. (1998). Analysis of the transcriptional activity of the 5′-flanking region of the rat catalase gene in transiently transfected cells and in transgenic mice. *J. Cell Biol.* **174,** 18–26.

Weindruch, R., and Walford, R. L. (1982). Dietary restriction in mice beginning at 1 year of age: Effect on lifespan and spontaneous cancer incidence. *Science* **215,** 1415–1418.

Weindruch, R., Walford, R. L., Fligiel, S., and Guthrie, D. (1986). The retardation of aging in mice by dietary restriction: Longevity, cancer, immunity and lifetime energy intake. *J. Nutr.* **116,** 641–654.

Wispé, J. R., Warner, B. B., Clark, J. C., Dey, C. R., Neuman, J., Glasser, S. W., Crapo, J. D., Chang, L.-Y., and Whitsett, J. A. (1992). Human Mn-superoxide dismutase in pulmonary epithelial cells of transgenic mice confers protection from oxygen injury. *J. Biol. Chem.* **267,** 23937–23941.

Yan, L.-J., Levine, R. L., and Sohal, R. S. (1997). Oxidative damage during aging target mitochondrial aconitase. *Proc. Natl. Acad. Sci. U.S.A.* **94,** 11168–11172.

Yan, L. J., and Sohal, R. S. (1998). Mitochondrial adenine nucleotide translocase is modified oxidatively during aging. *Proc. Natl. Acad. Sci. U.S.A.* **95,** 12896–12901.

24 Antioxidants in Senescence and Wasting

Wulf Dröge,* Volker Hack,* Raoul Breitkreutz,*
Eggert Holm,† Stefanie Holm,† and
Ralf Kinscherf*

*Deutsches Krebsforschungszentrum
Division of Immunochemistry
D-69120 Heidelberg, Germany
†Medical Clinic I
68167 Mannheim, Germany

I. INTRODUCTION: LOSS OF SKELETAL MUSCLE MASS AND MUSCLE FUNCTION AS A CORRELATE OF SENESCENCE AND WASTING

Senescence and wasting are largely autonomous biological processes associated with the increased probability of death. The processes of age- and disease-related wasting have, by these criteria, a biological function analogous to apoptotic cell death and may be advantageous for the species because diseased and/or nonreproductive food competitors are eliminated. For individual human subjects, however, senescence and wasting are often associated with psychological stress and financial burden. Its medical and social relevance may increase even further as the average life span is increasing progressively.

The hallmarks of senescence and wasting include the massive loss of body cell mass (bcm) and muscle function, decreased resistance to infections, frailty (increased probability of disability), and organ failure (Cohn *et al.*, 1980; Brennan, 1977; DeWys *et al.*, 1980; Strain, 1979; Pisters and Pearlstone, 1993; Long *et al.*, 1976; Ott *et al.*, 1993; Kotler *et al.*, 1985; Buchner and Wagner, 1992). Wasting is a common phenomenon in malignancies (Brennan, 1977; DeWys *et al.*, 1980; Strain, 1979; Pisters and Pearlstone,

1993), sepsis, trauma (Long et al., 1976), and certain infectious diseases, including HIV infection (Ott et al., 1993; Kotler et al., 1985). In the treatment of cancer patients, wasting is often a limiting factor that prevents the application of aggressive chemotherapy. In old age, the loss of skeletal muscle mass is associated with a compromised physical and social function (Brody, 1985; Lamberts et al., 1997; Campion, 1994). Almost half of the people over 85 years in the United States require assistance (Brody, 1985). It was shown that the loss of muscle strength in the elderly is mostly related to the loss of muscle mass (Fiatarone et al., 1994).

II. REDOX STATE AS A TARGET FOR THERAPEUTIC INTERVENTION

A. Plasma Redox State as a Correlate of Senescence and Wasting

1. Age-Related Shift in the Plasma Redox State of Healthy Subjects

An increasingly popular hypothesis states that senescence may result from the accumulation of oxidative damage (Harman, 1956; Shigenaga et al., 1994; Stadtman, 1992; Gilchrest and Bohr, 1997) and that dietary antioxidants may slow the degenerative process (Harman, 1956; Stadtman, 1992; Gilchrest and Bohr, 1997). In line with this hypothesis, vitamin E was shown to ameliorate age-related health problems (reviewed in Manton et al., 1997), and certain age-related degenerative processes were even found to be reversed by treatment with antioxidants (reviewed in Stadtman, 1992, Gilchrest and Bohr, 1997). However, the general use of antioxidants as dietary supplements is limited by the fact that there is little information about what type and quantity of antioxidant are needed, and how the individual can monitor that she or he has gotten enough.

A series of more recent studies may answer these questions, at least partly, by showing (i) that senescence and wasting are associated with an easily demonstrable change in the redox state of the blood plasma and (ii) that important consequences of changes in the redox state can be demonstrated in human subjects within a few weeks or months (Hack et al., 1998). In view of the relatively short observation periods, these findings may provide a quantitative guideline for a redox-oriented prophylactic therapy. Importantly, corresponding changes in bcm and redox state can be observed not only in pathological conditions, but also among seemingly healthy persons.

One of these studies, which included 205 healthy human subjects (Hack et al., 1998), revealed a significant age-dependent increase in the plasma cystine level and a decrease in the plasma thiol level indicative of an age-dependent shift to a more oxidized condition (Fig. 1). Using the bcm index (defined as bcm/height^2) as an indicator for the age-related loss of bcm, it was found that the bcm index was inversely correlated with the plasma cystine/thiol ratio (Fig. 2, left). Expectedly, the lowest bcm indices and highest cystine/thiol ratios were seen among subjects >75 years of age (Hack et al., 1998).

2. Changes of the Redox State and Body Cell Mass in Cancer Patients

A study of patients with different types of cancer (mean age 60.5 ± 1.1 years) revealed that these patients had, on average, lower bcm indices and higher cystine/thiol ratios than healthy subjects in the corresponding age range of 50–75 years, but similar to that of subjects >75 years of age without tumors (Hack et al., 1998). Cancer patients also showed a significant inverse correlation between the bcm index and the cystine/thiol ratio similar to the correlation seen in the group of healthy subjects (Fig. 2).

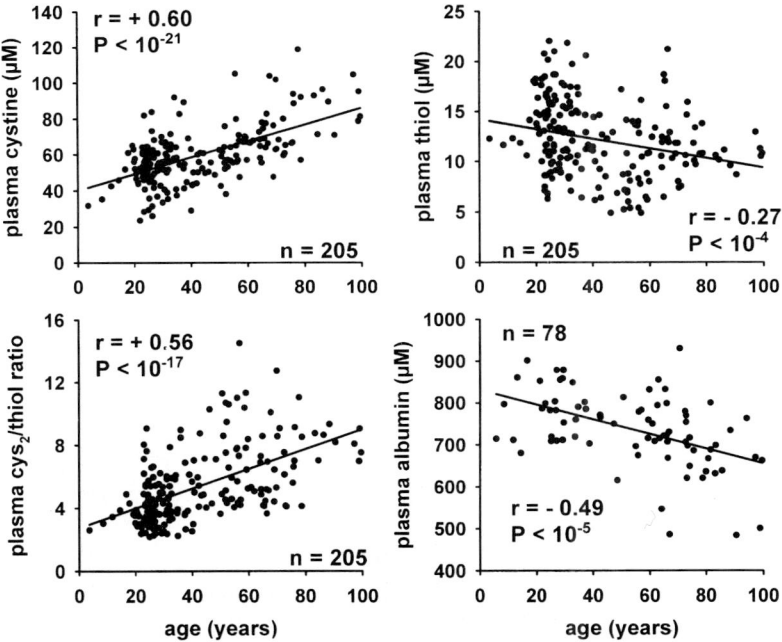

Figure 1 Correlation among plasma redox state, albumin level, and age in healthy persons. Postabsorptive plasma amino acid and acid-soluble thiol levels and plasma albumin concentrations have been determined in the plasma from the cubital vein of randomly selected healthy human subjects of both sexes. Each point represents an individual person (data from Hack et al., 1998).

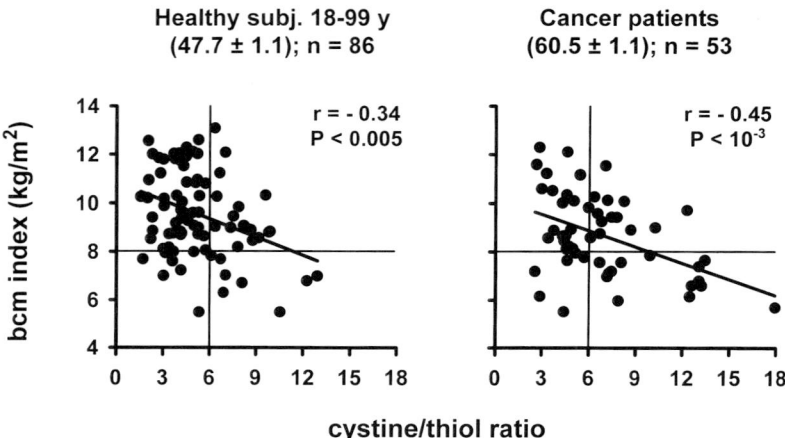

Figure 2 Correlation between body cell mass index and thiol redox state of healthy subjects and cancer patients. The bcm index was defined as the bcm/height^2 in analogy to the body mass index. Horizontal and vertical lines indicate the window that contains virtually all healthy young subjects <35 years old (data from Hack et al., 1998).

B. Plasma Albumin Level as a Correlate of Senescence and Wasting

1. Linkage between Albumin Level and Plasma Redox State

The plasma albumin level is another important parameter that is significantly correlated with the plasma redox state (Figs. 1 and 3) and with the bcm index of healthy subjects and cancer patients (Hack et al., 1998). In view of the relatively slow processes of senescence and wasting, it was important to note that the linkage between plasma albumin level and plasma redox state was demonstrable in longitudinal studies on the changes in albumin level and changes in the plasma cystine/thiol ratio during relatively short observation periods (Hack et al., 1998). In one of the most instructive longitudinal studies of this type, plasma cystine/thiol ratios and albumin levels were determined at 33 randomly chosen time points during a 2-year observation period in the peripheral blood from a single healthy individual in the sixth decade of life (Fig. 3). Observed data showed rather strong variations in both cystine/thiol ratios (range 3.3–9.4) and albumin levels (684–884 μM). The more detailed analysis revealed a significant correlation not only between albumin levels and plasma cystine/thiol ratios at given time points (Fig. 3, top), but also between the *changes* in the albumin level and the corresponding *changes* in the cystine/thiol ratio (Fig. 3, bottom). Similar correlations between changes in albumin levels and changes in the

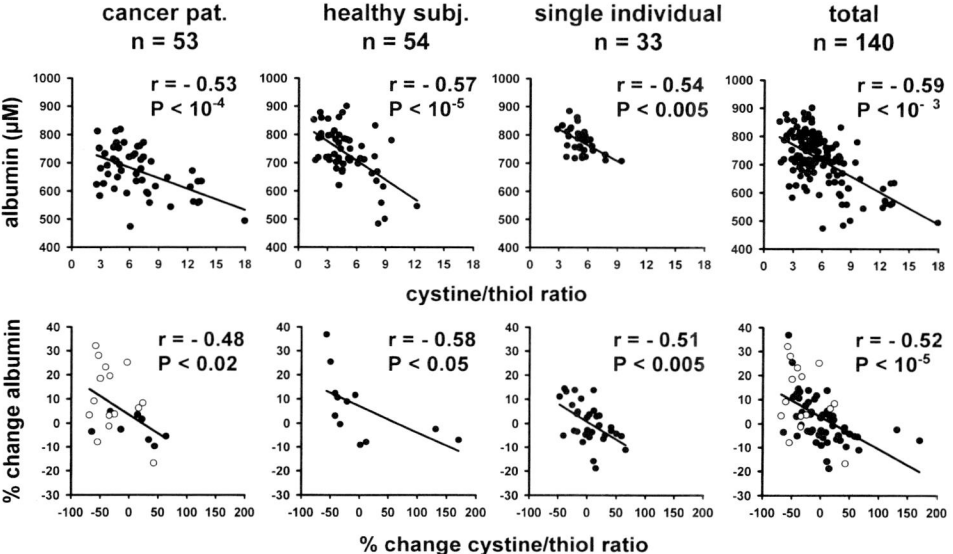

Figure 3 Correlation between plasma albumin level and plasma redox state of healthy subjects and cancer patients. The groups of subjects and the study design are described in Hack *et al.* (1998). Changes in albumin and cystine/thiol ratios have not been determined for all subjects under test. Open circles indicate NAC-treated patients.

cystine/thiol ratios were found in longitudinal observations of other healthy subjects and cancer patients (Fig. 3).

Plasma albumin is one of the major redox buffers in the blood plasma (Halliwell and Gutteridge, 1990). Moreover, the plasma albumin level decreases strongly in the course of senescence (Fig. 1) and wasting and has been used as a quantitative measure of cachexia in catabolic conditions (Rothschild *et al.*, 1988; Cooper and Gardner, 1989; Shibata *et al.*, 1991; Baumgartner *et al.*, 1996; Tayek, 1988; Naber *et al.*, 1997). In studies on healthy elderly subjects, a low plasma albumin level was correlated with a low 10-year survival rate (Shibata *et al.*, 1991) and with a loss of skeletal muscle mass (sarcopenia) (Baumgartner *et al.*, 1996). In cachectic patients, the plasma albumin level was found to be a strong predictor of survival and cost of treatment (Rothschild *et al.*, 1988). However, several attempts to increase the albumin level by nutritional therapy were not successful (Rothschild *et al.*, 1988; Tuten *et al.*, 1985, Erstad, 1992; Paluzzi and Meguid, 1987).

In principle, the albumin level can be regulated either at the level of albumin clearance (degradation) or at the level of albumin biosynthesis in the liver. Although a mechanistic link between plasma redox state and albumin biosynthesis is not formally excluded, available evidence favors the interpretation that the redox state affects the rate of albumin degradation rather than albumin synthesis. Human and bovine albumin contain a single unpaired cysteine residue (Cys-34) with antioxidative function (Halliwell and Gutteridge, 1990; Kuwata et al., 1994; Baumgartner et al., 1996; Finch et al., 1993). In the blood plasma, albumin occurs largely in its reduced form (mercaptoalbumin) and to a lesser extent in the oxidized form (nonmercaptoalbumin). The oxidized form consists mainly of a mixed protein–cysteine disulfide or protein–glutathione disulfide and increases in proportion with age (Halliwell and Gutteridge, 1990; Era et al., 1988, 1995). It has also been reported that redox processes mediate the conversion of albumin into an aged form with a three-fold higher catabolic rate (Kuwata et al., 1994).

2. Linkage between Plasma Albumin and Arginine Levels

Earlier experiments in rats have shown that the rate of albumin synthesis is regulated by arginine and its polyamine derivative spermine (Oratz et al., 1976). We, therefore, used the biparametric correlation analysis to determine whether and to what extent the albumin level may depend on the plasma arginine level in addition to the cystine/thiol ratio. The analysis of data from healthy subjects, cancer patients, and longitudinal measurements from a single healthy individual yielded very similar regression functions (Fig. 4). Exponential functions on the abscissa were obtained by biparametric analysis. These functions and the P values for the relative contribution of arginine and cystine/thiol ratio (Fig. 4, lower right) indicated that the contribution of arginine was not as strong as the contribution of the cystine/thiol ratio but was nevertheless significant ($P < 0.05$).

3. Consequences of a Decreased Albumin Level

The plasma albumin level is known to play an important role in the control of the oncoosmotic pressure and in the prevention of edema (Rothschild et al., 1972, 1988; Tayek, 1988; Erstad, 1992). A detailed analysis of plasma albumin levels and corresponding levels of relative body water as defined by the ratio of total body water/bcm among a group of cancer patients revealed that a decrease in plasma albumin level below 680 μM is associated with a strong increase in body water (Hack et al., 1998). A similar biphasic concentration dependency with a similar concentration threshold of about 680 μM albumin was found previously for the correlation between blood pressure and plasma albumin level (Hu et al., 1992). Abnormally low

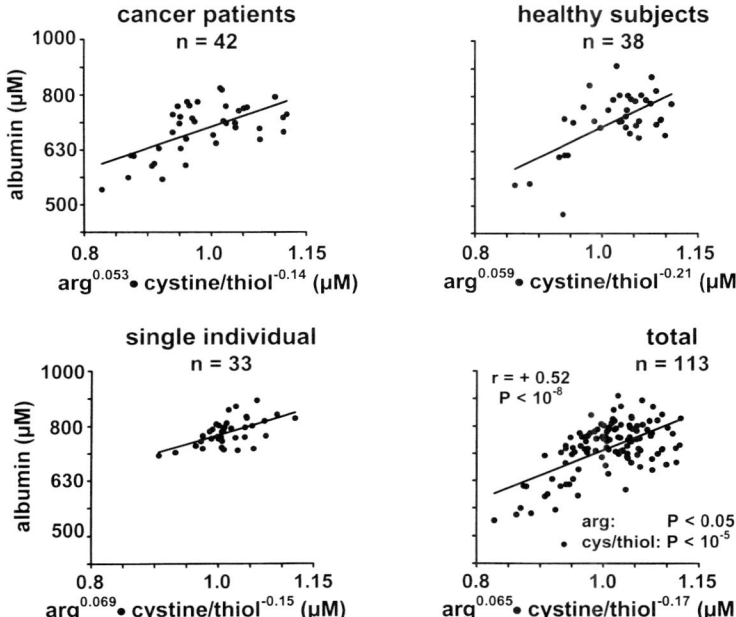

Figure 4 Linkage between albumin and plasma arginine levels. The units of the x axis are biparametric terms that were obtained by biparametric analysis of the correlations among plasma albumin concentrations, arginine levels, and cystine/thiol ratios. P values for the relative contribution of arginine and cystine/thiol ratios obtained from the total set of data are indicated in the lower right panel. The 33 measurements from the single individual were obtained after random time intervals during a 2-year observation period (see Hack *et al.*, 1998).

plasma albumin levels are among the major problems in critically ill patients but can be corrected in acute conditions by an infusion of freshly frozen plasma, human albumin, or synthetic colloids. This method is not applicable, however, for the long-term treatment of cancer patients and elderly subjects.

C. Therapeutic Intervention with an Antioxidant: Effect of *N*-Acetylcysteine (NAC) on bcm, Albumin Level, and Functional Activity of Cancer Patients and Healthy Human Subjects

1. A Placebo-Controlled Study on Healthy Subjects

Evidence for a cause/effect relationship between redox state and bcm was obtained in two independent studies on healthy human subjects and cancer patients, respectively, using NAC as an antioxidant. Healthy human subjects have been studied in a randomized placebo-controlled double-blind study on the effect of oral NAC treatment on the change in bcm. The volunteers were also subjected to a program of anaerobic physical exercise to generate a condition similar to that of cancer patients who are known to express a high rate of glycolytic activity in their muscle tissue (Hack et al., 1998). Physical exercise was already previously shown to cause the oxidation of glutathione in the blood (Sastre et al., 1992; Sen et al., 1994), and this oxidation was ameliorated by treatment with NAC (Sen et al., 1994). When the changes in bcm during a 5-week observation period, including a 4-week program of physical exercise, were analyzed separately in subgroups with a higher and lower than median cystine/thiol ratio (i.e., 6.29), it was found that NAC treatment of the subgroup with high cystine/thiol ratios induced a relative increase in bcm in comparison with the placebo group (see Fig. 5), whereas no such effect of NAC was seen in the subgroup with low baseline cystine/thiol ratios (Hack et al., 1998). Unexpectedly, NAC did not cause a significant increase in the intracellular glutathione level of the peripheral blood mononuclear cells (Hack et al., 1998).

2. Effect of NAC on bcm, Albumin Levels, and Functional Activity of Cancer Patients

In a study on cancer patients, NAC was used in combination with interleukin-2 (IL-2) treatment and was compared with IL-2 treatment alone or with conventional chemotherapy (Hack et al., 1998). Control patients with IL-2 treatment or standard therapy experienced a substantial loss in bcm indicative of the disease-related cachectic process. NAC-treated patients, in contrast, showed an improvement in four independent quantitative corre-

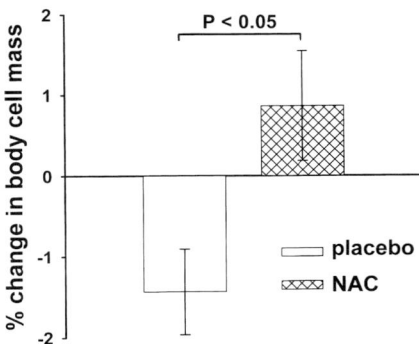

Figure 5 Effect of N-acetylcysteine (NAC) treatment on the bcm of healthy subjects. Changes in bcm (mean ± SEM) of healthy human subjects with baseline cystine/thiol ratios >6.28 during a 5-week observation period. Data are taken from Hack et al. (1998).

lates of the cachectic process, i.e., an increase in bcm (Fig. 6), an increase in plasma albumin, a decrease in the mean plasma cystine/thiol ratio (Fig. 7), and a decrease in the plasma glutamate level (Hack *et al.*, 1998). In addition, the NAC plus IL-2-treated group showed a substantial improvement of "functional capacity" as a measure of the patient's quality of life (Fig. 8). The "functional capacity index" has been defined by Ottery (1995) and is described briefly in the legend to Fig. 8.

III. WHAT DETERMINES THE PLASMA CYSTINE LEVEL?

A. An Increase in the Plasma Cystine Level and a Shift in Redox State May Be Caused by Skeletal Muscle Protein Catabolism and Export of Cystine

Mechanisms responsible for the age-related increase in plasma cystine levels and the corresponding shift in the plasma redox state are not known. It is a common paradigm that senescence is associated with an increased rate of oxidative processes and a decrease in antioxidative defense mechanisms.

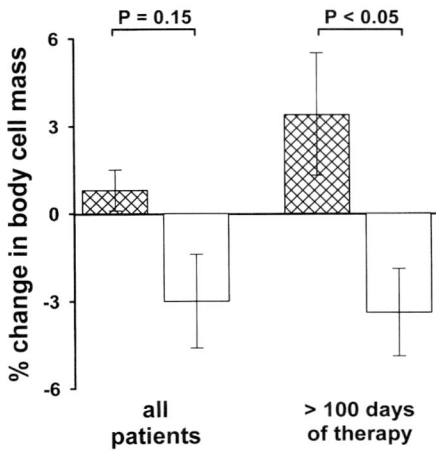

Figure 6 Effect of NAC treatment on the bcm of cancer patients. The group treated with NAC in combination with interleukin-2 has been compared with a combined group of patients who were treated either with IL-2 only or with conventional chemotherapy. The changes were defined by the first and last examination of each patient. The difference between treatment groups was statistically significant if patients with observation periods >100 days were compared. Data are taken from Hack *et al.* (1998).

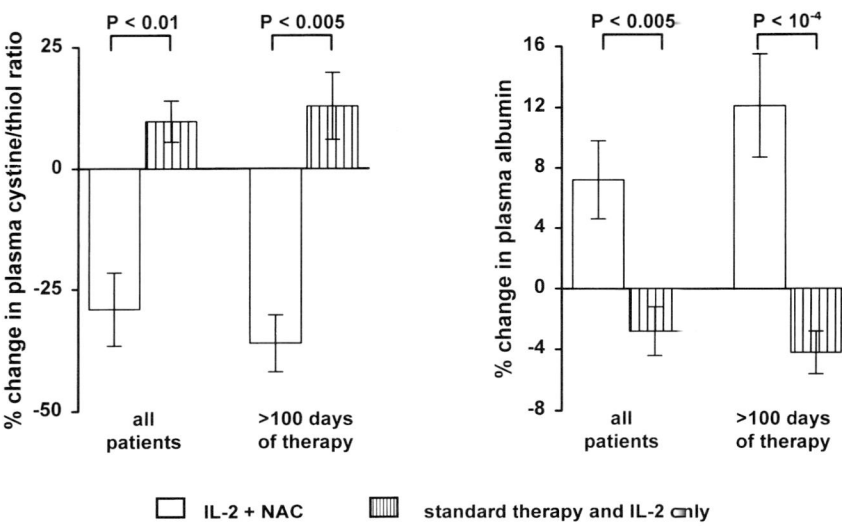

Figure 7 Effect of NAC treatment on the plasma cystine/thiol ratio and the plasma albumin level of cancer patients. Data are taken from Hack et al. (1998). For other details, see legend to Fig. 6.

However, the simplest and most trivial explanation for the increase in the plasma cystine level is that this phenomenon may result from an increased export of cystine from the skeletal muscle tissue as a consequence of an increased protein catabolism. This assumption is supported by several independent observations. First, the age-related increase in plasma cystine levels is associated with the corresponding increase in the plasma levels of most other protein-forming amino acids (Fig. 9). This nonspecific increase of almost all amino acids in the blood plasma would not be expected if the increase in the plasma cystine level would result directly from the oxidation of plasma cysteine by yet unknown oxidative processes. Second, amino acid exchange studies on young healthy subjects have shown that the plasma cystine level is strongly correlated with the rate of protein catabolism and the export of cystine from the skeletal muscle tissue (Hack et al., 1997). Third, another study on healthy human subjects with a 4-week program of anaerobic physical exercise has shown that *changes* in the plasma cystine level during the 5-week observation period were inversely correlated with *changes* in bcm (Kinscherf et al., 1996). Interestingly, the detailed analysis of this latter study revealed that this inverse correlation was seen only in the subgroup of persons with a lower than median cystine/thiol ratio (6.29) but not in the group with higher cystine/thiol ratios (Fig. 10). This study has been performed with two groups of subjects, i.e., one treated with NAC

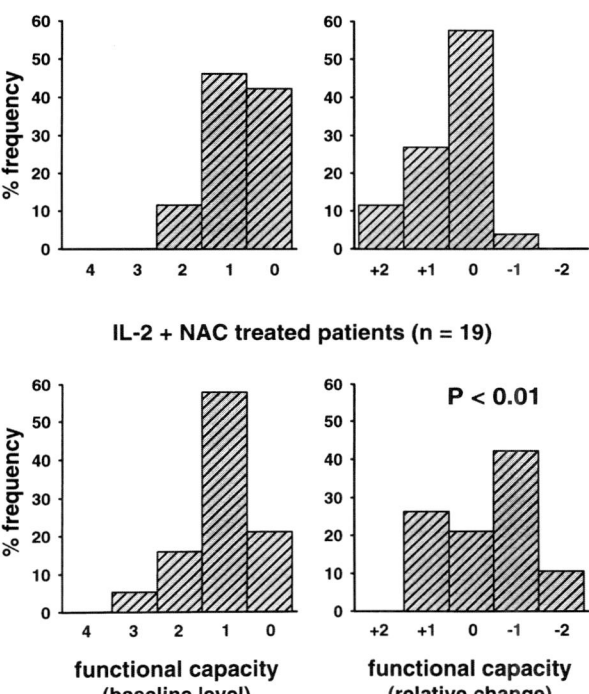

Figure 8 Effect of NAC treatment on the functional capacity of cancer patients (frequency histograms). The "functional capacity index" was defined according to Ottery (1995): 0, normal, no limitations; 1, not normal, but able to be up with fairly normal activities; 2, not feeling up to most things, but in bed less than half the day; 3, able to do little activity and most of the day in bed or chair; and 4, rarely out of bed. The scales of the x axis have been reversed to account for the fact that a higher "functional capacity index" means a lower quality of life according to this definition. Data are taken from Hack et al. (1998). For other details, see legend to Fig. 6.

and one with placebo. However, for the purpose of the analysis of Fig. 10, data of both groups were pooled because both groups showed similar correlations. The results of Fig. 10 are in agreement with the analysis of postabsorptive amino acid exchange rates of elderly subjects, which revealed that the release of tyrosine (i.e., an indicator of protein catabolism) was correlated with a more than proportional export of taurine and a less than proportional export of cystine if compared with the exchange rates of young subjects (Hack et al., 1997). This phenomenon may tend to slow a possible vicious circle of protein catabolism and an increase in plasma cystine.

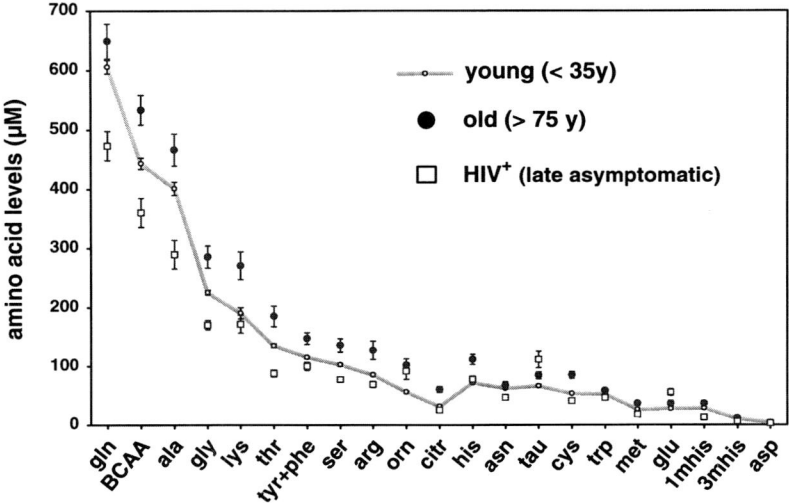

Figure 9 Plasma amino acid levels of healthy young subjects, elderly subjects, and HIV-infected patients. Plasma amino acid levels of a randomly selected group of persons <35 years old (○), a group of randomly selected elderly subjects >75 years old (●), and a group of HIV-infected patients of the late asymptomatic stage with CD4+ T-cell counts <400 mm^3 (□). This subgroup of patients was previously shown to have, on average, the lowest plasma cystine levels (Hack *et al.*, 1997). The groups of young and elderly healthy subjects are identical to the groups described in Hack *et al.* (1998).

B. What Determines the Rate of Skeletal Muscle Protein Catabolism in Subjects with Low Cystine/Thiol Ratios: Role of Ornithine in the Regulation of bcm

In view of the linkage between plasma cystine level and muscle protein catabolism in healthy young subjects, it was of interest to identify factors that control bcm and protein catabolism in persons with a low baseline cystine/thiol ratio. To test the hypothesis that the net protein catabolism in the skeletal muscle tissue may be limited in this case by the availability of one or several amino acids, we again analyzed data from the subgroups of healthy subjects with a lower or higher than median cystine/thiol ratio (see Section III,A) and determined the correlation between the change in bcm and baseline plasma amino acid levels. Data of the NAC group and placebo group (see Section III,A) were pooled again because both groups showed similar correlations. The strongest correlations obtained by this analysis were seen in the subgroup with low cystine/thiol ratios and specifically between the change in bcm and the baseline levels of ornithine (Fig. 11), histidine, arginine, cystine, and serine (not shown) in decreasing order of

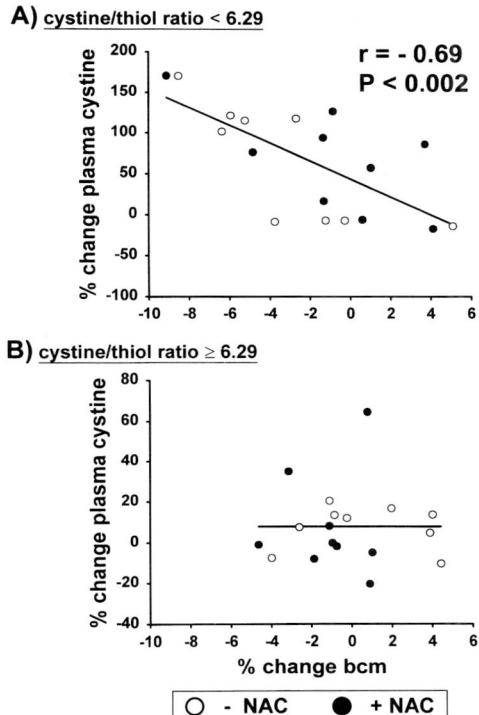

Figure 10 Correlations between changes in the plasma cystine level and changes in bcm of healthy subjects during a 5-week observation period, including a 4-week period of physical exercise, were determined independently for the subgroup of persons with lower (A) and higher (B) than median cystine/thiol ratios (6.29). Correlations obtained for persons with (●) and without (○) NAC treatment were essentially similar, and the results of these two treatment groups have therefore been pooled.

statistical significance. Multiparametric analysis of the correlation between changes in bcm and the baseline levels of these five amino acids finally revealed a significant correlation with ornithine only, whereas the other amino acids contributed only marginally. The group with higher than median cystine/thiol ratios showed no significant correlation between the change in bcm and any of the amino acids except for a marginally significant negative correlation between the change in bcm and 3-methylhistidine. This methylated amino acid is a well-known breakdown product of skeletal muscle fiber proteins. Taken together, the available information leads to the paradigm that the bcm of persons with a high cystine/thiol ratio may be determined largely by the redox state (see Section II), whereas the bcm of physically

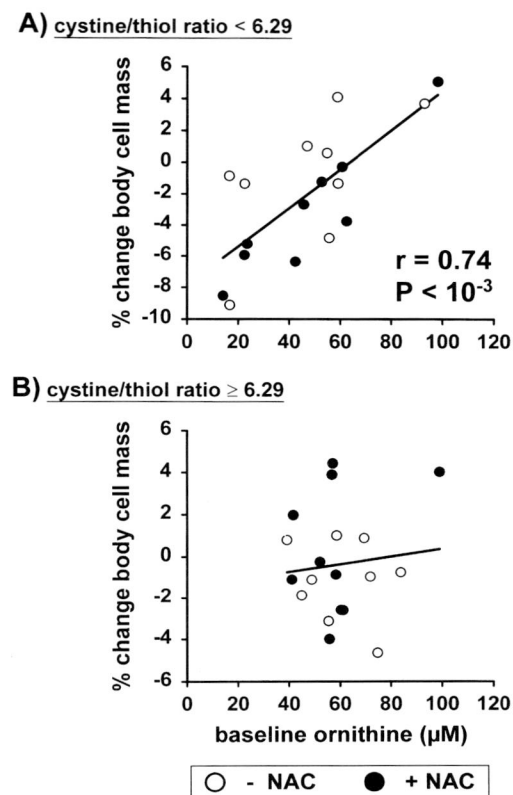

Figure 11 Correlation between the change in bcm and the baseline level of plasma ornithine of healthy subjects. Correlations were determined again separately for the two subgroups of persons with lower (A) and higher (B) than median plasma cystine/thiol ratios (6.29). Persons with (●) and without (○) NAC treatment showed essentially similar correlations, and data of these two treatment groups have therefore been pooled for the purpose of this analysis.

active healthy subjects with lower cystine/thiol ratios may be determined, at least partly, by the plasma ornithine level.

The hypothesis that the availability of ornithine may be a decisive factor in the regulation of bcm is supported further by a series of reports on the anabolic and anticatabolic effects of ornithine preparations in various catabolic conditions (Le Boucher et al., 1997; Le Bricon et al., 1994, 1997; Cynober, 1991, 1994; Vaubourdolle et al., 1990; Cynober et al., 1984a,b, 1987; Leander et al., 1985; Wernermann et al., 1987; Holm et al., 1991). Despite accumulating evidence on the limiting role of ornithine in catabolic conditions, it was an unexpected but potentially important finding to see that this amino acid may be limiting even in healthy physically active persons living on a standard Western diet.

It is reasonable to assume that other factors such as immobilization (lack of physical exericse), inadequate nutrition (starvation), and a decreased insulin responsiveness or other hormone-dependent mechanisms may also contribute decisively to the decrease in bcm and the corresponding increase in net protein catabolism in elderly subjects or cancer patients. In the context of antioxidant therapy, it is important to note, however, that all these processes may potentially contribute to a progressive increase in the plasma cystine level and to a corresponding shift in the redox state. It remains to be determined whether this shift may eventually become autonomous as a cause of skeletal muscle catabolism and ultimately trigger a vicious circle.

C. Factors That Determine the Clearance of Plasma Cystine

The plasma cystine level would not increase so markedly in response to an increased cystine export from muscle tissue if the clearance of cystine from the blood plasma were not exceptionally low. Indeed, the plasma cystine level shows, in contrast to most other protein-forming amino acids, on the average only moderate differences between the arterial and the venous blood and between the postprandial and the postabsorptive state (W. Dröge, unpublished observation; see also Clowes *et al.*, 1980; Bennegard *et al.*, 1984). Cystine is generally transported into cells and tissue by the membrane transport system x_c^-, which is driven by the glutamate antiport (i.e., by the relative concentration gradients of cystine and glutamate) rather than by the powerful Na^+ concentration gradient at the plasma membrane that drives most other amino acid transport activities (Makowske and Christensen, 1982; Takada and Bannai, 1984; Bannai, 1984; Watanabe and Bannai, 1987). The x_c^- transport system has been demonstrated in liver cells, fibroblasts, and macrophages but seems to be expressed only weakly on most other cells and tissues (Makowske and Christensen, 1982; Takada and Bannai, 1984; Bannai, 1984; Watanabe and Bannai, 1987).

IV. DISEASES AND CONDITIONS ASSOCIATED WITH RELATIVELY LOW PLASMA CYSTINE LEVELS (LOW CG SYNDROME)

A. Decreased Plasma Amino Acid Levels and Evidence for a Leakage of Cysteine in HIV Infection

In contrast to the relatively high plasma cystine levels in elderly subjects and cancer patients, HIV-infected patients have, on average, relatively low plasma cystine levels (Hack et al., 1997; Dröge et al., 1988b). This decrease was most pronounced in the late asymptomatic stage of HIV infection with CD4$^+$ counts <400 mm^3 (Hack et al., 1997) and is associated with the decrease of plasma levels of most amino acids with the notable exception of glutamate, ornithine, and taurine (Fig. 9). A relatively low mean plasma cystine level associated with an abnormally low mean plasma glutamine concentration and decreased immunological reactivity has also been found in several other diseases and conditions, including sepsis, trauma, Crohn's disease, ulcerative colitis, chronic fatigue syndrome, and, to some extent, in overtrained athletes (Dröge and Holm, 1997) and has been tentatively called "low CG syndrome."

Available evidence suggests that the immune system may be compromised in these conditions by more than one mechanism because deficiencies in cysteine, glutamine, and arginine may each have distinct effects on cells of the immune system (Dröge and Holm, 1997). In a more recent study in which 40 HIV-infected patients with CD4$^+$ T-cell counts <500 mm^{-3} undergoing state-of-the-art antiviral therapy were compared with 60 young healthy control subjects, the patients were found to have still relatively low plasma cystine, glutamine, and especially very low arginine levels (42 ± 2, 568 ± 14, and 50 ± 4 versus 45 ± 1, 605 ± 13, and 80 ± 3 μM, respectively), despite antiviral therapy and despite the relatively moderate mean virus load of 17,300 ± 4500 copies/ml (R. Breitkreutz, N. Pittack, and W. Dröge, unpublished observation). Unfortunately, published reports on sepsis, major injury, Crohn's disease, ulcerative colitis, chronic fatigue syndrome, and overtraining (reviewed in Dröge and Holm, 1997) did not indicate plasma arginine levels.

The decrease of the plasma cystine levels in HIV-infected patients (Hack et al., 1997; Dröge et al., 1988b) and the concomitant decrease of intracellular glutathione levels in HIV-infected patients (Eck et al., 1989; Roederer et al., 1991) were originally thought to be the consequence of oxidative processes in these patients. This interpretation was supported by reports on elevated levels of malondialdehyde, i.e., the end product of lipid peroxidation (Sonneborg et al., 1988; Revillard et al., 1992), decreased plasma vitamin A levels (Semba et al., 1993; Constans et al., 1995), decreased selenium levels (Constans et al., 1995; Beck et al., 1990; Cirelli et al., 1991), decreased GSH/GSSG ratios in peripheral blood mononuclear cells of HIV-infected patients (Aukrust et al., 1995; Walmsley et al., 1997), and decreased GSH/GSSG ratios in the skeletal muscle tissue of SIV-infected rhesus macaques (Gross et al., 1996). However, the paradigm of a generalized shift to more oxidative conditions in HIV-infected patients may have to be challenged in view of the more recent finding that some HIV-infected patients have relatively high concentrations of acid-soluble thiol in the arterial plasma indicative of a strongly reducing arterial plasma thiol redox state (Breitkreutz *et al.*, 1999). This surprising finding may need further investigation. If confirmed, this finding and the pattern of amino acids in Fig. 9 suggest the possibility that the organism may require a well-balanced intermediate plasma thiol redox state for optimal function and that deviations in either direction may be detrimental.

B. Therapeutic Intervention in HIV Infection with NAC or Other Cysteine Derivatives

To correct the low cystine level and to ameliorate the immunological and metabolic consequences of the cyst(e)ine and glutathione deficiency in HIV infection, we have proposed that HIV-infected patients may receive treatment with a cysteine derivative such as NAC (Dröge, 1989; Dröge *et al.*, 1992). Whether treatment with NAC will effectively slow the disease progression in HIV infection remains to be demonstrated in controlled clinical trials. However, longitudial observations on a small number of HIV-infected patients over a total period of more than 10 years have already shown that NAC treatment is capable of increasing not only plasma cystine, but also glutamine to levels that were even higher than the mean of healthy controls (Dröge and Holm, 1997; Dröge *et al.*, 1997). The $CD4^+$ T-cell numbers did not increase during NAC treatment, but remained essentially stable. Several preliminary studies by different laboratories on the therapeutic effect of NAC in HIV-infected patients also revealed encouraging results (Olivier, 1995; Akerlund *et al.*, 1996; Herzenberg *et al.*, 1997). Because two of these studies were not controlled rigorously (Olivier, 1995; Herzenberg *et al.*, 1997) and because the other study used only relatively low doses of NAC (Akerlund *et al.*, 1996), a well-controlled study of the therapeutic effect of adequate doses of NAC still needs to be done. There are strong reasons to believe, however, that the rationale for the use of NAC in HIV infection is entirely different from the rationale of antioxidant treatment of cancer patients as described earlier. In the latter case, NAC is administered as an antioxidant to ameliorate the consequences of the shift in the plasma redox state to more oxidative conditions. In HIV infection, NAC has been proposed as a tool to compensate for the loss of cyst(e)ine in these patients. Studies on SIV-infected rhesus macaques indicate that a considerable amount of cysteine may be lost by an increased conversion of cyst(e)ine into sulfate in the skeletal muscle tissue (Gross *et al.*, 1996). More recent studies on HIV-infected patients lend further support to this hypothesis (Breitkreutz *et al.*, 1999).

C. Putative Consequences of Cysteine Leakage in HIV Infection: "Push" and "Pull" Mechanisms in Catabolic Conditions

Because a decrease in the plasma cystine level in HIV-infected patients is commonly associated with a decrease in plasma glutamine and arginine levels and because treatment with NAC leads to an increase not only of the plasma cystine, but also of plasma glutamine and arginine levels (Dröge and Holm, 1997; Dröge et al., 1997), it has been suggested that the availability of cysteine may have a regulatory influence on the amino acid pool and specifically on the rate of hepatic amino acid catabolism. Tentatively, we have proposed that cysteine may regulate the rate of hepatic urea production by controlling the rate of carbamoyl-phosphate biosynthesis (Dröge and Holm, 1997; Dröge et al., 1997). This hypothesis is supported by several complementary studies in different experimental systems:

In SIV-infected rhesus macaques, increased hepatic concentrations and increased urea/glutamine ratios were found to be associated with decreased hepatic sulfate levels and decreased plasma cystine levels (Gross et al., 1996).

In weight-losing tumor-bearing mice, increased hepatic urea concentrations and increased urea/glutamine ratios were found to be associated with decreased hepatic sulfate and plasma cystine levels (Hack et al., 1996a). Individual urea/glutamine ratios were found to be correlated inversely with hepatic sulfate levels, and the administration of cysteine caused an increase in hepatic sulfate levels, a concomitant decrease in hepatic urea levels, and urea/glutamine ratios close to normal values (Hack et al., 1996a).

Treatment of normal mice with interleukin-6 (IL-6) induced a substantial decrease in hepatic sulfate levels and an increase in the urea/glutamine ratio within 4 hr, suggesting that cytokines may play an important role in the dysregulation of hepatic cysteine catabolism and urea production (Hack et al., 1996a).

Because the skeletal muscle protein catabolism and the corresponding loss of bcm in elderly persons and most cancer patients with high cystine/thiol ratios are, on average, associated with a relative increase in the plasma levels of most amino acids (see Fig. 9) and because asymptomatic HIV-infected patients with $CD4^+$ cells <400 mm^{-3} were found to have lower than normal plasma levels of most protein-forming amino acids, we have proposed to distinguish "push" and "pull" mechanisms in catabolic processes. It is our working hypothesis that the leakage of cysteine may be a common cause for a generalized amino acid deficiency characteristic of the "pull" mechanism (Dröge and Holm, 1997; Dröge et al., 1997).

V. PLASMA GLUTAMATE AND INTRACELLULAR GLUTATHIONE LEVELS

Despite the remarkable differences among HIV-infected patients, cancer patients, and elderly subjects with respect to general plasma amino acid levels, the plasma glutamate level is, on average, increased significantly in all these conditions. Significantly increased plasma glutamate levels were found in HIV-infected patients (Hack et al., 1997; Dröge et al., 1988b; Eck et al., 1989), in SIV-infected rhesus macaques (Eck et al., 1991), in cancer patients (reviewed in Dröge et al., 1988a), in elderly subjects (Hack et al., 1996b), and in tumor-bearing mice (Hack et al., 1996a). This phenomenon appears to have important implications for intracellular glutathione levels, at least in the skeletal muscle tissue. Amino acid exchange studies in the lower extremities of cancer patients have shown that the elevated venous glutamate levels result, at least partly, from a decreased glutamate transport activity into the skeletal muscle tissue (Hack et al., 1996b), and complementary studies on SIV-infected rhesus macaques and tumor-bearing mice have shown that the increase in the plasma glutamate level is indeed associated with a significant decrease in the intracellular glutamate level in the white sections of skeletal muscle tissue and with a concomitant increase in the liver (Gross et al., 1996; Hack et al., 1996a) (see Fig. 12). In all cases tested, the change in the intracellular glutamate level was associated with a corresponding change in intracellular glutathione (Fig. 12 and Gross et al., 1996; Hack et al., 1996a). The functional significance of these changes and their contribution to the plasma redox state remain to be shown.

VI. SUMMARY AND CONCLUSIONS

The conspicuous increase in the plasma cystine/thiol ratio in elderly persons and cancer patients indicates a shift of the plasma redox state to a more oxidized condition. The pathological consequences of this change in the redox state are not yet fully understood. A strong case can be argued for a causal relationship between the redox state and the plasma albumin level. First, albumin is known to have reactive cysteine residues, and there is evidence that the oxidation of albumin may increase the rate of albumin degradation. Second, there is a strong linkage between the plasma albumin level and the plasma cystine/thiol ratio in cancer patients and healthy subjects. Importantly, this linkage can be demonstrated even within relatively

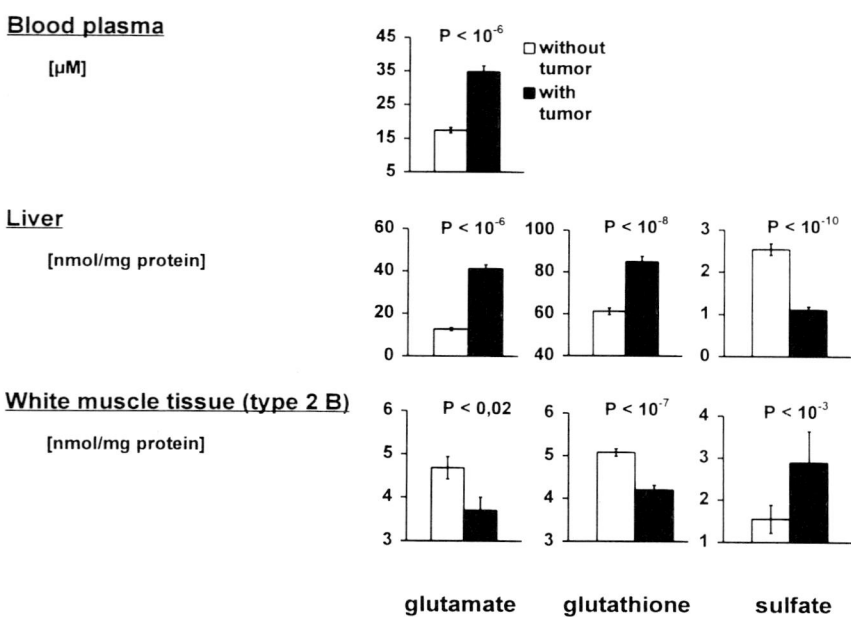

Figure 12 Glutamate, glutathione, and sulfate levels in tumor-bearing mice. The experimental system and the type of tumor (MCA-105 fibrosarcoma) have been described in Hack et al. (1996a). Data show the mean ± SEM. Analysis revealed a significant decrease of intracellular glutamate and glutathione levels and increased sulfate levels in skeletal muscle tissue of type 2B (white muscle sections). This pattern of tumor-mediated changes is virtually the mirror image of the changes seen in the blood plasma and in the liver and may be explained, at least partly, by the decreased membrane transport of glutamate from the blood into the skeletal muscle tissue.

short observation periods. Third, treatment of cancer patients with the antioxidant NAC was shown to cause, on average, a decrease in the plasma cystine/thiol ratio and a concomitant increase in the plasma albumin level. The plasma albumin level is commonly considered as a clinically important parameter. A decrease in plasma albumin was recognized as a quantitative measure of cachexia in various catabolic conditions, and it is well established that senescence is also associated with a decrease in the mean plasma albumin level. Whether the widely observed linkage between the decrease in albumin level and loss of bcm may reflect a cause and effect relationship is not known. There is good evidence, however, that the albumin level plays an important role in the control of oncoosmotic pressure and in the prevention of edema.

The conspicuous linkage between the plasma cystine/thiol ratio and bcm and the effect of NAC on bcm in cancer patients and healthy subjects are strongly indicative for a cause and effect relationship between redox state and bcm. However, the mechanism of this effect is again unknown. Whether this effect involves the increase of the plasma albumin level remains to be shown.

There is an increasing body of evidence suggesting that HIV-infected patients suffer from a massive loss of cysteine and, therefore, require reconstitution with a cysteine derivative. The therapeutic effects of NAC in cancer cachexia and HIV infection are, therefore, believed to involve different mechanisms, i.e., the antioxidant effect and the reconstituting effect, respectively.

Some of the concepts described in this chapter are still hypothetical, but the history of science has taught us that paradigms may be useful even if they have to be replaced eventually by better ones. Our knowledge about the consequences of the shift in the redox state in senescence and wasting and about the cysteine deficiency in HIV infection is admittedly still incomplete. Also, more work needs to be done to document the benefits and potential risks of a cysteine-based therapeutic intervention. Nevertheless, the albeit limited information that has been obtained so far should be weighed against the fact that we are dealing with the treatment of severely debilitating diseases and conditions for which no other satisfying treatments are available today. Should we keep these patients waiting until we know all the details? Among others, we are dealing with HIV-infected patients, who show progressive weight loss and disease progression despite antiviral therapy and low virus titers; patients with advanced malignancies for whom established therapies have failed; and last but not least with an increasing number of elderly persons who suffer from severely compromised physical and social activity.

The hypothesis that redox processes may contribute decisively to the phenomenon of senescence was first proposed in the 1950s. How many of us will see the benefits of this concept in our lifetime?

ACKNOWLEDGMENT

The assistance of Mrs. I. Fryson in the preparation of this manuscript is gratefully acknowledged.

REFERENCES

Akerlund, B., Jarstrand, C., Lindeke, B., Sönnerborg, A., Akerblad, A.-C., and Rasool, O. (1996). Effect of N-acetylcysteine (NAC) treatment on HIV-1 infection: A double-blind placebo-controlled trial. *Eur. J. Clin. Pharmacol.* 50, 457.

Aukrust, P., Svardal, A. M., Müller, F., Lunden, B., Berge, R. K., Ueland, P. M., and Froland, S. S. (1995). Increased levels of oxidized glutathione in $CD4^+$ lymphocytes associated with disturbed intracellular redox balance in human immunodeficiency virus type 1 infection. *Blood* **86**, 258.

Bannai, S. (1984). Transport of cystine and cysteine in mammalian cells. *Biochim. Biophys. Acta.* **779**, 289.

Baumgartner, R. N., Koehler, K. M., Romero, L., and Garry, P. J. (1996). Serum albumin is associated with skeletal muscle in elderly men and women. *Am. J. Clin. Nutr.* **64**, 552.

Beck, K. W., Schramel, P., Held, A., Jaeger, H., and Kaboth, W. (1990). Serum trace-element levels in HIV-infected subjects. *Biol. Trace Elem. Res.* **25**, 89.

Bennegard, K., Lindmark, L., Eden, E., Svaninger, G., and Lundholm, K. (1984). Flux of amino acids across the leg in weight-losing cancer patients. *Cancer Res.* **44**, 386.

Breitkreutz, R., Holm, S., Pittack, N., Beichert, M., Babylon, A., and Dröge, W. (1999). Peripheral cysteine catabolism and massive loss of sulphur in HIV-infection. Submitted for publication.

Brennan, M. F. (1977). Uncomplicated starvation versus cancer cachexia. *Cancer Res.* **37**, 2359.

Brody, J. A. (1985). Prospects for an ageing population. *Nature (London)* **315**, 463.

Buchner, D. M., and Wagner, E. H. (1992). Preventing frail health. *Clin. Geriatr. Med.* **8**, 1.

Campion, E. W. (1994). The oldest old. *N. Engl. J. Med.* **330**, 1819.

Cirelli, A., Ciardi, M., De Simone, V., Sorice, F., Giordano, R., Ciaralli, L., and Costantini, S. (1991). Serum selenium concentration and disease progress in patients with HIV infection. *Clin. Biochem.* **24**, 211.

Clowes, G. H. A., Jr., Heideman, M., Lindberg, B., Randall, H. T., Hirsch, E. F., Cha, C.-J., and Martin, H. (1980). Effects of parenteral alimentation on amino acid metabolism in septic patients. *Surgery* **88**, 531.

Cohn, S. H., Vartsky, D., Yasumura, S., Sawitsky, A., Zanzi, I., Vaswani, A., and Ellis, K. J. (1980). Compartmental body composition based on total-body nitrogen, potassium, and calcium. *Am. J. Physiol.* **239**, E524.

Constans, J., Peuchant, E., Pellegrin, J. L., Sergeant, C., Hamon, C., Dubourg, L., Thomas, M. I., Simonoff, M., Pellegrin, I., Brossard, G., Barbeau, P., Fleury, H., Clerc, M., Leng, B., and Conri, C. (1995). Fatty acids and plasma antioxidants in HIV-positive patients: Correlation with nutritional and immunological status. *Clin. Biochem.* **28**, 421.

Cooper, J. K., and Gardner, C. (1989). Effect of aging on serum albumin. *J. Am. Geriat. Soc.* **37**, 1039.

Cynober, L., (1991). Ornithine α-ketoglutarate in nutritional support. *Nutrition.* **7**, 313.

Cynober, L. (1994). Can arginine and ornithine support gut functions? *Gut* **35**, S42.

Cynober, L., Saizy, R., Nguyen Dinh, F., Lioret, N., and Giboudeau, J. (1984a). Effect of enterally administered ornithine α-ketoglutarate on plasma and urinary amino acid levels after burn injury. *J. Trauma* **24**, 590.

Cynober, L., Vaubourdolle, M., Doré, A., and Giboudeau, J. (1984b). Kinetics and metabolic effects of orally administered ornithine α-ketoglutarate in healthy subjects fed with a standardized regimen. *Am. J. Clin. Nutr.* **39**, 514.

Cynober, L., Lioret, N., Coudray-Lucas, C., Aussel, C., Ziegler, F., Baudin, B., Saizy, R., and Giboudeau, J. (1987). Action of ornithine α-ketoglutarate on protein metabolism in burn patients. *Nutrition* **3**, 187.

DeWys, W. D., Begg, C., Lavin, P. T., Band, P. R., Bennett, J. M., Bertino, J. R., Cohen, M. H., Douglass, H. O., Jr., Engström, P. F., Ezdinli, E. Z., Horton, J., Johnson, G. J., Moertel, C. G., Oken, M. M., Perlia, C., Rosenbaum, C., Silverstein, M. N., Skeel, R. T., Sponzo, R. W., and Tormey, D. C. (1980). Prognostic effect of weight loss prior to chemotherapy in cancer patients. *Am. J. Med.* **69**, 491.

Dröge, W. (1989). Metabolische Störungen bei HIV-Infektion. In "Project News," No. 2, p. 4. AIDS-Zentrum des Bundesgesundheitsamtes, Berlin.
Dröge, W., and Holm, E. (1997). Role of cysteine and glutathione in HIV infection and other diseases associated with muscle wasting and immunological dysfunctions. *FASEB J.* **11**, 1077.
Dröge, W., Eck, H.-P., Betzler, M., Schlag, P., Drings, P., and Ebert, W. (1988a). Plasma glutamate concentration and lymphocyte activity. *J. Cancer Res. Clin. Oncol.* **114**, 124.
Dröge, W., Eck, H.-P., Näher, H., Pekar, U., and Daniel, V. (1988b). Abnormal amino acid concentrations in the blood of patients with acquired immune deficiency syndrome (AIDS) may contribute to the immunological defect. *Biol. Chem. Hoppe-Seyler* **369**, 143.
Dröge, W., Eck, H.-P., and Mihm, S. (1992). HIV-induced cysteine deficiency and T cell dysfunctions - A rationale for treatment with N-acetyl-cysteine. *Immunol. Today* **13**, 211.
Dröge, W., Gross, A., Hack, V., Kinscherf, R., Schykowski, M., Bockstette, M., Mihm, S., and Galter, D. (1997). Role of cysteine and glutathione in HIV infection and cancer cachexia. Therapeutic intervention with N-acetyl-cysteine (NAC). *Adv. Pharmacol.* **38**, 581.
Eck, H.-P., Gmünder, H., Hartmann, M., Petzoldt, D., Daniel, V., and Dröge, W. (1989). Low concentrations of acid soluble thiol (cysteine) in the blood plasma of HIV-1 infected patients. *Biol. Chem. Hoppe-Seyler* **370**, 101.
Eck, H.-P., Stahl-Hennig, C., Hunsmann, G., and Dröge, W. (1991). Metabolic disorder as early consequence of simian immunodeficiency virus infection in rhesus macaques. *Lancet* **338**, 346.
Era, S., Hamaguchi, T., Sogami, M., Kuwata, K., Suzuki, E., Miura, K., Kawai, K., Kitazawa, Y., Okabe, H., Noma, A., and Miyata, S. (1988). Further studies on the resolution of human mercapt- and nonmercaptalbumin and on human serum albumin in the elderly by high-performance liquid chromatography. *Int. J. Pep. Protein Res.* **31**, 435.
Era, S., Kuwata, K., Imai, H., Nakamura, K., Hayashi, T., and Sogami, M. (1995). Age-related change in redox state of human serum albumin. *Biochim. Biophys. Acta* **1247**, 12.
Erstad, B. L. (1992). Serum albumin concentrations: Who needs them? *Ann. Pharmacother.* **26**, 1134.
Fiatarone, M. A., O'Neill, E. F., Ryan, N. D., Clements, K. M., Solares, G. R., Nelson, M. E., Roberts, S. B., Kehayias, J. J., Lipsitz, L. A., and Evans, W. J. (1994). Exercise training and nutritional supplementation for physical frailty in very elderly people. *N. Engl. J. Med.* **330**, 1769.
Finch, J. W., Crouch, R. K., Knapp, D. R., and Schey, K. L. (1993). Mass spectrometric identification of modifications to human serum albumin treated with hydrogen peroxide. *Arch. Biochem. Biophys.* **305**, 595.
Gilchrest, B. A., and Bohr, V. A. (1997). Aging processes, DNA damage, and repair. *FASEB J.* **11**, 322.
Gross, A., Hack, V., Stahl-Hennig, C., and Dröge, W. (1996). Elevated hepatic γ-glutamyl-cysteine synthetase activity and abnormal sulfate levels in liver and muscle tissue may explain abnormal cysteine and glutathione levels in SIV-infected rhesus macaques. *AIDS Res. Hum. Retroviruses* **12**, 1639.
Hack, V., Gross, A., Kinscherf, R., Bockstette, M., Fiers, W., Berke, G., and Dröge, W. (1996a). Abnormal glutathione and sulfate levels after interleukin-6 treatment and in tumor-induced cachexia. *FASEB J.* **10**, 1219.
Hack, V., Stütz, O., Kinscherf, R., Schykowski, M., Kellerer, M., Holm, E., and Dröge, W. (1996b). Elevated venous glutamate levels in (pre)catabolic conditions result at least partly from a decreased glutamate transport activity. *J. Mol. Med.* **74**, 337.

Hack, V., Schmid, D., Breitkreutz, R., Stahl-Hennig, C., Drings, P., Kinscherf, R., Taut, F., Holm, E., and Dröge, W. (1997). Cystine levels, cystine flux and protein catabolism in cancer cachexia, HIV/SIV infection and senescence. *FASEB J.* **11**, 84.

Hack, V., Breitkreutz, R., Kinscherf, R., Röhrer, H., Bärtsch, P., Taut, F., Benner, A., and Dröge, W. (1998). The redox state as a correlate of senescence and wasting and as a target for therapeutic intervention. *Blood* **92**, 59.

Halliwell, B., and Gutteridge, M. C. (1990). The antioxidants of human extracellular fluids. *Arch. Biochem. Biophys.* **280**, 1.

Harman, D. (1956). Aging: A therory based on free radical and radiation chemistry. *J. Geronterol.* **11**, 298.

Herzenberg, L. A., De Rosa, S. C., Dubs, J. G., Roederer, M., Anderson, M. T., Ela, S. W., Deresinski, S. C., and Herzenberg, L. A. (1997). Glutathione deficiency is associated with impaired survival in HIV disease. *Proc. Natl. Acad. Sci. U.S.A.* **94**, 1967.

Holm, E., Hess, Y., Leweling, H., Barth, H.O., and Hagmüller, E. (1991). Ornithine aspartate (OA) promotes amino acid (AA) retention in the peripheral tissue of patients with liver cirrhosis. A double-blind, randomized crossover study. *JPEN, J. Parenter. Enteral Nutr.* **15** (Suppl), 36S.

Hu, H., Sparrow, D., and Weiss, S. (1992). Association of serum albumin with blood pressure in the normative aging study. *Am. J. Epidemiol.* **136**, 1465.

Kinscherf, R., Hack, V., Fischbach, T., Friedmann, B., Weiss, C., Edler, L., Bärtsch, P., and Dröge, W. (1996). Low plasma glutamine in combination with high glutamate levels indicate risk for loss of body cell mass in healthy individuals: The effect of N-acetylcysteine. *J. Mol. Med.* **74**, 393.

Kotler, D. P., Wang, J., and Pierson, R. (1985). Body composition in patients with the acquired immunodeficiency syndrome. *Am. J. Clin. Nutr.* **42**, 1255.

Kuwata, K., Era, S., and Sogami, M. (1994). The kinetic studies on the intracellular SH, S-S exchange reaction of bovine mercaptalbumin. *Biochim. Biophys. Acta.* **1205**, 317.

Lamberts, S. W. J., van den Beld, A. W., and van der Lely, A.-J. (1997). The endocrinology of aging. *Science* **278**, 419.

Leander, U., Fürst, P., Vesterberg, K., and Vinars, E. (1985). Nitrogen sparing effect of ornicetil in the immediate postoperative state. Clinical biochemistry and nitrogen balance. *Clin. Nutr.* **4**, 43.

Le Boucher, J., Obled, C., Farges, M. C., and Cynober, L. (1997). Ornithine α-ketoglutarate modulates tissue protein metabolism in burn-injured rats. *Am. J. Physiol.* **273**, E557.

Le Bricon, T., Cynober, L., and Baracos, V. E. (1994). Ornithine α-ketoglutarate limits muscle protein breakdown without stimulating tumor growth in rats bearing Yoshida ascites hepatoma. *Metab. Clin. Exp.* **43**, 899.

Le Bricon, T., Coudray-Lucas, C., Lioret, N., Lim, S. K., Plassart, F., Schlegel, L., De Bandt, J. P., Saizy, R., Giboudeau, J., and Cynober, L. (1997). Ornithine α-ketoglutarate metabolism after enteral administration in burn patients: Bolus compared with continuous infusion. *Am. J. Clin. Nutr.* **65**, 12.

Long, C. L., Crosby, F., Geiger, J. W., and Kinney, J. M. (1976). Parenteral nutrition in the septic patient: Nitrogen balance, limiting plasma amino acids, and calorie to nitrogen ratios. *Am. J. Clin. Nutr.* **29**, 380.

Makowske, M., and Christensen, H. N. (1982). Contrasts in transport systems for anionic amino acids in hepatocytes and a hepatoma cell line HTC. *J. Biol. Chem.* **257**, 5663.

Manton, K. G., Corder, L. S., and Stallard, E. (1997). Monitoring changes in the health of the U.S. elderly population: Correlates with biomedical research and clinical innovations. *FASEB J.* **11**, 923.

Naber, T. H. J., de Bree, A., Schermer, T. R. J., Bakkeren, J., Bär, B., de Wild, G., and Katan, M. B. (1997). Specificity of indexes of malnutrition when applied to apparently healthy people: The effect of age. *Am. J. Clin. Nutr.* **65**, 1721.

Olivier, R. (1995). Flow cytometry technique for assessing effects of N-acetylcysteine on apoptosis and cell viability of human immunodeficiency virus-infected lymphocytes. In "Methods in Enzymology" (L. Packer, ed.), Vol. 251, p. 270. Academic Press, San Diego, CA.

Oratz, M., Rothschild, M. A., and Schreiber, S. S. (1976). Alcohol, amino acids, and albumin synthesis. II. Alcohol inhibition of albumin synthesis reversed by arginine and spermine. *Gastroenterology* **71**, 123.

Ott, M., Lembcke, B., Fischer, H., Jager, R., Polat, H., Geier, H., Recht, M., Staszeswki, S., Helm, E. B., and Caspary, W. F. (1993). Early changes of body composition in human immunodeficiency virus-infected patients: Tetrapolar body impedance analysis indicates significant malnutrition. *Am. J. Clin. Nutr.* **57**, 15.

Ottery, F. D. (1995). Supportive nutrition to prevent cachexia and improve quality of life. *Semin. Oncol.* **22**(Suppl 3), 98.

Paluzzi, M., and Meguid, M. M. (1987). A prospective randomized study of the optimal source of nonprotein calories in total parenteral nutrition. *Surgery* **102**, 711.

Pisters, P. W., and Pearlstone, D. B. (1993). Protein and amino acid metabolism in cancer cachexia: Investigative techniques and therapeutic interventions. *Crit. Rev. Clin. Lab. Sci.* **30**, 223.

Revillard, J. P., Vincent, C. M. A., Favier, A. E., Richard, M. J., Zittoun, M., and Kazatchkine, M. D. (1992). Lipid peroxidation in human immunodeficiency virus infection. *J. Acquired Immune Defic. Syndr.* **5**, 637.

Roederer, M., Staal, F. J. T., Osada, H., Herzenberg, L. A., and Herzenberg, L. A. (1991). CD4 and CD8 T cells with high intracellular glutathione levels are selectively lost as the HIV infection progresses. *Int. Immunol.* **3**, 933.

Rothschild, M. A., Oratz, M., and Schreiber, S. S. (1972). Albumin synthesis. *N. Engl. J. Med.* **286**, 748.

Rothschild, M. A., Oratz, M., and Schreiber, S. S. (1988). Serum albumin. *Hepatology* **8**, 385.

Sastre, J., Asensi, M., Gasco, E., Pallardo, F. V., Ferrero, J. A., Furukawa, T., and Vina, J. (1992). Exhaustive physical exercise causes oxidation of glutathione status in blood: Prevention by antioxidant administration. *Am. J. Physiol.* **263**, R992.

Semba, R. D., Graham, N. M. H., Caiaffa, W. T., Margolick, J. B., Clement, L., and Vlahov, D. (1993). Increased mortality associated with vitamin A deficiency during human immunodeficiency virus type 1 infection. *Arch. Intern. Med.* **153**, 2149.

Sen, C. K., Rankinen, T., Väisänen, S., and Rauramaa, R. (1994). Oxidative stress after human exercise: Effect of N-acetylcysteine supplementation. *J. Appl. Physiol.* **76**, 2570.

Shibata, H., Haga, H., Ueno, M., Nagai, H., Yasumura, S., and Koyano, W. (1991). Longitudinal changes of serum albumin in elderly people living in the community. *Age Aging* **20**, 417.

Shigenaga, M. K., Hagen, T. M., and Ames, B. N. (1994). Oxidative damage and mitochondrial decay in aging. *Proc. Natl. Acad. Sci. U.S.A.* **91**, 10771.

Sonneborg, A., Carlin, G., Akerlund, B., and Jarstrand, C. (1988). Increased production of malondialdehyde in patients with HIV infection. *Scand. J. Infect. Dis.* **20**, 287.

Stadtman, E. R. (1992). Protein oxidation and aging. *Science* **257**, 1220.

Strain, A. J. (1979). Cancer cachexia in man. A review. *Invest. Cell. Pathol.* **2**, 181.

Takada, A., and Bannai, S. (1984). Transport of cystine in isolated rate hepatocytes in primary culture. *J. Biol. Chem.* **259**, 2441.

Tayek, J. A. (1988). Albumin synthesis and nutritional assessment. *Nutr. Clin. Pract.* **3**, 219.

Tuten, M. B., Wogt, S., Dasse, F., and Leider, Z. (1985). Utilization of prealbumin as a nutritional parameter. *JPEN, J. Parenter. Enteral Nutr.* **9**, 709.

Vaubourdolle, M., Salvucci, M., Coudray-Lucas, C., Agneray, J., Cynober, L., and Ekindjian, O. G. (1990). Action of ornithine α-ketoglutarate on DNA synthesis by human fibroblasts. *In Vitro Cell Dev. Biol.* **26**, 187.

Walmsley, S. L., Winn, L. M., Harrison, M. L., Uetrecht, J. P., and Wells, P. G. (1997). Oxidative stress and thiol depletion in plasma and peripheral blood lymphocytes from HIV-infected patients: Toxicological and pathological implications. *AIDS* **11**, 1689.

Watanabe, H., and Bannai, S. (1987). Induction of cystine transport activity in mouse peritoneal macrophages. *J. Exp. Med.* **165**, 628.

Wernermann, J., Hammarqvist, F., von der Decken, A., and Vinnars, E. (1987). Ornithine α-ketoglutarate improves skeletal muscle protein synthesis as assessed by ribosome analysis and nitrogen use after surgery. *Ann. Surg.* **206**, 674.

Index

Aconitase, reactive nitrogen species, 154, *155*
Aging
  oxidative stress, 517–530
  premature, 427–432, *430*, *431*
AIDS, NF-κB, redox regulation, 205–206
Alpha-tocopherol, atherosclerosis, 359–380
Alzheimer's disease, 432–435, *434*
Antioxidant genes, Down's syndrome, reactive oxygen species in, 425–450
Antioxidant regulation, redox, cell adhesion processes, *282*, 282–283
  agonist-induced cell adhesion process, 278–283
Antioxidant therapy, redox signaling, clinical implications of, 337–556
Antioxidants
  senescence, 531–536
  wasting, 531–536
AP-1 pathway, singlet oxygen, 7–9, 10–12t
Apoptosis
  Bcl-2, 222–225, *225*
  Ca^{2+}, 111–114
  reactive nitrogen species, 167–168

Arthritis, rheumatoid, NF-κB, redox regulation, 207
Ascorbate, redox, cell adhesion processes, 281
Atherosclerosis
  alpha-tocopherol, 359–380
  redox, cell adhesion processes, 285
ATPase pumps, Ca^{2+}, 118–119

Bcl-2, 221–244
  antiapoptotic properties, antioxidants, 225–228
  apoptosis, 222–225, *225*
  functions, via antioxidant mechanism?, 231–236
  proapoptotic properties, antioxidants, 228–229
  proteins, 230t

Ca^{2+}
  apoptosis, 111–114

557

Ca^{2+} (continued)
  ATPase pumps, 118–119
  Na/Ca^{2+} exchanger, 120
  Ca^{2+} channels in plasma membrane, 120
  and cell death, 106–111
  cell death, 106–111
  in cell death, 106–111
  changes in intracellular concentration, 109–111
  endoplasmic reticulum, release from, 115–120, 117
  homeostasis, cell death, 111–113
  IP_3 activation, of Ca^{2+} release, endoplasmic reticulum, 109–110
  IP_3-sensitive channels, 118–119
  maintenance of, 106–109, 107
  mitochondrial, 110
  Na/Ca^{2+} exchanger, 120
  necrosis, 111–114
  oxidative stress, 111–114
  pools, release from, 121
  signaling by oxidative stress, 114–121
  study methods, 115
  targets of increase in, 111
  uptake of, extracellular, 110
Cadherins, redox, cell adhesion processes, 269
Calcium-dependent potassium channels, reactive nitrogen species, 158
Calcium release channel, reactive nitrogen species, 157
Cancer
  metastasis, NF-κB, redox regulation, 206–207
  prevention, lycopene, 409–424
  redox, cell adhesion processes, 284
Caspase-independent signaling to cell death, tumor necrosis factor toxicity, 247
Caspases, reactive nitrogen species, 160
Cell cycle control, p21 expression and, 313–316, 314, 315
Cell death, tumor necrosis factor toxicity, 245
Chemical oxidants, protein tyrosine phosphorylation, 136–141
Chromosome 21, altered redox state, 440–441
Cigarette smoke, redox, cell adhesion processes, 278

Cyclooxygenase, reactive nitrogen species, 162–163
Cytochrome P450, reactive nitrogen species, 152–153
Cytoprotective action, of thioredoxin, 300
Cytoskeleton, Rac, 63–66

Diabetes, redox, cell adhesion processes, 285
Dimerization hypothesis, protein tyrosine phosphorylation, 141–142
DNA transcription, mitochondrial, oxidant-mediated repression of, 451–478
Down's syndrome, antioxidant genes, reactive oxygen species in, 425–450

Endothelial cells, reactive nitrogen species, 164–166
Epidermal growth factor receptor, reactive nitrogen species, 159

Fe-S proteins, reactive nitrogen species, interaction with, 154–155
Firm adhesion, redox, cell adhesion processes, 271
Flavonoids, redox, cell adhesion processes, 281–282
Focal adhesion kinase, redox, cell adhesion processes, 273–276, 274, 275, 277

Gene expression, effects of singlet oxygen on, 6–7, 8
Glyceraldehyde-3-phosphate dehydrogenase, reactive nitrogen species, 159
Guanylyl cyclase, reactive nitrogen species, 151–152, 153t

Heavy metals, redox, cell adhesion processes, 278
Heme
  iron, reactive nitrogen species, interaction with, 151–153
  reactive nitrogen species, interaction with, 162–163

Hemoglobin, reactive nitrogen species, 162
HIV infection, peroxiredoxins, 381–408

Immune system, reactive nitrogen species, 167–169
Immunoglobulin superfamily, redox, cell adhesion processes, 267–268
Infectious diseases, redox, cell adhesion processes, 286–287
Inflammatory response, reactive nitrogen species, 166, 168–169
Inflammatory skin disorders, redox, cell adhesion processes, 286
Insulin gene promoter activity, suppression, by oxidative stress, 479–490
Integrins, redox, cell adhesion processes, 266
Intracellular messengers, reactive species as, 1–178
Ionizing radiation, protein tyrosine phosphorylation, 133–135
Ischemia-reperfusion injury, redox, cell adhesion processes, 284
Ischemic adaptation, redox regulation, 491–516

Kinase cascade, NF-κB, redox regulation, 211

L-selectin, redox, cell adhesion processes, 272–273, 274
Ligand-receptor complexes, redox, cell adhesion processes, 269
Lipoxygenases, 339–358
Long chain fatty acylation, reactive nitrogen species, 158–159
Lycopene, cancer prevention, 409–424

MAP kinases, 10–12t
  reactive nitrogen, 10–12t
Metastasis, redox, cell adhesion processes, 284
Mitochondrial DNA transcription, oxidant-mediated repression of, 451–478

Mitochondrial gene expression, oxidant-modulated gene, 28–29
mRNAs, oxidant-induced, 25t

N-methyl-D-aspartic acid receptor, reactive nitrogen species, 158
Na/Ca^{2+} exchanger, Ca^{2+}, 120
NF-κB
  activation pathways, 181–202
  antioxidants, 190–191
  functions, 181–184
  IL-1, 184–187, *185*
  prooxidant conditions, 187–190
  redox-regulating proteins, 190–191
  TNF, 184–187, *185*
  redox regulation, 203–220
  AIDS, 205–206
  cancer metastasis, 206–207
  kinase cascade, 211
  pathologies, 205–207
  radical oxygen intermediates, 208–210, *210*
  rheumatoid arthritis, 207
  signal transduction pathway, 207–214, *208–209*
  Trx-mediated redox, 212–214, *213, 214*
  Trx-mediated redox regulation, 211
Nitric oxide
  p21 induction, 328–330, *329*
  redox, cell adhesion processes, 278
Nitrosothiol, reactive nitrogen species, proteins inactivated by, 158–162
Nonphagocytic cells, Rac in, 54–57

Oxidant-modulated gene expression, 21–46
  analysis, 22–24
  examples, of human pathologies, 34t
  mitochondrial gene expression, 28–29
  mRNAs, oxidant-induced, 25t
  nuclear gene expression, modulation, 24–28, 25t
  oxidant stress-related pathologies, 33–37, 34t
  regulation, modes of, 31–33
  stable, *versus* transient gene expression, 29–31
Oxidant stress, pathologies, 33–37, 34t

Oxidative burst, in plants, 67–69
Oxidative stress
  apoptosis, 301–302, *302*
  Ca^{2+} homeostasis, apoptosis, 113–114
  physiological aging, 517–530
  response, thioredoxin, signaling, 297–310
Oxidative treatments, MAP kinases, 10–12t
Oxy R, reactive nitrogen species, 161–162

P-selectin, redox, cell adhesion processes, 272–273, *274*
p21
  induction
    oxidative stress, 318–325
    p53-dependent and -independent pathways, *314*, 316–318, *317*
  redox regulation of, 311–336
p53, redox regulation of, 302–303, *303*
Peroxiredoxins, 382–391, 383t, *384–385*, *386*
  cell signaling, HIV infection, 381–408
  major subfamilies of, 383t
Physiological aging, oxidative stress, 517–530
Proapoptotic properties, Bcl-2, antioxidants, 228–229
Protein tyrosine phosphorylation, 129–146
  chemical oxidants, 136–141
  dimerization hypothesis, 141–142
  ionizing radiation, 133–135
  redox, cell adhesion processes, 273–276, *274*, *275*, *277*
  signal pathways, 130–133, *131*
  ultraviolet radiation, 135–136

Rac
  cytoskeleton, 63–66
  development, 66–67
  in nonphagocytic cells, 54–57
  proteins, activation of, 50–52
  respiratory burst, in phagocytes, 50–54, *53*
  ROS production, 57–63, *58*, *60*, *61*
Radiation
  ionizing, protein tyrosine phosphorylation, 133–135
  ultraviolet, protein tyrosine phosphorylation, 135–136

Ras, reactive nitrogen species, 156–157
Reactive nitrogen, 147–178
  aconitase, 154, *155*
  AP-1, 161
  apoptosis, 167–168
  calcium-dependent potassium channels, 158
  calcium release channel, (Ryanodine receptor), 157
  caspases, 160
  cell proliferation, inhibition of, 164–166
  cellular events regulated by RNS/cGMP, 153t
  chemistry of, 148–149, *150*, 151t
  cyclooxygenase, 162–163
  cytochrome P450, 152–153
  cytotoxic mechanisms, 163
  differentiation, 167
  donors, commonly used, 151t
  endothelial cells, 164–166
  epidermal growth factor receptor, 159
  Fe-S proteins, interaction with, 154–155
  gene expression, 163–170
  glyceraldehyde-3-phosphate dehydrogenase, 159
  guanylyl cyclase, 151–152, 153t
  heme
    interaction with, 162–163
    iron, interaction with, 151–153
  hemoglobin, 162
  immune system, 167–169
  inflammatory response, 166, 168–169
  long chain fatty acylation, 158–159
  MAP kinases, 10–12t
  N-methyl-D-aspartic acid receptor, 158
  nervous system, 169–170
  nitrosothiol, proteins inactivated by, 158–162
  Oxy R, 161–162
  Ras, 156–157
  regulatory response, 166–167
  Ryanodine receptor, 157
  S-nitrosothiol formation, proteins activated by, 156–158
  signal transduction, 149–163
  SoxR, 154–155
  thiols, interaction with, 155–162
  tissue plasminogen activator, 157–158
  transcription factors, 160–162
  vascular smooth muscle, 164, *165*

vascular system, 164–167
zinc-finger proteins, 159–160
Reactive oxygen
  Down's syndrome, 425–450
  oxidative treatments, MAP kinases, 10–12t
  tumor necrosis factor toxicity, mediators of cell death by TNF, 248–259
Reactive species, as intracellular messengers, 1–178
  gene expression, effects of singlet oxygen on, 6–7, 8
  reactive oxygen, oxidative treatments, MAP kinases, 10–12t
  singlet oxygen
    AP-1 pathway, 7–9, 10–12t
    in biological systems, 3–6, 5
    signaling by, 3–20
  transcription factors, induction of, singlet oxygen, 9–14
Redox
  cell adhesion processes, 265–296
    antioxidant regulation, *282*, 282–283
      agonist-induced cell adhesion process, 278–283
    ascorbate, 281
    atherosclerosis, 285
    cadherins, 269
    cancer, 284
    cell adhesion molecules, 266–269
    cigarette smoke, heavy metals, and other oxidative environmental pollutants, 278
    diabetes, 285
    direct activation, cell adhesion processes, reactive oxygen species, 271–276, *272*
    effect of ·OH scavengers, iron chelators on adherence, *272*, *273*
    firm adhesion, 271
    flavonoids, 281–282
    focal adhesion kinase, 273–276, *274*, *275*, *277*
    heavy metals, 278
    immunoglobulin superfamily, 267–268
    infectious diseases, 286–287
    inflammatory skin disorders, 286
    integrins, 266
    ischemia-reperfusion injury, 284
    L-selectin, 272–273, *274*
    ligand-receptor complexes, 269
    metastasis, 284
    multistep model of cell adhesion, 270t, 270–271
    nitric oxide, 278
    P-selectin, 272–273, *274*
    L-selectin, and ICAM-1, 272–273, *274*
    protein tyrosine phosphorylation, 273–276, *274*, *275*, *277*
    redox imbalances, pathologies, 283–287
    rheumatoid arthritis, 285–286
    rolling, 270–271
    selectins, 268–269
    thiols, 279–280, *280*
    tocopherol, 280–281
    transmigration, 271
    ultraviolet radiation, 276–277
  imbalances, redox, cell adhesion processes, pathologies, 283–287
  regulation
    ion channels, 81–104
      glutathione-operated cation channel, vascular endothelial cells, 88–91, *90*, *92*, *93*
      patch clamp, 85–87, *86*
      redox modulation, of KC_{Ca} channels, 91–97, *96*, *97*
      ryanodine receptor Ca^{2+} channel, 97–101, *99*
    ischemic adaptation, 491–516
    p21, role of free radicals in, 318–330
  signaling, antioxidant therapy, clinical implications of, 337–556
Redox-sensitive molecular processes, 179–336
Regulatory response, reactive nitrogen species, 166–167
Respiratory burst, in phagocytes, 50–54, *53*
Rheumatoid arthritis
  NF-κB, redox regulation, 207
  redox, cell adhesion processes, 285–286
Rolling, redox, cell adhesion processes, 270–271
ROS, intracellular, tumor necrosis factor toxicity, mediators of TNF-mediated cell death, 250–255
Ryanodine receptor, reactive nitrogen species, 157

S-nitrosothiol formation, reactive nitrogen species, proteins activated by, 156–158
Selectins, redox, cell adhesion processes, 268–269
Senescence, antioxidants, 531–536
Signal pathways, protein tyrosine phosphorylation, 130–133, *131*
Signal transduction
  to gene expression, tumor necrosis factor toxicity, 247–248
  reactive nitrogen species, 149–163
Singlet oxygen
  AP-1 pathway, 7–9, 10–12t
  in biological systems, 3–6, *5*
  signaling by, 3–20
SoxR, reactive nitrogen species, 154–155
Stable, *versus* transient gene expression, 29–31

Thiol
  heme, reactive nitrogen species, interaction with, 162–163
  reactive nitrogen species, interaction with, 155–162
  redox, cell adhesion processes, 279–280, *280*
Thioredoxin, 298–299
  induction by oxidative stress, 299–300
  nuclear translocation of, 303–305
  oxidative stress response, signaling, 297–310
Tissue plasminogen activator, reactive nitrogen species, 157–158
TNF-R55 death-inducing signaling complex, tumor necrosis factor toxicity, 246
Tocopherol, redox, cell adhesion processes, 280–281
tPA, reactive nitrogen species, 157–158
Transcription factors
  induction of, singlet oxygen, 9–14
  reactive nitrogen species, 160–162
Transmigration, redox, cell adhesion processes, 271

Trx-mediated redox, NF-κB, redox regulation, 212–214, *213*, *214*
Tumor necrosis factor toxicity, 245–264
  caspase-independent signaling to cell death, 247
  cell death, 245
  intracellular ROS
    mediators of TNF-mediated cell death, 250–255
    signaling cascade to TNF cytotoxicity, 250–252, 251t
  iron chelators, 251t
  lipoxygenase pathway, ROS from, 253–255
  low molecular weight free radical scavengers, 251t
  mitochondrial origin, ROS of, 252–253, *254*
  reactive oxygen species, mediators of cell death by TNF, 248–259
  signal transduction to gene expression, 247–248
  TNF-induced oxidative burst, 249–250

Ultraviolet radiation
  p21 induction, 326–328
  protein tyrosine phosphorylation, 135–136
  redox, cell adhesion processes, 276–277

Vascular smooth muscle, reactive nitrogen species, 164, *165*
Vascular system, reactive nitrogen species, 164–167
Vitamin E, as antioxidant, 360–363

Wasting, antioxidants, 531–536

Zinc-finger proteins, reactive nitrogen species, 159–160